ALLE ZEIT WACH 1842

B.M. Johri · K.B. Ambegaokar · P.S. Srivastava

Comparative Embryology of Angiosperms Vol. 1

With 362 Figures
in two Volumes

Springer-Verlag
Berlin Heidelberg New York
London Paris Tokyo
Hong Kong Barcelona
Budapest

Prof. Dr. Brij M. Johri
Dr. Kunda B. Ambegaokar
Department of Botany
University of Delhi
Delhi 110007
India

Dr. Prem S. Srivastava
Department of Botany
Hamdard University
Hamdard Nagar
New Delhi 110062
India

ISBN 3-540-53633-7 (in 2 Bänden) Springer-Verlag Berlin Heidelberg New York
ISBN 0-387-53633-7 (in 2 Volumes) Springer-Verlag New York Berlin Heidelberg

Production Editor: Renate Münzenmayer
Typesetting: Macmillan, India
31/3145 – 5 4 3 2 1 0 – Printed on acid-free paper

Dedicated to

Professor Karl Schnarf
(1879–1947)

Professor Panchanan Maheshwari
(1904–1966)

Preface

The concept for the present book dates back to the early 1960s when Professor P. Maheshwari FRS (deceased May 1966) invited scientists to contribute to a multiauthored book – *Comparative Embryology of Angiosperms*. Soon thereafter, Professor Gwenda L. Davis (University of New England, Armidale, Australia) joined the University of Delhi as a Visiting Professor and announced that her book, *Systematic Embryology of Angiosperms*, would soon be published (it appeared in 1966). Therefore, Professor Maheshwari postponed his project for the time being.

The only book on embryology available until then (in english) was by Prof. Maheshwari, published in 1950 (McGraw-Hill, New York) although G. L. Davis' book (without illustrations) was also useful, but to a limited extent. The rapid increase in knowledge in the years that followed, with the advent of new techniques such as ultramicroscopy (TEM and SEM), histochemistry, and the study of organic and inorganic materials, provided impetus to the study of embryology, transgressing its traditional confinements to 'compound' microscopy. This resulted in the emergence of new ideas and methodologies, which helped in a better interpretation of data previously restricted to only the description of developmental aspects.

As new information on embryology accumulated, the need for a comprehensive, illustrated book on comparative embryology that could serve researchers as a reference work became imperative.

In 1979, when Dr. Klaus Peters of Springer-Verlag visited BMJ with an invitation to prepare a treatise on *Embryology of Angiosperms*, the book, *Experimental Embryology of Vascular Plants* (ed. B.M. Johri), was already in its final stage and was published by Springer-Verlag in 1982. The multiauthored book *Embryology of Angiosperms* (ed. B.M. Johri) was published by Springer-Verlag in 1984, and the outline of the book, *Comparative Embryology of Angiosperms*, was prepared simultaneously and given to Klaus Peters. Although a very large collection of relevant literature was available to us at the Department of Botany, University of Delhi, it took almost 10 years to survey the exhaustive literature and to complete the manuscript on Comparative Embryology.

This book endeavours to consolidate the scattered information on angiosperm embryology, to present a coordinated and coherent interpretation of significant data accumulated thus far, and to elucidate the interrelationship of taxa based on comparative embryological findings.

Comparative Embryology of Angiosperms is a review of the developmental processes leading to sexual reproduction in flowering plants, and provides illustrated source material on sexual reproduction, primitive and advanced embryological characters, and phylogenetic relationships. On the basis of embryological data and certain evidence from other areas of study, special emphasis has been placed on the relationship among and within the families and orders of angiosperms. For our work Engler's classification system for angiosperms proposed in *Syllabus der Pflanzenfamilien*, revised by Melchior (1964), has been adopted.

This treatise summarizes nearly 5000 publications (up to 1991) on angiosperm embryology; its resulting ultimate size prompted us to divide it into two volumes. For the convenience of the reader, the preface, acknowledgements and the table of contents are presented in both volumes. For economic reasons, the references for the entire work and the plant index are included only in Volume 2.

Each order begins with general features, followed by the embryology of its constituent families, culminating in critical taxonomic considerations on the basis of comparative embryologic data, and the placement of each family in six chosen systems of classification is discussed.

A comprehensive bibliography is given at the end of each family and includes the citations referred to in the text. Additional citations (not mentioned in the text) are provided for those who wish to obtain further information.

When appropriate, inaccuracies in observation and interpretation are pointed out, alternative interpretations offered, gaps in our knowledge highlighted, and prospects outlined.

Comparative Embryology of Angiosperms provides useful information to students and research workers involved in angiosperm embryology, taxonomists, plant breeders, agriculturists and horticulturists, and plant scientists interested in the evolutionary tendencies in flowering plants.

University of Delhi, July 1992

B.M. Johri
K.B. Ambegaokar
P.S. Srivastava

Acknowledgements

The preparation of *Comparative Embryology of Angiosperms* took almost a decade. We could not have accomplished this undertaking without the academic support and encouragement from our colleagues at the University of Delhi, and plant scientists in India and abroad. Over the last 55 years, B.M. Johri has had the privilege of discussions with embryologists on the reproductive biology of angiosperms, especially at national and international conferences. He is grateful to all of them.

We are indebted to fellow embryologists who generously provided relevant literature and advance information about ongoing research projects. To name a few: Prof. Martin Bopp (Heidelberg), Dr. F. Bouman and Dr. W. Verkerke (Amsterdam), the late Dr. R. Dahlgren (Copenhagen), Dr. M.I. Hakki (Berlin), Prof. U. Hamann (Bochum), Dr. R. Wunderlich (Vienna), Prof. M. Yakovlev and Prof. T. Batygina (Leningrad), Dr. H. Tobe (Chiba, Japan), Dr. P.H. Raven (St. Louis, USA), Prof. A.E. Cocucci (Cordoba, Argentina), the late Dr. Thanikoimani (Pondichery, India), Prof. K.K. Lakshmanan (Coimbatore, India), Dr. M.P. Nayar and Dr. B.D. Sharma (Botanical Survey of India, Calcutta), Prof. S. Chanda (Bose Institute, Calcutta) and our colleagues (at the University of Delhi) Prof. H.Y. Mohan Ram, Prof. S.C. Maheshwari, Prof. R.N. Kapil, Prof. N.N. Bhandari and Dr. Ashok Bhatnagar.

For advance information regarding the recently published (1989) Dedicatory Volume in honour of Prof. A. Lebégue, *Some Aspects and Orientations in Plant Embryology*, we wish to thank the editors, Prof. J. Paré and Prof. M. Bugnicourt (University of Picardie, Amiens, France).

For the translation of Russian literature we are obliged to Dr. Balram Sharma (Indian Agric. Res. Inst., New Delhi) and Dr. (Mrs.) Akhilesh Tiwari (University of Delhi); and to Dr. (Mrs.) Sheela Srivastava (University of Delhi) for German literature. Dr. M.A. Rau (Mysore) very kindly prepared a number of schematic figures on the development of endosperm in the Sympetalae. Dr. F. Bouman provided scanning electron micrographs of seeds which appear in this volume. Thanks are also extended to Mr. Dev Raj Khosla (Indian Natl. Sci. Acad., New Delhi) who prepared photocopies of literature and illustrations, so essential for checking the manuscript. Mr. S.B. Ambegaokar (Delhi) was kind enough to compare over 1800 pages of the final manuscript with the earlier draft. Dr. (Mrs.) Kapoor ably assisted us in the collection of literature and indexing the Contents and References. Dr. Kapoor and Dr. Manoj Kumar (University of Delhi) also helped in the preparation of plates.

We express our appreciation to Mr. G. A. Shastri and Ms. Sushil Kaur for easy access to the University Library, and to Dr. (Mrs.) Vatsala (Publ. Direc., CSIR, New Delhi), Dr. (Mrs.) Nina Dhiman and Ms. Binny Mathur (Editorial Assistants, Phytomorphology), and our research scholars Mr. N. Syed Abbas Ali and Ms. Babeeta Chrungu for their help in checking the periodicals. Mr. Krishan Lal, artist (University of Delhi), very competently designed and prepared several figures, and labelled all the illustrations. Mr. D. Daniel and Mr. S.K. Das (University of Delhi) attended to photographic work with professional competence.

We are indebted to Dr. Dieter Czeschlik, Life Science Editor, Springer-Verlag, Heidelberg, for his immense patience in waiting to see the completion of this work. Moreover, we would like to thank Mrs. Jean von dem Bussche for copyediting "Sympetalae" (pp. 615–829) in 1987; Dr. G. Forbath, Copy Editor, Ms. Carola Wissmeier, Secretary to Dr. D. Czeschlik, and Ms. Renate Münzenmayer, Production Editor, Springer-Verlag, for the meticulous handling of this publication. For the high quality of printing by MacMillan India, Bangalore, we have great appreciation.

We are obliged to *Janab* Hakim Abdul Hameed *Sahab* (Chancellor), Prof. M. Amin (Vice Chancellor) and Prof. Muhammed Iqbal (Dean, Faculty of Science) of the Hamdard University, New Delhi, for their interest and encouragement in this project.

Thanks are also due to Ms. Madhumati, Mr. Aniruddha, Ms. Deepshikha, and Mr. Deepak, research scholars at the Department of Botany, Hamdard University for cross-checking the text and references in the final proofs, and for preparing plant indices.

We would like to make special mention of Mr. R.K. Gupta (University of Delhi) for expert secretarial assistance, and typing twice the manuscript of about 2000 pages, and Mrs. Krishna Thakur (University of Delhi) for periodical secretarial assistance. Mr. Sukhbir Singh and Mr. Chandra Pal (University of Delhi) have been very helpful in attending to the day-to-day routine work during the preparation of this book.

The Indian National Science Academy, New Delhi, and Springer-Verlag, Heidelberg, provided partial financial support for this book.

We also wish to record gratitude to Mrs. Raj Johri (wife of B.M. Johri) and Dr. (Mrs.) Sheela Srivastava (wife of P.S. Srivastava) for their support, forbearance and interest during the long periods when *Comparative Embryology of Angiosperms* was in preparation. Dr. Lalit M.

Johri (son) and Mrs. Meera Johri (daughter-in-law) keenly followed the progress of this book and encouraged an expeditious completion. Once again, many thanks to all those who helped us in various ways.

May 28, 1992

B.M. Johri
K.B. Ambegaokar
P.S. Srivastava

Contents Volume 1

References see Vol. 2 p. 1013
Plant Index see Vol. 2 p. 1174

Contents Volume 2

II Dicotyledons: Sympetalae

III Monocotyledons

1 The Background

The history of "embryology" dates back to the age-old tradition of Arabs and Assyrians of dusting the male inflorescence of the date palm on to the female palm to obtain a good supply of dates. This religious ceremony indicated some sort of sex-blending. But, until the experiments of Camerarius (1694), no attention was paid to sexuality in plants. The only efforts till then were of Grew (1682) who treated the stamens as the male organs (of flowers) which induced the ovary to produce the fruit.

Camerarius (1694) observed that in mulberry, *Mercurialis annua*, *Ricinus* and *Zea mays*, the anthers are essential for preparing a young plant in seeds contained in the ovary. He was not aware of the exact role played by the anthers. His work was later confirmed by Kölreuter (1761) who gave a detailed account of the role of insects in pollination. While working on the interaction of pollen grains with the stigma of flowers, he was able to ascertain the specificity of pollen grains. He produced hybrids in *Dianthus*, *Hyoscyamus*, *Matthiola* and *Nicotiana*.

Meanwhile, the discovery of the microscope provided an insight into the then unknown world of microorganisms. The excited microscopists focussed their attention on whatever object they came across. The Italian mathematician, astronomer and good microscopist, Amici (1824), observed the splitting of a pollen grain of *Portulaca*, emergence of pollen tube, and entry of the pollen tube into the stigma. But he could not ascertain the fate of the pollen tube. Brogniart (1827) also examined several pollinated pistils. He observed the pollen tubes (spermatic tubules) but imagined that they burst and discharged the "spermatic granules". He also visualized the passing of spermatic granules through the style, and their reaching the placenta, as spermatozoids do in animals. Amici (1830), still doubtful about the process, continued his studies on *Portulaca oleracea*, *Hibiscus syriacus*, etc., and discovered that the pollen tube itself elongates and reaches the ovule.

The false notion of Schleiden (1837) – the propounder of the "cell theory" with Schwann (1838, 1839; see EJ Gardner 1978) – caused great confusion. According to him, the extreme tip of the pollen tube grows into a vesicle and then develops into an embryo. The embryo sac acts simply as an "incubator". However, Amici (1844, 1847), in spite of strong opposition, established the fact that it is not the tip of the pollen tube that develops into an embryo, but a "germinal vesicle" already present in the embryo sac in the ovule. The germinal vesicle becomes "fertilized" by the fluid from the pollen tube and develops into an embryo.

The well-illustrated works of Hofmeister (1849), finally confirmed Amici's observations and settled the relationship between pollen tube and the egg. He (Hofmeister 1847–1861) also provided detailed information on the formation of tetrads, organization of embryo sac, and development of the embryo in several genera. Hanstein (1870) traced the sequence of early cell divisions in the embryogeny of *Capsella* and *Alisma.*

Following Hofmeister, Strasburger (1879) presented a comprehensive account of development of the Polygonum type of embryo sac, its organization, and formation of secondary (fused polar nuclei) nucleus in *Polygonum divaricatum.* His most significant contribution is the discovery of syngamy (Strasburger 1884). He could illustrate the two male gametes in a pollen tube, their discharge in the embryo sac, and fusion of one of the male gametes with the egg cell (syngamy) in *Monotropa* and other plants. The fate of the other male gamete, its fusion with the two polar nuclei (triple fusion), was discovered by Nawaschin (1898). Polyembryony was also reported (Ernst 1918, Schnarf 1929), and the concept of "apomixis" had been put forward by Juel (1898, 1900) and Murbeck (1901).

By the end of nineteenth century, the major facts about the development of male and female gametophytes, endosperm and embryo became known (see Coulter and Chamberlain 1903).

2 Diagnostic Characters

A number of diagnostic features facilitate the study of comparative embryology, and are also significant for taxonomic considerations. These characters are mentioned below (see Tobe 1989).

Anther. Number of sporangia, developmental type of anther wall, endothecium fibrous or non-fibrous, tapetum secretory (glandular) or amoeboid (intrusive, invasive, plasmodial), cytokinesis simultaneous or successive, quadripartition by cell-plate formation or furrowing, tetrads, polyads, microspores, pollen grains shed singly, or clumped in dyads, tetrads, pollinia or massulae, at two- or three-celled stage, shape, ornamentation of exine, germinal furrow, pores and allied structures, special type of pollen as in Cyperaceae, Epacridaceae and Loranthaceae.

Ovule. Vascularized or non-vascularized, ortho-, hemiana-, amphi-, campylo-, ana-, circino-tropous or incompletely differentiated (reduced), ategmic, uni- or bitegmic, micropyle formed by one or both integuments, endothelium present or absent, tenui- or crassinucellate, formation of nucellar cap, operculum, epistase, hypostase, and ovular appendages.

Megasporogenesis. Reduction divisions, tetrad, dyad or coenomegaspore. If coenomegaspore, distribution of nuclei 2 + 2, 1 + 1 + 1 + 1, 1 + 3, and 1 + 2 + 1 (see Sects. 3.5, 3.6).

Female Gametophyte. Monosporic: Polygonum or Oenothera type; bisporic: Allium or Endymion type; tetrasporic: Adoxa, Plumbago, Penaea, Peperomia, Drusa, Plumbagella or Fritillaria type, variations in embryo sac development in the same taxon, micropylar, lateral or chalazal extension, embryo sac haustoria, synergid/s ephemeral or persistent, synergid haustoria, number of antipodals, ephemeral or persistent, uni- or multinucleate, antipodal haustoria, special features.

Pollination. Mode of pollination: anemophily, hydrophily, entomophily; structural adaptation of pollen and stigma, in two-celled pollen grain division of generative cell in pollen tube, on stigma or in the style, time lag between pollination and double fertilization, special mode of pollination as in *Ficus* and *Yucca*.

Fertilization. Entry of pollen tube in embryo sac: porogamous, mesogamous or chalazogamous; discharge of gametes, syngamy, triple fusion or fusion of sperm with secondary nucleus, persistent remnants of pollen tube, entry of more than one pollen tube and polyspermy, and fertilization of one or both synergids.

Endosperm. Nuclear, Helobial or Cellular type; stage of proembryo at wall formation in Nuclear (and Helobial) type, micropylar and chalazal haustoria, one or both aggressive, extension into circumjacent tissue, number of nuclei and nuclear behaviour, persistence of haustoria in developing seed, wall formation in haustoria, development of endosperm proper, reserve food, rumination due to integument/s or pericarp, special behaviour of peripheral layers of endosperm in consuming seed-coat and even pericarp (partially or completely).

Embryo. Stage of endosperm at division of zygote, quadrant, octant, derivation of embryonal organs, Crucifer, Asterad, Solanad, Caryophyllad, Chenopodiad and Piperad type of embryogeny, fully differentiated and organless (undifferentiated) embryo, delayed differentiation, suspensor absent, short or massive, uni- or multinucleate, ephemeral or persistent, suspensor haustoria, special feature/s.

Polyembryony. Cleavage and/or budding of zygotic embryonal mass, proliferation of suspensor, accessory embryo/s from haploid or diploid (fertilized) synergid, somatic (diploid) embryo from nucellar or integumentary cells, developmental pattern of accessory embryos.

Apomixis. Diplospory, apospory, auto- or pseudogamous development of endosperm, parthenogenetic development of egg cell in reduced and unreduced embryo sac, obligate and facultative apomict.

Seed-coat. Unspecialized, exo-, meso- or endotestal, exo- or endotegmic, testal-tegmic, "naked" seeds (without seed-coat).

Seed. Exalbuminous or albuminous, food reserve in endosperm, perisperm, and/or embryo, nature of food reserve, intraseminal leaf primordia, cotyledons equal or unequal, fleshy or thin, foliaceous, folded or convolute, embryo chloro- or non-chlorophyllous; appendages on seed; hairs, wing, aril, elaiosome.

When more than one type of development occurs in the same taxon, it is necessary to calculate their relative percentage. For example, on the basis of counts of several microscopic fields, mention the percentage of tetrahedral, isobilateral, decussate, linear and T-shaped microspore tetrads, types of embryo sac in ovules, and types of endosperm and embryogeny in developing seed (see Schmid 1982, Tobe and Raven 1987c).

The first comprehensive account of comparative embryology of angiosperms is by Schnarf (1931), followed by Davis (1966), B.M. Johri (1970a), Yakovlev (1981, 1983), Poddubnaya-Arnoldi (1982), and Batygina and Yakovlev (1985, 1987). The embryogeny has been discussed by Johansen (1950), and seed anatomy by Netolitzky (1926), Corner (1976) and Takhtajan (1985, 1988). Several examples of the role of embryological characters in taxonomic considerations have been cited by P. Maheshwari (1950, 1956, 1963b, 1964), B.M. Johri (1963b, 1967), Rau (1970), Palser (1975), Herr (1984), and Tobe and Raven (1983a, b, 1984b, c, d, e, f, 1987a, b, c).

R. Dahlgren (1975b, 1977a, 1983a) and R. Dahlgren and Clifford (1982) have designed a series of diagrams for comparative representation of several embryological characters in various families.

3 Developmental Aspects

3.1 Anther

A typical anther is tetrasporangiate with two microsporangia in each lobe (see Bhandari 1984). When transverse septa develop in each microsporangium (Sect. 18.6), it becomes polysporangiate, as in the Annonaceae. In the Onagraceae (Tobe and Raven 1986a, b), transverse as well as longitudinal septa develop in each microsporangium (Sect. 39.13). At dehiscence, the septa break down in each lobe and the mature anther becomes bilocular (see Engler 1876). In *Ficus ottoniifolia* (Verkerke 1986a), each of the four sporangia dehisces independently (Sect. 10.4).

In some taxa, as in Malvaceae (Sect. 35.5), the anthers are bisporangiate. *Arceuthobium minutissimum* (Bhandari and Nanda 1968a) shows a horseshoe-shaped archesporium around a central columella (Sect. 12.8).

Wall. The young anther is a homogeneous mass of parenchymatous cells covered by a distinct epidermis. It becomes four-lobed, and a single or a group of two or more hypodermal archesporial cells differentiates in each lobe. The archesporial cell/s divides periclinally separating the primary parietal layer (below the epidermis) and the primary sporogenous tissue. The latter may undergo mitotic divisions resulting in secondary sporogenous tissue. The primary parietal layer, by further periclinal divisions, gives rise to the anther wall (see Periasamy and Swamy 1964).

According to Bhandari and Khosla (1982), in *Triticale* the primary parietal layer divides periclinally and produces the endothecium, middle layer, and secretory tapetum. These three layers are concentric and surround the sporogenous tissue. The connective tissue does not contribute to the development of inner tapetum.

The middle layers are generally rich in reserve food. Their number varies, and may increase up to four or five layers, as in *Daphniphyllum himalayense* (A.K. Bhatnagar and Garg 1977) and *Nelumbo nucifera* (S.C. Gupta and Ahluwalia 1979). They are usually resorbed during development, but some of them may persist or develop fibrous thickenings as in members of the Liliaceae and Zingiberaceae.

On the basis of the formation of middle layers, Davis (1966) classifies the development of anther wall into four types: Basic, Dicotyledonous, Monocotyledonous and Reduced.

Basic Type. The primary parietal layer divides periclinally and produces two secondary parietal layers. Both the layers undergo further periclinal divisions, and the one towards the epidermis differentiates into endothecium followed by a middle layer. The other secondary parietal layer produces another middle layer, and the tapetum. The middle layers may undergo further divisions. Examples: Anacardiaceae, Lecythidaceae, Rhamnaceae, Vitaceae.

Dicotyledonous Type. Comparable to the "Basic" type, the outer secondary parietal layer produces the endothecium and middle layer; the inner layer functions directly as tapetum. Examples: Acanthaceae, Aristolochiaceae, Bignoniaceae, Convolvulaceae.

Monocotyledonous Type. The outer secondary parietal layer develops directly into endothecium, while the inner gives rise to the middle layer and tapetum. Examples: Alismataceae, Amaranthaceae, Aponogetonaceae, Cactaceae, Chenopodiaceae, Cruciferae.

Reduced Type. The primary parietal layer, by a periclinal division, produces the endothecium and tapetum. The middle layer is absent. "Reduced" type of development is reported in *Antigonon leptopus* (Agoram and Krishnamurthy 1980) and *Gaultheria* spp. (Chou 1952). Davis (1966) refers this type to the members of Lemnaceae and Najadaceae. B.L. Gupta (1935) confirms the absence of a middle layer in *Wolffia arrhiza*, S.C. Maheshwari (1954) in *W. microscopica*, Lawalrée (1952) in *Lemna minor*, and S.C. Maheshwari and Kapil (1963a) in *L. paucicostata* (Fig. 61.3.1G–L, N). Lodkina's (1977) report of a middle layer in *Lemna gibba*, *L. paucicostata*, *L. trisulca*, *Najas flexillus*, *N. graminea*, *N. marina* and *Wolffia microscopica* requires confirmation.

According to Davis (1966), the Basic type occurs in nine families, while the Dicotyledonous type in the majority of dicotyledonous families and one monocotyledonous family, Taccaceae. The Monocotyledonous type includes the majority of monocotyledonous families as well as several dicotyledonous families. Two different types of wall formation occur in only a few families: Combretaceae, Euphorbiaceae, Sterculiaceae and Thymelaeaceae. Davis (1966) mentions that variation occurs in development in the same species (*Stellaria media*, Caryophyllaceae). A.K. Bhatnagar and Kapil (1979) report Basic, Dicotyledonous and Monocotyledonous types in *Bischofia javanica*.

Varghese and Chowdhury (1972) question the validity of the terms suggested by Davis. They point out that these terms have only a descriptive value, and serve a limited purpose. In our opinion, judging from the variations in the development of anther wall, it does not serve any useful purpose to stick to the terminology of Davis.

The epidermal layer undergoes only anticlinal divisions, covers the inner growing tissue, and its function is protective. The epidermal cells may develop a cuticle on the outer tangential wall, or may contain tannin as in *Hibbertia* (Sastri 1958b). They develop fibrous thickenings in some members of Dilleniaceae and Zingiberaceae. The tapetum-like behaviour of the epidermal cells (Sect. 65.2) in *Zeuxine longilabris* (Karanth et al. 1979) is unique, and needs further study.

Endothecium. Closer inwards to the epidermis lies the endothecium, which is responsible for the dehiscence of the anther (see Chauhan 1979, Gerendy and French 1988). Its cells enlarge and develop callose thickenings on their inner tangential walls. These radiate outwards at the time of maturation of anther. The thickenings are usually absent in the region of dehiscence of the anther in cleistogamous flowers, and in some members of Dipterocarpaceae, Ericaceae and Hydrocharitaceae. Rarely, the endothecial thickenings extend on the dorsal side into the connective as in *Bupleurum tenue* (S.C. Gupta and M. Gupta 1964), and they extend and unite (in the region of the connective) surrounding the staminal bundle, as in *Eucommia ulmoides* (Eckardt 1963). Occasionally, the endothecium becomes biseriate. It is five or six layers thick in *Nicotiana tabacum* (Jose and Singh 1968) and *Voacanga foetida* (Maheswari Devi 1971).

From cleared anthers, sections and macerations, French (1985a, b) examined endothecial thickenings in Araceae, 106 representative species of 20 genera in Pathoideae and Monsteroideae, and representative species of 37 genera of Calloideae, Lasioideae and Philodendroideae. The wall thickenings in the endothecium are based on an annular–helical pattern. Variations in the thickenings are related to differences in cell shape, cell orientation, intergradation, presence of branched and reticulate thickenings, and various types of discontinuous thickenings. The pattern of wall thickenings can be used as a basis of classification of the Araceae.

The absence of endothecial thickenings is a derived condition often correlated with porate dehiscence (Davis 1966). However, Araceae, Philesiaceae and Rapateaceae have porate dehiscence and also endothecial thickenings (R. Dahlgren and Clifford 1982, see also Greenday and French 1988). Families with nonporate dehiscence and which lack endothecial thickenings include Hydrocharitaceae (Ernst-Schwarzenbach 1945), *Musa errans* (Juliano and Alcala 1933), *Oryza* (Juliano and Aldama 1937), and *Xyris* (see French 1985a). Untawale and Bhasin (1973) report annular and helical thickenings in *Musa.* Endothecial thickenings are also absent in some cleistogamous flowers as in *Amphicarpum* (Ritzrow 1908); in *Colocasia* (Pothoideae), *Cercestis* and *Rhektophyllum* (Lasioideae), and *Philodendron* (Philodendroideae) of the Araceae (French 1985b).

French (1986a) made a further study of endothecial thickenings in representative species from 43 genera of Colocasioideae, Aroideae and Pistioideae. All endothecial cells are anticlinally elongated with bars or reticulate thickenings on anticlinal walls and variably discontinuous thickenings on periclinal walls. Several genera show unusual and distinctive patterns. In *Stylochiton* (Aroideae), the endothecial cells are periclinally elongated and exhibit continuous helices and rings. Other genera with a similar pattern occur in Monsteroideae and Lasioideae. *Acorus* (Araceae) exhibits a unique pattern of stellate thickenings.

Endothecial thickenings in mature anther (ready to dehisce) should be studied from suitably stained dissected wholemounts and sections in various planes.

Tapetum. The tapetum is the innermost layer of the anther wall (see Warmke and Overman 1972) and originates from the derivatives of the primary parietal layer. Presumably, it differentiates from the sterilized sporogenous tissue in *Potamogeton crispus* (B.L. Gupta 1934). Lakshmanan and Rajeshwari (1979) report the origin of tapetum from peripheral sporogenous layers in *Syringodium isoetifolium.* Its dual origin has been observed in *Tarenna asiatica* (Periasamy and Parmeswaran 1965). Towards the outer side the tapetum is derived from the primary parietal layer, whereas on the inner side it develops from the cells of the connective. According to Periasamy and Swamy (1966), such a development occurs in all the angiosperms.

In *Annona squamosa* (Periasamy and Kandaswamy 1981), the origin of the tapetum is trimorphic. The archesporial tissue is uniseriate and alternate archesporial cells develop into sterile septa so that the microsporangium appears polysporangiate. The outer tapetum is derived from the primary parietal layer, the inner from the connective, and both are secretory. The septa in the anther locule are formed due to the sterility of the archesporial cells and develop into plasmodial tapetum (Fig. 18.6.1C–J).

Pacini et al. (1985) discuss the "form, function and possible phylogeny" of the tapetum in the Embryophyta.

The tapetum is mainly a nourishing tissue. In a young anther, the tapetal cells remain in continuity with the microspore mother cells through plasmodesmata through which the nutrients are provided to the latter, till the formation of meiocytes. These connections are plugged with callose thickenings, as soon as meiosis is completed.

The tapetal cells are uninucleate in the beginning, but in the majority of plants mitotic divisions of their nuclei are not accompanied by cytokinesis (see Cooper 1933). Thus, they become two- to four-nucleate or, at times, the number goes up to 16. Very often, due to fusion the nuclei become polyploid. Polyploidy may also be attained due to endomitosis, polyteny, or by formation of restitution nuclei (Figs 3.1.1, 19.2.1).

The tapetum may become irregularly biseriate, and in *Calotropis* it is multilayered. The tapetum is designated as amoeboid (invasive, intrusive, plasmodial) or secretory (glandular), according to its behaviour. The nature of tapetum is a diagnostic feature in the majority of families. However, both types may occur in the same family, in different genera. Rarely, in *Anemonopsis macrophylla* (Tobe 1974b), both types occur in different anthers of the same flower. The appearance of perforations in the tapetal cells of the anther is reported in *Acalypha cilliata* (P.K. Mukherjee 1964).

Amoeboid Type. The walls of tapetal cells break down, protoplasts enlarge and enter the anther sac, and by their fusion periplasmodium is formed around the microspore mother cells. The microspore mother cells, or microspores, or two-celled pollen grains remain embedded in the periplasmodium (Fig. 52.3.1A–E). Periplasmodial formation occurs at different developmental stages of microspores in different species. At the tetrad stage, the tapetal cells produce an enzyme callase which disorganizes the callose walls around the microspore tetrads and the microspores are released in the periplasmodium. The plasmodium is gradually absorbed by the developing pollen grains. In *Rhoeo spathacea* (Nanda and Gupta 1977), an extra-tapetal membrane develops on the outside of the periplasmodium, next to endothecium, and encloses the microspores and periplasmodium. Nanda and Gupta also report raphides in the periplasmodium. Later on the membrane breaks down, and the mature pollen grains are released. The development of an "extra-tapetal" membrane has been previously reported in some members of Compositae (J. Heslop-Harrison 1969, see also Gupta and Nanda 1972).

Pacini and Juniper (1983) studied ultrastructural details of the development of amoeboid tapetum in *Arum italicum*. The anther wall comprises the epidermis, fibrous endothecium, three middle layers, and two-layered tapetum. At the premeiotic phase of microspore mother cells, the tapetal cells are connected with each other by plasmodesmata, but these are absent in the tapetal walls facing the middle layer and the microspore mother cell (Fig. 3.1.2A). During the meiotic divisions of mother cells, the tapetum shows conspicuous changes. At leptotene (Fig. 3.1.2B), the plasmodesmata in the tapetal cells widen to form cytomictic channels; the tangential walls disappear at pachytene (Fig. 3.1.2C). During meiosis I, anaphase (Fig. 3.1.2D), the radial walls of tapetal cells also disappear and microtubules lie parallel to the inner radial walls of mother cells. The tapetal cytoplasm starts to intrude between the meiocytes; microtubules appear at that region (Fig. 3.1.2E).

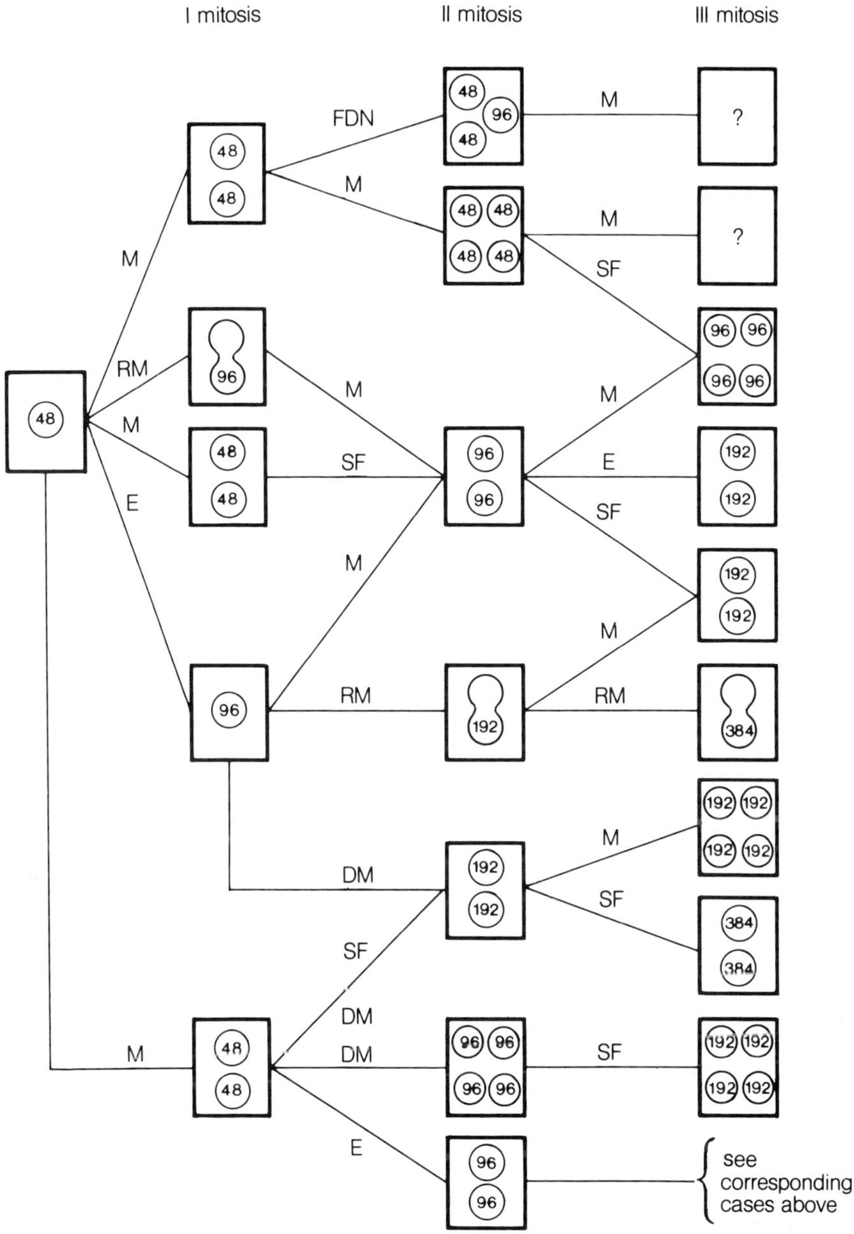

Fig. 3.1.1. Schematic representation of the methods of polyploidization in the cells of the internal anther tapetum of *Solanum tuberosum* (2n = 48). *DM* diplochromosome mitosis; *E* endomitosis; *FDN* fusion of daughter nuclei; *M* mitosis; *RM* restitutional mitosis; *SF* spindle fusion; *SF/DM* Spindle fusion in diplochromosome mitosis. (Adapted from Avanzi 1950, see D'Amato 1984)

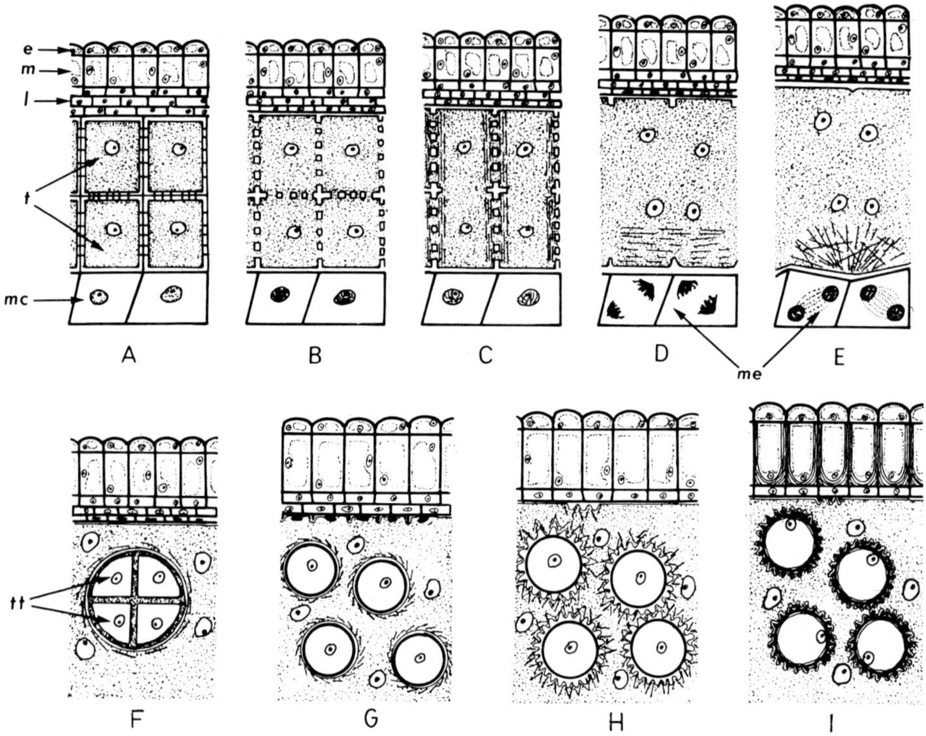

Fig. 3.1.2A–I. *Arum italicum*, amoeboid tapetum (semi-diagrammatic). **A** Premeiotic phase; the tapetal cells (*t*) are connected by plasmodesmata which are absent in the tapetal walls facing the middle layer (*l*) and microspore mother cells (*mc*); *e* epidermis; *m* mechanical layer/endothecium, **B** Leptotene (microspore mother cells), plasmodesmata widen. **C** Pachytene; tangential walls between tapetal cells have disappeared. Note microtubules (*thin lines*) along radial walls of tapetal cells. **D** Meiosis (*me*) I, anaphase; radial walls of tapetal cells have also disappeared. Microtubules lie parallel to the inner radial wall of meiocytes. **E** Meiosis I, telophase. Tapetal cytoplasm starts to intrude between meiocytes and at this site many microtubules appear in random orientation. **F** Tetrad (*tt*) embedded in multinucleate periplasmodium. Microtubules surround the tetrad. **G** Microspore (in periplasmodium) surrounded by microtubules; *dark line* exine. **H** Microspores, later stage. The tapetal plasmalemma facing microspores retracts leaving cone-shaped spaces; this retraction also occurs at the periphery of tapetum but is discontinuous. Microtubules are no longer parallel to exine surface. **I** Uninucleate pollen grain with intine and exine. Two middle layers are absorbed in the periplasmodium (**F–H**), thickenings appear in endothecium (**I**). (After Pacini and Juniper 1983)

Figure 3.1.2D and E show the formation of multinucleate periplasmodium in which the tetrads, microspores and uninucleate pollen grains are embedded (Fig. 3.1.2F–I). Microtubules surround the tetrad facing the callose wall and the microspore. The tapetal plasmalemma facing the microspores retracts from the exine surface leaving roughly cone-shaped spaces (Fig. 3.1.2H). Microtubules no longer lie parallel to the exine surface. By the localized retraction of the plasmamembrane, the spaces left on the microspore surface and tapetal periphery become filled with a material forming "spines" (Fig. 3.1.2I); the intine also develops.

The amoeboid type mostly occurs in the monocotyledons.

Secretory Type. The secretory or glandular tapetum is characteristic of most of the dicotyledons. The tapetal cells remain intact in situ till the maturation of pollen grains, and provide nutrition. A tapetal membrane develops on the inner tangential surface of tapetal cells (S.C. Gupta and Ahluwalia 1979, see also Shivanna et al. 1979). The tapetal cells develop pro-orbicules. The sporopollenin, a substance resistant to chemical decomposition, is produced in the mitochondria of tapetal cells (J. Heslop-Harrison 1963). The sporopollenin is deposited on the pro-orbicules. As the microspores are released in the anther locule, the inner periclinal and radial walls of the tapetal cells dissolve. According to Bhandari and Ram Kishori (1973), at the tetrad stage the protoplasts of the tapetal cells develop a membrane in *Nigella damascena*, and the orbicules become studded onto this membrane. Later, they are deposited on the tectum to form the exine-sculpture of pollen grains. Orel et al. (1977) describe the development of a sporopollenin film within the tapetal cells during the formation of microspore tetrads in five species of the Gramineae, and beet, potato and tomato.

The tapetum synthesizes certain substances which are released into the anther locule during maturation of pollen grains. These substances contain fibro-granular proteins and carotinoid-containing lipid droplets. The fibro-granular proteins enter the exine of pollen grain, and are stored in the region of bacula. These wall-bound proteins are extra-cellular and capable of recognizing the compatibility of stigma. The carotinoid-containing lipid granules spread over the exine and help in pollination. They are described as "pollen-kitt". The tapetum is a physiologically active layer of the anther wall (see Hesse 1981, Knox 1984, Shivanna and Johri 1985, Pacini et al. 1985).

The malfunctioning of tapetal cells has an adverse effect on the development of the endothecium and pollen fertility (Nanda and Gupta 1974, Vijayaraghavan and Ratnaparkhi 1979, see also Warmke and Overman 1972). The tapetal cells cause sterility of pollen grains when they become overstretched and form a persistent band around the pollen grains as in *Allium cepa* (Nanda and Gupta 1974) and *Belamcanda chinensis* (S.C. Gupta and Nanda 1973). The endothecial cells also remain small, and do not develop fibrous thickenings. The accumulation of a large quantity of micronutrients in the tapetum also affects the development of pollen grains. The malfunctioning and hypertrophied condition of tapetal cells exhibits callase activity which leads to the failure of formation of callose walls around the tetrads.

3.2 Microsporogenesis

Sporogenous Tissue. The sporogenous tissue may consist of a single vertical row of hypodermal cells as in some members of Cactaceae, Compositae, Cucurbitaceae and Malvaceae, or may undergo mitotic divisions to produce several rows. Each cell of the archesporium is potentially capable of developing into spores, but sometimes several cells take up the nourishing function and become sterile (see Dickinson 1982). The development of trabuculate partitions dividing the sporogenous tissue into groups occurs in *Hydrocera triflora* (Venkateswarlu and Laxminarayan 1957). In some members of Annonaceae, and in *Myristica malabarica* (N.C. Nair 1972), the microsporangium is septate. In *Cananga* (only in a few anthers), the partitions develop due to sterilization of sporogenous tissue, whereas in *Annona* only one microspore mother cell occurs in each compartment and the septa develop tapetum. *Goniothalamus* shows multilayered transverse septa, and each microspore mother cell has its own tapetum (Parulekar 1970). Several sterile transverse septa develop between sporogenous cells of *Parkia biglandulosa* (Dnyansagar 1954b). Each compartment contains several microspore mother cells nourished by its own tapetum.

Septate microsporangia also occur in Balsaminaceae, Convolvulaceae, Gentianaceae, Plumbaginaceae, Rhizophoraceae (Lersten 1971), Flacourtiaceae, Myrsinaceae and Sterculiaceae (Endress and Vosser 1975), and Onagraceae (Tobe and Raven 1986a).

An abnormal behaviour of sporogenous cells (in an anther) is reported in *Leptomeria billardierii* (Ram 1959c). Some of the cells enlarge, the nucleus divides, the daughter nuclei divide again, and again for the third time. Thus, eight nuclei are produced, and these organize more or less like an embryo sac. This feature has not been reported so far in any other angiosperm (see B.M. Johri and Ambegaokar 1984).

Microspore Mother Cells. The microspore mother cells are usually uninucleate. However, occasionally, polynucleate condition has been observed in *Santalum obtusifolium* (S.P. Bhatnagar 1966), triticales (Kempanna et al. 1971) and in *Secale-Agropyrum* hybrids (Poddubnaya-Arnoldi and Makhaline 1969). Perhaps, this condition is due to the absence of cytokinesis. The microspore mother cells are interconnected, and are also connected with the tapetal cells through plasmodesmata. As they enter the phase of meiotic divisions, they lose contact with the tapetal cells, become enclosed in thick callose walls, but remain interconnected through bigger channels.

Meiosis. The microspore mother cells undergo meiotic divisions. Meiosis I is a heterotypic division, and reduction occurs in the number of chromosomes. Meiosis II is a homotypic division, only mitosis occurs. Cytokinesis follows meiotic divisions of the microspore mother cells. The interconnections between the microspore mother cells become obliterated, and the tetrads become independent.

Cytomixis. The migration of nuclear material from cell to cell is designated as cytomixis (Gates 1911). It takes place at all stages of meiotic divisions in pollen mother cells. The presence of cytoplasmic links in between the meiocytes provides evidence for migration of nuclear materials. Anucleate, polyploid and aneuploid pollen mother cells are the consequence of cytomixis. In case of complete migration, the donor cell becomes empty and the recipient polyploid (see Whelan 1974).

Cytomixis in pollen mother cells has been observed throughout the year, both in fixed and fresh flower buds of *Ervatamia divaricata.* It occurs to the maximal degree from March to June; the temperature fluctuation affects cytomixis. Sometimes, most of the pollen mother cells of an anther are involved in cytomixis while the development in the adjacent anthers is normal; direct fusion of two pollen mother cells also occurs. The adjacent pollen mother cells may be linked by two or more cytoplasmic channels of varying breadth. Cytomixis is more common at prophase I but occurs almost at every stage up to tetrad formation.

The extrusion of chromatin and cytomixis in pollen mother cells affect the viability of gametes. It is associated with high pollen sterility, and usually there is no seed formation.

In *Nigella sativa* (A.K. Datta and Biswas 1984), cytomixis plays a significant role in the production of aneuploid gametes, and as a result thereof trisomic and other aneuploids may appear in the subsequent generations.

Sinha (1985) observed the migration of chromatic material in 11%–18% of cells at metaphase I in *Bacopa monnieri, Centranthera tranquobaria, Limnophila indica, Lindernia ciliata, L. parviflora, L. verbaenifolia, Mazus pumilus* and *Sutera dissecta.* There is 26%–40% cytomixis in *Dopatrium junceum, Lindenbergia indica* and *L. crustacea* at metaphase I, 30% at anaphase I and II in *Lindenbergia indica,* 15% in *L. crustacea* and 20% in *D. junceum.*

The flow or migration of chromatin material from one pollen mother cell into another is unidirectional. A pollen mother cell donates its chromatin material to one cell and might receive chromatin simultaneously from another cell, forming a chain.

In *Rauwolfia serpentina* (N. Banerjee and Sharma 1988), the frequency of cytomixis showed significant increase during monsoon months (9.62%). It was predominant in prophase I to early metaphase I. The regular occurrence of cytomixis in all the individuals of a particular population of *R. serpentina* in successive years suggests a genetic control, and its manifestation depends on the physiological state of the tissue.

Cytokinesis. Cytokinesis is either simultaneous or successive. The former is characteristic of the dicotyledons, and the latter of the monocotyledons. The occurrence of both types in different genera of the same family is also reported. However, the occurrence of both types of cytokinesis, as in *Catharanthus pusillus* (Bhasin 1971), *Conyza stricta* (H.P. Sharma and Murty 1978) and *Vernonia cinerea* (Dakshini and Dadlani 1978), is rare.

In the simultaneous type, the first meiotic division of microspore mother cell is not accompanied by wall formation. The two-nucleate condition is immediately followed by meiosis II. The callose wall intrudes through centripetal furrows, unites in the centre, and separates the microspores. Quadripartition is also brought about by the laying down of cell plates after meiosis II. In contrast, in the successive type, the first meiotic division separates two dyads, a cell plate develops between the two nuclei and extends to the sides centrifugally. Each dyad undergoes meiosis II, resulting in a tetrad. The cell plate formation takes place as after meiosis I (see Longly and Waterkeyn 1979).

Tetrads. The arrangement of microspores in a tetrad exhibits five patterns. Generally, they are tetrahedral and isobilateral. In addition, they may be linear, T-shaped and decussate. All five types are reported in *Aristolochia elegans* (Fig. 21.2.1G–M), and *Sparganium erectum* (Fig. 62.3.1H). Linear tetrads also occur in *Halophila ovata* (Kausik and Rao 1942). The microspores form uniseriate chains and the individual tetrads are separated by sterile cells (Fig. 52.4.7C). The variation in tetrad formation depends more or less on the shape of microspore mother cells (Chennaveeraiah and Mahabale 1962), and the orientation of spindles during meiosis II. In *Typha*, Sahasi and Ikuse (1974) attribute the formation of various types to rotation and sliding of the microspores.

The microspores (in a tetrad) are interconnected and surrounded by a thick callose wall. The microspores in a massula of Orchidaceae are also interconnected. The massulae of *Zeuxine strateumatica* (Vijayaraghavan and Shukla 1980) are held together by elastic, refractive viscin threads. These threads are PAS-positive and probably contain elastin, but are not autofluorescent.

In *Allium tuberosum* and *Cyclamen persicum* (Bhandari et al. 1981), the callose is deposited within the primary wall around each microspore mother cell. The primary wall persists until the late tetrad stage. In *A. tuberosum*, the primary wall dissolves simultaneously with the callose and the microspores are released. Bhandari et al. (1981) suggest that the primary wall assists the callose in preventing the entry of macromolecules, and isolating the microspores.

Polyspory. Occasionally, the microspores in a tetrad undergo further divisions and polyspory occurs producing polyads. The number of microspores goes up to eight in *Hyphaene* (Mahabale and Cheenaveeraiah 1957), 11 in *Cuscuta reflexa* (Fig. 48.4.1A–E) and 22 in *Thunbergia mysorensis* (Khaleel and Boraiah 1973). Compound pollen grains occur in several families (Fig. 48.4.1L–S). In *Brachycome ciliaris* (Davis 1964a), the compound pollen grains become four-nucleate, due to failure of cytokinesis (see B.M. Johri and Tiagi 1952).

In Podostemaceae (Sect. 28.2), both simultaneous (*Indotristicha* and *Terniola*) and successive (*Dicraea*) cytokinesis follows meiotic divisions in microspore mother cells. The microspores separate in pairs, and form "double pollen grains" in *Dicraea* (Fig. 28.2.1A), *Griffithella* (Razi 1949), *Hydrobryum*, *Podostemon* (Magnus 1913), *Angolaea*, *Mniopsis* and *Lophogyne* (Went 1910, 1912). Double pollen grains are also reported in *Bulbostylis* and *Cyperus* (C.K. Shah 1962), *Cuscuta* (Fedortshuk 1931, B.M. Johri and Tiagi 1952), *Marianthus* (Sundri and Narayana 1977) and *Strychnos* (Mohrbutter 1936). The pollen grains adhere together to form a pollinium as in some members of Asclepiadaceae (Fig. 47.7.1A, D), Mimosaceae (Fig. 26.19.1A) and Orchidaceae. The microspores change their shape and size as the pollinium rounds off.

In Orchidaceae, the microspores of a tetrad may become free from one another, or adhere together to form a compound pollen grain. In tribes

Neottieae and Ophrydeae, the compound grains adhere together to form a massula (Fig. 65.2.2L).

Due to callase activity, the callose wall around the tetrad is digested, and the microspores are released in the anther locule except in some families where they remain in permanent tetrads as in Ericaceae, Juncaceae (Fig. 55.2.1A–D), Monimiaceae, Phylidraceae (Fig. 54.18.1A–G), Winteraceae, etc.

3.3 Male Gametophyte

Two- and Three-Celled Pollen Grains. The development of male gametophyte starts soon after the uninucleate pollen grains are produced. However, overwintering may occur in the species of colder regions. Vacuolation occurs in the pollen grain and its nucleus takes up a wallward position. The mitotic division is always unequal separating a larger vegetative cell from the smaller generative cell (see Steffen 1963a, Hageman 1981). The mitotic spindle is either symmetrical or asymmetrical. It may be conical or flattened. A curved cell-plate separates the two nuclei, and extends up to the intine. The curved wall grows between the intine and plasmalemma of the generative cell. The generative cell gets separated from the wall of the pollen grain, and comes to lie in the cytoplasm of vegetative cell.

The generative cell may divide before, or after the dehiscence of anther. Meanwhile, the vegetative cell continues to grow, and accumulates reserve food in the form of starch and fat. It becomes rich in proteins, and its nucleus may undergo endoduplication and become highly polyploid. The generative cell also elongates. This is the two-celled stage of pollen grain at which shedding occurs in the majority of angiosperms.

When the first mitotic division occurs in a microspore within a tetrad, the position of the generative cell is usually predetermined due to certain genetic factors, and this feature is of taxonomic significance. The generative cell may be cut off towards the centre (proximal pole) (Fig. 54.18.1B, C), or the outer wallward side (distal pole) (Fig. 55.2.1D). The polarity can be ascertained only when the pollen grains are attached in a tetrad. As the microspores in a tetrad, or a massula, are interconnected, the mitotic divisions usually occur simultaneously.

In several species, the generative cell divides when the pollen grains are still within the anther locule, and the two male gametes are formed. The division is mitotic, and cytokinesis occurs by cell-plate formation. The male gametes are oval or spindle-shaped and invested by a sheath of cytoplasm. Then the pollen grains are liberated at the three-celled stage.

If the pollen grains are dispersed at the two-celled stage, the division of generative cell may occur after the deposition of pollen on the stigma, or in the pollen tube (Fig. 3.3.1A–C) while traversing the style or, rarely, after reaching the embryo sac.

The formation of supernumerary sperm nuclei, through the division of sperm cells, is an abnormality and is known as polyspermy.

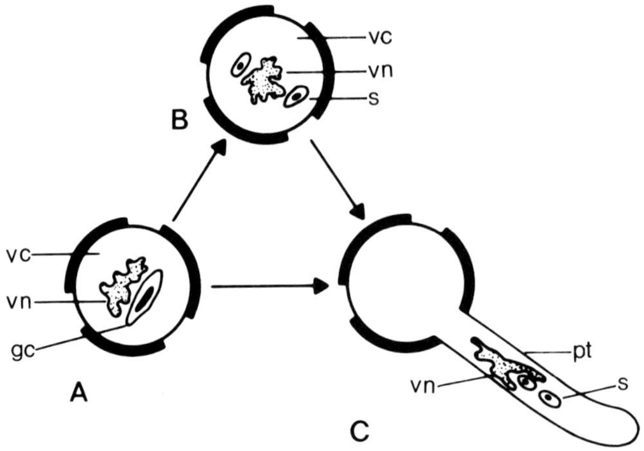

Fig. 3.3.1A,B. Bi- and tri-celled pollen grains; *vc* vegetative cell; *vn* vegetative nucleus; *gc* generative cell. *s* sperm cell. **C** Germinated pollen grain with vegetative nucleus and two sperm cells in pollen tubes (*pt*). (Adapted from Dexheimer (1970)

Usually, the pollen grains are disseminated singly, but they may remain attached in dyads (Fig. 28.2.1A), tetrads (Fig. 55.2.1A–D), pollinia (Fig. 26.19.1A), or massula (Fig. 65.2.1L).

The occurrence of supernumerary nuclei in pollen grains is not very common, e.g., in Pittosporaceae (Sundri and Narayana 1977, L.L. Narayana and Sundri 1981, 1983).

The cytoplasm of two- or three-celled pollen grain is rich in carbohydrates, lipids, proteins and amino acids (see H.G. and I. Baker 1979). It also contains enzymes for pollen germination and tube elongation. The enzymes are mostly hydrolytic; their nature depends on the type of reserve food in the pollen grains and the tissue through which the tube has to grow. The two- and three-celled pollen grains exhibit physiological differences in their behaviour (B.M. Johri and Shivanna 1977). The two-celled pollen grains remain viable for a longer time, and show a high percentage of germination in vitro, as compared to the three-celled grains.

Pollen Grains in Cyperaceae and Epacridaceae. Cyperaceae (Fig. 63.2.1A–H) present a peculiar type of pollen development. The meiotic divisions of microspore mother cells are followed by simultaneous cytokinesis in *Carex*, *Cyperus*, *Kyllinga*, *Lipocarpha* and *Scirpus*. Of the four nuclei thus formed (Fig. 63.2.1A–D), three occupy wallward position while the fourth lies in the centre. Walls develop in between the three nuclei, and the large functional microspore is separated from the rest (Fig. 63.2.1E). The three nuclei may undergo a further division, but they soon degenerate. However, the functional microspore remains enclosed in the wall of the tetrad, and is often described as pseudomonad.

Epacridaceae (Fig. 42.6.1A–J) exhibits another noteworthy example with a variety of pollen types (Smith-White 1948, 1955a, Venkata Rao 1961b). Venkata Rao (1961b) presumes that there is evidence of parallelism between the development of male and female gametophytes of angiosperms in various patterns of pollen development in the Epacridaceae. He compares the full tetrad type with all four functional microspore nuclei to the tetrasporic embryo sac, the Pentachondra type with two functional microspore nuclei to the bisporic type, and the Styphellia type to the monosporic type of embryo sac. This comparison is of no consequence.

Pollen Embryo Sacs. An abnormal behaviour of microspore nuclei is also exhibited in the formation of pollen embryo sacs. Stow (1930, 1934) observed binucleate, four-nucleate and well-developed eight-nucleate "embryo sacs" in the anthers of *Hyacinthus orientalis*. Vijayaraghavan and Ratnaparkhi (1977) report that 10% of anthers of *Heuchera micrantha* show a tendency towards the formation of pollen-embryo sacs. The elongated microspores with well-organized eight-nucleate embryo sac-like structures simulate the Polygonum, Allium or Adoxa type of female gametophyte. In addition, enlarged microspores show nuclear arrangements resembling the Penaea or Plumbago type. An increase in the number of nuclei in a microspore also occurs in *Capsicum annuum* var. *acuminatum* but they do not get organized (Dharamdhaj and Prakash 1978).

Pollen Grains in Aquatic Plants. Filiform (thread-like) pollen grains (Fig. 52.9.1D–F) occur in marine plants (sea grasses) of Hydrocharitales – *Halophila*, *Thalassia*, and Zosterales – *Amphibolis*, *Cymodocea*, *Diplanthera*, *Halodule*, *Phyllospadix*, *Posidonia*, *Syringodium* and *Zostera* (Fig. 52.8.1A–I). The microspores are arranged linearly parallel to the locule in *Diplanthera* (Lakshmanan 1968), *Halodule* (Yamashita 1976b) and *Syringodium* (Lakshmanan and Rajeshwari 1979). The microspores elongate after they are released from the tetrad, become filamentous and 1–5 mm long. The pollen grains lack a well-defined exine, and there is no specialization for shedding. These long rope-like masses are carried in the water currents, and may adhere to the bibrachiate styles of the submerged stigmas of female flowers (Fig. 52.9.1A–E).

The pollen grains of sea-grasses exhibit three different adaptation to the environment. The spherical grains of Hydrocharitaceae are embedded in slime (Fig. 52.4.2A, B), they may be ellipsoidal and are released in tetrads within a slime tube as in *Halophila*, or may be filiform.

Pollen Wall. The microspores synthesize part of their walls from the substances produced by their own cytoplasm while in tetrad condition (Heslop-Harrison 1968a, b, c, 1971, Banerjee and Barghoorn 1971b). The initial wall termed primexine is of cellulose, and is discontinuous opposite the endoplasmic reticulum. The gaps oriented parallel to plasmalemma mark the position of germ-pores. The cytoplasm of the microspore also synthesizes precursors of sporopollenin. They are polymerized and deposited on the lamellae in such a way that various patterns of exine of mature pollen grains become predetermined. The wall is made up of radially-elongate rod-like structures, the baculae, their lower ends spread out forming a continuous foot layer. The upper ends of baculae may also expand, and give rise to a roof, the tectum; such pollen grains are tectate.

The uninucleate pollen grain is densely cytoplasmic with a prominent nucleus. It enlarges due to increase in cytoplasm and vacuolation. In a free state, the pollen grain initiates formation of intine, and the innermost layer of exine (see Godwin 1968). In the formation of intine, the proteins synthesized by the cytoplasm of microspores are incorporated. The plasmalemma of the pollen cytoplasm develops radially oriented cytoplasmic tubules containing proteins, in the developing intine. As the intine grows, these tubules with proteins are cut off from the plasmalemma by a layer of intine which is free from tubules. Thus, the intine proteins are the products of the pollen grain cytoplasm (gametophytic). The intine proteins are concentrated more in the region of germ-pores in aperturate pollen grains, whereas they are distributed throughout in the non-aperturate pollen grains of monocotyledons (Knox 1971, 1984, Shivanna 1978, Chanda et al. 1979, Shivanna and Johri 1985, see also Stanley and Linskens 1974, Walker 1974a, b, Walker and Doyle 1975).

The proteins synthesized by the tapetum (sporophytic) penetrate through the micropores of the tectum and are stored in the spaces between the baculae, whereas the lipid substances spread over the tectum layer.

The exine is differentiated into two layers, the outer sculptured layer, the sexine, and inner non-sculptured layer, the nexine. The sexine comprises the baculae with tectum in tectate grains; in non-tectate grains the baculae may remain separate or join to form various patterns. The exine is variously described according to its ornamentation, such as psilate (smooth), scabrate (very fine projections), verrucate (warty), echinate (spiny), reticulate (network-like), etc.

The morphology of pollen grain is usually considered on the basis of its shape, size, thickness and ornamentation of exine. The most important feature is the form of aperture, their number, distribution and position (designated NPC: N = number, P = position, C = character). The aperture is designated as "treme" (Erdtman 1969). When the apertures are regularly distributed over the surface, the arrangement is called "nomotreme", irregular distribution is "anomotreme", and absence of apertures is referred to as "atreme" or inaperturate.

The number of apertures varies from zero to eight (N_0–N_8) and, accordingly, mono-, bi-, tri-, tetra- and so on is prefixed to the character of pollen grain.

The position of aperture may be distal (outwards in a tetrad), i.e. "anatreme", or the apertures may be distributed on the equatorial plane at equal distance as in "zonotreme".

The two basic forms of apertures (morphoform) are the porate and colpate. The apertures have a circular outline in porate form; in colpate grains there are elongated furrows (colpi) along the axis. The compound forms are designated colporate or pororate. In a colporate pollen grain the outer region bears a furrow with a pore in the centre.

In a very useful and informative discussion, Blackmore et al. (1987) emphasize ontogeny and phylogeny of all aspects of microsporogenesis, male gametophyte and pollen grains on the basis of 26 character states.

3.4 Ovule

The ovule, or the megasporangium, is attached to the placenta of the ovary by a stalk, the funicle. It is a mass of potentially fertile tissue called nucellus, enveloped by one or two integuments (Fig. 3.4.1A–F; see also Bouman 1974).

Types of Ovules. According to the position of the ovule, in relation to hilum, the mature ovules are classified as ortho-, ana-, hemiana-, campylo- and amphitropous (Figs. 3.4.1A–F, 3.4.2A–H). In orthotropous form, the micropyle lies in the line of hilum, above it, whereas it is reversed in the anatropous type. The curvature is much less than the anatropous form in campylotropous type (see Bocquet 1959). The ovule is at right angles to the funiculus in hemianatropous type. The amphitropous form occurs when the ovule, as well as the embryo sac, curves like a horseshoe. In Aristolochiaceae (*Bragantia*- N.C. Nair and Narayanan 1961), Cactaceae and Plumbaginaceae the ovule becomes anatropous due to unequal growth of the axis of funiculus, then curves further till the micropyle faces upward and develops into circinotropous type.

The ortho-, ana-, hemiana- campylo- and amphitropous (Fig. 3.4.2A–E) forms of ovules are considered as basic forms. The rest of the forms have been derived by various degrees of curvature as in circinotropous form (Fig. 3.4.2F–H).

The position of the ovule is also described according to its orientation. The raphe may be dorsal (abaxial: away from the axis) or ventral (adaxial: towards the axis). The micropyle may point upward (epitropous), downward (hypotropous) or to a side (lateral: pleurotropus). The position of raphe is a diagnostic feature in the families Buxaceae, Euphorbiaceae and Theacae. Variations have been observed in orientation of ovules in the same ovary. In *Pittosporum floribundum* (Sheela and Narayana 1966), the ovules are hypotropous in the upper region of ovary and epitropous in the lower region; the raphe is dorsal.

According to Bor (1978), the ovule consists of two separate developmental units: the funicular or basiovular unit which gives rise to raphe, chalaza and outer integument (together forming testa), and the nucellar unit from which the inner integument is derived.

Integument/s. The ovules generally have one or two integuments and are designated uni- and bitegmic. The nucellus may be massive or scanty and is designated crassi- and tenui-nucellate.

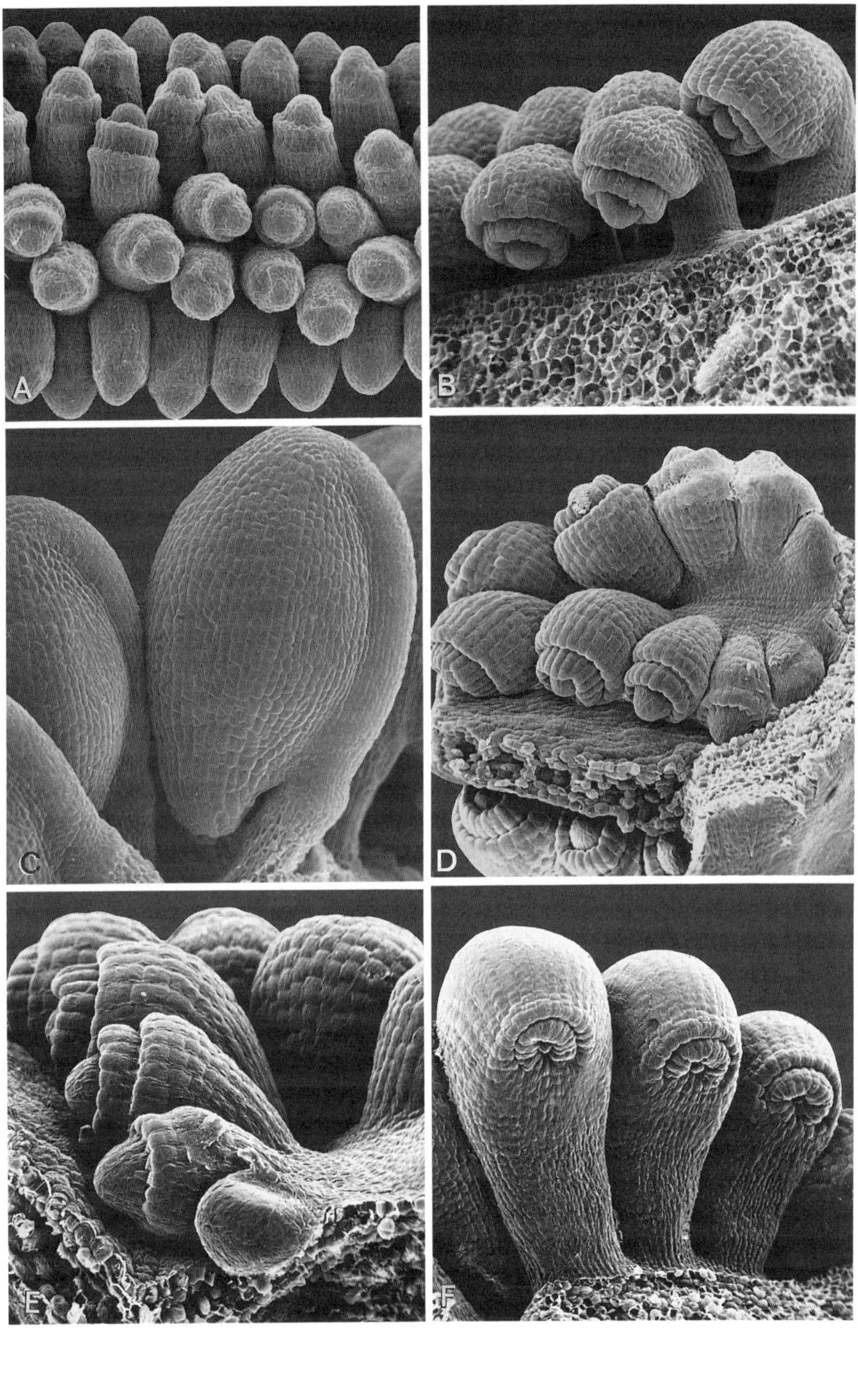
A
B
C
D
E
F

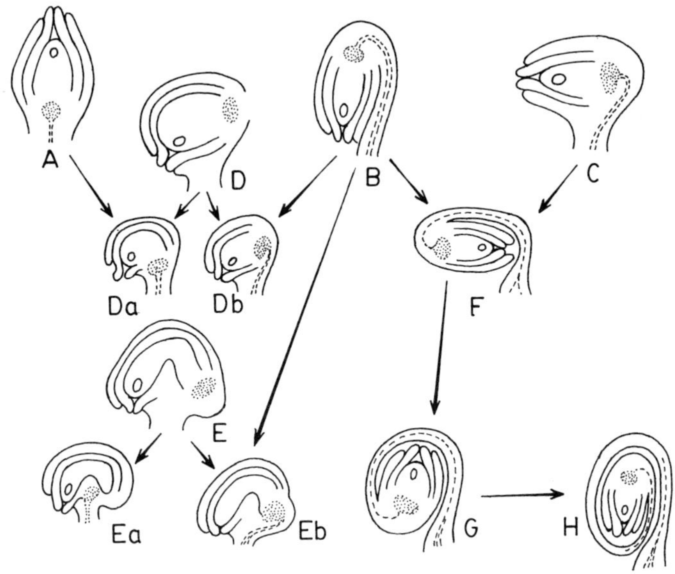

Fig. 3.4.2A–H. Relationship between different types of ovule. **A** Orthotropous. **B** Anatropous. **C** Hemianatropous. **D** Campylotropous. **Da** Ortho-campylotropous. **Db** Ana-campylotropous. **E** Amphitropous. **Ea** Ortho-amphitropous. **Eb** Ana-amphitropous. **F** Hypertropous. **G** Ortho-circinotropous. **H** Ana-circinotropous. (After Savchenko 1973, see Poddubnaja-Arnoldi 1976)

W.R. Philipson (1977) classifies the ovules into four types: Bitegmic crassinucellate (Type A), Bitegmic tenuinucellate (Type B), Unitegmic crassinucellate (Type C), and Unitegmic tenuinucellate (Type D).

Ovules of Type A have been observed in Magnoliidae, Hamamelidae, Caryophyllidae, and most members of Dilleniidae and Rosidae (subclasses of Cronquist 1968). In a number of families of Dilleniidae, the ovules are of Type B, and in a few families of Type D. A large number of families of the Rosidae have ovules of Type C and D. In the subclass Asteridae, most members have ovules of Type D.

The families which consistently have ovules of Type B form a subclass, the Thealean-Primulalean complex. For the families with ovules of Type C and D, W.R. Philipson suggests the term "Unitegminae".

The transition from ovules of Type A to Type D probably took place in two steps, since a few families under the Unitegminae have crassinucellate ovules: Araliaceae and Umbelliferae. A gradation from Type A to C, with some examples of D, occurs in the Celastrales.

Fig. 3.4.1A–F. SEM micrographs of developing ovules. **A** *Passiflora racemosa*, ovular primordia with initiation of anatropous curvature, and inner and outer integuments. **B, C** *P. verspitilio*, overgrowth of nucellus beyond integuments (**B**), and mature anatropous (**C**) ovule. **D–F** *Agrostemma gracile*, ovular primordia with initiation and development of integuments (**D, E**), and mature (**F**) ovules. (After Bouman 1984)

Type D has remained very stable and is known in about half of the dicotyledons. This type is of much significance as a diagnostic feature in systematics.

In angiosperms, the inner integument is always dermal except in *Codiaeum variegatum* and *Euphorbia milii* (Bor and Bouman 1974), *E. geniculata* (Bor and Kapil 1975), and *Chrozophora obliqua* (Bor and Kapil 1976). The outer integument is generally subdermal, but its dermal origin has been observed in Dichapetalaceae – *Dichapetalum mombattense* (Boesewinkel and Bouman 1980), Erythroxylaceae – *Erythroxylum coca* (Boesewinkel and Geenen 1980), Geraniaceae – *Erodium, Geranium, Monsonia, Pelargonium* and *Sarcocaulon* (Boesewinkel and Been 1979), Linaceae – *Linum usitatissimum* (Boesewinkel 1980a), and Rutaceae – *Ruta graveolens* and *Zanthoxylum simulans* (Boesewinkel 1977). Both the integuments are of dermal origin in Centrospermae, Parietales and Capparales.

Bouman and Calis (1977) suggest the origin of unitegmic condition due to integumentary shifting. In *Eranthis hiemalis*, the inner integument of the pseudo-crassinucellate ovule originates by the periclinal divisions of the dermal cells. The outer integument develops subdermally, proximal to the inner one. Due to activity in the subdermal initials, the inner integument is shifted upward; its growth is arrested. The incision between the outer and inner integument ends at or above the level of nucellar apex. Thus, a single massive integument surrounds the nucellus. In *Delphinium nudicaule*, the joint activity of the initials of outer and inner integument gives rise to a single, mainly subdermal, massive integument. The activity of the dermal initials of the inner integument is arrested.

The formation of unitegmic ovule by integumentary shifting has also been observed in *Helleborus niger* (Bouman and Calis 1977). Here the integument appears bifid in transmedian longitudinal sections of older ovules. In the same way, unitegmic ovules are also formed in *Gentiana cruciata* (Bouman and Schrier 1979).

In Polygonaceae, Trapaceae and some members of Euphorbiaceae, the nucellus produces a long beak which projects through and beyond the micropyle (Fig. 3.4.3A). In Commelinaceae (Chikkannaih 1964), the apical region of the nucellus is broad, and there is no well-organized micropyle.

In some taxa, an additional integument develops between the outer and the inner integument but usually it remains arrested (Fig. 3.4.3B).

When the integument/s do not develop, the ovule is designated ategmic (Fig. 3.4.3C).

In bitegmic ovules the two integuments arise independently at the base of the ovule. They overtop the nucellus leaving a narrow opening called the micropyle.

The opening of the inner integument is called endostome, and that of the outer integument exostome. Thc exostome and the endostome may not lie in the same alignment producing a zigzag micropyle. In the majority of families with bitegmic ovules, the inner integument overgrows the outer and forms the micropyle. The unitegmic ovules with a massive integument have a long, narrow micropyle. The unitegmic ovules of Sabiaceae (*Meliosma* spp. – M.V.S.

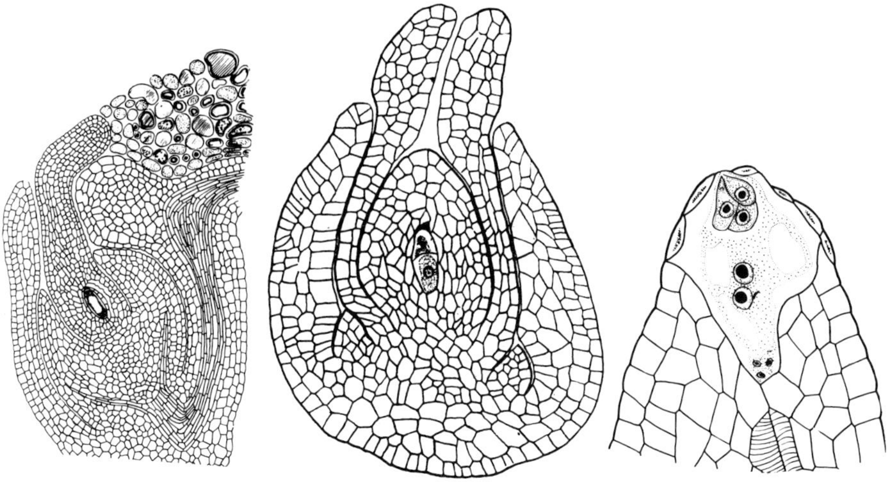

Fig. 3.4.3A. *Baliospermum montanum*, longisection of ovule showing nucellar beak, obturator, and vascular strand in inner integument. (After P.N. Rao 1970). **B** *Cananga odorata*, longisection of ovule with poorly developed middle integument. (After Corner 1949). **C** *Exocarpus bidwillii*, longisection of ategmic "ovule". (S.P. Bhatnagar and Joshi 1965)

Raju 1952a) and Orchidaceae (*Gastrodia elata* – Abe 1976) also do not develop a micropyle.

Sunder Rajan (1976) considers the micropyle formed by both the integuments as primitive. He compares this condition with six other primitive characters occurring in different families: (1) trees or shrubs, (2) actinomorphic flowers, (3) free petals, (4) axile placentation, (5) crassinucellate ovule, and (6) Nuclear type of endosperm. The author does not cite the examples of families, but relies on the data provided by Davis (1966). This information is not at all helpful in judging the primitiveness of a family.

An apparently poorly developed middle integument occurs in *Canangium*, *Mezzettia*, *Xylopia* (Corner 1949), *Cananga* (Fig. 3.4.3B), *Myristica* (N.C. Nair 1972), *Nelumbo nucifera* (S.C. Gupta and Ahluwalia 1977) and *Sarcandra* (Vijayaraghavan 1964).

The aril is often considered as the third integument which, in our opinion, is not justified at all. The aril arises from the funicle, at the base of the integuments, and from the funicle and exostome in *Myristica* (Fig. 18.8.2A–C). It grows during post-fertilization stages and surrounds the ovule as in Aizoaceae, Crossosomataceae, or merely remains as a vestige at the base as in Bixaceae. It is thin-walled and contains oil and starch in Passifloraceae. It is fimbriate in Crossosomataceae, vascularized in Myristicaceae and Paeoniaceae, and with coenocytic cells in Dilleniaceae (*Acrotrema* – Fig. 22.2.2A–I).

The bitegmic ovules of annonaceous genera and *Myristica* (N.C. Nair 1972), where both the middle integument and aril are present, have four integuments.

The number of integuments indicates the evolutionary status of the family. The bitegmic condition occurs usually in dicotyledonous Polypetalous families, whereas the unitegmic condition is associated with most of the Sympetalous families. Exceptions occur in both cases (see de Boer and Bouman 1972).

In the ovules of Anacardiaceae, the bitegmic condition occurs in *Rhus* and *Lannea* (Kelkar 1958a, b) due to the splitting of the integumentary primordia. In *Pistacia* spp. (Grundwag 1976), the ovule is initially unitegmic but, at a later stage, the integument splits close to the micropyle forming a partial second integument. According to P. Maheshwari (1950), the unitegmic condition may arise due to elimination of one integument, or the fusion of two separate primordia. The Santalales and Rafflesiaceae depict the former feature, whereas the accretion of integuments is exhibited in Anacardiaceae, Balsaminaceae, Betulaceae, Haloragaceae, Lecythidaceae, Marcgraviaceae, Rosaceae (*Sanguisorba*), and others.

Van Heel (1970a, b, 1971a, b, c, 1973, 1976) attributes an evolutionary significance to the formation of distally lobed integuments. The bitegmic ovule of *Exochorda racemosa* shows a three- to four-lobed outer integument and a bilobed inner integument; while the single integument in *Juglans regia* may exhibit four to seven, entire or fringed, lobes in the distal region (see also Leroy 1955).

In the ovules of Centrospermales and Cactales, there is a prominent air-space between the outer and inner integument at the base, and it is a characteristic feature of these orders.

The occurrence of stomata on the integuments is a common feature in several families such as Malvaceae, Sapotaceae and Umbelliferae. In Amaryllidaceae and Moringaceae, the integuments become chlorophyllous in the young seed.

Generally, the vascular supply in an ovule terminates at the chalaza and branches there. The occurrence of integumentary vascular system is treated as a primitive feature. The bitegmic ovules of Magnoliaceae (*Magnolia*) and Euphorbiaceae (*Dalechampia*) have an independent vascular system in each integument. The outer integument is vascularized in several families: Berberidaceae, Betulaceae, Boraginaceae, Compositae, Cyrillaceae, Leguminosae, Moringaceae, Ranunculaceae, Sterculiaceae. In *Hyphaene indica* (Mahabale and Chennaveeraiah 1957), 16–17 vascular bundles extend in the outer integument. Rarely, the bundles occur only in the inner integument (*Baliospermum* – Fig. 3.4.3A; *Croton, Jatropha, Ricinus*), whereas in Myristicaceae both outer and inner integument as well as the aril are vascularized.

The development of operculum occurs in the members of Lemnaceae (S.C. Maheshwari 1956b), and Araceae (S.C. Maheshwari and Khanna 1956). The inner integument proliferates at its micropylar tip, and forms a cap-like structure. Its cells are pigmented, and their walls are thickened.

Endothelium. In the ovule, where the nucellus is ephemeral, the innermost layer of the integument in contact with the embryo sac develops into integumentary tapetum or endothelium (Fig. 3.4.4A–E). The cells elongate radially, become richly cytoplasmic, and perform the function of transport of nutrients to the embryo sac.

The flow of nutrients from the integumentary cells to the embryo sac depends on the chemical nature of the intervening cell walls of endothelial cells, and the embryo sac. The flow may be restricted to the micropylar, sub-micropylar, or chalazal region (see Kapil and Tiwari 1978a).

In some members of Scrophulariaceae, the unequal radial elongation of the endothelial cells causes rumination in the endosperm. In *Celsia coromandeliana* (Kapoor et al. 1975), some cells develop thickenings on the inner tangential and radial walls, and protect endosperm and embryo.

The cells of the endothelium may become multinucleate (Fig. 3.4.4B), or the ploidy level of the nuclei increases due to endomitosis. In *Pedicularis palustris* (Steffen 1955), the nuclei become 32-ploid. The endothelium is multilayered (Fig. 3.4.4C) in the chalazal region of the embryo sac in *Tithonia rotundifolia* (Pullaiah 1978a), and is supposed to be a characteristic feature of the tribe Heliantheae. In *Melampyrum pratense* (D. Raju and Arekal 1977), the inner epidermis of the integument becomes obliterated in the micropylar region, during maturation of the embryo sac. Finally, the integumentary tapetum develops from the hypodermal layer. The endothelium is a common feature in unitegmic, tenuinucellate ovules of the Sympetalae. However, in Acanthaceae and Loganiaceae the endothelium develops only in some members. Its absence in Diapensiaceae is a taxonomic feature.

In *Flaveria repanda* (S. Misra 1964), *Melampodium divaricatum* (Maheswari Devi and Pullaiah 1976) and sunflower (Fig. 3.4.4D), the endothelial cells divide in all planes and produce a multilayered, massive structure which destroys the embryo sac. Perhaps, this may be due to the accumulation of nutritive substances (see Savchenko 1960).

In *Utricularia flexuosa* (R. Khan 1963), the tangential cells separating the endothelial cells disorganize, and their nuclei accumulate in a nodule which bulges into the degenerating embryo sac (Fig. 3.4.4E). Rarely, the biseriate endothelium of *Chelone glabra* (Arekal 1963a) also develops such nodules.

Nucellus. Enclosed within the integuments lies the reproductive homogeneous tissue, the nucellus. An archesporial cell, or a group of cells, differentiates in the hypodermal region of the nucellus in tenuinucellate ovules. The archesporial cell/s cut/s off a parietal cell in crassinucellar ovules. The parietal cell may remain undivided as in *Flagellaria indica* (Subramanyam and Narayana 1972), or may undergo divisions to produce a parietal tissue. The cells of the nucellar epidermis divide periclinally, and add to the parietal tissue. The megaspore mother cell becomes deep-seated either due to the growth of parietal tissue and derivatives of nucellar epidermis, or due to the tissue produced by the nucellar epidermis alone. In the latter case, the ovule is termed as pseudo-crassinucellar (see Davis 1966) as in *Cordia* (Tasneem 1966) and *Foeniculum vulgare* (S.C. Gupta 1964, 1970).

The occurrence of twin nucelli within a common integument is not very uncommon.

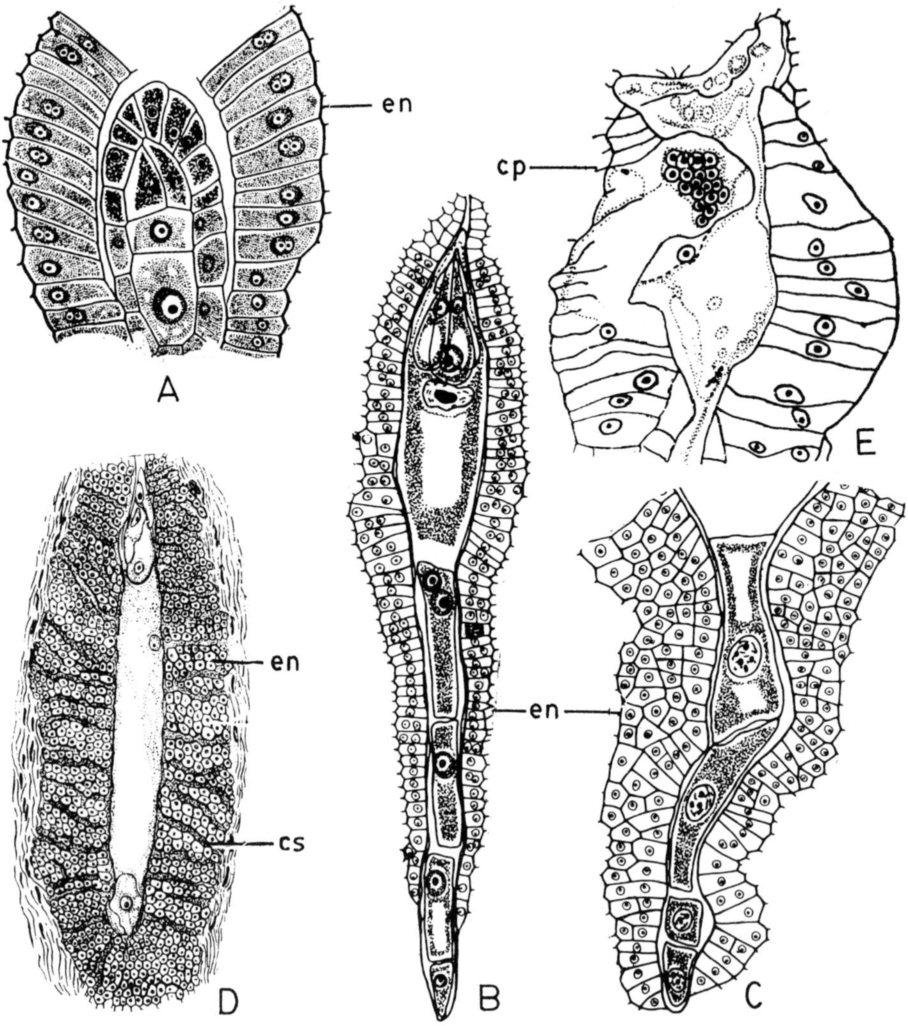

Fig. 3.4.4A–E. Endothelium. **A** *Volutarella ramosa*, T-shaped tetrad of megaspores, endothelium (*en*) with prominent uninucleate radially elongate cells. (After P.K. Deshpande 1964a), **B, C** *Tithonia rotundifolia*. **B** Mature embryo sac with multinucleate endothelial cells. Of the four antipodals one is binucleate; **C** Chalazal portion of embryo sac invested by multilayered endothelium, antipodal cells polyploid. (After Pullaiah 1978a). **D** *Helianthus annuus*, part of ovule at mature embryo sac, endothelium highly proliferated and interrupted by cuticular septae (*cs*). (After Savchenko 1969). **E** *Utricularia flexuosa*, portion of ovule with degenerated embryo sac. Due to break down of endothelial cells a coenocytic pouch (*cp*) has been formed. (After Khan 1963)

Sometimes, the derivatives of nucellar epidermis form a nucellar cap as in Lemnaceae (S.C. Maheshwari 1954). In *Elettaria* (Panchaksharappa 1962b) and *Hedychium* (Sachar and Arora 1963), the cells of the nucellar epidermis below the micropyle elongate radially and form a nucellar pad. Such a radial elongation of nucellar cells also occurs in Aizoaceae and Cactaceae. In Leitneriaceae, Polygonaceae, Trapaceae, and some members of Euphorbiaceae, the nucellus develops a beak which projects through the micropyle. The nucellar cap, nucellar pad, and nucellar beak are features of taxonomic importance in several families.

Vascular elements have been observed in the nucellus of *Acalypha* (Sect. 29.9), *Casuarina* (Sect. 4.2), *Duabanga* (Sect. 39.7) and several other taxa.

The nucellus is generally consumed during the growth of the embryo sac, but in Centrospermales (Fig. 16.6.1A) and some members of Proteaceae, Piperaceae and Zingiberaceae it stores food and persists in the seed as perisperm. In Podostemaceae (Figs 28.2.1D, E), and *Cenarrhenes nitida* (Proteaceae, Venkata Rao 1967), the nucellar cells below the embryo sac break down forming a pseudo-embryo sac which provides nourishment to the developing gametophyte and embryo.

In the ovules of *Cabomba caroliniana* (Ramji and Padmanabhan 1965), *Halophila ovata* (Lakshmanan 1963a), and *Vateria indica* (Fig. 22.11.1G–I), the nucellar growth in height is due to the presence of a basal meristem.

The nucellar tissue, at the base of the embryo sac, in *Quercus gambelii* (Brown and Mogensen 1972), grows upward and forms a column in the embryo sac. This structure is termed postament.

A rare feature in *Halophila ovata* (Fig. 52.4.3C), *Tamarix* (Fig. 37.15.2A–C) and *Pandanus parvus* (Cheah and Stone 1975; see also Fagerlind 1940b) is that the walls of nucellar cells at the base of the embryo sac break down and their nuclei migrate into the embryo sac.

The nucellar cells become multinucleate in the chalazal region in *Samadera indica* (N.C. Nair and Joseph 1957); the nuclei fuse and exhibit polyploidy. They reach up to 256n in *Trapa natans* (Nägl 1962, Trela-Sawicka 1978).

Hypostase and Epistase. Anatomical differentiation occurs during the formation of hypostase and epistase. Some of the nucellar cells, at the base of the embryo sac, become specialized and delimit the downward growth of the embryo sac. This structure is called hypostase. The cells are usually sclerenchymatous, or suberized (*Cannabis*), or they may remain thin-walled and become secretory. According to Venkata Rao (1963), the hypostase acts as a connecting link between the vascular strand and the embryo sac, and helps in the transportation of nutrients. Autoradiographic studies have shown that the hypostase is not concerned in the conduction of nutrients (see Kapil 1974). The hypostase probably serves to maintain the water balance in seed during dormancy (Johansen 1928a, b; see also van Tieghem 1901). It persists in seed in several plants. Some of the cells at the nucellar tip become thick-walled, and differentiate as epistase. In *Rhamnus wightii* and *Zizyphus glabrata* (see N.C. Nair 1970b), these cells are filled with dark contents.

In *Turnera fournieri* (Tiwari 1983), a thick-walled cup-shaped hypostase differentiates at the megaspore mother-cell stage; apparently devoid of cytoplasm. At the periphery the cells are partially thick-walled but completely thick-walled in the centre. The funicular provascular strand of the ovule runs almost up to the level of the hypostase. The cell walls of the hypostase predominantly consist of callose with a considerable amount of pectin; lignin is scanty. Presumably, the callose serves as a "molecular filter" which regulates the flow of nutrients to the embryo sac, and prevents the inflow of the unwanted large molecules. The occurrence of callose in the hypostase has also been detected in *Antirrhinum* (Rodkiewicz 1967), *Pentstemon* and *Digitalis* (Kuran 1972).

Obturator. An obturator is an outgrowth of the placenta, funicle, integument or style which bridges the nucellus or the micropyle to the adjoining integuments. It is presumed that it guides the course of pollen tube from the style or the ovary wall to the micropyle. It is either in the form of thin-walled compact tissue, or of loose hairy outgrowths. The cells may be glandular. Usually, in the majority of families, it develops from the funicle. The obturator is of placental origin in *Nelumbo*, and stylar in Thymeleaceae. Occasionally, both placental and funicular obturator occur in Aizoaceae. In Urticaceae, the inner integument develops hairy outgrowths which penetrate into the style as in *Leucosyke* and *Myriocarpa* (Fagerlind 1944b).

Incompletely Differentiated (Reduced) Ovules. The reports on "naked" ovules in the Apocynaceae, Olacaceae and Rubiaceae are inaccurate (see Eames 1961). In these families, there is a tendency for the reduction of nucellus.

Guignard (1917a, b) suggested a series of reduction of nucellus in unitegmic, tenuinucellate ovules of Apocynaceae (Sect. 47.6). In *Cerbera*, the nucellus is a single layer of cells which invests the megaspore tetrad, and is distinct from the integument. The nucellus collapses on the sides of the megaspore mother cell in *Vinca*, whereas in *Apocynum* it is represented by only three or four cells above the megaspore mother cell, and is continuous with the epidermis of the integument.

A similar series of reduction of nucellus can be traced in Rubiaceae (Fig. 47.8.2A–F). In *Phyllis*, the nucellus is a uniseriate layer, the lateral nucellar cells have disappeared in *Bouvardia* and *Vaillantia*, and the nucellus is represented by only a couple of cells below the micropyle. A few radially elongate cells in the micropylar region above the archesporial tissue represent the nucellus in *Rubia*, and only one or two cells in *Oldenlandia*. In *Houstonia*, the nucellar cells are not distinguishable from the integument.

Maheswari Devi (1971) reports reduction of nucellus in four genera of the tribe Plumiereae: *Catharanthus*, *Holarrhena*, *Rauwolfia*, and *Voacanga*.

Ovary-Ovule Complex. There is an ovary-ovule complex in the Loranthaceae (Sect. 12.8), Santalaceae (Sect. 12.6) [except *Comandra* (M. Ram 1957), *Iodina* (S.P. Bhatnagar and Sabharwal 1968) and *Thesium* (S.P. Bhatnagar and Agarwal 1961) where the ovules are unitegmic)], and Opiliaceae (Sect. 12.4), and a distinct tendency of reduction in the number of ovules and placenta. *Exocarpus* (M. Ram 1959a) reaches an extreme, and there is no distinction between the placenta and the ovule. Reduction also occurs in the integuments, and nucellus. In *Leptomeria* (M. Ram 1959b), the integument is more or less suppressed, and the nucellus is represented by a single cell.

In the Loranthaceae (Fig. 12.8.1$A_{1,2}$– $F_{1,2}$), the "mamelon" represents the placenta, the ovules of normal structure have not been observed. In *Helixanthera ligustrina* (P. Maheshwari and Johri 1950a), the placenta is absent. In *Amyema* (Dixit 1958a), the placenta may be absent or a rudimentary placenta may develop due to the elongation of sporogenous cells. *Helicanthes elastica* (B.M. Johri et al. 1957) shows a conspicuous conical unlobed placenta. The mamelon is basally three-lobed in an unilocular ovary in *Macrosolen cochinsinensis* (P. Maheshwari and B. Singh 1952) and *Peraxilla tetrapetala* (S. Prakash 1960a). The ovary is three-chambered with a basally three-lobed placenta in *Nuytsia floribunda* (R. Narayana 1958b), and four-chambered with a completely four-lobed placenta in *Lysiana exocarpi* (R. Narayana 1958a).

The maximum reduction is reported in *Balanophora* (Fagerlind 1945a) where the entire gynoecium is affected (Sect. 13.2). The ovular locule and stylar canal do not differentiate, and the archesporial cell differentiates in the centre of the basal region of a simplified carpel (Eames 1961). Ontogenetically, this simplified carpel is considered as the nucellar tissue (Arekal and Shivamurthy 1979).

3.5 Megasporogenesis

Archesporium and Sporogenous Tissue. The hypodermal archesporium is usually one- or two-celled, but multicellular archesporium develops in some families like Casuarinaceae, Rosaceae and Sterculiaceae. In *Casuarina equisetifolia* (Swamy 1948a), vascular elements differentiate in the centre of sporogenous tissue (from the procambial strand). A few tracheids have been observed among the sporogenous cells in the mamelon of *Amyema miquelii* (Dixit 1954).

In crassinucellar ovules, the archesporial cell cuts off a parietal cell which is not the case in tenuinucellar and pseudo-crassinucellar ovules. Generally, a single archesporial /sporogenous/ megaspore mother cell is functional.

Dyad, Tetrad and Coenomegaspore. The megaspore mother cell (Fig. 3.5.1A) has a cellulosic wall and is connected with the circumjacent cells through plasmodesmata. During megasporogenesis, callose (Rodkiewicz 1970) is deposited on the megaspore mother-cell wall, plasmodesmatal connections become obliterated, and the megaspore mother cell becomes isolated from the adjoining tissue (see Kapil and Bhatnager 1981). It undergoes meiotic divisions. The heterotypic division results in the formation of a dyad (Fig. 3.5.1B). The homotypic division occurs in each dyad cell and a linear or T-shaped tetrad is produced (Fig. 3.5.1C, E). Suppression of homotypic division in one of the dyads results in the formation of a triad (Fig. 3.5.1G) as in Cactaceae,

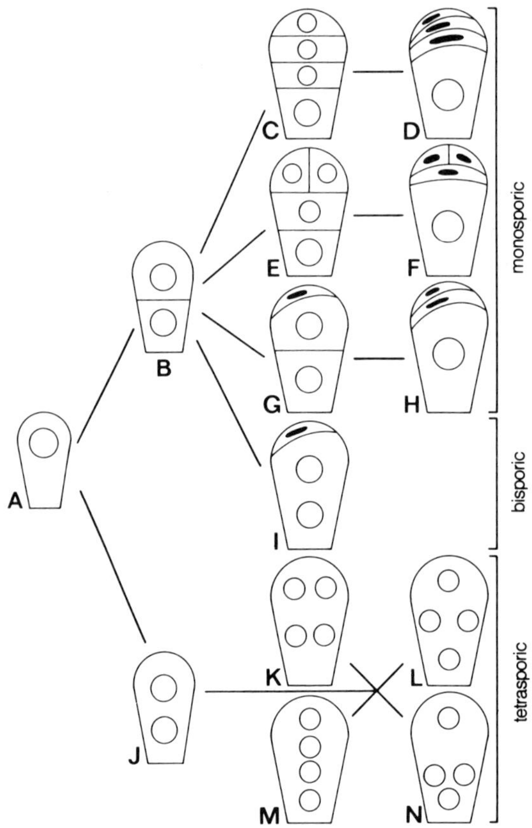

Fig. 3.5.1A–N. A Megaspore mother cell. **B** Dyad following heterotypic division (meiosis I). **C** Tetrad following homotypic division (meiosis II). **D** Three megaspores degenerate, chalazal functional. **E** T-shaped tetrad. **F** Chalazal megaspore functional, three degenerated megaspores. **G** Two megaspores with degenerated dyad. **H** Functional megaspore with degenerated dyad and one megaspore: **I** Lower dyad binucleate, upper degenerated. **J** Meiosis I without wall formation. **K–L** Meiosis II without wall formation. 2 + 2 (**K**), 1 + 1 + 1 + 1 (**L**), 1 + 2 + 1 (**M**) and 1 + 3 (**N**) arrangement of nuclei in coenomegaspore. **A–H** Monosporic, **I** bisporic, and **J–N** tetrasporic type of development of embryo sac. (After Bouman 1984)

Connaraceae and Stachyuraceae. In Crassulaceae, isobilateral tetrads have also been observed; this appears doubtful. Rembert (1971) observed ten patterns of megaspore tetrads in Leguminales, seven in monosporic type (including triad) and three in bisporic type of the embryo sac.

The absence of wall formation in megaspore mother cell, following meiosis I and II, leads to the development of a four-nucleate coenomegaspore (Fig. 3.5.1I–N; see also Schnarf 1936, Smith 1973b, Swamy 1974b, Willemse 1981).

During megasporogenesis, a notable feature is the development of callose thickenings around the non-functional megaspores. In the Polygonum type of embryo sac, the callose appears first in the chalazal region of megaspore mother cell, spreads over its entire surface, and isolates it from the circumjacent maternal tissue. Then the callose is retained only around the micropylar dyad, or around three upper megaspores so that the functional chalazal megaspore does not have any callose. The reverse condition has been noted in the Oenothera type. Its appearance first in the micropylar region, isolation of the megaspore mother cell for a short period, and later on disappearance from the micropylar dyad and the functional micropylar megaspore denote that the absence of callose is necessary for further growth of a megaspore. The distribution of callose only on transverse walls has been observed in homeopolar tetrad where any one of the megaspores develops. In the bisporic Endymion type of embryo sac of *Oenothera lamarckiana*, callose disappears from the micropylar dyad.

The functional activity of the megaspore and dyad depends on its larger size, viability for a longer period, and the occurrence of maximal organelles. In *Epipactis*, the chalazal megaspore has a porous wall which allows the inflow of metabolites, whereas the callosic walls in three chalazal megaspores, and chalazal dyad result in the loss of activity of megaspore dyad.

The accumulation of plastids containing starch in a megaspore mother cell, dyad, or a megaspore denotes its functional activity (see K.V.O. Dahlgren 1927a).

The chalazal megaspore is functional in the majority of taxa (Fig. 3.5.1C–F), and the development of embryo sac is of the Polygonum type. The micropylar megaspore functions in Balanophoraceae, Onagraceae, and Theaceae (*Camellia*). Occasionally, the epichalazal megaspore (one above the chalazal) is functional in *Cannabis sativa* (Mohan Ram and Nath 1964). At times, all the four megaspores of a tetrad develop concurrently and four binucleate megaspores (two-nucleate embryo sacs) are formed as in *Casuarina equisetifolia* (Swamy 1948a), and *Iphigenia indica* (Sulbha 1954a).

The formation of megaspore haustoria in *Sedum chrysanthum* (Fig. 25.5.2A–D) and *S. sempervivoides* (Mauritzon 1933a) is an unusual feature.

3.6 Female Gametophyte

Types of Embryo Sac

Depending on the number of megaspore nuclei taking part in its formation, the embryo sac is designated as Mono-, Bi- and Tetrasporic (Fig. 3.5.1A–N). P. Maheshwari (1937, 1941, 1946, 1947, 1948), B.M. Johri (1963a, b), and Willemse and van Went (1984) have reviewed different types of embryo sacs in angiosperms.

Monosporic Type. The monosporic type is represented by the Polygonum and Oenothera types. The chalazal megaspore of a tetrad develops into an embryo sac in the former, while the micropylar develops in the latter. The other three megaspores degenerate (Fig. 3.6.1).

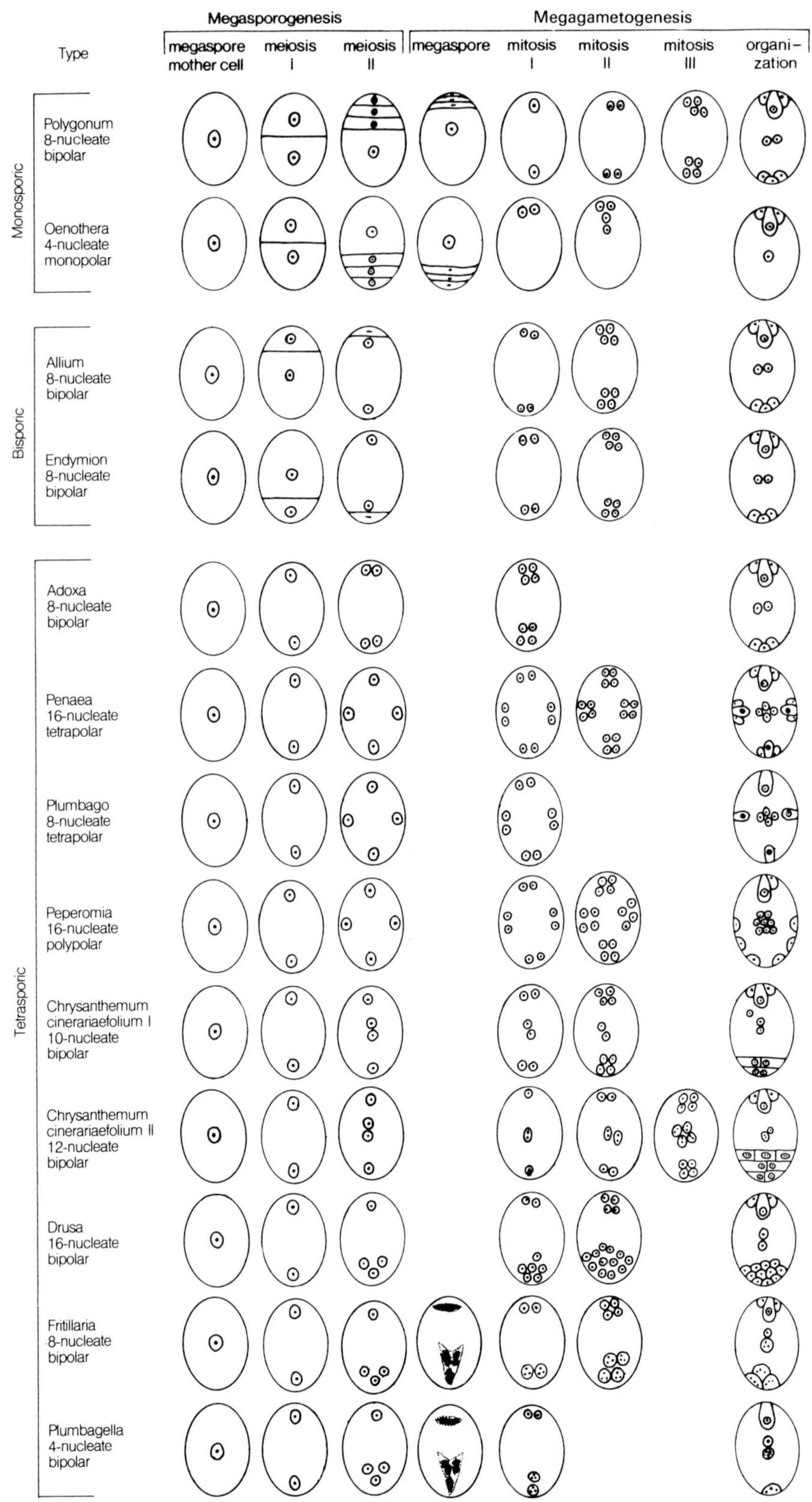

Megasporogenesis
Megagametogenesis
Type
megaspore mother cell
meiosis I
meiosis II
megaspore
mitosis I
mitosis II
mitosis III
organi-zation
Monosporic
Polygonum 8-nucleate bipolar
Oenothera 4-nucleate monopolar
Bisporic
Allium 8-nucleate bipolar
Endymion 8-nucleate bipolar
Tetrasporic
Adoxa 8-nucleate bipolar
Penaea 16-nucleate tetrapolar
Plumbago 8-nucleate tetrapolar
Peperomia 16-nucleate polypolar
Chrysanthemum cinerariaefolium I 10-nucleate bipolar
Chrysanthemum cinerariaefolium II 12-nucleate bipolar
Drusa 16-nucleate bipolar
Fritillaria 8-nucleate bipolar
Plumbagella 4-nucleate bipolar

Polygonum Type. In the Polygonum type of embryo sac, the nucleus of the chalazal megaspore undergoes three consecutive mitotic divisions. Concurrently, the embryo sac grows along the micropylar-chalazal axis, the nuclei are separated to the poles by a central vacuole, and polarity is established at the binucleate stage. During maturation, the three nuclei from the micropylar quartet differentiate as egg apparatus consisting of the egg cell and two synergids. The fourth one is the upper polar nucleus. The three nuclei of the chalazal quartet organize into antipodal cells except in some taxa like *Cassytha*, *Scilla*, *Thismia*, *Xylopia* and *Xyris* where they degenerate before organization into cells. The fourth nucleus functions as the lower polar nucleus which usually moves upward and fuses with the upper polar nucleus to form a secondary nucleus. However, variations occur in the behaviour of the polar nuclei. They may fuse in the micropylar region of the embryo sac, in the centre, or chalazal region. Likewise, they may fuse before fertilization, or may remain appressed till fertilization.

Sometimes, the chalazal nuclei of the four-nucleate embryo sac divide earlier than the micropylar resulting in a six-nucleate embryo sac as in *Macrosolen cochinchinensis* (P. Maheshwari and Singh 1952); subsequently the micropylar nuclei also divide. Due to the suppression of division of chalazal nuclei, the embryo sac may be six- or seven-nucleate as in Alismataceae, Butomaceae and Lythraceae. A rare condition of the absence of central vacuole, during development and maturation of the embryo sac, has been noted in *Aneilema paniculatum* (Chikkannaiah 1969). This needs confirmation.

In *Balanophora* (Fagerlind 1945a, Arekal and Shivamurthy 1978, Engell 1979), the egg apparatus is always organized in the lower pole of the gametophyte, while the nuclei in the upper pole fuse successively (Sect. 13.2).

Oenothera Type. In the Oenothera type, the lower three megaspores degenerate (Fig. 39.13.2B–E), and the nucleus of the functional micropylar megaspore undergoes mitotic division. A notable feature is that at the two-nucleate stage of embryo sac the nuclei do not migrate to the two poles, but stay at the micropylar pole and undergo one more mitotic division. These four nuclei organize into the egg apparatus and upper polar nucleus (Fig. 3.6.1). This feature is characteristic of the Onagraceae. In the mono- and bisporic type of embryo sac in *Schisandra chinensis*, Swamy (1964a) describes four-nucleate embryo sacs with a similar organization. The primary chalazal nucleus is absent in both cases.

Bisporic Type – Allium and Endymion type. Bisporic development occurs when either the chalazal or micropylar dyad undergoes the second meiotic division. The former condition occurs in the Allium type (Fig. 3.6.1), the latter in the Endymion type. The non-functional dyad, in both types, degenerates. The functional dyad undergoes two mitotic divisions forming an eight-nucleate embryo sac. The organization is similar to that in the Polygonum type.

Fig. 3.6.1. Schematic representation to show the development of different types of embryo sacs. (See Willemse and van Went 1984)

Tetrasporic Type. In the tetrasporic type, cytokinesis does not follow meiotic divisions in the megaspore mother cell, and a coenomegaspore with four haploid nuclei is formed (Fig. 3.5.1A–N).

The variations in the arrangement of the four megaspore nuclei in a coenomegaspore lead to the formation of different types of embryo sacs (Fig. 3.6.1): (1) Two megaspore nuclei lie at each pole (Adoxa type), (2) four megaspore nuclei lie in a crosswise manner – one at each pole and two placed laterally (Penaea, Plumbago and Peperomia type), (3) one megaspore nucleus lies at each pole, and the other two remain in the centre (Chrysanthemum cineraraefolium type), (4) the arrangement of the nuclei is 1 + 3, i.e. one at the micropylar pole and three at the chalazal pole (Drusa, Fritillaria and Plumbagella type).

Adoxa Type. The nuclei (2 + 2) of the coenomegaspore undergo one mitotic division to form an eight-nucleate embryo sac. The organization is similar to that of the Polygonum type (Fig. 3.6.1).

Penaea Type. The nuclei (1 + 1 + 1 + 1) of the coenomegaspore undergo two mitotic divisions forming 16 nuclei, in four quartets. A single nucleus from each quartet migrates to the centre, and these four nuclei behave as polars. The three nuclei at the micropylar pole organize into the egg apparatus; the three nuclei in the other groups also organize into cells but do not function as egg-apparatus (Fig. 3.6.1).

Plumbago Type. The arrangement of nuclei in the coenomegaspore is crosswise (1 + 1 + 1 + 1). The nuclei undergo only one mitotic division forming eight nuclei in four groups. One nucleus from each group migrates to the centre and functions as polar nucleus. The nucleus at the micropylar pole organizes as the egg cell; a synergid is absent. The remaining three nuclei also organize into cells but soon degenerate (Fig. 3.6.1).

Peperomia Type. The arrangement of the four nuclei in the coenomegaspore is again crosswise (1 + 1 + 1 + 1) or irregular. Each nucleus undergoes two mitotic divisions forming four quartets, or the nuclei are irregularly distributed in the embryo sac. The two nuclei at the micropylar pole differentiate as the egg cell and a synergid. Eight nuclei migrate to the centre and unite to form an octoploid secondary nucleus. The remaining six nuclei organize as peripheral cells at the chalazal end, and function as antipodals (Fig. 3.6.1). This behaviour is characteristic of *Peperomia pellucida* (Fagerlind 1939a). Occasionally, the egg and one synergid organize at the micropylar pole, and 14 nuclei fuse in the centre to form the secondary nucleus. This happens in *P. hispidula* (D.S. Johnson 1914).

Chrysanthemum cinerariaefolium Type. The nuclei of the coenomegaspore show 1 + 2 + 1 arrangement. Two variations occur in this type, due to different behaviour of the two centrally located megaspore nuclei. In the first case, these two nuclei do not undergo any division, while the nuclei at the poles divide twice and the embryo sac becomes ten-nucleate. The three nuclei of the micropylar quartet organize into the egg apparatus, the fourth one, along with two central nuclei, behave as polar nuclei. The four nuclei at the chalazal pole are cut off as antipodal cells (Fig. 3.6.1).

In the second case, the two centrally-situated megaspore nuclei fuse. The latter and two nuclei at the poles undergo two mitotic divisions resulting in three quartets, two with haploid nuclei, one at each pole, and the third with diploid nuclei in the centre. The three nuclei of the micropylar quartet differentiate as egg apparatus. The fourth one and one diploid nucleus from the central quartet function as polar nuclei. The remaining seven nuclei (three diploid and four haploid) organize as antipodals (Fig. 3.6.1).

Drusa Type. The arrangement of nuclei in a coenomegaspore is 1 + 3. Each nucleus undergoes two mitotic divisions to form a 16-nucleate embryo sac. The three nuclei at the micropylar pole differentiate as egg apparatus, the remaining one migrates to the centre as the polar nucleus. Eleven antipodal cells are cut off and the remaining one behaves as the lower polar nucleus (Fig. 3.6.1).

Fritillaria Type. The arrangement of the nuclei in a coenomegaspore is again 1 + 3. These nuclei undergo two mitotic divisions. During the first division, the three chalazal spindles fuse to produce two triploid daughter nuclei. A haploid egg apparatus differentiates from the micropylar quartet. The remaining haploid nucleus recedes to the centre as the upper polar nucleus. One of the triploid nuclei from the chalazal quartet behaves as lower polar nucleus; the other three nuclei are cut off as triploid antipodal cells (Fig. 3.6.1).

Bambacioni (1928a), for the first time, discovered that in *Fritillaria* the spindles of the three chalazal nuclei fuse during the mitotic division resulting in two triploid nuclei. This phenomenon is referred to as the "Bambacioni effect".

The Fritillaria type in *Clintonia* is noteworthy. Pahuja and Kumar (1970) reinvestigated the development in *C. undensis* var. *alpina.* The arrangement of four megaspore nuclei in a coenomegaspore is 1 + 3. The micropylar nucleus undergoes two successive mitotic divisions to produce a micropylar quartet which later organizes into the egg apparatus and upper polar nucleus. The three chalazal megaspore nuclei fuse and degenerate. Thus, the antipodals and the lower polar nucleus are absent.

Pahuja and Kumar (1970) designate this reduced Fritillaria type as biphasic (two mitotic divisions after the meiotic divisions of megaspore mother cell), and pseudomonosporic (single megaspore nucleus functional).

Plumbagella Type. The arrangement of megaspore nuclei in a coenomegaspore is 1 + 3, and a single mitotic division leads to the formation of a mature four-nucleate embryo sac. The three chalazal spindles fuse during division resulting in two triploid daughter nuclei. A single nucleus at the micropylar pole differentiates as the egg cell, the other behaves as the upper polar nucleus. The lower polar nucleus from the chalazal pole is triploid. A single triploid antipodal cell is cut off (Fig. 3.6.1).

Variations in the Types of Embryo Sacs

The development of more than one type of embryo sac in a family, or genus, or even in a species, is not uncommon. Kapil and Prakash (1966) recorded five different types (Polygonum, Endymion, Drusa, Penaea and Adoxa) of development in *Delosperma cooperi* (Aizoaceae). Dharamadhaj and Prakash (1978) observed twin embryo sacs, one monosporic and the other bisporic, in the same nucellus of an ovule of *Capsicum annuum* var. *acuminatum*. However, according to them, the formation of the monosporic type is more frequent in summer, and that of the bisporic type in winter. The temperature may have a definite role in determining the type of embryo sac. These observations support the results obtained by Hjelmqvist and Grazi (1965).

Fagerlind (1944a) classified tetrasporic embryo sacs into ten categories. The criteria used are: (1) number of mitotic divisions (phases) intervening between meiosis in megaspore mother cell and organization of mature female gametophyte, (2) number of poles (or groups) at which the nuclei are arranged in a coenomegaspore, and (3) presence or absence of fusion of nuclei during mitotic divisions.

Thus, the development in the Adoxa and Plumbago type is monophasic (only one mitotic division occurs), but the arrangement of nuclei is bipolar (two groups) in Adoxa and tetrapolar (four groups) in the Plumbago type. Similarly, it is biphasic (two mitotic divisions) in the Penaea type, with tetrapolar arrangement of nuclei.

Swamy and Krishnamurthy (1975) discuss a new approach to the classification of embryo sacs. According to them, in the previous system, the number of megaspore nuclei involved in the constitution of the female gametophyte, and the number of divisions intervening between the megaspore mother cell and the mature embryo sac are emphasized. They consider the development of egg apparatus from a single primary micropylar nucleus as more "stable and stereotyped". The development at the chalazal pole of the embryo sac is more variable, and has less functional significance.

In their proposed system, the successive steps involved in the ontogeny of embryo sac are: (1) attainment of haploidy, (2) establishment of polarity leading to the formation of egg apparatus mother nucleus, (3) elaboration phase including steps leading to the formation of egg apparatus, and (4) mature organization. They recognize four basic types of embryo sac development, acknowledging the time of polarity as the guiding factor. The proposed system consists of "Supra-Homotypic" and "Homotypic" categories (Fig. 3.6.2).

Supra-Homotypic Category. Polarity is established after one more division following the homotypic division, i.e. at the two-nucleate stage of embryo sac, and a single primary antipodal nucleus is present – Polygonum type. The Oenothera type is similar but the primary antipodal nucleus is absent.

Homotypic Categories. *Category I.* Characterized by the establishment of polarity immediately following homotypic division. A single primary antipodal nucleus is present – Allium, also Endymion type. *Category II.* Polarity is established following homotypic division, and three primary antipodal nuclei are present – Penaea type; it also gives rise to the Drusa (even Peperomia),

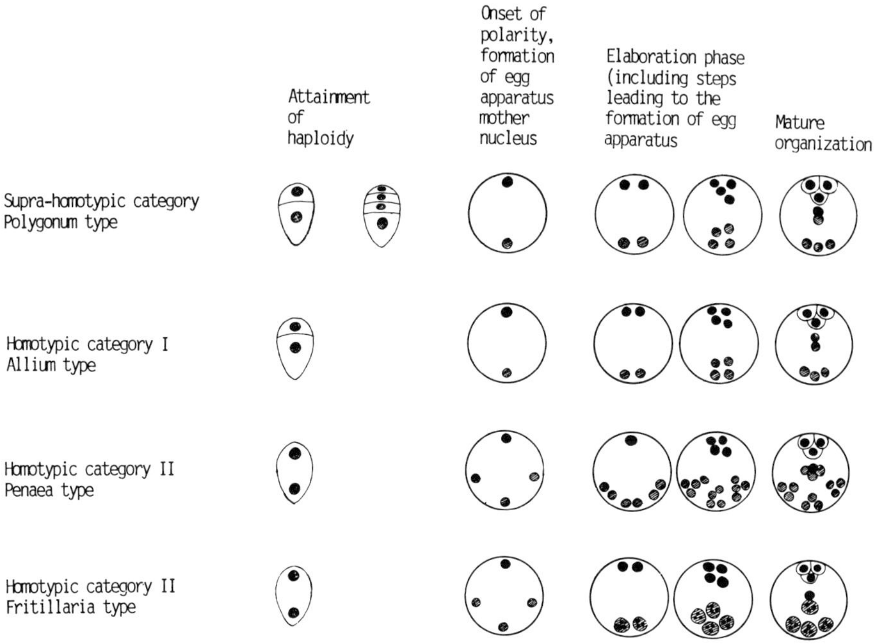

Fig. 3.6.2. Schematic representation of ontogenetic sequence of embryo sac types, based on the onset of polarity (micropylar nuclei *black*, chalazal *shaded*) after attaining haploidy. (After Swamy and Krishnamurthy 1975)

Plumbago and Adoxa types. The Fritillaria type, also the Plumbagella type, belong to Category II; the chief difference is that there is nuclear fusion and triploid nuclei are formed in the chalazal region.

While discussing the "Supra-Homotypic" and "Homotypic" categories, Swamy and Krishnamurthy (1975) state that at the two-nucleate stage of embryo sac the micropylar nucleus should be designated "egg apparatus mother nucleus", and the chalazal one as "primary antipodal nucleus". Since the micropylar and chalazal nuclei also give rise to upper and lower polar nuclei, Nagendran and Arekal (1976) emphasize that the upper nucleus be designated as "micropylar quartet mother nucleus", and the lower as "chalazal quartet mother nucleus".

The classification of embryo sacs proposed by Swamy and Krishnamurthy, and the nomenclature suggested by them and Nagendran and Arekal for the micropylar and chalazal nuclei of two-nucleate embryo sacs are of no special significance, and confusing. The conventional classification and nomenclature are much simpler, and convincing.

3.7 Mature Embryo Sac

In the majority of angiosperms, the embryo sac remains confined to the nucellus (within the limits of integuments) during its growth and maturation. But in *Pentaphragma horsfieldii* (Kapil and Vijayaraghavan 1962) and *Torenia violacea* (Yamazaki 1954), it becomes extra-micropylar at the four-nucleate stage. In *Lepidagathis trinervis* (Ambegaokar 1959) and *Nelsonia campestris* (Mohan Ram and Masand 1963), the apical end grows beyond the micropyle in the schizogenous cavity in the funicular region, and the egg apparatus differentiates in this cavity. An extreme condition is observed in *Exocarpus* spp. (Ram 1959a), where the embryo sac grows in the placenta (in the absence of a true ovule) and beyond it so that the egg apparatus becomes exposed (Fig. 3.4.3C).

Loranthaceae show a different pattern. Several embryo sacs grow concurrently up to varying lengths in the stylar and stigmatic region (Fig. 12.8.2F). The embryo sac exhibits maximum length in *Moquiniella rubra* (B.M. Johri and Raj 1969) where it grows up to the tip of stigma as in *Helixanthera lingustrina* (Fig. 12.8.2C), but (in *Moquiniella*) it curves backward and becomes inverted J-shaped (Fig. 12.8.2D).

The embryo sac develops micropylar and chalazal extensions, as in Acanthaceae, Santalaceae, Stylidiaceae and Opiliaceae. A singular condition is reported in *Exocarpus*, and finger-like outgrowths are produced in the apical and middle region of the embryo sac (Fig. 12.6.3A–C). The apex of the embryo sac extends in the stylar canal and broadens laterally. It develops several tubular processes which grow downward from the micropylar region of the embryo sac along the side of the placenta (Fig. 12.6.3A). Some of these haustorial processes extend upward beyond the level of the egg apparatus (Fig. 12.6.3B), and may reach the outer epidermis of the style. The development of a branched micropylar caecum in *Nuytsia floribunda* (R. Narayana 1958b) is also rare (Fig. 12.8.2E).

The mature embryo sac usually shows three-celled egg apparatus, three antipodal nuclei or cells, and two polar nuclei or secondary nucleus (Masand and Kapil 1966, B.M. Johri and Bhatnagar 1973). The synergids are absent in the Plumbago and Plumbagella types, and antipodals in the Oenothera type (Fig. 3.6.1). The number, as well as the ploidy-level, of the antipodal cells varies.

The mature embryo sac in Podostemaceae exhibits a different pattern (Fig. 28.2.1D, E), and a detailed description is given along with the family (Sect. 28).

The egg apparatus consists of the egg cell and two synergids (see Favre Duchartre 1975). It remains within the limits of the embryo sac wall except in Compositae, Ericaceae and Stylidiaceae where the synergids project into the micropyle. The egg cell shows a distinct polarity with its nucleus embedded in the dense cytoplasm at the lower end; the upper portion is occupied by a vacuole. In *Plumbago capensis* (Cass 1972), at the micropylar end the egg cell shows conspicuous wall ingrowths, the filiform apparatus. In the absence of synergids, the egg cell has taken over the function of the synergid, in addition to its gametic function.

The egg cell of the aposporous (nucellar) embryo sac of *Cortaderia jubata* (M.N. Philipson 1978b), rarely exhibits an extension (haustorium) which enters the micropyle, adjoining the synergid haustorium.

The accumulation of starch grains in the embryo sac is a common feature in several families, such as Acanthaceae, Aizoaceae, Bruniaceae, Cactaceae,

Crassulaceae, Frankeniaceae, Portulacaceae and Tiliaceae. In *Blepharis maderaspatensis* (Phatak and Ambegaokar 1963), the starch grains in the mature embryo sac are in such abundance that they mask the nuclei. The starch is gradually consumed, and disappears during post-fertilization stages. Compound starch grains have been reported in *Cansjera rheedii* (Swamy 1960). Fat globules occur in Sonneratiaceae.

Multiple Embryo Sacs. The concurrent development of two megaspore mother cells leads to the development of twin tetrads as in *Turnera ulmifolia* (Vijayaraghavan and Kaur 1966), and twin embryo sacs as in *Linum perenne* (D. Rao 1966). Rarely, mono- and bisporic gametophytes may develop side by side as in *Capsicum annuum* var. *acuminatum* (Dharmadhaj and Prakash 1978).

Six embryo sacs develop concurrently in *Cassytha filiformis* (Sastri 1962), ten or more in *Pterospermum suberifolium* (Venkata Rao 1952b), and 20–25 in *Casuarina equisetifolia* (Swamy 1948a).

Synergids. The synergids are pyriform and "hooked", or the tip is prolonged into a "beak". The cell wall is present only in the micropylar region. Similar to the egg, they also exhibit polarity. The nuclei lie in the upper narrow region, while the lower region is occupied by a vacuole. They show a characteristic structure, the filiform apparatus, at the tip. The apical "caps" have been reported in *Helianthemum vulgare* (Kapil and R. Maheshwari 1964).

The synergids are haustorial in Compositae – *Calendula* and *Ursinea* (K.V.O. Dahlgren 1924), *C. fulgida*, *C. maritima*, *C. persica*, *C. tunetana* (Ahlstrand 1979d) and *Osteospermum sinuatum* (Fig. 3.7.1A), *Cotula* (Davis 1961c), *Elephantopus* (Pullaiah 1979a) and *Ursinia anthemoides* (see Poddubnaya-Arnoldi 1976); Crassulaceae – *Sedum sempervivoides* (Mauritzon 1933a); Santalaceae – *Quinchamalium chilense* (Fig. 3.7.1B–D); Gramineae – *Cortaderia jubata* (Fig. 3.13.2D) and *C. selloana* (M.N. Philipson 1977, 1978b).

The role of synergids in the process of fertilization has attracted much attention. They secrete chemotropic substances and guide the growth of pollen tube; one of the synergids serves as a landing site for the pollen tube and provides nutrition (see Vijayaraghavan and Bhat 1983).

Following two successful pollinations, the sections of 75 ovules of *Hordeum vulgare* (cv Bonus) were examined to determine the pattern of early synergid degeneration (Mogensen 1984). The degeneration of both synergids is equally as likely as the degeneration of only one synergid. When only one synergid degenerates, it is mostly located towards the placenta (proximal).

Antipodals. The antipodals show much variation in number, size, ploidy-level, and may be ephemeral or persistent. Normally, there are three antipodal cells, but in a large number of plants they undergo secondary multiplication; the largest number is reported in Gramineae. The antipodals do not organize into cells in *Cassytha*, *Hydnora*, *Scilla*, *Thismia* and *Xyris*. The antipodal cells become polyploid, due to endomitosis or polyteny. They may be multinucleate and multinucleolate.

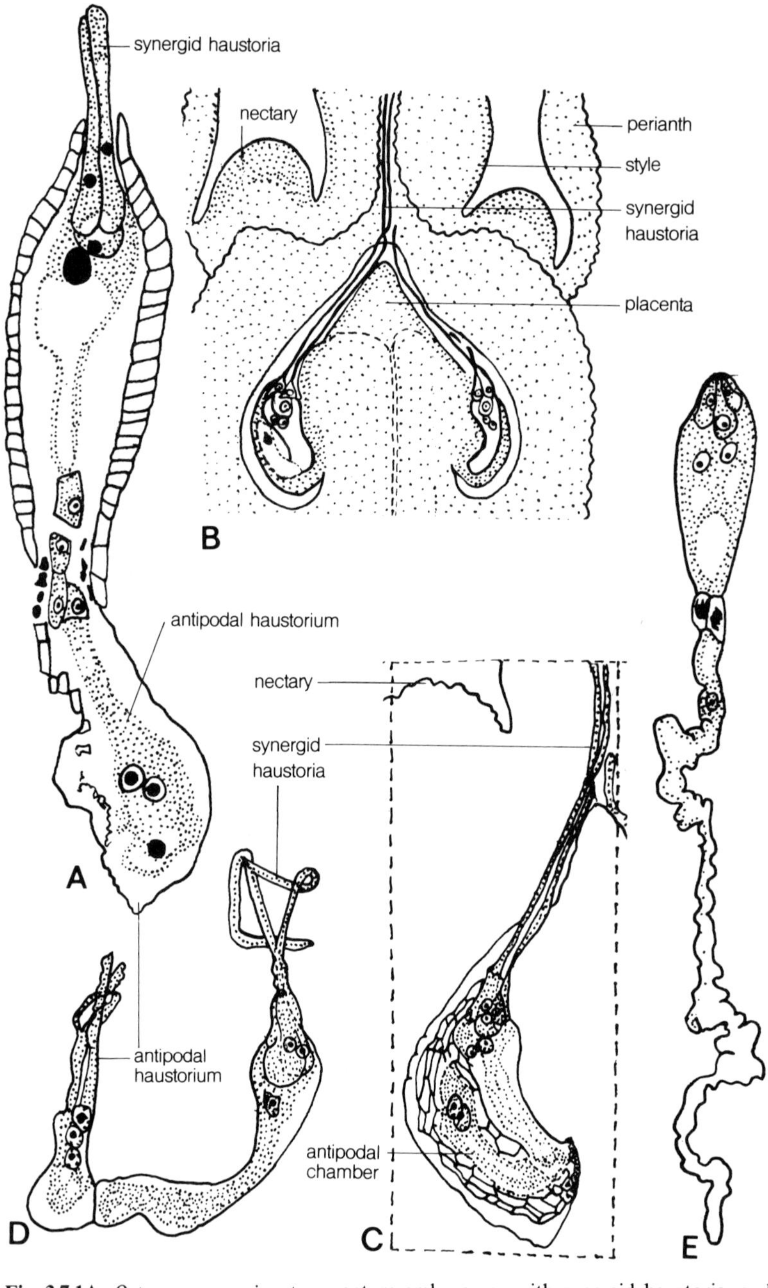

Fig. 3.7.1A. *Osteospermum sinuatum*, mature embryo sac with synergid haustoria, and an antipodal haustorium. (After Ahlstrand 1979d). **B–D** *Quinchamalium chilense*. **B** Longisection of lower portion of gynoecium; **C** Embryo sac from **B** (*left*). **D** Mature embryo sac (from dissection). Note extensive synergid haustoria in **B–D**, and branched antipodal haustorium in **D**. (After Agarwal 1962). **E** *Galium mollugo*, mature embryo sac with much-elongated basal antipodal cell. (After Fagerlind 1937)

At times, the antipodal cells enlarge considerably as in *Aconitum napellus* (Osterwalder 1898), *Argemone* (Sachar 1955) and *Ranunculus tripartitus* (T.K. Patel and Cook 1972), and become haustorial. In *Zea mays* (Diboll and Larson 1966), the antipodals are separated from one another by incomplete walls, and those adjacent to chalaza show conspicuous wall ingrowths. In *Grindelia* and *Prionopsis*, Howe (1975) observed that one or both antipodal cells grow laterally into the integuments. These outgrowths invade the endothelial cells, and occasionally approach the surface of the ovule. Howe's observations need confirmation. Their role in the nutrition of the embryo sac is undisputable.

A unique condition has been observed in *Quinchamalium chilense* (B.M. Johri and Agarwal 1965). The three nuclei at the chalazal end of the embryo sac are contained in an antipodal chamber which develops into a prominent haustorium (Fig. 3.7.1C, D). The haustorium grows through the funiculus, and branches profusely as it reaches the tip of placenta.

Osteospermum sinuatum (Fig. 3.7.1A) shows a conspicuous antipodal haustorium. In *Gallium mollugo* (Fig. 3.7.1E), the basal antipodal cell forms a long tubular haustorium.

The antipodals exhibit an unusual feature in the hybrids and polyploid varieties of Gramineae. According to Beaudry (1951), the antipodals alter the nutrient supply causing the break down of endosperm in a cross between *Elymus virginicus* L. × *Agropyron repens* (L.) Beauv. In an octoploid variety of *Hierochloë odorata*, Weimarck (1967a) noticed a degenerated mature embryo sac with five multinucleate antipodals. He considered it as an abnormality due to degeneration. B.M. Johri and Ambegaokar (1975) made a comparative study of the development of antipodals in relation to development of endosperm in three varieties of hexaploid triticales (a cross between *Triticum* and *Secale cereale*). The variety T 6437 shows maximum shrinkage of seeds due to failure of the development of endosperm. In this variety, the antipodals exhibit maximal size with hypertrophied nuclei, and they persist for a much longer time as compared to the varieties Triticale-1 and Bronco 90, which show normal development of endosperm.

Central Cell. The central cell is separated from the egg cell and synergids by plasma membranes, and from the antipodals by thin walls often traversed by plasmodesmata. The outer wall of the central cell may show transfer cell-like wall ingrowths. Thus, it is well-located for absorption, conduction and secretion of nutrients. It also serves as a repository for starch, lipids and proteins (Kapil and Bhatnagar 1983). During post-fertilization stages, the wall labyrinth of the central cell increases and new organelles appear.

Normally, there are two polar nuclei in the central cell, except in the Oenothera, Penaea, Plumbago, Peperomia and Chrysanthemum cinerariaefolium types (Fig. 3.6.1). When there are more than two, all of them come together and fuse simultaneously to form a secondary nucleus. Sometimes, the two polar nuclei remain distinct till triple fusion. The latter is said to be responsible for the stimulation of metabolic activity within the central cell.

3.8 Pollination

During pollination, the pollen grains are brought over to the stigmatic surface by various agencies. The role of style as a "sense-organ" is discussed by Gilissen (1976). Pollination in land plants has been studied for a much longer period than in aquatic plants. This aspect has received some attention in the description of the families (see also E.G. Williams et al. 1986).

The primitiveness of wind-pollination has been a matter of much discussion. There is no doubt, however, that the aquatic habit of higher plants and hydrophily are derived conditions. Most aquatic plants are still pollinated in the air, like other terrestrial plants: *Nymphaea*, *Alisma* and *Hottonia* are good examples of entomophily, *Potamogeton* or *Myriophyllum* for anemophily, and *Lobelia dortmanna* for autogamy. Some aquatics have taken advantage also of the water for pollination (Faegri and van der Pijl 1971).

Water-pollination may take place either on the surface of water (ephydrophily), a further step in the development from entomophily or anemophily. Not all pollinations taking place in water are hydrophilous. Many small, autogamous terrestrial plants may flower even when submerged, and the autogamous mechanism functions even under these circumstances, generally in an air bubble confined within a flower (Faegri and van der Pijl 1971).

Ephydrophily is unique among abiotic pollinations in that it takes place in a two-dimensional medium. Compared with a three-dimensional medium of anemophily or hyphydrophily, this makes possible a great saving of pollen.

In the classical example of *Vallisneria*, the entire male flower is released instead of individual pollen grains; the pollen itself therefore does not touch the water surface. Small depressions in the surface around the floating female flowers cause the male flower to slide down and the anthers attach themselves to the stigma. This effective mode of pollination results in the reduction of the number of pollen grains (ca 72) in a male flower. A similar mechanism is also reported in *Hydrilla* (explosive anthers), and *Lemna trisulca* where the whole plant rises up to the surface.

Water may come into the picture as a pollinating agent also in terrestrial plants, e.g. in rain-pollination. During rain the flowers fill with water up to a certain level. Pollen grains float on the surface of this water, and eventually reach the stigmas that are exposed at the level of water. Primarily, rain-pollination is autogamous, but a mechanism of allogamy is also conceivable with rain-splash, even if it is not very effective. So far, no flower is known which is dependent exclusively on rain-pollination.

It is essential for all forms of hydrophily that the pollen is not killed by immersion in water. Ephydrophily and rain-pollination presume water-repellent pollen. In hyphydrophilous plants pollen grains endure wetting.

Cox and Knox (1986, see also Knox and Ducker 1976) have given an excellent account for four taxa. The marine sea-grasses *Halodule pinifolia* and *Halophila ovalis*, and fresh-water aquatics *Lepilaena cylindrocarpa* and *Ruppia polycarpa* are surface-pollinated. The pollen floats on the surface upon release from the anthers. The individual pollen grains join together in large floating rafts that function as search vehicles. Initially, each search vehicle represents the products of a single anther but subsequently through collisions even larger search vehicles of mixed parentage are formed. Pollen and stigma differ significantly between the marine and freshwater types (Tables 3.8.1). In the freshwater genera *Lepilaena* and *Ruppia*, the pollen is ellipsoidal or spherical while the floating stigmas present a small, round target (the indusiate stigma plus the radius of surface depression in the surface plane). In the marine genera *Halophila* and *Halodule*, however, the pollen is filiform as are floating stigmas which present a linear target of decreasing length as the tide rises. In all types they are either long and filamentous or borne on long filamentous peduncles which (after fertilization) retract the developing fruit beneath the surface.

In *Lepilaena* and *Ruppia*, anthesis occurs under water or at the surface. If the *Ruppia* inflorescence is under water, a gas bubble propels the mass of pollen to the water surface. There the pollen grains line up side by side forming search vehicles. Surface and water currents propel these vehicles, and on encountering the stigma, they adhere. In *Lepilaena*, the boat-like anthers are released under water, float to the surface, and dehisce. These search vehicles drift through the influence of wind and waves, some are trapped in the slight depressions in the water surface formed by the floating tripartite stigmas of the apocarpous female inflorescences. In both these freshwater systems, search vehicles which fail to encounter stigmas fuse into a yellow hydrophobic surface film that accumulates towards the edge of the pond.

In *Halodule*, the tiny stamens become erect and dehisce when exposed to the air at low tide. The anthers express the filiform pollen as a flocculent mass. As the tide comes in, the pollen grains form search vehicles of large diameter which float on the surface. The long linear stigmas also float and pollination occurs through the collision of a search vehicle with the floating stigmas. In *Halophila*, the staminate flowers produce three anthers which are adnate until just before dehiscence at low tide. At dehiscence, the pollen tetrads (which remain held in their mucilaginous tubes) are released as flocculent masses that rapidly expand into search vehicles. The long floating filaments of the three branched stigmas form linear targets for the search vehicles.

After a detailed consideration of four systems of aquatic pollination, pollination in a few land plants may be taken up. In grasses (Gramineae) and sedges (Cyperaceae), pollination is typically anemophilous. Pant et al. (1982a, b) observed entomophilous pollination in *Dichanthium annulatum*, *Cyperus rotundus* and *Scirpus lacustris*. The inflorescences of these plants are regularly visited for foraging by two types of insects. Those which collect pollen grains on their hind-legs – *Apis indica*, *Micrapis florea*, *Megapis dorsata* and *Melipona laeviceps*, and those which collect pollen grains on their hind-legs, thorax and abdomen – *Nomia andrenina*, *N. westwoodii*, *Lasioglossum* (*Ctenonomia*) *albescens*, *Halictus* (*Seladonia*) *lucidipennis* and *Evylaeus* sp.

Some of the insects are pollen specific, while others visit different flowers for pollen and nectar.

Armbruster (1984) has drawn attention to the fact that instead of the usual nectar, some flowers produce resin, e.g. *Dalechampia* (Euphorbiaceae) and *Clusia* (Guttiferae). Bees that pollinate these flowers collect and use the resin in their nest construction. Floral resins are slow in hardening as compared to

Table 3.8.1. Characteristics of surface-pollinated genera (Adapted after Cox and Knox 1986)

	Cymodoceaceae *Halodule* n = 114	Hydrocharitaceae *Halophila* n = 5	Zannichelliaceae *Lepilaena* n = 172	Ruppiaceae *Ruppia* n = 300
Habitat	Marine	Marine	Freshwater	Freshwater
Pollen Morphology				
Shape	Filiform	Ellipsoidal[a]	Spherical	Boomerang
Length (μm)	1000	80	50	140
Search Vehicle Diameters				
Median (μm)	5000	4690	338	1361
Range	1000–22000	1320–7520	118–2020	220–1817
Transport	Floats	Floats	Floats	Floats
Stigma Characteristics				
Shape	Filiform	Filiform	Indusiate	Indusiate
Length (μm)	22000	13000	776	1222
Placement	Floats	Floats	Floats	Floats

[a] Embedded in long mucilaginous tube.

many other plant resins. This facilitates collection and storage by bees, and probably enhances the attractiveness of the flowers. Floral resins are presumably of great utility and dependability and may be an especially important resource to certain bees, which are major pollinators of many species of tropical plants. Bees appear to forage floral resin using a strategy of energy efficiency: large bees collect resin only from sources with a large amount of resin; small bees collect resin from sources with a small as well as a large amount of resin.

Resin secretion in flowers may have primarily arisen as a defence against herbivores, and secondarily as an "award" to the pollinator. These bees may also collect some pollen, besides resin. This aspect of floral resin attracting pollinators requires further study.

Ant pollination systems are remarkably rare. Beattie et al. (1984) point out that the pollen exposed to ants for brief periods exhibits reduced viability, reduced per cent of germination, and shorter pollen tubes relative to control pollen. Pollination with ant-borne pollen on *Lycopersicon peruvianum* resulted in lower seed-set than with control pollen. This disruption of pollination processes could exert a powerful selection pressure against the evolution of ant-pollination systems. It is suggested that the nest-building and brood-rearing habit of ants requires that they secrete large amounts of antibiotics to combat pathogenic microorganisms. It is these secretions that disrupt pollen function; bees and wasps exhibit very different nesting behaviour. The eggs, larvae and pupae lie exposed in the nest, unprotected by either a wax cell as in bees, or a paper one as in wasps. Consequently, there are no chemical barriers to their co-evolving with flowers as pollinators.

Intracarpellary Pollen Grains. In *Butomopsis lanceolata* (B.M. Johri 1936b, see also B.M. Johri and Ambegaokar 1984), occasionally the pollen grains were observed in the stylar canal. In one instance a pollen grain had found its way into the ovary and germinated on the surface of an ovule. The occurrence of intrastylar and intraovarian pollen grains is a rare feature, and has been observed in *Amianthium muscaetoxicum* (Eunus 1951a), *Annona* (Adatia 1946), *Bootia cordata, Butomus umbellatus, Hydrocleis nymphoides, Limnocharis emarginata* and *Trillium* (B.M. Johri and Bhatnagar 1957), *Erythronium americanum* (Haque 1951), *Fritillaria roylei* and *Lilium tigrinum* (Vasil and M.M. Johri 1964), and *Ottelia alismoides* (Islam 1950).

Pollen Germination. When the pollen germinates, mostly it produces a single tube. Polysiphonous condition refers to the emergence of several pollen tubes (Fig. 3.8.1A). Sometimes, three to five tubes emerge and one or two tubes may branch (Fig. 3.8.1B, C). The generative cell and tube nucleus migrate into separate branches (Fig. 3.8.1C). In *Lilium regale* (Fig. 3.8.1D), the pollen tube becomes coiled and produces one or two side branches. The main tube continues to elongate even after the generative cell and tube nucleus migrate into a branch. Branched pollen tubes are common in the Casuarinaceae (Fig. 4.2.1B) and Betulaceae (Figs. 9.2.1, 9.2.2A).

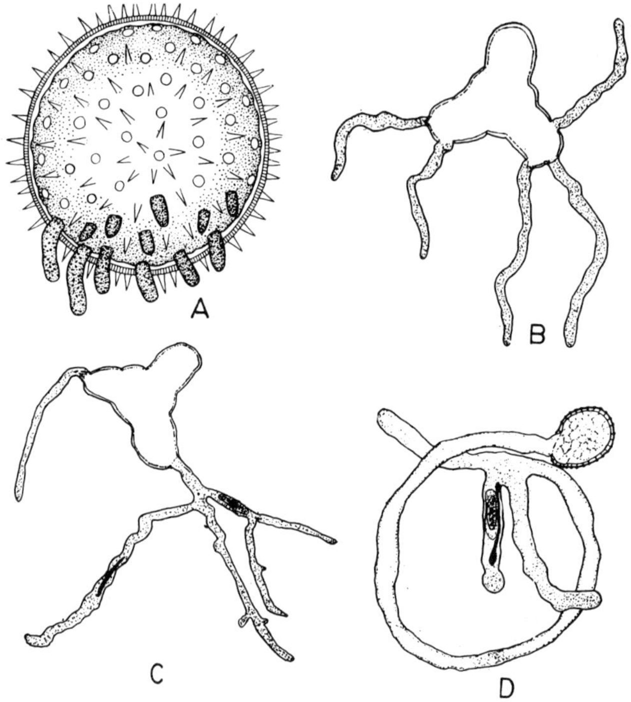

Fig. 3.8.1A–D. Polysiphonous germination of pollen grains. **A** *Lavathera* sp., emergence of many tubes. (After Poddubnaya-Arnoldi 1976). **B, C** *Clarkia elegans*. **B** Pollen with five unbranched tubes; **C** Pollen with one un-branched tube, and second tube branched with generative cell and tube nucleus in different branches. **D** *Lilium regale*, the pollen tube has coiled; the generative cell and tube nucleus are in the same branch. (After Johnston 1959)

Usually, the pollen grains germinate singly (Fig. 65.2.1A). In taxa in which pollen grains are shed in tetrads (Fig. 65.2.1H–K), or pollinia (Fig. 26.19.1B), or massulae (Fig. 65.2.1M), pollen germination occurs in composite units. All the four microspores in a tetrad may germinate synchronously; the outermost microspores in a massulae germinate first and then the others.

Post-Pollination Events. Using a semi-vivo pollination technique, G.B. and D.L. Mulcahy (1985) draw attention to the influence of the ovary on the growth of pollen tube.

Pollination on cut styles of *Nicotiana elata* and *Petunia hybrida* showed poorer growth of pollen tubes as compared to growth on styles intact with the ovary. The presence of other floral parts has no effect on the growth of pollen tubes.

Following a detailed study of six species of the Umbelliferae, eight of Cruciferae, six of *Solanum* and ten unrelated species, Cruden and Lyon (1985) emphasize that there is a positive correlation between stigma depth – approximate distance a pollen tube has to grow to reach the exogenous source of nutrients in the transmitting tissue – and pollen grain size, and *not* between pollen grain size and stylar length. All these pollen grains have a starchy food reserve. Absence of a correlation between pollen grain size and stylar length is inconsistent with the hypothesis that pollen grains must contain sufficient nutrients to sustain the growth of the tube up to the ovule. In fact, as early as 1830, Amici had suggested that the pollen tubes, during their growth, obtain nutrients from the transmitting tissue. In recent years, many experimental studies have demonstrated the utilization of stylar nutrients by the growing pollen tubes.

Pollen-Pistil Interaction. The surface of the stigma is the landing and germinating platform for pollen grains, and it also participates in the recognition and acceptance/rejection (compatibility/incompatibility) reactions (for stigma characteristics, see J. Heslop-Harrison 1981).

Pollen grains have a sticky or oily surface in many plants. They adhere to the exudate of wet stigmas, or to the papillae of dry stigmas. The stigmatic exudate protects the stigma against excessive desiccation, and provides a suitable medium for pollen germination. The papillae of dry stigmas are covered with a pellicle of external proteins which are hydrophylous (Shivanna 1978). Following pollen adhesion to the papillae, they become hydrated, and the wall-bound proteins of pollen grain are released. The cuticle of stigma is dissolved, and pollen tubes enter the transmitting tissue of the stigma. The transmitting tissue in the style provides mechanical and chemical guidance for growing pollen tubes as well as the nutrients (see Kroh and Helsper 1974). In the ovary, transmitting tissue (the obturator) provides additional guidance for the growth of pollen tubes.

Based on physiological studies, the entire mechanism from the germination of pollen grain up to the discharge of male gametes in the embryo sac has been divided into several subsystems. These proceed effectively only when the interaction of pollen tube with the tissues of pistil is healthy. Otherwise, the process may terminate at several subsystems. The following flow chart (adapted after E.G. Williams et al. 1982, Knox 1982, 1984, Shivanna 1982,

Compatible Pollination	Incompatible Pollination
Pollen lands on Stigma Pollen hydration	
1. Pollen grains germinate and produces tubes	Do not germinate
2. Tubes grow through the transmitting tissue of the stigma and enter style	Tubes abort on stigmatic surface
3. Tubes grow through stylar canal/transmitting tissue	Tubes abort in the upper part of style
4. Tubes ėnter ovary	Tubes fail to enter ovary
5. Tubes enter micropyle of ovules	Tubes fail to enter micropyles
6. Tubes enter embryo sacs	Tubes do not enter embryo sacs
7. Tubes discharge male gametes in the embryo sac	Tubes fail to release male gametes
8. Fertilization occurs	Fertilization fails

Shivanna and Johri 1985, Knox et al. 1986) indicates the arrest points (*right*) in incompatible crosses compared to normal pollination (*left*).

In a pollen tube the cytoplasm and most of the organelles are located in the tip-region. As the tube grows and elongates, callose plugs develop on the inner side of the tube in a centripetal manner in the region away from the tip so that the backward flow of cytoplasm in the tube is checked. The growth of the tube is unidirectional through the transmitting tissue, guided by the architecture of the transmitting tissue.

In compatible pollinations, the pollen tubes effectively utilize the nutrients of the transmitting tissue (see Sassen 1974).

More than one tube can reach the embryo sac, but only one tube penetrates the degenerated synergid via the filiform apparatus. Sperm cells, vegetative nucleus and tube cytoplasm with many amyloplasts are discharged into this synergid.

Usually, the pollen tube passes through the micropyle to the embryo sac and exhibits porogamy. But in Anacardiaceae (*Pistacia*), Betulaceae (Figs. 9.2.1, 9.2.2A), Casuarinaceae, and Juglandaceae the pollen tube penetrates the chalaza, often branches there, and one of the branches passes upward through the nucellus and penetrates through the micropyle (extra-vesicular chalazogamy). In *Corylus* (see Swamy 1948a), and rarely in *Eucommia ulmoides* (Eckardt 1963), the pollen tube enters the embryo sac through the chalazal end, traverses through it and then reaches the egg cell (intra-vesicular chalazogamy). In mesogamous mode, the pollen tube penetrates through the integuments and reaches the micropyle. Reayat Khan (1970) coined the term "exogamy" for fertilization in Lentibulariaceae. The apex of the embryo sac is extraovular and fertilization occurs in the ovarian cavity. In Loranthaceae, true ovules are absent. In many taxa the tip of embryo sacs grow into the stylar canal and stigma, and the pollen tubes meet the embryo sacs in this region.

3.9 Double Fertilization

After pollination, one of the synergids shows signs of degeneration. Synergid degeneration far ahead of the pollen tube penetration into the ovary (Jensen et al. 1983) is another evidence of this interdependence, or reciprocal influence of the two gametophytes well before they come into contact with each other. The pollen tube penetrates the embryo sac, and enters the degenerated synergid through the filiform apparatus. It bursts in the chalazal region of the degenerated synergid. A coiled persistent pollen tube inside the embryo sac has been observed in *Passiflora calcarata* (M.V.S. Raju 1956b). The contents of the pollen tube – two male gametes, sometimes vegetative nucleus, and cytoplasm – are liberated through a pore which is either terminal (*Petunia hybrida* – van Went 1970, *Plumbago zeylanica* – Russell 1980a, b), or subterminal (*Gossypium hirsutum* – Jensen and Fisher 1968).

Double fertilization in angiosperms (see Steffen 1963b, Korobova 1974, Battaglia 1981, Belyaeva 1981, van Went and Willemse 1984) follows pollination. In this process one of the male gametes (sperm) fuses with the egg cell (syngamy), while the other male gamete fuses with the secondary nucleus (Fig. 3.9.1) or the two polar nuclei (triple fusion) (Figs. 3.9.2A–D, 3.9.3A).

Rarely, more than one pollen tube may enter the embryo sac (for pollen-ovule ratio see Cruden 1977, Cruden and Jensen 1979). As many as three tubes and sperm cells have been reported in *Cuscuta reflexa* (Fig. 3.9.3B).

Frequently, two darkly staining structures have been observed in the discharged pollen tube, or in the degenerated synergid, or lying close to the egg apparatus. Both structures are of nuclear material as revealed by ultrastructural studies and radioautographs. These so-called "X"-bodies may represent the degenerated vegetative nucleus of the pollen tube and degenerated nucleus of the synergid.

The additional pollen tubes, on entering the embryo sac, may fertilize the intact synergid leading to the formation of a proembryo (Fig. 3.12.1G).

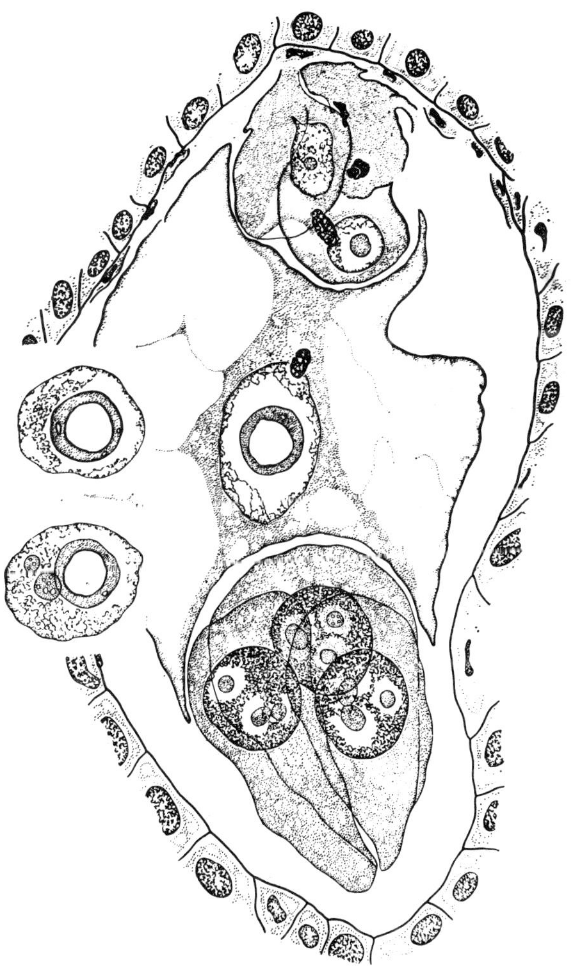

Fig. 3.9.1. *Anemone nemorasa*, double fertilization. *Inset*, zygote nucleus and primary endosperm nucleus. (After Yoffe 1969)

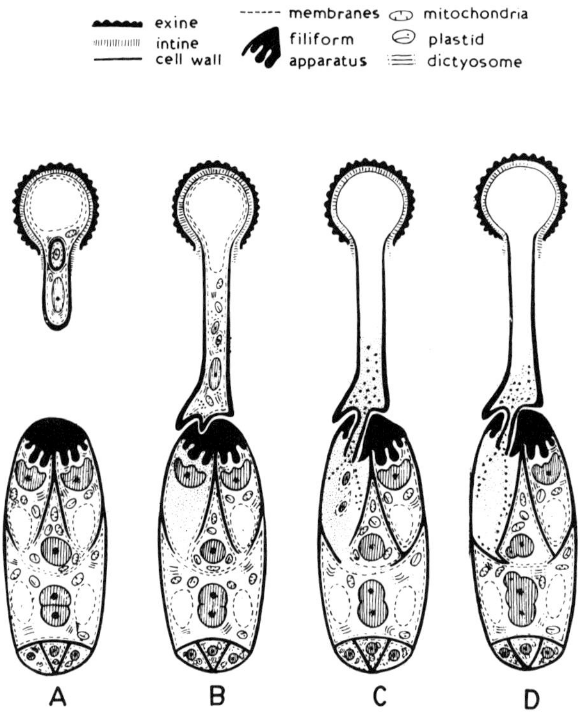

Fig. 3.9.2. Diagrammatic representation of male and female gametophyte. **A** Pollen tube and embryo sac. **B** Entry of pollen tube. **C** Discharge of male gametes inside the degenerated synergid. **D** Syngamy and triple fusion. There is no wall in the chalazal region of egg cell and synergids. Note the distribution of mitochondria, plastids and dictyosomes in the male and female gametophyte. (After Cocucci 1969)

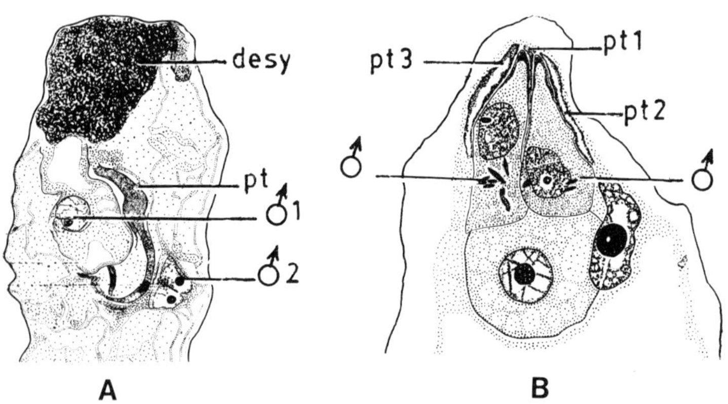

Fig. 3.9.3A, B. Fertilization. **A** *Camellia sinensis*, micropylar portion of embryo sac to show syngamy and triple fusion; *desy* degenerated synergid; *pt* pollen tube; (After Sethi 1965). **B** *Cuscuta reflexa*, upper part of fertilized embryo sac with zygote and primary endosperm nucleus; there are remnants of three pollen tubes, seven (or eight?) sperms are lying in one synergid and three in the other. (After B.M. Johri and Tiagi 1952)

The synergid bursts and the male gametes enter the egg cell, or the central cell through the bridges fromed due to the fusion of plasma membranes (see Kapil and Bhatnagar 1975).

The egg cell is said to have an attraction zone (Belyaeva 1977). As the sperms emerge from the synergid, one after the other, the first one is attracted towards the egg cell. As soon as it comes in contact with the wall of the egg cell, the attraction ceases. The second sperm is carried to the secondary nucleus.

Our information on the movement of male gametes to the egg and polars or secondary nucleus is very meagre. There is no certainty that the amoeboid movement of the male gametes, or the streaming of cytoplasm of the central cell, brings them in contact with the egg and polars/secondary nucleus.

At the time of fertilization, the male gamete is at the late telophase of mitosis (i.e. in the form of a spireme) initiated at spermatogenesis, while the egg nucleus is in the "aging" or "maturing" phase (Steffen 1951, Gerassimova-Navashina 1960, 1982).

Attention is invited to the detailed studies of Russell (1980a, b, 1982, 1983, 1984) and Russell and Cass (1981, 1983, 1988) on fertilization in *Plumbago* and *Plumbagella.*

Persistent Pollen Tubes. After fertilization, the remnants of the pollen tube persist for some time in the micropyle as well as in the embryo sac, or only in the embryo sac. There are several instances of the remnants persisting in the mature seed (see P. Maheshwari 1950, B.M. Johri and Ambegaokar 1984). From time to time it has been argued that after the discharge of gametes the persistent pollen tubes have a haustorial role, absorb nutrients from the circumjacent tissue, and pass on to the embryo until it is able to establish its own source of nutrition (see Longo 1903, Cook 1909, Venkata Rao 1952a, D. Singh 1963). Werner (1915) and Täckholm (1915) mention that in the Onagraceae the pollen tube becomes greatly broadened in the micropyle, and often sends out branches into the outer integument and nucellus. At the same time, the pollen tube grows towards the embryo sac destroying the cells in its path. The tube is recognizable until the embryo sac has enlarged considerably. Werner and Täckholm presume that the pollen tube absorbs nourishment from the adjoining tissue and transmits it to the embryo. In *Oxybaphus* (Cooper 1949), the pollen tube remains recognizable throughout the development of seed and probably transports nutritive materials from the secretory cells of the funiculus to the embryo.

Tilquin et al. (1983) refer to haustorial pollen tubes in *Fuchsia boliviana.* The pollen tubes do not collapse (disappear) after fertilization, but become even more conspicuous and persist for a considerable time, and remain recognizable in the mature seed. Tilquin et al. assume that after fertilization the pollen tube absorbs food materials still present in the style and partly also from the central region of the ovary and conveys these to the embryo until it can obtain nourishment from the endosperm. They strengthen their argument from the well-known fact that the pollen grain does not contain adequate reserve food to sustain the total growth and elongation of the tube which absorbs nourishment from the transmitting tissue of the style.

Indra and Krishnamurthy (1985) emphasize that in *Ottelia alismoides* the pollen tube forms a bulla (balloon-like swelling) as it enters the micropyle. It grows into a synergid and discharges the two male gametes and some cytoplasm. Soon after fertilizations, the micropyle in the region of outer integument becomes extremely wide; now the actual micropyle is said to be formed only by the inner integument. The "bulla" formation occurs at two or more sites between the obturator (glandular tissue) and the synergid.

Indra and Krishnamurthy further state: "There is a substantial increase in cytoplasmic content of the pollen tube in the post-fertilization stages especially in the 'bulla' regions as indicated by all staining procedures". The persistent pollen tube finally bursts at the bulla above the inner integument. The pollen tube is enucleate, and the increase in cytoplasm is due to absorption of reserve food material from outside the ovule (stylar glandular transmitting tissue) and transmission to the embryo sac. Indra and Krishnamurthy, using histochemical techniques, found only indirect evidence of the absorption of nutrients from outside the ovule and their transmission to the embryo sac.

The widening of the micropyle (exostome) is very doubtful and unconvincing. P. Maheshwari and Johri (1950b) also examined post-fertilization development in *O. alismoides* but did not observe either the widening of the micropyle or the formation of the bulla.

Briggs (1989) studied early post-fertilization development in *Swainsona greyana*. He reported that the pollen tube and the penetrated synergid persist throughout proembryonic development until the heart-shaped stage. After fertilization, the pollen tube swells, forms a bulla within the nucellar cap cells and stains for insoluble carbohydrates and proteins. Two days after anthesis, wall ingrowths develop along the suspensor cell wall in contact with the penetrated synergid before developing along all of the outer walls of the basal suspensor cells. A pollen tube-synergid-proembryo nutrient pathway is postulated.

Herscovitch and Martin (1989, 1990) made a histochemical and ultrastructural study of pollen-pistil interaction in *Grevillea banksii*; Martin and Herscovitch (1989) reported the behaviour of the pollen tube in *G. banksii* and *G. sericea*. In *G. banksii*, after fertilization of the ovules, the pollen tube branches near or at the micropyle. The enucleate branches invade the ovary wall where further branching occurs and the branches ramify intercellularly in the ovarian tissue. Branching was not observed in the styles of excised flowers, or in the ovular tissue. Martin and Herscovitch's conclusion on the haustorial function of pollen tube branches, which invade the ovary wall, is based on circumstantial evidence. They used fluorescence microscopy after decolourizing aniline blue staining of thick sections or entire ovaries macerated in 8 N KOH for 8 h or more. This is the first and only report that the pollen tubes branch and "parasitize" the ovary wall.

P. Maheshwari (1950) and B.M. Johri and Ambegaokar (1984) discussed the role of persistent pollen tubes and emphasized that these are merely the remnants of a "dead" structure which are not at all concerned in absorbing and transmitting the nutrients.

This controversy could be resolved by examining live dissected material with a phase contrast microscope. Zhou and Yang (1982, 1985) have developed a technique for enzymatic isolation of live pre- and post-fertilized embryo sacs. Success would depend on the proper selection of the material at the right stage.

Dissections, tran- and longisections, fluorescence, electron and interference microscopy, and histochemical studies to follow up the food reserve in the ovular and ovarian tissue from the time of entry and branching of the pollen tube up to the cessation of growth of the pollen tube would be very rewarding.

Another approach is to label the metabolites in emasculated flowers with $^{14}CO_2$, pollinating the pistil with normal (unlabelled) pollen, and then to demonstrate the ^{14}C-labelled metabolites in the pollen tube branches in the ovular/ovarian tissue. This would be a convincing demonstration that (during post-fertilization development) the persistent tubes absorb labelled metabolites from the adjoining tissue, and actually have a haustorial function.

The labelling can be achieved (see Coe 1954) by feeding a leafy branch with emasculated flowers (ovules at the mature embryo sac stage) with $^{14}CO_2$ for 36 h or more. Due to photosynthetic activity, the metabolites in the stigma, style, ovary and ovules will become labelled with ^{14}C. By autoradiography of sections of frozen/fixed material (following pollination and fertilization), labelled compounds in the bulla, unbranched/branched pollen tubes in the ovular/ovarian tissue can be detected.

So far, none of the investigators have used these procedures to demonstrate the haustorial function of the pollen tube (see B.M. Johri 1992). Assumptions are essentially based on indirect and circumstantial evidence.

3.10 Endosperm

Endosperm, a product of double fertilization, develops only in angiosperms. A male gamete fuses with the polar nuclei, or the secondary nucleus, and the primary endosperm nucleus is produced. The division of the primary endosperm nucleus, and repeated divisions of the daughter nuclei lead to the formation of endosperm tissue (Swamy and Ganapathy 1957a, Bhatnagar and Sawhney 1981, Vijayaraghavan and Prabhakar 1984, Bhojwani and Bhatnagar 1986).

The primary endosperm nucleus is not formed in Podostemaceae. In Trapaceae (Sect. 39.3) and Orchidaceae (Sect. 65.2), either the division of the primary endosperm nucleus is suppressed, or only a few divisions occur but all the daughter nuclei degenerate.

The nutritive role of endosperm has long been recognized. It nurses the embryo from the proembryo stage till it becomes self-sufficient, and regulates the pattern of development (Krishnamurthy 1988). The endosperm contains growth regulators, and coconut milk, sometimes the extract of corn endosperm at milk stage, is added to the nutrient medium for growing tissues in cultures. In *Salix* (Håkansson 1954), the endosperm is ephemeral and provides only growth regulators.

The peripheral cells of endosperm develop phellogen in the "naked" seeds of *Crinum defixum* (B.S.M. Dutt 1962), while the inner cells are chlorophyllous and store starch. The mature endosperm stores carbohydrates, proteins or fatty food reserve. The chemical composition of the storage products forms the basis for taxonomic classification of monocotyledons (see Huber 1977).

The developing endosperm derives nutrients from the food reserves stored in the nucellus and integuments. The development of chalazal, micropylar or secondary endosperm haustoria leads to partial or entire absorption of integuments in several families. The peripheral layers of the endosperm become cutinized, and serve as protective layers as in *Andrographis* spp. (Mohan Ram 1960b, Mohan Ram and Masand 1963), *Elytraria acaulis* (B.M. Johri and Singh 1959) and *Haplanthus tentaculatus* (Phatak and Ambegaokar 1961a).

Depending on its mode of development, the endosperm is said to be of the Nuclear (Figs. 3.10.1A–D, 3.10.2A–F), Helobial (Fig. 3.10.1E–L) and Cellular (Fig. 3.10.1M–O) type.

Nuclear Type. The primary endosperm nucleus divides followed by free-nuclear divisions (Fig. 3.10.1A–D). The free nuclei either aggregate in the micropylar or chalazal region, or in both, or may become distributed in a film of cytoplasm around the central vacuole. At times, some of the nuclei may fuse to form polyploid nuclei as in *Stemona tuberosa* (Swamy 1964b).

The endosperm remains free-nuclear throughout in *Floerkea proserpinacoides* (P. Maheshwari and Johri 1956) and *Oxyspora* (Subramanyam 1951a). In the majority of plants, however, it does become cellular. Wall formation occurs at different stages of development in different species, as early as the four-nucleate endosperm in *Coffea arabica* (Mendes 1941), or even after the formation of 1000 free nuclei as in *Anemone* (Bhandari 1968). Wall formation is centripetal or it is initiated at the micropylar end and proceeds to the chalaza (Fig. 3.10.3D). It commences from the chalazal end and proceeds upward in *Carya* (McKay 1947), *Elatine* (Frisendahl 1927), *Frankenia* (Walia and Kapil 1965) and *Pistacia* (Grundwag 1976).

The single layer of endosperm in *Begonia crenata* (Swamy and Parmeswaran 1960) undergoes periclinal divisions; the outer layer develops into an aleurone layer while the inner produces the endosperm tissue. The peripheral layer of the cellular endosperm becomes meristematic in Gramineae, and contributes cell layers to the inner endosperm. Later, the peripheral layer develops into an aleurone layer.

Fig. 3.10.1A–O. Types of endosperm. **A–D** *Scoliopus bigelvii*, Nuclear type. (After Berg 1959). **E–L** Helobial type. **E–J** *Philydrum lanuginosum*. (After Hamann 1966b). **K–L** *Sparganium erectum*. **K** Lower part of embryo sac; *ME* micropylar endosperm chamber; *CE* chalazal endosperm chamber; *Ant* antipodals; **L** Chalazal chamber and remnants of antipodals (*AntR*). (After M. Muller-Doblies 1969a). **M–O** Cellular type. **M** *Arisaema wallichianum*, note the large chalazal haustorium corresponding to the chalazal chamber in Helobial type. (After S.C. Maheshwari and Khanna 1956); **N** *Spirodela polyrhiza*. (After S.C. and N. Maheshwari 1963); **O** *Lemna paucicostata*, the entire endosperm is cellular and a chalazal haustorium is absent. (After S.C. Maheshwari and Kapil 1963b)

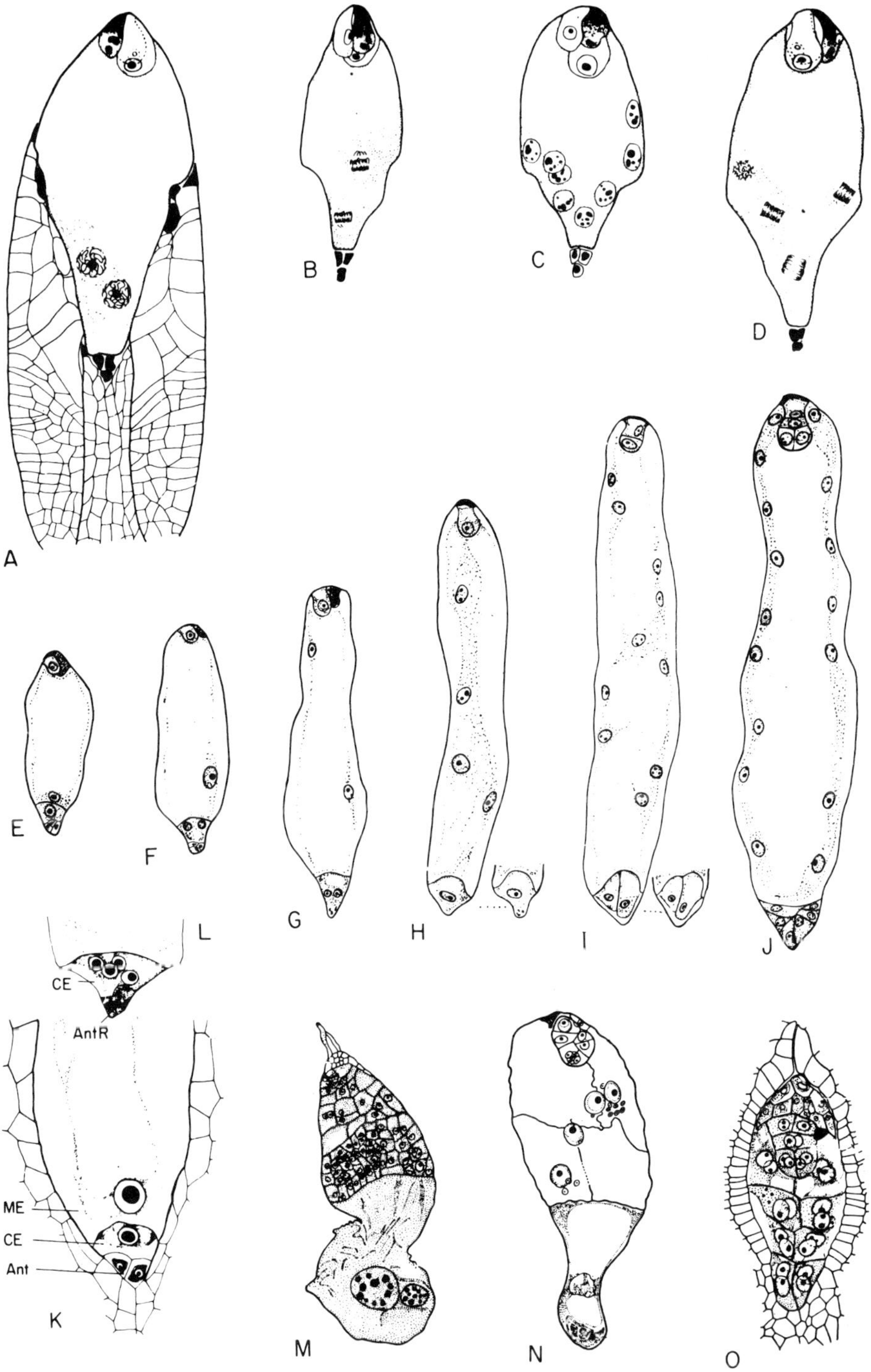
A
B
C
D
E
F
G
H
I
J
L
CE
AntR
ME
CE
Ant
K
M
N
O

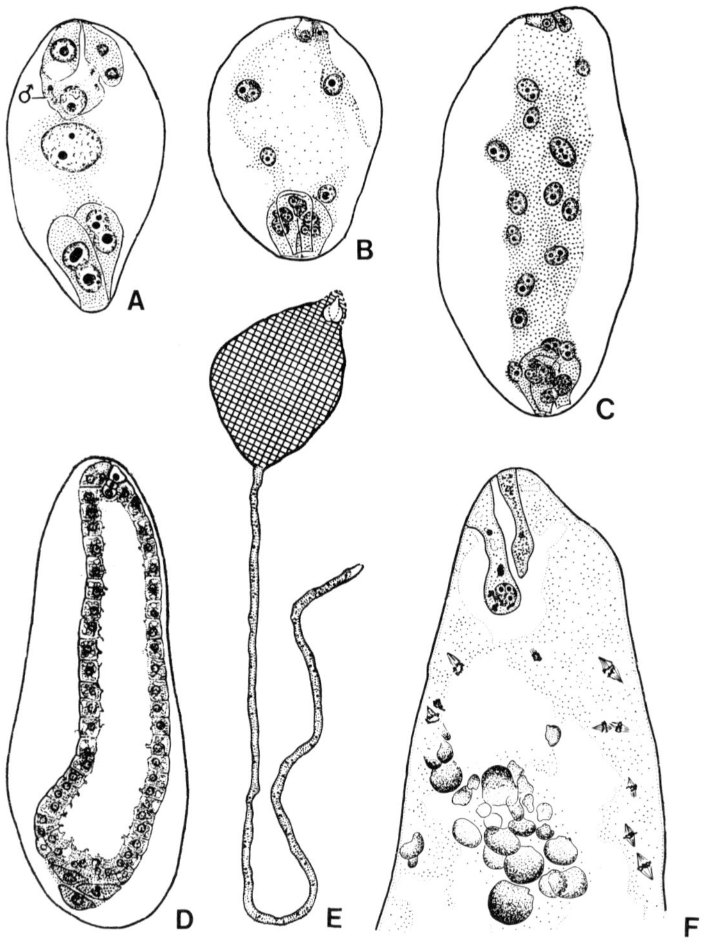

Fig. 3.10.2A–F. Nuclear endosperm. **A–D** *Caltha palustris*. **A** Fertilized embryo sac, syngamy has not yet been completed; **B** Four-nucleate endosperm, the antipodals have become polyploid; **C** 16-nucleate endosperm, the zygote has not yet divided; **D** Centripetal wall formation at three-celled proembryo. (After Kapil and Jalan 1962). **E** *Echinocystis lobata*, the endosperm proper has become cellular and there is a long coenocytic chalazal haustorium, embryo dicotyledonous (dissected wholemount). (After Chopra and Seth 1977). **F** *Foeniculum vulgare*, micropylar portion of embryo sac with zygote, and endosperm nuclei in division. Note enucleate endosperm vesicles. (After S.C. Gupta 1964)

The development of micropylar haustorium, as in *Frankenia hirsuta* (Walia and Kapil 1965), is a rare feature. It becomes cellular at a very late stage.

The formation of chalazal haustorium has been noted in Anacardiaceae (*Pistacia*), Cucurbitaceae (Fig. 3.10.2E), Euphorbiaceae, Leguminoseae and Proteaceae. Streaming movement of the protoplasm has been observed in the vermiform appendage of *Grevillea robusta* (Kausik 1941a), and coenocytic haustorium of *Trichosanthes anguina* (Chopra 1955). The length of the chalazal haustorium is maximal (19,000 μm) in *Sechium edule* (D. Singh 1964a). The development of unicellular finger-like appendages all over the surface of the endosperm in *Lomatia polymorpha* (Venkata Rao 1962) is unique. It increases the absorptive surface.

Another noteworthy feature in the Nuclear type of endosperm is the formation of enucleate (Fig. 3.10.2F) and nucleate cytoplasmic nodules. Enucleate nodules have been reported in *Passiflora* (M.V.S. Raju 1956b) and *Foeniculum vulgare* (S.C. Gupta 1964). According to Al-Attar (1977), the formation of such nodules distributes the endosperm nuclei to the periphery and prevents their accumulation around the zygote. Otherwise, it causes hindrance to the growth of the embryo. Nucleate nodules have been reported in *Capsella bursa-pastoris* (P. Maheshwari and Sachar 1954b). Both enucleate and nucleate nodules occur in *Carica papaya* (Dathan and Singh 1976). D. Singh (1964b) states that the nodules are much better developed in plants at higher altitudes, and is a possible method of filling up the vacuoles of the endosperm. However, R.C. Misra (1966) and Prasad (1969, 1974) attribute nutritional significance to them since the nodules persist for only a short duration in the developing endosperm.

Helobial Type. While reviewing the Helobial type of endosperm, Swamy and Parmeswaran (1963) suggest certain norms: (1) The primary endosperm nucleus lies in the chalazal region of the embryo sac, (2) it divides in situ, (3) the micropylar chamber is larger than the chalazal and separated either by a wall or a membrane, (4) the nucleus of the micropylar chamber moves away from the wall, undergoes free-nuclear divisions, and later the tissue becomes cellular, (5) the nuclear division in the chalazal chamber is strictly free-nuclear and it becomes haustorial, (6) if the chalazal chamber becomes coenocytic, the few initial divisions are synchronous; however, the chalazal chamber lags behind in the number of nuclei and their rate of formation, and (7) if wall-formation occurs in the chalazal chamber, it is significantly belated not only in its own ontogeny but also as compared with the time of initiation of cells in the micropylar chamber (see Fig. 3.10.1E–M).

After the formation of micropylar and chalazal endosperm chambers, usually the nucleus of the micropylar chamber divides first. However, there are exceptions, and the nucleus of the chalazal chamber may divide earlier than that of the micropylar as in *Phylidrum* (Kapil and Walia 1965). The free-nuclear divisions in the micropylar and chalazal chamber may or may not be synchronous.

Asynchronous free-nuclear divisions in the chalazal chamber have been noted in *Sparganium* spp. and *Typha* sp. (Asplund 1972, 1973). Wall formation in the micropylar as well as the chalazal chamber is not a rare feature, e.g. *Trillium undulatum* (Swamy 1948c), *Lindmania pendulifolia* (Lakshmanan 1967), *Phylidrum lanuginosum* (Kapil and Walia 1965), *Sparganium* spp. and *Typha* (Asplund 1972, 1973). In the latter four genera, the chalazal chamber becomes coenocytic by free-nuclear divisions, or cellularization occurs. In addition, in Philydraceae, Bromeliaceae, Sparaganiaceae and Typhaceae, wall formation is initiated first in the chalazal and then in the micropylar chamber.

Swamy and Parmeswaran (1963), Swamy and Krishnamurthy (1973) and Swamy (1974a) discuss several instances of Helobial type of endosperm in dicotyledonous families, and draw attention to their either Nuclear or Cellular ontogeny. According to Swamy and Parmeswaran, the formation of true Helobial endosperm is confined only to the monocotyledonous families; if it occurs in dicotyledonous families, it is only a modification of the Cellular or Nuclear type of endosperm. However, typical Helobial type of endosperm has been observed in dicotyledonous families, such as Santalaceae and Saxifragaceae.

Cellular Type. The division of the primary endosperm nucleus is followed by wall formation. The subsequent divisions may be followed by transverse walls resulting in a linear row of cells as in *Cercidiphyllum* (Swamy and Bailey 1949) and *Sarracenia* (Shreve 1906), or the walls may be crosswise as in *Parrotiopsis* (Kapil and Kaul 1972).

The majority of sympetalous families are characterized by the formation of either micropylar, or chalazal, or both types of endosperm haustoria. The sequence of planes of wall formation, following the first few divisions, exhibits several subtypes of development. These are discussed along with the respective families. Rau has drawn up a comparative schematic representation of endosperm types in sympetalous families: Labiatae, Verbenaceae (Fig. 48.10.2A–F), Scrophulariaceae (Fig. 48.15.6A–K_4), Gesneriaceae (Fig. 48.22.1A–K), and Campanulaceae (Fig. 51.2.2A–F_1).

Usually, for obtaining nutrients and transporting them to the developing endosperm, the haustoria grow in the direction of the tissue rich in food reserves. Their hypertrophied, often polyploid nuclei indicate high metabolic activity. The order Santalales presents extreme cases of development of the haustoria. In *Opilia*, the long tubular chalazal haustorium penetrates the pedicel of the flower, whereas in *Cansjera rheedii* (Swamy 1960) it branches profusely and grows towards the base of the ovule. In addition, secondary branches also develop (Fig. 12.4.2A, B). In *Iodina rhombifolia* (Bhatnagar and Sabharwal 1968), the aggressive chalazal haustorium branches extensively at the free end (Fig. 12.6.7A, B).

Both, the inter- and intracellular growth of the haustorium have been observed in *Nothapodytes foetida* (Swamy and Ganapathy 1957b). The basal endosperm cell develops chain-like connections with the chalazal tissue (through the perforations in the intervening walls), and absorbs nutrients (Fig. 33.13.1A–C). The outline of the nucleus appears most unusual, and exhibits a high-ploidy level.

The development of micropylar and chalazal haustoria is noteworthy in Loasaceae (Fig. 37.19.1A–G), Scrophulariaceae (*Pedicularis sylvatica* – Fig. 48.15.3A, B), Gesneriaceae (*Klugia notoniana* – Fig. 48.22.3A–C), and Orobanchaceae (*Aeginetia indica* – Fig. 48.24.1H, I, J, L, M).

In *Rhinanthus* spp. (B. Tiagi 1972), secondary haustoria develop from the central cellular endosperm after the inactivation of aggressive micropylar and chalazal haustorium. The formation of cambium-like strips in the central cellular endosperm is unique (for further details, refer to Sect. 48.15).

The asymmetrical development of the central cellular endosperm, formation of "basal apparatus", ruminations, and the development of primary and secondary haustoria characterize the endosperm in Acanthaceae (Sect. 48.19).

The formation of a "composite" endosperm in Loranthaceae is unknown elsewhere in angiosperms. As several embryo sacs grow concurrently in an ovary, each develops endosperm which fuse to form a "composite" structure (Fig. 12.8.3A, B).

Which of the types – Nuclear, Helobial, Cellular – is primitive? One can derive the Helobial from the Nuclear, and Cellular from the Helobial type. These three types are randomly distributed in the primitive and advanced angiosperms, as is also true of several other embryological characters. Generally, the Nuclear type is common in the Polypetalae, Cellular in the Sympetalae, and Helobial in the Monocotyledons. In Saxifragaceae (Sect. 26.7), all the three types occur in different taxa, but both Cellular and Helobial types may occur in *Boykinia*, *Mitella*, *Ribes* and *Tiarella* (Rau 1970).

Mitrasacme alsinoides (Fig. 47.2.2A–G) shows a Cellular–Nuclear–Cellular type of development. In a uniseriate four-celled endosperm, nuclear divisions occur in the micropylar and chalazal cell. Finally, the entire endosperm becomes cellular.

As the Nuclear type of endosperm has been observed in a larger number of taxa than the other two types, it may be considered to be primitive.

What is the morphological nature of the endosperm? It is 2n in *Butomopsis* (Fig. 53.3.2F, G), 3n in most plants, 5n in *Plumbago* (Sect. 44.2) and *Fritillaria* (Sect. 54.2), and up to 15n in *Peperomia* (Sect. 20.3). Whatever the ploidy-level of the endosperm, the development always conforms to one of the three types, and its main function is to nourish the embryo during development and maturation, and during germination (in albuminous seeds).

When the primary endosperm nucleus is 2n (upper polar + sperm cell), it is truly sporophytic and genetically identical with the zygote but the two behave quite differently – one produces the endosperm and the other the embryo. This also disproves Sargant's (1900) view that the participation of the lower polar causes imbalance and the primary endosperm nucleus (2n) is unable to produce the embryo. Now we know that it is not two polars but up to 14 which are concerned in the formation of the primary endosperm nucleus. Therefore, the only interpretation could be that the endosperm is a sui generis tissue.

In the absence of endosperm, the embryo usually aborts, unless nutrition can be obtained otherwise as in Trapaceae (Fig. 39.3.2N, O). Krishnamurthy (1988) emphasizes that the endosperm exercises a hormonal control on the growth and differentiation of embryo.

Balanophora (Fig. 13.2.3A, B) is so far the only example where, at the time of germination of the fruit on the host rootlet, some of the endosperm cells produce tubular slender sticky extensions for anchorage to the young rootlet.

The endosperm has long been presumed to be a "dead" tissue. B.M. Johri and Bhojwani (1965) clearly demonstrated, for the first time, that in (in vitro) cultures the endosperm of *Exocarpus cupressiformis* (Santalaceae) generates shoot buds.

Ruminate Endosperm. The ruminate condition of endosperm occurs during the maturation of seed so that the surface of endosperm becomes uneven. The ruminations are caused due to localized activity of the seed-coat, or endosperm. On morphological basis, Periasamy (1962a) distinguishes seven types: Annona (Fig. 3.10.3A–C), Passiflora (Fig. 3.10.3D), Myristica (Fig. 3.10.3E), Spigelia (Fig. 3.10.3F), Verbascum (Fig. 3.10.3G), Coccoloba (Fig. 3.10.3H, I) and Elytraria (Fig. 3.10.3K). In all these types, except Elytraria, irregularities occur in the growth of integument/s which bring about the rumination of endosperm. In the Elytraria type, during the development of seed, localized regions in the peripheral layers of cellular endosperm show active growth so that it becomes ruminate (Fig. 3.10.3K). In this process the seed-coat is consumed by the endosperm, and only the remnants persist.

An excessive elongation and enlargement of cells of the seed-coat has been observed in the Annona type (Fig. 3.10.3A–C). In Annonaceae, the cells of the outer, inner, or middle (Fig. 3.10.3C) integument elongate and cause ruminations. They are not vascularized. In some members of Annonaceae, the ruminations are plate-like and lie one above the other (Fig. 3.10.3A).

In the Passiflora type, the ovules are bitegmic, and the inner epidermis of testa and outer epidermis of tegmen show unequal radial elongation of cells which cause ruminations (Fig. 3.10.3D). In the Verbascum type, the ovules are unitegmic, and the cells of the persistent endothelium elongate unequally (Fig. 3.10.3G).

In *Degeneria* (Swamy 1949a), *Tiliacora* (Sastri 1954a), and some members of the Myristicaceae (Fig. 3.10.3E), Leeaceae and Vitaceae, the ingrowths develop due to the meristematic activity of localized areas of tegmen. The position of the meristematic zones is not constant except in Vitaceae and Leeaceae. In Myristicaceae, the ingrowths develop due to localized meristematic activity in testa and the massive chalaza (Periasamy 1962a, see also N.C. Nair 1972). According to Mohana Rao (1974b), the inner layers of the inner integument produce the ingrowths which become vascularized.

The ovules are bitegmic in *Coccoloba*, and unitegmic in *Spigelia*. In the Coccoloba (Fig. 3.10.3H, I) and Spigelia (Fig. F) type, the mature seed-coat comprises only the epidermis of testa. The epidermal cells of testa and of the massive chalaza, during the growth of seed, develop irregular infoldings due to cell enlargement. The endosperm tissue grows in between the infoldings, and ruminations are produced.

The Spigelia type of ruminate endosperm is common in Rubiaceae (Fig. J), but the condition in *Borreria hispida* (Periasamy and Vaidyanathan 1980) is rather rare. In the pachychalazal seed, the epidermal layer (of seed-coat) along the boundary between the strophiole and ovular part forms two inward extensions into the mature endosperm making it three-lobed and ruminate along the ventral face.

The endosperm also consumes the strophiolar tissue so that the strophiole is reduced to only a few cell layers at the region immediately adjoining the septum. The growth of the endosperm into the strophiole also makes it ruminate.

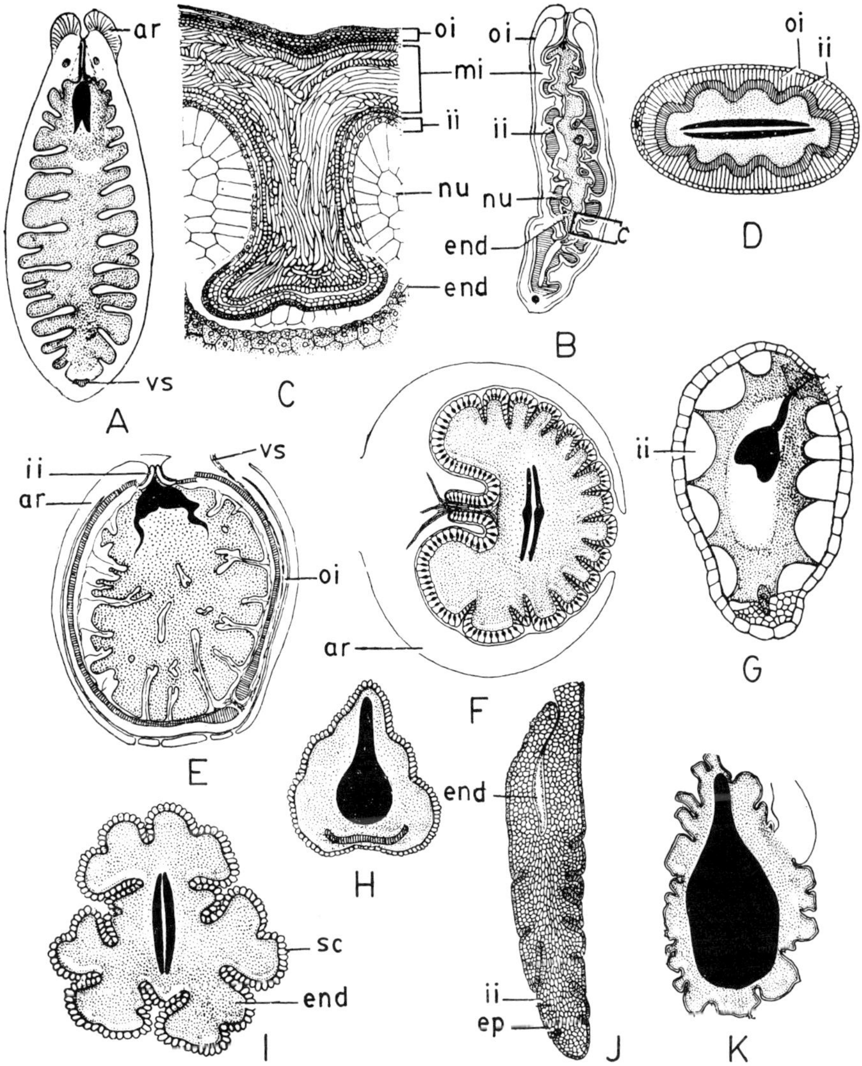

Fig. 3.10.3A–K. Ruminate endosperm. **A** *Annona squamosa*, longisection of mature seed with aril (*ar*) and vascular strand (*vs*); embryo *black*. **B, C** *Cananga odorata*. **B** Trans-median longisection of immature seed with outer (*oi*), middle (*mi*) and inner integument (*ii*), endosperm (*end*) and nucellus (*nu*); **C** *Inset C* from **B**. (After Periasamy 1962a). **D** *Passiflora calcarata*, transection of mature seed with inner (*ii*) and outer integument (*oi*). (After M.V.S. Raju 1956b). **E** *Myristica fragrans*, longisection of mature seed with aril (*ar*), inner integument (*ii*), and vascular strand (*vs*). (After Voigt 1888). **F** *Ervatamia heyneana*, transection of mature seed with aril. (After Periasamy 1962a). **G** *Tetranema mexicana*, longisection of mature seed with inner integument (*ii*). (After Krishna Iyengar 1942a). **H, I** *Coccoloba uvifera*. **H** Longisection of mature seed; **I** Transection of seed *end* endosperm; *sc* seed-coat (exotesta). **J** *Psychotria congesta*, longisection of immature seed. (**H–J** after Periasamy 1962a). **K** *Andrographis serpyllifolia*, longisection of mature seed. (After Mohan Ram 1960b)

In *Bentinckia codapanna* (Padmanabhan and Regupathy 1981), the endocarp is closely appressed to the thin seed-coat. The endocarp, the seed-coat and the outer layers of endosperm, as a unit, produce sinuous inpushings which impart a semblance to ruminate endosperm. The undulations originate in the fruit-wall even before the endosperm becomes cellular.

According to van Heel (1970a, 1971a, b), the labyrinth seed is an extreme form of rumination; the seed cut in any plane exhibits ruminations. In *Kingiodendron pinnatum*, the testa first encroaches on the endosperm and then on cotyledons. It penetrates in the lobes and crevices of the cotyledons and makes the seed ruminate giving it a labyrinth appearance. Other examples of labyrinth seeds are met with in Convolvulaceae, Hernandiaceae and Anacardiaceae (see Boesewinkel and Bouman 1984).

3.11 Embryo

The process of syngamy results in the formation of zygote. The zygote undergoes, apparently, a resting period. In some species it shrinks as in *Syringodium* (Lakshmanan and Rajeshwari 1979). The polarity becomes more pronounced, and differentiation occurs in the apical and basal region of the zygote. The apical region is an active centre of protein synthesis and cell division, while the basal part accumulates osmotically active substances. In sympetalous families like Acanthaceae, Gesneriaceae, and several others the micropylar part of the embryo sac elongates considerably. The apical region of the zygote elongates into a long tube-like structure so that the tip with the nucleus comes to lie in the central part of cellular endosperm. The zygote, in *Acrotrema arnottianum* (Swamy and Periasamy 1955), exhibits the unique development of a mantle (Fig. 22.2.1A, B). As the growth proceeds, the zygote emerges out of the mantle.

Early Embryogeny. Usually, a transverse division occurs in the zygote, and separates a spherical terminal cell *ca* from the basal cell *cb* (see Sivaramkrishna 1978). The division of zygote may rarely be vertical as in Loranthaceae (Sect. 12.8), or vertical/oblique as in *Quercus gambelii* (Brown and Mogensen 1972). The next division may be transverse in both the terminal (*ca*) and basal (*cb*) cell, and a linear tetrad is produced (Fig. 3.11.5). The daughter cells of *ca* are designated *l* and *l′*. The derivatives of *cb* are designated *m* and *ci* (see Fig. 3.11.5). The term proembryo is applied from the two-celled condition till the initiation of organs in the embryonal mass.

In the formation of a T-shaped type of tetrad, a vertical division occurs in *ca*, and the two juxtaposed hemispherical cells form the tier *q* (Fig. 3.11.1A). *q* gives rise to a quadrant by vertical divisions in *l* and *l′*, or by vertical division in each of the two juxtaposed cells, perpendicular to the previous one (Fig. 3.11.1B). The vertical walls do not intersect, but a small intercept is left between them.

Subsequent to the formation of a quadrant, the proembryo develops into an octant. The formation of octant is of two types (Fig. 3.11.2A–F). In the first

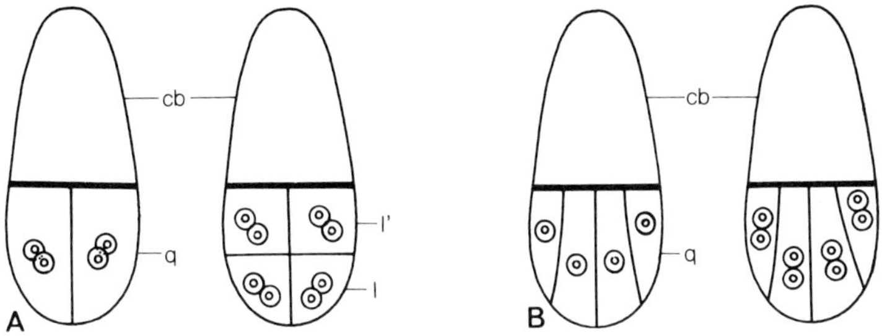

Fig. 3.11.1A, B. Quadrants (four cells at distal end of proembryo) produce octants (eight cells) either in two tiers (**A**), or one tier (**B**). When in two tiers, apical tier is designated *l* and the basal tier *l'*. (After Natesh and Rau 1984)

instance, in a linear type of tetrad, one more vertical division occurs in each cell of the quadrant, at right angles to the previous one, forming two superposed tiers (*l*, *l'*) of four cells each (Fig. 3.11.2A, B). In a T-shaped tetrad, similarly, the two tiers *l* and *l'* develop by a transverse division in each cell of *q* (Fig. 3.11.2C, D). In the second instance, instead of undergoing a transverse division, the four cells of tier *q* divide vertically in such a way that each quadrant is divided into unequal cells – one adjoining the axis and the other towards the periphery (Fig. 3.11.2E, F). They form together a single tier of eight cells.

The variations in the octants occur due to the plane of formation of wall, following nuclear division in each cell of a quadrant. The walls are laid down in three planes: transverse, vertical or longitudinal, and oblique (periclinal to curved wall). According to Takao (1976), theoretically, the octant can show 15 patterns, but in *Impatiens textorii* he observed only five: (1) One of the cells of the quadrant divides in vertical plane, and the other three in transverse plane. Then the octant is composed of one pair of juxtaposed cells, and three pairs of superposed cells. (2) If two cells of the quadrant divide vertically, and the other two transversely, the octant shows two pairs of juxtaposed and two pairs of superposed cells. (3) An octant consists of three pairs of juxtaposed cells, and one pair of superposed cells. (4) All the four pairs of cells in an octant are juxtaposed. (5) In an octant two pairs of cells are juxtaposed, the other two cells of the quadrant divide by oblique walls so that the daughter cells are arranged circumaxially.

The development of proembryo up to an octant stage and the formation of two types of octant are similar in dicotyledons and monocotyledons (Swamy 1962, see also Guignard and Mestre 1969a, Guignard 1975a). The morphogenetic potential of the cells is described later (see also D'Amato 1977).

Embryogenic Classification. The system of classification by Souèges (1937a–f) and Mestre (1967) is mainly based on the origin of four-celled proembryo, and organ differentiation from each cell. The rate at which these cells divide varies in different species, but the planes of their divisions and the disposition of cells formed are specific and predetermined. The four-celled proembryo has a

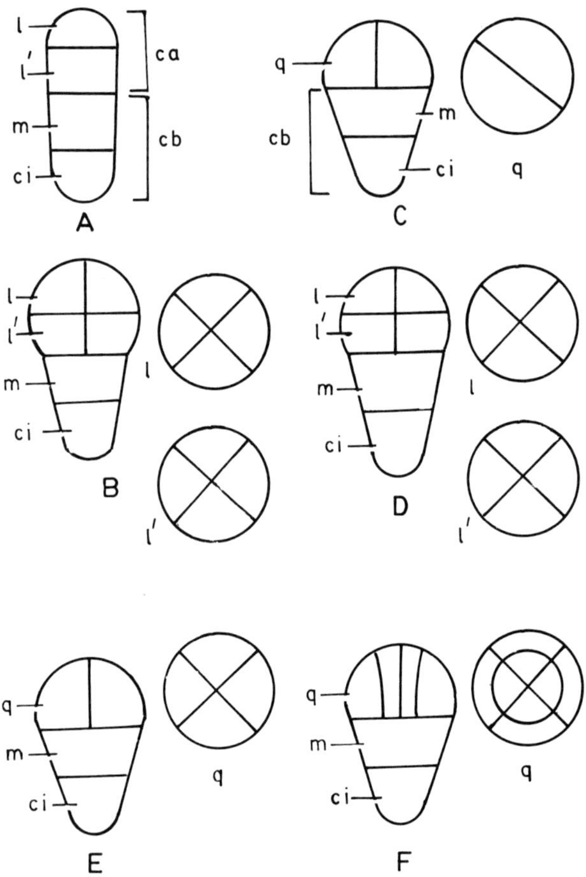

Fig. 3.11.2 A–F. Types of octant formation during early embryogeny. (After Swamy 1962)

specific, predetermined potential blue-print of the mature embryo (see Brogniart 1827, Crété 1963a, Guignard 1975b, Lebégue 1989).

A transverse division of the zygote gives rise to the terminal cell (*ca*) and the basal cell (*cb*). The destiny of these two cells is represented by Megarchetypes – *pvt* stem tip, *phy* hypocotyledonary part, *icc* root cylinder, *iec* hypophysis (root epidermis), *co* root cap, *s* suspensor, *pco* cotyledonary region proper:

Megarchetype I: cb = pvt + phy + icc + iec + co + s
ca = pco

Megarchetype II: cb = phy + icc + iec + co + s
ca = pco + pvt

Megarchetype III: cb = $\frac{1}{2}$ phy + icc + iec + co + s
ca = pco + pvt + $\frac{1}{2}$ phy

Megarchetype IV: cb = iec + co + s
ca = pco + pvt + phy + icc

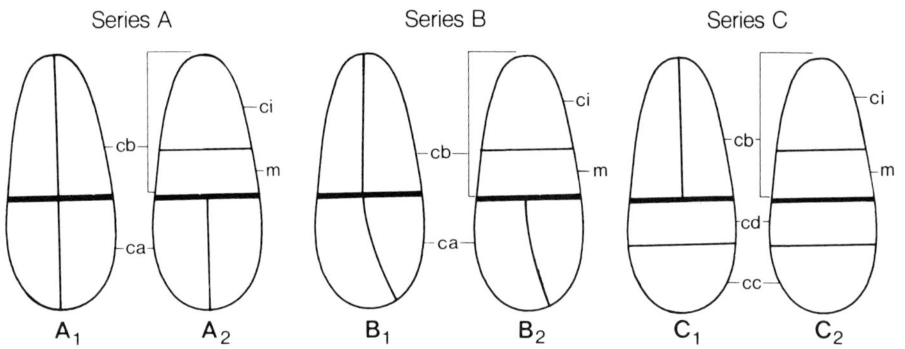

Fig. 3.11.3. Proembryonal tetrads, classification based primarily on the plane of division in apical cell; vertical in *Series A*, oblique in *Series B*, and transverse in *Series C*. Further subdivision takes into account the mode of division of basal cell which may be either vertical. (A_1, B_1, C_1) or transverse (A_2, B_2, C_2). (After Natesh and Rau 1984)

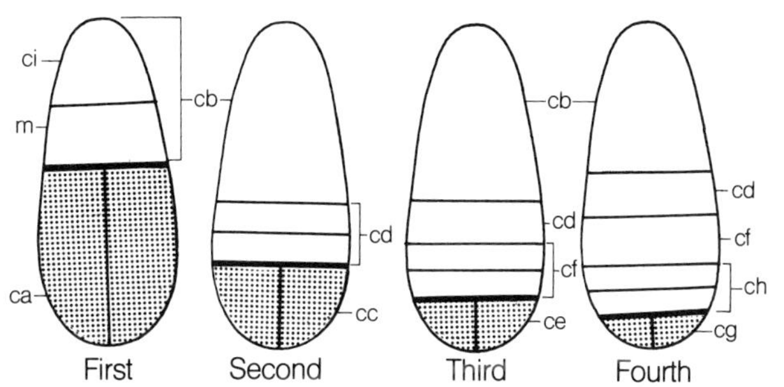

Fig. 3.11.4. Proembryonal tetrads of Series C to show tetrads of first, second, third and fourth order, respectively. The potential embryonal pole (*shaded*) shifts progressively to the extreme chalazal end. (After Natesh and Rau 1984)

Megarchetype V: cb = co + s
ca = pco + pvt + phy + icc + iec

Megarchetype VI: cb = s
ca = pco + pvt + phy + icc + icc + co

The Megarchetypes are divided into Series A, B and C, depending on the variation in the plane of division of the terminal cell (*ca*) of the two-celled proembryo. When the division is in vertical plane, the tetrad is referred to Series A, when oblique, to Series B, and a division in transverse plane to Series C (Fig. 3.11.3).

When the terminal cell (*ca*) and a part of the basal cell (*cb*) give rise to embryo proper, it is referred to period I. This potentiality shifts from the basal cell (*cb*) to the terminal cell (*ca*) in Period II, and then to the apical derivative of the terminal cell (*cc*) in Period III, and so on (Fig. 3.11.4). Thus, there is a progressive series of shift in potentiality to the apical derivatives.

In the Series C, the laws governing the embryogeny are applicable to the apical derivatives (see Crété 1963a).

Simplified System of Mestre (1967). In view of the fact that, very often, many irregularities occur in the sequence of divisions during early stages, Mestre formulated a simplified sytem. According to him, only the potentialities of the first two cells, apical and basal, are of relevance and he recognizes only three Megarchetypes:

Megarchetype A: ca = cotyledons only
cb = stem + root

Megarchetype B: ca = stem, cb = root (in general)

Sub-type B_1: ca = stem (major part of hypocotyl)
cb = minor part of hypocotyl and root

Sub-type B_2: ca = entire hypocotyl (stem)
cb = root

Megarchetype C: ca = stem and root
cb = only suspensor, i.e. the basal cell has no contribution in the development of new plant

Johansen's System. Schnarf (1929) and Johansen (1945) recognize six principal types of embryos among dicotyledons (Fig. 3.11.5). P. Maheshwari (1950) distinguished them as follows:

I. The terminal cell of the two-celled proembryo divides by a longitudinal wall:

i) The basal cell has only a minor role, or none, in the subsequent development of the embryo — Crucifer or *Onagrad* type

ii) The basal and terminal cells both contribute to the development of the embryo — Asterad type

II. The terminal cell of the two-celled proembryo divides by a transverse wall.

1. The basal cell plays only a minor role, or none, in the subsequent development of the embryo:

i) The basal cell usually forms a suspensor of two or more cells — Solanad type

ii) The basal cell does not divide further, and the suspensor, if present, is always derived from the terminal cell — Caryophyllad type

2. The basal and terminal cells both contribute to the development of the embryo — Chenopodiad type

III. The zygote divides by a vertical wall — Piperad type

Embryogenic types, as suggested by Souèges, Johansen, and P. Maheshwari are compared in the following Table:

The derivatives of the two-celled proembryo, *ca* (apical cell) and *cb* (basal cell), lead to six types of embryogeny (Fig. 3.11.5).

Table 3.11.1. Types of Embryogeny in Angiosperms

R. Souèges (1939–1951)	D.A. Johansen (1950)	P. Maheshwari (1950)
1) Period I: Series 'A' and 'B'– Group I. Megarchetypes IV–VI	Onagrad (includes Myosurus, Juncus, Capsella, Veronica, Mentha, Trifolium, Ruta and Lilium variations)	Crucifer/Onagrad
2) Period I: Series 'A' and 'B'– Groups I and II. Megarchetypes I–III	Asterad (includes Muscari, Geum, Polygonum, Urtica, Lamium, Oxalis, Senecio and Poa variations)	Asterad
3) Period I: Series 'C'– Group III (mostly). Megarchetypes IV – V.	Solanad (includes Hyoscyamus, Nicotiana, Papaver, Sherardia and Linum variations)	Solanad
4) Period I: Series 'C'– Groups III–VIII. Megarchetypes III, V.	Chenopodiad (includes Chenopodiad, Polemonium and Myosotis variations)	Chenopodiad
5) Period II (also a few species in period III and IV). Series 'A', 'B', 'C'– Groups IX–XVI. Megarchetypes I–VI	Caryophyllad (includes Sagina, Vaccaria, Hypecoum, Corydalis, Sedum, Medicago, Drosera, Fumaria, Myriophyllum, Ruppia, Sagittaria, Zannichellia and Aglaonema variations)	Caryophyllad
6) —	Piperad	Piperad

1) Crucifer (Onagrad) Type. The general plan of development is given below:

The apical cell *ca* (Fig. 3.11.6A) divides vertically; the two daughter cells undergo another vertical division, at right angles to the previous one, to form a quadrant *q* (Fig. B). A transverse division gives rise to two superposed tiers of an octant designated *l* and *l′* (Fig. C, D). Periclinal division in each cell of an octant demarcates the protoderm from the axial zone (Fig. E, F). Further divisions in the cells of protoderm are always anticlinal, and they keep pace with the growing axial zone. The cells of the axial zone divide vertically, separating the central cells, the plerome initials (*pl*) from the outer periblem (*pc*) initials (Fig. F–I). The axial tissue divides in both the planes. Vertical divisions are more frequent in the plerome, and the tiers *l* and *l′* give rise to the cotyledons and stem tip.

The basal cell *cb* divides transversely to produce *m* and *ci* (Fig. B). A transverse division in *m* produces tiers *d* and *f*, and *ci* gives rise to *n* and *n′* (Fig. E–G). The tiers *m* and *ci* contribute to the suspensor. The lowermost cell of the suspensor functions as hypophysis (Fig. H, I). It divides transversely, and both the cells undergo crosswise vertical divisions so that eight cells are produced in two tiers (Fig. I, J). The upper four cells produce root cap and root epidermis, while the lower four give rise to root cortex.

2) Asterad Type. The development is according to the following plan.

The terminal cell *ca* of the two-celled proembryo (Fig. 3.11.7A) undergoes a vertical division (Fig. B). The two sister cells undergo one more vertical

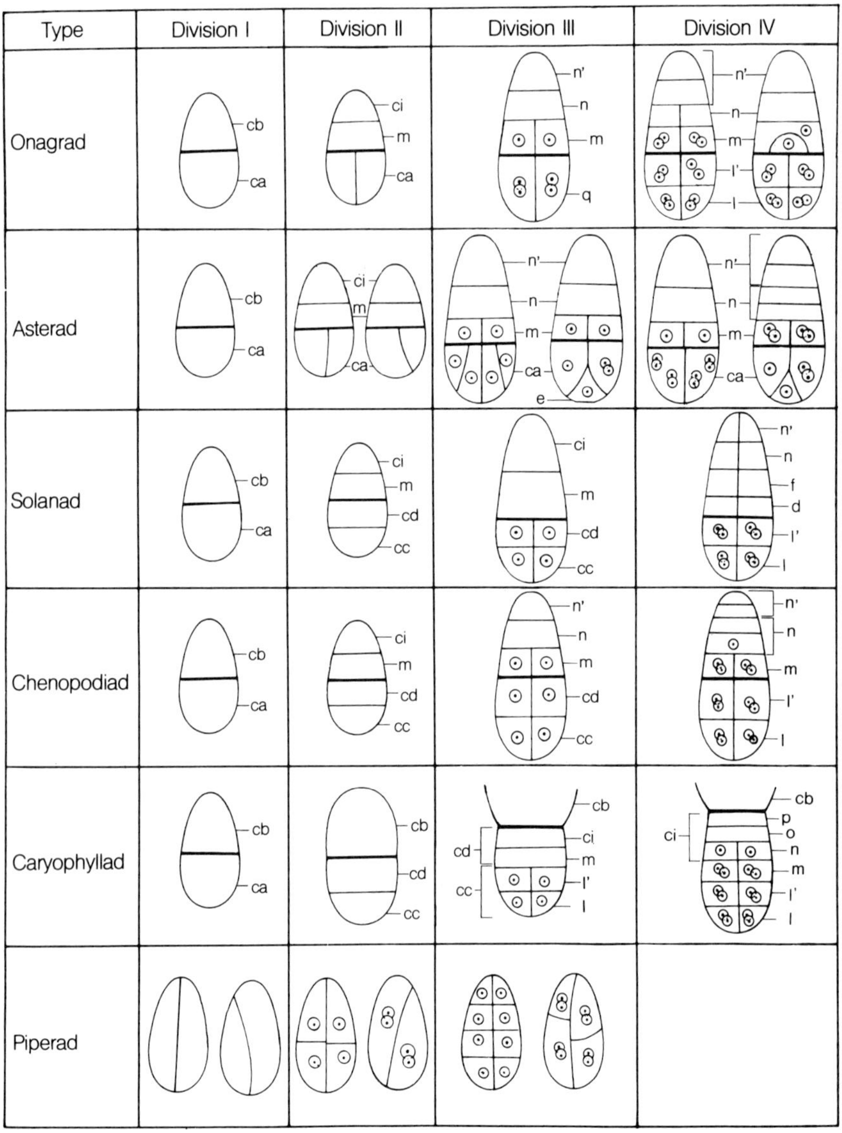

Fig. 3.11.5. Schematic representation of main types of embryogeny; based on Schnarf (1929) and Johansen's (1950) system of classification. (After Natesh and Rau 1984)

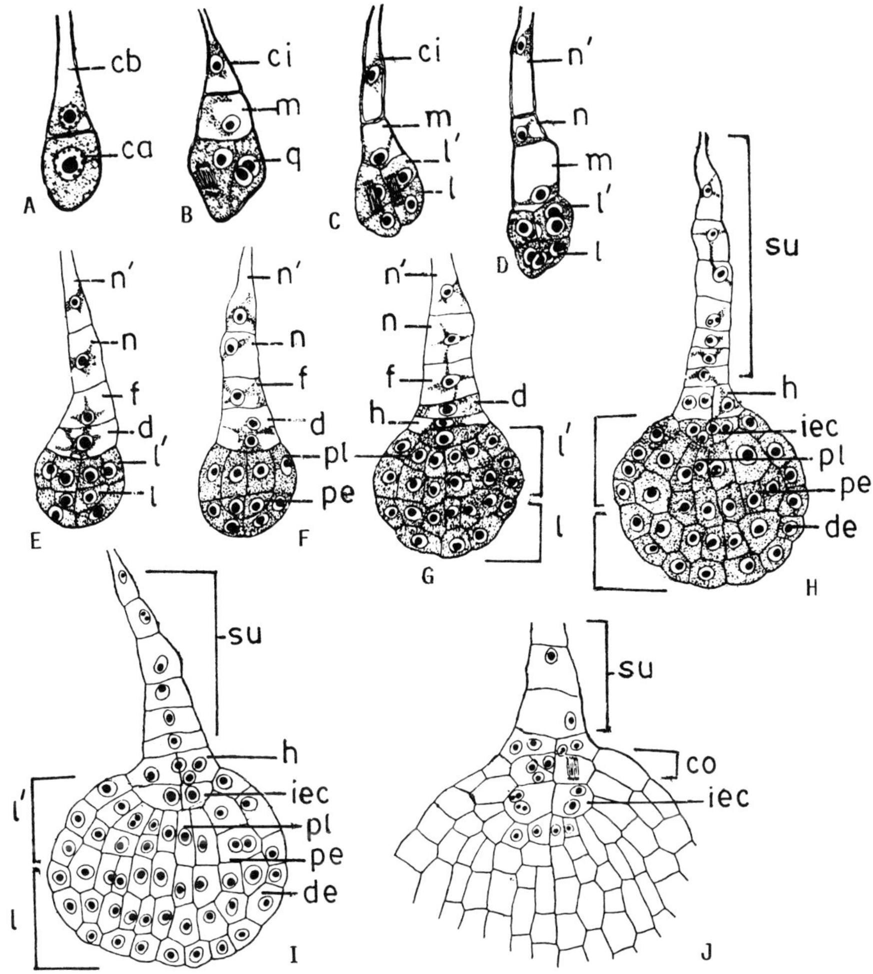

Fig. 3.11.6A–J. *Lophanthus chinensis*, embryogeny of Onagrad type; *su* suspensor. (After Crété 1942)

division, in a plane at right angles to the previous one forming a quadrant *q* (Fig. C–F). Each cell of a quadrant divides by an obliquely oriented vertical wall producing two unequal cells, one towards the central axis and the other towards the periphery. Thus, in a single tier of an octant, an axial group of four cells is separated from the peripheral four cells (Fig. G, H). Further divisions in *q* give rise to the initials of cotyledons and stem tip (Fig. G–K).

The basal cell *cb* divides transversely and produces tiers *m* and *ci*. The tier *ci* divides transversely, and produces *n* and *n′* (Fig. C). The tier *n* divides vertically (Fig. G–I), while *n′* divides transversely to give rise to *o* and *p* (Fig. G, H). Of the derivatives of *cb*, *m* gives rise to hypocotyl and radicle, *n* to the remaining

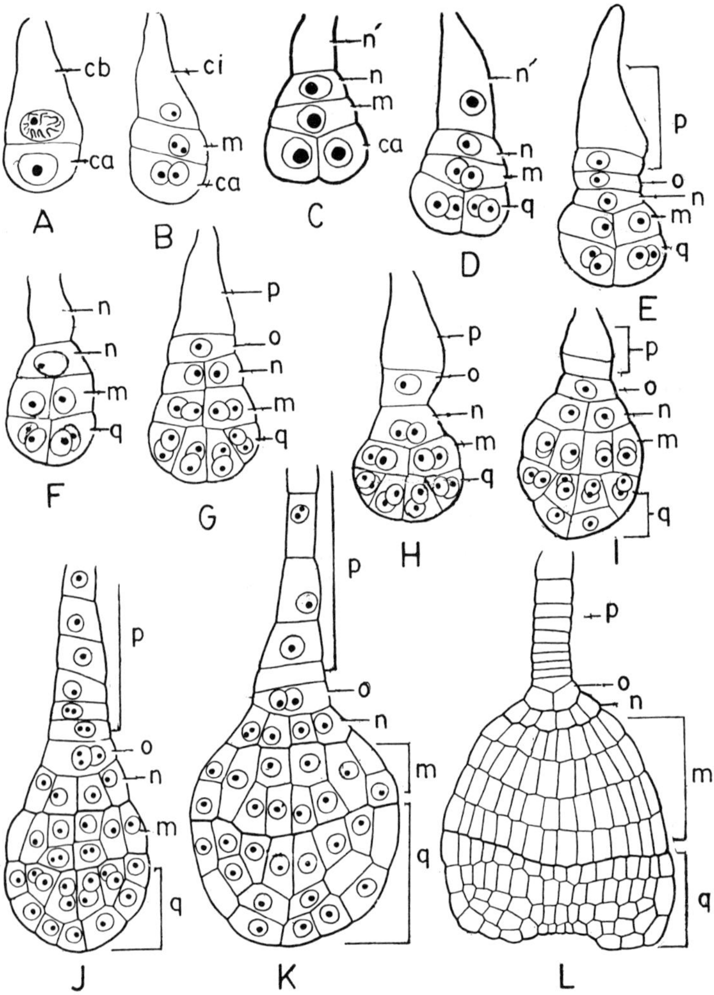

Fig. 3.11.7A–L. *Bidens biternata*, embryogeny of Asterad type. (After P.K. Deshpande 1964b)

part of root tip, and *o* and *p* are the initials of root cap and suspensor, respectively (Fig. G–L).

3) Solanad Type. The development is according to the following scheme:

The cells *ca* and *cb* of the two-celled proembryo (Fig. 3.11.8A) undergo transverse divisions to form a four-celled linear proembryo – these cells are designated *l*, *l′*, *m* and *ci* (Fig. B). The tiers *l* and *l′* divide vertically to form a quadrant (Fig. C). Another vertical division in each cell produces an octant (Fig. D). The tier *l* gives rise to the cotyledons and shoot apex, and *l′* to

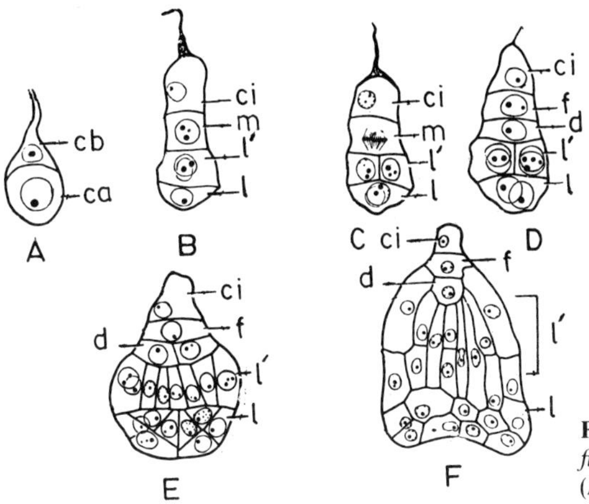

Fig. 3.11.8A–F. *Pentaphragma horsfieldii*, embryogeny of Solanad type. (After Kapil and Vijayaraghavan 1965)

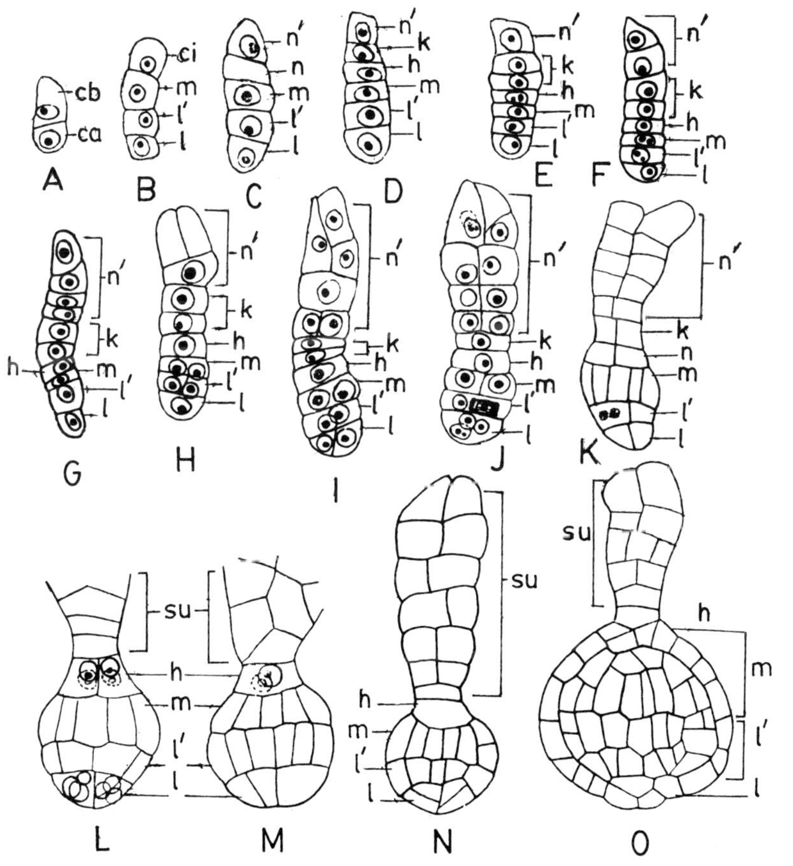

Fig. 3.11.9A–O. *Chenopodium murale*, embryogeny of Chenopodiad type. (After Mahabale and Solanky 1954c)

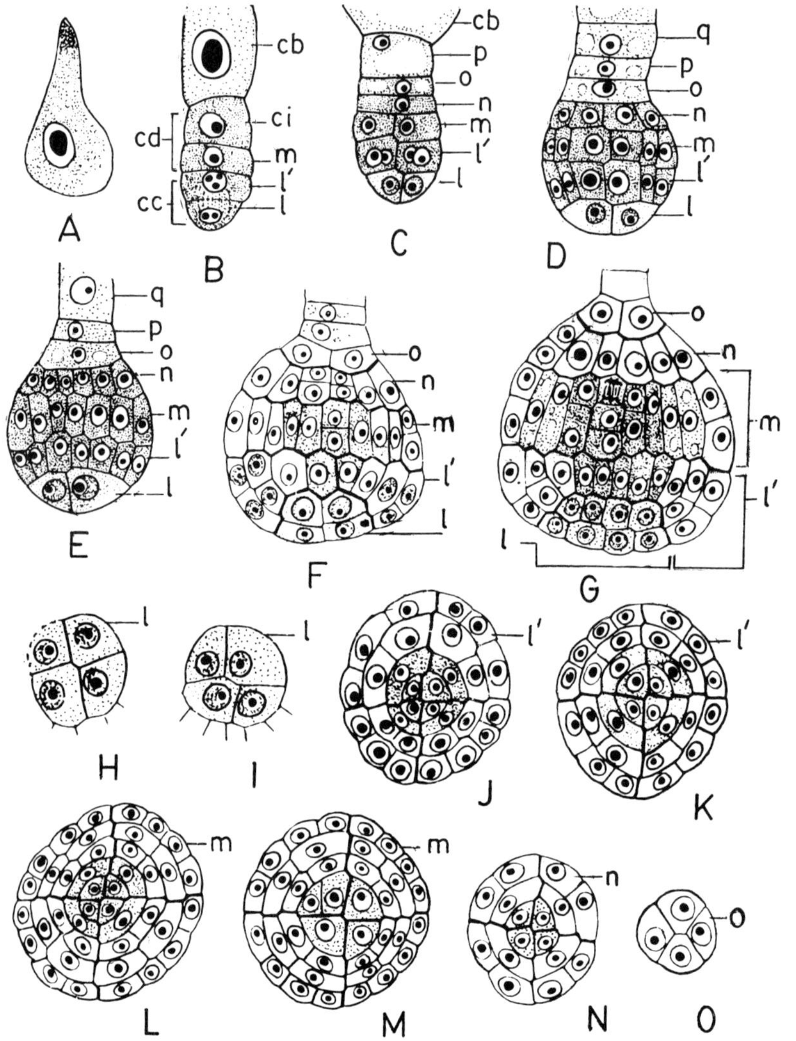

Fig. 3.11.10A–Q. *Stellaria media*, embryogeny of Caryophyllad type. (After Ramji 1975)

hypocotyl, and periblem and plerome of the root. *m* produces the tiers *d* and *f* by a transverse division. The tier *d* is the initial of root tip, and *f* and *ci* are the initials of suspensor (Fig. D–F).

4) Chenopodiad Type. The development is according to the following plan:

Transverse divisions in tiers *ca* and *cb* produce a four-celled proembryo: *l*, *l'*, *m* and *ci* (Fig. 3.11.9A, B). *ci* divides transversely to produce *n* and *n'*, and

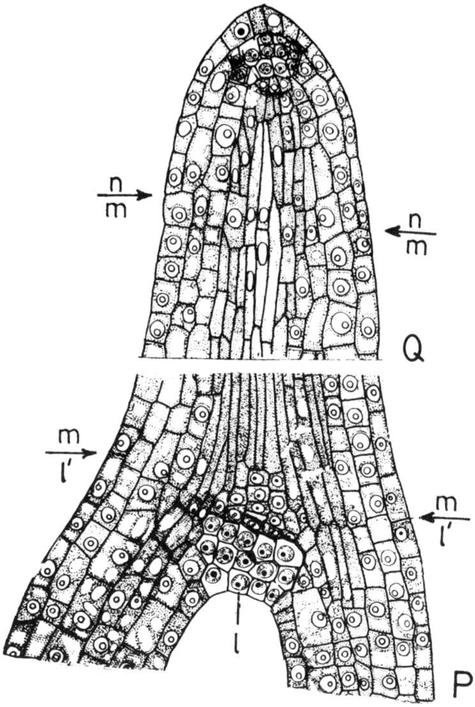

Fig. 3.11.10P, Q (*Contd.*)

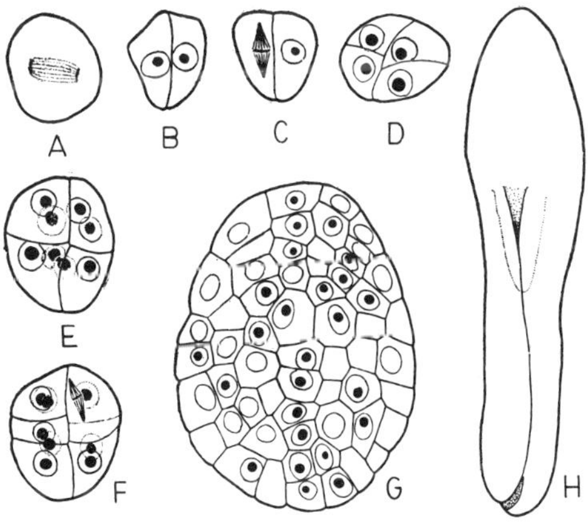

Fig. 3.11.11A–H. *Comandra umbellata*, embryogeny of Piperad type. (After M. Ram 1957)

by another transverse division *n* gives rise to *h* and *k* (Fig. C, D). Further divisions in *n'* produce a multicellular suspensor (Fig. E–O). The embryo proper develops from the tiers *l*, *l'*, *m* and *h* (Fig. E–O). In *Chenopodium murale* (Mahabale and Solanky 1954c), a vertical division occurs in *l'* followed by a vertical wall in *l* forming a quadrant (Fig. H, I); *m* also divides vertically (Fig. H–J).

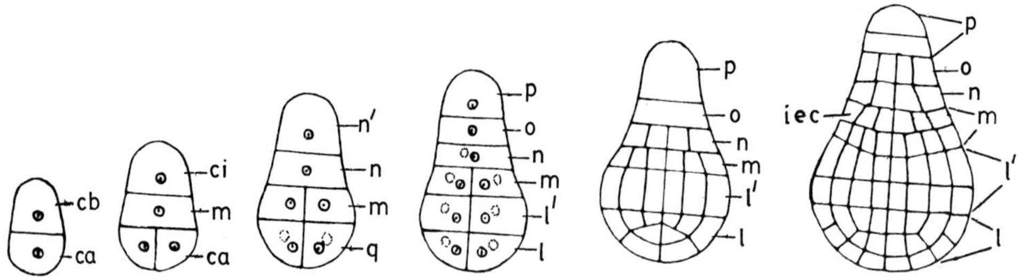

Polygonad type

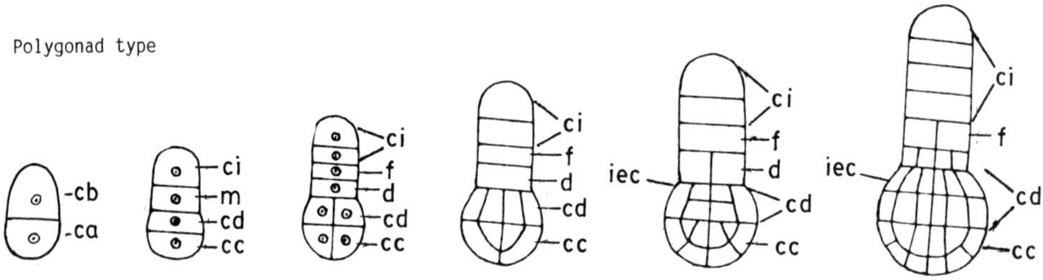

Solanad type

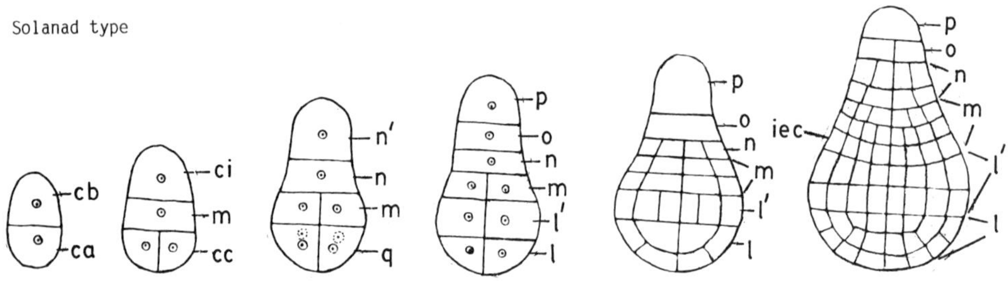

Geraniad type

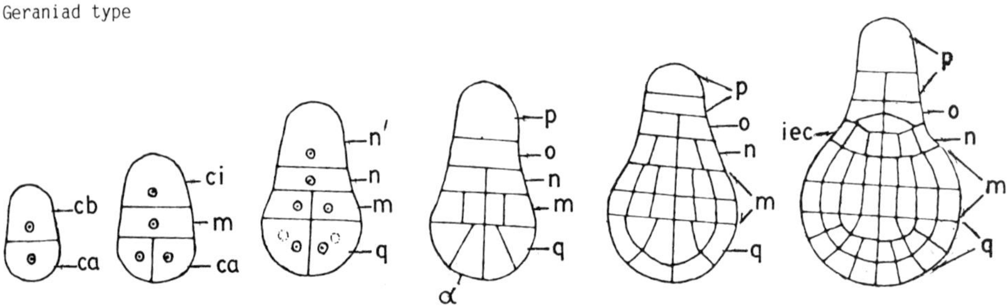

Asterad type

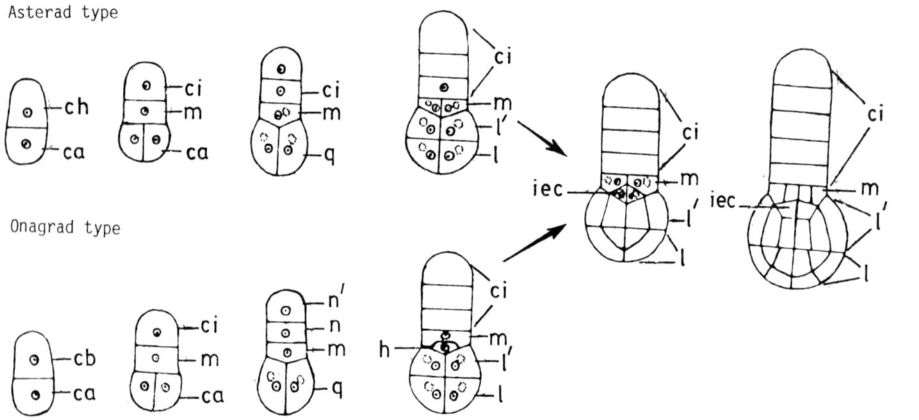

Onagrad type

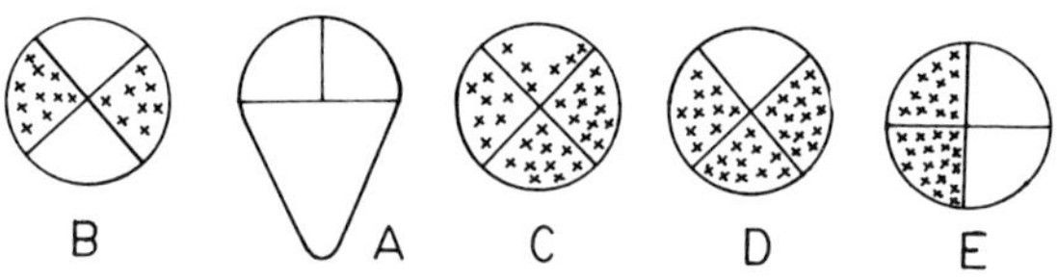

Fig. 3.11.13A–E. Proembryo at quadrant stage, initials of cotyledon *cross-marked*. **A** Vertical section of proembryo. **B** Transection of terminal tier of dicotyledonous proembryo; two opposite sectors denote loci of cotyledons. **C–E** Transections of terminal tier of monocotyledonous proembryo; there is a single locus of cotyledon. **C** Condition in Philydraceae; **D** Pontederiaceae, Sparganiaceae and Iridaceae; **E** Hydrocharitaceae, Potamogetonaceae and Amaryllidaceae. (After Lakshmanan 1972)

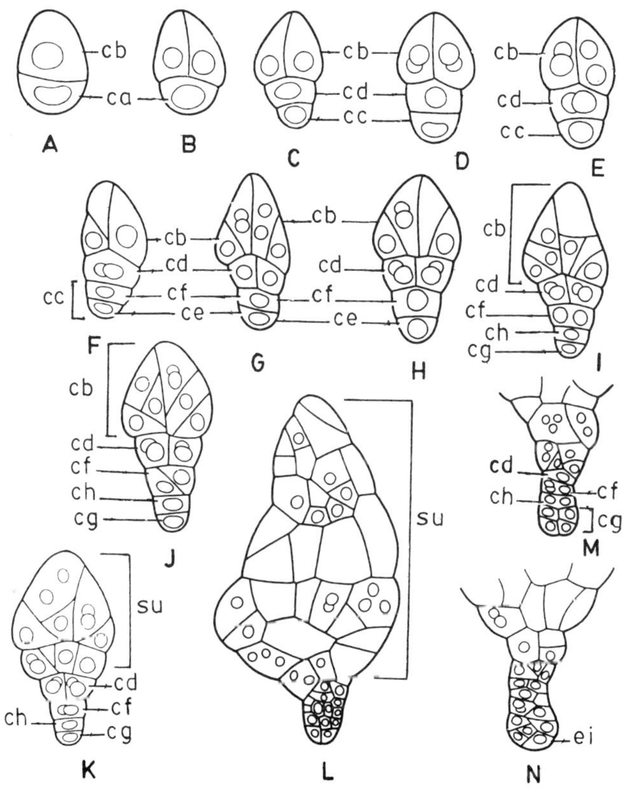

Fig. 3.11.14A–N. *Tulipa kolpakowskiana*, early embryogeny. **A** Two-celled proembryo. **B** Vertical division in *cb*. **C** Tetrad. **D** Second vertical division in *cb*. **E** Later stage, vertical division in *cb* and *cd*. **F** Transverse division in *cc* to form *ce* and *cf*. **G, H** Later stages. **I–K** Transverse division in *ce* to form *cg* and *ch*, vertical divisions in *cf* and *cd*. **L** Proembryo with a well-developed suspensor (*su*). **M** Octant (*cg*). **N** Proembryo with epiphysis initial (*ei*). (After Haccius and Hausner 1972)

Fig. 3.11.12. Developmental types of embryo: Polygonad, Solanad, Geraniad, Asterad and Onagrad. (After Yamazaki 1982)

After the formation of quadrant and octant, the protoderm is first demarcated in tier *m*, then in *l'*, and finally in *l* (Fig. K–O).

The derivatives of tier *l* give rise to the two cotyledons and shoot tip, and of *l'* to the lower and *m* to the upper part of hypocotyl. *h* functions as hypophysis, and gives rise to root-tip. Thus, the embryo proper develops from the derivatives of both *ca* and *cb*.

5) Caryophyllad Type. The general plan of development is given below:

The zygote (Fig. 3.11.10A) divides transversely into the apical cell *ca* and basal cell *cb*. A transverse division in *ca* produces tiers *cc* and *cd*; by one more

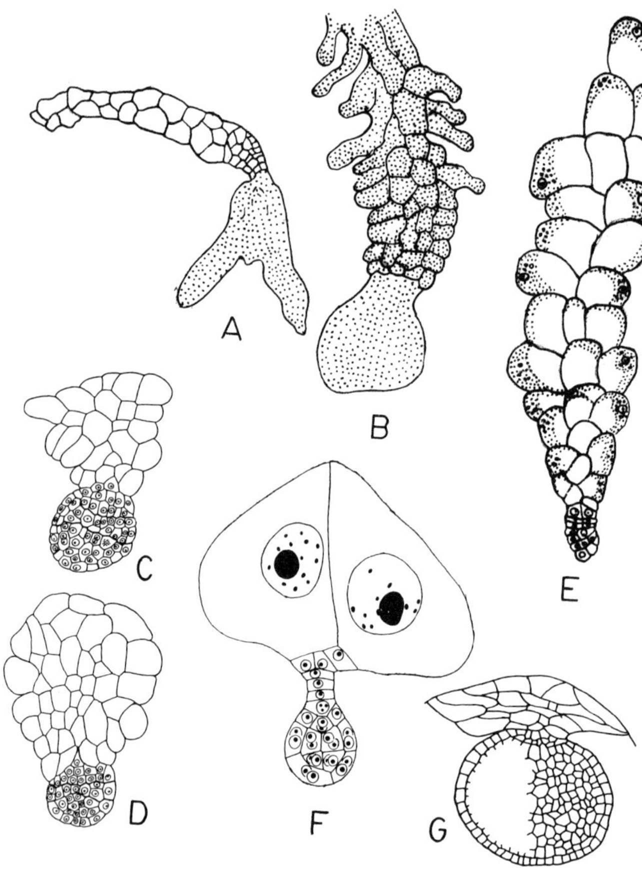

Fig. 3.11.15A–G. A *Stylosanthes mucronata*, dicotyledonous embryo with a long, multicellular suspensor. **B** *Crotalaria stricta*, the prominent tubular suspensor cells (*middle region*) remain in intimate contact with endosperm cells. (After Anantaswamy Rau 1953). **C, D** *Sesbania grandiflora*, multicelled conspicuous suspensor. (After Anantaswamy Rau 1951d). **E** *Rubia cordifolia*, proembryo with vesicular suspensor cells. (After Venkateswarlu and Rajeswara Rao 1958). **F** *Haloragis micrantha*, proembryo with two juxtaposed hypertrophied suspensor cells. (After Nijalingappa 1975). **G** *Glottidium vesicarium*, globular proembryo with fan-shaped suspensor. (After N. Prakash and Herr 1979)

division in the same plane *cc* gives rise to *l*, *l′*, and *cd* to *m* and *ci* (Fig. B). The tier *ci*, by two successive transverse divisions, produces tiers *n*, *o* and *p* (Fig. C).

The tier *cb* remains undivided and becomes vesicular (Fig. C) in *Stellaria media* (Ramji 1975).

The tier *l* gives rise to epicotyl (shoot apex), *l′* to cotyledons, *m* to hypocotyl, and *n* to the tip of hypocotyl. The derivatives of the tier *o* form the root-cap, and of *p* and *cb* give rise to the suspensor (Fig. B–Q). Further details are given under "Morphogenetic Concept".

Thus, the embryo proper is derived from the daughter cells of *ca* only; in addition, the tier *ca* also contributes to the formation of suspensor. It may be noted that the tiers *m*, *n*, *o* and *p* are the derivatives of *cb* in the Crucifer, Asterad, Solanad and Chenopodiad type. However, in the Caryophyllad type they are the derivatives of *cd*, a daughter cell of *ca*.

6) Piperad Type. The division of the zygote is invariably vertical as in Loranthaceae, Piperaceae, and occasionally in some other taxa. The division is obliquely vertical in *Scabiosa succia* (Souèges 1937d). *Comandra umbellata* (Ram 1957) is an exception in Santalaceae. The zygote divides vertically (Fig. 3.11.11A), and the two daughter cells (Fig. B, C) undergo transverse divisions to form a quadrant (Fig. D) and an octant (Fig. E). Subsequent divisions give rise to a globular proembryo (Fig. F, G) which differentiates into a dicotyledonous embryo (Fig. H). The suspensor is absent.

In contrast, in Loranthaceae the vertical division of the zygote is followed by a series of transverse divisions producing a long, biseriate suspensor (Sect. 12.8). In *Balanophora*, the first two or three divisions in the zygote are vertical (Zweifel 1939).

Yamazaki's Classification. Yamazaki (1974, 1982) proposes a system of classification of angiosperms on the basis of patterns of differentiation of initials of the root cortex (*iec*). He emphasized that it is a constant feature not only in the families, but even in the orders. He recognised five types of embryogeny (Fig. 3.11.12): Polygonad type, Solanad type, Geraniad type, Asterad type, and Onagrad type – Ruta variation and Onagrad variation.

1) Polygonad Type. The cell *ca* of the two-celled proembryo gives rise to shoot apex and cotyledons. The cell *m*, daughter cell of *cb*, produces two tiers. The central cells of the upper tier produce hypophysis (*iec*), while the tier *m* gives rise to hypocotyl.

2) Solanad Type. The terminal cell *ca* divides transversely to form tiers *cc* and *cd*. The derivatives of *cc* form the initials of shoot apex and cotyledons. *cd* forms two tiers. The central cells of the upper tier produce *iec*, whereas the lower tier produces the entire hypocotyl.

3) Geraniad Type. The cell *ca* gives rise to shoot apex, cotyledons and hypocotyl. The tier *m*, daughter cell of *cb*, produces three or more cell layers. *iec* is derived from the central cells of the lowermost tier of *m*. The upper tiers of *m* give rise to the root-cap, and a part of the suspensor.

4) Asterad Type. The terminal cell *ca* directly gives rise to tier *q* which develops into shoot apex and cotyledons. *cb* divides transversely to produce *m* and *ci*; *ci* gives rise to *n* and *n'*, and *n'* to *o* and *p*. The tier *n* (or *m*) consists of one layer of cells, and the central cells give rise to the hypophysis initial (*iec*). Sometimes, the tier *m* is two- or three-layered, and the hypophysis initial is derived from the lower tier, while the upper one produces the root-cap. The embryogeny then conforms to the Calendula variation.

5) Onagrad Type. The cell *ca* and its derivatives form the initials of shoot apex, cotyledons, and hypocotyl. In the Ruta variation the cell *m*, lower daughter cell of *cb*, undergoes two longitudinal divisions, followed by a transverse wall. The lower four cells form the initials of the root cortex.

In the Onagrad variation, the cell *m* cuts off a lenticular cell *h*, which divides twice longitudinally to produce the four-celled hypophysis.

Lastly, Yamazaki mentions the "Irregular" type where the segmentation and differentiation of the root cortex is irregular, and difficult to follow. The zygote may divide transversely or longitudinally.

In our opinion, there is no merit in Yamazaki's concept of the differentiation of initials of root cortex (*iec*) as a basis of classification of angiosperm families and orders. The embryogenic classification, as discussed by Natesh and Rau (1984), is certainly the most valid system and has been widely used for over four decades since Schnarf (1939) and Johansen (1945) proposed the six basic types.

The embryogenic classification, according to these five types (except Piperad), is inadequate to classify various types of variations which have been described as "unclassified or abnormal" (P. Maheshwari 1950) or "irregular" (Crété 1963a). Swamy and Padmanabhan (1962) point out that variations occur even in the laying down of walls at different stages within the same species.

These embryogenic classifications consist of a hierarchy of types and subtypes which have been widely criticized. First, most investigators have not studied sufficient material to emphasize the range of variation. Second, when more than one type of embryogeny has been observed, their relative percentage has never been studied. Third, the sequence of early cell divisions on which the categories are based vary a good deal in the proembryos of the same species; therefore, the categories are no more valid (see Wardlaw 1955, Swamy and Padmanabhan 1962, Ramji 1975, Periasamy 1977, Lersten 1983).

Padmanabhan (1967, 1968), Ramji (1975), and Swamy and Krishnamurthy (1980) emphasize the importance of organization of meristems and organogenesis. The entire developmental sequence is possibly controlled by physical, biochemical and genetic factors. The zygote exhibits distinct polarity, and the terminal pole is far more active. According to Swamy and Krishnamurthy (1980), the formation of a partition wall in the zygote depends on the "principle of minimal area and distribution of dictyosomes. During embryo development, there occurs a gradient of decreasing growth activity from the shoot pole to the suspensor/root pole". The systems based on cell lineages fail to explain these features.

Morphogenetic Concept. Padmanabhan (1967, 1968) and Ramji (1975) interpret embryogeny on the basis of morphogenic phases. Ramji (1975) distinguishes four phases: (1) filamentous phase following transverse divisions in zygote and its derivatives (Fig. 3.11.10A, B), (2) onset of vertical divisions (three-dimensional phase) (Fig. C), (3) periclinal phase (Fig. D, F), and (4) block-meristem phase (Fig. F–O).

1) Filamentous Phase. In *Stellaria media* (Ramji 1975), the zygote (Fig. 3.11.10A) divides transversely forming *ca* and *cb*. *ca*, by three successive divisions in the same plane, gives rise to *l*, *l'*, *m* and *ci* (Fig. B); *cb* remains undivided. Thus, there is much more activity (cell divisions) at the terminal pole. Ramji (1975) also draws attention to the differential staining reaction of the protoplast. This is much more intense in *ca* than *cb*.

2) Onset of Vertical Divisions or Three-Dimensional Phase. Vertical walls are laid in tiers *l'*, *m* and then in *l* (Fig. C). The vertical walls in *l'* and *m* do not coincide, or are perpendicular to each other so that a small sector of the original wall (intercept) is left at the point of contact. During later stages, this intercept helps in determining the bilateral symmetry of the proembryo.

3) Periclinal Phase. Periclinal divisions parallel to the surface of the embryo occur in *m* and *l'* separating a group of axial cells in the centre (Fig. D). The periclinal division in *n* occurs when the tiers *m* and *l'* become six-celled (Fig. E). The activity of the cells of tier *l* is very slow as they are the potentials of the shoot-apex, which grows vigorously during the germination of seed. The four cells of tier *l* (Fig. F) divide by oblique transverse walls, separating the outer cover of four protoderm cells, and inner group of four cells. This results in separation of a continuous protodermal layer around the central axial zone. The increase in the number of cells in the axial zone is due to vertical divisions. The central zone of four axial cells along with the encircling layer gives rise to procambium; the cells lying between this zone and the protoderm give rise to cortex (Fig. G; for further details of each tier, apex upward, see Fig. H–O).

4) Block Meristem Phase or Organ Indication Phase. The outward globular appearance (Fig. F, G) of the proembryo is retained with the increase in the number of cells. The cells of the central axial zone divide in a tangential plane, whereas those of outer layers undergo divisions in a radial plane.

The cells of the subprotodermal layer of *l'*, on either side of the apex of epicotyl, divide anticlinally and differentiate as initials of cotyledons, whereas those in the centre elongate to form the rib meristem, later producing pith of the cotyledonary node. Tier *m* gives rise exclusively to the hypocotyl. The three zones – protoderm, initials of cortex, and axial region – are well-demarcated due to differential staining of the protoplasts. Frequent divisions in transverse plane (Fig. G) take place in central cells, which develop into the procambium. Tier *n* forms the tip of the hypocotyl, which is the potential root.

Figures P and Q show longisections of apices of epicotyl and hypocotyl. (The author includes radicle in hypocotyl.) The cortical cells developed from the tier *l′* are in continuation with those produced by tier *m* (Fig. P). In the region *n* the subcortical layer is developed by the periclinal division of protoderm cells, and the remaining two layers are in continuation with *m*. Thus, the differentiation of cortical tissue in the tiers *m* and *n* differs (Fig. P, Q).

Periasamy (1977) proposes a new system of classification of embryos based on morphogenetic significance. Accordingly, he classifies 45 species of dicotyledonous and monocotyledonous embryos into five groups. Periasamy emphasizes that the external surface of the embryo and its derivatives towards the inner side are subject to two different environments. The criteria used for classification are: (1) number of cell tiers that comprise the proembryo at the time when the first series of internal cells are segmented, (2) relative location of the tiers with varying cell number, and (3) terminal or subterminal location of the tier in which the first internal cells are segmented.

The embryogeny in *Paeonia* shows a totally different type of development (see Sect. 22.3).

Dicot and Monocot Embryos. The similarity of development between the dicotyledonous and monocotyledonous type of embryos has been traced by Lakshmanan (1972). Apparently, the external appearance of the proembryo up to the globular stage is similar in both. Swamy and his co-workers (see Lakshmanan 1972) have demonstrated that the epicotyl and cotyledon arise from one and the same terminal tier (*q*) in monocotyledons. The epicotyl shows belated development, while the cotyledon overgrows in early stages. Consequently, the epicotyl assumes a lateral position. The subterminal tier and its derivatives contribute to the hypocotyledonary region.

Lakshmanan (1972) also discusses the differences between the ontogeny of the two types of embryos (Fig. 3.11.13A–E). The main difference lies in the number of sectors of a quadrant, and their position in a quadrant, which contribute to the formation of cotyledon(s). In a dicotyledonous embryo, the two opposite cells of a terminal quadrant give rise to two cotyledons (Fig. B). In monocotyledons, the number of sectors of a quadrant involved in the formation of a single cotyledon varies (Fig. C–E). "It is practically almost all the four (except a few cells derived from one of the quadrant cells) in the Philydraceae, three cells of the quadrant in the Pontederiaceae, Sparganiaceae, and Iridaceae, and adjacent two cells in the Hydrocharitaceae, Potamogetonaceae, and Amaryllidaceae that constitute the cotyledonary locus" (Lakshmanan 1972).

Guignard and Ly-Thi-Ba (1989) have reviewed the embryogeny of the Helobiae.

There is a single cotyledonary locus in monocotyledons, and two in dicotyledons. Likewise, the morphogenetic potential of a shoot apex differentiates at the end of the first vertical division in the terminal cell in monocotyledons, and at the end of the second vertical division in dicotyledons (see also Swamy and Krishnamurthy 1980).

It is customary to classify monocotyledonous proembryos, like those of dicotyledons, according to the embryogenic systems of classification by Souèges (see Crété, 1963a), and Johansen (1950). However, Figure 3.11.13A–E clearly indicates the difference between the cotyledonary loci of the two.

In taxa such as Hydrocharitales, Aponogetonales, Potamogetonales, Najadales, Commelinales, Xyridales, Zingiberales, Liliales, Amaryllidales, Orchidales, Graminales, etc., equal volumes of the shoot apex are involved in the organization of the cotyledon (single) and epicotyl, whereas in Alismataceae, Juncaceae, Pontederiaceae, Typhales, etc., a larger volume of the shoot apex takes part in the development of the cotyledon (Swamy 1979).

The embryogeny in *Triticum* (Batygina 1969) exhibits the orientation of oblique walls at the onset of division of the zygote (Fig. 58.2.3A, B). They are disposed in such a way that a continuous dorsiventral symmetry is maintained during the entire development of the embryo (Fig. C–T). The vegetative point (shoot apex) originates from the apical cell *ca* and is terminal (Fig. N), the scutellum initial originates from the cell *ca* as well as from *cb* (*m* and *n*) and is apical–lateral in origin (Fig. N). The region of intensive colour and high mitotic activity of cells denotes the origin of meristems during organogenesis. The dorsiventral symmetry of the embryo (Fig. C–T), inception of oblique walls (Fig. A–H, B′–F′), displacement of tiers *m* and *ci*, and early differentiation of scutellum led Batygina (1969) to propose a new type, the Graminad type.

Juguet (1989) discusses the problems of total number of cells in a proembryo when the symmetry becomes established in the monocots and dicots. According to him, the dicot embryo has an "axial" and the monocot embryo a "bilateral" symmetry. The cotyledon and the future stem-tip both develop in lateral–apical position but become respectively terminal and lateral. Juguet re-emphasizes "that the quarrel between "terminalists" and "lateralists" is a futile one". We support this view, but consider the cotyledon/s as lateral and the shoot apex as terminal structures, both in monocots and dicots.

Delayed Embryo Differentiation. Haccius (1971) and co-workers (Haccius and Hausner 1972, Haccius and Bhandari 1975, Haccius and Philip 1979) draw attention to "embryogenesis with delayed embryo-differentiation". The embryogeny in *Tulipa* (Haccius and Hausner 1972) is a good example, and shows the formation of a vertical wall in the apical cell at the end of the fifth or sixth cell generation, initiating the globular stage of the proembryo. The major portion of the derivatives of *ca* and all of *cb* do not contribute to the formation of the embryo proper.

After the first cell generation, the zygote produces two superposed cells *ca* and *cb* (Period I, see Fig. 3.11.4). In the second generation (Period II), *ca* produces tiers *cc* and *cd*, in the third (Period III) *ca* gives rise to *ce* and *cf*, in the fourth (Period IV) *ce* to *cg* and *ch*, and the fifth and sixth generations continue similarly.

In *Tulipa kolpakowskiana* (Haccius and Hausner 1972), a vertical division occurs first in the basal cell *cb* of the two-celled proembryo (Fig. 3.11.14A–C). Successive vertical divisions occur in each of these two sister cells up to the end of the third cell generation (Fig. D–H). The derivatives of *cb* undergo divisions in all planes and the basal portion becomes massive (Fig. I–L).

The tier *ca* undergoes a transverse division forming *cc* and *cd* (Fig. C–E). The tier *cc* divides transversely to form *ce* and *cf*, while *cd* undergoes a vertical division (Fig. F, G). At the end of the fourth cell generation (Fig. H–J), *ce* produces two superposed daughter cells, *cg* and *ch*, whereas *cf* and the derivatives of *cd* divide vertically. The fifth cell generation begins with the transverse division of *cg* (Fig. M), *ch* divides vertically. The vertical wall is laid down in the terminal cell mostly at the end of the fifth or sixth cell generation (Fig. L); M and N are later stages. The embryogeny of *Tulipa* can be classified to Period V or VI of Souèges system. It cannot be referred to Johansen's system, as in this classification the provision is only up to Period II (Caryophyllad type). It may be noted that, during early embryogeny, vigorous growth is restricted to the basal region only and the differentiation of the embryo is delayed. The derivatives of *cb* and a considerable portion of the cells derived from *ca* do not contribute to the formation of the embryo proper.

Suspensor. The suspensor is usually uniseriate with uninucleate cells in young proembryos. However, the occurrence of massive suspensor with multinucleate cells is not uncommon (Fig. 48.4.4J–O; see also Raghavan 1986a, b). The highly coiled suspensor of multinucleate cells (Fig. 24.3.1A–F) in *Isomeris arborea* was incorrectly interpreted as endosperm nodules by Billings (1937). During the earlier stages of development the suspensor cells exhibit endomitotic polyploidization in *Trapa natans* (Nägl 1962). The highest ploidy-level noted is 256×. The polyploid level corresponds to 1024× in *Tropaeolum majus*, and 409× in *Phaseolus coccineus* (Nägl 1962). The occurrence of giant chromosomes in *Phaseolus* (Nägl 1962) and *Gagea lutea* (Hasitschka-Jenschke 1959, 1962) indicates a high metabolic activity of suspensor cells.

In *Stylosanthes mucronata* (Fig. 3.11.15A), at the dicotyledonous stage of the embryo the suspensor is long and multicellular. In *Crotalaria stricta* (Fig. 3.11.15B), the peripheral suspensor cells in the middle region of a long multicellular suspensor develop tubular projections which penetrate into the endosperm tissue. The suspensor is fairly broad, multicellular and conspicuous in *Sesbania grandiflora* (Fig. 3.11.15C, D). The suspensor cells in *Rubia cordifolia* (Fig. 3.11.15E) become prominently vesicular. The two juxtaposed cells (daughter cells of *cb*) enlarge and their nuclei become hypertrophied in *Haloragis micrantha* (Fig. 3.11.15F). The suspensor is fan-shaped in *Glottidium vesicarium* (Fig. 3.11.15G).

A four-celled suspensor with coenocytic cells is reported in *Lathyrus latifolius* (Nägl 1962), and a broad and conspicuous suspensor in *Phaseolus coccineus* (Yeung and Clutter 1978).

The extensive suspensor haustorium in Podostemaceae (Figs. 28.2.2K–N, 28.2.3A–C), Trapaceae (Fig. 39.3.1N, O) and Orchidaceae (Fig. 65.2.3E–H) serves the function of absorption and transport of nutrients in the absence of endosperm tissue. In *Dicraea stylosa* (Fig. 28.2.3A–C), the binucleate basal cell of the two-celled proembryo develops conspicuous tubular haustorial branches. During later stages, the suspensor pushes the embryo into the pseudo-embryo sac which provides nutrients in the absence of endosperm.

The development of haustoria (Sect. 26.5) from the suspensor in Crassulaceae (Mauritzon 1933a, Subramanyam 1963, 1967) is well known. The hypha-like branches of the haustorial micropylar cell of the suspensor grow into the integument and the raphe, and in some members become extraovular and penetrate the funicle. The Orchidaceae (Sect. 65.2) show much variation in the organization of the suspensor (Swamy 1942a). A well-developed branched suspensor persists in the globular incompletely differentiated embryo in the mature seed (Fig. 65.2.3H) of *Dendrobium nobile* (Poddubnaya-Arnoldi 1967).

The role of suspensor has been revealed by ultra-structural, histochemical and in vitro cultural studies. P. Schulz and Jensen (1969) observed that in *Capsella* the octant proembryo has a six-celled suspensor. The wall of the micropylar terminal vesicular cell bears projections on its inner surface. The wall ingrowths are more concentrated at the micropylar tip and are similar to "transfer cells" (see Gunning and Pate 1969). The latter perform the function of transport of nutrients from the adjoining tissue to the developing embryo.

The histochemical studies in *Brassica campestris* (Malik and Vermani 1975) also support these observations. Malik and Vermani further suggest that the suspensor has a secretory function. Bhalla et al. (1979) experimented with enzymatic reactions occurring in the suspensor haustorium (in the carpellary region) and the suspensor of *Tropaeolum majus*. Their data indicate that the suspensor haustorium absorbs nutritive material from the adjoining carpellary tissue and the suspensor transports it to the young proembryo.

The suspensor has an important role in embryogenesis by acting as a site for the synthesis of growth regulators required by the embryo (Alpi et al. 1975). While experimenting with the major role of gibberellins in characterizing the physiological role of the suspensor, Cionini et al. (1976) demonstrated that GA_3 (10^{-8} to 10^{-6} M) replaces the function of the suspensor (in *Phaseolus coccineus*) when suspensor-deprived embryos of heart-shaped and early cotyledonary stages (0.5–1.5 mm long) are cultured in vitro.

Chlorophyllous Embryos. Yakovlev and Zhukova (1980) review the occurrence of chlorophyll in the embryos of 438 species, belonging to 224 genera of 72 families. The seeds of 1094 species of 666 genera of 182 families have been investigated so far. The greening of embryos is a genetic feature (Yakovlev and Zhukova 1975), and in several species the chlorophyll disappears in the mature embryo.

Chlorophyllous embryos invariably occur in Brassicaceae (Cruciferae), Convolvulaceae, Dipsacaceae, Papilionaceae, Plumbaginaceae, Rhamnaceae and Tiliaceae, whereas the families Betulaceae, Bignoniaceae, Caprifoliaceae, Fagaceae, Oleaceae, and Saxifragaceae have totally non-chlorophyllous embryos. The rest of the 58 families, such as Anacardiaceae, Apocynaceae, Malvaceae, Myrtaceae, Scrophulariaceae and others, have both chlorophyllous and non-chlorophyllous embryos. The monocotyledonous families rarely show green embryos.

The chlorophyllous embryo may be considered as a diagnostic feature, e.g. closely related families Capparaceae and Cruciferae have chlorophyllous embryos. On this basis, Celastraceae (green embryo) can be distinguished from Aquifoliaceae (non-green embryo), Dipsacaceae from Morinaceae, and Plumbaginaceae from Primulaceae.

3.12 Polyembryony

Polyembryony is of common occurrence in angiosperms. In most cases, the additional embryos do not develop up to maturity and degenerate early. Polyembryony has been classified by Schnarf (1929), P. Maheshwari and Sachar (1963), Yakovlev (1967), and Bouman and Boesewinkel (1969). The classification suggested by P. Maheshwari and Sachar (1963) is followed in the text (see also Lakshmanan and Ambegaokar 1984).

True Polyembryony. 1) Embryos arising within an embryo sac, either by cleavage of the zygotic proembryo, or from the synergid/s. 2) Embryos arising from the tissues lying outside the embryo sac, i.e. nucellus and integument/s (adventive embryony).

False Polyembryony. This includes twin nucelli formed due to cleavage of the nucellar primordium (*Pterocarya fraxinifolia* – Bouman and Boesewinkel 1969), or the development of two or more embryo sacs within the same nucellus either due to multicellular archesporium, or development of more than one megaspore from a single tetrad, or concurrent development of sporic and aposporic embryo sacs.

Cleavage Polyembryony. Polyembryony may be due to proliferation, or budding, or splitting of the zygotic proembryo. All these have been observed in *Eulophia epidendraea* (Fig. 65.2.3E, F). Narayanan (1968) reported cleavage polyembryony in 86 percent of ovules of *Streblus taxoides*. Two to four embryos are produced due to cleavage of the zygotic proembryo, and the twin embryos develop up to the dicotyledonous stage (Fig. 3.12.1A). Mathew and Chaphekar (1977) also report cleavage of the zygotic proembryo in *Stachyurus chinensis* (Fig. 3.12.1B, C).

The proliferation of proembryonal and suspensor cells giving rise to additional embryos is a common feature in the members of Acanthaceae (Fig. 3.12.1D–F), Actinidiaceae, Santalaceae, Zygophyllaceae and others.

Synergid Polyembryony. In several families the persistent synergid develops into an embryo, e.g. Amaryllidaceae, Apocynaceae (Plumiereae), Burmanniaceae, Cucurbitaceae, Stachyuraceae, Tamaricaceae and some others. In most cases the synergid does not get fertilized. B.M. Johri and Kak (1954) reported the presence of two pollen tubes in the embryo sac of *Tamarix ericoides* (Fig. 3.12.1G). The synergid embryo appears to have developed after fertilization. Similar fertilized synergids have also been reported in Alismataceae (B.M. Johri 1936a).

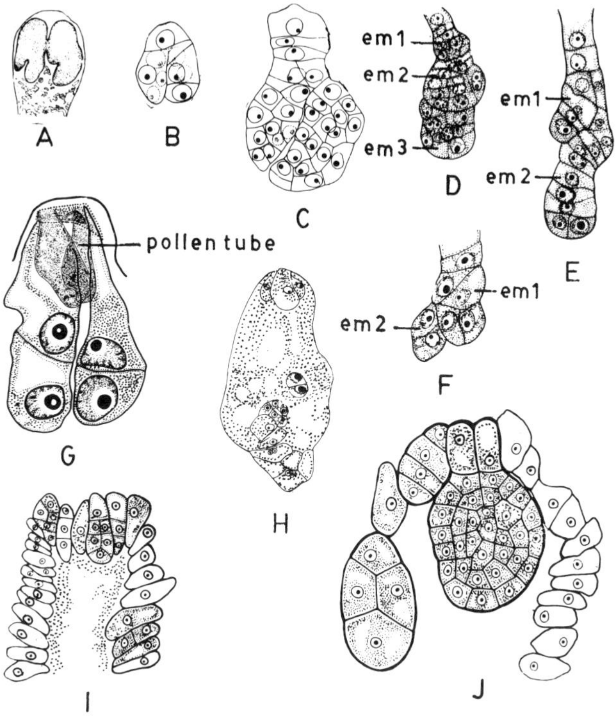

Fig. 3.12.1A–J. Polyembryony. **A** *Streblus taxoides*, longisection of seed, twin dicotyledonous embryos result from cleavage of zygotic embryo. (After Narayanan 1968). **B, C** *Stachyurus chinensis*, budding/cleavage of proembryo. (After Mathew and Chaphekar 1977). **D** *Barleria prionites*, additional proembryo *em 1* formed by proliferation of a suspensor cell (?), and *em 3* by proliferation of proembryo *em 2*. (After Phatak and Ambegaokar 1956). **E** *Sericocalyx scaber*, additional proembryo *em 1* formed by budding of a suspensor cell. (After Phatak and Ambegaokar 1961b). **F** *Lepidagathis trinervis*, additional proembryo *em 2* formed by proliferation of proembryo *em 1*. (After Ambegaokar unpublished). **G** *Tamarix ericoides*, portion of embryo sac showing two-celled zygotic and two-celled synergid proembryo; note remnants of two pollen tubes. (After B.M. Johri and Kak 1954). **H** *Paspalum scrobiculatum*, embryo sac showing antipodal (nucellar?) proembryo; note the adjoining antipodal cells. (After Narayanaswami 1954). **I, J** *Melampodium divaricatum*, endothelium with some cells developing into proembryonal structures. (After Maheswari Devi and Pullaiah 1976)

Sachar (1955) reported synergid polyembryony in *Argemone mexicana*. The synergids are usually ephemeral, but one of them may persist and develop into a proembryo. This proembryo is smaller, has an independent suspensor, and lies juxtaposed to the zygotic proembryo.

In *Dioscorea composita*, the zygotic and synergid proembryos develop until a late stage. They follow a uniform pattern of development up to organogenesis in *Nechamandra alternifolia* (Lakshmanan 1963b).

The development of haploid proembryo from the synergid, along with the diploid zygotic proembryo, is reported in several taxa: *Lilium martagon*, *Arabis lyallii*. The suspensor of the zygotic as well as of the synergid proembryo of

Table 3.12.1. *Capsicum frutescens*, multiple seedlings in reciprocal crosses of two varieties (After Morgan and Rappleye 1954)

Varieties	Twins				Triplet
	Unattached			Conjoined	Unattached
	Sexual–sexual 2n 2n	Maternal–sexual n 2n	Maternal–maternal n n	Sexual–sexual 2n 2n	Sexual–sexual–sexual 2n 2n 2n
Non-pungent, non-elongated, dark-green immature fruit × pungent, elongated yellow immature fruit	72	11	1	7	1
Pungent, elongated, yellow, immature fruit × non-pungent, non-elongated, dark-green immature fruit	10	4	0	3	0
Total 109	82	15	1	10	1

A. lyallii elongates into similar filamentous structures with a small embryonal mass at the apical end.

In *Cuscuta*, polyembryony is rare, but in *C. reflexa* (B.M. Johri and Tiagi 1952) a two-celled synergid proembryo has been observed along with the bicelled zygotic proembryo. Accessory pollen tubes and supernumerary male gametes have also been observed in the embryo sac (Fig. 3.9.3B). However, it could not be determined whether the synergid proembryo is haploid or diploid.

In *Pennisetum squamulatum* (Sindhe et al. 1980), the diploid synergid proembryo develops along with the zygotic proembryo. The suspensor of the zygotic proembryo is typically multicellular as in the Gramineae, and that of the synergid is partly uniseriate and partly biseriate.

In *Azadirachta indica* (N.C. Nair and Kanta 1961), at times one or both synergids develop into haploid proembryos. In *Carthamus tinctorius* (Maheswari Devi and Pullaiah 1977), the triplets develop due to fertilization of the egg and both synergids.

Multiple seedlings arise in *Capsicum frutescens* (2n = 24) from individual polyembryonic seeds, following reciprocal crosses of two varieties (Table 3.12.1).

The haploid members of twin seedlings arose from reduced cells of the female gametophyte, presumably the synergid (the antipodals degenerate prior to fertilization). All diploid members of the multiple seedlings were of sexual origin (adventive embryos do not develop in this sexual species). Diploid–diploid twins could arise from twin embryo sacs. The conjoined diploid–diploid twins presumably originated from the cleavage of zygotic embryo. Haploid–haploid twins could originate from the synergids, if the zygotic embryo degenerates. Most of the unattached 2n-2n twins developed within a single embryo sac by fertilization of a synergid and the egg. It is likely that the pollen tube had supernumerary sperm cells, or more than one pollen tube entered the embryo sac. The origin of the unattached triplet could not be traced. The possibilities are: (a) polyspermy can lead to fertilization of the egg and both the synergids, and (b) fertilization of the egg in three embryo sacs in the same nucellus could also lead to a 2n-2n-2n triplet – rather rare. The study of polyembryony in hybrids and their n/2n constitution needs better attention.

Antipodal Polyembryony. The formation of antipodal embryos is not so frequent. These have been observed in *Alangium lamarkii* (Gopinath 1943), *Paspalum scrobiculatum* (Fig. 3.12.1H), *Rudbeckia sullivanti* (Battaglia 1955), *Ulmus americana* (Shattuck 1905), *U. capensis* (Guignard and Mestre 1966) and *U. glabra* (Ekdahl 1941). In *Solidago canadensis*, Pullaiah (1978b) observed two proembryos developing at the antipodal end but these did not mature to full size.

In our opinion, the development of embryos from antipodals needs confirmation. When the antipodal cells divide, the configuration of cells may simulate embryogenic patterns. Since advanced stages have never been recorded, it is not at all appropriate to interpret such cell assemblages as antipodal embryos. We would rather interpret them as internal divisions in antipodal cells – a feature already known in Dipsacaceae (Fig. 50.5.7C–G). This is probably also true of endothelial (Fig. 3.12.1I, J), and integumentary proembryos in *Spiranthes cernua* (Swamy 1948d).

Adventive Polyembryony. The development of proembryo/s from the integument/s is less frequent: *Spiranthes cernua* (Swamy 1948d). Either a single cell, or a group of two or three cells of the inner epidermis (endothelium) of the inner integument proliferate to produce the additional embryo/s (Fig. 3.12.1I, J). Integumentary embryos are known in *Carthamus tinctorius* (Maheswari Devi and Pullaiah 1977), *Datura* (Satina et al. 1950), and *Spiranthes* (Swamy 1948e).

The formation of embryos from the nucellar cells is a common feature. The cells of the nucellus, in the micropylar region, lying on the periphery of the embryo sac proliferate and nucellar embryos are initiated. As they grow, they come to lie in the embryo sac and attain various stages of development. In *Opuntia dillenii* (P. Maheshwari and Chopra 1955), the egg degenerates, several nucellar embryos develop, and one of them reaches up to the mature stage. The nucellar embryos are frequent in Anacardiaceae (mango), Cactaceae (*Mammillaria* – Fig. 17.2.1A–D), Combretaceae, Gramineae, Malpighiaceae, Rutaceae, Urticaceae and others. Polyembryonate seeds have been reported in Myrtaceae.

False polyembryony is reported in Loranthaceae (Sect. 12.8), Gramineae (Sect. 58.9) and some other taxa.

3.13 Gametophytic Apomixis

Apomixis, believed earlier to be a rare phenomenon, has now been observed frequently in families like Compositae, Gramineae, Rosaceae and many others. As described by Winkler (1908), apomixis is a process in contrast to amphimixis (sexual reproduction) where the sporophyte is produced from the gametophyte without sexual fusion.

Classification. The classification of apomixis is rather diverse. P. Maheshwari (1950) divides it into four classes: (1) non-recurrent apomixis, (2) recurrent apomixis, (3) adventive embryony, and (4) formation of vegetative propagules. The formation of vegetative propagules is merely a form of vegetative reproduction.

Battaglia (1963) adopted several terminologies, and defined apomixis as the production of sporophyte from the gametophyte without sexual fusion. He classifies apomixis into two types: recurrent and non-recurrent. Unlike Fagerlind (1940a), Gustafsson (1946, 1947a, b), P. Maheshwari (1950) and Battaglia do not include (and rightly so) adventive embryony and vegetative reproduction in "apomixis".

Solntzeva (1969) accepts Battaglia's classification, but includes adventive embryony in apomixis. The vegetative reproduction is totally excluded. While defining apomixis, Solntzeva emphasizes: (1) apomixis is a special and peculiar mode of seed reproduction; (2) this mode of reproduction implies the use of morphological structures inherent in seed reproduction – flower, ovule, embryo sac, embryo, seed and fruit; (3) in this mode of reproduction disturbances may occur both in sporogenesis and in sexual process, and (4) production of a new organism takes place through the formation of an embryo. In her scheme of classification of apomixis in angiosperms, she describes the cytological disturbances which occur during sporogenesis and sexual process.

Davis (1966, 1972) also follows Battaglia's classification. J. Heslop-Harrison (1972) does not recognize the non-recurrent type as a part of apomixis as it includes the process of meiosis in the formation of reduced embryo sacs.

Types of Apomixis. In a "non-recurrent" type, or in the formation of a reduced embryo sac, the megaspore mother cell undergoes regular meiotic divisions. The megasporogenesis and gametogenesis are normal, and the components of the embryo sac carry a haploid set of chromosomes. Syngamy does not occur, and the haploid egg cell gives rise to the embryo by parthenogenesis, pseudogamy, or androgenesis.

In parthenogenesis, the egg cell divides without pollination. Depending on the reduced and unreduced embryo sac, parthenogenesis is referred to as haploid or diploid, respectively. Parthenogenesis can also be induced experimentally as an effect of temperature.

In pseudogamy, pollination occurs but the male gamete disintegrates either inside or outside the egg cell. It provides only a stimulus as in Compositae and Rubiaceae. The development of haploid embryo from the sperm nucleus, i.e. androgenesis, is reported in *Crepis* (Gerassimova 1936). Here the egg nucleus degenerates, and the sperm nucleus functions in the cytoplasm of the egg cell.

The embryo may develop from some other cell of the reduced embryo sac, such as the synergid. It is termed "haploid apogamy".

The second type of apomixis, "recurrent", occurs in a non-reduced (diploid) embryo sac. It is effected by (1) aneuspory, (2) gonial apospory, (3) somatic apospory, and (4) synkaryogenesis.

Nogler (1984) critically analyzes the phenomenon of gametophytic apomixis, and suggests appropriate terminologies. The following account is based on his review.

Apomixis or asexual (agamic) reproduction by seeds, i.e. agamospermy (seed apomixis) covers two aspects.

1) Adventive Embryony. One or more embryos develop from the nucellar cells (Sect. 3.12). Only the endosperm is produced in the embryo sac.

2) Gametophytic Apomixis. The embryo sac develops (diplospory) from an unreduced generative cell (sporogenous or megaspore mother cell) either directly by mitosis (Antennaria type), or indirectly by modified meiosis resulting in an unreduced restitution nucleus (Taraxacum and Ixeris type). Unreduced embryo sac can also result from somatic cells of the nucellus (apospory).

In diplospory and apospory, there is no meiosis and fertilization of the egg does not occur. The unreduced egg develops parthenogenetically. The formation of endosperm depends on the fertilization of polar nuclei/secondary nucleus.

When both egg cell and polar nuclei are not fertilized, the development of embryo and endosperm is autonomous.

Diplospory (Generative Apospory)

Antennaria Type. The megaspore mother cell becomes vacuolated (as in functional megaspore), undergoes three successive mitoses producing an unreduced eight-nucleate embryo sac: *Antennaria, Erigeron, Eupatorium, Hieraceum, Poa alpina, P. nervosa* and others.

Taraxacum Type. The megaspore mother cell does enter meiosis but forms a meiotic restitution nucleus. The second meiotic division leads to an unreduced dyad (instead of a reduced tetrad). Following three successive mitoses, the lower dyad gives rise to an unreduced eight-nucleate embryo sac: *Taraxacum, Erigeron, Chondrilla* and others.

Ixeris Type. As in Taraxacum type, the meiotic restitution nucleus undergoes a division but is not followed by a wall. The coenomegaspore contains two unreduced (instead of four reduced) nuclei. Two successive mitotic divisions produce an unreduced eight-nucleate embryo sac: *Ixeris dentata, Rudbeckia, Erigeron.*

Female Meiosis in Diplosporous Apomicts

Various possibilities of meiosis in megaspore mother cell in diplosporous apomicts are shown in Fig. 3.13.1A–E. During normal meiosis, a linear tetrad develops, the upper three megaspores degenerate, and the lower is functional (Fig. 3.13.1A). Disturbed meiosis results in irregular distribution of chromosomes leading to aneuploidy (Fig. 3.13.1B) and the embryo sac is always diploid. The meiotic restitution nucleus undergoes meiosis II followed by a wall, and the lower unreduced dyad forms a Taraxacum type of embryo sac (Fig. 3.13.1C). If the wall is not formed after meiosis II, the embryo sac is of the Ixeris type (Fig. 3.13.1D). If meiosis does not occur at all, the embryo sac is of the Antennaria type (Fig. 3.13.1E).

Apospory (Somatic Apospory)

The unreduced embryo sacs develop from the centrally and/or laterally placed nucellar cells, or cells in the chalazal region (Fig. 3.13.2A–D), or even from the cells of the inner integument. The aposporous embryo sac initials enlarge, become vacuolate, compress and resorb the circumjacent tissue, and develop into one or more embryo sacs.

Cortaderia jubata (M.N. Philipson 1978b) is an excellent example of somatic apospory (Fig. 3.13.2D) and non-pseudogamy. In this species pollen grains are not produced; the embryos develop in the aposporous embryo sacs,

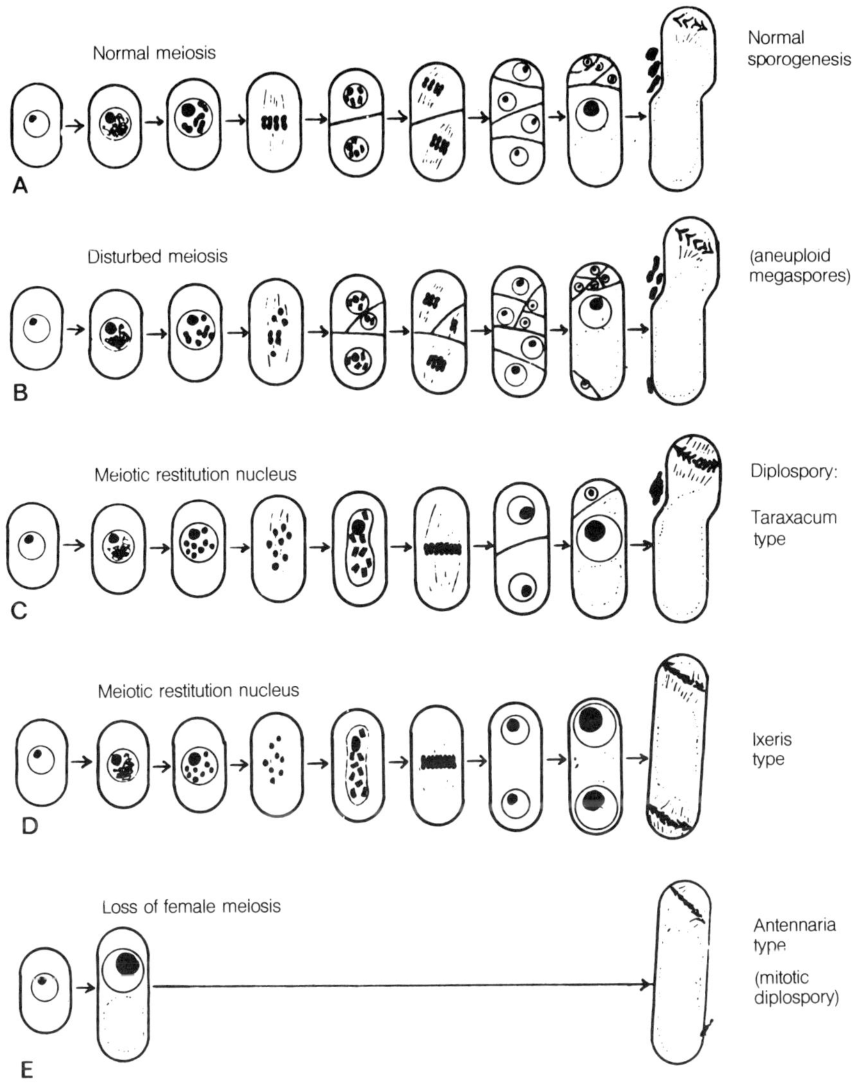

Fig. 3.13.1A–E. Female meiosis in diplosporous apomicts. (After Nogler 1984, modified from Fagerlind 1944a)

produced from the nucellus; the embryos are well developed by the time the inflorescence emerges.

Hieraceum Type. The aposporous embryo sac is bipolar, and is widely distributed: some Compositae, *Hypericum perforatum*, *Poa pratensis*, *P. granitica*, *Elatostema acuminatum* and others.

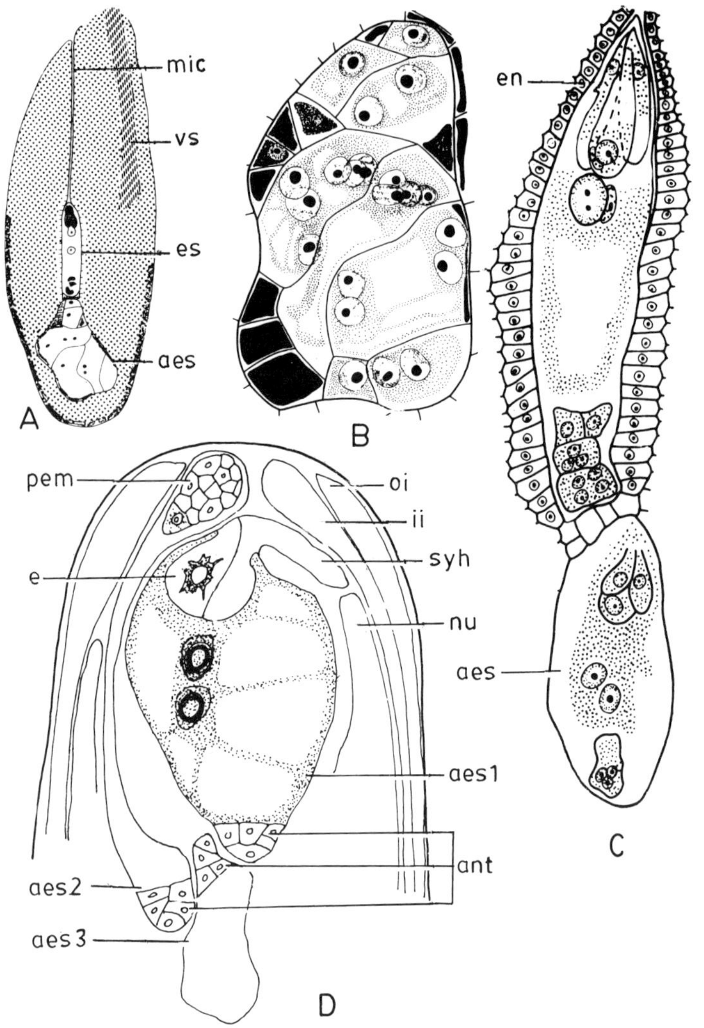

Fig. 3.13.2A–D. Somatic apospory. **A**, **B** *Cliftonia monophylla*. **A** Longisection of ovule with normal embryo sac (*es*); several aposporous embryo sacs (*aes*) develop in the chalazal region. *mic* micropyle. *vs* vascular supply; **B** Aposporous embryo sacs in the chalazal region (of another ovule). (After Vijayaraghavan and Dhar 1978). **C** *Melampodium divaricatum*, normal embryo sac with endothelium (*en*), and an aposporous embryo sac at chalazal end. (After Maheswari Devi and Pullaiah 1976). **D** *Cortaderia jubata*, longisection of young bitegmic seed to show three somatic embryo sacs: *aes 1*, *aes 2*, *aes 3*. *aes 1* contains the undivided egg (*e*), a synergid with its haustorium (*syn*) deflected from the micropyle, two polar nuclei, and five antipodals (*ant*); *aes 2* contains an inverted (?) globular proembryo (*pem*) and five antipodals; there were also a few endosperm cells which are not shown; *aes 3* has four antipodals at the micropylar end (reversed polarity), the rest of the embryo sac is empty; *oi*, *ii* outer and inner integument; *nu* nucellus. (After M.N. Philipson 1978b)

Panicum Type. The unreduced embryo sac is monopolar. The nucleus of the aposporous initial undergoes two successive mitotic divisions, and of these four nuclei three organize into egg apparatus and the fourth is the single polar. The lower polar nucleus and the antipodals are absent: common in Panicoideae and Andropogoneae.

Female Meiosis in Aposporous Apomicts

In aposporous apomicts, the sexual and aposporous embryo sacs often develop concurrently (Fig. 3.13.2A–C). However, the megaspore mother cell before undergoing normal meiosis, or later, or even the early embryo sacs, may remain arrested and degenerate. Thus, only the aposporous embryo sacs develop. When the aposporous initials take the lead, the chances of development of sexual embryo sacs decline since the megaspore mother cell must undergo meiosis and in the meantime the aposporous embryo sac is already well advanced.

The diplosporous and aposporous embryo sacs usually have the same organization as sexual eight-nucleate embryo sacs but various abnormalities may occur.

Endosperm and Embryo in Apomicts

When pollination does not occur, the endosperm develops from the polar nuclei without fertilization. This mode is referred to as "autonomous" and the egg develops parthenogenetically: many Compositae, *Burmannia coelestis*, *Elatostema*, *Cortaderia jubata* and others.

When fertilization of polar nuclei is necessary, the development of endosperm is said to be "pseudogamous": most apomictic Rosaceae (except *Alchemilla*, *Aphanes*), apomictic Gramineae (except *Calamagrostis*, *Cortaderia jubata*), *Zephyranthes texana*, *Ranunculus auricomus*.

The apomicts in the same genus show either autogamy or pseudogamy; exceptions do occur. The ploidy-level of endosperm is more variable in pseudogamous than in autogamous apomicts.

In many pseudogamous apomicts, the development of embryo begins only after the formation of endosperm as in sexual plants: *Rubus*, *Hypericum perforatum*, *Ranunculus auricomus*, *Panicum maximum*, *Pennisetum ciliare*.

In *Poa*, *Parthenium* and *Tripsacum*, also pseudogamous apomicts, precocious development of embryo occurs in most ovules while still in the bud; the endosperm develops only later after the fertilization of polar nuclei.

In autonomous apomicts both types of development of embryo occurs, but precocious embryony prevails: *Hieraceum*, *Taraxacum*. The endosperm may be initiated before anthesis, and in *Crepis* and *Cortaderia jubata* the egg cell divides only after the endosperm formation has begun.

Hemigamy

In certain pseudogamous apomicts, while one male gamete fertilizes the polar nuclei, the other male gamete enters the egg cell, does not fuse with the nucleus, and degenerates as in *Atamosco* (*Zephyranthes*). In *Rudbeckia*, the male gamete may even divide once (Fig. 51.9.4A–G) or several times before the daughter nuclei degenerate, or even participate in the formation of the embryo (especially the suspensor) which can become a chimera (Solntzeva 1974, P.N. Rao and Narayanan 1980). Hemigamy is not a prerequisite for embryo differentiation, but only an "additional feature in certain pseudogamous apomicts" (Coe 1953).

In "androgenesis" (male parthenogenesis), after the entry of the male gamete in the egg cell, the egg nucleus degenerates and the embryo originates from the male gamete.

Savidan (1983) experimented to determine how far the sexuality in facultative apomicts can be modified following hybridization. He observed that, in *Panicum maximum*, apomixis is associated with sexual reproduction in biotypes from South Africa and the Antilles. Three of the hybrids exhibited high percentage of sexuality, while the others appeared to be obligate or nearly so.

The average percentage of sexual reproduction in apomictic hybrids was 5.6, as compared to 8.1 in apomictic forms. He concluded that there is no increase of sexuality in apomicts due to hybridization.

Further information on apomixis is given along with the families where it has been reported.

3.14 Seed

The act of fertilization brings about several changes in the structure of ovule, and a seed is produced (see Linskens 1973). The outer and inner integument undergo structural modifications, and are termed testa and tegmen, respectively. They together constitute the seed-coat. The aril, if present, may grow and envelope the seed partially, or completely. The nucellus is usually consumed during development. But, if it persists as in Centrospermales, it is called perisperm (Fig. 16.6.1A). As the seed gets detached from the funicle, a scar is left on its surface, known as hilum. In anatropous ovules it leaves a longitudinal ridge, termed raphe. The endosperm and embryo occupy the major portion of seed. The endosperm is a storage tissue and, according to its presence or absence, the seeds are called albuminous (endospermous) or exalbuminous (non-endospermous). In a ripe seed, in the majority of families, the embryo is fully differentiated into the radicle, plumule, and cotyledons, and maintains its dormancy. In some members of Ranunculaceae, Burmanniaceae (Fig. 54.16.1J), Orchidaceae (Fig. 65.2.3H), and some parasitic plants such as *Orobanche* (see Sect. 48.24) the embryo is apparently undifferentiated.

Embryo. Martin (1946) studied the size of embryo in relation to the endosperm, and differences in size, shape and position of the embryo in the seeds of 1287 genera of angiosperms. The information is very revealing. He used as measurement quarter units of a circle. When the embryo is smaller than a quarter, it is designated "small", a quarter as "quarter", more than two quarters as "half", and more than three units as "dominant". The embryo is called "total" when it occupies the entire seed, as in exalbuminous seeds.

The embryo in a mature seed is also classified according to its position and shape into three divisions and a number of subdivisions (Martin 1946, see also Radford et al. 1974). R. Dahlgren and Clifford (1982) give a good account of two divisions and subdivisions, especially with reference to the monocotyledons (see Fig. 3.14.1A–Y).

1) Basal division. Embryo usually relatively small and restricted to lower part of seed (except certain grasses), seeds medium to large.

1a) Rudimentary. Embryo small, globular to ovate-oblong, (not truly rudimentary), relatively undifferentiated, massive starchy or non-starchy endosperm: most Magnoliales, Laurales (increasing embryo volume, reduction in endosperm), Piperales (perisperm partly takes over the function of endosperm), *Erythronium*, *Ptychosperma.*

1b) Broad. Embryo as wide as or wider than long, basal, globular or lenticular, in copious endosperm, peripheral or nearly so: *Aneilema*, *Mayaca*, Eriocaulaceae, Rapateaceae, Xyridaceae, *Nymphaea*, *Houttuynia.*

1c) Capitate. Embryo more or less basal, distally expanded, in copious endosperm: *Dioscorea*, *Strelitzia*, *Commelina*, *Tradescantia*, most Cyperaceae.

1d) Lateral. In relation to endosperm and ovule symmetry, "basal or basolateral", discoid or lenticular, surrounded by copious endosperm: Poaceae.

2) Axile Division. Embryo small to total, central (axile), straight or variously curved, seeds minute to large.

2a) Linear. Axial embryo much larger than broad, straight, curved or coiled, cotyledons not expanded, endosperm present or absent: several genera of Hydrocharitaceae, Alismataceae, *Zosteraceae*, *Najas*, *Uvularia*, *Calochortus*, *Erythronium*, *Lilium*, *Sparganium* and many others.

2b) Minute. Axial embryo in minute seed, less than 0.2 mm long, minute and undifferentiated to almost total size of seed: Burmanniaceae, Orchidaceae.

The cotyledons show much variation: a) foliate embryo with expanded and thick cotyledons (in an axile position) bent upon the hypocotyl, b) foliate embryo with cotyledons usually thin and extensively expanded and folded in various ways, c) axial embryo usually erect with thick cotyledons which overlap and encase the somewhat short hypocotyl, the endosperm is scanty or absent, d) cotyledons narrow or expanded, perisperm central or lateral; such a peripheral embryo is large and elongate, arcuate, annular, spirolobal (incumbent cotyledons folded once) or straight, and e) foliate erect embryo with cotyledons thin to thick and slightly expanded to broad (Martin 1946, Radford et al. 1974).

Embryos with a swollen hypocotyl or radicle are referred to as "macropodous" and are associated with exalbuminous seeds: Najadales, Zos-

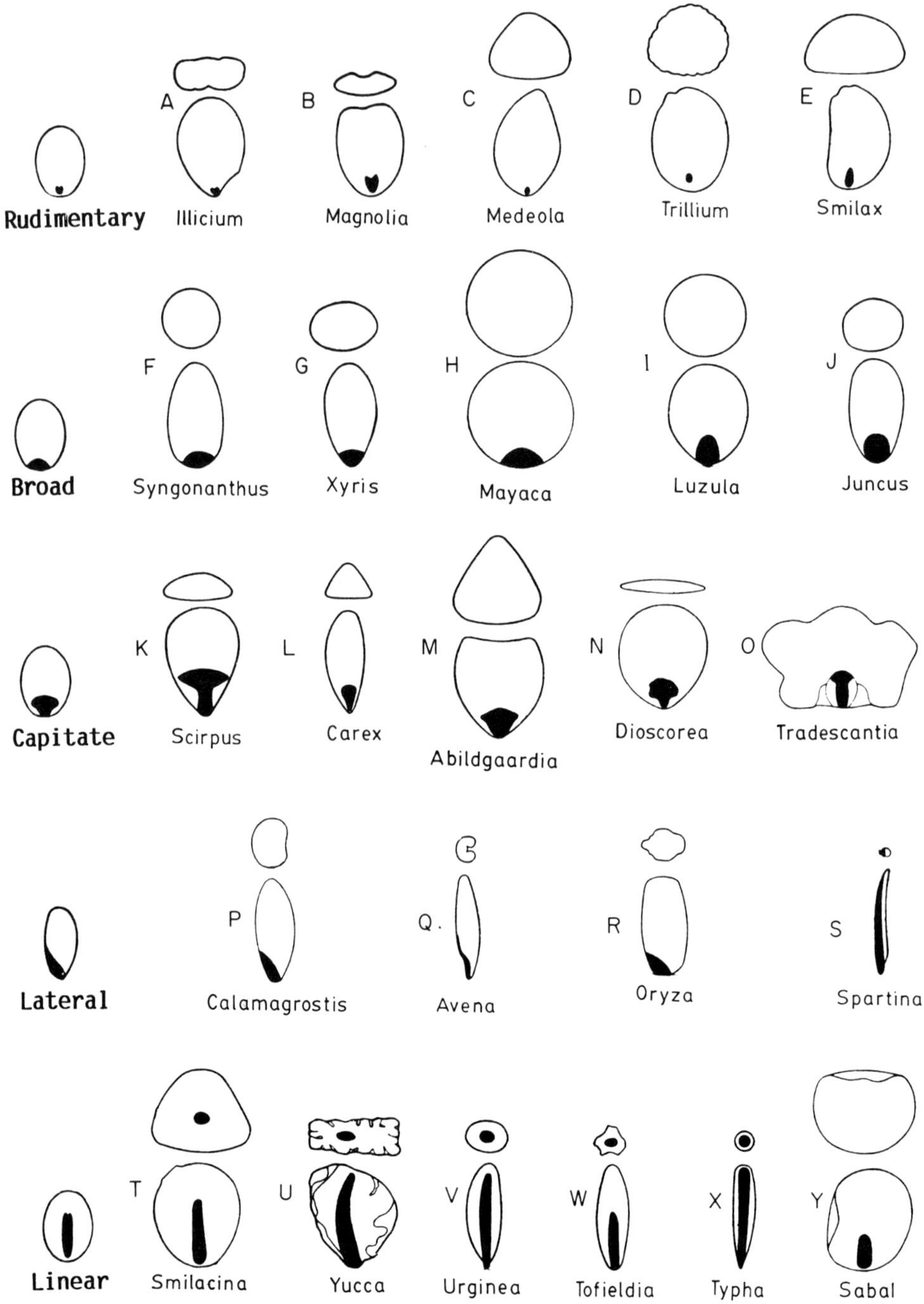

Fig. 3.14.1A–Y. Embryo types, tran- and longisections; an example is cited under each figure. **A–E** Rudimentary. **F–J** Broad. **K–O** Capitate. **P–S** Lateral. **T–Y** Linear. (After Martin 1946, see R. Dahlgren and Clifford 1982)

terales (except Juncaginaceae and Schaeuchzeriaceae) and Hydrocharitales (Thallassioideae and Halophiloideae of Hydrocharitaceae) – all submerged aquatics; not known in dicots (R. Dahlgren and Clifford 1982).

Classification of Seed. During the post-fertilization growth of ovules, sometimes the chalazal region shows conspicuous changes, especially in the ovules with late development of the endosperm. The cells, between the integument and nucellus, undergo intercalary growth pushing the integument in the micropylar region. The chalaza becomes massive (Fig. 3.14.2A–D), and the seed is said to be pachychalazal. In a perichalazal seed the growth is unilateral, and the chalaza exhibits bilateral symmetry (Fig. 3.14.2E–J). A peculiar feature, the formation of chalazosperm, has been noted in the seed of Cynastraceae. The chalazosperm (Fig. 3.14.5E) serves as a storage tissue, in the absence of nucellus and endosperm in the seed.

The development of ingrowths from the chalaza, inner and outer integument, and unequal growth of the peripheral endosperm leads to a ruminate condition of the endosperm (Fig. 3.10.3A–K). This is a diagnostic feature for considerations of systematic assignment.

Morpho-anatomical study of seed is often the basis of classification. The shape, size, colour, nature of surface, and position of hilum (in legumes) is taken into consideration. R.C. Misra (1970) classified the seeds of *Plantago* on the basis of external morphology. Clifford and Smith (1969), Chowdhury and Buth (1970), Ambegaokar (1976), and Behl and Tiagi (1976) used the above criteria, along with internal morphology, to classify legumes. On a similar basis the

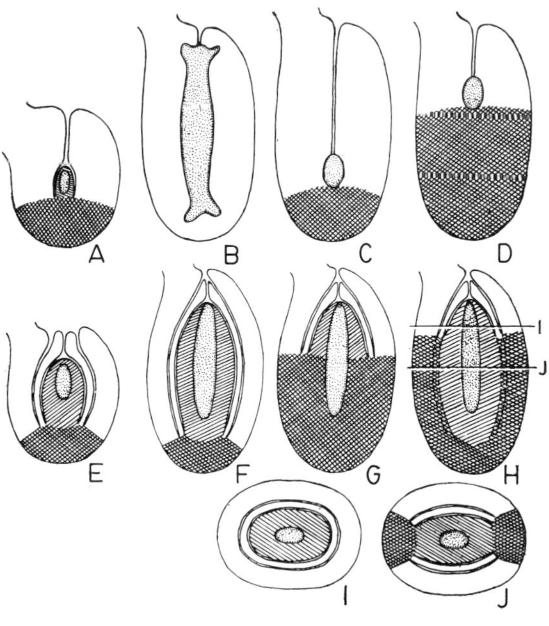

Fig. 3.14.2A–J. Variations in the growth of chalaza during seed development (diagrammatic). All ovules anatropous (chalaza-*cross hatched*). Endosperm and embryo sac *stippled*. Nucellus *striped*. **A–D** Unitegmic ovule, median longisections. **A** Ovule at fertilization; **B** Immature seed with endosperm; **C, D** Immature seed with normal (**C**) and massive (**D**) chalaza. **E–J** Bitegmic ovules; **E–H** median longisections; **I, J** cross-sections of **H** at respective levels; **E** Ovule at fertilization; **F–H** Immature seeds with normal (**F**), massive (**G**) and perichalaza (**H**). (After Periasamy 1962a)

seeds of Cruciferae have been classified by Martin (1946) and Vaughan and Whitehouse (1971), and of Convolvulaceae by Sampathkumar and K.R. Ayyangar (1982).

The sequential study of the developmental changes in the structure of ovule, after fertilization till the formation of mature seed, provides several important features for phylogenetic classification.

Vascular Supply. The vascular supply of the seed is a diagnostic feature. Usually, a well-differentiated raphal bundle terminates at the chalaza (Fig. 3.14.3A). In Annonaceae, Cucurbitaceae, Vitaceae, Polygalaceae, the single bundle may extend in the median plane into the outer integument (postchalazal bundle/s) to the micropyle (Fig. 3.14.3B). In Meliaceae, Connaraceae,

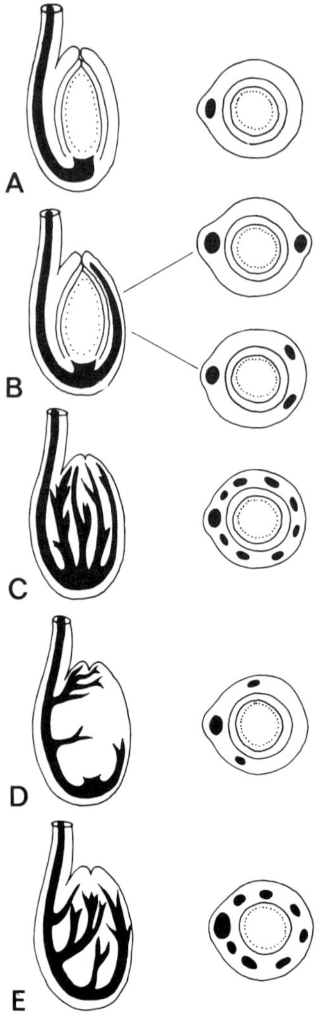

Fig. 3.14.3A–E. Types of seed vasculature: *left row* side views, *right row* transverse sections. **A** Seed with well developed raphal bundle. **B** One postchalazal bundle in the median plane, or two postchalazal bundles. **C** Many postchalazal bundles. **D** *Pre-raphe bundles.* **E** Pachychalazal bundles. (After Boesewinkel and Bouman 1984)

Dichapetalaceae, a number of post-chalazal bundles run up to the micropyle in the outer integument (Fig. 3.14.3C). In Meliaceae (*Dysoxylon*, *Trichilia grandifolia*) and Connoraceae, in addition to the raphal, chalazal and testal bundles, pre-raphal bundles run from the beginning of raphe in the direction of the micropyle (Fig. 3.14.3D). In Meliaceae, Tropaeolaceae, *Glycosmis*, a network of vascular bundles originates from the chalaza and (by intercalary growth) replaces the seed-coat partly or wholly (Fig. 3.14.3E). This condition is termed pachychalaza.

The shape of the seed mostly depends on the shape of the ovule. In the majority of angiosperms, the ovule is anatropous and gives rise to a normal seed (Fig. 3.14.4E). All the other types of seeds can be derived from the anatropous seed as shown in Figure 3.14.4A–N.

The "hilar" seed in some Meliaceae, *Erythrina* and *Aesculus* has an enlarged hilum (Fig. 3.14.4C) and a greater/smaller part of the surface of seed-coat consists of the extended hilum.

In Meliaceae, Sapindaceae, Lauraceae and Tropaeolaceae, which have pachychalazal seeds, the place of attachment of the integument/s is extended (Fig. 3.14.4F, I, K). Below the place of attachment, there is intercalary growth so that the endosperm and embryo are contained in the pachychalaza.

In *Ricinus* and *Myristica*, the zone below the nucellus and above the attachment of outer integument proliferates and results in a short inner and a long outer integument (Fig. 3.14.4F, G). In these endopachychalazal seeds, the seed-coat may remain either endotestal or exotegmic.

In Annonaceae and Meliaceae, in the perichalazal seeds, the extension of the chalaza takes place only in the median plane of the seed (Fig. 3.14.4J) and one of the vascular bundles extends around the periphery from funicle to the micropyle.

Campylotropous seeds are reported in Papaveraceae, Capparidaceae, Cactaceae, Leguminosae, Geraniaceae, Malvaceae and other taxa. The anti-raphal side is strongly extended (Fig. 3.14.4L–N). Mostly, from the chalazal or raphal direction, or inner integument at the raphal side, pressure is exerted on the embryo sac so that it becomes U- or V-shaped. The U- or V-shaped embryo becomes twice as large as the seed.

The obcampylotropous seed is the converse of the campylotropous seed. The raphal side extends, and the antiraphe remains small (Fig. 3.14.4B) as in Leguminosae and Vitaceae.

The orthotropous seed in one-seeded fruits in Piperaceae, Polygonaceae, Urticaceae and Juglandaceae lacks the raphe, and the hilum is situated opposite the micropyle (Fig. 3.14.4H).

Corner (1976), in his well-illustrated book *Seed Anatomy of Dicotyledons*, elaborately discusses the bearing of seed anatomy on the phylogeny of the orders, families, and genera. He defines several terminologies which have been followed in the text.

Seed-Coat. According to Corner (1976, see also Rauh et al. 1975), during development the cells of the testa and tegmen may undergo periclinal divisions so that the number of cell layers increases, i.e. they become multiplicative.

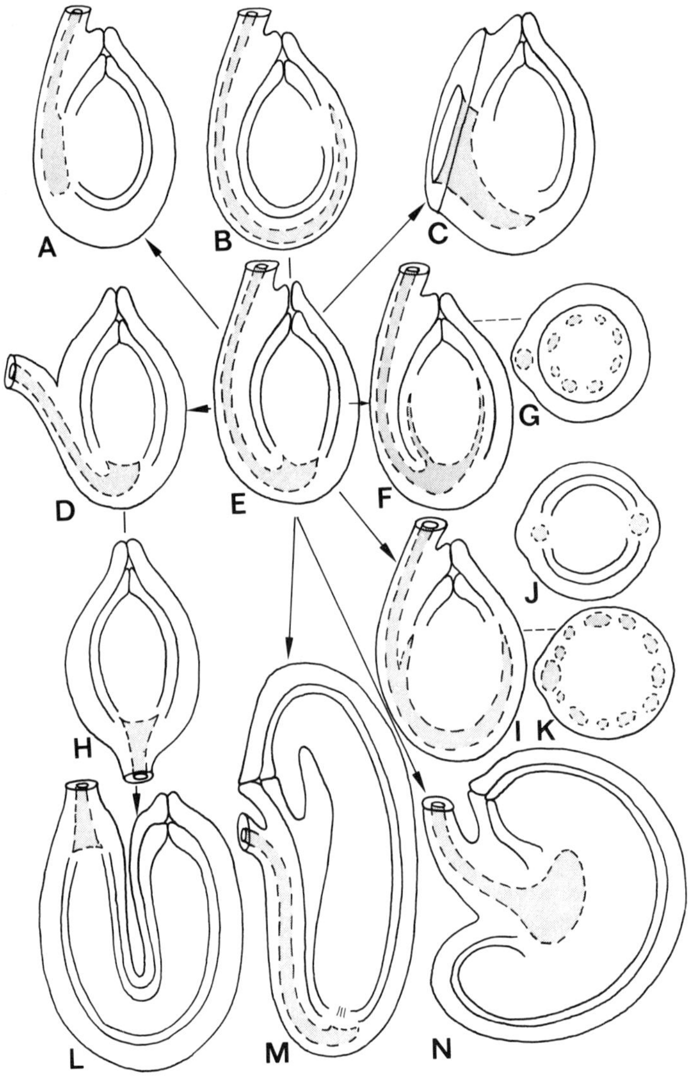

Fig. 3.14.4A–N. Possible derivation of seed types from normal anatropous seed; mostly longisections, **G**, **J** and **K** transections. **A** Chalaza shifted in raphal direction. **B** Obcampylotropous seed, chalaza shifted in antiraphal direction. **C** Hilar seed. **D** Hemianatropous seed. **E** Normal anatropous seed. **F** Pachychalazal seed with short inner integument. **G** Transection. **H** Orthotropous seed. **I** Pachychalazal seed with short inner and outer integuments. **J** Perichalazal seed in transection. **K** Transection of **I**. **L–N** Different types of campylotropous seeds. (After Boesewinkel and Bouman 1984)

Sometimes, some of the cell layers collapse during maturation, or one of the integuments may collapse and degenerate. The seed is then designated as "testal" (Annonaceae – Sect. 18.6, Cucurbitaceae – Sect. 38.2), or "tegmic" (Podostemaceae – Sect. 28.2) as the case may be. In unitegmic ovules, the seed is described as testal (sympetalous families). The integument, i.e. testa (or tegmen), differentiates into three regions: outer epidermis – exotesta (or exotegmen); middle region – mesophyll; inner epidermis – endotesta (or endotegmen).

In the seed-coat, a protective covering, has the mechanical tissue either in the outer epidermis, the mesophyll, or inner epidermis of the testa. Accordingly, the seed is designated "exotestal" (Leguminosae – Sect. 26.19), "mesotestal" (Theaceae – Sect. 22.12), or "endotestal" (Magnoliaceae – Sect. 18.2, Geraniaceae – Sect. 29.4). Likewise, the disposition of mechanical elements in the outer and inner epidermis of the tegmen forms "exotegmic" (Malvaceae Sect. 35.5, Sterculiaceae – Sect. 35.7), and "endotegmic" (Zygophyllaceae – Sect. 19.6) seeds. The mechanical tissue comprises lignified cells, compact palisade-like cells (Leguminosae), fibres, or tracheidal thickenings (see R.P. Singh 1985).

The seed of pomegranate (*Punica granatum*) is mesotestal. The exotesta is a thin-walled, palisade layer, and is pulpy. It is called sarcotesta. In *Magnolia*, the outer part of testa becomes fleshy, and develops into "sarcotesta" (Sect. 18.2). The sarcotestal seeds are dispersed by animals.

Generally, the seed contains endosperm of uniform surface. However, localized peripheral growth of endosperm, ingrowths of the inner and/or outer integument, chalaza, and even pericarp leads to a condition when the endo sperm is called ruminate (Fig. 3.10.3).

Takhtajan (1985, 1988) has further added to our knowledge of the seeds of monocots and dicots.

Appendages. The modifications of the seed which help in dispersal are epidermal hairs (coma), aril, arillode, caruncle, strophiole, elaiosome, wing, and jaculator (Figs 3.14.5 A–F, 3.14.6 A–F, 3.14.7 A–J). According to Pijl (1955), these appendages are classified on the basis of site of origin. Bor (1978) states that "in an anatropous ovule, hilum, raphe, chalaza and exostome are merely topographical zones of the testa. The accessory seed appendages arising from the surface have a common origin".

The epidermal cells of the testa often develop into long, unicellular, hair-like, thin or thick-walled structures, either all over the surface, or in patches, or at localized sites as in Acanthaceae, Asclepiadaceae, Malvaceae, and Salicaceae. These seeds are called "comose".

In euphorbiaceous seed the free-end of the testa proliferates, and develops a tissue on the side of or all round the funicle. It differentiates into a thick, whitish caruncle (Fig. 3.14.5A).

In Linaceae, the tip of the testa develops an outgrowth which grows downward and covers the seed partially. It is termed an arilloid or false aril (Eames 1961). Kapil and Vasil (1963) regard it as an excessively developed caruncle.

The aril may be thin- or thick-walled and, usually, arises from the funicle, outer integument, or both (Fig. 3.14.5B). In a nutmeg seed it develops from the funicle, and exostome (Mohana Rao 1974b). The orthotropous ovule of *Hitchenia caulina* (Panchaksharappa 1962b) shows a similar feature. Accord-

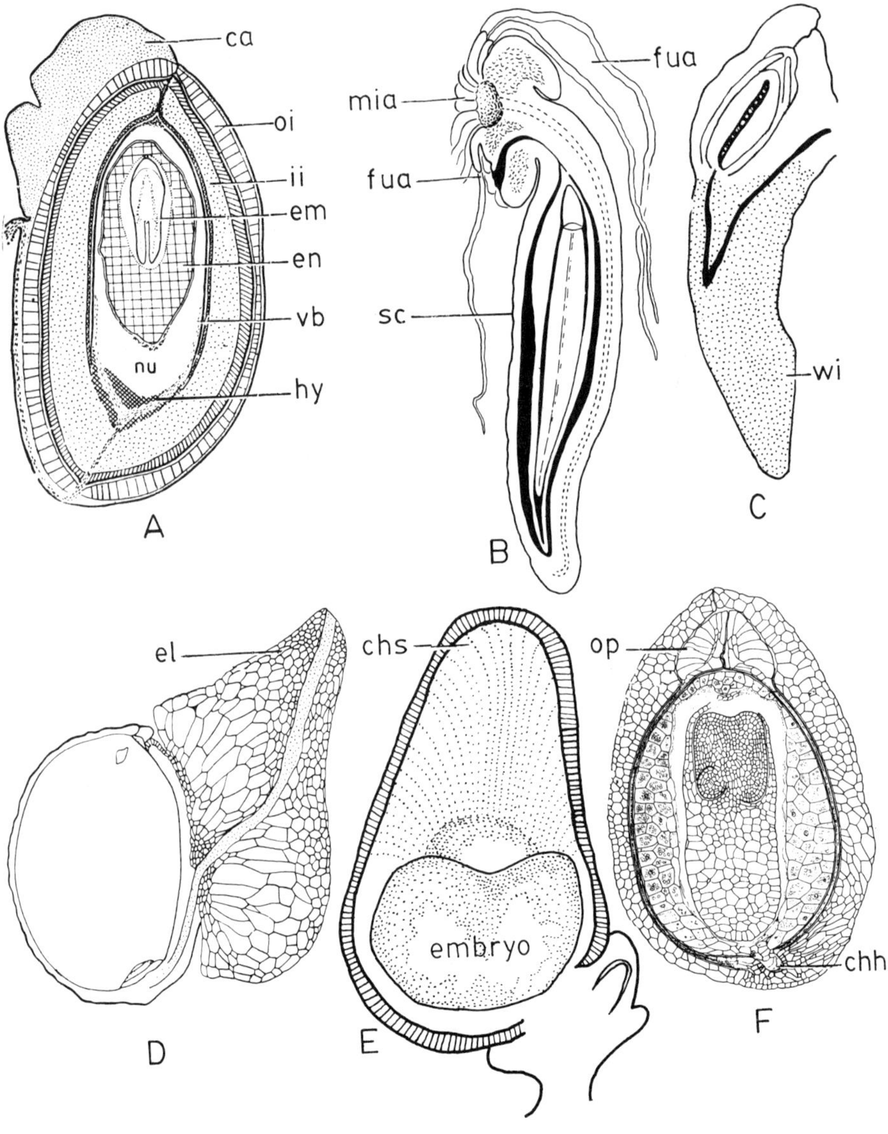

Fig. 3.14.5. A *Acalypha indica*. longisection of seed with caruncle (*ca*); outer integument (*oi*); inner integument (*ii*); embryo (*em*); endosperm (*en*); vascular bundle (*vb*); nucellus (*nu*) and hypostase (*hy*). (After B.M. Johri and Kapil 1953). **B** *Sarawakodendron* sp., longisection of seed in side view; *fua* funicular aril; *mia* micropylar aril; *sc* seed. (After Corner 1976). **C** *Cercidiphyllum japonicum*, longisection of ovule, the triangular extension of raphe and chalaza later develop into a thin membraneous wing (*wi*). (After Swamy and Bailey 1949). **D** *Trillium ovatum*, seed with elaiosome (*el*). (After Berg 1958). **E** *Cynastrum* sp., mature seed to show chalazosperm (*chs*). (After Melchior 1964). **F** *Lemna pausicostata*, longisection of seed to show operculum (*op*), endosperm with chalazal haustorium (*chh*), and embryo (*centre*). (After S.C. Maheshwari and Kapil 1964)

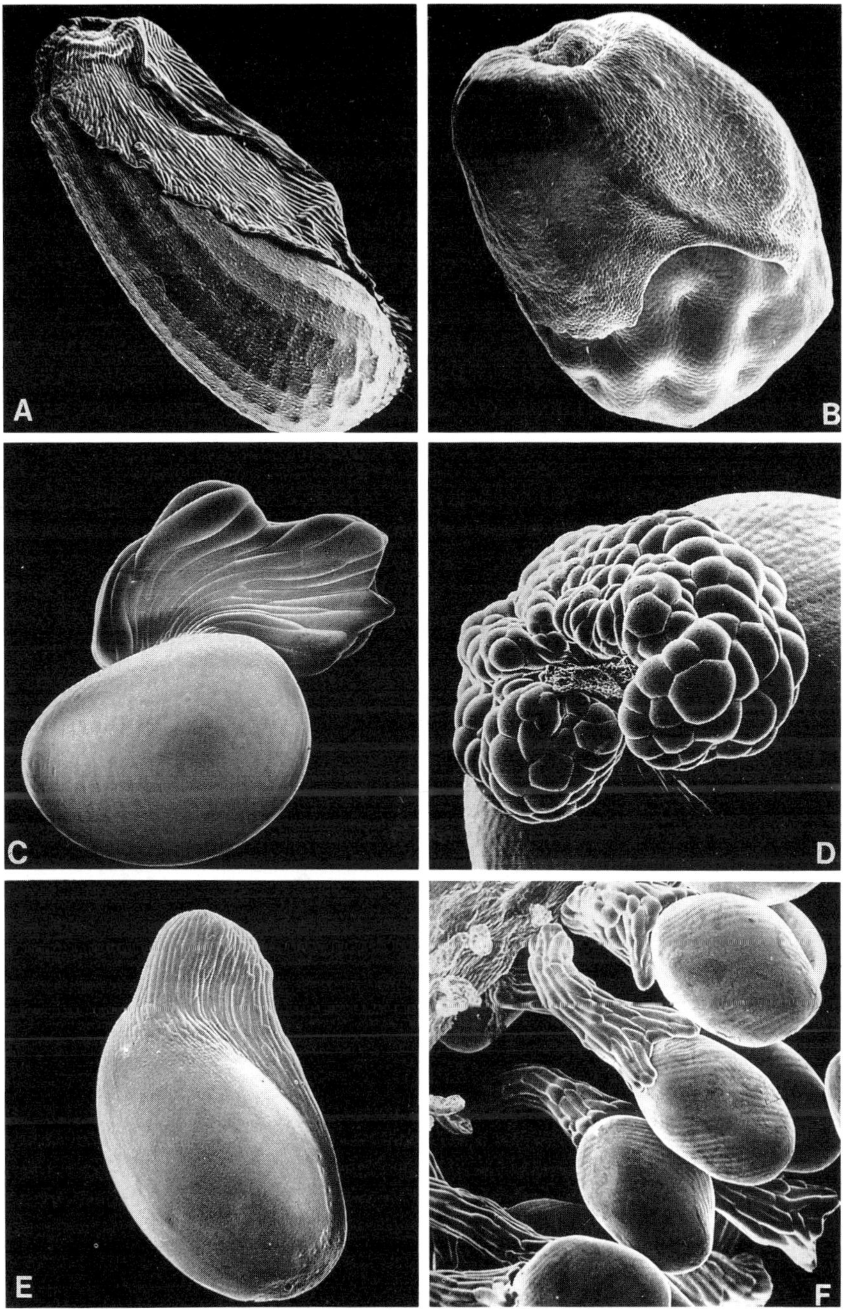

Fig. 3.14.6A–F. Types of seed arils, SEM micrographs. **A** *Turnera ulmifolia*, raphal axis. **B** *Passiflora suberosa*, funicular outgrowth. (After Kloos and Bouman 1980). **C** *Chelidonium majus*, strophiole. **D** *Chionodoxa luciliae*, caruncle. **E** *Viola odorata*, Exotestal elaiosome. **F** *Episcia melittifolia*, funicular aril. (After Boesewinkel and Bouman 1984)

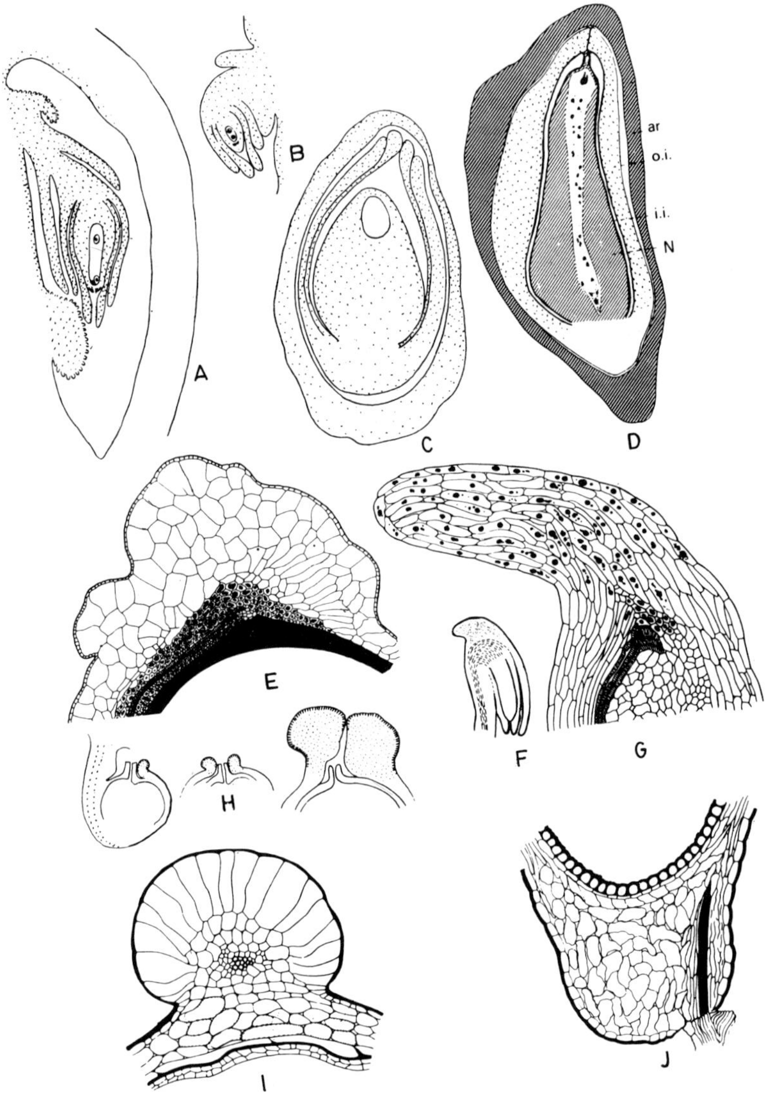

Fig. 3.14.7A–J. Aril and strophiole. **A** *Asphodelus fistulosus*, longisection through ovary showing an ovule with aril and obturator. **B** *Bulbine annua*, longisection of bitegmic ovule with aril. **C** *Eremurus himalaicus*, ovule surrounded by aril. (After Stenar 1928). **D** *Kniphofia praecox*, longisection of ovule at young proembryo with nuclear endosperm. Note the aril (*ar*), outer (*oi*) and inner integument (*ii*), and nucellus (*N*). (After Schnarf and Wunderlich 1939). **E** *Iris ruthenica*, upper part of seed with strophiole. **F, G** *Galanthus nivalis*. **F** Ovule; **G** Strophiole. **H** *Scilla* sp. to show development of strophiole on the outer integument at the micropyle. (After Bresinsky 1963). **I** *Scoliopus bigelovii*, lower part of seed with strophiole. (After Berg 1959). **J** *Luzula campestris*, strophiole on seed. (After Bresinsky 1963)

ing to V. Puri (1970), the morphological nature of arils is variable. Mohana Rao (1974b) states: "the single anatomically and functionally similar structure arises from two morphologically different sites". Mohana Rao (1974b) also mentions that it is superfluous to classify various outgrowths from the functional, and ecological point of view.

Various types of aril are shown in Figure 3.14.6A–F, and aril and strophiole in Figure 3.14.7A–J.

In the seeds of *Cercidiphyllum*, the chalaza and raphe together produce a triangular outgrowth which develops into a membranous wing in mature seed (Fig. 3.14.5C). A wing also develops in Hamamelidaceae, but due to the elongation of the tip of the outer integument.

The elaiosome arises as an outgrowth of raphe, or hilum. It contains food material, and attracts ants which help in seed dispersal (Amaryllidaceae, Aristolochiaceae, Liliaceae – *Trillium*; Fig. 3.14.5D). Excessive chalazal growth in the seeds of *Cynastrum* produces a tissue known as chalazosperm (Fig. 3.14.5E) which stores food reserves.

Several acanthaceous genera are characterized by the development of a jaculator or retinaculum. It is a hook-like outgrowth of the funiculus, and ejects the seed out of the capsule.

Sometimes, during post-fertilization, the cells of tegmen in the micropylar region undergo excessive expansion and develop into an operculum (Fig. 3.14.5F) as in *Lemna paucicostata* (S.C. Maheshwari and Kapil 1964). In the mature seed, rest of the tegmen collapses and is represented by two refractive cuticular thickenings which enclose the endosperm and embryo. In a dormant seed the operculum fits at the top as a stopper and, during germination, it is thrown off so that the embryo pushes out.

3.15 Nutrition of the Anther, Ovule and Seed

Anther. Initially, the floral vascular supply transports nutrients to the various parts. As soon as the tapetum differentiates in the microsporangium, it provides nutrition, or the nutrition passes through it (Pacini et al. 1985), for the development of microspore mother cells, tetrads/microspores/pollen grains – including "clumped" pollen grains: dyads (Fig. 28.2.1A), tetrads (Fig. 18.5.1A, C, D), pollinia (Fig. 47.7.1A–D) and massulae (Fig. 65.2.1L, M).

The absorbing external surface of each grain is progressively reduced. In the massulae, the pollen has no direct contact with the tapetum. The inner exine walls, or sometimes the intine itself, or joined grains are generally reduced or even discontinuous so that the nutrients may pass easily through and ensure the nutrition of the inner spores/pollen. This also ensures a synchronous pollen development.

According to Pacini et al. (1985), all the substances produced by the secretory tapetum – starch and/or proteins in plastids, soluble polysaccharides in the vacuoles – reach the microspores/pollen grains via the locular fluid. In the periplasmodial tapetum, the cytoplasm adheres closely to the microspores. In the secretory tapetum, the cells are polarized both in structure and function;

this is not so in the amoeboid tapetum. The inner and outer tangential faces of secretory tapetum differ in their capacity to release substances.

The locular cavity increases in size during anther development, the secretory tapetum continuously supplies locular fluid until it starts to degenerate. The enzyme callase, produced by the tapetum, depolymerizes the callosic walls of tetrads and the microspores are released. The tapetum then becomes very active and has a multiple function. From the second meiotic division onwards, PAS-positive material present in tapetal vacuoles passes into the locule.

The tapetum also produces: (a) sporophytic proteins and enzymes which are deposited next to the aperture regions or in the interaperturate areas; (b) tryphine which, in contrast to pollenkitt, comprises a mixture of hydrophobic and hydrophilic substances often containing cytoplasmic elements (degenerated organelles); (c) pollenkitt, a hydrophobic oily layer containing mainly lipids and carotenoids.

The tapetal cell organelles (mostly plastids, perhaps other organelles too) form lipid masses which are deposited outside the mature pollen grains just before they are released.

In polysporangiate microsporangia of Annonaceae (Fig. 18.6.1C–J), Onagraceae (Fig. 39.13.1A–I), and some other families, each sublocule has its independent tapetum which enhances the nutritive function (Tobe and Raven 1986a). It is well known that when the tapetum disfunctions, fertile pollen grains are not produced.

Ovule. The funicular vascular supply may not extend beyond the chalaza, branch at the base of nucellus, extend into one or both integuments and may even branch there. The euphorbiaceous ovule is a good example (Landes 1946). With the development of the ovule into a seed, the vascular supply becomes very pronounced (Figs 3.14.4A–E, 3.14.5A–N). Besides the vascular supply, there are several other strategies to provide the nutrition, e.g. by various types of haustoria.

Megaspore Haustoria. In Crassulaceae, each megaspore of a linear tetrad produces a tubular haustorium, apically from the micropylar megaspore and laterally from the other megaspores (Fig. 26.5.2A–D). These haustoria intertwine, extend up to the nucellar epidermis, and contain starch.

Embryo Sac Haustoria. The apical extension of the embryo sac, and even the chalazal one are common in Acanthaceae (Sect. 48.19). In *Exocarpus* (Fig. 12.6.3A–C), several tubular haustoria directed upward develop from the micropylar end; some haustoria grow towards the chalazal region. Sometimes, flap-like haustoria, directed towards the chalaza, grow all round the middle region of the embryo sac.

In Podostemaceae (Fig. 28.2.1D, E), the central cells of the nucellus break down, a plasmodium (pseudo-embryo sac) is formed, and after fertilization the embryo grows in this pseudo-embryo sac (Fig. 28.2.3A–C).

The numerous embryo sacs in *Utricularia* derive nutrition from the placental and chalazal nutritive tissue. The endothelium is also very conspicuous and the nutrients pass through it (Figs. 48.25.3, 48.25.4).

The endothelium (see Kapil and Tiwari 1978) is a characteristic feature of the Sympetalae (Fig. 3.4.4A–E), and is actively concerned in the transport of nutrients.

In *Helianthus annuus*, the endothelium is many-layered and interrupted by transverse cuticular septa (Fig. 3.4.7D). Its role in the efficient transport of nutrients should be carefully examined.

The occasional reports of the differentiation of endothelium in some monocots appear doubtful (see Swamy and Krishnamurthy 1970), and deserve to be examined histochemically.

Synergid and Antipodal Haustoria. Synergid haustoria are reported in Compositae (Sect. 51.9). The most conspicuous synergid haustoria develop in *Quinchamalium chilense* (Figs. 12.6.5A–C, 12.6.6A–C), and the tip extends up to one-third of the basal region of the style. *Cortaderia* (Fig. 3.13.2D; see also Fig. 58.2.1), an apomict, shows lateral synergid haustoria which contain starch.

In *Q. chilense* (Fig. 12.6.5A–E), the antipodal nuclei do not organize into cells. The tubular antipodal chamber functions as a haustorium.

The basal antipodal cell in Rubiaceae (Fig. 47.8.3F, G) elongates a great deal and becomes tortuous.

The frequent occurrence of polyploid nuclei and multi-nucleate antipodal cells indicates a high metabolic activity (see D'Amato 1984). As early as 1898, in *Aconitum napellus*, Osterwalder observed giant antipodal cells with the tip reaching up to the egg apparatus.

Haustorial Pollen Tube. For almost nine decades, the controversy about the haustorial role of the persistent remnants of the pollen tube has been going on (Longo 1903, Venkata Rao 1952a, Tilquin et al. 1983, Indra and Krishnamurthy 1985, Briggs 1989).

The latest report on haustorial pollen tubes, which parasitize the ovary wall during post-fertilization development in *Grevillea banksii*, was given by Martin and Herscovitch (1989).

P. Maheshwari and Johri (1950b) and B.M. Johri and Ambegaokar (1984) have critically examined this assumption and conclude that the "dead" remains of the pollen tube do not have any haustorial role (see also Sect. 3.9).

Endosperm Haustoria. In most plants, the endosperm is the main nutritive tissue and supports the development of the embryo. The haustoria are the most spectacular structures and augment the nutritive function.

A prominent chalazal haustorium, which invades the adjoining tissue, develops in Leguminosae and Cucurbitaceae (Fig. 38.2.2A–I). The endosperm is of the Nuclear type; the haustorium may remain free-nuclear or become partially or wholly cellular. It becomes coiled in Proteaceae (Figs 11.2.2D–H, K, L, 11.2.3A–C); in Caryophyllaceae (Fig. 16.10.2B, C) superficial hairy growths occur. In Icacinaceae (Fig. 33.13.1A, B), the chalazal haustorium shows many interesting features.

In the Helobial type, the haustoria are not very common. In *Monochoria* (Ono 1928), the bilateral "saccate" nuclear micropylar haustoria are very conspicuous (see Sect. 54.13).

In the Cellular type, only a chalazal haustorium may develop: Olacaceae (Fig. 12.2.1A, B), Opiliaceae (Figs 12.4.1A–F, 12.4.2A, B), Santalaceae (Figs 12.6.4D, E, 12.6.7A, B), Myzodendraceae (Fig. 12.7.1G, H, J) and others. In Olacaceae, Opiliaceae and Myzodendraceae, the haustorium invades the placenta and extends as far as the pedicel.

Both micropylar and chalazal haustoria develop in Balsaminaceae (Fig. 31.11.2D, E, G–I), Cyrillaceae (Fig. 33.2.1C–F), Loasaceae (Fig. 37.19.1A–G), Hydrophyllaceae (Fig. 48.5.1E, F), Verbenaceae (Fig. 48.8.3A–C), Labiatae (Fig. 48.10.3), Scrophulariaceae (Figs. 48.15.3A, B, 48.15.4O–T), Acanthaceae (Fig. 48.19.3H, I), Gesneriaceae (Fig. 48.22.3A–C), Lentibulariaceae and others.

Both micropylar and chalazal haustoria may or may not branch. In many cases, the chalazal haustorium is not at all aggressive and collapses. The micropylar haustoria invade the exotesta as in *Aeginetia* (Fig. 48.24.1E, H, I, J, L–N).

Suspensor Haustoria. The suspensor haustorium is very aggressive and invades the integument (Figs. 26.5.1G, 26.5.3A–E). The uppermost cell of the suspensor enlarges considerably, and it develops many haustorial extensions (Figs. 28.2.1F, 28.2.2N, 28.2.3A–C). The suspensor may comprise biseriate elongated multinucleate cells as in Leguminosae (Sect. 26.19), or a uniseriate row of multinucleate bulbous cells (Figs. 48.4.4H–P, 48.4.7A–D, N, O). The suspensor cells may invade the endosperm (Figs. 47.8.4A, B). In *Plumbago*, the group of suspensor cells is much larger than the proembryo (Fig. 44.2.6F–K). There are many other variations. The most unusual filamentous highly coiled suspensor of uniseriate multinucleate cells is reported in *Isomeris* (Fig. 24.3.1A–F′).

Haustorial Cotyledons. In Hippocastinaceae (Fig. 31.7.1B–E), one cotyledon remains enclosed in the liquid endosperm to absorb nutrients for the growth and maturation of the other cotyledon. In *Yucca*, the cotyledon remains permanently embedded in the perisperm to absorb nutrients for germination and establishment of the seedling.

Transfer Cells. The cells actively concerned in short-distance transport of nutrients develop wall ingrowths to increase the absorptive surface (see Pate and Gunning 1972, B.M. Johri and Ambegaokar 1984). The wall ingrowths are simple and papillate, or reticulate or labyrinth when branched. An important feature is that in transfer cells the plasma membrane follows the contour of the wall ingrowths, and the middle lamella is absent.

The embryo sac wall, synergids, antipodals, endosperm, perisperm, embryo suspensor, and even pericarp have transfer cell ingrowths.

After fertilization, wall ingrowths are initiated at the micropylar region of the embryo sac. They increase in size and persist up to the heart-shaped embryo. These ingrowths are especially concentrated around the embryo. At the chalazal end, the wall ingrowths are usually short-lived.

The central cell is the largest cell of the embryo sac, and its wall usually has tuberculate wall ingrowths adjacent to the nutrient-rich nucellar cells. In *Helianthus annuus*, the micropylar part of the central cell has wall ingrowths which project towards the synergids. In this taxon, the synergids are reported to lack the filiform apparatus; which requires confirmation (see Berger and Erdelská 1973, Vijayaraghavan et al. 1988).

Bhandari and Mal (1989) report a prominent wall ingrowth, in *Scilla siberica*, close to the chalazal end of embryo sac; it separates a small chalazal and a large micropylar chamber. This ingrowth is interrupted by one or more gaps through which the metabolites, and even nuclei, can pass from one chamber to the other. The authors compare it to a transfer cell.

A filiform apparatus in the synergids has been known since long. The ingrowths of the wall of the synergid are synonymous with the filiform apparatus (see Willemse and van Went 1984). It is said to be actively concerned in a secretion into the micropyle which possibly directs the tip of the pollen tube into the synergid. Probably, the filiform apparatus in the two synergids have different and specific functions, and the one which receives the pollen tube is in a degenerate condition as compared to the other healthy synergid.

The tetrasporic eight-nucleate embryo sac of *Plumbago* lacks synergids, and the filiform apparatus has been demonstrated in the apical region of the egg cell (Cass 1972, see also Russell 1989). This is likely to be the case in *Plumbagella*.

The persistent antipodal cells may show massive and highly branched wall ingrowths, and tuberculate wall projections. The wall ingrowths proliferate after fertilization in *Nigella damascena*, *Papaver somniferum*, *Scilla sibirica* and *Zea mays*. During the free-nuclear endosperm, wall ingrowths occur all over the antipodal walls except in a small area at the chalazal region, in *Aconitum vulparia* (see Vijayaraghavan et al. 1988). The major pathway for the flow of nutrients into the embryo sac is the chalazal region, and the wall ingrowths are involved in translocation.

The inner and outer epidermis of the endosperm show cell wall ingrowths, and are actively concerned in nutritive processes. Brisson and Peterson (1975) demonstrated wall ingrowths on the inner wall of the endosperm haustoria of *Vaccinium macrocarpum*. This may be a common feature in other endosperm haustoria, and needs to be investigated.

The suspensor anchors the proembryo/embryo to the embryo sac, and the circumjacent tissue (endosperm) is nutritionally rich. The cells of the suspensor show prominent ingrowths, especially on the side walls; there are no ingrowths on the periclinal walls. The uppermost basal cell, especially large in some taxa as in *Capsella bursa-pastoris*, shows conspicuous wall ingrowths and is actively concerned in transporting the nutrients to the embryo (Gunning and Pate 1974, Vijayaraghavan et al. 1988).

Wall ingrowths also occur on the epidermis of cotyledons, epidermis of endosperm facing the perisperm as in *Mesembryanthemum*. Aleurone transfer cells have been observed at the placenta of grass seeds.

In *Vicia*, transfer cells occur in endothelial cells, epidermis of endosperm, suspensor cells, and cotyledonary epidermis facing the endosperm (Gunning and Pate 1974).

In *Pappophorum subbulbosum* (Poaceae; Rost et al. 1984), transfer cells occur in the outer pericarp epidermis (continuous across the placental pad), and discontinuous layer of the inner pericarp epidermis (absent across the placental pad). From the inside, several partially collapsed cell layers of the nucellar projection occur which have wall-ingrowths. The aleurone is a continuous uniseriate layer around the entire caryopsis (except the placental pad where it is crushed and contains remnants of wall-ingrowths). At maturity, the caryopsis-coat comprises the layers of pericarp, integument and nucellus. Numerous other examples of transfer cells are cited and illustrated in various families.

According to Buell (1952a), there are two periods of accumulation of food reserve (fats, starch, possibly proteins) in the ovary wall, placenta and ovules (chiefly nucellus) about the time of fertilization: (a) during the development of the embryo sac, and (b) from the globular proembryo up to the maturation of seed.

At first, the food reserve in the nucellus is mobilized, followed by the food reserve in the integuments. During its entry into the ovule, the cells adjoining the micropyle nourish the pollen tube, and then the filiform apparatus in the synergid. To begin with, the proembryo is nourished by the suspensor. The globular proembryo and, up to the maturation of the embryo, the endosperm provides the nutrients. Additional supply of nutrients becomes available from the circumjacent tissue through the development of various types of haustoria, their disposition, and duration of aggressive activity. The aggressive activity, and the duration thereof, vary from plant to plant, and also depend on the metabolic status of the parent tissue. The branching, or labyrinth condition, of wall-ingrowths is directly correlated with the absorbtive activity.

Nutritive Role of Seed-Coat. With elaborate physiological and biochemical studies, Murray (1987) has demonstrated the dynamic nutritive role of testa and tegmen in legume seeds. In *Pisum sativum*, the post-fertilized embryo sac enlarges, the central vacuole fills with a solute-enriched fluid. This fluid is considered to arise as a secretion from the tegmen, and is distinct from the endosperm. The secretion of sucrose by tegmen and testa is essentially irreversible in vivo. The flow of nutrients through the maternal tissues, and the relationship of seed-coat metabolism to the nutrition of the developing embryo, have been reported. Essentially, all the metabolites acquired by the developing embryo must first pass through a variable number of seed-coat cells. This involves the "unloading" of solutes from the phloem sieve elements of seed-coat veins, and the processing of certain compounds remained in seed-coat cells prior to secretion. The composition of the embryo sac reservoir of nutrients alters predictably as the seed develops. Further investigations on the nutritive role of the seed-coat for the developing embryo, especially in ex-albuminous seeds, would be very significant. The accumulation of reserve food in the seed-coat has been observed in many taxa, e.g. in *Butomopsis lanceolata* (B.M. Johri 1936b) and in *Indotristicha ramosissima* (Fig. 28.2.1F, G).

Finally, the food reserves accumulate in the endosperm in albuminous seeds, sometimes in the perisperm, and in the embryo in exalbuminous seeds. Jacobsen (1984) gives a good account of the reserve food in seeds, and their mobilization during germination.

3.16 Comparative Studies

The classificatory systems of angiosperms by Bentham and Hooker (1965a, b, c), Hutchinson (1973), Takhtajan (1980), Cronquist (1981), and R. Dahlgren (1980b) are frequently used for comparative "Taxonomic Considerations". Attention is also invited to the updated classification of angiosperms by G. Dahlgren (1989).

Other classifications/relationships mentioned in the following pages are those of Baillon (1874, 1892), Benson (1957, 1962), Bessey (1897, 1915), Corner (1981), Cronquist (1957, 1965, 1968, 1981, 1983), Dumortier (1829), Ehrendorfer (1973), Endlicher (1836–1840), Good (1957), Gunderson (1943, 1950), Hallier (1905, 1912), de Jussieu (1789), Kuijt (1969), Lawrence (1951), Netolitzky (1926), Radford et al. (1974), Rendle (1959, 1971), Schürhoff (1926), A.C. Smith (1972), Stebbins (1974a, b), Thorne (1963, 1968, 1976, 1977, 1983) and Wettstein (1935).

Robyns (1972) proposes a new system of orders and families of Sympetalae; Yamazaki (1974) does so on the basis of embryology. W.R. Philipson (1975) draws attention to evolutionary lines within the dicotyledons. The systematics, phylogeny and evolution of the monocotyledons have been highlighted by Ankerman (1927), Deyl (1955), Doyle (1973), Fritsh (1932) and Huber (1977).

Brewbaker (1967) points out the phylogenetic significance of two- and three-celled pollen grains. Erdtman (1945, 1952) refers to the systematic significance of dyads and tetrads of microspores, and the importance of pollen morphology in taxonomy. Canright (1963) uses pollen morphology as a basis of phylogeny of some Ranalian families. Walker and Skvarla (1975) propose a new concept in the evolutionary morphology of angiosperms on the basis of primitively columella-less pollen.

Sporne (1969) considers the ovule as an indicator of the evolutionary status of angiosperms. W.R. Philipson (1974, 1977) uses ovular morphology as a basis of classification of dicotyledons. Bouman (1971b) points out the usefulness of developmental and structural studies of integuments for consideration of taxonomic and phylogenetic problems. Sunder Rajan (1976) considers the ovular micropyle as an indicator of primitiveness in angiosperms. Bocquet and Bersier (1960) report the systematic value of teratological ovular development.

Tobe (1989, see also Cocucci 1973a, 1983, Hamann 1977, di Fulvio 1981) emphasizes the role of embryology for a systematic and evolutionary study. Schnarf (1933), Kapil (1962, see also Kapil and Bhatnagar 1980), P. Maheshwari (1964), B.M. Johri (1967), Yakovlev (1974), Palser (1975), Poddubnaya-Arnoldi (1982) and Herr (1984) also draw attention to the importance of embryology as a basis for taxonomic considerations. Fagerlind (1944a) refers to the phylogeny of female gametophyte on the basis of tetrasporic embryo sacs (see also Battaglia 1989). A.C. Joshi (1938d) states that the distribution of Fritillaria type of embryo sac indicates the mono- or polyphyletic origin of angiosperms; a wholly incorrect assumption. Other embryological characters used for phylogenetic relationships and classification are "events leading up to fertilization" (Gerasimova-Navashina 1961), embryo development (Yakovlev 1969, Periasamy 1977, Yamazaki 1982), and seed structure (Wunderlich 1967, Vaughan 1968, Corner 1976, Barthlott 1981).

Single embryological characters are not as useful for phylogenetic and evolutionary considerations as a group of characters. Anatomical (Solereder 1908, Eames 1951, Metcalfe 1961, Metcalfe and Chalk 1950), and chaemotaxonomic data (Hegnauer 1963, 1964, 1969, 1973, Frohne and Jensen 1973, Gibbs 1974, Jensen et al. 1975, R. Dahlgren 1983a, b) provide significant support to embryological data.

Most of the above characteristics, and several others, are appropriately discussed in respect of taxonomic considerations. There is no doubt that a comparative analysis of cumulative data is useful in the delimitation of species, genera, families and orders.

I Dicotyledons: Archichlamydeae

4 Casuarinales

The order Casuarinales comprises the monotypic family Casuarinaceae.

4.1 General Features

The order Casuarinales exhibits the following embryological features: (1) acolpate, three-porate pollen grains, two-celled at dispersal stage; (2) orthotropous, bitegmic and crassinucellar ovule with multicellular archesporium, and micropyle formed by inner integument; (3) sporogenous tissue contributes to the development of conducting strand, and formation of a massive parietal tissue; (4) concurrent development of several Polygonum type of embryo sacs with chalazal caecum; (5) persistence of pollen tube and its profuse branching at the chalaza; (6) extravesicular chalazogamy; (7) Nuclear type of endosperm; (8) Onagrad type of embryogeny; (9) exalbuminous seeds with protective crystalliferous, sclerotic pericarp.

4.2 Casuarinaceae

The anther wall comprises an epidermis, fibrous endothecium, one or two middle layers, and tapetum with two-nucleate cells in *Casuarina equisetifolia* (Swamy 1944a).

The acolpate, three-porate pollen grains with thick exine are two-celled at the dispersal stage (Moseley 1948).

The bicarpellary ovary contains one fertile orthotropous, bitegmic and crassinucellate ovule. The inner integument is crushed except in the apical region where it organizes the micropyle. The nucellus projects through the micropyle in *Casuarina montana* (Swamy 1948a). The presence of nucellar tracheids has been noted.

A procambial strand differentiates in the chalaza and extends upward as vascular strand, up to the sporogenous mass (Swamy 1944a). Some of the centrally situated sporogenous cells seem to elongate and contribute to the extension of the strand into the nucellus. In spite of such diversion of a considerable part of the archesporium, as many as 20–25 embryo sacs develop in the same nucellus.

The hypodermal archesporium is multicellular. Each cell cuts off a parietal cell. The parietal tissue is five to six cells thick. Meiosis occurs in the megaspore mother cells, producing linear tetrads. Due to concurrent development of megaspores up to two- to four- or eight-nucleate stage, nearly 20–25 embryo sacs are organized. The chalazal megaspore develops into a Polygonum type of

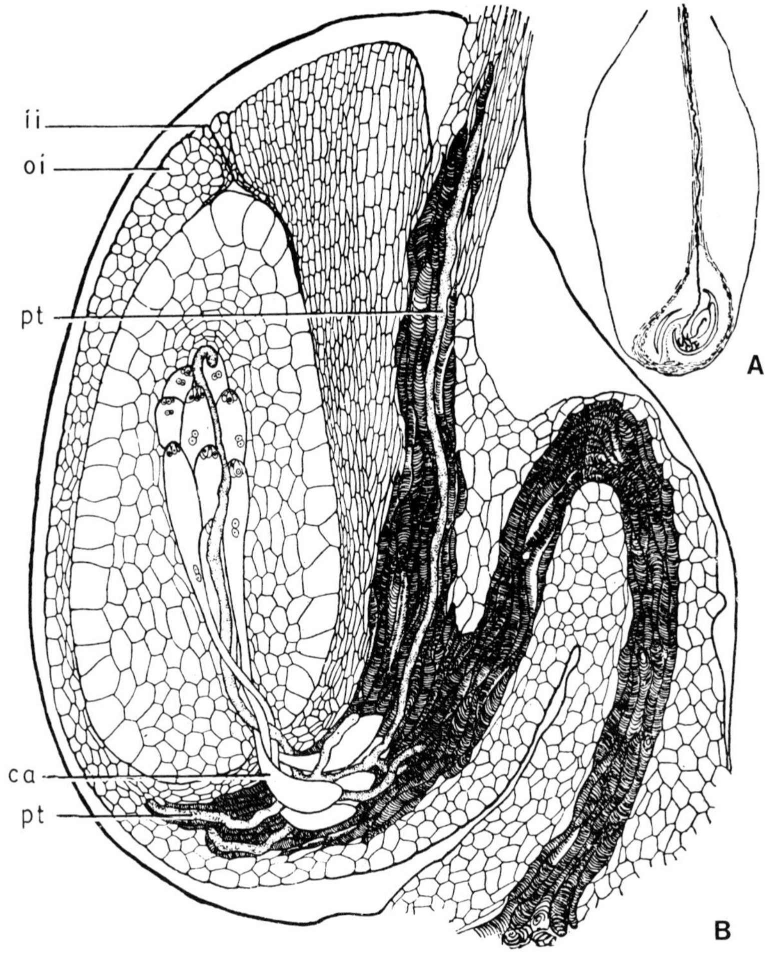

Fig. 4.2.1A, B. *Casuarina equisetifolia.* **A** Longisection of ovary, *central black line* denotes pollen tube. **B** Longisection of ovule to show mature embryo sacs (nine), and path of pollen tube (*pt*). The pollen tube branches in the chalazal region; *ii* inner integument; *oi* outer integument; *ca* embryo sac caecum. (After Swamy 1948a)

embryo sac. The chalazal end of gametophytes form caeca with terminal pouches extending up to the funiculus (Fig. 4.2.1B). The synergids are hooked, and the antipodals lie in caeca. In *C. montana* (Swamy 1948a), rarely the persistent synergid becomes binucleate. This is doubtful.

The growth of the pollen tube continues intercellularly, for about 45–60 days. It grows downward, along the vascular strand, to the chalaza and branches there profusely. Generally, the branches end blindly and some of them may even reach up to the outer epidermis of outer integument, in the basal region. Ultimately, the branch containing the male gametes makes its way upward, between the chalazal caeca of the embryo sacs, passes through the nucellus, and penetrates the embryo sac at a point adjacent to the egg apparatus (Fig. 4.2.1A, B). The mode of fertilization is chalazogamous, and such a penetration of the pollen tube into the embryo sac at the egg apparatus is described as extravesicular (Swamy 1948a).

The development of endosperm is of the Nuclear type, wall formation is initiated at the micropylar end and proceeds centripetally as far as the chalaza. The tissue in the micropylar region is compact. During its growth the embryo consumes the entire endosperm.

The zygote undergoes a transverse division when the micropylar half of the embryo sac contains cellular endosperm. A further transverse division produces a linear row of three cells. The terminal cell of the three-celled proembryo develops into the embryo proper, the middle and basal cell into the suspensor. As the basal cell enlarges, its nucleus becomes hypertrophied. The embryogeny appears to conform to the Onagrad type. Polyembryony due to fertilization of multiple embryo sacs is common. Parthenocarpy has also been reported (Swamy 1948a).

The exalbuminous seed contains an erect, spathulate, large embryo with a well-differentiated radicle and two cotyledons. The testa shrivels, or its outer epidermis may develop thick outer walls. The sclerotic pericarp with crystalliferous tissue gives protection to the dry fruit.

TC Frye (1903), MF Moseley Jr (1948), BGL Swamy (1944a, 1948a), M Treub (1891)

4.3 Taxonomic Considerations

Casuarinaceae. Casuarinaceae show gymnospermous morphological characters. The flowers are extremely reduced, naked and monoecious. The male flower is represented by a single stamen, and the female flower by a tricarpellate pistil.

Several contradictory opinions have been expressed concerning the systematic position of Casuarinaceae. Engler and Prantl (1931), Wettstein (1935), Hjelmqvist (1948) and Rendle (1959) treat it as the most primitive taxon of dicotyledons. Wettstein and Rendle presume its origin from Ephedraceae, and Wettstein retains it in Amentiferae. Hallier (1912), Bessey (1915), Hutchinson (1967) and Tippo (1938) recognize it to be an advanced member. Hutchinson (1973) places it with amentiferous orders, and derives the entire group from the

Hamamelidales. Tippo (1938) and Cronquist (1968, 1981) support this view. Moseley (1948) considers it as "moderately specialized", and on the basis of its anatomy and floral morphology suggests its derivation from Hamamelidaceae as a result of reduction of floral organs during the course of evolution. R. Dahlgren (1977a, 1980b, 1981) places the order Casuarinales in the superorder Hamamelidanae.

The embryological features, such as three-four-five-colpate pollen grains, persistent endosperm, Solanad type of embryogeny, and winged seed with massive sclerotic testa of Hamamelidaceae, differ from those of Casuarinaceae. The wood structure is similar in both families.

Eames (1961) assigns Casuarinaceae to Betulaceae, on the basis of pollen structure, two-trace stamen, and other floral characters. Both the families exhibit true chalazogamy, Polygonum type of embryo sac, and Nuclear endosperm. However, unitegmic ovule with single archesporium, persistent endosperm, and Asterad type of embryogeny in Betulaceae do not occur in Casuarinaceae.

Casuarinaceae resemble Betulaceae and Hamamelidaceae in the belated development of ovule, massive sporogenous tissue giving rise to multiple embryo sacs, and delay in syngamy in relation to anthesis.

Melchior (1964) places Casuarinales near Juglandales. Both the orders show several common embryological features such as three-porate pollen, orthotropous crassinucellate ovule, parietal tissue, Polygonum type of embryo sac, chalazogamy and delayed fertilization, Nuclear endosperm, and a single-layered seed-coat.

Casuarinaceae are undoubtedly angiosperms; the simplicity of flower is due to reduction. The bitegmic ovule with multicellular archesporium, and Polygonum type of embryo sac are markedly angiospermous. The double-trace stamen is a primitive character, and occurs in Proteaceae and Betulaceae. The pollen grains show affinity to Betulaceae and Myricaceae. Chalazogamy has little or no phylogenetic significance. Eames (1961) considers Casuarinaceae as a highly specialized family, derived from primitive stock, and which has remained isolated.

5 Juglandales

The order Juglandales comprises two families: Myricaceae and Juglandaceae.

5.1 General Features

This order is characterised by (1) three-porate pollen grains, (2) basal, orthotropous, unitegmic ovules with massive nucellus, (3) single archesporial cell and well-developed parietal tissue, (4) Polygonum type of embryo sac; occasionally tetrasporic in Juglandaceae, (5) delayed fertilization and chalazogamy in Juglandaceae, (6) Nuclear endosperm, (7) Asterad type of embryogeny, and (8) single-layered seed-coat.

5.2 Myricaceae

The embryology has not been fully investigated.

The ovule is basal, orthotropous, unitegmic and crassinucellate in *Myrica*, and pendulous and anatropous in *Canacomyrica*. The nucellus is elongated, and the embryo sac is located in its upper region. The growth of the integument is arrested, occasionally, and then the micropyle does not organize in *Myrica gale* (Håkansson 1955). The vascular supply branches at the chalaza and passes into the integument.

The archesporial cell cuts off a parietal cell, the latter produces six to eight layers of parietal tissue. Meiosis in the megaspore mother cell results in a linear tetrad. The chalazal megaspore develops into a Polygonum type of embryo sac. The polar nuclei fuse before fertilization, and antipodals are ephemeral. A hypostase of lignified cells develops. Twin embryo sacs may occur.

The growth of pollen tube is slow, and fertilization is delayed. The entry of pollen tube is porogamous.

The development of endosperm precedes the division of zygote by about 4 weeks. Yen (1950) reported the period as 6 to 7 weeks. The development of endosperm is of the Nuclear type. A noteworthy feature is the formation of nucleated cytoplasmic nodules in *Myrica gale* (Håkansson 1955). Wall formation is initiated at the 32-nucleate stage of endosperm; at first cells develop in the micropylar region and proceed towards the chalaza. The entire endosperm becomes cellular. It is generally ephemeral, or may persist as a single layer with oil and aleurone grains as storage products. Its main role is to produce growth substances (Håkansson 1955).

The exalbuminous seed is small, and contains a straight embryo. The seed-coat is represented by the single outer epidermal layer of thick-walled cells.

Å Håkansson (1955), TK Yen (1950)

5.3 Juglandaceae

The anther wall comprises the radially elongate epidermis, an endothecium, two to four middle layers, and glandular tapetum with multinucleate cells. The microspore mother cells undergo meiotic divisions, and tetrahedral and decussate tetrads are produced. Cytokinesis is simultaneous.

The three- to seven-porate pollen grains are two-celled at the dispersal stage. The formation of male gametes occurs in pollen grains before germination, in *Carya illinoinensis* (Solntzeva and Ogebuadze 1974). The male nuclei are spherical in the beginning, become curved afterwards, and remain surrounded by a thin film of cytoplasm (Finn 1925).

The ovule is orthotropous, unitegmic and crassinucellate with lobulate exostome. The single, deeply situated archesporial cell cuts off a parietal cell, and the parietal tissue is well developed. The cells of the nucellar epidermis divide and add to the parietal tissue. In *Pterocarya fraxinifolia*, Bouman and Boesewinkel (1969) report twin nucelli in an ovule due to cleavage of nucellar primordium. The megaspore mother cell undergoes meiotic divisions, producing a linear or T-shaped tetrad. Twin tetrads develop in *Juglans* (Langdon 1934). The chalazal megaspore produces a Polygonum type of embryo sac (Fig. 5.3.1A–D). Twin and triple embryo sacs (per ovule) are formed in *Carya illinoinensis* (Ogebuadze 1977). There are three small antipodals. Karsten (1902) reported tetrasporic type of development in *J. cordiformis*, which needs confirmation.

Extravesicular chalazogamy (see Swamy 1948a) occurs in *Juglans cinerea*, *J. nigra* (Nawaschin and Finn 1913) and *J. regia* (Nast 1941). However, in

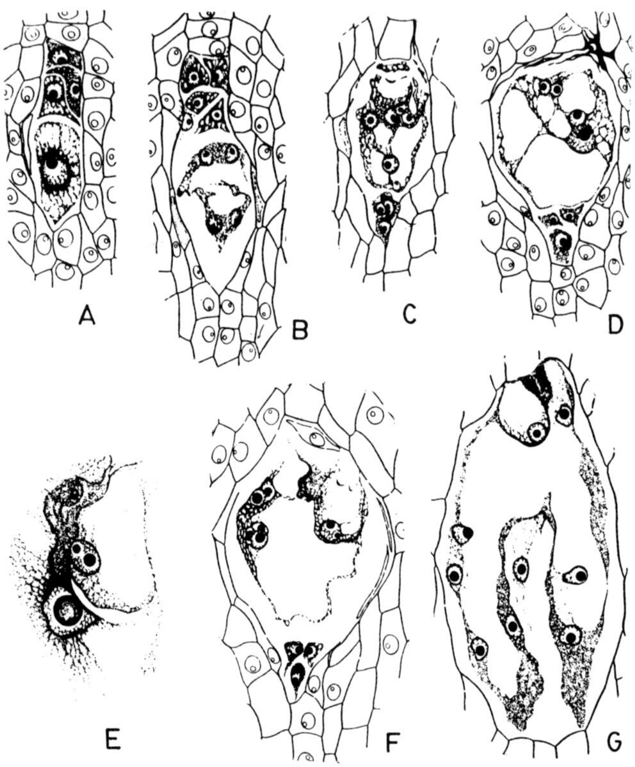

Fig. 5.3.1A–G. *Juglans mandshurica*. **A** Tetrad of megaspores, chalazal functional. **B** Four-nucleate embryo sac, four megaspores persist at the micropylar end. **C** Eight-nucleate embryo sac. **D** Embryo sac prior to fertilization, the polar nuclei have fused. **E** Zygote and primary endosperm nucleus, a sperm (or vegetative?) nucleus is still present in the discharged pollen tube. **F** Embryo sac with zygote, nuclear endosperm, and three antipodal cells. **G** Later stage; remnants of pollen tube persist in **F** and **G**. (After Langdon 1934)

Hicoria (J.G. and N.C. Woodroof 1927) porogamy is reported. As many as two (Langdon 1934) or five (Nawaschin and Finn 1913) pollen tubes enter the embryo sac in *Juglans*. Fertilization is delayed. Figure 5.3.1E shows the zygote and primary endosperm nucleus.

The development of endosperm is of the Nuclear type, and a large number of free nuclei are produced around the central vacuole (Fig. 5.3.1F, G). Wall formation begins from the micropylar end in *Juglans* (Nast 1941), and from the chalazal (needs reinvestigation) end in *Carya* (McKay 1947). The endosperm is reduced to a single layer of cells, and stores fatty substances.

According to Langdon (1934), the division of the zygote is transverse. The apical cell produces cotyledons and stem tip, whereas the basal cell gives rise to hypocotyl, root axis, root-cap, and a multicellular suspensor. Ogebuadze (1977) reports Asterad or Piperad type of embryogeny in *Carya*.

In *Carya* (Solntzeva and Ogebuadze 1974), the zygote undergoes a resting period of 1.5–2 months. A period of 2–2.5 months elapses between the division of zygote and initiation of differentiation of organs. Complete ripening of fruit begins 5 to 6 months after fertilization.

Twin nucelli are reported in *Pterocarya fraxinifolia* (Bouman and Boesewinkel 1969). This is due to the cleavage of nucellar primordium after the initiation of integument. Such nucelli contain normal embryo sacs.

The seed is large and the massive embryo has contorted cotyledons. The seed-coat is highly vascularized, and reduced to the outer epidermal layer of large, radially elongate, thin-walled cells. Stomata occur in the epidermal layer.

F Bouman and FD Boesewinkel (1969), WW Finn (1925), WA van Heel (1976), G Karsten (1902), LM Langdon (1934), JF Leroy (1955), JG Luza and VS Polito (1988), JW McKay (1947), CG Nast (1941), SG Nawaschin (1895a), SG Nawaschin and WW Finn (1913), NV Ogebuadze (1977), MP Solntzeva and NV Ogebuadze (1974), DE Stone and CR Broome (1971), BGL Swamy (1948a), DR Whitehead (1965), JG Woodroof (1924), JG and NC Woodroof (1927), NC Woodroof (1928)

5.4 Taxonomic Considerations

Myricaceae, Juglandaceae. Bentham and Hooker (1965c) place Myricaceae and Juglandaceae in Unisexuales, Engler and Diels (1936) place Myricaceae in Myricales, and Juglandaceae in Juglandales, and Hutchinson (1973), R. Dahlgren (1977a, 1980b) and Cronquist (1968, 1981) agree with this arrangement. Hjelmqvist (1948) considers Juglandaceae to be advanced over Myricaceae, and includes both the families in the order Juglandales. Hutchinson (1973) maintains that the family Myricaceae is derived from Rosalean stocks ancestral to the Hamamelidales. Bessey (1915) considers the origin of Juglandaceae through Sapindales. R. Dahlgren (1977a) places both the orders, Juglandales and Myricales, near Casuarinales in his superorder Rosiflorae. Juglandaceae bear affinity with Betulaceae and Fagaceae in floral and gross anatomical characters.

In addition to the general features of the order, Juglandaceae is characterized by the occurrence of chalazogamy, and Asterad or Piperad type of embryogeny. In Myricaceae the mode of fertilization is porogamous. Nucleate cytoplasmic nodules occur in the endosperm, and the endosperm is ephemeral.

Both the families show close affinities, and the order is homogeneous.

The features of Sapindales, comparable to Juglandales are: two-celled pollen grains, crassinucellar ovules, Polygonum type of embryo sac, chalazogamy in *Anacardium* and *Pistacia* (Anacardiaceae), and Nuclear endosperm with cytoplasmic nodules in Anacardiaceae. Cronquist (1968, 1981) negates the relationship, as the wood of Juglandaceae is primitive.

For the relationships of Juglandales with Fagales, refer to Section 9.4.

6 Balanopales

6.1 Balanopaceae

The monotypic family Balanopaceae represents the order Balanopales. The embryology has not been investigated.

7 Leitneriales

The order Leitneriales comprises two families: Leitneriaceae and Didymelaceae.

7.1 General Features

Leitneriales show: (1) secretory anther tapetum, (2) three- to six-colporate pollen grains with reticulate exine, (3) amphitropous, bitegmic crassinucellar ovule with a slender nucellar beak, (4) unicellular archesporium with a massive parietal tissue, (5) Polygonum type of embryo sac with ephemeral antipodal cells, (6) Nuclear endosperm, (7) persistent massive endosperm, and (8) perisperm in seed.

7.2 Leitneriaceae

Pfeiffer (1912) gives a good account of the embryology.

The anther wall comprises persistent epidermis, fibrous endothecium, an ephemeral middle layer, and glandular tapetum. The microspore mother cells undergo meiotic divisions. The tetrads are tetrahedral, isobilateral and decussate.

The pollen grains are three- to six-colporate with reticulate surface.

The ovule is amphitropous, bitegmic and crassinucellate in *Leitneria floridana*. Both the integuments take part in the formation of micropyle. A single archesporial cell differentiates, and cuts off a parietal cell which produces a massive parietal tissue. The nucellar epidermal cells, by periclinal divisions, produce a slender nucellar beak. The nucellus is persistent. The megaspore mother cell, by meiotic divisions, gives rise to a linear tetrad. The chalazal megaspore develops into a Polygonum type of embryo sac. The synergids are non-vacuolate (requires confirmation), and the antipodals are evanescent.

The development of the endosperm is of the Nuclear type, the free nuclei lie peripherally. After 18 days (of its formation) the endosperm starts becoming cellular. In the chalazal region the endosperm is composed of loosely packed large cells with irregular nuclei. The endosperm occupies the entire seed, and is persistent and starchy.

The zygote divides at the 35-nucleate stage of the endosperm. The divisions are oblique, and cell formation is irregular. The proembryo, before the differentiation of cotyledons, is pear-shaped with a massive suspensor. The mature embryo is large and straight, with two thin and broad cotyledons.

The seed is endospermic and perispermous. The outer epidermis of the testa is thin-walled. The mesophyll cells are thick-walled with reticulate striations. The inner epidermis of the tegmen comprises thick-walled cells with reticulate markings.

WM Pfeiffer (1912)

7.3 Didymelaceae

The embryology has not been studied in detail.

There is a brief account of *Didymeles medagascariensis* by Leandri (1937, see also Yakovlev 1981). The anther is tetralocular (bilocular according to Hutchinson 1973) and extrose. The ovule is hemianatropous and bitegmic. The exostome forms a tube which is broad in the apical region. The seed is exalbuminous (copious endosperm according to Hutchinson 1973) with a large embryo which has flat-concave cotyledons and a short radicle.

H Baillon (1892), J Hutchinson (1973), J Leandri (1937), MS Yakovlev (1981)

7.4 Taxonomic Considerations

Leitneriaceae. Bentham and Hooker (1965c) place Leitneriaceae in Unisexuales, and Bessey (1915) in Ranales. Engler and Prantl (1931), Hutchinson (1973), and R. Dahlgren (1977a) consider it as a monotypic family of Leiteneriales. Hutchinson traces its ancestry in Rosales. Hjelmqvist (1948) includes this family in the Amentiferae, in the line of Myricaceae. Cronquist (1968, 1981) and others derive it, by reduction, from Hamamelidales.

The embryological features of Leitneriaceae resemble those of Myricaceae except in the structure of ovule and seed-coat, and the occurrence of endosperm and perisperm in seed. The presence of starchy endosperm, and reticulate thickenings on the walls of testa and endotegmen suggest that this family may not be related directly with other Amentiferae (see also Thorne 1973). In *Leitneria* the division of zygote is followed by a transverse or longitudinal wall. Further information about its embryogeny may be helpful in determining the systematic position.

Leitneriaceae show affinities with Geraniales in having tricolporate pollen grains, bitegmic crassinucellar ovule, Polygonum type of embryo sac with ephemeral antipodals, porogamous entry of pollen tube, Nuclear endosperm, and albuminous seed with perisperm.

The mature pollen grains, ovule, development of embryo sac and endosperm, and mesotestal seed-coat are comparable to that in Hamamelidales.

8 Salicales

8.1 General Features

The representative monotypic family Salicaceae shows:

(1) secretory anther tapetum, (2) two-celled pollen grains, (3) uni- or bitegmic crassinucellate ovule, formation of parietal tissue, (4) Polygonum type of embryo sac, (5) porogamous entry of pollen tube, (6) Nuclear endosperm, and (7) Asterad or Onagrad type of embryogeny. A central cavity is present in the endosperm of *Salix*, and the exalbuminous small seed is adapted to wind dispersal.

8.2 Salicaceae

The anther wall comprises the epidermis, fibrous endothecium, ephemeral middle layers, and glandular tapetum of binucleate cells. There are one to four middle layers in *Salix* (Chamberlain 1897), and two in *Populus* (Nagaraj 1952).

The pollen grains are inaperturate and granulose in *Populus*, and three-colpate with reticulate exine in *Salix*. They are two-celled at the dispersal stage.

The ovule is anatropous and crassinucellar. In *Populus deltoides* (Nagaraj 1952), it is initially bitegmic but, in later stages, only the outer integument can be distinguished. The ovules are unitegmic in *P. tremuloides* (Nagaraj 1952) and *Salix*. The archesporium is single-celled in *P. tremuloides* and *Salix* (P. Maheshwari and Roy 1951), but five- to six-celled in *P. deltoides*. The functional archesporial cell cuts off a parietal cell. The megaspore mother cell undergoes meiotic divisions, tetrads are T-shaped, and the chalazal megaspore functions.

In a short paper, P. Maheshwari and Roy (1951) contradict Chamberlain's observations (1897) on *Salix*, and point out that the embryo sac is of the Polygonum and not Allium or Adoxa type. The micropylar tip of embryo sac comes in contact with the integument. The synergids and antipodals persist up to fertilization.

Fertilization is porogamous, and is not "delayed" as in other amentiferous families.

The endosperm is of the Nuclear type. Multinucleate cytoplasmic nodules are formed in *Salix cinerea* (Håkansson 1954). Wall formation in endosperm occurs in the micropylar and chalazal region at the heart-shaped embryo in *Salix*. In *Populus* the endosperm completely fills the central cavity, whereas a distinct cavity persists in *Salix*. According to Håkansson (1954): "the endosperm of *Salix* has no storage function, being of ephemeral nature. It has certain physiological functions during this period. The formation of growth substances, inciting the growth of the fertilized ovules is the function. When no endosperm is formed, the ovules remain small. A second function of *Salix* endosperm is probably a certain influence on the development of the embryo".

The embryogeny is of the Asterad type Polygonum-variation in *Salix* (Souèges 1923f), and Onagrad type in *P. deltoides*. Synergid embryos are reported in *Salix* (Chamberlain 1897), but this needs confirmation.

The seed is small with a straight embryo, bears a plumage of aseptate hairs at the base (from the placenta), and is adapted to wind dispersal. The testa comprises only the outer epidermal layer of the integument, with the outer and radial walls thickened.

CJ Chamberlain (1897), Å Håkansson (1954), P Maheshwari and SK Roy (1951), M Nagaraj (1952), R Souèges (1923f), H Takeda (1936)

8.3 Taxonomic Considerations

Salicaceae. Bentham and Hooker (1965c) place Salicaceae among anomalous families of Monochlamydeae. Hallier (1912) and Bessey (1915) include it in Caryophyllales. Melchior (1964), Hutchinson (1967, 1973), R. Dahlgren (1977a) and Cronquist (1968, 1981) treat it as a lone member of Salicales. Hutchinson considers it as a descendant of hamamelidaceous ancestors, Cronquist (1981) of violaceous ancestors, whereas Hjelmqvist (1948) regards it as an advanced order within the Amentiferae.

Salicaceae share several embryological features with Hamamelidaceae, such as anatropous crassinucellate ovule, Polygonum type of embryo sac, Nuclear endosperm, and seeds adapted to wind dispersal (comose in the former and, occasionally, winged in the latter).

Ephemeral endosperm occurs in Myricaceae, and unitegmic crassinucellar ovule and Nuclear type of endosperm are similar to those in Juglandaceae. However, the primitive features of Juglandaceae, such as presence of integumentary vascular bundle, massive nucellus and delayed fertilization, are absent in Salicaceae. Moreover, reduced flower, scanty nucellus, seed adapted to wind dispersal, and rapid germination of seed indicate its advanced nature (in the Amentiferae). The small comose seed with a well developed exotesta is comparable to that in Tamaricaceae (Corner 1976).

9 Fagales

Fagales comprises two families: Betulaceae and Fagaceae.

9.1 General Features

As members of a common order Fagales, Betulaceae and Fagaceae show some similarities: (1) inferior ovary with one to many ovules, anatropous crassinucellar ovule, (2) Polygonum type of embryo sac with ephemeral antipodal cells, (3) delayed fertilization, and Nuclear type of endosperm. The differences are more pronounced: (1) The Betulaceae have (1) three- to four-porate pollen grains, (2) unitegmic ovules, (3) single-celled archesporium, (4) chalazogamy, (5) massive persistent endosperm, and (6) Asterad type of embryogeny. *Alnus* exhibits unreduced parthenogenesis, and antipodal, synergid and nucellar polyembryony. On the other hand, the Fagaceae show: (1) three- to six-colporate pollen grains, (2) bitegmic ovule with multicellular archesporium, (3) postament and chalazal caecum of embryo sac, (4) porogamous fertilization, (5) endosperm with a large central vacuole, (6) embryogeny of either Piperad or Asterad type, and (7) exalbuminous seed.

9.2 Betulaceae

The embryology has not been fully studied.

The pollen grains are three-, four- or five-porate.

The ovule is anatropous, unitegmic and crassinucellar with a short straight micropyle. The hypodermal archesporial cell probably cuts off a parietal cell, producing parietal tissue. Nawaschin (1894) traced the origin of this tissue to nucellar epidermis which Håkansson (1957) doubted. The megaspore mother cell undergoes meiotic divisions resulting in a linear tetrad. The chalazal megaspore develops into a Polygonum type of embryo sac. The polar nuclei fuse before fertilization; the antipodals are ephemeral.

The sporogenous tissue is massive in *Carpinus betulus*, and several young embryo sacs occur in Coryloideae (Endress 1977a). The embryo sac develops a long chalazal caecum in *Corylus*. The fusion of the polar nuclei occurs in the caecum, whereas the antipodals degenerate in situ.

Extravesicular chalazogamy (see Swamy 1948a) is reported in *Alnus glutinosa*, *A. viridis*, *Betula alba* and *B. ermani* (Fig. 9.2.1). Intravesicular chalazogamy (Benson 1894) occurs in *Carpinus betulus* and *Corylus avellana*.

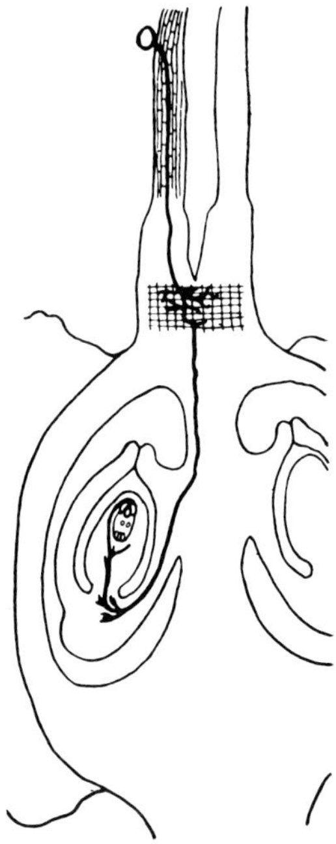

Fig. 9.2.1. *Betula ermani*, longisection of gynoecium with unitegmic ovule at mature eight-nucleate embryo sac. The pollen grain germinates on the style, the tube passes through the ovary, enters the ovule at the chalaza, and then moves upward in the nucellus up to the egg apparatus. The pollen tube gives out branches on its way. (See Yakovlev 1981)

The pollen tube penetrates the embryo sac at the chalazal end, and then reaches the egg apparatus (Swamy 1948a). This needs confirmation. Branching of the pollen tube occurs in *Betula alba* (Benson 1894), *Betula* sp., *Carpinus* and *Corylus* (Nawaschin 1895b). In *B. costata* (Fig. 9.2.2A), the pollen tube enters at the chalaza and puts out branches which traverse through the nucellus. There is again profuse branching at the micropylar end, and further branching occurs inside the embryo sac. The entry of pollen tube at the chalazal end of the embryo sac is reported in *Alnus incana* (Fig. 9.2.2D). A period of 3 to 4 months elapses between pollination and fertilization in *A. glutinosa* (Benson 1894).

The development of the endosperm is of the Nuclear type. Later the endosperm becomes cellular, crushes the nucellus, and fills up the chamber completely (Håkansson 1957). The development is similar in *Alnus rugosa* (R.H. Woodworth 1930a) and *Betula verrucosa* (Håkansson 1957). The massive endosperm fills the seed, and contains aleurone grains and oil as reserve food.

The embryogeny conforms to the Asterad type.

As the viability of pollen grains is very low, and as meiotic divisions do not occur during the formation of the embryo sac, 50% of the seeds contain diploid egg-embryos. Thus, *Alnus* exhibits unreduced parthenogenesis. In addition, polyembryony due to the activity of antipodals, synergids, nucellus, and endosperm is reported in *Alnus rugosa* (R.H. Woodworth 1930a). The concurrent development of multiple embryo sacs also gives rise to additional embryos. Nucellar, synergid and zygotic polyembryony also occurs in *A. incana*, *B. cycoviensis*, *B. pendula* and *Duschekia kamschatica* (Fig. 9.2.2B–G). The presence of four to seven cotyledonary buds on certain embryos suggests

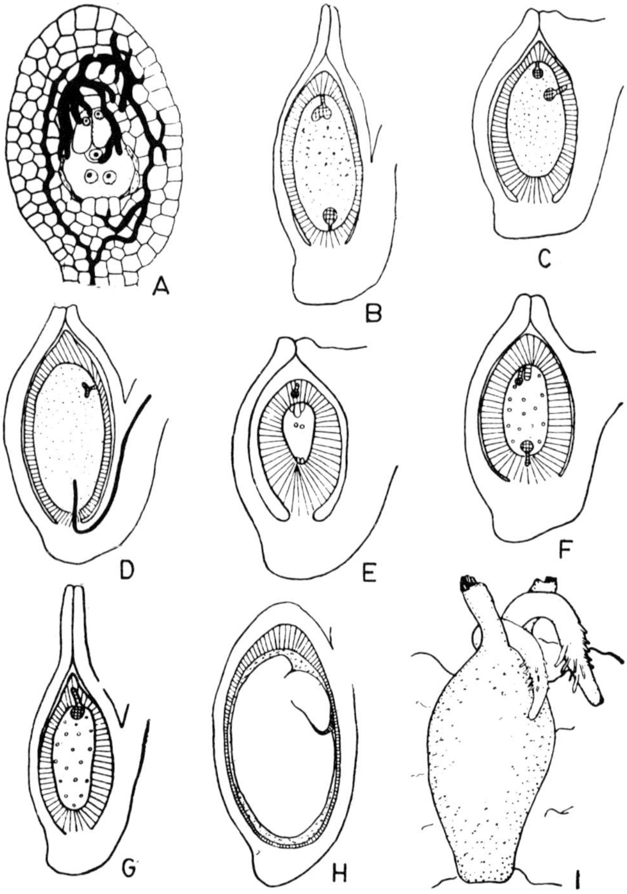

Fig. 9.2.2A–I. **A** *Betula costata*, **B**, **G** *Duschekia kamschatica*, **C** *B. pendula*, **D**, **F** *Alnus incana*, **E** *B. cycoviensis*, **H** *B. davurica*, **I** *D. fruticosa*. Nucellus striped in **B–H**. **A** Longisection of nucellus with branched pollen tube. **B**, **C** Longisection of young seeds with proembryo from the zygote and antipodal (**B**), and zygote and nucellus (**C**). Endosperm *dotted in* **B–D**, **H**. **D** Chalazogamous entry of pollen tube, note a nucellar embryo. **E** Mature embryo sac with a nucellar proembryo above the egg apparatus. **F** Proembryos from zygote, synergid, and nucellus (chalazal end). *Empty circles* denote endosperm nuclei. **G** Nucellar proembryo and endosperm. **H** Mature embryo in seed (longisection). **I** Germinated fruit with two embryos. (See Yakovlev 1981)

fusion of embryos forming a composite mass. P. Maheshwari (1950) suggested that some of the conclusions need confirmation, particularly the formation of embryos from endosperm.

The seed is small and may be pachychalazal. The unitegmic condition must be the result of connation of two integuments (Corner 1976).

The seed of *Betula davurica* (Fig. 9.2.2H) contains a large embryo with scanty endosperm. Two embryos may mature in *Duschekia fruticosa* (Fig. 9.2.2I).

The passage of the pollen tube through the style, its branching in the stylar region, entry through the funiculus, entry into the embryo sac at the chalazal end, branching at the chalazal end of nucellus, branching in the nucellus, at its micropylar end, inside the embryo sac, process of fertilization, parthenogenesis, polyembryony, and structure of the seed (persistence of nucellus as perisperm, if any) require thorough reinvestigation.

M Benson (1894), EJH Corner (1976), PK Endress (1977a), Å Håkansson (1957), AD Macdonald and DH Mothersill (1987), P Maheshwari (1950), JC Mestre (1964), SG Nawaschin (1894, 1895b), BGL Swamy (1948a), A Szwabowicz (1976), RH Woodworth (1929, 1930a, b), MS Yakovlev (1981)

9.3 Fagaceae

The anther wall comprises four layers; the innermost is the tapetum with two- to multinucleate cells. Simultaneous cytokinesis results in the formation of tetrahedral tetrads in *Quercus* (Stairs 1964).

The pollen grains are three- to six- or seven-aperturate (colpate), and two-celled at the dispersal stage.

The ovule is anatropous, bitegmic, unitegmic in *Nothofagus* (Poole 1952), and crassinucellar. The outer integument is extremely long in Fagaceae (*Fagus sylvatica*), and perhaps partially serves as an obturator (see Endress 1977a). In *Fagus* (Hjelmqvist 1957a), the inner integument is reduced, thinner and irregularly folded with papillate elevations on the surface. A 20- to 60-celled archesporium differentiates in the hypodermal region, only one cell functions. True parietal cells are not produced in *Quercus robur* (Hjelmqvist 1953). In *Fagus* a parietal cell is produced giving rise to several cell layers. Meiotic divisions in megaspore mother cell result in a linear or T-shaped tetrad. The chalazal megaspore develops into a Polygonum type of embryo sac. Adoxa type of development is reported in *Quercus alba* (Turkel et al. 1955).

In *Quercus* the embryo sac develops a caecum reaching up to the base of nucellus. Another interesting feature is the development of a central resistant column of cells below the antipodals, in continuity with the vascular trace. It protrudes into the embryo sac and persists until late stages of the embryo. It is referred to as "postament" (Poole 1952, Brown and Mogenson 1972).

The polar nuclei fuse prior to fertilization. The antipodals are inconspicuous and ephemeral.

Fertilization is porogamous. The formation of zygote is delayed, as compared to the primary endosperm nucleus. Two months elapse after pollination to complete the nuclear fusion of gametes (Hjelmqvist 1953).

After fertilization, the primary endosperm nucleus migrates into the caecum while the antipodals remain in situ at the apex of postament, where they degenerate. In *Q. alba* the antipodals move to the chalazal end of the caecum.

The development of endosperm is of the Nuclear type. The aggregation of nuclei in the chalazal region of the embryo sac and their fusion is very common. In *Quercus robur* (Conrad 1900), wall formation is initiated in the micropylar region after the formation of hundreds of nuclei and differentiation of dermatogen in the globular proembryo. The cellular endosperm occupies the upper one-third portion of the embryo sac, while a large vacuole occupies the lower part. The endosperm is consumed by the developing embryo.

In *Q. coccinea* (Stairs 1964), the embryo requires more than two months for its complete development. Johansen (1950) and Davis (1966) refer the embryogeny to the Onagrad type. Hjelmqvist (1957a) suggests that due to oblique wall formation and variability in cleavage patterns, the embryogeny (as a whole) in Fagales cannot be assigned to any type so far described.

In *Q. gambelii* (Brown and Mogensen 1972), the division of zygote is followed by an oblique or nearly vertical wall, producing a small apical and a large basal cell. As the division of zygote is followed by almost vertical partition, the embryogeny can be assigned to the Piperad type. If it is considered that this division is followed by an oblique wall, and as apical and basal cell both contribute to the formation of embryo proper, it can be assigned to the Asterad type. Brown and Mogensen (1972) assigned the embryogeny to a variation of Piperad or Asterad type rather than the Onagrad type. Hjelmqvist (1953) presumed the varying positions of additional embryos to be due to fertilization of multiple embryo sacs.

The seed is exalbuminous, has a large straight embryo with thick or plicate cotyledons, and is enclosed in a nut. The testa is thin-walled, and contains branches of vascular bundle. The tegmen gets crushed.

RC Brown and HL Mogensen (1972), AH Conrad (1900), GL Davis (1966), PK Endress (1977a), H Hjelmqvist (1953, 1957a), DA Johansen (1950), HL Mogensen (1970a), M Okamoto (1989), AL Poole (1952), AP Singh and HL Mogensen (1975, 1976), GR Stairs (1964), HS Turkel, AL Rebuck and AR Grove Jr (1955)

9.4 Taxonomic Considerations

Betulaceae, Fagaceae. Bentham and Hooker (1965c) combine Betulaceae and Fagaceae in Cupiliferae, and Hallier (1912) in Amentiferae. Wettstein (1935), Engler and Diels (1936), Rendle (1959) and Hutchinson (1973) include them in Fagales. Hutchinson assigns the genus *Corylus* to the family Corylaceae. Hjelmqvist (1948) subdivides the Fagales into Fagales and Betulales, within Amentiferae. Bessey (1915) assigns the two families with Juglandaceae, and treats the three as derived from sapindaceous ancestors. Tippo (1938) derives them from hamamelidaceous stocks, and considers Betulaceae as more primitive. Cronquist (1968, 1981) postulates their derivation from ancestors similar to Hamamelidales.

Betulaceae and Juglandaceae share several similar features, such as unitegmic ovule with single archesporium, chalazogamy, delayed fertilization, Nuclear endosperm, and Asterad type of embryogeny. The occurrence of unitegmic ovules in Betulaceae, Juglandaceae and *Nothofagus* (Fagaceae) is a derived condition due to reduction of bitegmic ovules.

The multicellular archesporium and delayed syngamy in Fagaceae indicate primitiveness. Exalbuminous seeds with a large embryo and contorted cotyledons, and vascularized testa, occur in Juglandaceae and Fagaceae. All the three families exhibit primitive embryological features.

While dealing with the general similarities of inflorescence, flower, pollen, fruit and seed in the Hamamelidales-Fagales group, Endress (1977a) emphasizes embryological features, such as belated development of ovule, and delayed syngamy in relation to anthesis. Coryloideae in Betulaceae, and *Nothofagus* in Fagaceae show close affinity to Hamamelidales during earlier stages of development (up to fertilization). The similarities are: number of integuments (except *Nothofagus*), outer integument partially functioning as obturator, massive sporogenous tissue in the ovule (Coryloideae), development of multiple embryo sacs (Coryloideae), and Polygonum type of embryo sac.

10 Urticales

The Urticales comprise five families: Rhoipteleaceae, Ulmaceae, Eucommiaceae, Moraceae and Urticaceae.

10.1 General Features

The Urticales show (1) secretory anther tapetum, (2) simultaneous cytokinesis following meiotic divisions of pollen mother cells, (3) two- to six-porate pollen grains with simple exine, two-celled at the dispersal stage, (4) anatropous (orthotropous in Urticaceae), bitegmic, crassi-nucellar ovule, weakly-crassinucellar and unitegmic in Eucommiaceae, both crassi- and tenuinucellar in Moraceae, (5) Polygonum type of embryo sac in all members except Ulmaceae where it is of the Adoxa type or an intermediate type between Adoxa and Drusa, (6) distinct hypostase, (7) antipodals small and ephemeral, they proliferate in Ulmaceae, Moraceae and Urticaceae, (8) Nuclear type of endosperm except in Eucommiaceae where it is Cellular; chalazal haustorium develops in Moraceae and Urticaceae, (9) embryogeny of the Asterad type in Moraceae and Urticaceae, and of the Onagrad or Solanad type in Ulmaceae, (10) polyembryony frequent, (11) apospory in Urticaceae, (12) seeds exalbuminous except in Ulmaceae, and (13) seed-coat unspecialized and either the testa, or tegmen, or both, become crushed; remain thin and membraneous if they persist.

10.2 Rhoipteleaceae

The embryology has not been investigated. The anther has four lobes which are connected into two thecae. The pollen grains are triangular.

The ovule is hemianatropous, bitegmic, attached to the septum by a short ascending funicle.

The seed is exalbuminous with a straight embryo which has a slender radicle and thick cotyledons.

J Hutchinson (1973), MS Yakovlev (1981)

10.3 Ulmaceae

The anther is tetrasporangiate. The anther wall comprises an epidermis, fibrous endothecium, two ephemeral middle layers in *Ulmus* (Shattuck 1905) and three or four in *Holoptelea* (Capoor 1937b), and a glandular tapetum with binucleate cells. The tapetal cells are multinucleate in *Zelkova serrata* (Yamada and Yoshida 1979). Partial sterilization of sporogenous tissue, which probably provides nourishment to the remaining cells, occurs in *Holoptelea*. The microspore mother cells undergo meiotic divisions giving rise to tetrahedral and decussate tetrads. Cytokinesis is simultaneous.

The pollen grains are two-celled at the dispersal stage. In *Holoptelea*, division of the generative cell occurs on the stigma.

The ovule is anatropous, bitegmic and crassinucellar with a single hypodermal archesporial cell. Both the integuments form the micropyle in *Celtis* and *Ulmus*, while in *Holoptelea* and *Zelkova* only the inner one. The nucellus develops a beak, and the hypostase is well developed.

The hypodermal archesporial cell cuts off a parietal cell (Shattuk 1905). On the other hand, R.I. Walker (1938, 1950) reported that the single hypodermal archesporial cell functions directly as megaspore mother cell in *Ulmus americana*, *U. fulva*, *U. glabra* var. *camperdownii*, *U. pumila* and *U. racemosa*. The megaspore mother cell undergoes meiotic divisions and a coenomegaspore of four linearly arranged nuclei is produced. The nuclei undergo one further division to produce eight-nucleate, seven-celled gametophyte. This happens in 40%–60% of ovules of *Ulmus*, while in the remaining ovules the four megaspore nuclei assume a 1 + 3 position. The micropylar nucleus undergoes two more mitotic divisions, while three chalazal nuclei undergo one or more divisions. Mature gametophytes with four to 12 antipodals have been observed. According to R.I. Walker (1950), the development of female gametophyte in *Ulmus* is transitional between the Adoxa type and a modified Drusa type.

The Adoxa type is also reported by Leliveld (1935) and D'Amato (1940).

Zelkova serrata (Yamada and Yoshida 1979) has a single ovule in the ovary. The megaspore mother cell undergoes meiosis producing a coenomegaspore which shows 1 + 1 + 1 + 1, 1 + 2 + 1, 2 + 2 arrangement of nuclei, or even a 1 + 3 arrangement. The two lower nuclei may divide earlier than the upper two or all the four nuclei may divide simultaneously and an eight-nucleate embryo sac is formed. If some of the nuclei divide again, the embryo sac has more than eight nuclei and one to four nuclei in the chalazal end degenerate. The organized embryo sac has three-celled egg apparatus, two polars, and a variable number of antipodal cells (some nuclei do not organize into cells). The mature embryo sacs are of the Adoxa type, sometimes Drusa type with a variable number of antipodal nuclei (always less than 11). In our opinion a reinvestigation may be very useful.

In *Celtis occidentalis* (Jönsson 1879/80) and *Holoptelea integrifolia* (Capoor 1937b), the gametophyte is of the Polygonum type. Supernumerary antipodals occur in *Holoptelea*.

Capoor (1937b) reported some abnormalities in *Holoptelea*. In several cases, due to the elongation of funiculi, the ovules grow through the style reaching as far as the stigma. Therefore, the stylar canal remains completely open.

In *Ulmus* the pollen tubes branch, anastomose, and wander profusely through the funiculus, integuments and, occasionally, in the nucellus. Fertilization is porogamous, and the pollen tubes persist up to the two-celled proembryo (Shattuck 1905). Several features reported in *Ulmus* require reinvestigation.

The development of endosperm is of the Nuclear type, and the nuclear divisions are mostly synchronous. Later, it becomes cellular.

The development of embryo is of the Onagrad type in *Holoptelea*, and Solanad in *Ulmus*.

In *U. americana*, the presence of antipodal embryos, extra embryos at the micropylar region, and pseudopolyembryony due to the presence of two embryo sacs in an ovule are common (Shattuck 1905). Ekdahl (1941), and Guignard and Mestre (1966) reported antipodal embryos in *U. campestris*, and *U. glabra*, respectively.

The seed is small and exalbuminous with straight embryo; the cotyledons are fleshy and the radicle short. The testa is crushed in *Ulmus*. It persists in *Holoptelea*; the thick-walled outer epidermis contains brown pigment. The tegmen is crushed.

SP Capoor (1937b), F D'Amato (1940), WC Dickison and EM Swetrer (1970), I Ekdahl (1941), JL Guignard and JC Mestre (1966), H Hjelmqvist and F Grazi (1965), B Jönsson (1879/80), JA Leliveld (1935), CH Shattuck (1905), T Takaso (1987), T Takaso and H Tobe (1990), H Tobe (1987), RI Walker (1938, 1950), H Yamada and O Yoshida (1979)

10.4 Eucommiaceae

The anther wall comprises persistent epidermis, endothecium, one or two ephemeral middle layers, and secretory tapetum with two- to six-nucleate cells in *Eucommia ulmoides* (Tang 1962). During early development of the anther, the epidermal cells between the theca become detached. The endothecium is extensively developed. In *E. ulmoides* it is two- to three-layered, extends to the connective, and surrounds the staminal vascular bundle on either side (Eckardt 1963). The microspore mother cells undergo meiotic divisions, and tetrahedral tetrads are produced. Cytokinesis is simultaneous.

The pollen grains are tricolporoidate, and shed at two-celled stage.

The ovule is anatropous, weakly crassinucellar and unitegmic with a long micropylar canal. The cells of the nucellar epidermis divide periclinally to form a small nucellar cap. The multicellular archesporium cuts off parietal cells. The development of the female gametophyte is of the Polygonum type. One of the embryo sacs develops to maturity, while the others may develop up to four-nucleate stage and collapse.

Fertilization is porogamous. In two cases intravesicular chalazogamy has been observed in *Eucommia* (Eckardt 1963). Branching of the pollen tube is also reported.

The development of endosperm is of the Cellular type.

In *Eucommia ulmoides* the zygote remains in resting condition even up to 6 weeks; the embryogeny has not been studied. Tang (1962) reports suspensor polyembryony.

The seed is albuminous with a straight embryo. The seed-coat is membraneous and thin-walled.

T Eckardt (1957, 1963), H Harms (1930), SH Tang (1962), O Tippo (1940)

10.5 Moraceae

The reproductive biology of *Ficus* is highly specialized and is dealt with separately. The following account refers to other members of the Moraceae.

The wall of the tetrasporangiate anther comprises heavily cutinized epidermis, fibrous endothecium, three to four ephemeral middle layers, and secretory tapetum. According to S. Kumar (1978), in *Morus* spp. the development of anther wall conforms to the "Basic type" (see Davis 1966). Resin occurs in the endothecial cells, and connective of *Cannabis sativa* (Mohan Ram and Nath 1964). The tapetal cells are two- to multinucleate, and show nuclear fusions. It becomes irregularly biseriate. The microspore mother cells undergo meiotic divisions, simultaneous cytokinesis, and produce tetrahedral and decussate, rarely isobilateral, tetrads.

The pollen grains are biporate with a smooth exine, and two-celled at the dispersal stage. S.P. Singh (1954) observed pollen grains in the ovary wall of *Morus*.

The ovule is hemianatropous, anatropous, bitegmic and crassinucellate with a micropyle formed by the inner integument. The cells of the nucellar epidermis produce a nucellar cap. The corky hypostase is well developed.

The archesporium is unicellular, except in *Morus* where it is multicellular but only one cell functions. It cuts off a parietal cell, the parietal tissue is well developed. The megaspore mother cell undergoes meiotic divisions, gives rise to a linear or T-shaped tetrad, and the chalazal megaspore is functional. Frequently, the sub-chalazal megaspore develops in *Cannabis sativa* (Mohan Ram and Nath 1964). The development of the female gametophyte is of the Polygonum type. The synergids are usually small except in *Artocarpus* (Subba Rao 1940a). The antipodals are ephemeral in *Dorstenia contrayerva* and *D. drakena*. Modilewski (1908b) reports proliferation of antipodals, and parthenogenesis in *D. contrayerva* and *D. drakena*.

In *Morus alba* (S.P. Singh 1954), very few embryo sacs reach maturity, and generally parthenocarpic fruits are produced.

The endosperm is of the Nuclear type, but later becomes cellular. In *Artocarpus*, free-nuclear condition persists until a late stage of embryogeny (Davis 1966). In *Cannabis*, enucleate cytoplasmic vesicles develop in the nuclear endosperm which ultimately becomes cellular (Mohan Ram and Nath 1964). M. Ram (1960) had pointed out earlier that a chalazal free-nuclear haustorium also develops. Mohan Ram and Nath state that enucleate vesicles also occur in the chalazal haustorium later which becomes cellular with binucleate vacuolate cells (see also M. Ram 1960).

The embryogeny conforms to the Asterad type in *Streblus* (Anantaswamy Rau 1942) and *Dorstenia* (Padmanabhan 1961b). Tricotyly is reported in *Morus multicaulis* (R.M. Datta 1941). Cleavage and nucellar polyembryony occur in *S. taxoides* (Narayanan 1968), and the supernumerary embryos are always confined to the micropylar part of the embryo sac.

Srivastava and Tandon (1985) point out that in *Morus indica* and *M. laevigata* adventive polyembryony originates from cells of the ovary and nucellus. The adventive embryo initially has a well-defined nucleus and dense cytoplasm. These embryos grow only up to the four- to eight-celled stage and then degenerate. Of the 65 adventive embryos studied, 30 were observed in open-pollinated flowers, 23 in intra- and interspecific crosses, and 12 in unpollinated (isolated) female flowers. Possibly, the pollen grains stimulate the formation of adventive embryos.

Gonial apospory is reported in several species of *Dorstenia* (see Davis 1966).

The seeds are small to large, albuminous or exalbuminous, enclosed in the woody endocarp. The embryo is straight or curved, cotyledons flat and thin or thick or convolute (Corner 1976).

The outer epidermis of outer integument and the inner epidermis of inner integument form the seed-coat in *Broussonetia* and *Morus*. In *Artocarpus* (Subba Rao 1940a), the outer integument alone forms the seed-coat.

Ficus

For a proper growth and development of the syconium (Fig. 10.5.1), *Ficus* has a mutualistic symbiotic dependence on a sycophilous wasp. *F. religiosa* (B.M. Johri 1989, B. M. Johri and Konar 1955, 1956, Galil and Eisikowitch 1968a) and *F. ottoniifolia* (Verkerke 1986a) have been studied in detail.

At the apex of the syconium is an ostiole with inwards directed bracts on the inside (Fig. 10.5.2A–C).

F. religiosa and *F. ottoniifolia* are monoecious, and each syconium contains male and female flowers (Figs. 10.5.3A–D, 10.5.4A–D, 10.5.5A–D). On the inner fleshy wall of the syconium, *F. religiosa* (B.M. Johri and Konar 1956) has 11–19 pedicellate male flowers around the closed ostiole, each flower has three hooded perianth lobes and a stamen; 105–113 long-styled or seed flowers and 80–81 short-styled or gall flowers, each pedicellate flower has five perianth lobes, and an uniovulate ovary with a laterally attached style with papillate

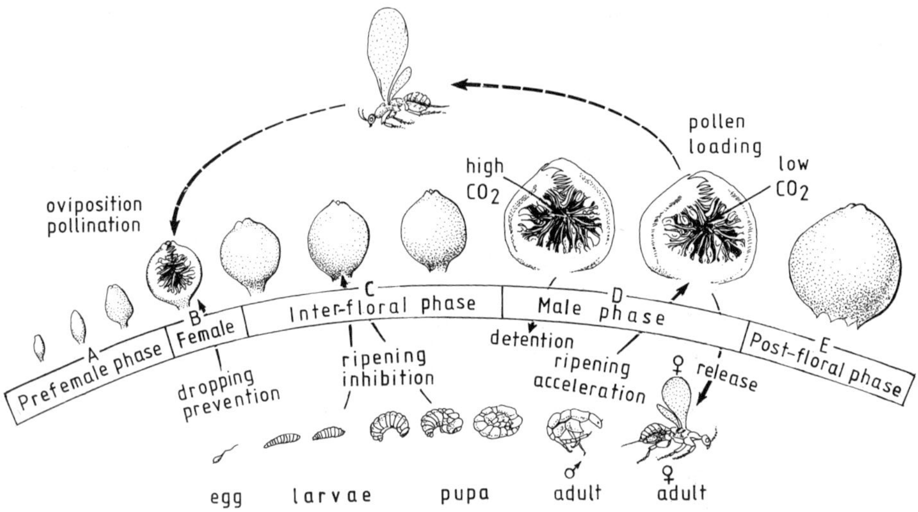

Fig. 10.5.1. Concurrent symbiotic developmental cycle of fig syconium and wasp, phases **A**–**E**. The female wasp loaded with pollen (phase **D**) leaves the old syconium, and enters a young syconium (phase **B**). Following oviposition, pollination occurs. The development of the wasp is also shown. (Adapted after B.M. Johri and Konar 1956, Galil 1977)

stigma. In *F. ottoniifolia* (Verkerke 1986a), there are 60 pedicellate male flowers, each with two cucullate perianth segments and one stamen (Figs. 10.5.3A, 10.5.5D). There are 1200–1400 female flowers, each flower has three to four membraneous perianth segments, a uniovulate ovary and a single terminal style (Fig. 10.5.3A, D). The seed flowers may be sessile, shortly pedicellate and sometimes slightly stipitate, and have a style up to 1800 μm long (Fig. 10.5.3A). During phase A, the stigma of a gall flower develops two unequal lobes (Fig. 10.5.4C, D). The transmitting tissue bordering the stylar canal consists of two layers of cytoplasm-rich cells.

The dithecous anther (Fig. 10.5.5A–D) undergoes normal development. Abnormal pollen grains develop due to arrested cytokinesis after meiosis. The pollen grains are shed at the two-celled stage, and the generative cell divides in the pollen tube.

The ovule is hemianatropous, bitegmic and crassinucellate (Figs. 10.5.8A, C, G, 10.5.9A, B, 10.5.10A–C). The outer integument remains short and the inner alone forms the micropyle. The apical cells of the nucellar epidermis divide and produce a conical cap. The development of the embryo sac is of the Polygonum type, and shows usual organization. The polar nuclei fuse before fertilization.

Grover (1969) reports that both Polygonum and Allium type of embryo sacs develop in *F. carica*. The Allium type of development needs confirmation.

The most interesting part of the study of figs is the pollination, concurrent development of the sycophilous chalcid wasp, emergence of the pollen-bearing wasp, and its entrance into a second generation young syconium (Fig. 10.5.1).

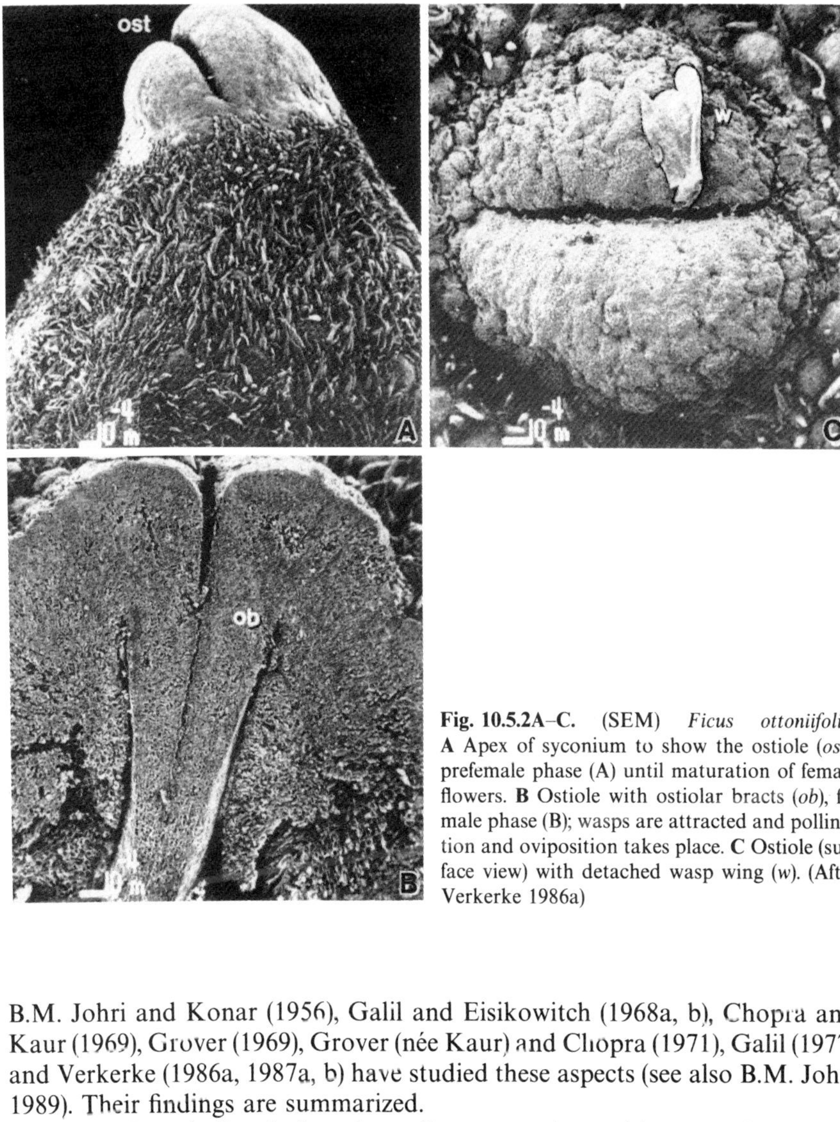

Fig. 10.5.2A–C. (SEM) *Ficus ottoniifolia.* **A** Apex of syconium to show the ostiole (*ost*), prefemale phase (A) until maturation of female flowers. **B** Ostiole with ostiolar bracts (*ob*), female phase (B); wasps are attracted and pollination and oviposition takes place. **C** Ostiole (surface view) with detached wasp wing (*w*). (After Verkerke 1986a)

B.M. Johri and Konar (1956), Galil and Eisikowitch (1968a, b), Chopra and Kaur (1969), Grover (1969), Grover (née Kaur) and Chopra (1971), Galil (1977) and Verkerke (1986a, 1987a, b) have studied these aspects (see also B.M. Johri 1989). Their findings are summarized.

It was long believed that the pollen vector (agaonid wasp *Blastophaga psenes* in *Ficus carica*) effected pollination passively by rubbing against the open anthers on its way out of the old syconium and touching the receptive stigmas on entering the new one. It has now been discovered that pollination is not always as simple as that. Several fig pollinators have specific organs which serve to transport pollen from old to young syconia. The loading and unloading through these organs entails very characteristic behavioural patterns not encountered in other pollination syndromes.

In *Ficus sycomorus* and *F. religiosa* (Galil 1977), and many other species, the number of male flowers is comparatively small and the wasps do not

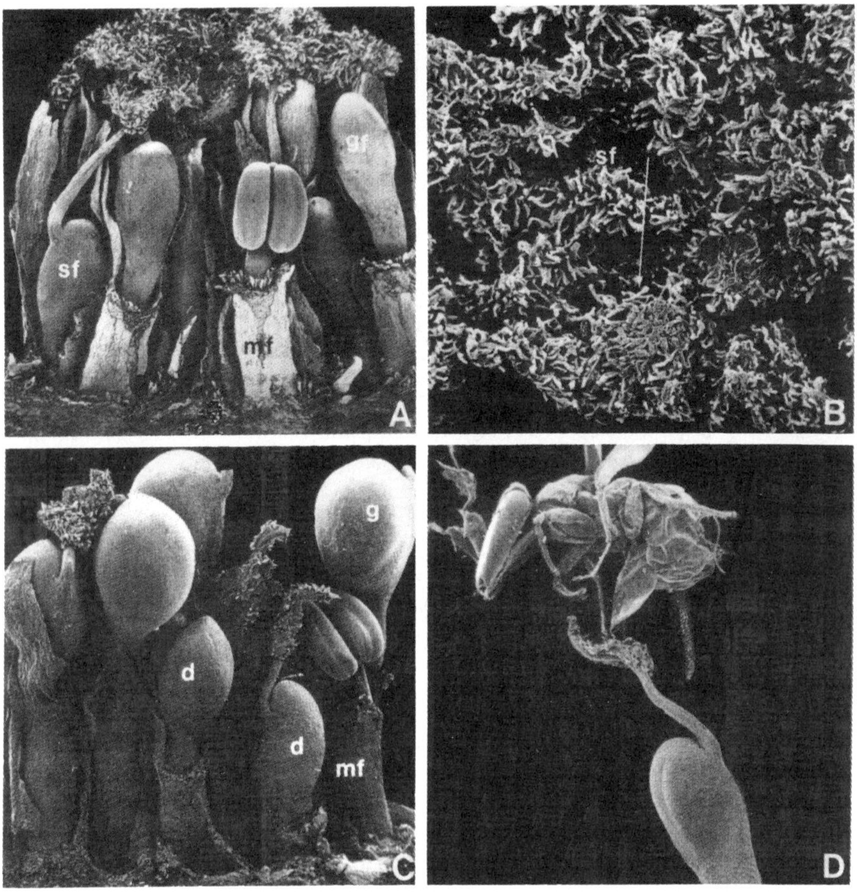

Fig. 10.5.3A–D. (SEM) *Ficus ottoniifolia*, flowers at the time of fertilization, phase B; perianth partly removed in **A**, **C**, **D**. **A** Syconium wall lined with seed (*sf*), gall (*gf*) and male flowers (*mf*). Note the synstigma. **B** Synstigma (surface view), note crushed stigmatic papillae of gall flower (*arrow*); *sf* seed flower stigma. **C** Syconial wall lined with male and gall (*g*) flowers, and drupes (*d*). **D** Gall flower, stigma penetrated by ovipositor of the wasp (*Courtella gabonensis*). (After Verkerke 1986a)

become dusted with pollen on their way out. Sporadically, many pollen grains have been observed on their bodies as they leave the fig. Many of the fig wasps [*Ceratosolen arabicus* (Figs. 10.5.6A, 10.5.7C), pollinator of *F. sycomorus*, and *Blastophaga quadraticeps* (Figs. 10.5.6B, C, 10.5.7A, B) of *F. religiosa*] have thoracic pollen pockets closed from the outside by a movable lid. Many other wasp species of the *Urostigma* section have, in addition to open pockets, coxal corbiculae in the form of longitudinal depressions along the coxae (proximal joints) of the forelegs, which also serve to transport pollen. The corbiculae are fringed on one side by a row of long and stiff hairs (comb) which aid in the loading and unloading of pockets.

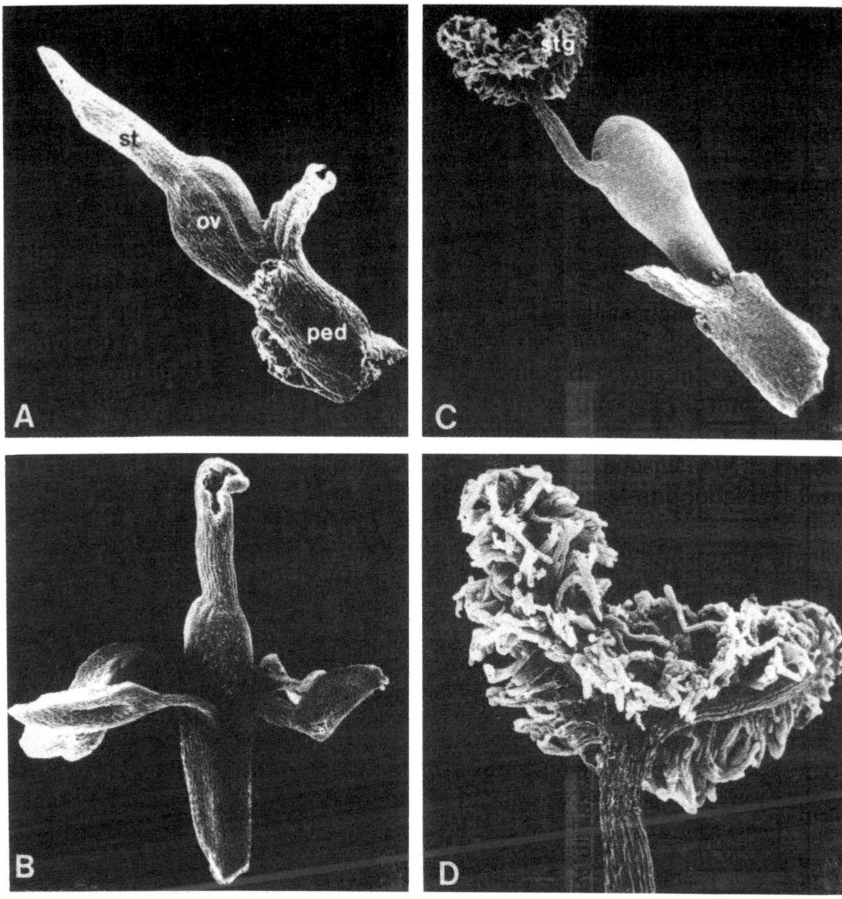

Fig. 10.5.4A–D. (SEM) *Ficus ottoniifolia*, prefertilization development of gall flower, perianth partly removed. **A**, **B** Prefemale phase until maturation of female flower. Note the pedicel (*ped*), ovary (*ov*) and style (*st*). **C** Female phase, wasps are attracted and pollination and oviposition takes place; *stg* stigma, **D** Stigma from **C**. (After Verkerke 1986a)

Before leaving the fig, the female wasp approaches the anthers and with the aid of the arolia (lobes or pads) of the forelegs, raises pollen into its pockets which open in response to touch. After entering the receptive fig (Fig. 10.5.6D), the above movements are repeated in reverse order at the end of each act of oviposition. Pollination proper is effected by striking the tarsi of forelegs one against the other, or by delicate touching of the stigmas.

Not all agaonid wasps have pollen pockets or corbiculae. The pollen vector of the common fig, like those of a whole group of species belonging to the subgenus *Ficus*, has no specific pollen containers. By way of compensation, however, the plants concerned produce an abundance of pollen which ensures passive coating of the wasps with pollen as they pass between the numerous stamens.

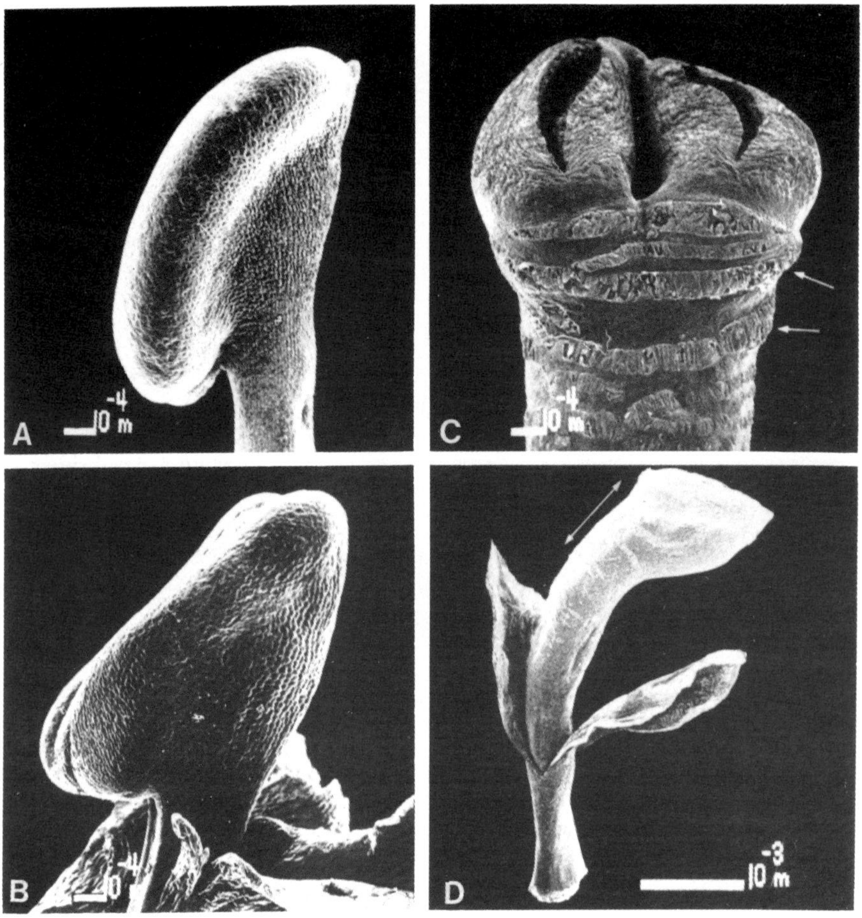

Fig. 10.5.5A–D. (SEM) *Ficus ottoniifolia*, development of male flower, perianth partly removed; **A, B, D** lateral view, **C** ventral view. **A** Anther oblique, phase A–C. **B** Apical thickening of filament initiates tilting of anther, early phase D. **C, D** Ventral filament elongation (*double arrow*) causes both tilting of anther and opening of stomium (**C**), and tanniferous epidermal cells become disrupted (*arrow*). (After Verkerke 1986a)

Blastophaga psenes, pollinator or *F. carica*, is an exception since the pollen is stored in the intersegmental and pleural invaginations on the body of the wasp.

In *F. religiosa*, the syconia are inhabited exclusively by *Blastophaga quadraticeps* (Agaonidea, Agaoninae). Galil and Eisikowitch (1969) mention five developmental phases of the syconia (Fig. 10.5.1A–E).

The female flowers mature at phase B and vary in style length – some sessile, or almost sessile with a comparatively long style, or the flowers are pedicelled with short styles. The long curved stigmas are situated approximately at the same distance from the receptacular wall. They are interlaced and

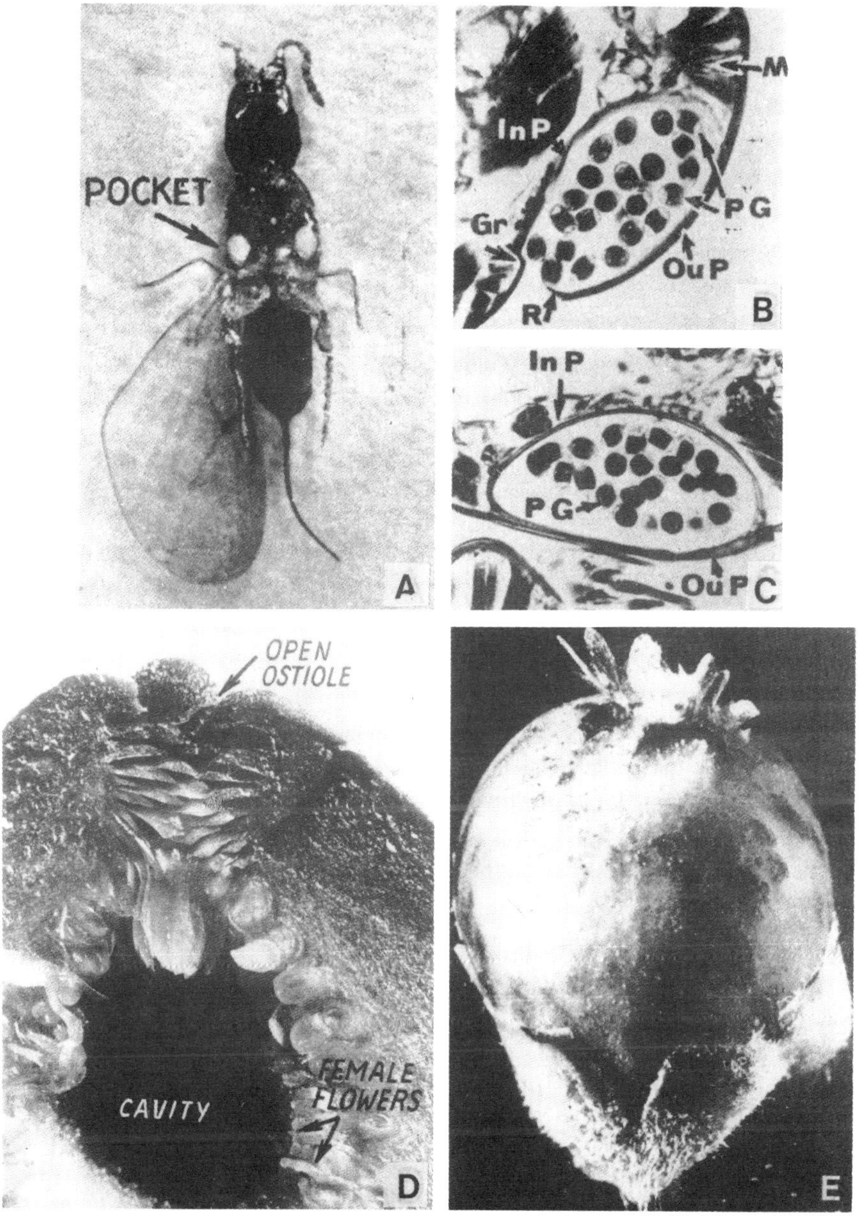

Fig. 10.5.6. **A** *Ceratosolen arabicus*, ventral view of female to show thoracic pollen-pocket (*arrow*). **B**, **C** *Blastophaga quadraticeps*. **B** Transection through slightly open pollen-pocket close to metathorax muscles (*M*) of insect. It contains pollen grains (*PG*), and has an opening (*R*) near groove (*Gr*) bound by inner (*In P*) and outer plate (*Ou P*); **C** Longisection of closed pollen-pocket containing pollen grains (*PG*), and bound by inner (*In P*) and outer plate (*Ou P*). **D** *Ficus* sp., longisection through receptive syconium, female-phase with open ostiole and female flowers lining the central cavity. **E** *F. religiosa*, syconium with a tuft of detached wings (at the ostiole) of *B. quadraticeps*. (After Galil 1977)

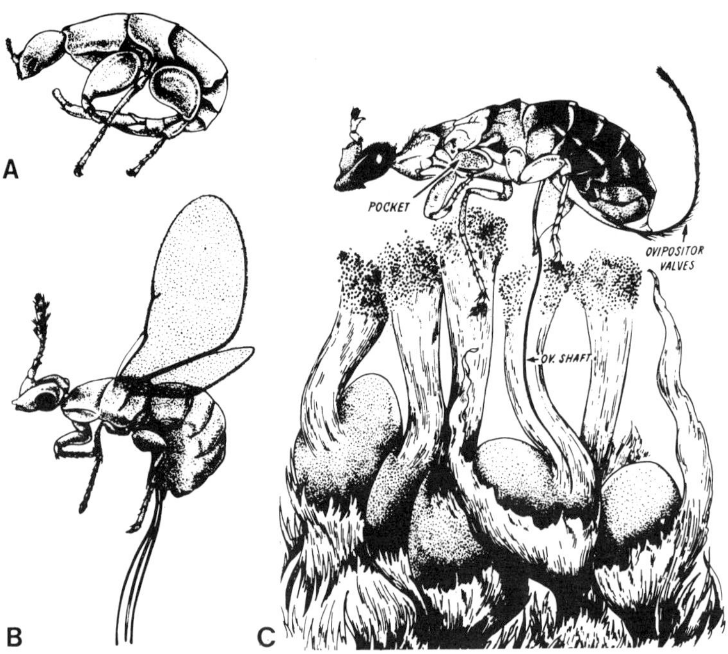

Fig. 10.5.7A, B. *Blastophaga quadraticeps*, male (**A**) and female (**B**), specific pollinators of *Ficus religiosa*. **C** *Ceratosolen arabicus*, wingless female with broken antennae, ovipositing the pistils of female flowers during pollination (pollen is contained in a pocket), by inserting ovarian shaft (*ov. SHAFT*). The ovipositor valves facilitate the process. (After Galil 1977)

stuck together forming a cup-shaped chamber connected by the styles with the ovaries below (Fig. 10.5.3A). This is the compound "synstigma" (Fig. 10.5.3B). It prevents the wasps within from slipping down between the styles and ovaries. Consequently, the fig-wasps trapped in the stigmatic chamber can oviposit only through the synstigma and styles.

At phase D, the synstigma shrinks and withers so that the way from the ovaries into the cavity of the syconium opens. The male flowers mature and the stamens (one per flower) project into the cavity. Most anthers dehisce (Fig. 10.5.5C) by narrow slits (some do not open at all) but the pollen does not get released.

The male wasps now emerge from the galls. Copulation takes place through a hole in the gall (while the female wasp is still within). On emerging from the gall, the female does not leave the syconium immediately. After stretching her body and wings, she approaches male flowers pushing her head into the open anther, or even into a closed anther, thereby causing it to open. Thus, the female becomes coated with pollen. The wasp even eats some pollen since pollen grains have been detected in its alimentary canal.

Now the wasp leaves the syconium through the ostiole. Its scales loosen and the wasp crawls through the slits. If the scales remain adpressed, one to four exit holes are made by the males, in the zone of the ostiolar scales. The syconium wall is thin and the male wasps bite their way through the wall. Only afterwards do the females appear in the outlet and fly away. The wasp-galls are now perforated and easily distinguishable from the fertile ovaries containing seeds.

The ripe anthers "attract" the young female wasps immediately after they emerge from the gall. After once visiting the anther, the female wasp would not be attracted again. The attractant is not known. The male wasps have no role in pollination and are not attracted to the anthers.

The pollen-loaded females, on escape from the older syconia, are attracted to the young second generation syconia (Fig. 10.5.1). At this time the ostiolar scales somewhat loosen, and the female wasp forces its way into the syconium (Fig. 10.5.6D, E). During this process, it looses the wings and flagellae of the antennae (Fig. 10.5.7C). Generally, only one wasp enters the syconium; more than one wasp may also enter. In some figs there is a one-to-one relationship, i.e. only the wasp of a particular species is associated with each fig species. In other fig species, more than one species of wasps may be concerned.

Once having succeeded in entering the syconial cavity, she wanders about on the synstigma and pollination occurs. The wasp stumbles and falls down between attempts to oviposit (Fig. 10.5.3D). The hypogonium guides the ovipositor into the stylar canal of the gall flower (Fig. 10.5.7C). When the ovipositor reaches the curved part of the stylar canal, it forces its way between the raphal strand and nucellus. The ovipositor punctures the tissues. Following "recognition", an egg is deposited between the cytoplasm-rich cells at the junction of the integuments and the nucellus (Figs. 10.5.8A, 10.5.9A). Near the wasp egg there are some drops of a secreted fluid. On withdrawal of the ovipositor, the injured tissues become restored to their original condition, due to growth.

The pollen germinates on the stigmatic papillae (Fig. 10.5.8C, D). The tube grows through the style, the generative cell divides to form two male gametes, and double fertilization is brought about as usual (see Chopra and Kaur 1969, Verkerke 1987b). The ovule may not have been fertilized at the time of oviposition. After oviposition the synstigma and style turn brown, and the stigma falls apart opening up the central syconial cavity.

Usually, gall and some seed-flowers are oviposited but both receive pollen tubes and double fertilization occurs. Oviposition and fertilization are followed by rapid growth of the ovaries and the syconium.

Galil and Eisikowitch (1968a) have experimentally demonstrated that the pollen-loaded wasps with amputated ovipositor do bring about pollination, but the wasp-egg cannot be deposited. Therefore, purely seed syconia were obtained. It has been concluded that the growth of syconium is dependent on pollination and not on the oviposition of the wasp-egg.

In both gall and seed flowers, the endosperm and embryo develop, and in gall flowers the wasp larva grows concurrently (Fig. 10.5.8A–M). The larva migrates from the nucellus into the embryo sac (Figs. 10.5.8G–M, 10.5.9B) causing abortion of the young embryo. The endosperm continues to develop and the free-nuclear endosperm becomes cellular (Figs. 10.5.8E–M, 10.5.9E,

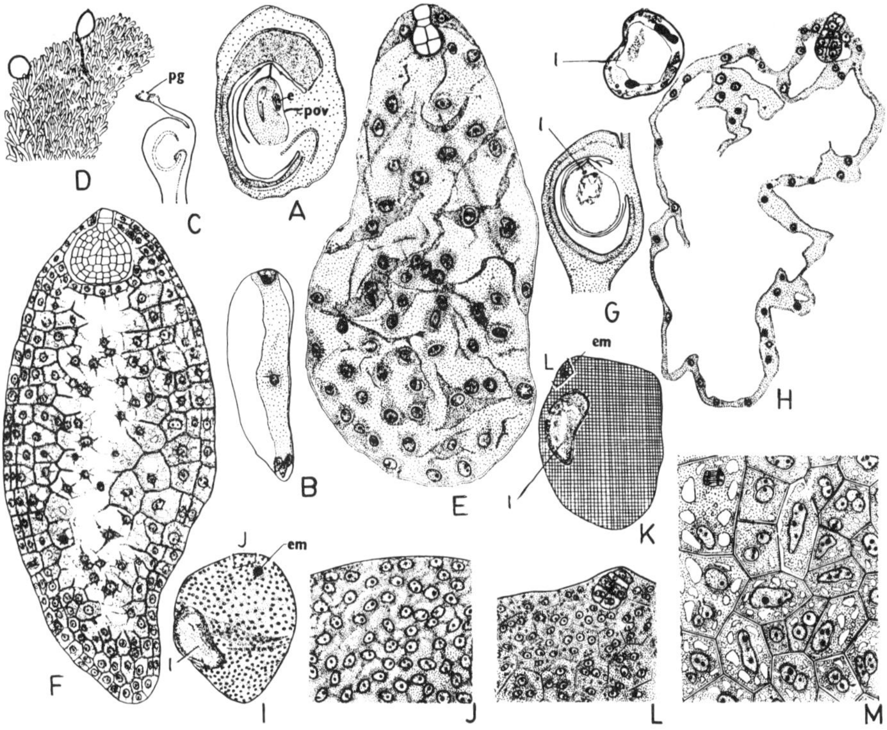

Fig. 10.5.8A–M. *Ficus religiosa.* **C–E** dissected wholemounts, rest sections; **E, F** from seed flowers, others from gall flowers. **A** Longisection ovary of gall flower after oviposition (*pov* path of oviposition) and deposition of egg (*e*) of *Blastophaga quadraticeps*, between nucellus and inner integument (diagrammatic). **B** Embryo sac (from **A**) with egg (synergids not seen), secondary nucleus and antipodals. **C** Short-styled pollinated (*pg*, pollen grains) female flower. **D** Germinated pollen grain on stigma, from **C**. **E** Free-nuclear endosperm and young proembryo, from long-styled flower. **F** Later stage, endosperm partly cellular and late globular proembryo. **G** Gall flower, longisection of ovary with larva (*l*) of wasp. **H** Larva, free-nuclear endosperm and globular proembryo from **G**. **I** Free-nuclear endosperm, proembryo (*black*) and larva. **J** *Inset* from **I**, free-nuclear endosperm. **K** Advanced stage, cellular endosperm, proembryo (*em black*) and larva (diagrammatic). **L** *Inset* from **K**, note wall formation in endosperm and proembryo. **M** Later stage, several cells of the endosperm are multinucleate and others have polyploid nuclei of irregular shape. (After B.M. Johri and Konar 1956)

10.5.10B, G, H). The larva consumes the endosperm gradually and almost none of it is left during metamorphosis of the insect. Most of the ovular tissue is resorbed; remnants of nucellus and outer epidermis of the outer integument persist. Ultimately, the wasp fills the gall almost completely.

In *F. religiosa* (B.M. Johri and Konar 1956), the endosperm nuclei do not divide synchronously in seed flowers. About 60 free nuclei are present at the quadrant proembryo. Centripetal walls are laid down at the early globular proembryo. The endosperm cells are usually uninucleate, occasionally binucleate and the nuclei may fuse. The peripheral cells elongate tangentially and divide periclinally recalling cambial activity (Fig. 10.5.8F). At the heart-shaped embryo, the cellular endosperm fills the embryo sac.

In gall flowers, the earlier stages of endosperm development are comparable to those in seed flowers. In Figure 10.5.8G, H, the wasp larva has digested the nucellus and lies close to the embryo sac containing the free-nuclear endosperm and quadrant proembryo. Later stages are shown in Figure 10.5.8I–L. At the octant proembryo, cleavage furrows followed by wall formation produce multinucleate vacuolated cells; the nuclei fuse forming irregular polyploid masses (Fig. 10.5.8L, M). Further nuclear divisions may or may not be accompanied by wall formation. As the larva grows, it encroaches on the endosperm. Finally, the proembryo aborts, and the entire endosperm is consumed so that the drupelet contains only the wasp.

The development of embryo in a seed flower is quite normal, including tricotyly which is also known in other plants, such as *Magnolia grandiflora*, and *Opuntia dillenii* (P. Maheshwari and Chopra 1955). This is the normal condition in *Degeneria vitiensis* (Swamy 1949a).

In gall flowers, the growth of the proembryo beyond the octant stage is arrested, but the endosperm continues to develop. Obviously, it is not for lack of nutrition that the embryo fails to develop, but the *Blastophaga* may secrete some hormone/enzyme which inhibits the growth of the embryo.

Verkerke (1986a) has studied the fig–fig wasp symbiosis in *Ficus ottoniifolia* for which the pollinator is *Courtella gabonensis* (Fig. 10.5.3D) or *C. camerunensis*. The pollen reaches maturity only 4 days after the organization of the Polygonum type of embryo sac. It remains enclosed in the anther for 6 weeks. Although not mentioned, the large number of male flowers indicates that the female wasp gets dusted with pollen while forcing its way out through the ostiole. Prefertilization gall flowers are shown in Figure 10.5.4A–D. In galls the enlarged cells of the outer epidermis of the pericarp and the collapsed layers of the underlying tissues persist at maturity (Fig. 10.5.9A–F). In "seeds" the pericarp differentiates into epicarp, mesocarp and sclerotic endocarp; a few layers of outer integument persist, while the inner integument collapses (Fig. 10.5.10A–H). In main features, the development of endosperm, embryo, seeds/fruit and syconium is similar to that in *F. religiosa* (B.M. Johri and Konar 1956).

Finally, the wasp exits from the gall (Fig. 10.5.1).

The gynodioecious *F. aspergifolia* (Verkerke 1987a) has (a) figs which contain both male and female gall flowers, and (b) figs with female flowers which produce only seeds. The ovules in the two types of female flowers differ in relation to the different functions, and the author uses the term "ovule dimorphism".

In *F. religiosa* (B.M. Johri and Konar 1956), in the drupelet the seed shows an exotesta (outer epidermis of outer integument), a few collapsed layers of inner integument and a conspicuous endotegmen, and cellular endosperm with reserve food. The curved embryo has only a few layers of endosperm around it except for a broad zone on the funicular side. The pericarp can be distinguished into a few layers of epicarp, a single-layered thick-walled mesocarp, and

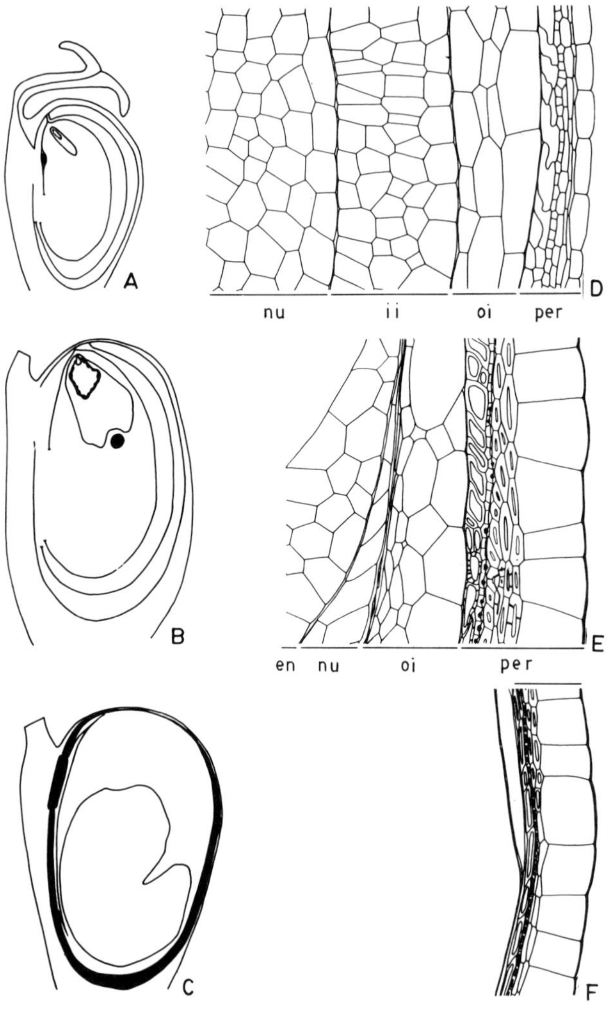

Fig. 10.5.9A–F. *Ficus ottoniifolia*, gall flower. **A–C** Longi- and **D–F** transections. **A** Carpel, 3 days after fertilization. The wasp egg (*black*) has been laid between the cytoplasm-rich raphal cells and the nucellus. Note asymmetric growth of nucellus. **B** Ovary, 9 days after fertilization. The embryo sac contains a globular proembryo and free-nuclear endosperm. The larva (*black*) lies below the embryo sac. **C** Ovary, 30 days after fertilization. **D** Pericarp (*per*), outer (*oi*) and inner (*ii*) integuments, and nucellus (*nuc*) from **A**. **E** The inner layers of cells of pericarp have become sclerified (*black* region in **C**), the inner integument has collapsed, and nucellus is followed by cellular endosperm (*end*), from lateral region of **C**. **F** Except for a part of the pericarp, all the other tissues have collapsed, from apical region of **C**. (After Verkerke 1986a)

a sclerotic endocarp. The wall of the syconium is parenchymatous with a few sclereids in the outer region. It becomes pulpy in a ripe fig.

The ripe syconia with seeds (protected by the stony endocarp) are eaten by birds. The seeds are not damaged at all while passing through the alimentary

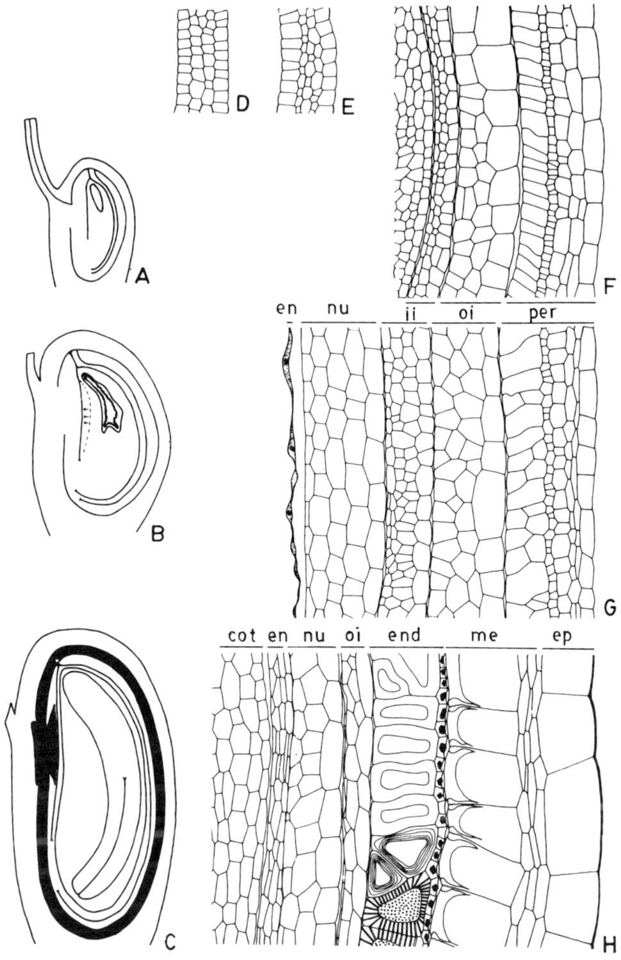

Fig. 10.5.10A–H. *Ficus ottoniifolia*, seed flower. **A**, **B** Longisection of seed carpels. **A** At mature embryo sac when pollination occurs (phase B). **B** 3 days after fertilization when the embryo sac contains globular proembryo and free-nuclear endosperm; the nucellar epidermal cells elongate (*arrows*). **C** Longisection of drupe with dicotyledonous embryo (end of phase **C**). **D**–**H** Cross-sections. **D**, **E** Pre-fertilization ovary wall. **F** Pericarp (*per*), outer (*oi*) and inner (*ii*) integuments and part of nucellus from **A**. **G** *Per*, *oi*, *ii*, nucellus (*nu*) and free-nuclear endosperm (*en*), from **B**. **H** Epicarp (*ep*), mesocarp (*me*) and endocarp (*end*) followed by *oi*, *nu*, *en* and cotyledonary (cot) tissue, from **C**. (After Verkerke 1986a)

canal, and retain 100% viability. This also accounts for the widespread distribution of *Ficus*.

In *F. sycamorus* (see Galil 1977), which does not set seed in the Middle East, the developing larvae of *Sycophaga sycomori* exert a direct influence on the development of the syconium. The occupied syconia do not drop and also resist premature ripening, which enables the contained wasp brood to complete its developmental cycle.

After impregnation of the females in the gall, the male wasps tunnel through the syconial wall. As carbon dioxide (10% in *F. religiosa*) escapes, and the internal atmosphere equilibrates with the external, the females widen the fertilization holes and emerge from the galls. Generally, high carbon dioxide content (Fig. 10.5.1, phase D) in the internal atmosphere of fruits inhibits ripening, whereas its depletion promotes ripening (see Galil 1977).

The relationship between the sycophilic wasp, developing in the ovary, and the syconium is not merely a dietary one, for the insect becomes involved in certain vital processes of the fig. This is an instance of coevolution where both – the fig and the wasp – have become highly specialized. This mutualistic symbiosis must have become established in ancient times (see Murray 1985).

The fig–fig-wasp biology has been fully investigated but there are still many unanswered questions which deserve attention.

M Anantaswamy Rau (1942), I Banerji and A Hakim (1954), RN Chopra and H Kaur (1969), EJH Corner (1976), BC Das (1961), RM Datta (1941), GL Davis (1966), NK Dwivedi (1990), J Galil (1977), J Galil and D Esikowitch (1968a, b, 1969), J Galil and Y Snitzer-Pasternak (1970), H Grover (1969), H Grover (née Kaur) and RN Chopra (1971), BM Johri (1989), BM Johri and RN Konar (1955, 1956), H (Harjinder) Kaur (1970a), S Kumar (1978), P Maheshwari and RN Chopra (1955), J Modilewski (1908a, b), HY Mohan Ram and R Nath (1964), MG Murray (1985), CR Narayanan (1968), G Neeman and J Galil (1978), D Padmanabhan (1961b), M Ram (1960), SP Singh (1954), BL Srivastava and S Tandon (1985), AM Subba Rao (1940a), BGL Swamy (1949a), O Tippo (1938), W Verkerke (1986a, 1987a, b, 1990), RC Verma, A Sarkar and BC Das (1986)

10.6 Urticaceae

Simultaneous cytokinesis accompanies meiosis in microspore mother cells. The tetrads are tetrahedral and decussate.

The male gametes are rod-shaped in *Urtica* (Strasburger 1910). The pollen grains are two-celled at the dispersal stage.

The ovule is basal, orthotropous, bitegmic and crassinucellar with the inner integument forming the micropyle. By periclinal divisions the cells of the nucellar epidermis add to the parietal tissue. The archesporial cell cuts off a parietal cell. The parietal tissue is well developed. The megaspore mother cell undergoes meiotic divisions. The chalazal megaspore of a linear tetrad develops into a Polygonum type of embryo sac. The polar nuclei fuse before fertilization. The antipodal cells multiply in *Urtica*. A hypostase is conspicuous. An obturator develops from the outer integument in *Elatostema machaerophyllum*, and from the inner epidermis of the ovary wall in *E. penduncul osum* (Fagerlind 1944b).

The development of endosperm is of the Nuclear type, which later becomes cellular. The chalazal region of endosperm develops into a short haustorium in *Urtica cannabina* and *U. pilulifera*. It is richly cytoplasmic with hypertrophied nuclei (see Schnarf 1931).

The embryogeny conforms to the Asterad type. Supernumerary embryos are produced from synergids, antipodal cells and nucellus. Sometimes, multiple embryo sacs fuse followed by polyembryony. If the developing embryos lie close together, they may fuse. Unreduced parthenogenesis occurs in apomictic species of *Elatostema* (Fagerlind 1944b).

Somatic apospory of Hieracium type occurs in *E. acuminatum* and *E. penninerve*; aneuspory of Taraxacum type in *E. eurynchum* (2% embryo sacs), *E. machaerophyllum* (88% embryo sacs) and *E. peltifolium* (Fagerlind 1944b). Gonial apospory has also been noted in *E. eurynchum* (98% embryo sacs), *E. machaerophyllum*, *E. rostratum* and *E. sinuatum* (Fagerlind 1944b).

The testa and tegmen become crushed, and there is no definite seed-coat.

G Bentham (1880), F Fagerlind (1944b), GG Franchi and E Pacini (1980), FM Jarrett (1976), N Prakash, LR Bohm and J Maze (1979), K Schnarf (1931), M Sharma (1964), R Souèges (1920c, 1921), E Strasburger (1910)

10.7 Taxonomic Considerations

The order Urticales is considered as phylogenetically advanced, and as the terminal taxon of an evolutionary line from which Fagales have also been derived (Lawrence 1951). Engler and Diels (1936) include Ulmaceae, Rhoipteleaceae, Moraceae and Urticaceae in Urticales. Eucommiaceae has been placed in Rosales by Harms (1930), near Hamamelidaceae. Tippo (1940) and Wettstein (1935) assign it to Urticales. Hutchinson (1973) includes all the families in Urticales, except Rhoipteleaceae which has been placed in Juglandales. He assigned the genus *Cannabis* to a separate family, Cannabaceae (Cannabinaceae). Bentham and Hooker (1965c) place it in Urticaceae, whereas Engler and Prantl (1931) and Melchior (1964) assign it to Moraceae.

Ulmaceae. The development of embryo sac of the tetrasporic Adoxa type or of an Intermediate type between Adoxa and Drusa marks out the family Ulmaceae from the rest. The persistent remnants of the pollen tube, Onagrad or Solanad type of embryogeny, and exalbuminous seed with curved embryo and folded cotyledons are additional features in which the family Ulmaceae differs from the other members. Otherwise, secondary multiplication of antipodals, Nuclear endosperm, and polyembryony occur in Moraceae and Urticaceae. Rest of the features are in general conformity with those of the order.

Eucommiaceae. Ulmaceae and Eucommiaceae have unilocular, bicarpellary ovary with axile placentation, a single anatropous, pendulous ovule with nucellar cap or beak, and samaroid fruit, but the seed is exalbuminous in Ulmaceae. In addition, Eucommiaceae show several noteworthy features. The fibrous thickenings in the endothecium develop extensively. The thickenings are also present in the region of the connective, and around the staminal vascular bundle. The unitegmic ovule, a rare occurrence of intravesicular chalazogamy, branched pollen tubes, Cellular endosperm, and a long resting period of zygote for about 6 weeks distinguish the family and makes its assignment to Urticales anomalous. Unitegmic ovules with multicellular archesporium and intravesicular chalazogamy occur in Fagales (Betulaceae –

Corylus), and there is a long resting period of zygote in Juglandales (Juglandaceae – *Carya illinensis*).

Eucommiaceae are considered to be advanced derivatives of Hamamelidales, or of stocks ancestral to them. The reduced flowers, presence of secretory ducts, Cellular endosperm, and samaroid fruit of *Eucommia* adapted to wind pollination are some of the features allied to Hamamelidaceae.

C.C. Berg (1977) considers the position of Eucommiaceae in Urticales doubtful. Cronquist (1968, 1981) and R. Dahlgren (1975a, 1981) erect a separate order, Eucommiales. Due to the presence of iridoid substances, R. Dahlgren (1975a) suggests its position near Cornales (see also Jensen et al. 1975).

Moraceae, Urticaceae. Moraceae and Urticaceae show a close alliance. The two-celled pollen grains, hemianatropous bitegmic ovules with prominent nucellar cap, Polygonum type of embryo sac, secondary multiplication of antipodals, Nuclear endosperm with chalazal haustorium, Asterad type of embryogeny, nucellar polyembryony, parthenocarpy, and albuminous seeds are the similarities. But the occasional formation of an Allium type of embryo sac, unusual type of pollination and seed development in *Ficus*, tricotyly, cleavage polyembryony, and seed with curved embryo and a definite seed-coat distinguish Moraceae. In Urticaceae the embryo is straight, and a definite seed-coat is absent.

The presence of unreduced parthenogenesis, antipodal, synergid and nucellar polyembryony, polyembryony due to multiple embryo sacs, gonial and somatic apospory in *Elatostemma* spp. are outstanding features comparable to those of *Alnus* (Betulaceae – Fagales).

Mohan Ram and Nath (1964) made a detailed study of *Cannabis sativa*, and support its assignment to an independent family Cannabinaceae, erected by Rendle (1959) and Hutchinson (Cannabaceae) (1973). The family segregates itself from Moraceae in the (1) absence of milky latex, (2) resin in endothecial cells and connective of anther, (3) corky hypostase, (4) frequent development of Polygonum type of embryo sac from the epichalazal megaspore, (5) tubular chalazal endosperm haustorium, and (6) enucleate vesicles in the nuclear endosperm. The position of Cannabinaceae is rather isolated. Anatomical and cytological data (Tippo 1938, Raven 1975) also support this view.

Cronquist (1968, 1981) and Takhtajan (1969, 1980) place the order Urticales near Hamamelidales, whereas Thorne (1968), Stebbins (1974a), R. Dahlgren (1975a, 1981) and C.C. Berg (1977) place it close to Malvales. The phytochemical data (Meeuse 1975) as well as the morphology of inflorescence (Endress 1977a), fruit and seed do not support any relationship of Urticales with Hamamelidales.

11 Proteales

The order Proteales comprise a single family, Proteaceae.

11.1 General Features

The characteristic features of Proteales are: (1) secretory anther tapetum, (2) simultaneous cytokinesis of pollen mother cells, (3) bitegmic, crassinucellate ovules with well-developed parietal tissue, a nucellar cap, hypostase, and epistase, (4) Polygonum type of embryo sac with ephemeral antipodals, (5) porogamous mode of fertilization, (6) Nuclear endosperm, and (7) Asterad type of embryogeny.

The occurrence of varying degrees of development of the endosperm haustoria in Grevilloideae, and their total absence in Proteoideae is noteworthy. The nucellus becomes the source of nourishment in exceptional cases, as in *Cenarrhenes nitida*. In this species pseudo-embryo sacs also develop. The nucellus persists as perisperm (endosperm scanty) in *Helicia*, *Lambertia*, etc. The seed structure is elaborately developed. The seeds are winged in some members. Perichalazal growth occurs in *Helicia*, *Macadamia* and some other taxa.

11.2 Proteaceae

The anther is tetralocular. The hypodermal archesporial cells cut off primary sporogenous cells, and primary parietal cells. The anther wall (Fig. 11.2.1A) comprises an epidermis, fibrous endothecium, two or three ephemeral middle layers, and secretory tapetum. Uninucleate vacuolate tapetal cells occur in *Macadamia ternifolia* (Kausik 1938b), and binucleate in *Hakea* (Venkata Rao 1969a) and *Lambertia* (Venkata Rao 1968). The tapetum is two-layered (Fig. 11.2.1A) on the side of the connective in *Lomatia tinctoria* (Venkata Rao 1962). Following meiosis in microspore mother cells (Fig. 11.2.1B), simultaneous cytokinesis occurs by furrowing, and pollen tetrads are tetrahedral and decussate.

The pollen grains are two-celled when shed, and have two to three or seven apertures (poroid) (Fig. 11.2.1C–E).

The ovule is anatropous, hemianatropous, or orthotropous, bitegmic and crassinucellar. The inner integument grows before the outer one covering the nucellus and leaving a wide micropyle (Fig. 11.2.1F–P). The archesporium is multicellular except in *Hakea* (Venkata Rao 1969a) where it is one-celled. Only one cell is functional, and a parietal cell is cut off. The apex of the nucellus is glandular, and a nucellar cap develops in *Lomatia*. In the chalazal region of the embryo sac a zone of meristematic tissue develops above the pad of tannin-filled cells. Epistase and hypostase are present. Meiotic divisions occur in megaspore mother cell, and a linear, T-shaped or, rarely, inverted T-shaped tetrad is produced. The chalazal megaspore is functional. The female gametophyte (Fig. 11.2.2A) is of the Polygonum type in *Protea lepidocarpon* (Ballantine 1909). The three ephemeral antipodals are inconspicuous. The synergids are broad at the base, and hooked.

Venkata Rao (1967) reported two types of mature ovules. Those of *Cenarrhenes*, *Conospermum*, *Protea*, *Stirlingia* and *Synaphea* have a massive nucellus comprising large, thin-walled, highly vacuolate parenchyma, and

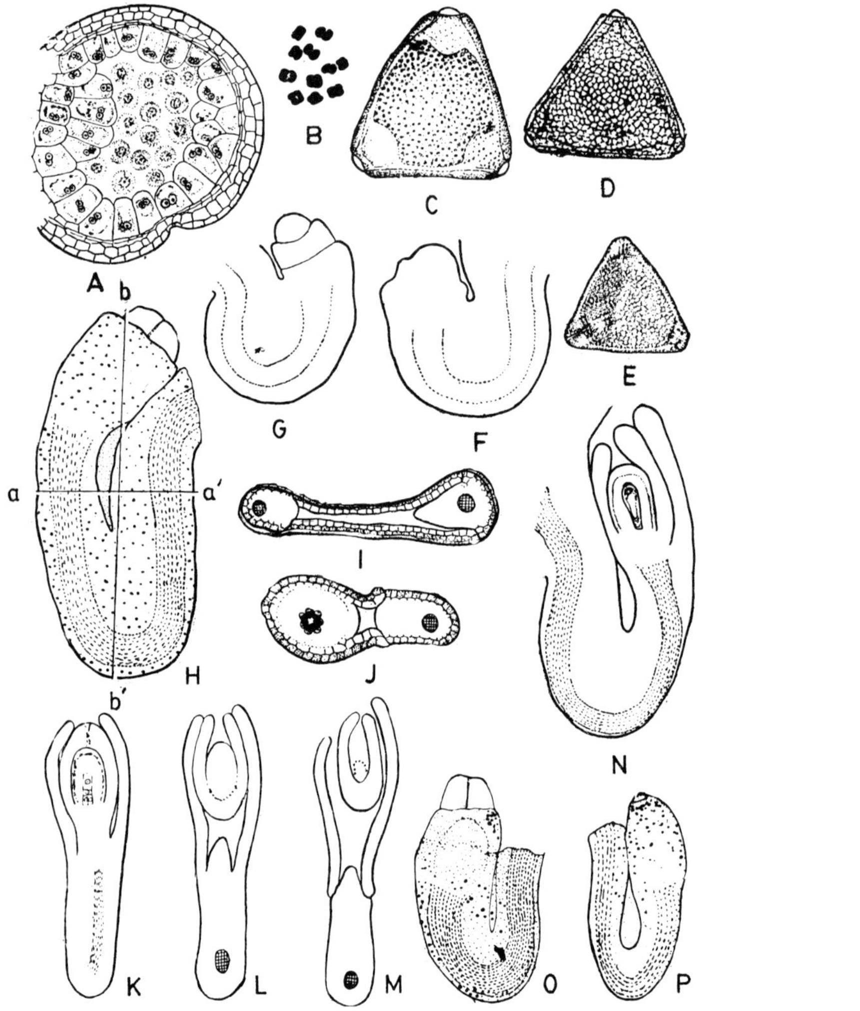

Fig. 11.2.1.A *Lomatia polymorpha*, transection of young anther lobe at microspore mother cell, the binucleate secretory tapetal cells are two-layered at places. **B** *L. tinctoria*, diakinesis showing 11 bivalents. **C–E** Poroid pollen grains of *L. salaifolia* (**C**), *L. polymorpha* (**D**) and *L. hirsuta* (**E**). The exine is ornamented. **F–M** *L. polymorpha*; **F–H** Stages in the development of ovule. **I, J** Cross-sections of ovule along the plane *a-a'* marked in **H**; **K–M** Longisections of ovule along plane *b-b'* marked in **H. N** *L. tinctoria*, median longisection of ovule cut parallel to the flat surface, the outer integument is absent on the funicular side. Note the conspicuous horseshoe-shaped growth of funicle in **H, K–N**. It develops into the seed-wing. **O, P** Ovule of *L. hirsuta* and *L. dentata* (**P**). (After Venkata Rao 1962)

small embryo sac. Starch grains are present in *Protea* and *Stirlingia*. In *Agastachys*, *Bellendena*, *Franklandia* (Venkata Rao 1965), *Persoonia*, and most of the members of Grevilleoideae, the ovules have scanty nucellus confined to the sides of the embryo sac which elongates and extends up to the zone of tannin-bearing cells in the chalaza. *Isopogon trilobus* shows a prominent parietal tissue.

A unique feature is described in the ovule of *Cenarrhenes nitida*, belonging to Persoonieae. Before fertilization, the central region of nucellus disorganizes resulting in a cavity which becomes filled with nutritive fluid. The chalazal end of the embryo sac comes in contact with the lower part of the cavity. This is somewhat akin to the pseudo-embryo sac of Podostemaceae, and the nucellus serves as a source of nutrition. The cells of the hypostase are absorbed.

Fertilization is porogamous. The cells of the glandular epistase help in directing and nourishing the pollen tube. In *Franklandia* the remnants of pollen tube persist up to the globular proembryo.

The development of endosperm is of the Nuclear type. Generally, free-nuclear divisions occur in the endosperm, and wall formation commences from the micropylar end (Fig. 11.2.2B–P). Cytokinesis occurs in varying degrees, in different genera. The endosperm in *C. nitida* is an exception. The embryo sac does not enlarge after fertilization; the free-nuclear endosperm is scanty, and probably does not become cellular. The nucellus persists as perisperm, and is the chief source of nutrition.

In *Bellendena* and *Persoonia* the elongation of the embryo sac (by digesting the nucellus) is slow. The entire endosperm becomes cellular, and haustorial activity has not been noticed.

Grevilleoideae show a rapid elongation of the embryo sac. The nucellus is absorbed.

In *Knightia* and *Orites revoluta*, there is no sharp demarcation between the upper cellular and lower free-nuclear region of endosperm. The entire endosperm becomes cellular, and is afterwards absorbed. The endosperm of *Brabejum* (Jordaan 1946), *Macadamia* (Kausik 1938b) and Branksieae is similar to that of *Knightia*, except that the nuclear region is haustorial.

In *Orites accularia* and *O. revoluta*, the nuclear region is narrow and swollen at the distal end. Secondary haustoria arise from the lower part of the cellular region. In *O. diversifolia*, the nuclear region is thrown into worm-like folds. It becomes partly or wholly cellular in mature seed.

The chalazal nuclear region is haustorial in *Lambertia* (Venkata Rao 1968), whereas some secondary haustorial outgrowths are present on the sac-like distal end in *Stenocarpus*. Finger-shaped hypha-like outgrowths, all over the surface of endosperm in *Lomatia* (Venkata Rao 1962), are noteworthy (Fig. 11.2.3A–C). The chalazal nuclear part becomes coiled and appears like an appendage; ultimately it is absorbed.

In *Buchinghamia* several haustorial lobes arise from the cellular region, and increase the absorptive surface.

The endosperm is highly evolved in *Grevillea* (Kausik 1938c, 1939a, 1941a, 1942) and *Hakea* (Venkata Rao 1969a). In *Grevillea* the upper cellular region shows haustorial activity in some species, otherwise it is even in outline. The lower region develops into a coenocytic, haustorial, "vermiform" appendage (Fig. 11.2.2C, E, G, H, K, L). Streaming movement of protoplasm has been noticed in the coenocytic region (Kausik 1941a). According to Venkata Rao (1967), the nuclear region of the endosperm in *G. pinnatifida* forms multinucle-

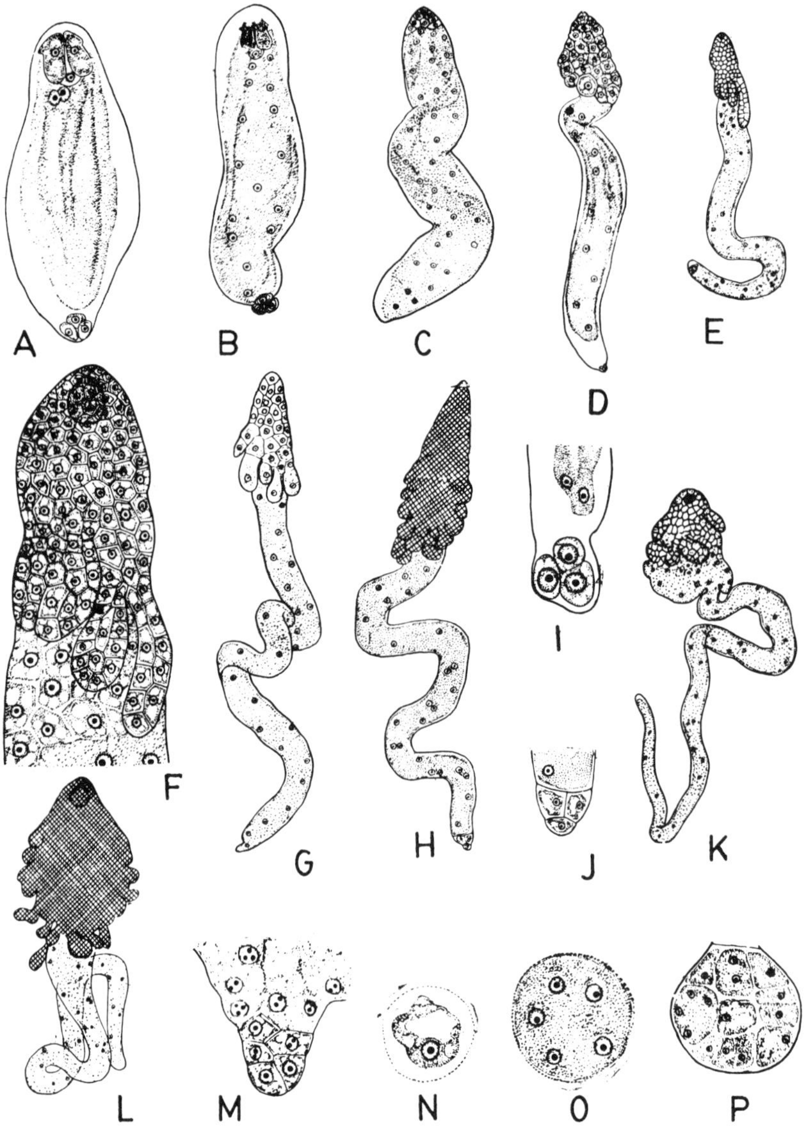

Fig. 11.2.2A–P. *Grevillea banksii*. **A** Mature embryo sac (from dissection). **B** Zygote with remnants of synergids and antipodals, and nuclear endosperm. **C, D** Endosperm showing cytokinesis from the micropylar end. **E–H** Cellular region of endosperm with secondary haustorial lobes at globular proembryo. Chalazal free-nuclear endosperm forms a "vermiform appendage". **I, J** Persistent antipodals. **K, L** Embryo sac with heart-shaped embryo, lobed cellular endosperm, and coiled free-nuclear chalazal appendage. **M** Secondary haustorial lobe. **N–P** Development of secondary haustorial lobe in cellular region. (After Venkata Rao 1966)

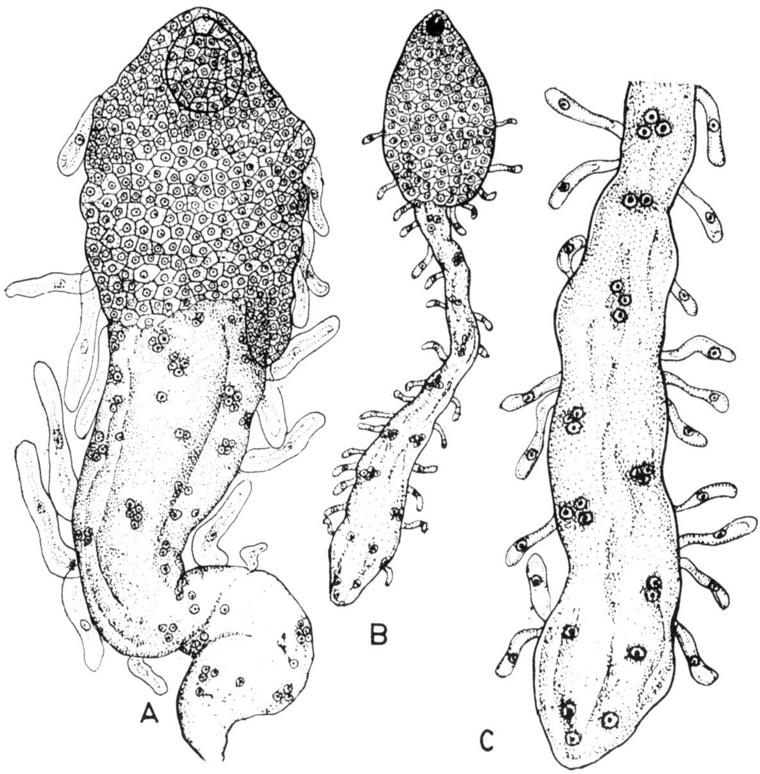

Fig. 11.2.3.A *Lomatia polymorpha*, endosperm (dissected wholemount) with coenocytic chalazal haustorium, uninucleate hairy projections occur on the surface of cellular and haustorial region. **B, C** *L. tinctoria*. **B** Same as **A**; **C** Coenocytic chalazal haustorium with one-celled projections on the surface. (After Venkata Rao 1962)

ate cells. In *G. alpina* and *G. bankskii*, finger-like and sac-like secondary haustoria grow at the base of the cellular endosperm (Fig. 11.2.2E–H, K, L), and later on the secondary haustoria coalesce. The secondary haustoria in *G. banksii* become coenocytic, but do not coalesce (Fig. 11.2.2M–P). A wide sac-like haustorium with secondary haustoria develops in *Hakea*. Later, this part becomes cellular and stores food.

A progressive evolution in the development of haustorium has been observed.

The embryogeny conforms to the Asterad type-Penaea variation (Johansen 1950) in *Franklandia* (Venkata Rao 1965) and *Lomatia* (Venkata Rao 1962). The mature embryo shows foliaceous cotyledons with prominent basal lobes.

The seed is small with a thin seed-coat, or is large with a massive seed-coat as in *Macadamia*, and is exalbuminous except in *Bellendena*. The structure of the seed-coat varies in different genera. The outer epidermis of the testa may consist of thin-walled, papillose cells, or cells with thickened, pitted outer walls as in *Grevillea*, or longitudinally elongate fibres as in *Hakea*, or it may get crushed. The mesophyll is sclerotic in *Macadamia*, but perishes in other genera. The wing, when present, develops from the testa. The tegmen consists of lignified fibers, or fibrous cells. In indehiscent fruits, it is undifferentiated and gets crushed.

In *Macadamia* (Venkata Rao 1970), the growth of integuments is arrested after fertilization, but the growth of chalaza is very conspicuous. The sclerotic pachychalazal zone forms the seed-coat.

A winged seed of *Lomatia tinctoria* is shown in Figure 11.2.4A. The wing is formed by a horseshoe-shaped vascularized extension of the funicle (Fig. 11.2.1H, K–N). In surface view the immature wing shows polygonal richly cytoplasmic cells, but at maturity the walls become thick and the cells lose contents (Fig. 11.2.4B, C). The exotesta is tanniniferous, and the endotesta becomes thick-walled (Fig. 11.2.4D, E). Only the two inner layers of tegmen and a few layers of nucellar tissue persist in the seed.

Banksia ornata is a common dominant in heaths in South-eastern Australia, which are frequently swept by fires which kill plants. Regeneration is only from seed, suggesting that the future of the species is almost completely fire-dependent. Other fire-dependent species are *B. serrata* and *B. serratifolia*.

The multiple fruit consists of a number of follicles attached to a central rachis. The junction between the valves of the follicle has a layer of inter-digitating cells.

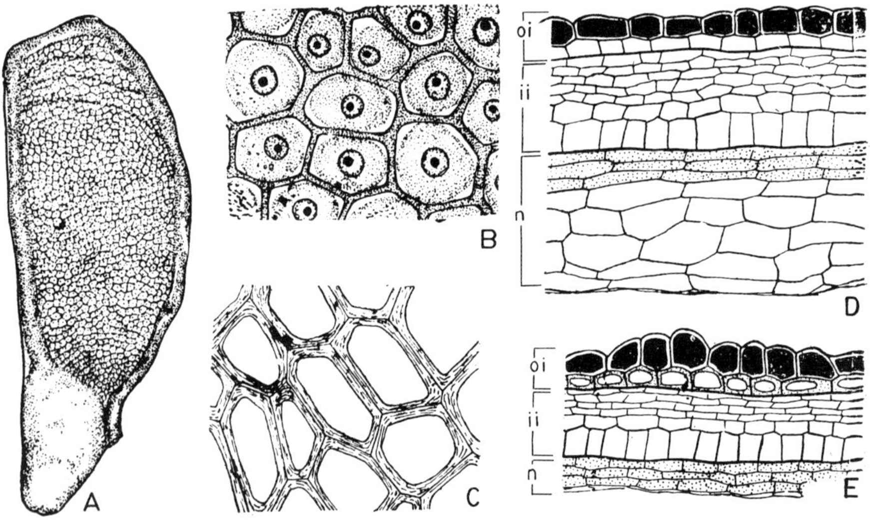

Fig. 11.2.4A–C. *Lomatia tinctoria.* **A** Seed with a conspicuous funicular wing (cf. Fig. 11.2.1H, K–N); **B, C** Surface view of young (**B**) and mature (**C**) seed-wing. **D–E** *L. polymorpha*, development of seed-coat. **D** Immature seed, the outer (*oi*) integument is two-layered and the inner (*ii*) seven-layered, followed by several layers of nucellus (*n*); the outer layers of nucellus are richly cytoplasmic; **E** Mature seed, the exotesta is tanniniferous, and the endotesta is thick-walled. Only the endotegmen, hypodermis, and a few layers of nucellar tissue persist at maturity, the rest of the layers collapse. (After Venkata Rao 1962)

AJ Ballantine (1909), P Brough (1933), DA Johansen (1950), PG Jordaan (1946), SB Kausik (1938a, b, c, 1939a, 1940a, 1941a, 1942), ARH Martin and JC Herscovitch (1989), A Messeri (1928), FG Smith (1968), C Venkata Rao (1960, 1961a, 1962, 1963, 1964, 1965, 1966, 1967, 1968, 1969a, b, c, 1970), AB Wardrop (1983)

11.3 Taxonomic Considerations

Bentham and Hooker (1965c) include the family Proteaceae in the Series Daphnales along with Thymelaeaceae and Elaeagnaceae. Engler and Prantl (1931), Rendle (1971) and Melchior (1964) place the Proteales between Urticales and Santalales. Hutchinson (1973) sought its phyletic relationship with Thymelaeaceae. Takhtajan (1969, 1980) and Cronquist (1981) relate it to Elaeagnaceae. R. Dahlgren (1981) erects the superorder Proteiflorae with the order Proteales, and family Proteaceae, and places it between the superorders Fabiflorae and Myrtiflorae.

Proteaceae seem to be a primitive monochlamydous family. It shows primitive features such as a double-trace stamen as in Casuarinaceae and Betulaceae.

The family Proteaceae resembles Thymelaeaceae and Elaeagnaceae. All the three families have several common features: secretory anther tapetum, simultaneous cytokinesis of pollen mother cells, anatropous, bitegmic, crassinucellar ovule with Polygonum type of embryo sac, Nuclear endosperm, and Asterad type of embryogeny. Moreover, Elaeagnaceae show ephemeral antipodals, and chalazal endosperm haustorium, as in Proteaceae.

In Elaeagnaceae the seed is exotestal with outer epidermis forming a lignified palisade layer. In Proteaceae and Thymelaeaceae it is the outer epidermis of tegmen that comprises the lignified cells. A good deal of variation occurs in the structure of seed in Proteaceae. By evaluating the seed structure, Corner (1976) rejects its alliance with Thymelaeaceae and Elaeagnaceae.

Venkata Rao (1967) wrongly emphasizes the resemblance of Proteaceae to Santalaceae and Loranthaceae. To some extent, the development of chalazal endosperm haustorium may be compared with that in the Santalaccae, but the embryological differences are more marked.

12 Santalales

The Santalales comprise two suborders, Santalineae with six families: Olacaceae, Dipentodontaceae, Opiliaceae, Grubbiaceae, Santalaceae and Myzodendraceae; and Loranthineae with a single family: Loranthaceae and subfamilies Loranthoideae and Viscoideae.

12.1 General Features

The Santalales show the following general features: (1) secretory anther tapetum, (2) two- or three-celled pollen grains, (3) anatropous, uni- or bitegmic tenuinucellate ovule; very often the distinction between the nucellus and integument is absent, (4) development of caecum at micropylar and chalazal end of the embryo sac; in Loranthoideae the micropylar tip of the embryo sac extends to varying length in the stylar canal and stigma, (5) Polygonum type of embryo sac with ephemeral antipodal cells; Allium type in Olacaceae and Viscoideae, (6) occurrence of synergid and antipodal haustoria, (7) Cellular type of endosperm; Helobial type in Olacaceae and Santalaceae, (8) development of micropylar and chalazal haustoria, (9) composite endosperm in Loranthoideae, (10) the division of zygote is transverse; it may be vertical in Santalaceae and Viscoideae; in Loranthoideae it is always vertical, and (11) the albuminous seed lacks seed-coat.

In this order the placental-ovular complex presents a rather confusing picture. The ovules are referred to as bitegmic, unitegmic, and ategmic (undifferentiated).

Ategmic ovules are reported in Olacaceae (*Aptandra, Liriosma, Olax imbricata, Ptychopetalum* and *Ximenia*), Opiliaceae (*Agonandra, Opilia* and *Schoepfia*), Santalaceae (*Buckleya, Chloretum, Leptomeria, Mida, Osyris, Quinchamalium, Santalum* and *Scleropyrum*), and Myzodendraceae (*Myzodendron*).

12.2 Olacaceae

The anther wall comprises an epidermis, fibrous endothecium, two or three ephemeral middle layers, and a secretory tapetum. The endothecium becomes irregularly biseriate in *Strombosia* (Agarwal 1961b). The tapetal cells are two- or four-nucleate. They are uninucleate in *Ximenia americana* (Sankara Rao and Shivaramaiah 1977). The microspore mother cells undergo meiotic divisions resulting in tetrahedral, isobilateral and decussate tetrads. Cytokinesis is simultaneous.

The three-colporate pollen grains are two- or three-celled at the dispersal stage in *Olax stricta* and *Strombosia*, and three-celled in *O. scandens* and *O. wightiana* (Agarwal 1963a). They are triporate in *S. ceylanica* (Shamanna 1961).

In *Olax* the ovary is trilocular below and unilocular above. The ovules are pendulous on the placenta and directed downward (Agarwal 1963a).

The ovule is anatropous and tenuinucellar. A single cell above the archesporial cell represents the nucellus in *Anacolosa frutescens* (Fagerlind 1947a). There is a tendency towards reduction of the nucellus which C.F. Reed (1955) has tried to correlate with the degree of parasitism. The ovules are bitegmic in *Coula, Heisteria, Macrotheca* and *Minquartia*. A single massive integument with a long micropyle develops in *Anacolosa* (Fagerlind 1948), *Olax scandens, O. stricta* and *O. wightiana* (Shamanna 1954, S. Agarwal 1963a). The ovules are undifferentiated (ategmic) in *Aptandra, Liriosma, Olax imbricata, Ptychopetalum petiolatum* and *Ximenia americana* (Sankara Rao and Shivaramaiah 1977).

The archesporial cell functions directly as megaspore mother cell. In *X. americana* there are two to three archesporial cells, but only one functions and produces a parietal cell. The parietal tissue is two to three layers thick. The megaspore mother cell undergoes meiosis and produces a linear or T-shaped tetrad. The chalazal megaspore develops into a Polygonum type of embryo sac. The development is of the Allium type in *Olax wightiana* (Shamanna 1954, Agarwal 1963a). The tip of the embryo sac extends up to the base of stigma in *Aptandra*, and only up to the base of style in *Olax scandens* and *O. wightiana*, and *Chaunochiton*, *Ptychopetalum* and *Schoepfia* (Fagerlind 1948). The embryo sac remains confined to the ovule in *Anacolosa*, *Olax stricta* and *Strombosia* (Fig. 12.2.1A). A lateral caecum of the embryo sac ramifies through the funiculus, and reaches up to the apex of the placental column in *Olax*. A similar lateral haustorium develops in some ovules of *Ximenia*.

The synergids are hooked. The antipodals are ephemeral. They organize into cells in *Strombosia* and *Ximenia*, and are haustorial in *Ximenia* (Sankara Rao and Shivaramaiah, 1977). In *Olax* all the three ephemeral nuclei remain in one cell, or there are two cells one of which is binucleate. The embryo sac contains starch grains in *Strombosia* and *Ximenia*.

The endosperm is of the Cellular type in *Strombosia* (Agarwal 1963b), the chalazal tip develops into a uninucleate haustorium extending up to the apex of placenta. The development is of the Helobial type in *Olax* (Agarwal 1963a). The chalazal chamber becomes tetranucleate, haustorial, and elongates downward through the ovule, grows in the ovarian cavity, penetrates the base of the

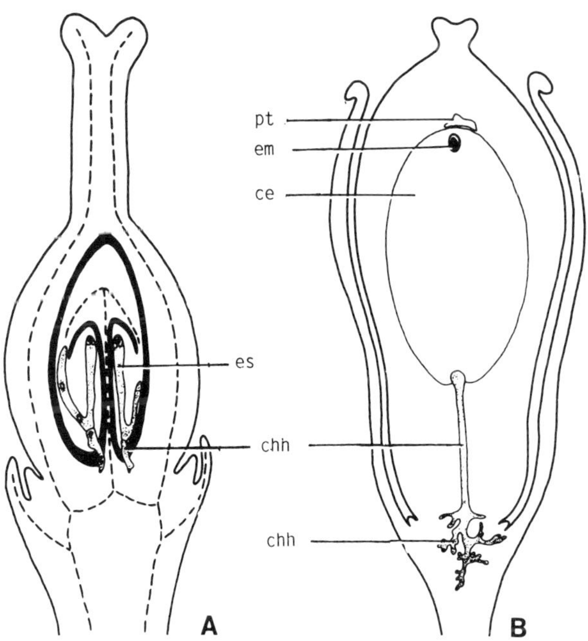

Fig. 12.2.1A,B. *Olax stricta.* **A** Longisection of ovary to show two embryo sacs (*es*) with chalazal haustorium (*chh*) extending up to the receptacle. **B** Longisection of ovary at globular proembryo (*em*), remnants of pollen tube (*pt*), cellular endosperm (*ce*), and chalazal endosperm haustorium extending from the base of endosperm up to receptacle of young fruit. (After Agarwal 1963a)

ovary, and branches profusely in the pedicel (Fig. 12.2.1B). Repeated nuclear divisions occur in the micropylar chamber, and most of the nuclei migrate into the lateral caecum. Wall formation takes place in the micropylar chamber and lateral caecum, producing endosperm proper. The endosperm is persistent and copious, and contains oil or starch (in *Coula*) as storage product.

The zygote divides transversely, and shows a conspicuous multiseriate suspensor in *Strombosia.*

The developing endosperm consumes the integument, and therefore the seed-coat is absent. The fruit is a pseudo-drupe. The drupe-like or nut-like fruits in some members are the largest in this alliance of families.

S Agarwal (1961b, 1963a, b), F Fagerlind (1947a, 1948), BM Johri (1962a, b), BM Johri and SP Bhatnagar (1960), M Ram (1970a), CF Reed (1955), K Sankara Rao and G Shivaramaiah (1977), S Shamanna (1954, 1961)

12.3 Dipentodontaceae

The embryology has not been investigated.

12.4 Opiliaceae

The anther wall comprises an epidermis, fibrous endothecium, one or two ephemeral middle layers, and secretory tapetum. The tapetal cells are binucleate in *Opilia* (Shamanna 1955) and multinucleate in *Cansjera* (Swamy 1960). Microspore mother cells undergo meiotic divisions, followed by simultaneous cytokinesis resulting in tetrahedral tetrads.

The three-colporate pollen grains are two-celled at the dispersal stage.

In Opiliaceae there is a tendency for the reduction of the placental-ovular complex. The ovule (directed downward) is borne on a central placental column in *Rhopalopilia* (Fagerlind 1948) and *Opilia* (Swamy and Dayanand Rao 1963). The placental column is rather small in *Cansjera* (Swamy 1960). It shows further reduction, and in *Agonandra* there is an undifferentiated central mound bearing laterally placed embryo sacs (Fagerlind 1948).

The ovule is anatropous, unitegmic and tenuinucellar with a long micropylar canal. The nucellus is extremely reduced. The nucellar epidermis, integument and micropyle are short-lived in *Cansjera* and *Opilia.* The archesporial cell functions as megaspore mother cell. The latter undergoes meiotic divisions and produces a linear tetrad. The chalazal megaspore gives rise to a Polygonum type of embryo sac. The synergids are hooked, and the antipodals are ephemeral. Compound starch grains occur in the embryo sac of *Cansjera.* The mature embryo sac is characteristically bent like the letter U, in which the chalazal arm is much longer. With the initiation of caecum, the antipodal nuclei migrate into the "pouch" and degenerate. The enlargement of the micropylar part of embryo sac crushes the integument and the egg apparatus becomes exposed. In *Opilia* the chalazal caecum develops into a long, tubular structure reaching up to the level of thalamus, leaving the antipodals in situ.

The development of endosperm is of the Cellular type. In *Opilia* the division of the primary endosperm nucleus occurs at the junction of a cyst-like micropylar chamber and a long tubular chalazal caecum (Fig. 12.4.1A, B). It is followed by wall formation. The micropylar chamber develops into the endosperm proper, while the uninucleate, tubular chalazal chamber develops

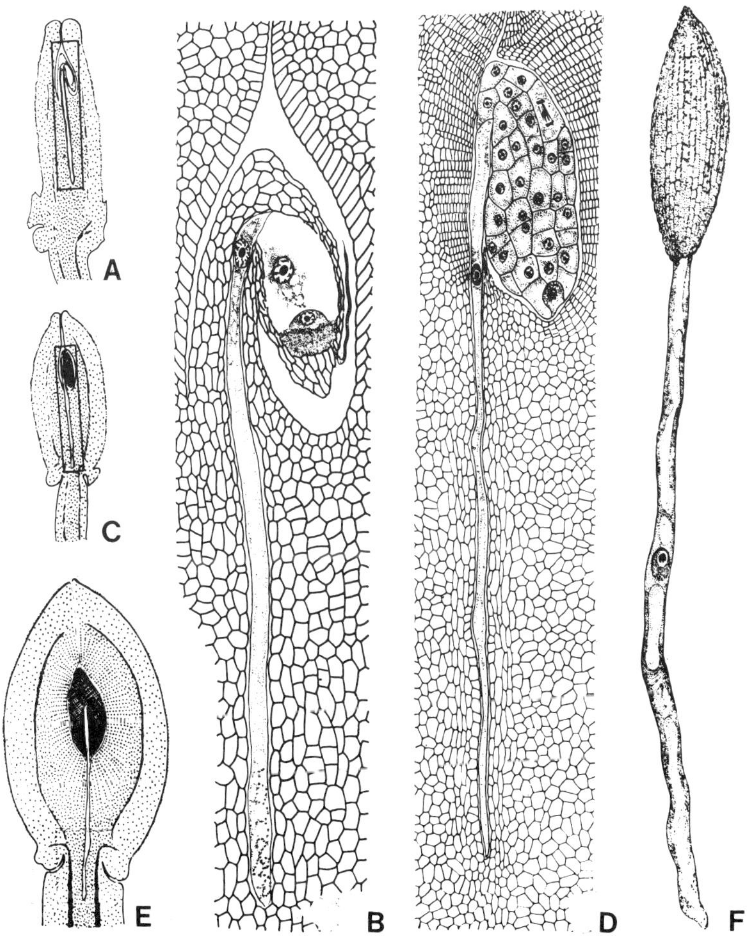

Fig. 12.4.1A–F. *Opilia amentacea.* **A** Median longisection of ovary, outline diagram for **B**. **B** *Inset* from **A**, the embryo sac contains zygote, uninucleate micropylar endosperm chamber, and uninucleate tubular chalazal haustorium. **C** *Outline* for **D**, endosperm *black*. **D** *Inset* from **C**, embryo sac with zygote, cellular endosperm, and uninucleate chalazal haustorium which reaches up to the base of ovary. **E** Longisection of fruit. The chalazal tubular haustorium penetrates as far as the pedicel, endosperm *black*. The ovary wall shows parenchymatous and sclerotic zones. **F** Cellular endosperm and associated uninucleate chalazal haustorium, dissected from young fruit. (After Swamy and Dayanand Rao 1963)

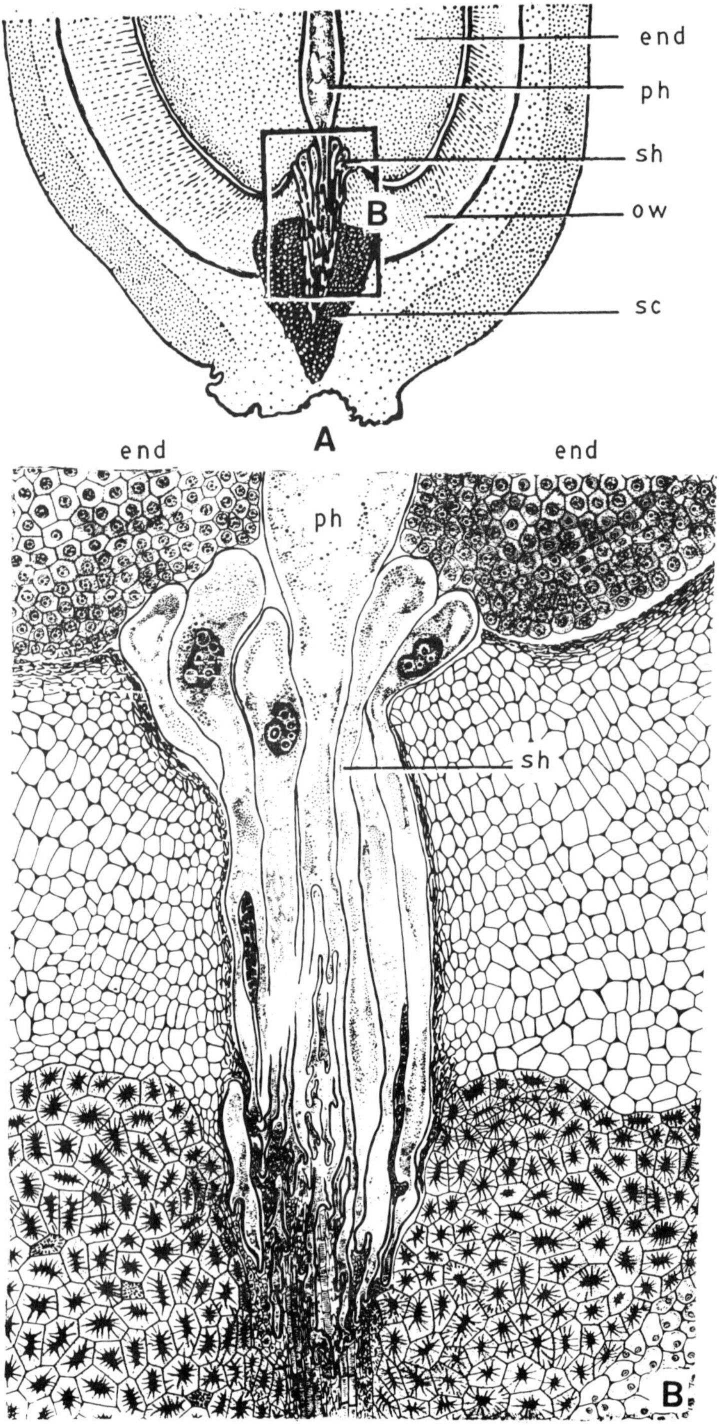
end
ph
sh
B
ow
sc
A
end
end
ph
sh
B

into a haustorium. It elongates parallel to the axis, and penetrates the pedicel (Fig. 12.4.1C–F).

In *Cansjera*, during the maturation of the embryo sac, the solid basal region of the ovary exhibits considerable growth due to meristematic activity. This region also shows a peripheral and axial vascular system. The division of primary endosperm nucleus is followed by a wall resulting in a micropylar and a chalazal chamber (caecum). As the Cellular endosperm consumes the wall of the ovule, the endosperm comes to lie in contact with the inner wall of the ovary. The unicellular chalazal chamber grows intercellularly, and its lower end extends downward up to the axial vascular system. It looks spindle-shaped, and its nucleus becomes hypertrophied. When the chalazal chamber extends up to the point of separation of the peripheral and axial vascular supply, it develops three or four branches which again branch and grow towards the pedicel (Fig. 12.4.2A). The endosperm cells adjoining the primary endosperm haustorium (chalazal chamber) also develop into tubular, branched secondary haustoria. The ultimate branches of both primary and secondary haustoria penetrate and ramify in the axial vasculature in the basal part of the ovary. These haustoria destroy the primary walls of the adjoining cells.

The lateral growth of the haustoria is limited by the development of a zone of sclereids in the basal region of the ovary (Fig. 12.4.2B).

The zygote divides transversely. A suspensor is present.

The seed is albuminous. The seed-coat is thin-walled and gets crushed.

F Fagerlind (1948), M Ram (1970b), S Shamanna (1955), BGL Swamy (1960), BGL Swamy and J Dayanand Rao (1963)

12.5 Grubbiaceae

The embryology has not been fully worked out.

The pollen grains are three-colpate and two- or three-celled at the shedding stage.

Fagerlind (1947d) studied *Grubbia tomentosa*. The ovule is anatropous, unitegmic and tenuinucellate with a long micropylar canal. The sporogenous cell functions as megaspore mother cell. The latter undergoes meiotic divisions, and produces a linear tetrad. The chalazal megaspore develops into a Polygonum type of embryo sac which elongates and comes to lie in the micropylar canal. The presence of an endothelium is reported.

The massive cellular endosperm, at the cotyledonary stage of the embryo, shows chalazal and micropylar haustoria. The epidermis of the endosperm is cuticularized. The reserve food in endosperm and embryo is proteinacous. The embryo occupies the entire length of seed.

Fig. 12.4.2A, B. *Cansjera rheedii*. **A** Longisection of lower portion of fruit to show endosperm (*end*), primary (*ph*) and secondary haustoria (*sh*); *ow* ovary wall; *sc* sclerenchymatous pad. **B** *Inset* from **A**. (After Swamy 1960)

S Carlquist (1977, 1978), F Fagerlind (1947d), BM Johri and SP Bhatnagar (1960), M Ram (1970b, c), Ph van Tieghem (1897)

12.6 Santalaceae

The anther wall comprises an epidermis, fibrous endothecium, two to three middle layers, and multinucleate tapetum. A tapetum with uninucleate cells is reported in *Osyris* (P.C. Joshi 1960). The epidermis of the anther wall is persistent and cutinized in *Santalum*. In *Comandra* the endothecial thickenings are feebly developed, whereas in *Thesium* they show a stellate pattern. The tapetum is multilayered in *Osyris* and *Santalum*. Meiotic divisions in the microspore mother cells followed by simultaneous cytokinesis due to furrowing result in tetrahedral, isobilateral, linear and T-shaped tetrads.

Pollen grains are two-celled at the shedding stage in *Comandra*, *Exocarpus* (M. Ram 1957, 1959a), *Leptomeria* (M. Ram 1959b), *Osyris* (P.C. Joshi 1960), *Thesium* (S.P. Bhatnagar and Agarwal 1961) and *Quinchamalium* (B.M. Johri and Agarwal 1965). They are three-celled in *Santalum* (Paliwal 1956), *Leptomeria* (M. Ram 1959b) and *Mida* (S.P. Bhatnagar 1960).

Cytomixis was observed between the microspore mother cells. Polynucleate pollen mother cells are reported in *Santalum obtusifolium* (S.P. Bhatnagar 1966). Sometimes, the number of microspores in a tetrad may be fewer or more than four.

An interesting feature in *Leptomeria billardierii* (M. Ram 1959c) is the formation of eight-nucleate embryo sac-like structures, originating from the sporogenous cells in the anther.

The ovule may be pendulous, straight, hemianatropous, anatropous or orthotropous. There is no distinction between the nucellus and integument, except in *Comandra*, *Iodina* and *Thesium* where the ovule is unitegmic. In *Exocarpus* there is no organized placenta, and the ovule does not differentiate (see Fagerlind 1959, S.P. Bhatnagar and Joshi 1965).

The archesporial cell functions directly as megaspore mother cell. The latter undergoes meiotic divisions and produces a linear tetrad. Usually, the chalazal megaspore functions and gives rise to a Polygonum type of embryo sac, except in *Buckleya* (B. Raj 1964) and *Iodina* (S.P. Bhatnagar and Sabharwal 1966, 1968) where it is of the Allium type. Endymion type of development is reported in *Buckleya* (B. Raj 1964).

The embryo sac remains confined to the ovule in *Choretrum* (see S.P. Bhatnagar 1970), *Iodina* (S.P. Bhatnagar and Sabharwal 1966), *Osyris* (Paliwal 1956, P.C. Joshi 1960), *Scleropyrum* (L.N. Rao 1942) and *Thesium* (Paliwal 1956, S.P. Bhatnagar and Agarwal 1961).

The formation of extra-ovular prolongations of the embryo sac is a noteworthy feature of the family. In *Comandra umbellata* (M. Ram 1957), a lateral caecum originates in the micropylar region of the embryo sac, on the funicular side of the ovule, and grows into the long and twisted placenta (Fig. 12.6.1A, B). Even at the four-nucleate stage the chalazal end of the embryo sac elongates, and grows beyond the ovule. Finally, the caecum comes

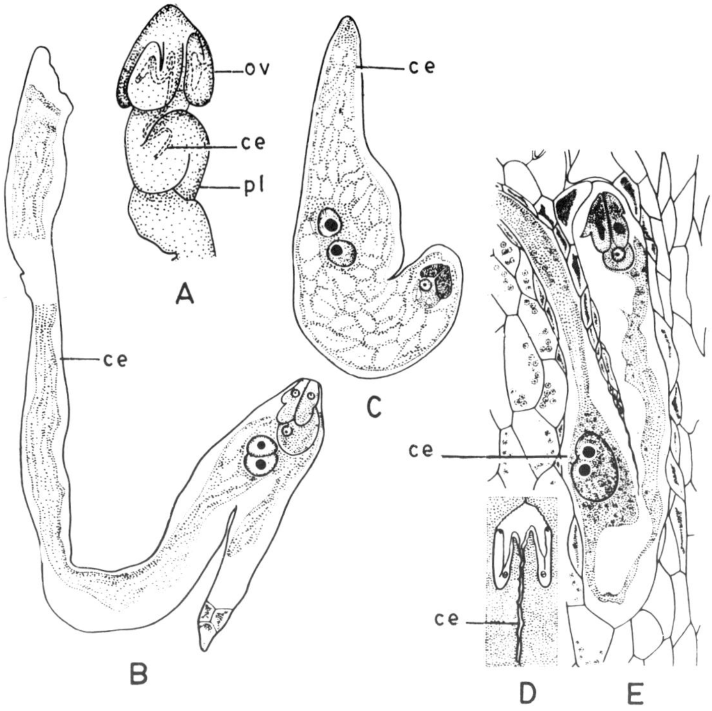

Fig. 12.6.1A, B. *Comandra umbellata* (wholemount). **A** Placenta (*pl*), *dotted lines* indicate the extension of the embryo sac; *ce* caecum; *ov* ovule. **B** Mature embryo sac with a well-developed lateral caecum. (After M. Ram 1957). **C–E** *Leptomeria cunninghamii.* **C** Embryo sac with caecum (wholemount). **D** Portion of ovary and placenta to show extension of embryo sac (*ce*). **E** Portion of embryo sac from **D**, with egg apparatus and polar fusion nucleus. (After M. Ram 1957)

to lie in the ovarian cavity (Fig. 12.6.1C–E) in *Leptomeria* (Ram 1959b), *Mida* (S.P. Bhatnagar 1960), and *Quinchamalium* (Agarwal 1961a, B.M. Johri and Agarwal 1965).

The placental column is straight and elongated with pendulous ovules (Fig. 12.6.2A) in *Santalum album* (Sedgley 1982), and the embryo sac develops a chalazal caecum which grows into the placenta leaving the antipodals in situ. The development of endosperm is of the Helobial type. Free-nuclear divisions occur in the micropylar chamber followed by cellularization. The chalazal chamber remains uninucleate throughout (Fig. 12.6.2B–D).

In *Exocarpus*, the development of the embryo sac is of the Polygonum type with an eight-nucleate gametophyte. The embryo sac develops lateral finger-like haustorial extensions in *E. menziesii* (Fig. 12.6.3A). In a mature embryo sac the growth in the apical region is rather extensive (Fig. 12.6.3B). In the fertilized embryo sac of *E. strictus* (Ram 1959a), the haustoria are produced in the middle region (Fig. 12.6.3C).

Due to the apical growth of the gametophyte, the epidermal cells of the "placenta" are crushed and the embryo sac becomes exposed (Fig. 12.6.4A–C).

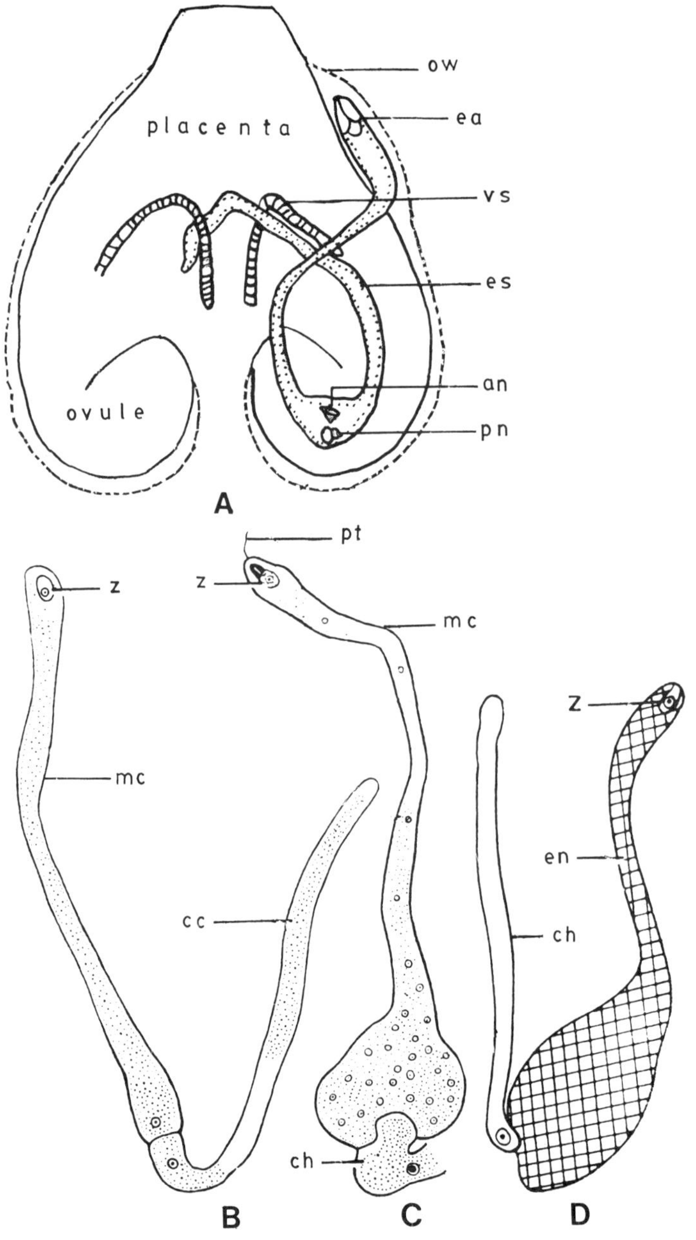

Fig. 12.6.2A. *Santalum acuminatum*, longisection (semi-diagrammatic) of placenta with one embryo sac; *ow* inner ovary wall; *ea* egg apparatus; *vs* vascular supply; *es* embryo sac; *an* antipodals; *pn* polar nuclei. (After Sedgley 1982). **B–D** *S. album*. **B** Embryo sac with zygote (*z*) and two-celled endosperm-micropylar (*mc*) and chalazal chamber (*cc*). **C** Upper part of embryo sac with zygote (*z*), and coenocytic micropylar chamber (*mc*) and portion of uninucleate chalazal haustorium (*ch*); *pt* remnants of pollen tube. (After S.P. Bhatnagar 1959). **D** Later stage, cellular endosperm (*en*) and uninucleate chalazal haustorium. (After B.M. Johri and Bhatnagar 1969)

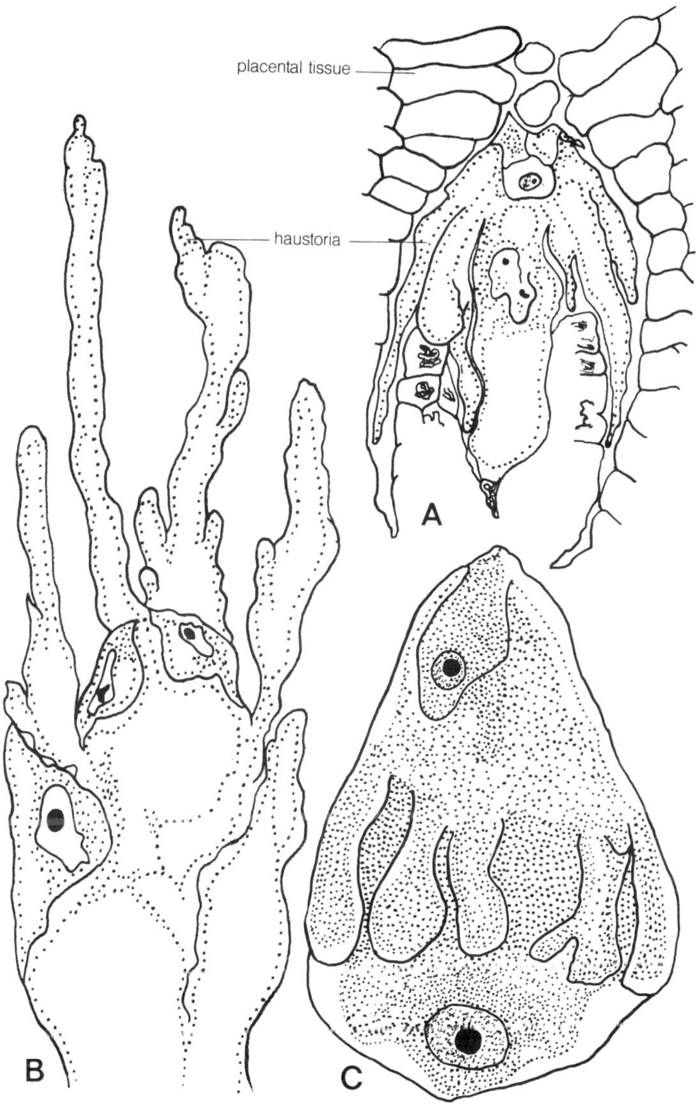

Fig. 12.6.3A, B. *Exocarpus menziesii.* **A** Organized embryo sac with finger-like haustoria growing downward; **B** Later stage. There is profuse development of haustoria growing upward. (After Fagerlind 1959). **C** *E. sparteus*, embryo sac with haustoria from the middle region. (After M. Ram 1959a)

The development of endosperm and young embryo is shown in Figure 12.6.4B–H.

The synergids are generally hooked, except in *Quinchamalium* where the tips elongate and protrude beyond the embryo sac, even before fertilization. They form extensive haustoria (Figs. 12.6.5A–D, 12.6.6A–C) penetrating up to

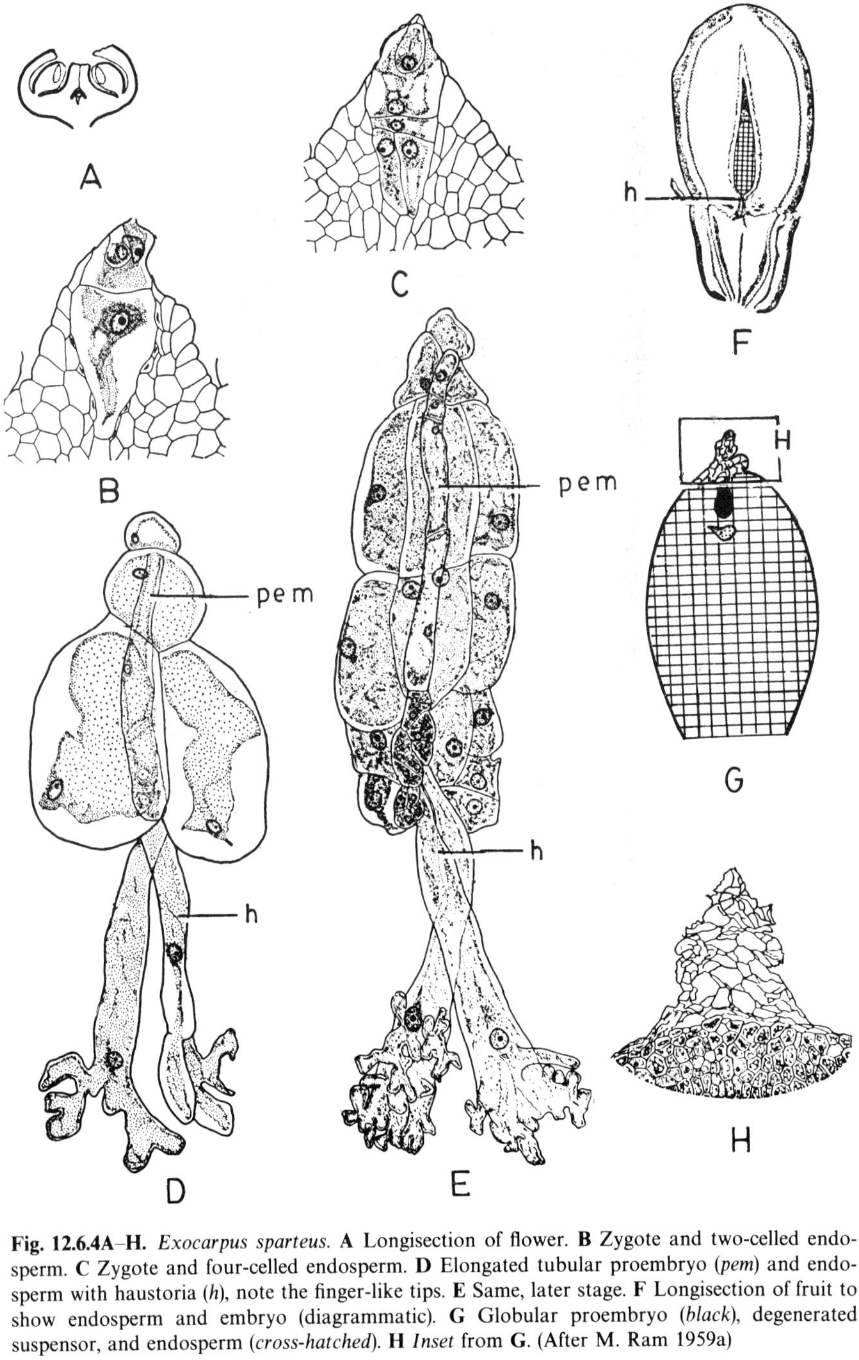

Fig. 12.6.4A–H. *Exocarpus sparteus.* **A** Longisection of flower. **B** Zygote and two-celled endosperm. **C** Zygote and four-celled endosperm. **D** Elongated tubular proembryo (*pem*) and endosperm with haustoria (*h*), note the finger-like tips. **E** Same, later stage. **F** Longisection of fruit to show endosperm and embryo (diagrammatic). **G** Globular proembryo (*black*), degenerated suspensor, and endosperm (*cross-hatched*). **H** *Inset* from **G**. (After M. Ram 1959a)

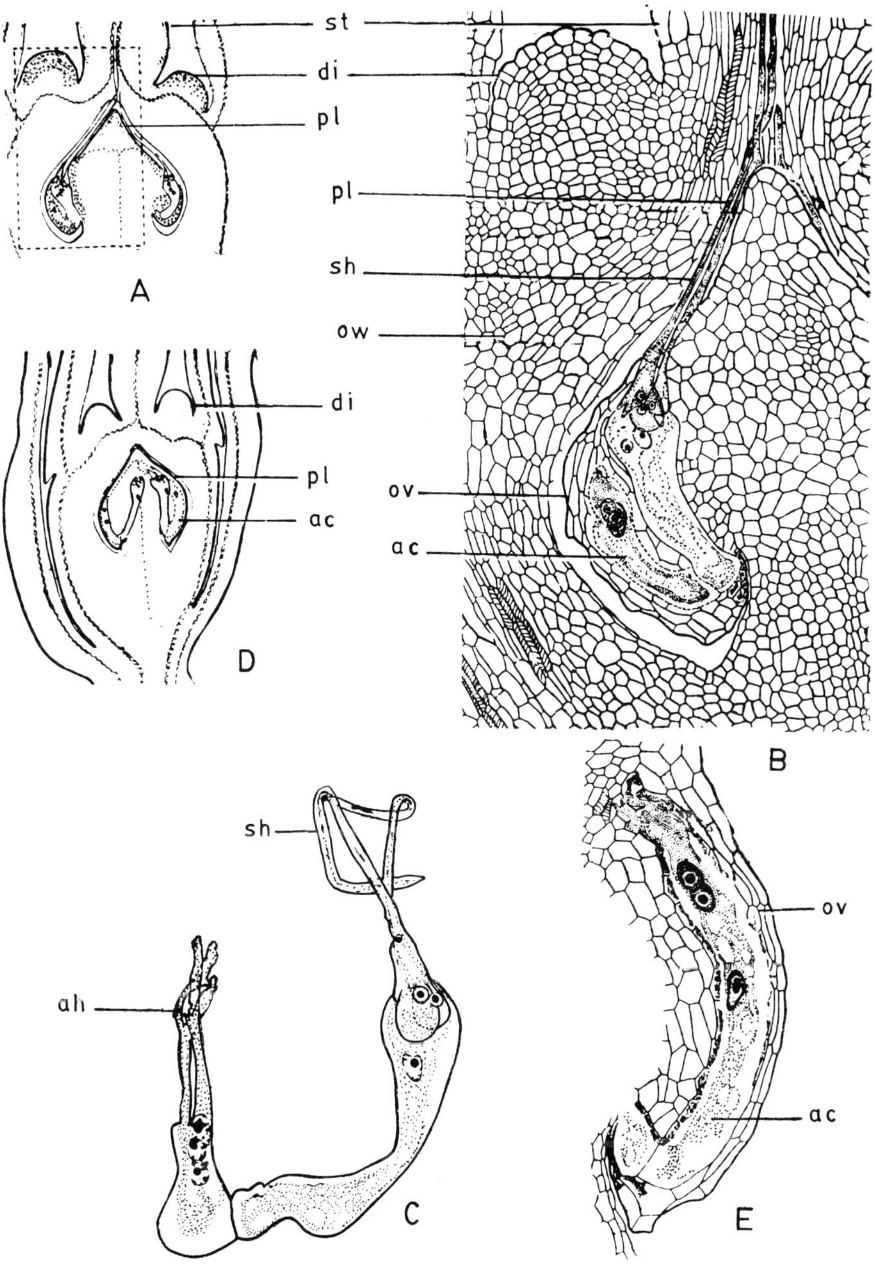

Fig. 12.6.5A–E. *Quinchamalium chilense.* **A** Portion of ovary (*l.s.*) to show conical placenta (*pl*) bearing two ovules; *st* style; *di* disc. **B** *Inset* from **A**. The tip of embryo sac has become extra-ovular and grown along the outer margin of the placenta (*pl*). The synergid haustoria (*sh*) reach up to the base of style (*st*), and the antipodal chamber (*ac*) extends into the funiculus; *ow* ovary wall; *ov* ovule. **C** Mature embryo sac to show extensive synergid (*sh*) and antipodal (*ah*) haustoria (wholemount). **D** Longisection of ovary showing two embryo sacs. **E** Antipodal chamber (*ac*) from **D**. (After Agarwal 1962)

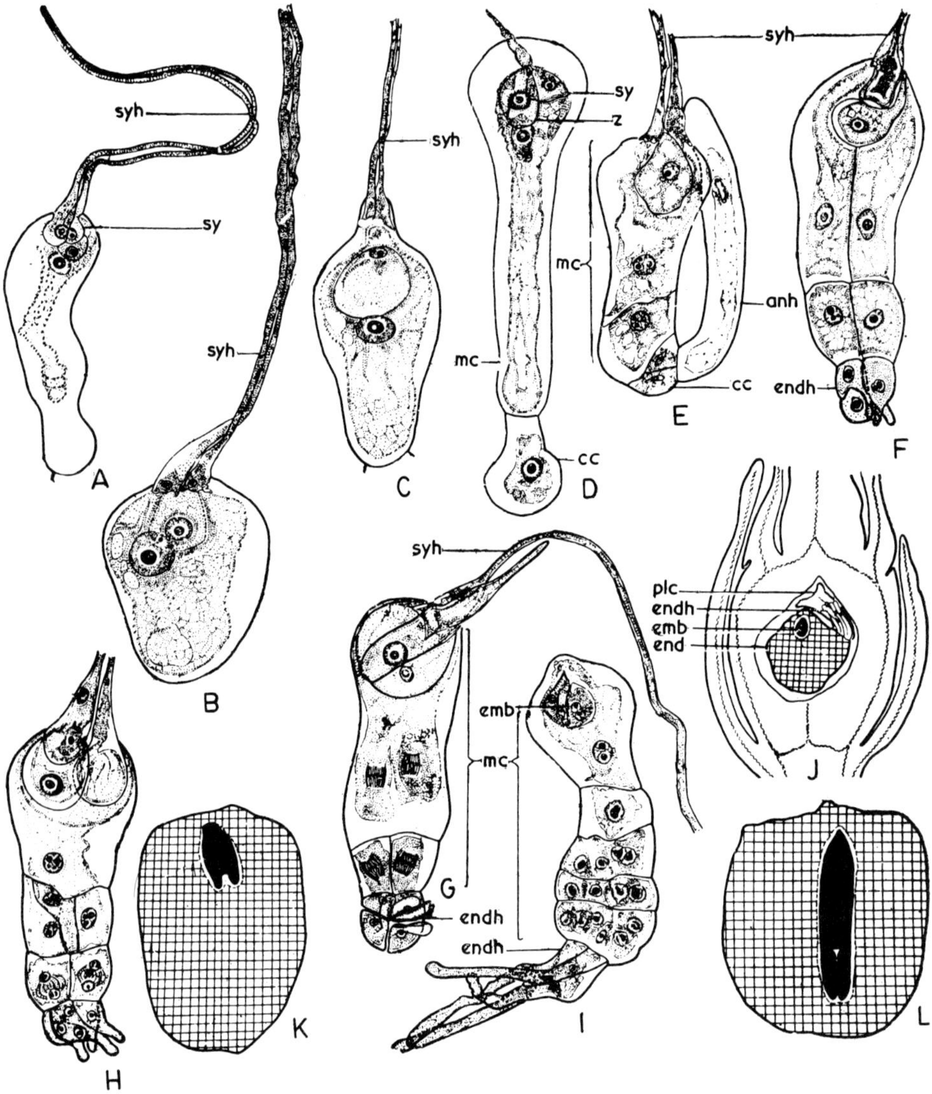

Fig. 12.6.6A–L. *Quinchamalium chilense.* (**E**, **J**–**L** from microtome sections, rest from whole-mounts. The remnants of antipodal haustoria have been omitted in **A**–**D**.) **A**–**C** Fertilized embryo sac with synergid haustoria (*syh*). The nuclei of synergids (*sy*) were not seen in **C**. **D** Two-celled endosperm; *z* zygote; *mc* micropylar chamber; *cc* chalazal chamber. **E** Two-celled micropylar chamber, the chalazal chamber has divided into four cells. Note the persistent antipodal haustorium (*anh*). **F**–**I** Progressive development of endosperm (*end*) and chalazal endosperm haustoria (*endh*). **J** Longisection of ovary to show endosperm at globular proembryo (*emb*). The endosperm haustoria extend through the funicle; *plc* placenta. **K**, **L** Endosperm at early and late dicotyledonous embryo. (After Agarwal 1962)

one-third the length of style (Agarwal 1961a, 1962). In *Comandra* the synergids are ephemeral.

The polar nuclei fuse before fertilization. The entry of pollen tube is porogamous. Persistent pollen tubes have been observed in *Santalum* (S.P. Bhatnagar 1959).

The formation of a chalazal caecum, leaving the antipodals in situ, is also common. The three antipodals are generally ephemeral, but persist in *Mida* and *Santalum*. In *Quinchamalium* a three-nucleate antipodal chamber elongates and develops into a haustorium (Fig. 12.6.5E) which invades the placental tissue. In *Scleropyrum*, L.N. Rao (1942) reported supernumerary free antipodal nuclei and their migration into the chalazal chamber. A lateral caecum, extending along the vascular supply, is reported in *Comandra*.

The development of endosperm is of the Cellular type, except in *Mida* where it is Helobial. Generally, the division of primary endosperm nucleus results in the formation of micropylar and chalazal chamber. The former

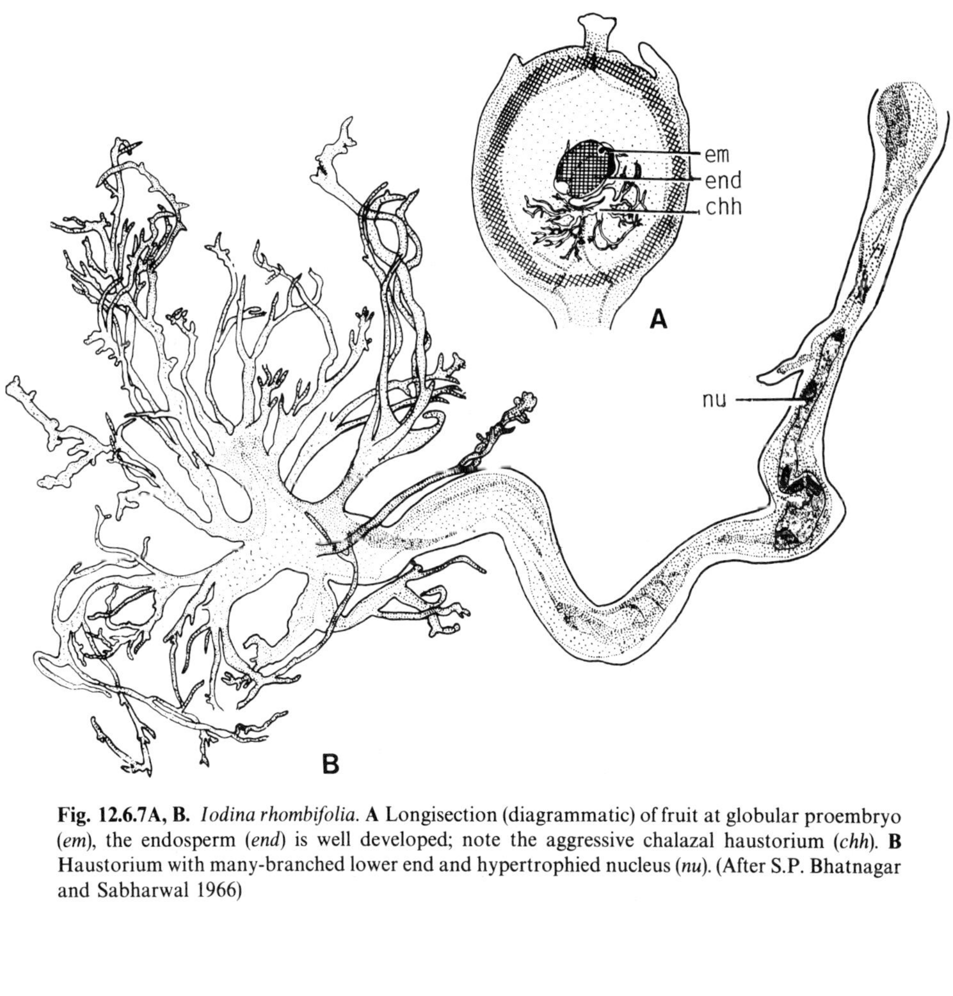

Fig. 12.6.7A, B. *Iodina rhombifolia.* **A** Longisection (diagrammatic) of fruit at globular proembryo (*em*), the endosperm (*end*) is well developed; note the aggressive chalazal haustorium (*chh*). **B** Haustorium with many-branched lower end and hypertrophied nucleus (*nu*). (After S.P. Bhatnagar and Sabharwal 1966)

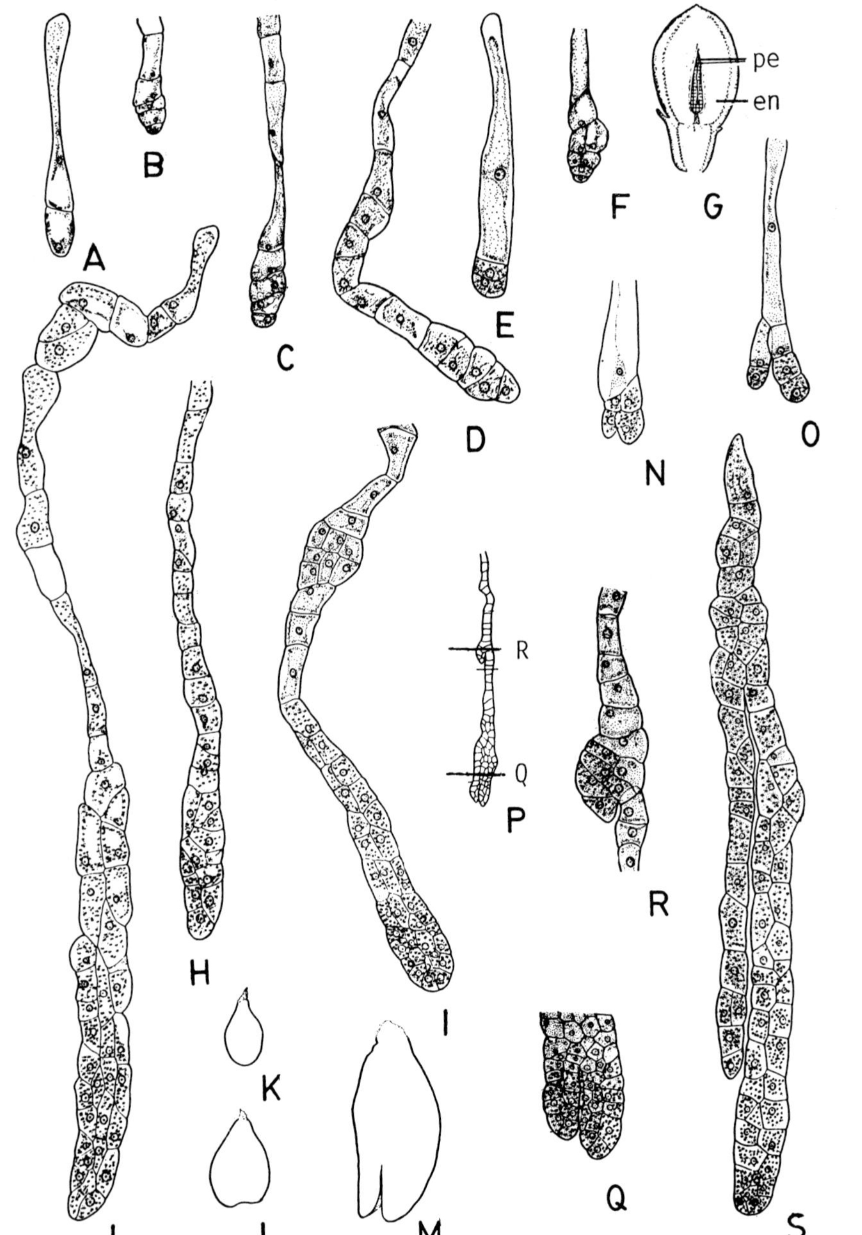

Fig. 12.6.8A–S. *Exocarpus sparteus*, embryogeny. **A** Two-celled proembryo. **B–D** Uniseriate proembryos. **E** Transverse division in terminal cell. **F** Multicelled apical region. **G** Longisection (diagrammatic) of fruit showing endosperm (*en*) and proembryo (*pe*). **H** Proembryo from **G**. **I**, **J** Progressive stages of proembryo, terminal region shows small richly cytoplasmic cells, and suspensor comprises elongated vacuolate cells. **K–M** Globular (**K**), heart-shaped (**L**) and dicotyledonous (**M**) embryo (diagrammatic). **N–S** Polyembryony. **Q**, **R** Portions marked *Q* and *R* in **P**. **S** Both proembryos are long and filamentous. (After Ram 1959a)

develops into the endosperm proper, and the latter forms a haustorium (Fig. 12.6.6D). In *Mida* the chalazal chamber forms a uninucleate haustorium. Free-nuclear divisions in the micropylar chamber are followed by wall formation.

The lateral caecum acts as a primary haustorium in *Comandra*. It cuts off a small cell and its derivatives produce secondary haustoria. Three to four cells, formed in the same manner in *Mida*, elongate and function as secondary haustoria. In addition, the micropylar chamber of endosperm produces numerous finger-like processes towards the base.

The chalazal chamber divides by a vertical wall in *Exocarpus* (Fig. 12.6.4C). In *E. sparteus* the daughter cells produce numerous lobed outgrowths at the base (Fig. 12.6.4D, E). In *E. cupressiformis* the haustorium becomes multicelled.

In *Quinchamalium* four cells at the base of endosperm develop into independent haustoria (Fig. 12.6.6E–I). They elongate and extend along the course of antipodal haustorium of the embryo sac, reaching up to the apex of placenta (Agarwal 1962). The endosperm proper is well developed (Fig. 12.6.6J–L).

A profusely branched, aggressive, coralloid haustorium with a highly hypertrophied nucleus is reported in *Iodina* (Fig. 12.6.7A, B). The nucleus increases to almost 20-times its original size. In *Thesium* (Erbrich 1965) the nucleus becomes highly polyploid (12 n-384) due to endomitosis.

The zygote divides transversely, except in *Comandra* and *Quinchamalium* where the division is vertical. A suspensor may or may not develop. Further divisions are variable. Figure 12.6.8 shows the development of embryo (A–M), and cleavage (N–S) and suspensor polyembryony (P, R) in *E. sparteus*.

The additional embryo in *Mida*, *Santalum* and *Quinchamalium* arises from the zygote of twin embryo sacs.

The pericarp directly surrounds the endosperm which consumes the endocarp, and the stony mesocarp gives protection. In *Quinchamalium*, as the ovary wall is consumed, the persistent perianth functions as "pericarp". In *Iodina* the persistent perianth becomes an additional protective layer.

S Agarwal (1961a, 1962), SP Bhatnagar (1959, 1960, 1965a, 1966, 1970), SP Bhatnagar and S Agarwal (1961), SP Bhatnagar and PC Joshi (1965), SP Bhatnagar and G Sabharwal (1966, 1968), P Erbrich (1965), F Fagerlind (1959), F Gagnepain and E Boureau (1947), GS Iyengar (1937), BM Johri (1962a, b), BM Johri and S Agarwal (1965), BM Johri and SP Bhatnagar (1960, 1969, 1970), PC Joshi (1960), RL Paliwal (1956), R Pilger (1935), B Raj (1964, 1969), M Ram (1957, 1959a, b, c), LN Rao (1942), M Sedgley (1982), FH and EC Smith (1942), HU Stauffer (1957), Ph van Tieghem (1896)

12.7 Myzodendraceae

The unigeneric monotypic family, Myzodendraceae, has fascinated botanists for their curious structure and reproductive biology. *Myzodendron* is essentially a dioecious genus; Skottsberg (1913) reports bisexual flowers. The staminate flowers, without a calyx and corolla (Hutchinson 1973), have two or three stamens. The bisporangiate (monothecous) anther (Fig. 12.7.1A, B) dehisces by terminal slits.

The pistillate flower has a three-lobed stigma on the top of the superior (not inferior as generally reported) ovary. The calyx is adnate to the ovary and free at the top. There are no petals and staminodes. Within the ovary, there are three reduced ovules at the apex of a short placental column (Fig. 12.7.1A, G–I), each hanging down into a pocket of the ovarian cavity. The upper part of each pocket becomes confluent with the others so that the ovary is trilocular below and unilocular above (Kuijt 1969).

The embryology has not been studied in sufficient detail.

The pollen grains are spheroidal, spinuliferous, and polyporate. The 3 to 19 apertures are scattered randomly over the entire surface. The ektexine is represented only by spines and a thin granular layer between them. Endexine is thick in the region bordering the pore, and extends into the pore (Feuer 1981).

The tenuinucellate ovule is highly reduced and undifferentiated (ategmic), and is represented only by the nucellus (Fig. 12.7.1C–F). With the growth of the endosperm, the scanty nucellus is completely crushed (Fig. 12.7.1G–I). This organization is not very different from that in Santalaceae and Olacaceae.

The archesporial cell functions directly as megaspore mother cell. The latter undergoes meiotic divisions, and the chalazal megaspore (Fig. 12.7.1C) of a linear or T-shaped tetrad develops into a Polygonum type of embryo sac (Fig. 12.7.1D, E). The polar nuclei fuse before fertilization; the three antipodals are ephemeral.

In *Myzodendron punctulatum* (Skottsberg 1913), the chalazal end of the embryo sac forms a prominent tubular caecum which extends into the placental column (Fig. 12.7.1F).

The development of the endosperm is of the Cellular type. The first cross-wall in the caecum is formed following the division of the primary endosperm nucleus. With the development of the endosperm proper, uniseriate cells are also formed in the haustorium (caecum; Fig. 12.7.1H, I). The haustorium branches, extends into the funicle and reaches the pedicel of the flower through the placenta (Fig. 12.7.1H, J). The development of endosperm is comparable to that in Olacaceae.

According to Johnson (1889), at first the chalazal haustorium consists of a uniseriate row of cells. Due to the dissolution of the walls, the haustorium becomes coenocytic. Skottsberg (1913) reports only a uninucleate chalazal haustorium which branches at the tip (see M. Ram 1970d). In view of the contradictory reports, a detailed reinvestigation is necessary.

The development of embryo has not been studied. It is reported to lack a suspensor. Figure 12.7.2 shows a globular proembryo (A), mature embryo (B), and disposition of embryo in the endosperm with a prominent haustorium (C).

Only one of the three ovules in an ovary matures into a seed, and it is lifted up by the elongation of the placental column.

The fruit is a three-angled or winged small achene. The naked (without testa) seed has a fleshy chlorophyllous endosperm, and a fleshy central embryo (Fig. 12.7.2C).

The seedling has a curious structure. It shows a small thick-walled cup

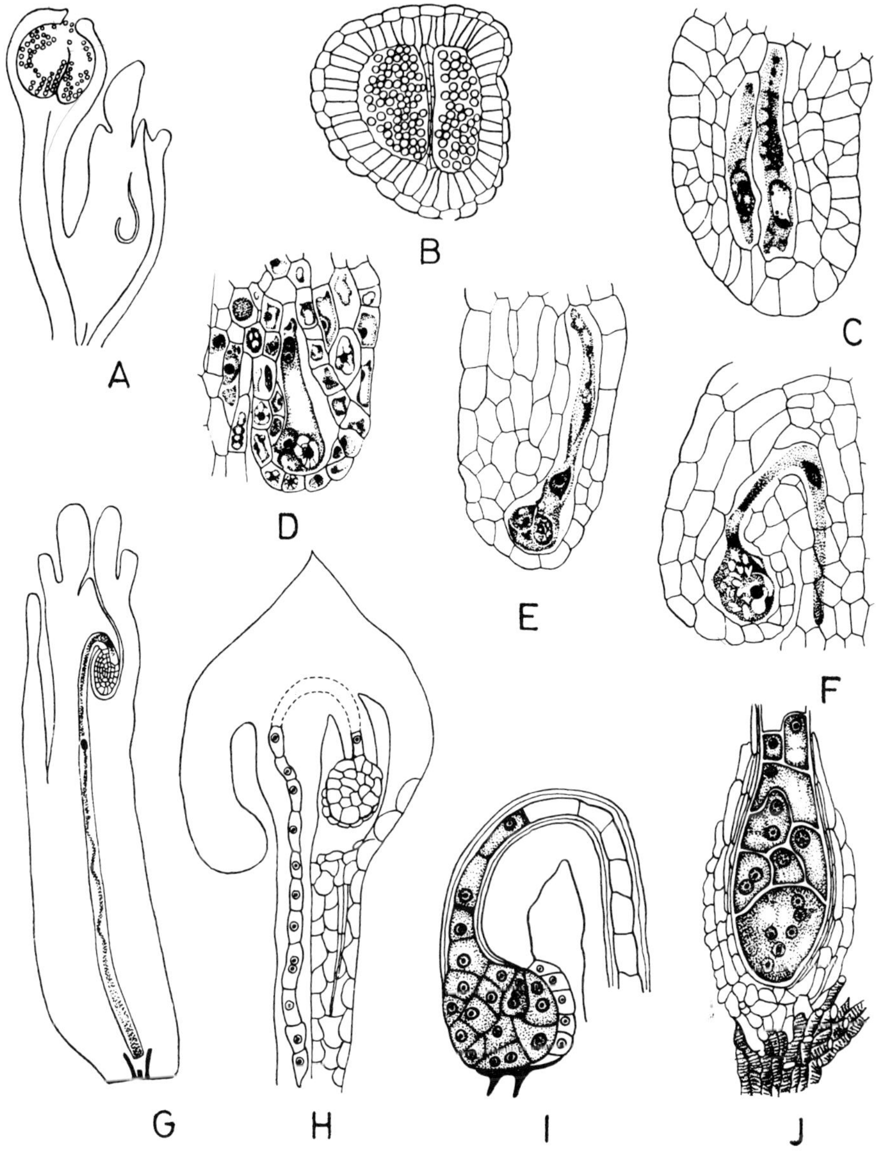

Fig. 12.7.1.A *Myzodendron linearifolium*, longisection of flower; stamen with bisporangiate anther, and gynoecium with superior ovary and undifferentiated (ategmic) pendulous ovule (nucellus only). **B** *M. recurvum*. Anther with pollen, the wall shows the epidermis and endothecium. **C** *M. quadriflorum*. Nucellus with two functional chalazal megaspores. **D–J** *M. punctulatum*. **D** Embryo sac with egg apparatus and polar nuclei, antipodals have degenerated; **E** Later stage, secondary nucleus lying close to egg apparatus; **F** Fertilized embryo sac with zygote, primary endosperm nucleus in chalazal caecum; **G** Zygote, endosperm proper, and unseriate endosperm cells in caecum/haustorium; **H** Gynoecium, the haustorium extends through the placenta up to the base of the pedicel; **I** Placenta, later stage of endosperm than in **G**; **J** Tip of chalazal haustorium close to the base of vascularized pedicel, the endosperm cells are uni-, bi- and multinucleate. (**A–F**, **I** After Skottsberg 1913, **G**, **H** and **J** after Johnson 1889)

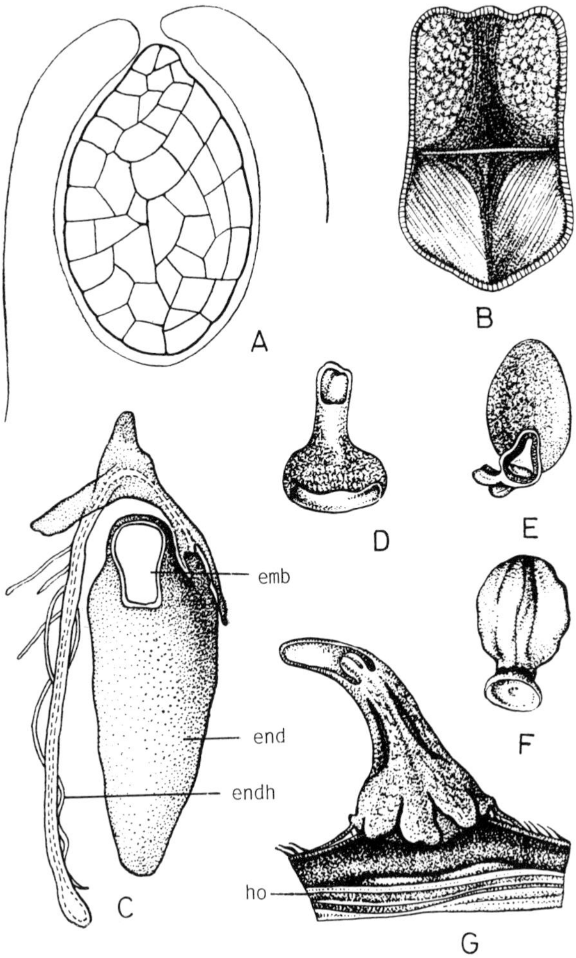

Fig. 12.7.2A–C. *Myzodendron punctulatum*, **A**, **B** Globular proembryo (**A**) and differentiated (mature?) embryo (**B**); **C** Embryo (*emb*) and endosperm (*end*) with conspicuous haustorium (*endh*). **D–G** *M. brachystachyum*. **D** Germinated embryo with perforated trumpet-shaped radicular end, and cotyledonary tube (hypocotyl) with first pair of leaves at the apex of notched tube; **E** Fruit, embryo as in **D**; **F** Fruit, embryo at later stage than in **E**. Note the cup-shaped adhesive membrane; **G** Seedling attached to host (*ho*) stem. The shoot has not yet emerged from the cotyledonary tube. (**A–C** After Johnson 1889, **D–G** after Hooker 1846)

(radicular pole) with a short stem (Feuer 1981). The stem is really the hypocotyl with a tubular slightly notched tip which completely encloses the first two foliage leaves (Fig. 12.7.2D, E, G). The cotyledonary tube (formed by the fusion of the two cotyledons) remains in the endosperm for some time. The cotyledonary tube is eventually ruptured when the foliage leaves emerge (Kuijt 1969).

On germination on the host surface, the radicle emerges from the achene, quickly expands, a thin adhesive membrane extends radially, and penetration follows from the axial part of the haustorial disc (Fig. 12.7.2E, F). An early stage of the host and semi-parasite is shown in Figure 12.7.2G.

TB Batygina and MS Yakovlev (1987), S Feuer (1981), JD Hooker (1846), J Hutchinson (1973), T Johnson (1889), BM Johri and SP Bhatnagar (1960), J Kuijt (1969), M Ram (1970d), C Skottsberg (1913), JC Willis (1973)

12.8 Loranthaceae: Loranthoideae

The representatives of Loranthoideae have been studied at the University of Delhi for over 25 years. The chief events in their life history are shown in Figure 12.8.7. B.M. Johri (1987a, see also S.P. Bhatnagar and Johri 1983) has reviewed the reproductive biology of mistletoes (Loranthaceae and Viscaceae).

The anther wall comprises the epidermis, fibrous endothecium, one or two ephemeral middle layers, and glandular tapetum of binucleate cells. The fibrous thickenings in the endothecium are absent in *Amylotheca dictyophleba* (B. Raj 1970a), *Lepeostegeres gemmiflorus* (Dixit 1958b) and *Macrosolen cochinchinensis* (P. Maheshwari and Singh 1952). In *Loranthus* (*Hyphear*) *europaeus* (Pisek 1924), and at times in *Scurrula*, the tapetum is amoeboid. The microspore mother cells undergo meiotic divisions, simultaneous cytokinesis, and produce tetrahedral and decussate tetrads. Because the microspores are triradiate, they are so arranged in a tetrad that all the three tips of each microspore touch the tips of the other microspores.

Polyspory with five or six spores has been observed in *Lysiana exocarpi* (R. Narayana 1958a).

Sterile transverse partitions in pollen sacs have been reported in some members: *Amyema gravis*, *Elytranthe* (Schaeppi and Steindl 1942), *Helixanthera ligustrina* (B.M. Johri and Bhatnagar 1972), *Taxillus cuneatus* (R. Narayana 1955) and *Tapinostemma acaciae* (B.M. Johri and S. Prakash 1965). Finally, the septae break down and the microsporangium becomes continuous.

The pollen grains are triangular with three slightly curved concavo-convex arms. Sometimes, the pollen may be four-armed as in *Barathranthus* (S. Prakash 1963), *Helicanthes* (B.M. Johri et al. 1957), *Lepeostegeres* (Dixit 1958b), *Lysiana* (R. Narayana 1958a) and *Moquiniella* (B.M. Johri and Raj 1969). Spherical pollen grains have been observed in *Atkinsonia* (S. Prakash 1961), *Tupeia* (van Tieghem 1895, Danser 1933, Smart 1952) and *Ixocactus* (Kuijt 1967, 1969). Usually, the pollen grains are tricolpate, but in *Ixocactus* they are tetracolpate (see B.M. Johri and Bhatnagar 1972). They are two celled at the dispersal stage.

B.M. Johri (1956) reports that in *Helixanthera ligustrina* some of the microspore mother cells fail to undergo meiosis, and form large restitution nuclei. Then an exine develops (within the original wall of the mother cell) simulating a pollen grain. In a few cases such pollen grains also showed a pollen tube. Other microspore mother cells in these microsporangia developed normally.

Ovary-Ovule Complex

B.M. Johri and Bhatnagar (1972) discuss the morphology, structure and disposition of placenta (mamelon), from four-lobed condition in four-chambered ovary (*Lysiana*) to complete absence of the placenta (*Helixanthera*). B.M. Johri and Ambegaokar (1984) and B.M. Johri (1987b) reconsider the status of the ovary-ovule complex under six categories (Fig. 12.8.1A_1–F_2).

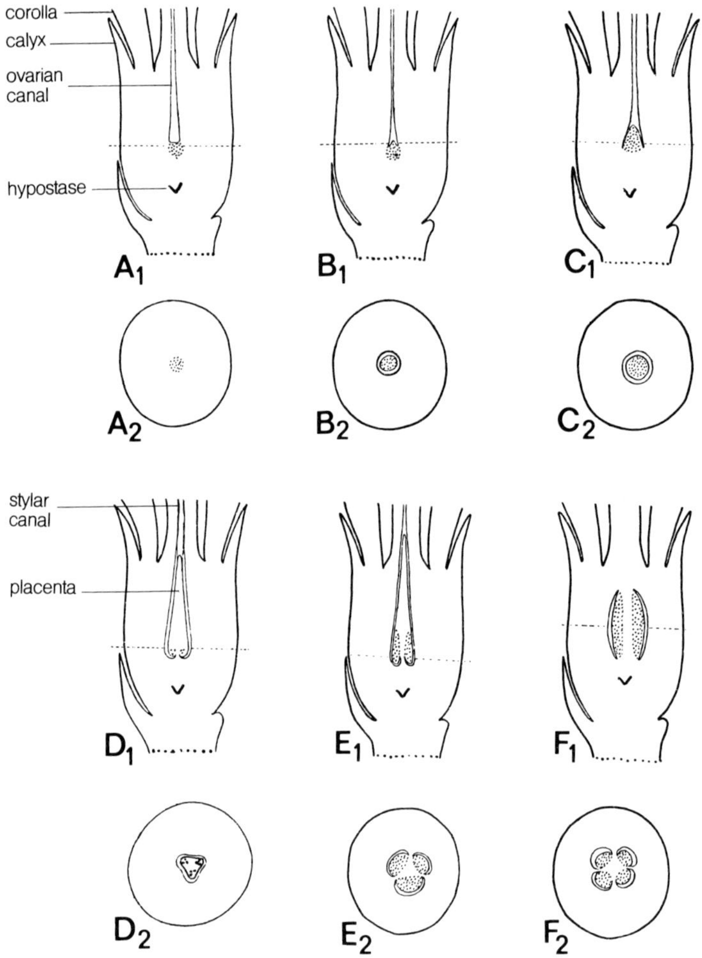

Fig. 12.8.1A_1–F_2. Loranthoideae, longi- and transections (diagrammatic) of ovaries, level of transections indicated by a *dotted line* in longisections. **A_1, A_2** *Helixanthera ligustrina*, sporogenous tissue (*stipped*) at the base of stylar canal. **B_1, B_2** *Amyema preissii*, sporogenous tissue lifts the base of the ovarian canal. **C_1, C_2** *Helicanthes elastica*, sporogenous tissue in an unlobed mamelon. **D_1, D_2** *Macrosolen cochinchinensis*, sporogenous cell/s in each of the three basal lobes of mamelon in an unilocular ovary. **E_1, E_2** *Nuytsia floribunda*, sporogenous tissue in each lobe of the mamelon in a trilocular ovary. **F_1, F_2** *Lysiana exocarpi*, sporogenous tissue in each lobe in a tetralócular ovary. (Modified after R. Narayana 1958a, Kapil and Vasil 1963)

1) Placenta absent, archesporium differentiates in the subepidermal region below the base of the stylar canal: *Helixanthera ligustrina* (Fig. 12.8.1A_1, A_2), *Dendrophthoe falcata* (B. Singh 1952a), *Barathranthus axanthus* (S. Prakash 1963), *Moquiniella rubra* (B.M. Johri and Raj 1969) and *Struthanthus vulgaris* (Venturelli 1981a).

2) Placenta transitory, archesporium differentiates as in category 1, but the base of stylar canal is pushed up as a result of elongation of sporogenous tissue: *Dendrophthoe pentandra*, *Scurrula montana* (Schaeppi and Steindl 1942), *D. neelgherrensis* (R. Narayana 1956), *Amyema miquelii* (Fig. 12.8.1B_1, B_2), *Tolypanthus involucratus* (Dixit 1961) and *Struthanthus vulgaris* (Venturelli 1981a). In *A. miquelii* (Dixit 1954), the mamelon shows tracheidal thickenings in some cells. In *A. miquelii* (Dixit 1954), the mamelon shows tracheidal thickenings in some cells.

3) Placenta prominent and unlobed, free from the carpellary wall, ovary completely unilocular, archesporium differentiates in the entire subepidermal region of the placenta: *Amyema gravis* (Schaeppi and Steindl 1942), *Helicanthes elastica* (Fig. 12.8.1C_1, C_2), *Tapinostemma acaciae* (B.M. Johri and Prakash 1965).

4) Placenta three-lobed at the base only, free from the ovary wall, ovary completely unilocular, archesporium differentiates in each lobe of placenta: *Macrosolen cochinchinensis* (Fig. 12.8.1D_1, D_2).

5) Placenta three-lobed at the base, fused with the ovary wall in between the lobes, free at the tip, ovary trilocular below and unilocular above, archesporium differentiates in each lobe: *Nuytsia floribunda* (Fig. 12.8.1E_1, E_2), *Lepeostegeres gemmiflorus* (Dixit 1958b), *Amylotheca dictyophleba* (Raj 1970a), *Tripodanthus acutifolius* and *T. flagellaris*.

6) Placenta four-lobed throughout, fused with the ovary wall in between the lobes, except at the tip where it is free, ovary tetralocular, archesporium differentiates in each lobe: *Lysiana exocarpi* (Fig. 12.8.1F_1, F_2), *Peraxilla tetrapetala*.

In contrast, *Phthirusa pyrifolia* (Kuijt and Weberling 1972) lacks the ovarian cavity which is characteristic of the family. The archesporium differentiates just above the collenchymatous pad (hypostase). Such a condition is not known in any other member of Loranthaceae, and *Phthirusa* needs to be reinvestigated.

Cocucci (1983, see also Cocucci and Venturelli 1982) does not accept the interpretation of mamelon as placenta. He refers to the ovules as "discrete" and "collective". Discrete ovules are equated to the lobes of mamelon, and unlobed mamelon with a massive sporogenous tissue to a collective ovule. In our opinion the equivalence of mamelon to a placenta is fully justified.

Distinct ategmic ovules comprise *only* the nucellus (Fig. 12.8.1$D_{1,2}$, $E_{1,2}$, $F_{1,2}$). The concept of a collective ovule is erroneous, since in the Loranthaceae there are no indications as if a number of ategmic ovules had fused to form a common structure, the so-called collective ovule.

The archesporium is uni- or multicellular (*Amyema*, *Lysiana*, *Peraxilla*), function/s directly as megaspore mother cells which undergo meiotic divisions and produce linear tetrads. Sometimes ⊥-shaped tetrads occur in *Amyema preissii* (Dixit 1958b). The chalazal megaspore develops into a Polygonum type of embryo sac (Figs. 12.8.2B, C, E, 12.8.7J).

Sometimes, the subchalazal megaspore develops (*A. preissii*). An ovary may show archesporial and megaspore mother cells, tetrads (*A. preissii*), and two- and four-nucleate embryo sacs (*L. exocarpi* – R. Narayana 1958a).

The polar nuclei fuse below the egg apparatus as in *Tapinostemma acaciae* (B.M. Johri and S. Prakash 1965). The antipodals may be ephemeral or persistent (Figs. 12.8.4F, 12.8.5D). Occasionally, only two cells organize, one binucleate.

The formation of thick-walled saucer-shaped (collenchymatous pad) or tubular (collenchymatous tube) hypostase is a characteristic feature of Loranthoideae (Figs. 12.8.1 longisections, 12.8.2A, F, 12.8.5B, G, 12.8.6A, F, 12.8.7K′–P).

The elongation of embryo sacs is noteworthy (Fig. 12.8.2F). The lower end of the embryo sac extends up to the base (caecum) of hypostase, leaving the antipodals in situ (Fig. 12.8.4F). In *Nuytsia floribunda* (R. Narayana 1958b), the lower end of the embryo sac pierces the hypostase and grows beyond it. The upper end extends up to the base of style in *Amylotheca*, *Atkinsonia* (Fig. 12.8.2A), *Elytranthe*, *Macrosolen* (P. Maheshwari and Singh 1952) and *Peraxilla*, and slightly above the base of style in *Lepeostegeres* (Dixit 1958b). It grows to almost half- or two-thirds the height of style in *Amyema* (Dixit 1958a), *Dendrophthoe* (B. Singh 1952a), *Helicanthes* (B.M. Johri and Agrawal 1954), *Lysiana* (R. Narayana 1958a), *Nuytsia* (R. Narayana 1958b), *Struthanthus* (Venturelli 1984c), *Tolypanthus* (Dixit 1961) and *Tripodanthus* (Venturelli 1983). It reaches the stigma in *Barathranthus* (S. Prakash 1963), *Helixanthera hookeriana* (Schaeppi and Steindl 1942), *Scurrula* (Agrawal 1954), *Tapinostemma* (B.M. Johri and S. Prakash 1965) and *Tupeia* (Smart 1952). In *Helixanthera ligustrina* (Fig. 12.8.2C), it comes to lie just below the stigmatic epidermis (Fig. 12.8.2F). After reaching the apex of the stigma in *Moquiniella* (B.M. Johri and Raj 1965, 1969), the tips of embryo sacs curve backward into the style (Fig. 12.8.2D). Some of the embryo sacs, however, remain arrested in the style (Fig. 12.8.2F). In different genera of Loranthoideae, the tips of embryo sacs grow to varying heights in the style and stigma (Fig. 12.8.2F). There is no correlation in the disposition and structure of the mamelon and

Fig. 12.8.2. A *Atkinsonia ligustrina*, longisection of ovary, mature embryo sacs. **B** *Amyema miquelii*, tip of Polygonum type of embryo sac with egg apparatus and upper polar nucleus (at two-thirds the height of style). The tip of embryo sac in *Tripodanthus* also reaches up to two-thirds the height of the style. **C** *Helixanthera ligustrina*, longisection of stigma with tips of five unfertilized and fertilized embryo sacs, one reaches as far as the stigmatic epidermis. One of the embryo sacs shows two-celled proembryo. **D** *Moquiniella rubra*, longisection of stigma with tips of four embryo sacs (es_1–es_4), three reach up to the stigma and curve backward so that the egg apparatus/proembryo acquires reverse polarity. Two embryo sacs show zygote and one four-celled proembryo. **E** *Nuytsia floribunda*, tip of embryo sac in the stylar tissue. Note the lateral caecum. **F** Longisection of carpel. The names of different taxa indicate the height to which the tip of the embryo sac reaches in the style and stigma. (**A** After S. Prakash 1961, **B**, **F** after B.M. Johri and Bhatnagar 1972, **C** after P. Maheshwari and Johri 1950a, **D** after B.M. Johri and Raj 1965, **E** after R. Narayana 1958b)

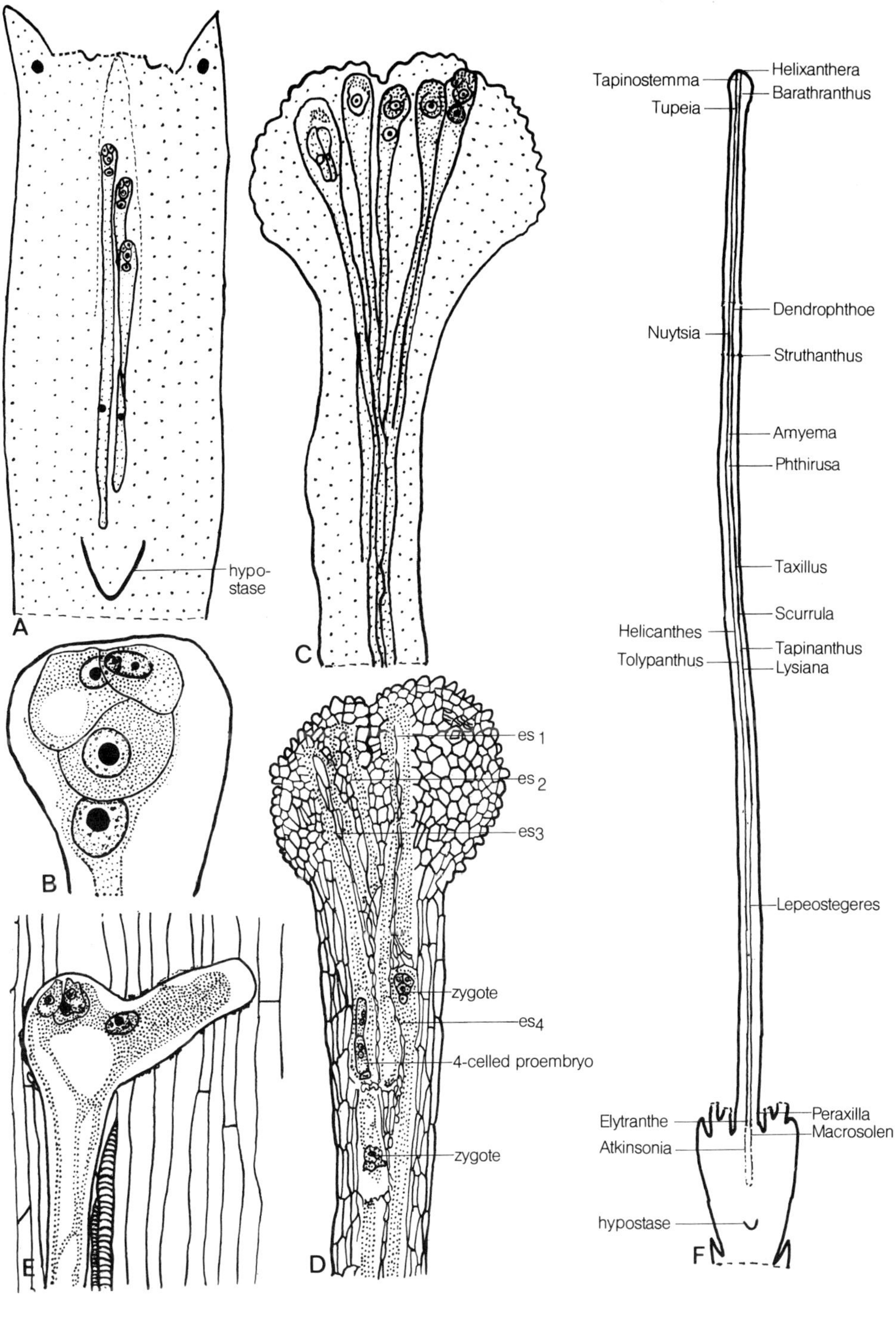
hypo-
stase
A
B
C
es 1
es 2
es3
zygote
es4
4-celled proembryo
zygote
D
E
Tapinostemma
Tupeia
Helixanthera
Barathranthus
Dendrophthoe
Nuytsia
Struthanthus
Amyema
Phthirusa
Taxillus
Scurrula
Helicanthes
Tapinanthus
Tolypanthus
Lysiana
Lepeostegeres
Elytranthe
Peraxilla
Macrosolen
Atkinsonia
hypostase
F

length of the embryo sac. This feature is not known in any other angiosperm. The significance, evolutionary or functional, is not clear.

The formation of six-nucleate embryo sacs, due to earlier division of the two chalazal nuclei of a four-nucleate embryo sac, is quite common. Several embryo sacs develop simultaneously in the same ovary. They digest the contents of the adjoining tissue rich in starch. In *Nuytsia* (Fig. 12.8.2E) and *Atkinsonia* (S. Prakash 1961), a lateral caecum develops at the tip of embryo sac, invading the circumjacent tissue.

The primary endosperm nucleus travels down to the lower part of the embryo sac. Its division is followed by a wall (P. Maheshwari and Singh 1952). The development of endosperm is of the Cellular type. The endosperm formation occurs in several embryo sacs in the same ovary (Fig. 12.8.3A); due to the growth of individual endosperm the ovarian tissue is destroyed. The endosperms from different embryo sacs fuse to form a composite structure (Fig. 12.8.3B).

The collenchymatous pad (hypostase) is often shifted to a lateral position due to the downward growth of the endosperm, as in *Tolypanthus lagenifer* (Dixit 1961).

In *Tapinostemma* (B.M. Johri and Prakash 1965), the basal end of the endosperm produces finger-like haustorial processes (Fig. 12.8.4F). These processes invade the adjoining tissue, and sometimes penetrate even through the thick-walled cells of the hypostase (Fig. 12.8.4I). Such a development has not been observed in any other member of Loranthoideae.

The only exception is *Struthanthus vulgaris* (Fig. 12.8.5A, C–G, I), where a composite structure is not formed and the endosperm develops from only one

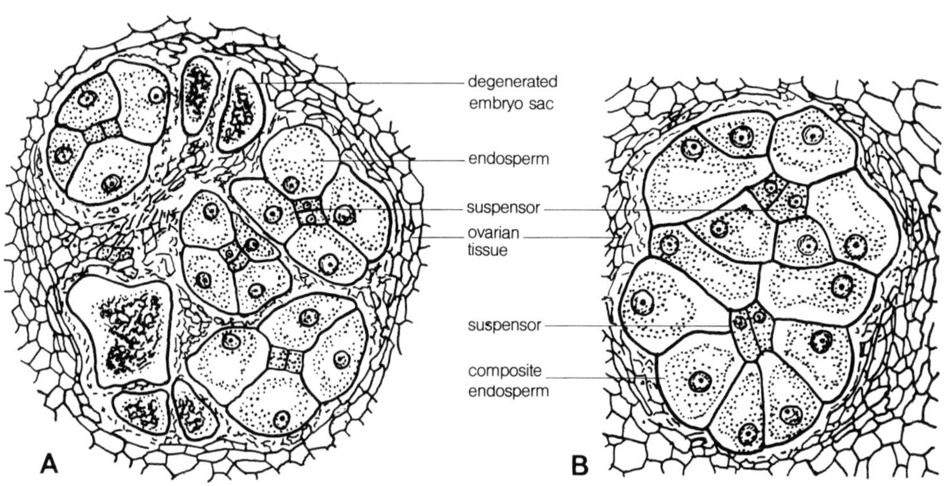

Fig. 12.8.3A, B. *Tolypanthus involucratus*. **A** Transection of ovary with four embryo sacs each containing four-seriate endosperm and biseriate suspensor (*t.s.*) of proembryos. Five embryo sacs have degenerated. **B** Later stage, composite endosperm with two biseriate suspensors. (After Dixit 1961)

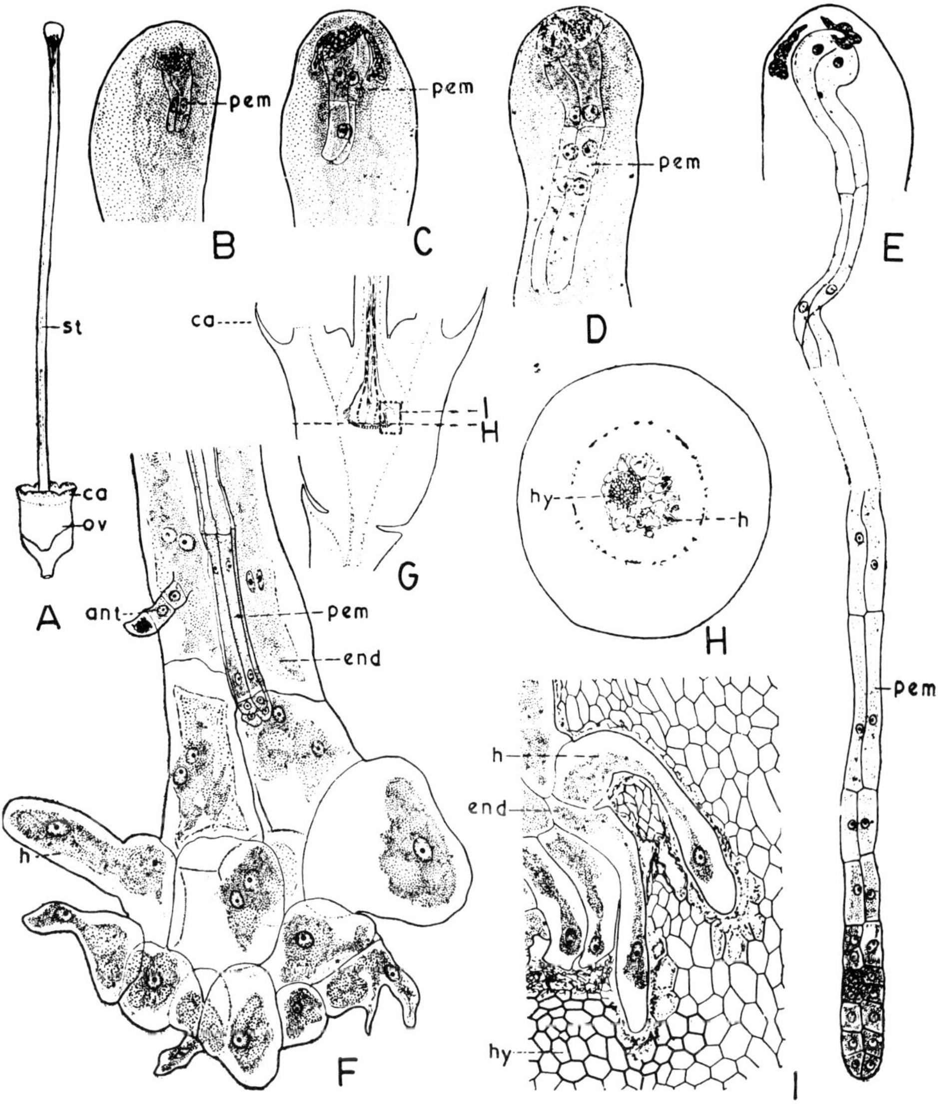

Fig. 12.8.4A–I. *Tapinostemma acaciae.* (**G–I** from microtome sections, others dissected wholemounts.) **A** Gynoecium (semi-diagrammatic); the tips of embryo sacs extend into the stigma (not shown); *st* style; *ca* calyx; *ov* ovary. **B–D** Tips of embryo sacs from stigma, two-, four- and eight-celled proembryos (*pem*). **E** Biseriate proembryo with a long suspensor and embryonal mass at the tip. **F** Lower part of endosperm (*end*), cells with haustorial (*h*) processes. The tip of the proembryo is in the upper middle part of the endosperm; *ant* persistent antipodal cells. **G** Longisection of ovary to show lower end of several embryo sacs. **H** Cross-section of ovary at level *H* marked in **G. I** *Inset I* from **G**, haustorial (*h*) extensions of endosperm cells; *hy* hypostase. (After B.M. Johri and Prakash 1965)

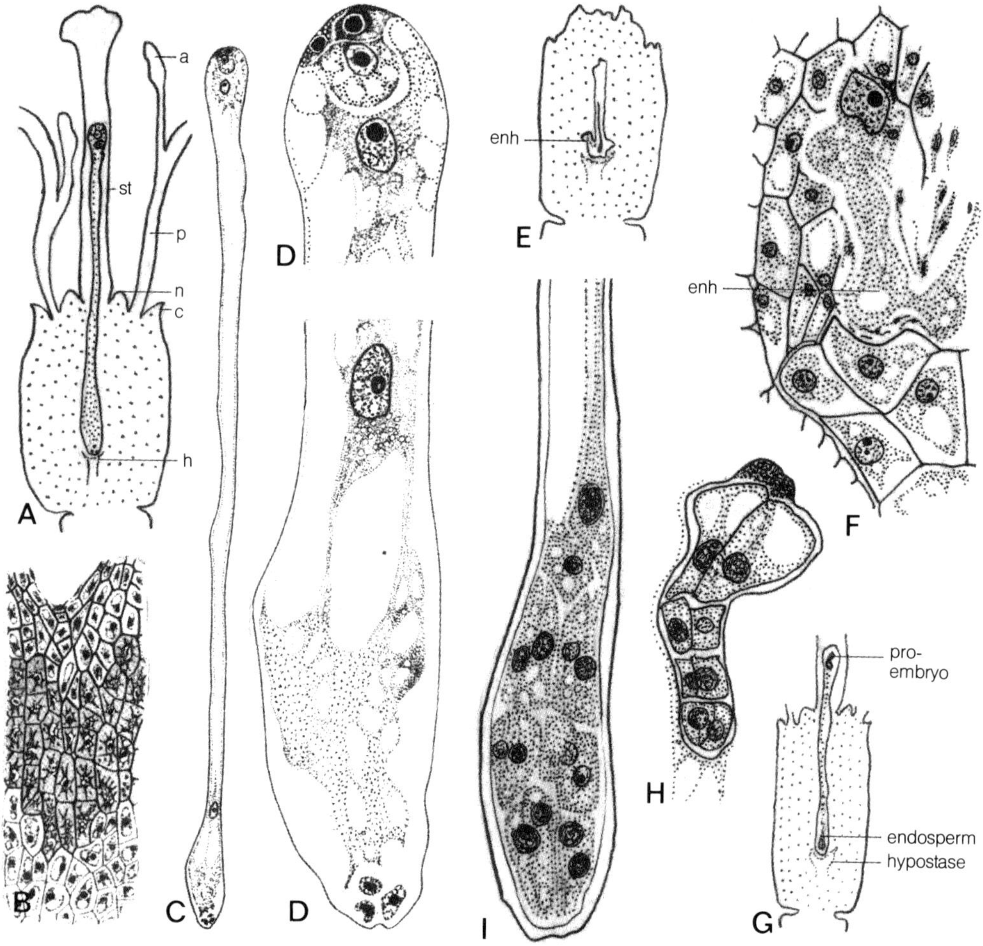

Fig. 12.8.5A–I. *Struthanthus vulgaris.* **A** Longisection of flower to show elongation of embryo sacs in style; *a* anther of epipetalous stamen; *st* style; *p* petal; *n* nectary; *c* calyculus; *h* hypostase. **B** Hypostase with thick-walled cells, from **A**. **C** Eight-nucleate embryo sac from **A**. **D**, **D′** Micropylar (**D**) and chalazal (**D′**) region from **C**. **E** Ovary with an embryo sac containing endosperm (formed from a *single* embryo sac) with lateral haustorium (*enh*). **F** Endosperm haustorium from **E**. **G** Portion of ovary with style to show an embryo sac with biseriate proembryo, and cellular endosperm. **H** Proembryo from **G**. **I** Endosperm from **G**. (After Venturelli 1981a)

embryo sac as in other angiosperms. This feature needs to be re-examined even though the author confirms that her findings are correct (pers. comm. 1987).

The division of zygote is vertical (Fig. 12.8.4B) (Griffith 1843), and transverse divisions follow to form a long biseriate proembryo (Figs. 12.8.4B–E, 12.8.5G, H). The terminal tier of the proembryo undergoes repeated divisions forming an embryonal mass. Several proembryos develop simultaneously in the style, but only a few grow down to the ovary. Normally, a single embryo reaches maturity.

With continued growth of the suspensor, it elongates, grows through the endosperm, and reaches the ovary. Figure 12.8.3A shows four suspensors (t.s.), one in each endosperm, B shows two suspensors in a composite endosperms. In Figure 12.8.4F, a proembryo has entered the composite endosperms (l.s.). The endosperm and the proembryo/s grow concurrently.

When the embryonal mass reaches the base of the endosperm, it gradually develops into a globular, heart-shaped and dicotyledonous embryo (Figs. 12.8.6A, D, 12.8.7M–P). With the progressive growth and upward shift of the embryo, the suspensor/s become coiled.

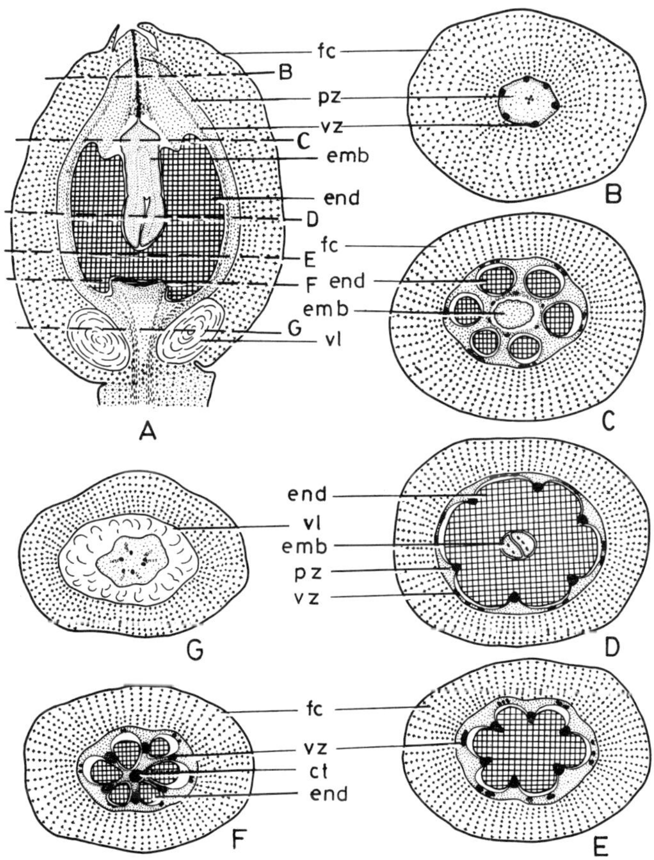

Fig. 12.8.6A–G. *Lepeostegeres gemmiflorus.* **A** Longisection of fruit, the pericarp shows fleshy coat (*fc*), parenchymatous zone (*pz*), vascular zone (*vs*), and viscid layer (*vl*) restricted only to the base. The endosperm (*end*) encloses the embryo (*emb*) except at the radicular end. The *black thick line* in the middle and at the base of endosperm represents the collenchymatous tube (*ct*, *black* central circle in **F**). **B–G** Transections of fruit at respective levels marked in **A**. **B** Pericarp; **C** Six apical prongs (*cross-hatched*) of endosperm, and radicular end of embryo (*emb*); **D** Six-lobed endosperm with vascular supply in between each lobe. Note the two cotyledons in the middle. **E, F** Cross-sections at levels below the embryo; **F** shows six separate lobes of endosperm; **G** Note the circular viscid layer (*vl*). (After Dixit 1958b)

In a mature embryo, the radicular end (= hypocotyl; a true radicle is absent) projects beyond the endosperm (Figs. 12.8.6A, C, 12.8.7P). The radicular end is generally studded with papillate projections and hairs (*Lysiana exocarpi* – R. Narayana 1958a). *Amylotheca dictyophleba, Amyema pendula, Dendrophthoe neelgherrensis* and *Helicanthes elastica* also show similar outgrowths (see B.M. Johri and Bhatnagar 1972).

The mature embryo may show two separate cotyledons, or the cotyledons may become closely appressed to each other (except in the region of plumule). Such a condition has been designated as "pseudomonocotyledonous", and is reported in *Amyema, Barathranthus, Dendrophthoe, Helicanthes, Helixanthera, Lysiana, Scurrula, Tapinostemma* and *Tolypanthus* (see B.M. Johri and Bhatnagar 1972).

Maturity of the endosperm is accompanied by several conspicuous features. At the base it grows peripherally around the hypostase. The lateral growth brings about a lobed condition due to the growth of endosperm around the vascular strands. At the apex, the endosperm grows in between the vascular strands. Thus, the mature endosperm looks like a flower vase with a number of lobes and apical prongs corresponding to the number of vascular strands (Fig. 12.8.6A–G). The endosperm cells contain tannin and starch.

Venturelli (1981a) has formulated a reproductive calendar (as in Brazil) for *Struthanthus vulgaris* (Fig. 12.8.7A–P). In an elegant study, she emphasizes the correlation between the size of bud, flower and ovary, and temporal development of microsporogenesis (Fig. 12.8.7A–E), megasporogenesis (Fig. 12.8.7G–I), male (Fig. 12.8.7E, F) and female gametophyte (Fig. 12.8.7J, K), pollination (Fig. 12.8.7E′, K′), fertilization (Fig. 12.8.7K, L), endosperm and embryo (Fig. 12.8.7M–P).

The male and female buds appear in January, bloom in April, followed by entomophilous pollination in April and fertilization in May and June. The endosperm, embryo and fruit develop from July to November.

The fruit is a pseudodrupe in *Atkinsonia* and *Gaiadendron*, in other members it is a pseudoberry. The seed is "naked" since it lacks a testa, and the endosperm is directly surrounded by the pericarp (Figs. 12.8.6A–E, 12.8.7M–P).

The pericarp is distinguishable into (a) an outer fleshy zone followed by (b) the viscid zone, (c) a parenchymatous zone, and (d) a vascular zone (Fig. 12.8.6A–G). In *Lepeostegeres* (Fig. 12.8.6A, G) and *Peraxilla* (B.M. Johri and Bhatnagar 1972), the viscid zone is restricted to the base of the fruit only.

The birds pick the fruit, try to crush it and in this process the viscin (from viscid layer enclosing the endosperm and the embryo) sticks to the beak. In its efforts to get rid of the tenaciously sticking viscin, the bird causes wounds on the bark and the endosperm sticks to the bark. On germination, the radicular end produces a discoid hold-fast and establishes itself on the host. The plumular leaves also develop, and finally the seedling.

Further studies of the uninvestigated genera, especially from Africa, Australia and South America, would be very rewarding.

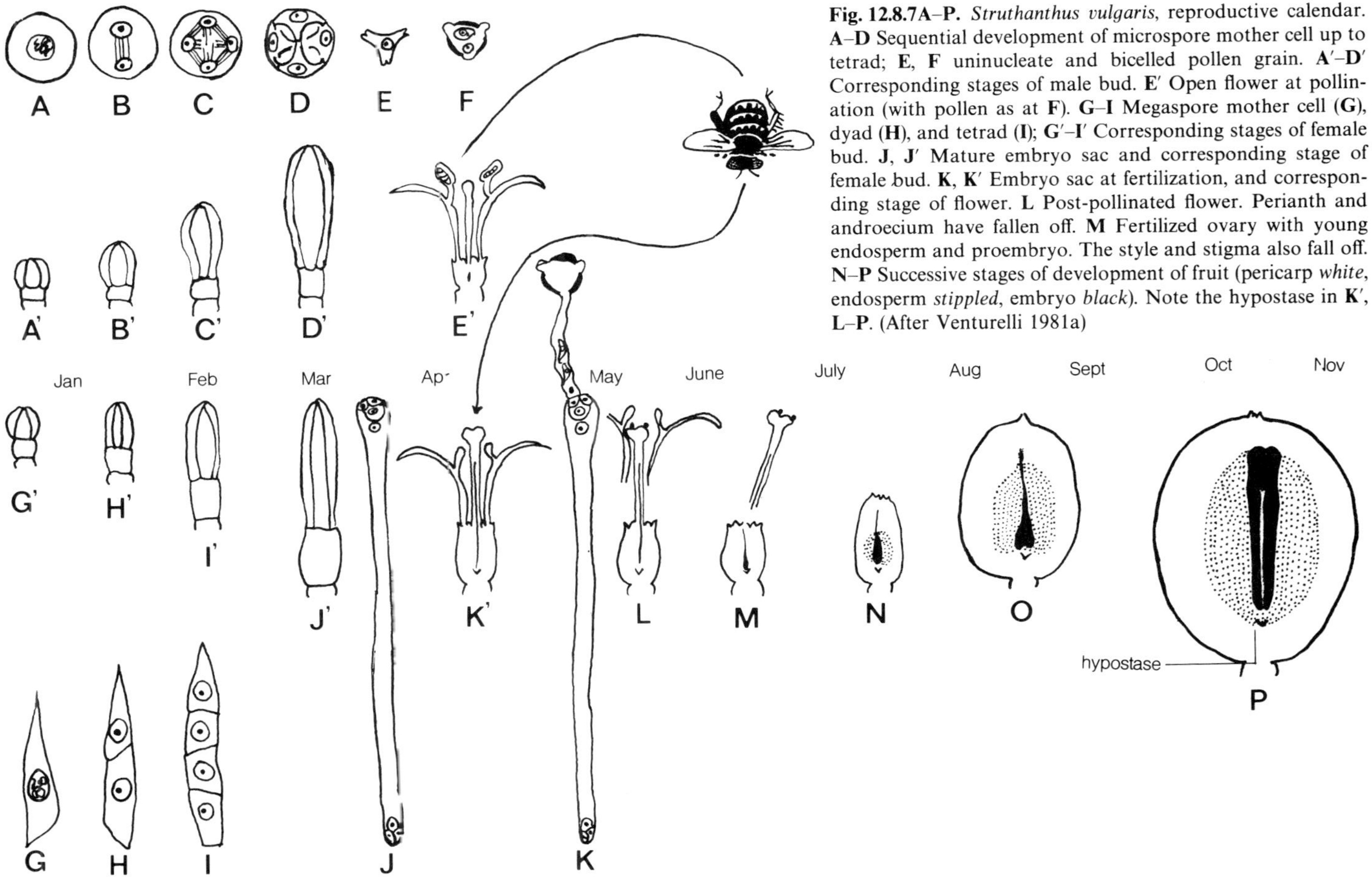

Fig. 12.8.7A–P. *Struthanthus vulgaris*, reproductive calendar. **A–D** Sequential development of microspore mother cell up to tetrad; **E, F** uninucleate and bicelled pollen grain. **A′–D′** Corresponding stages of male bud. **E′** Open flower at pollination (with pollen as at **F**). **G–I** Megaspore mother cell (**G**), dyad (**H**), and tetrad (**I**); **G′–I′** Corresponding stages of female bud. **J, J′** Mature embryo sac and corresponding stage of female bud. **K, K′** Embryo sac at fertilization, and corresponding stage of flower. **L** Post-pollinated flower. Perianth and androecium have fallen off. **M** Fertilized ovary with young endosperm and proembryo. The style and stigma also fall off. **N–P** Successive stages of development of fruit (pericarp *white*, endosperm *stippled*, embryo *black*). Note the hypostase in **K′**, **L–P**. (After Venturelli 1981a)

JS Agrawal (1954), BA Barlow (1983), SP Bhatnagar and S Chandra (1968), SP Bhatnagar and R Garg (1984), SP Bhatnagar and BM Johri (1983), AE Cocucci and M Venturelli (1982, 1984), BH Danser (1933), SN Dixit (1954, 1958a, b, 1961), W Griffith (1843), BM Johri (1956, 1987a, b), BM Johri and JS Agrawal (1954), BM Johri and KB Ambegaokar (1984), BM Johri and SP Bhatnagar (1972), BM Johri and S Prakash (1965), BM Johri and B Raj (1965, 1969), BM Johri, JS Agrawal and S Garg (1957), RN Kapil and IK Vasil (1963), J Kuijt (1967, 1969), J Kuijt and F Weberling (1972), P Maheshwari and BM Johri (1950a), P Maheshwari and B Singh (1952), P Maheshwari, BM Johri and SN Dixit (1957), R Narayana (1954, 1955, 1956, 1958a, b), A Pisek (1924), S Prakash (1960b, 1961, 1963), B Raj (1970a, b), H Schaeppi and F Steindl (1942, 1945), B Singh (1952a), C Smart (1952), Ph van Tieghem (1895), M Treub (1883), M Venturelli (1980, 1981a, b, 1983, 1984a, b, c)

Loranthaceae: Viscoideae

Bhandari and Vohra (1983) and B.M. Johri (1987b) have reviewed the embryology of Viscoideae.

In *Arceuthobium minutissimum* (Bhandari and Nanda 1968a), the structure of the anther is rather unique. There is a single annular pollen sac forming a continuous ring around a central column of sterile cells called "columella" (Fig. 12.8.8A–F).

The anther wall comprises an epidermis, fibrous endothecium, one or two ephemeral middle layers, and glandular tapetum. In *A. minutissimum* the endothecium is absent, and fibrous thickenings develop in the epidermis which is designated epithecium (Fig. 12.8.8G–L). Simultaneous cytokinesis follows meiotic divisions in the microspore mother cell, and the tetrads are tetrahedral and decussate in *Korthalsella opuntia* (Bhandari and Chaube 1969).

The spherical pollen grains are deeply trilobed, and two-celled at the dispersal stage (Fig. 12.8.8M–P).

The central ovarian papilla (mamelon = placenta) develops from the base of the ovary (Fig. 12.8.9A), and contains one or more archesporial cells. The true ovules, with nucellus and integuments, do not differentiate. The archesporial cells function directly as megaspore mother cells, and the development of the embryo sac is of the Allium type (Fig. 12.8.9B–L). In *Arceuthobium douglasii*, Bhandari and Nanda (1968b) refute the earlier report of tetrasporic type (Jones and Gordon 1965). Bisporic Endymion type is reported in *Ginalloa* (Rutishauser 1937), *Korthalsella* (Rutishauser 1937, Schaeppi and Steindl 1945,

Fig. 12.8.8A–P. *Arceuthobium minutissimum.* **A** Transection of perianth lobe bearing stamen primordium at archesporial stage. **B** Transection of perianth lobe bearing sessile stamen with horseshoe-shaped archesporium. **C–E** Stamen has differentiated into columella (*col*) and horseshoe-shaped archesporium surrounding it. **F** Transection of stamen to show horseshoe-shaped archesporium and sterile central columella. **G, I, K, M** Longisection of male flowers to show stamens at various stages of development. **H** *Inset H* from **G**. Some cells of the outermost archesporial layer have formed parietal layer. The tapetum has already differentiated on the inner side. **J, L** *Inset J* and *L* in **I** and **K**. The anther wall shows epidermis, two- to three-layered glandular tapetum and sporogenous cells. Note the disintegration of wall layer in **L. N** Longisection of dehisced anther from **M**. The epidermis has a cuticle and fibrous bands (exothecium). Note also the remnants of degenerated columella. **O, P** Uni- and bi-celled pollen grain. (After Bhandari and Nanda 1968a)

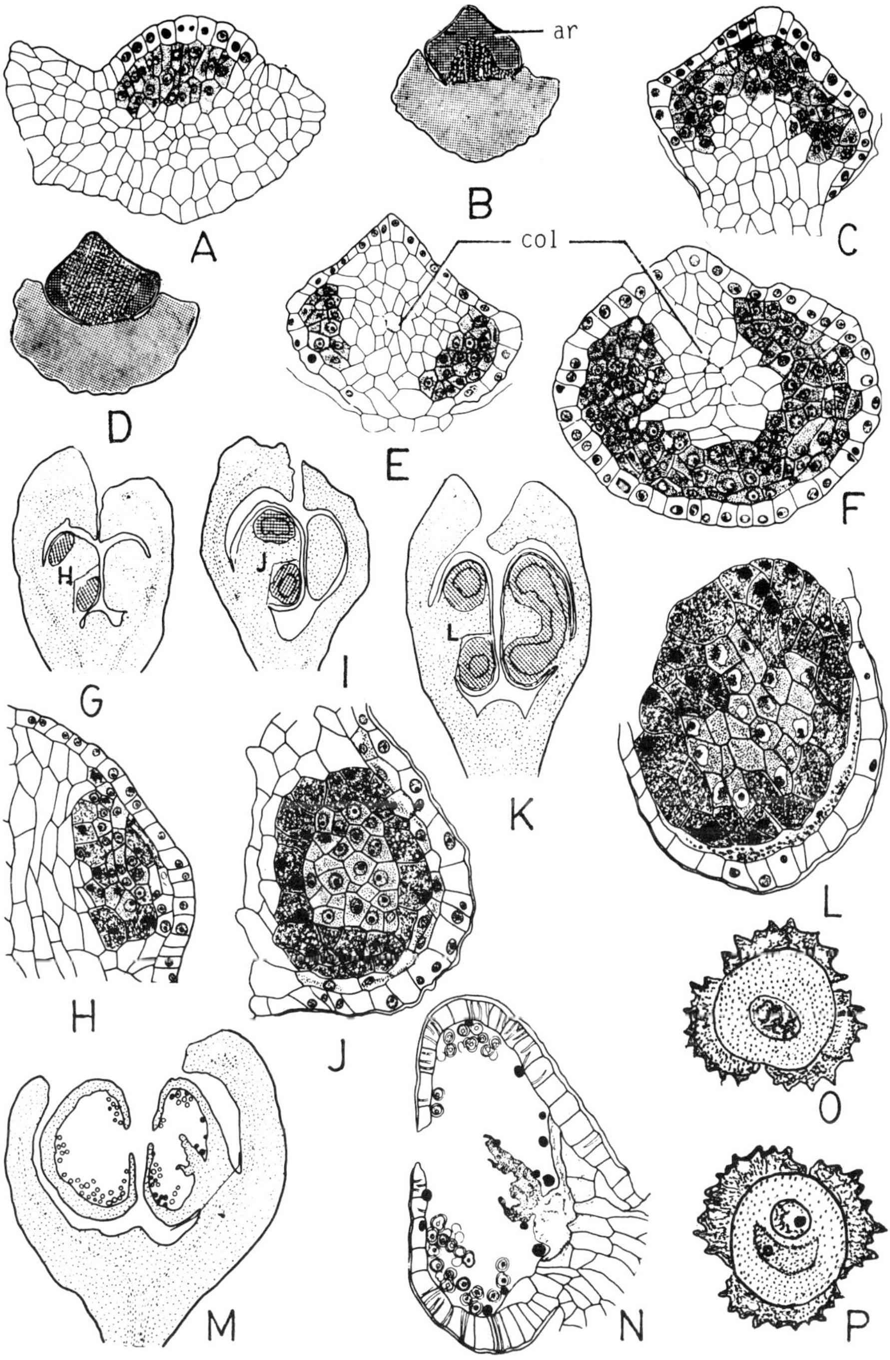
A
B
ar
C
D
E
col
F
G
H
I
J
K
L
M
N
O
P

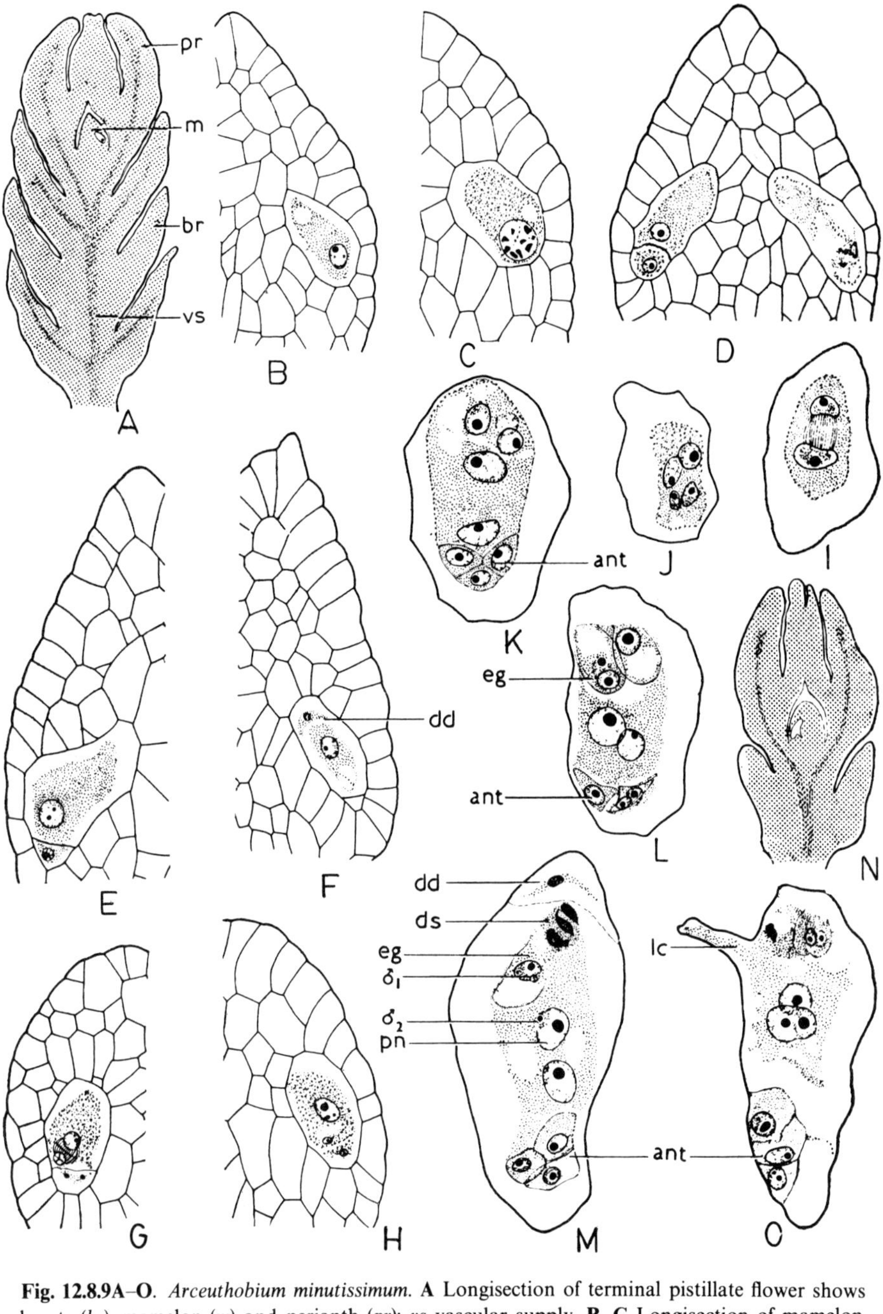

Fig. 12.8.9A–O. *Arceuthobium minutissimum.* **A** Longisection of terminal pistillate flower shows bracts (*br*), mamelon (*m*) and perianth (*pr*); *vs* vascular supply. **B**, **C** Longisection of mamelon with megaspore mother cell. **D** Mamelon with megaspore mother cell (*right*) and dyad (*left*). **E**, **F** Dyad. In **E** the chalazal dyad cell is functional, and in **F** micropylar; *dd* degenerated dyad. **G** Nucleus of lower dyad cell in division; the upper dyad shows two degenerated nuclei. **H** Lower dyad cell still uninucleate; the upper dyad cell has divided transversely but both the nuclei have

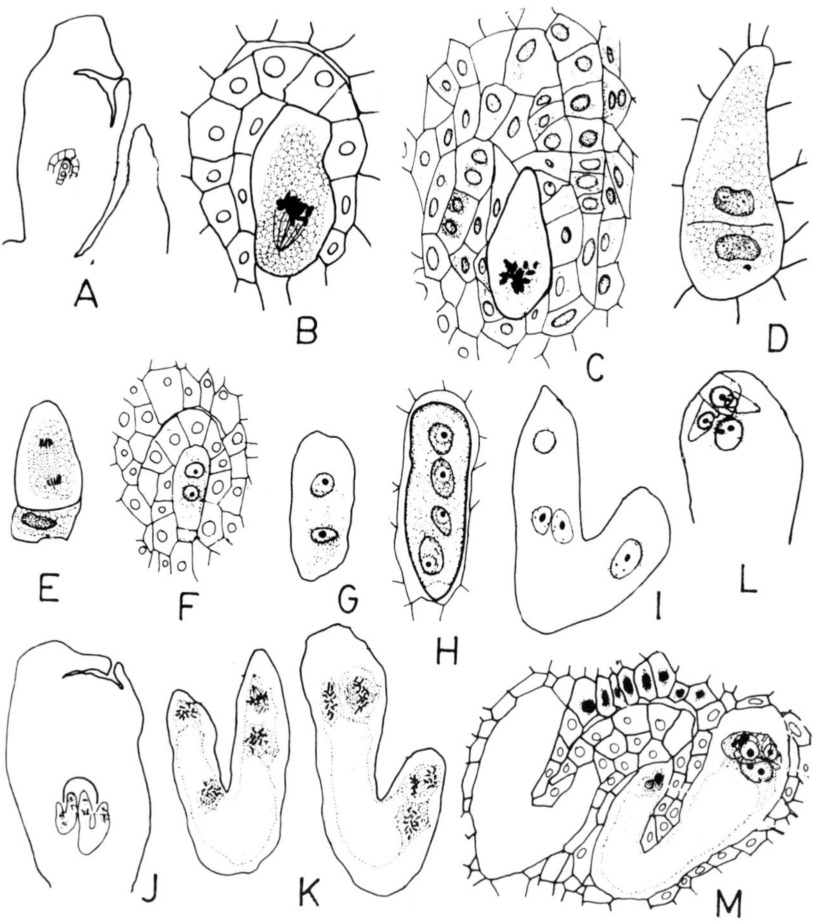

Fig. 12.8.10A–M. *Korthalsella japonica*. **A** Longisection of female flower at binucleate embryo sac. **B**, **C** Megaspore mother cell in metaphase I. **D** Dyad. **E** Micropylar dyad at anaphase. **F** Binucleate embryo sac with "ovular epidermis", from **A**. **G**, **H** Two- and four-nucleate embryo sac. **I** Four-nucleate embryo sac with the chalazal region curving laterally and upward. **J** Female flower with two curved four-nucleate embryo sac, nuclei in division. **K** Embryo sacs from **J**. **L** Micropylar portion of embryo sac with egg apparatus and secondary nucleus. **M** Mature embryo sac (*right*). (From Yoshida and Kawaguchi 1970)

Correa 1958, Yoshida and Kawaguchi 1970), and *Viscum* (Schaeppi and Steindl 1945).

In *Arceuthobium* and *Viscum* the embryo sac is straight, or slightly curved, and in *A. minutissimum* (Bhandari and Nanda 1968a) the fertilized embryo sac develops a lateral caecum (Fig. 12.8.9N–O).

degenerated. **I–K** Two-, four- and eight-nucleate bisporic embryo sac; *ant* antipodals. **L** Organized embryo sac; *eg* egg cell. **M** Embryo sac at fertilization; *ds* degenerated synergid; *pn* polar nuclei. **N** Longisection of pistillate flower at fertilization. **O** Embryo sac from **N** to show syngamy and triple fusion, and lateral caecum (*lc*). (After Bhandari and Nanda 1968a)

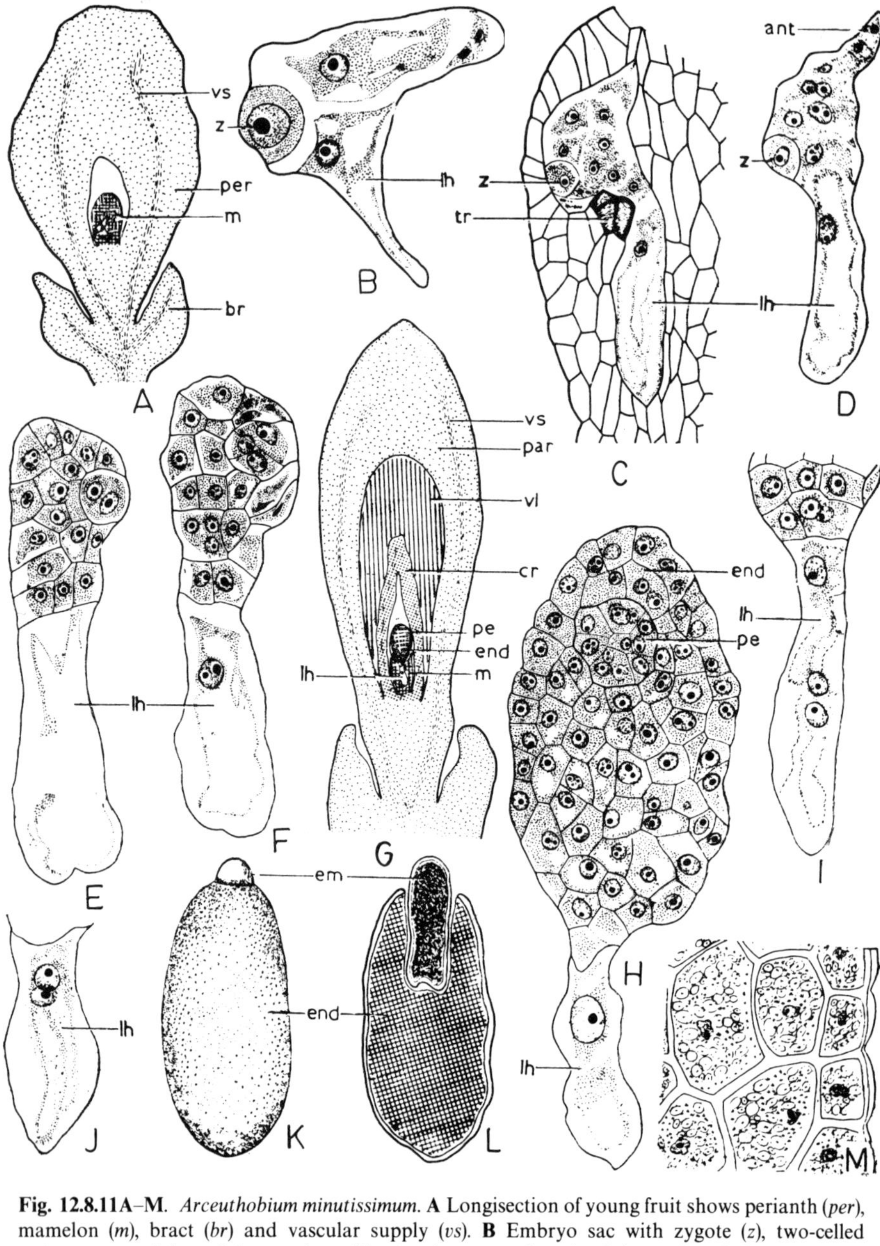

Fig. 12.8.11A–M. *Arceuthobium minutissimum.* **A** Longisection of young fruit shows perianth (*per*), mamelon (*m*), bract (*br*) and vascular supply (*vs*). **B** Embryo sac with zygote (*z*), two-celled endosperm, and lateral haustorium (*lh*); the degenerated antipodals persist. **C–F** Progressive development of endosperm. Note tracheids (*tr*) in the mamelon in **C**. The proembryo is two-celled in **E** and three-celled in **F**. **G** Longisection of fruit at preglobular proembryo (*pe*) with endosperm (*end*) and lateral haustorium (*lh*). The pericarp shows exocarp, vascular supply (*vs*), parenchymatous zone (*par*), viscid layer (*vl*) and crest (*cr*). **H** Endosperm, proembryo, and lateral haustorium from **G**. **I, J** Endosperm haustorium, three-celled in **I** and binucleate in **J**. **K, L** Mature endosperm encloses the embryo; **K** wholemount, **L** longisection almost at the same stage. Note the protruding radicular portion of embryo and absence of haustorium. **M** Mature endosperm tissue with starch-packed cells, the epidermis is covered with a cuticle. (After Bhandari and Nanda 1968a)

The bisporic embryo sac is U-shaped in *Dendrophthora*, *Ginalloa*, *Korthalsella* (Fig. 12.8.10A–M) and *Phoradendron*. At the four-nucleate stage it elongates and the basal end curves upward (carrying the nuclei) and penetrates the carpellary tissue (Fig. 12.8.10I–M). The chalazal end, thus, comes to lie at a higher level than the original micropylar end, and the egg apparatus differentiates at the upper pole. Generally, only one embryo sac develops in each papilla.

The development of endosperm is of the Cellular type (Fig. 12.8.11A, B). The division of the primary endosperm nucleus results in an unequal micropylar and chalazal chamber. The endosperm proper develops either from the micropylar chamber, or from both the chambers (Fig. 12.8.11C–H). In *Arceuthobium* (Bhandari and Nanda 1968a), the caecum contains one to three nuclei and behaves as a haustorium (Fig. 12.8.11H–J). The endosperm

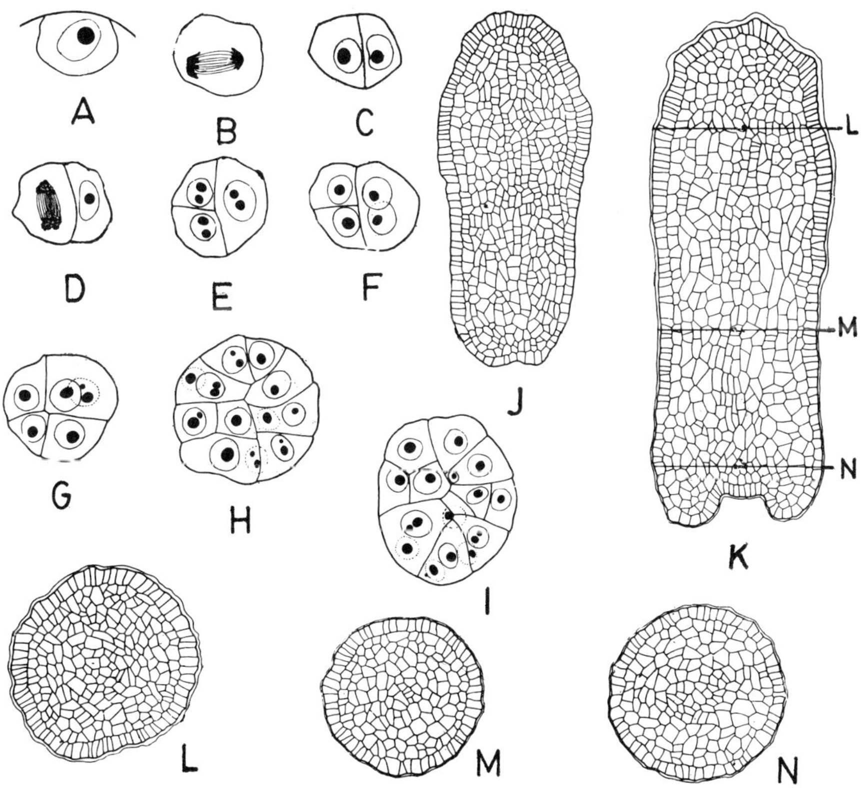

Fig. 12.8.12A–N. *Arceuthobium minutissimum.* **A** Zygote. **B, C** Vertical division in zygote. **D–I** Stages in the development of globular proembryo. **J** Longisection of bipolar torpedo-shaped proembryo; there is a continuous protoderm, and cotyledonary primordia have been initiated. **K** Longisection of mature embryo with thick cuticle, continuous protoderm, well-developed radicle, small cotyledons and shoot-apex, and absence of suspensor. **L–N** Cross-sections at levels *L*, *M* and *N* marked in *K*. (After Bhandari and Nanda 1968a)

is massive, and stores reserve food (Fig. 12.8.11K–M). The cells contain chlorophyll.

The division of zygote is followed by a transverse wall, except in *Arceuthobium* (Bhandari and Nanda 1968a, Correa 1958) and *Korthalsella* (Correa 1958) where it is obliquely vertical (Fig. 12.8.12A–C). Bhandari and Chaube (1969), and Yoshida and Kawaguchi (1970) describe the formation of a transverse wall in the zygote of *K. japonica* and *K. opuntia*. The suspensor does not differentiate. According to Cohen (1963), the embryogeny conforms to the Penaea variation of Asterad type in *A. americanum*, *A. campylopodum* and *A. douglasii*. Bhandari and Nanda report Dendrophthora variation of Piperad type in *A. minutissimum* (Fig. 12.8.12A–N).

The mature embryo is reduced. Only traces of vestigial cotyledons, as slight depressions on the cotyledonary pole, are reported in *A. pusillum* (Tainter 1968). In *A. minutissimum* (Cohen 1963, Bhandari and Nanda 1968a), *K. opuntia* (Rutishauser 1937) and *Viscum* the cotyledons are rudimentary.

The fruit is baccate, and the pericarp can be distinguished into four or five zones. In *Arceuthobium* it shows a leathery exocarp followed by parenchymatous zone, vascular zone, crest of viscid layer, and remnants of endocarp.

NN Bhandari and RK Chaube (1969), NN Bhandari and K Indra (1969), NN Bhandari and K Nanda (1968a, b), NN Bhandari and SCA Vohra (1983), SP Bhatnagar and S Chandra (1968), DM Calder (1983), LI Cohen (1963), JP Correa (1958), JD Hooker (1856), BM Johri (1962a, 1987a, b), BL Jones and CC Gordon (1965), J Kuijt (1955, 1969), HJ Lam (1948), R Narayana (1956), A Rutishauser (1935, 1937), H Schaeppi and F Steindl (1945), H Schulle (1933), FH Tainter (1968), HH York (1913), O Yoshida and H Kawaguchi (1970)

12.9 Taxonomic Considerations

Bentham and Hooker (1965c) group Loranthaceae, Santalaceae and Balanophoraceae in Santalales. Hutchinson (1973) placed Loranthaceae, Grubbiaceae, Santalaceae, Myzodendraceae and Balanophoraceae in Santalales, and Olacaceae, Opiliaceae, Octoknemataceae, Aptandraceae, Dipentodontaceae and Medusandraceae in Olacales.

Olacaceae. Kuijt (1969) regards Olacaceae – or some part thereof – as the starting point of four separate evolutionary lines: (a) Santalaceae, (b) Myzodendraceae could have only been derived from Olacaceae, (c) Viscaceae may be traceable to santalacean plants similar to present-day epiphytic parasites such as *Phacellaria*, and (d) Loranthaceae could not have evolved from Santalaceae. The flower of Loranthaceae is dichlamydous, and can scarcely have a monochlamydous ancestor in Santalaceae. Loranthaceae, therefore, represent the fourth evolutionary line originating in the Olacaceae, possibly through a *Chaunochiton*-like species.

Van Tieghem (1896), C.F. Reed (1955) and Hutchinson (1959, 1973) raise the family Olacaceae to an ordinal rank, Olacales. B.M. Johri and Bhatnagar (1960) compare the embryological features of Olacaceae and Santalaceae (Table 12.9.1), and conclude that both the families show marked similarities

Table 12.9.1. Embryology of Olacaceae and Santalaceae (After B.M. Johri and Bhatnagar 1960)

	Olacaceae	Santalaceae
Habit	Autotrophic or root parasite	Root parasite
Flower	Unisexual or bisexual, mono- or dichlamydeous, disc present	Unisexual or bisexual, monochlamydeous, disc present
Anther tapetum	Glandular, cells 2- or 4-nucleate	Glandular, cells uni- or multinucleate
Pollen	Tricolpate, 2- or 3-celled	Tricolpate, 2- or 3-celled
Ovary	Superior, inferior or semi-inferior, unilocular, sometimes chambered at base and unilocular above	Superior, semi-inferior or inferior, unilocular or 2–5-chambered at base and unilocular above
Ovule	1–5, borne on central placenta, anatropous, differentiation into nucellus and integument lacking (*Olax imbricata*), unitegmic (*Olax wightiana*, *Strombosia*), or bitegmic (*Coula*, *Ximenia*)	1–5, pendulous, borne on central placenta, hemianatropous or anatropous, differentiation into nucellus and integument lacking (*Santalum*), or unitegmic (*Thesium*); ovule absent in *Exocarpus*.
Embryo sac	Polygonum or Allium type, micropylar end may or may not extend beyond the ovule, chalazal caecum is formed	Polygonum, tip of embryo sac may or may not extend beyond the ovule, chalazal caecum is formed
Endosperm	Cellular and chalazal haustorium unicellular in *Strombosia*; Helobial in *Olax* and chalazal chamber haustorial	Cellular, or Helobial (*Santalum*, *Mida*), chalazal haustorium uni- or multicellular, secondary haustoria develop in *Osyris* and *Iodina*
Division of zygote	Zygote divides transversely; suspensor multiseriate	Zygote divides transversely or longitudinally (*Comandra*), suspensor present or absent

except for the presence of calyx and bitegmic ovules in some cases in Olacaceae. Even the comparison of the placental-ovular complex of Olacaceae and Santalaceae (Fagerlind 1947a, 1948) and floral anatomy (F.H. and E.C. Smith 1942) indicate their close alliance. B.M. Johri and Bhatnagar emphasize that there is no justification to raise Olacaceae to ordinal rank as suggested by Van Tieghem, C.F. Reed and Hutchinson.

The genus *Octoknema* has been given a family status: Octoknemataceae. Only scanty data are available about their embryology. According to Fagerlind (1948), the structure of wood, pollen grains, ovary, placenta, and development of the embryo sac resemble the members of Olacaceae. Fagerlind (1948), Stauffer (1957) and B.M. Johri and Bhatnagar (1960) do not feel the necessity of assigning it to a separate family. Its merger with Olacaceae appears to be more appropriate.

Opiliaceae. The spherical pollen grains, Polygonum type of embryo sac, and unitegmic ovule in Opiliaceae resemble those of *Thesium*, whereas the undifferentiated ovule of *Agonandra* shows similarity with that of *Exocarpus*. The chalazal caecum penetrating the thalamus in *Opilia*, and the much-branched chalazal haustorium with its secondary haustoria in *Cansjera* (Swamy 1960), are comparable to the condition in Santalaceae. Fagerlind (1948) suggests its inclusion as the tribe Opilieae under Santalaceae (see also B.M. Johri and Bhatnagar 1960).

Grubbiaceae. Endlicher (1840) and de Candolle (1857) established Grubbiaceae as a distinct family; Bentham and Hooker (1965c) do not recognize the family status and reduce it to the tribe Grubbieae under Santalaceae.

Van Tieghem (1897) studied the structure of flower and fruit of *Grubbia*. The bicarpellate ovary with axile placentation, anatropous, unitegmic ovule, and integument developing into seed-coat (as in Grubbiaceae) has no parallel in the families of Santalineae. Consequently, he refers the family to Hamamelidales, with which we do not agree.

In Grubbiaceae the ovule has a thick integument with a long micropyle; the embryo sac lies in the micropyle. The endosperm develops micropylar and chalazal haustoria in *Grubbia* (Fagerlind 1947b); the peripheral layer of endosperm is thickly cuticularized and the food reserve is proteinaceous. The embryo occupies the entire length of seed. In view of these features, Fagerlind (1948) relates Grubbiaceae to Ericaceae. B.M. Johri and Bhatnagar (1960) point out that similar features also occur in Olacaceae and Santalaceae (see Schulle 1933, Pilger 1935, M. Ram 1957, 1970a, S.P. Bhatnagar 1959). Therefore, there is no justification for relating Grubbiaceae to Ericaceae.

The pollen grains are different from those of Santalaceae. The occurrence of an endothelium, integument developing into seed-coat, and micropylar endosperm haustorium in *Grubbia* warrant the erection of a separate family, the Gubbiaceae, in the Santalineae.

Santalaceae. Apart from the general features of the order Santalales, Santalaceae exhibit several interesting features: (1) polynucleate pollen in *Santalum obtusifolium* (S.P. Bhatnagar 1966), (2) embryo sac-like structures from male archesporial/sporogenous cells in *Leptomeria billardierii* (M. Ram 1959c), (3) unitegmic (*Comandra*, *Iodina* and *Thesium*), or undifferentiated ovule (*Exocarpus*), (4) Polygonum, Allium and Endymion type of embryo sac; formation of extra-ovular extensions of embryo sac, apical and lateral caeca in *E. menziesii* (Fagerlind 1959), (5) antipodal and synergid haustoria in *Quinchamalium*, (6) Cellular, and Helobial (*Mida* and *Santalum*) type of endosperm, (7) profusely branched haustoria of *Iodina* with hypertrophied nucleus, (8) transverse division of zygote; vertical in *Comandra* and *Quinchamalium*, (9) cutinized epidermis of endosperm with proteinaceous food reserve, (10) absence of seed-coat, and (11) embryo occupying major part of the seed.

The position of *Exocarpus* has been controversial. Gagnepain and Boureau (1947) suggest that *Exocarpus* should be assigned to a separate family, Exocarpaceae, and placed near the Taxaceae under gymnosperms. Lam (1948) includes *Exocarpus* in Protangiospermae along with *Casuarina* and *Salix*. M. Ram (1959a) investigated the embryology of *Exocarpus cupressiformis*, *E. sparteus* and *E. strictus* and emphasizes their affinities. The flower has a typical angiospermous structure. The anther wall develops normally, pollen grains are two-celled, there is no differentiation of ovule, the archesporium differentiates at the apex of the "placenta", and development of the embryo sac is of the Polygonum type. The Cellular type of endosperm develops a chalazal haustorium. The division of zygote is transverse, and the embryo shows a suspensor and two cotyledons. The seed is enclosed in pericarp. These features indicate its alliance with Santalaceae. It differs from the latter in having a straight, instead of a U- or N-shaped embryo sac, and a long, proliferated suspensor (see B.M. Johri 1963b, P. Maheshwari and Kapil 1963). The embryological features are not comparable to those of *Casuarina* or *Salix*. Therefore, *Exocarpus* should be assigned to Santalaceae.

Myzodendraceae. The genus *Myzodendron* has been assigned to the tribe Osyrideae in Santalaceae. Rendle (1959) confirmed its inclusion in Santalaceae. The common features between *Myzodendron* and Santalaceae are: (1) receptacular ovary, (2) three pendulous and undifferentiated ovules on a free central placenta, (3) Polygonum type of embryo sac, and (4) Cellular type of endosperm with a conspicuous chalazal haustorium. *Myzodendron* differs from Santalaceae in the structure of flower, pollen grains, and presence of rudimentary testa. Embryologically, *Myzodendron* (Kuijt 1969) shows clear affinities with Olacaceae. The ovules are undifferentiated, a parenchyma layer covers the embryo sac, and the antipodal end of embryo sac forms a long tubular haustorium which invades the placental column. The first cross-wall, isolating the haustorium, is probably formed after the division of the primary endosperm nucleus [cf. *Cansjera* (Opiliaceae), Swamy 1960]. The haustorium reaches up to the base of flower, and ramifies in that region. The features mentioned here are scarcely different from that of the Olacaceae. Kuijt emphasizes that *Myzodendron* could have been derived directly from Olacaceae. The erection of a separate family Myzodendraceae appears justified and is supported by studies on pollen morphology (Feuer 1981). A more detailed account would be helpful to determine its systematic assignment.

After considering the comparative embryology of the order Santalales,

B.M. Johri and Bhatnagar (1960) state that the suborder Santalineae should include four families – Olacaceae, Grubbiaceae, Santalaceae and Myzodendraceae – and the family Santalaceae should comprise six tribes: Comandreae, Thesieae, Osyrideae, Santaleae, Opilieae and Anthoboleae.

Loranthaceae. The suborder Loranthinae includes a single family, Loranthaceae. P. Maheshwari et al. (1957), B.M. Johri and Bhatnagar (1960), S.P. Bhatnagar and Johri (1983), Bhandari and Vohra (1983) and B.M. Johri (1987b, see also 1963b) discuss the affinities of the two subfamilies, Loranthoideae and Viscoideae. The Loranthoideae are characterized by the presence of calyculus, triradiate pollen grains, except in *Atkinsonia* and *Tupeia*, Polygonum type of embryo sac, in some members extension of female gametophyte into the style and stigma, formation of composite endosperm, and division of zygote followed by a vertical wall. The embryo has a long, tortuous suspensor. The viscid zone lies outside the vascular zone in the pericarp. In contrast, the Viscoideae usually do not have a calyculus (except in male flowers of *Viscum articulatum* and *V. orientale*), the pollen grains are spherical, and the U-shaped embryo sac is bisporic (Allium type). The endosperm develops from a single embryo sac. The zygote divides transversely, except in *Arceuthobium minutissimum* (Correa 1958) and *Korthalsella* (Rutishauser 1935, Correa 1958), and the embryo usually lacks a suspensor. The viscid zone lies inward to the vascular zone (see B.M. Johri 1963b, P. Maheshwari and Kapil 1966).

In view of the foregoing account, the elevation of Loranthoideae and Viscoideae to the family rank as Loranthaceae and Viscaceae (Table 12.9.2) is justified (B.M. Johri and Bhatnagar 1960, B.M. Johri 1963b, see also Kuijt 1969, Bhandari and Vohra 1983). Van Tieghem (1896) erected two independent orders, Loranthales and Viscales. B.M. Johri and Bhatnagar (1960) do not consider it necessary.

Cocucci (1983, see also Cocucci and Venturelli 1982) emphasizes the importance of gynoecium, mamelon, ovules (lobes of mamelon) and embryo in the taxonomy of Loranthaceae (Fig. 12.9.1). The primitive condition of gynoecium occurs in *Tripodanthus* where the ovary contains discrete ovules, i.e. they are ategmic but with well-differentiated nucellus (megasporangium) (Fig. 12.9.1). The two diverging tendencies concern the orientation of ovules met with in the tribes Lorantheae and Lysianeae.

In the first category (Elytrantheae and Lysianeae), the nucellar axis (micropylar-chalazal axis) is so oriented that the chalaza points towards the apex, becoming more parallel to the longitudinal axis of the ovary. The nucellus (lobe of mamelon) fuses with the carpellary tissue. The "compitum" (mamelon) enlarges and develops into amyliferous tissue in which the embryo sacs lie: *Peraxilla* → *Lepeostegeres* → *Lysiana* (Fig. 12.9.1).

In the second category (Tapinostemmeae and Lorantheae), the nucellar axis is erect and the chalaza points downward, i.e. towards the inferior position (Fig. 12.9.1). The ovary becomes unilocular, and the ovules unite to form a "collective" ovule (mamelon without lobes). The next step is the reduction and sinking of the collective ovule till the disappearance of the mamelon: *Tapinostemma* → *Helicanthes* → *Amyema* → *Moquiniella* (Fig. 12.9.1).

Table 12.9.2. Embryology of Santalaceae, Loranthoideae (Loranthaceae) and Viscoideae (Viscaceae) (After Bhandari and Vohra 1983)

	Santalaceae	Loranthaceae	Viscaceae
Pollen grains	Spherical or oblong, 2- or 3-celled	Triradiate, spherical in *Atkinsonia*, *Ixocactus*, *Oryctanthus* and *Tupeia*; 2-celled	Spherical, 2-celled
Ovary	Superior, semi-inferior, or inferior, unilocular or 2–5-chambered at base and unilocular above	Inferior, unilocular, often 3- or 4-chambered	Inferior, unilocular
Ovule	1–5, borne on central placenta, hemiana- or anatropous, uni-, bi- or ategmic (undifferentiated), ovule does not at all differentiate in *Exocarpus*	Ovule does not develop, placenta (mamelon) unlobed and conical (*Helicanthes*), 3- or 4-lobed (*Lysiana*), free or attached to ovary wall (*Lepeostegeres*)	Placenta (mamelon) conical (basal attachment), bilobed in Phoradendreae, absent in *Viscum*
Embryo sac	Monosporic, Polygonum type, may remain confined to the ovule or the tip may grow out towards the stylar canal	Monosporic, Polygonum type, multiple embryo sacs except in *Phthirusa pyrifolia* and *Struthanthus vulgaris*, tip remains limited to the ovary, or reaches up to different heights in the style and stigma. The base elongates up to the hypostase, in *Nuytsia* beyond the hypostase	Bisporic, Allium type (*Arceuthobium*, *Eubrachion*, *Viscum*), or Endymion (*Korthalsella*, *Phoradendron*), embryo sac usually straight but U-shaped in *Arceuthobium* and *Korthalsella*, lower end bends, extends beyond the papilla, and proceeds upward towards the carpellary tissue
Caecum	Chalazal caecum common, micropylar caecum reported in *Comandra*	Chalazal caecum common (antipodals left in situ), micropylar caecum in *Atkinsonia* and lateral in *Nuytsia*	Lateral caecum in *Arceuthobium* and *Korthalsella*
Endosperm	Cellular or Helobial type (*Mida*, *Santalum*), chalazal haustorium develops, secondary haustoria in *Comandra*, *Iodina*, *Mida*; non-chlorophyllous	Cellular, composite structure except in *Phthirusa* and *Struthanthus*, chalazal haustorium reported in *Tapinostemma*; non-chlorophyllous	Cellular endosperm develops individually in each embryo sac, non-aggressive chalazal haustorium in Phoradendreae; chlorophyllous
Embryo	Division of zygote transverse, vertical in *Comandra*, massive suspensor in *Mida*, small in *Leptomeria*, and absent in *Osyris* and *Thesium*	Division of zygote vertical, suspensor biseriate, long and tortuous	Division of zygote transverse, vertical in *Arceuthobium* and *Korthalsella*, suspensor absent in *Eubrachion ambiguum*
Testa	Integuments become crushed during endosperm development, testa absent	Absent; ovules lack differentiation of integument	Absent; ovules lack integument
Pericarp	Parenchymatous epicarp and stony mesocarp persist, parenchymatous endocarp consumed by endosperm	Leathery epicarp followed by viscid zone (mesocarp) which is external to vasculature of ovary, endocarp parenchymatous	Leathery or parenchymatous epicarp, viscid zone (mesocarp) internal to vasculature of ovary, endocarp parenchymatous

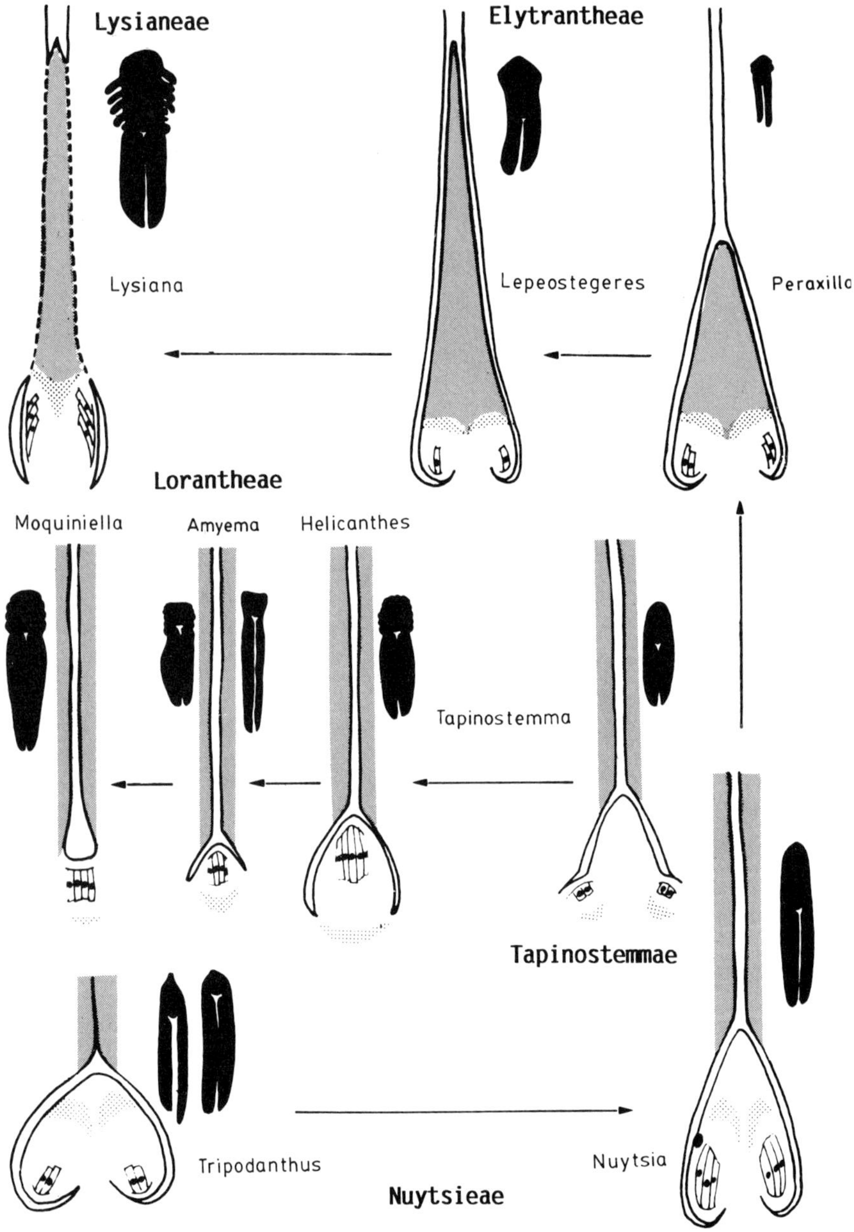

Fig. 12.9.1. Loranthaceae (Loranthoideae), evolutionary tendencies based on "gynoecium organization and embryo specialization". Embryo *black*, amyliferous tissue in style *shaded*, placenta *dotted*. (After Cocucci and Venturelli 1982)

With the above consideration Cocucci (1983) also correlates the condition of cotyledons in the mature embryo, i.e. free or fused. *Peraxilla* to *Lysiana* the cotyledons are free (in *Lysiana* the hypocotyl shows projections), and *Tapinostemma* to *Moquiniella* the cotyledons are partially fused. The primitive genus *Tripodanthus* shows both types of embryos – free as well as fused cotyledons (Fig. 12.9.1).

According to Danser (1933), who revised Engler and Prantl's (1897) classification of Loranthaceae, there are only three tribes: Elytrantheae, Nuytsieae and Lorantheae (see also Melchior 1964, B.M. Johri and Bhatnagar 1972).

1) Elytrantheae: Ovary multilocular, placenta with three to six basal lobes, each lobe represents only the nucellus (ategmic ovule). The placenta is amyliferous and provides nutrition to the developing megagametophyte. The style lacks amyliferous tissue. The tip of embryo sacs extends only up to the basal part of style, e.g. *Lepeostegeres*, *Macrosolen*, *Peraxilla*.

2) Nuytsieae: Ovary trilocular at the base, placenta with three basal lobes, each lobe represents only the nucellus. The style has amyliferous tissue and provides nourishment to the developing megagametophyte. The tips of embryo sacs reach up to two-thirds the height of styles, e.g. *Nuytsia*.

3) Lorantheae: Ovary unilocular, placenta unlobed when present, the tip of embryo sacs extends up to the middle of style or as far as the stigma, e.g. *Helicanthes*, *Helixanthera*.

On the basis of (a) uni-/multilocular ovary, (b) superior/inferior placenta, free/fused mamelon with tissue of stylar canal, (c) discrete/"collective" ovule, Cocucci (1983) suggests two additional tribes: Lysianeae and Tapinostemmae (Fig. 12.9.1).

4) Lysianeae: Ovary multilocular with "collective" ovule. The mamelon is fused with the tissue of stylar canal. The tip of embryo sacs reaches up to the middle of style, e.g. *Lysiana*.

5) Tapinostemmae: Ovary unilocular with discrete ovules. Mamelon is present. The ovules are free from one another, but remain imbedded in carpellary tissue. The tip of embryo sacs extends up to the stigma, e.g. *Tapinostemma*.

We do not see any justification for the erection of the additional tribes, Lysianeae and Tapinostemmae, since they do not show distinctive features. *Lysiana* is well placed in the tribe Elytrantheae, and *Tapinostemma* in Lorantheae.

Taxonomic significance cannot be attached to the presence or absence and structure of the placenta (B.M. Johri and Ambegaokar 1984, see also B.M. Johri 1987a). In the tribe Elytrantheae, the placenta is four-lobed in a tetralocular ovary (*Lysiana*, *Peraxilla*), or three-lobed in a unilocular ovary (*Macrosolen*). In Nuytsieae the placenta is three-lobed in a trilocular ovary (*Nuytsia*). In Lorantheae the placenta may be absent, rudimentary (*Amyema*,

Dendrophthoe), or well developed, unlobed, and free from the carpellary wall (*Helicanthes*, *Lepeostegeres*). Therefore, Cocucci's (1983) emphasis on mamelon as a taxonomic character is not justified at all.

The differences between the two suborders Santalineae and Loranthineae are well-marked. In Santalineae the ovules are distinct, except in *Agonandra* (Opiliaceae) and *Exocarpus* (Santalaceae). The development of the embryo sac is of the Polygonum type, and the embryo sac may develop micropylar and chalazal haustoria. The development of endosperm is either Cellular or Helobial. The zygote divides transversely except in *Comandra*. During seed development, the integuments are consumed, and ultimately there is no testa. The pericarp does not show the presence of a viscid layer. On the other hand, in Loranthineae true ovules are absent; in an inferior ovary a mamelon may develop in some members. Development of the embryo sac is of the Polygonum type in Loranthaceae, and Allium type in Viscaceae. In Loranthaceae the embryo sacs extend to varying lengths in the stylar canal and stigma, and micropylar and chalazal caeca may be formed in some members. The development of endosperm is always Cellular, and the endosperm is composite in Loranthaceae. The zygote divides vertically in Loranthaceae, and in *Arceuthobium* and *Korthalsella* of Viscaceae, otherwise a transverse division occurs in the members of Viscaceae. As true ovules do not differentiate, the testa is absent. A viscid layer develops in the pericarp (see B.M. Johri and Bhatnagar 1960).

In view of such marked differences, in our opinion Santalinae and Loranthineae should both be raised to ordinal rank – Santalales and Loranthales (Table 12.9.2). Santalales would comprise the families Olacaceae, Opiliaceae, Grubbiaceae, Santalaceae and Myzodendraceae; Loranthales would comprise Loranthaceae and Viscaceae.

13 Balanophorales

Balanophorales include only one family, the Balanophoraceae.

13.1 General Features

Balanophorales show: (1) absence of fibrous thickenings in endothecium, (2) three-celled pollen grains, (3) highly reduced ovule, (4) micropylar megaspore of an inverted T-shaped tetrad, or a triad, produces a Polygonum type of embryo sac; Allium in *Corynaea*, (5) organization of egg apparatus and upper polar nucleus in the chalazal region of the embryo sac, and successive fusion of four nuclei in the micropylar region in *Balanophora*, (6) Cellular type of endosperm, (7) Piperad type of embryogeny; absence of suspensor, (8) undifferentiated, minute embryo in the so-called "seed", and (9) unusual mode of germination of seed.

13.2 Balanophoraceae

The family comprises herbaceous fleshy root parasites devoid of chlorophyll and stomata. The flowers are mostly male and female, rarely hermaphrodite. The anther wall comprises the persistent epidermis, a fibrous endothecium does not differentiate, the parietal layers are ephemeral, and the epidermis abuts directly on the tapetum in *Balanophora* (Fagerlind 1945a).

In *Corynaea crassa* var. *sprucei* (Engell 1979), the male flower has three perianth lobes. The three stamens form a synandrium which encloses a sterile lobe in the centre. In the synandrium (Engell 1979), there are three, four, five or six compartments. Beneath the elongated epidermal cells there are one to three middle layers, and a secretory tapetum of uni- or binucleate cells; fibrous endothecium and stomium do not develop. A multilayered epidermis covers the compartments as a cap. Cytokinesis is successive, and the tetrads are tetrahedral. In *Helosis* the tapetum is plasmodial (Umikar 1920), and cytokinesis simultaneous.

The family is highly multipalynous with colpate or porate forms. Sometimes, even inaperturate forms occur. The pollen grains are three-celled at the dispersal stage. The male flower, perhaps, opens by throwing off the epidermal cap. In *Helosis* (Umikar 1920), the cap does not develop; this genus is considered as an apomict.

The female flowers of *Balanophora* and *Corynaea* have a rather different structure, and have been treated separately.

In *Balanophora abbreviata* (Shivamurthy and Arekal 1980), the female floral primordia develop on the inflorescence axis above the region of male flowers, while in *B. fungosa* spp. *indica* (Shivamurthy and Arekal 1980) they develop on the surface of the female inflorescence. They appear as finger-like protuberances amidst the spadicles. The mature flower is extremely reduced in size and structure, archegonium-like without vasculature and perianth. Floral development exactly simulates ontogenetic events normally noted in the nucellus of an angiosperm ovule. It is concluded that this female flower is ontogenetically a nucellus but, functionally, a gynoecium.

Concerning *Corynaea crassa*, Engell (1979) states: "A nucellus primordium develops in the bottom of the ovary and in this nucellus probably two archesporial cells will be differentiated. From these two cells a crassinucellate ovule will develop with two deep-lying sporogenous cells and without development of an integument". He calls this condition of ovule as ategmic and crassinucellate. This is not a correct interpretation. The ovule is indeed undifferentiated.

The archesporial cell is enclosed in a long, beaked "placenta". The megaspore mother cell undergoes meiotic divisions; the chalazal dyad is much smaller than the micropylar. Both the dyads undergo meiosis II and an inverted T-shaped tetrad develops (Fig. 13.2.1A–E) in *Balanophora abbreviata*, *B. dioica*, *B. fungosa* subsp. *indica* var. *indica* (Arekal and Shivamurthy 1978), *B. elongata* (Fagerlind 1945b), and *B. globosa* and *B. thwaitesii* (Teryokhin and Yakovlev 1967). A triad has been observed in *B. abbreviata*. The uppermost megaspore develops further, and produces an embryo sac of the Polygonum

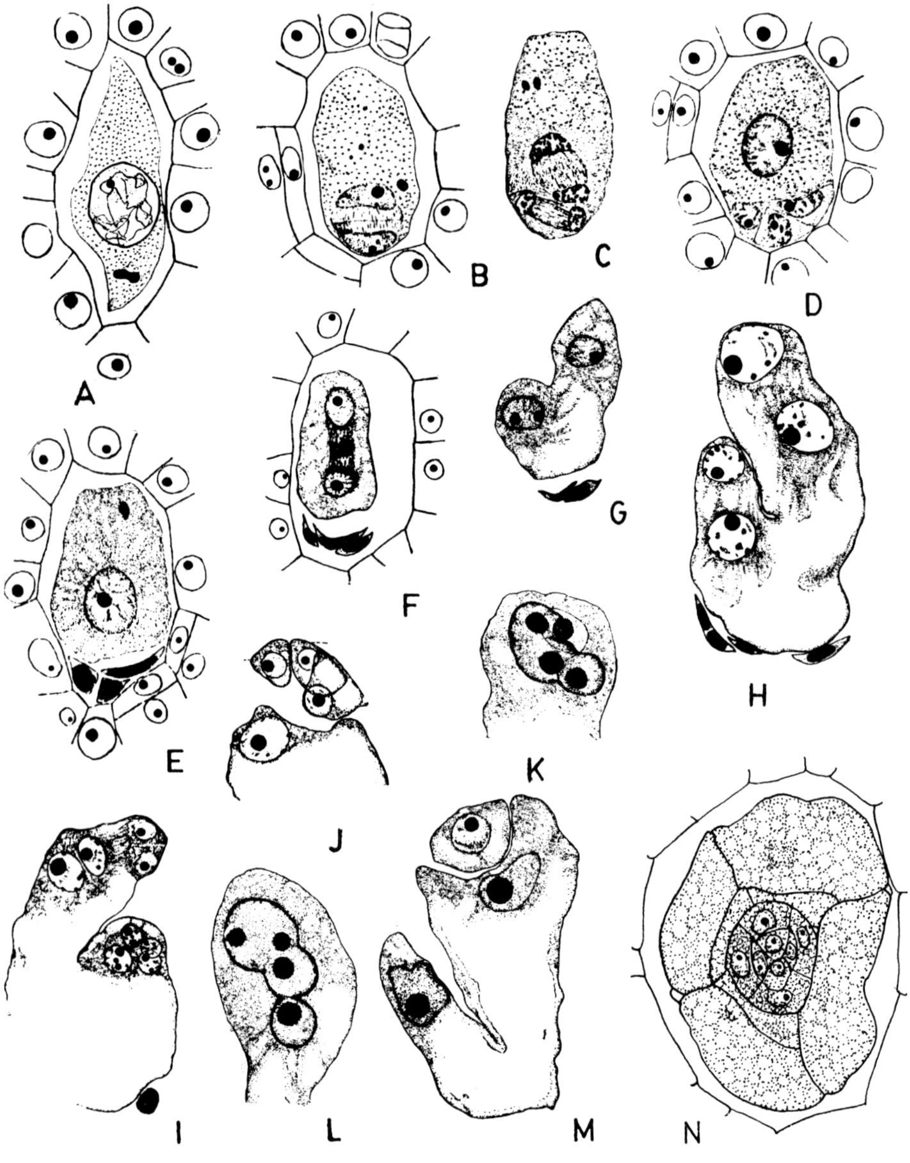

Fig. 13.2.1A–N. *Balanophora dioica.* **A** Megaspore mother cell in prophase. **B** Telophase, meiosis I. **C** Telophase, meiosis II. **D, E** Megaspore tetrad, upper megaspore functional. **F** Functional megaspore nucleus at late telophase. **G** Binucleate embryo sac, chalazal end curves upward. **H** Four-nucleate embryo sac. The chalazal end is larger with prominent nuclei. Degenerated megaspore nuclei are at the base. **I** Eight-nucleate, unorganized embryo sac. **J** Upper portion of embryo sac with egg cell, two synergids, and upper polar nucleus. **K, L** Fusion of antipodal nuclei. **M** Embryo sac with zygote, primary endosperm nucleus (?), and fused antipodal. **N** Transection of embryo sac to show cellular endosperm enclosing proembryo. (After Ekambaram and Panje 1935)

type (Fig. 13.2.1F–L). Similar observations have been reported in *B. dioica* (Ekambaram and Panje 1935) and *Langsdorffia hypogea* (Fagerlind 1945c).

During its development, the embryo sac shows "reversed polarity". At the two-nucleate stage a tubular process originates at the basal end in *Balanophora* (Fig. 13.2.1G), and at the apical end in *Langsdorffia.* This caecum grows rapidly, and comes to lie at a higher level than the original upper micropylar end of the embryo sac. The chalazal nucleus passes into the caecum. Figure 13.2.1H shows a four-nucleate embryo sac. During the differentiation of eight nuclei (Fig. 13.2.1I), four nuclei in the caecum develop into the egg apparatus and upper polar nucleus (Fig. 13.2.1J), those in the morphologically micropylar end fuse successively to form an irregularly lobed nucleus (Fig. 13.2.1K–M). In *B. dioica* (Ekambaram and Panje 1935), this nucleus assumes a more or less irregular outline. During endosperm formation it degenerates with the collapse of the antipodal arm. Thus, the lower polar nucleus is absent, and also the antipodal cells (Arekal and Shivamurthy 1978). However, the fusion of two polar nuclei in the embryo sac has been reported in *B. abbreviata, B. indica* (Zweifel 1939) and *B. thwaitesii* (Teryokhin and Yakovlev 1967). The mature embryo sac is U- or J-shaped (Fig. 13.2.1I, M). Bisporic Allium type is reported in *Ditepalanthus afzelli* and *Helosis cavennensis* (Fagerlind 1938a). However, Chodat and Bernard (1900) described Endymion type in *Helosis.* This needs reinvestigation.

In *Corynaea crassa* var. *sprucei* (Engell 1979), the flowers as well as the inflorescences are unisexual (Fig. 13.2.2A); sometimes both male and female flowers develop on the same branch (Fig. 13.2.2B). The young flowers are embedded in paraphyses or trichomes which function as nectaries.

The insects, perhaps, perforate the anthers and pollinate the flower.

The female flowers of *Corynaea* are pedicillate and devoid of perianth lobes (Fig. 13.2.2B). The gynoecium is bicarpellate and forms a compressed ovary with two styles. One, sometimes two, archesporial cells develop in the "placenta" but only one functions. The micropylar dyad degenerates, and the development of the embryo sac is of the Allium type. The mature embryo sac consists of the egg apparatus, a central cell with secondary nucleus, and three antipodal cells (Fig. 13.2.2C, D1–D3). The placenta degenerates, and the embryo sac becomes "naked".

The development of endosperm is of the Cellular type. The division of primary endosperm nucleus is followed by a wall separating a small micropylar from a large chalazal chamber. The micropylar chamber gives rise to the endosperm proper, while the chalazal develops into a large, uninucleate haustorium. The latter is crushed by the developing endosperm tissue.

The embryogeny conforms to the Piperad type. The first, as well as the second division is vertical in *Balanophora* (Zweifel 1939). In *B. dioica* (Ekambaram and Panje 1935) the third division is periclinal in all the four quadrants (Fig. 13.2.1N). The embryo consists of four central and four peripheral cells. The nuclei are all arranged in the equatorial region of the embryo. When the embryo lies deep in the endosperm, the next division in the embryo is transverse, but takes place only in the peripheral cells. The fully formed embryo consists of 12 cells. Treub (1898) reported five-celled embryo in

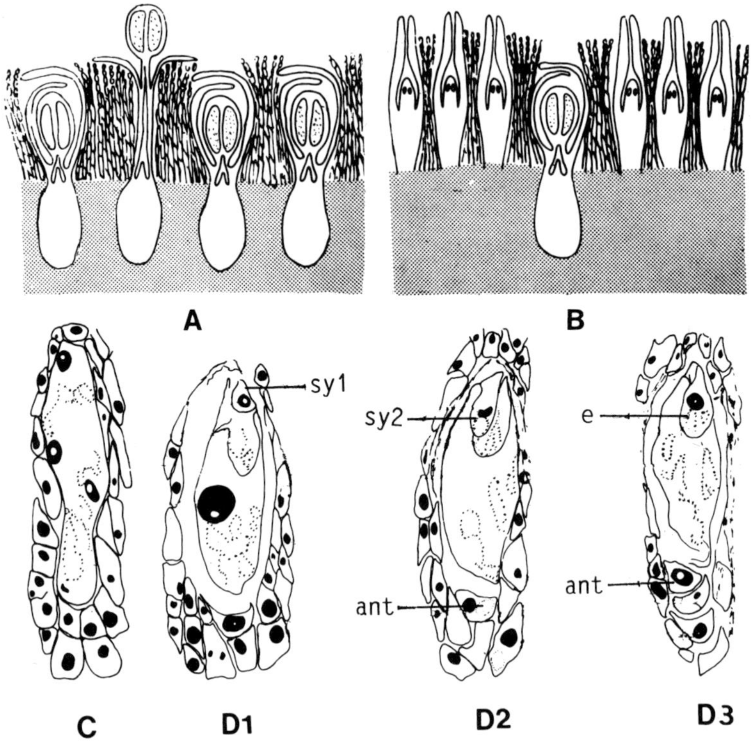

Fig. 13.2.2A–D. *Corynaea crassa* var. *sprucei*. **A** Schematic drawing of one tall and three short male flowers and paraphyses. Tall flower with unfolded perianth on the top of paraphyses. Short flowers with folded perianth. Three stamens with fused anthers and filaments; there is a small sterile cone above the receptacle in each flower. **B** Monoecious side-shoot with a single male flower in the centre, and some six pedicellate, naked (without perianth) flowers with bicarpellate gynoecium surrounded by paraphyses. **C** Four-nucleate embryo sac. **D1–D3** Successive serial longisections of a mature embryo sac; **D1** Egg cell, one synergid (*sy1*), and secondary nucleus; **D2** Egg cell, another synergid (*sy2*), central cell, and one antipodal (*ant*); **D3** Egg cell (*e*), central cell, and three antipodals. The nucellus has degenerated. (After Engell 1979)

B. elongata, while Ernst (1914) reported six-celled condition in the same species. The embryo in the mature seed is unorganized. A suspensor is absent. Teryokhin and Yakovlev (1967) describe embryogenesis as of "non-oriented" zygote cleavage type. The development of embryo requires further investigation. Polyembryony has been reported in *B. globosa* and *Helosis guianensis*.

Fig. 13.2.3A–D. *Balanophora abbreviata*, germination of "seed" and establishment of parasite on host rootlet; longisection, semidiagrammatic. **A** Germination of "fruit" on host rootlet; note the anchoring endosperm tubules (*et*), embryo (*em*), endosperm (*end*), fruit wall (*fw*), host-root cortex (*hrc*), and host-root central cylinder (*hrcl*). **B** Germination of "fruit" shows entry of primary

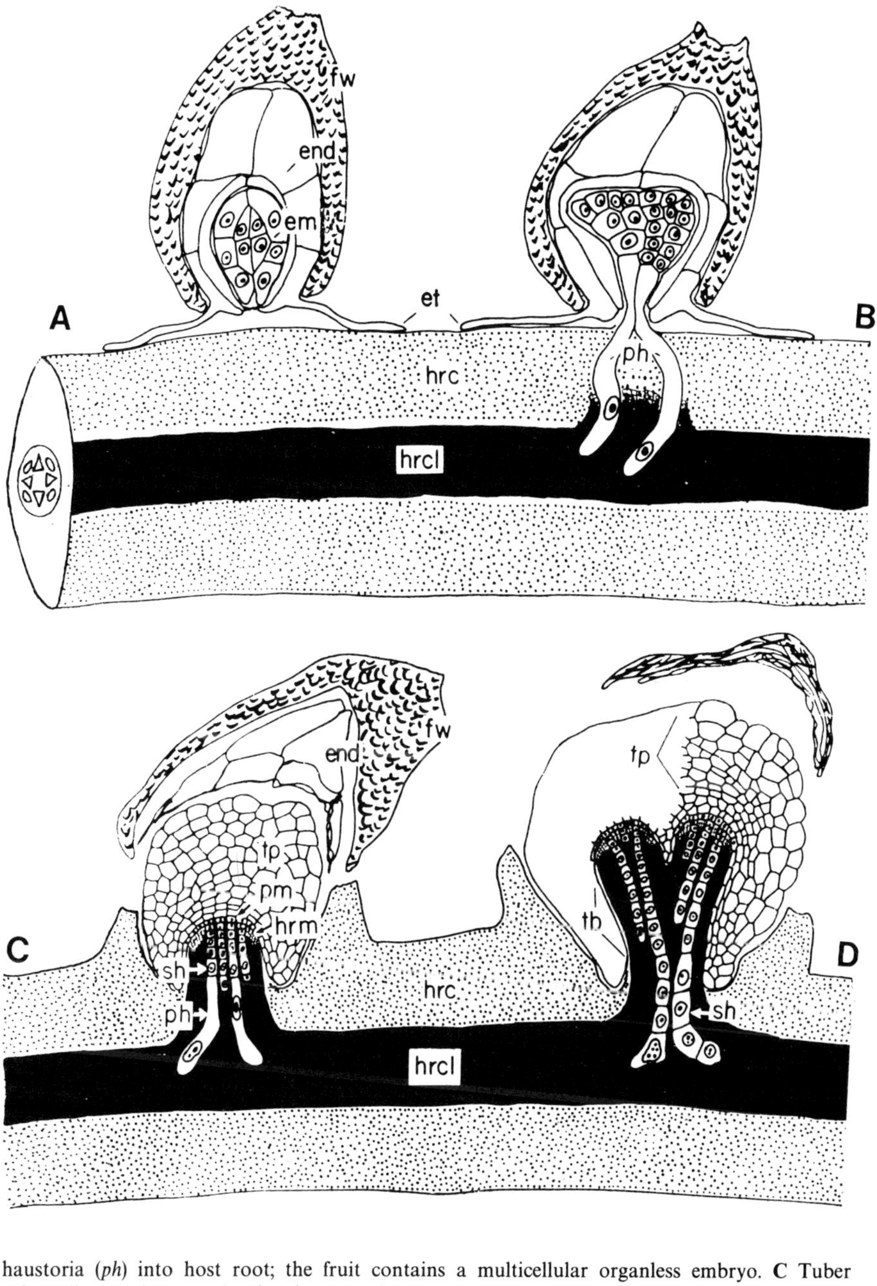

haustoria (*ph*) into host root; the fruit contains a multicellular organless embryo. **C** Tuber primordium on host rootlet showing organization of secondary haustorial cells and tuber bundle; *hrm* host-root meristem; *pm* meristem of parasite; *sh* secondary haustorium; *tp* tuber parenchyma. **D** Tuber primordium, slightly later stage, to show forked tuber bundle (*tb*). (After Arekal and Shivamurthy 1976)

In aposporous embryo sacs of *Balanophora globosa* (Ernst 1913), the lower polar nucleus of the embryo sac degenerates, and the upper polar nucleus alone gives rise to endosperm. *B. globosa* and *B. japonica* (Kuwada 1928, Zweifel 1939, Fagerlind 1945b) exhibit Taraxacum type of apospory. The first meiotic division of megaspore mother cell ends in the formation of a restitution nucleus. Meiosis II results in a dyad of unreduced cells, and the lower one develops into an eight-nucleate embryo sac.

The germination of "seed' in *Balanophora*, as described by Arekal and Shivamurthy (1976), is unique and needs special mention. Typical seeds are absent, and the ripe fruit comprises an undifferentiated embryo, few-celled endosperm, and sclerotic pericarp. These fruits, in the vicinity of host rootlets, break open. The endosperm cells produce tubular, slender, sticky extensions for anchorage of the fruit to the young host-rootlets (Fig. 13.2.3A, B). The undifferentiated embryo develops four to five tubular processes (as primary haustorium) which penetrate the vascular tissue of the host (Fig. 13.2.3B). It also forms a tuber by repeated divisions, and emerges out of the pericarp above the ground. The cells of the central cylinder of the host-rootlet become active in the region of contact (Fig. 13.2.3C). A lateral branch of the central cylinder of host enters the tuber, and branches dichotomously (Fig. 13.2.3D). The tuber develops secondary haustorium in the form of chains of cells which penetrate the central vascular cylinder of the host, and absorb food material. The tuber grows, and develops flowering scapes.

GD Arekal and GR Shivamurthy (1976, 1978), R Chodat and Ch Bernard (1900), AJ Eames (1961), T Ekambaram and RR Panje (1935), K Engell (1979), A Ernst (1913, 1914), F Fagerlind (1938a, 1945a, b, c, 1959), JD Hooker (1856), Y Kuwada (1928), LN Rao (1932), GR Shivamurthy and GD Arekal (1980), ES Teryokhin and MS Yakovlev (1967), M Treub (1898), O Umikar (1920), R Zweifel (1939)

13.3 Taxonomic Considerations

Balanophoraceae. Kuijt (1969) points out that Hooker's (1856) emphasis on Halorrhagaceae as possible relatives of Balanophoraceae, especially *Gunnera*, may be nearest the mark. Juel (1910) confirms this notion, but Harms (1935) does not refer to it in his treatment of Balanophoraceae. This may be due to the fact that in *Genera Plantarum* Hooker himself abandoned the idea.

Bentham and Hooker (1965c), Engler and Prantl (1931), Wettstein (1935), and Hutchinson (1973) group the family Balanophoraceae in Santalales, with Santalaceae and Loranthaceae. Balanophoraceae show distinct affinities with both the families. The common features with Santalaceae are: parasitic mode, extreme reduction of ovule, monosporic eight-nucleate or, sometimes, Allium type of female gametophyte, and Cellular type of endosperm with chalazal haustorium. The Balanophoraceae resemble the Loranthaceae-Viscoideae (*Korthalsella*) in reversed polarity of the embryo sac, vertical division of zygote, and absence of suspensor. The testa is absent in Santalales, and an undifferentiated embryo in the mature "seed" generally occurs in parasites.

For further discussion, see Section 39.19, Cynomoriaceae.

The presence of synandrium, extreme reduction of gynoecium without any histological differentiation, and the peculiar mode of germination of "seed" may be some of the reasons for erection of a separate order Balanophorales (see Melchior 1964). This order is placed between Santalales and Medusandrales; the affinities with Santalales are referred to above but we do not have any embryological data for Medusandrales.

With our present knowledge it does not seem possible to discuss the taxonomic assignment any further. We must await more detailed information on taxa which can be considered as relatives of this family (see also Kuijt 1969).

14 Medusandrales

This order is represented by a single family Medusandraceae; the embryology has not been investigated.

15 Polygonales

The Polygonales includes a single family, Polygonaceae.

15.1 General Features

The Polygonaceae are: (1) highly multipalynous with colpate, colporate or porate pollen grains, the number and distribution of pores vary widely, the exine is elaborately reticulate, granulose or spinuloid, (2) the pollen grains are three-celled at dissemination, (3) the ovules are orthotropous, bitegmic, and crassinucellate with a prominent nucellar beak, (4) multicellular archesporium, (5) Polygonum type of embryo sac (antipodals ephemeral), (6) persistent remnants of pollen tube, (7) Nuclear type of endosperm, ruminate in *Coccoloba*, (8) Asterad type of embryogeny, and (9) straight embryo in an albuminous seed with unspecialized seed-coat.

15.2 Polygonaceae

The anther is tetrasporangiate, and the anther wall comprises an epidermis, fibrous endothecium, ephemeral middle layer, and secretory tapetum. The development of anther wall is of the Reduced type (Davis 1966) in *Antigonon leptopus* (Agoram and Krishnamurthy 1980), otherwise the development is of the Monocotyledonous type (Maheswari Devi and Manorama 1985). The tapetal cells become two- to four-nucleate. Simultaneous cytokinesis in the microspore mother cells follows meiosis, and the tetrads are tetrahedral and isobilateral.

The family is highly multipalynous, with colpate, colporate or porate pollen grains. The bronchi of exine may be well organized and small. Pollen polymorphism has been observed in *Polygonum* spp. (Maheswari Devi and Manorama 1984). The largest pollen grain is three-nucleate and embryo sac-like. The pollen grains are two-celled at the dispersal stage in *Rumex uesicarius*. In *Polygonum alatum*, *P. capitatum*, *P. chinense* and *P. molle* the pollen grains are three-celled, and triaperturate with reticulate exine (Maheswari Devi and Manorama 1985).

The ovule is orthotropous, bitegmic (or unitegmic through connation), and crassinucellar with a hypostase. The inner integument forms the micropyle. A nucellar break is prominent, and projects beyond the micropyle in *Polygonum persicaria* (Souèges 1919). The hypodermal archesporium is multicellular (Dudgeon 1918), but only one cell functions. It cuts off a parietal cell. The parietal tissue is five to six layers thick. The megaspore mother cell undergoes meiotic divisions giving rise to a linear or T-shaped tetrad. The chalazal megaspore is functional.

Development of the female gametophyte is of the Polygonum type. The synergids are hooked in *Rumex*. The three antipodals become multinucleate, but are ephemeral. The first clear account of the Polygonum type of embryo sac development was reported in *Polygonum diuericatum* (Strasburger 1879).

Fertilization is porogamous, and the pollen tube is persistent and unbranched. The synergids remain intact during the entry of the pollen tube (Mahony 1935). This needs confirmation.

Development of the endosperm is of the Nuclear type. Wall formation is initiated at the micropylar end at the globular proembryo in *Polygonum pensylvanicum* (Neubauer 1971). A meristematic layer differentiates at the periphery in the cellular endosperm. The cells in the central region become multinucleate. After 10 days of its formation the meristematic layer develops into an aleurone layer. The nucellus persists as a thin covering. The chalazal region of endosperm remains free-nuclear in *Fagopyrum esculentum* (see Davis 1966). This aspect requires further study.

The embryogeny conforms to the Asterad type (Neubauer 1971, B.A. Rao and Mukherjee 1974) in *Polygonum*. According to Guignard (1966), the embryogeny in *Rumex aquaticus* resembles that of *P. persicaria*.

The mature seed is pachychalazal with ruminate endosperm in *Antigonon leptopus* (Agoram and Krishnamurthy 1980). The ruminations conform to the "Coccoloba" type (Periasamy 1962a). The invaginations of seed-coat are restricted to the portion of seed derived from the chalazal region. They develop first at the chalazal end, and proceed upward.

The seed contains a small and straight embryo with starchy endosperm. The seed-coat is unspecialized with the outer epidermis of testa showing reticulate thickenings in *Rumex*.

In *Coccoloba uvifera* (Periasamy 1964a), during post-fertilization development, the young seed becomes tightly enclosed by the ovary wall. In a bitegmic young seed the endosperm is surrounded by a well-developed nucellus

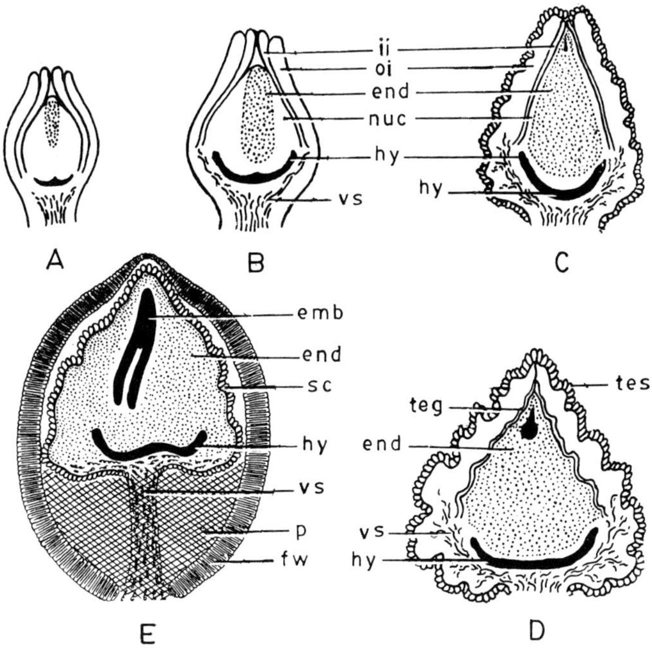

Fig. 15.2.1A–E. *Coccoloba uvifera*, development of seed (semidiagrammatic). **A** Longisection of ovule, endosperm *dotted*. **B** Later stage; *ii* inner integument; *oi* outer integument; *end* endosperm; *nuc* nucellus; *hy* hypostase; *vs* vascular supply. **C** Early globular proembryo (*black*), nucellus mostly consumed by endosperm, outer integument overarches the inner integument, and epidermis of the testa has become wavy. **D** Globular proembryo, nucellus completely consumed, tegmen (*teg*) wavy, and epidermis of testa (*tes*) thrown into folds. **E** Longisection of mature fruit to show dicotyledonous embryo (*emb*), ruminate endosperm (*end*), and excessive growth of chalaza; *sc* seed-coat; *p* pericarp; *fw* fruit wall. (After Periasamy 1964a)

(Fig. 15.2.1A); the chalazal tissue becomes meristematic, increases in breadth, and assumes a shallow cup-like appearance. A hypostase also develops (Fig. 15.2.1B). The vascular supply branches below the chalaza, and the branches reach up to the base of integuments. During its growth the endosperm consumes the nucellus (Fig. 15.2.1C, D).

The outer epidermis of the outer integument divides only anticlinally, the cells enlarge, become papillate, and accumulate dark-brown contents. The inner layers of the outer integument become meristematic and develop into irregular (mostly three) lobes. The chalazal growth is also massive and irregular. The two-layered inner integument shows undulations (Fig. 15.2.1D).

At first the epidermis of the outer integument becomes irregular, then this irregularity proceeds inward and the entire seed-coat assumes undulated appearance (Fig. 15.2.1D).

As the endosperm becomes cellular and grows, it consumes the nucellus and integument except for the lobed outer epidermis of outer integument which alone persists and renders the endosperm ruminate (Fig. 15.2.1E). The epidermis of the fruit wall develops into a palisade-like layer of radially elongate sclereids with heavily lignified walls (Fig. 15.2.1E). During the maturation of fruit, tissues of the fruit wall disorganize except for the epidermis.

S Agoram and KV Krishnamurthy (1980), HR Bhargava and RH Sawhney (1958), J Brandbyge (1989), GL Davis (1966), Y Doida (1960), W Dudgeon (1918), K Engell (1973), JL Guignard (1966), E Kirsten (1973), H Maheswari Devi and K Manorama (1984, 1985), KL Mahony (1935), PK Mukherjee (1972), BF Neubauer (1971), K Periasamy (1962a, 1964a), BA Rao and PK Mukherjee (1974), VS Rao (1936b), BS Siva Rao (1952), R Souèges (1919), E Strasburger (1879)

15.3 Taxonomic Considerations

Bentham and Hooker (1965c) place the family Polygonaceae in Curvembryae, and Bessey (1915) in Caryophyllales. Hallier (1912) recognizes the affinities between Caryophyllales and Polygonales, and includes them in his Centrospermae. Engler and Diels (1936) and Hutchinson (1973) treat Polygonaceae as a separate order, Polygonales. Hutchinson (1973) considers it as a reduced degraded type of Caryophyllales. Polygonales have their ancestry in Caryophyllales (Lawrence 1951).

Cronquist (1968, 1981) places Polygonales prior to Plumbaginales, and states that both are related to Caryophyllales. All the three orders contain anthocynins, and have food storage tissue in seed. R. Dahlgren (1981) places Plumbaginales (position uncertain) and Polygonales in superorder Polygoniflorae. The relationship of Polygonaceae with Plumbaginaceae is discussed in Section 44.3.

Polygonaceae. Polygonaceae show closest affinity with Caryophyllaceae, and in both the families the pollen grains are colpate, colporate or porate. The bi-, tri- or tetracarpellary unilocular ovary has a basal ovule, occasionally also in Caryophyllaceae. Further resemblance is a bitegmic, crassinucellar ovule with multicellular archesporium, Polygonum type of embryo sac with ephemeral antipodal cells, Nuclear type of endosperm, and albuminous seed. However, the Asterad type of embryogeny, small straight embryo in mature seed, absence of perisperm, and unspecialized seed-coat differ from those of Caryophyllaceae. The latter family has campylotropous ovule, Caryophyllad type of embryogeny, and curved embryo (around the perisperm) in mature seed. The seed is exotestal with reticulate markings on the exotesta. The mesophyll is unspecialized, and the tegmen is crushed in Caryophyllaceae.

On the whole, the embryological data support the phyletic position assigned to the order.

16 Centrospermae

The order Centrospermae comprises 12 families, distributed over four suborders: Phytolaccineae – Phytolaccaceae, Gyrostemonaceae, Achatocarpaceae, Nyctaginaceae, Molluginaceae, Aizoaceae; Portulacineae – Portulacaceae, Basellaceae; Caryophyllineae – Caryophyllaceae; Chenopodiineae – Dysphaniaceae, Chenopodiaceae, Amaranthaceae. The family Didieraceae is appended to it.

16.1 General Features

The Centrospermae show: (1) secretory anther tapetum, (2) simultaneous cytokinesis in pollen mother cells, (3) three-celled pollen grains, two- to three-celled in Nyctaginaceae, (4) ana-campylotropous, bitegmic, crassinucellar ovule with a characteristic air-space (between the integuments) at the base of ovule, (5) curved nucellus producing a nucellar cap, persists as mealy perisperm, (6) uni- or multicelled archesporium, (7) massive parietal tissue, (8) Polygonum type of embryo sac, mono-, bi- and tetrasporic embryo sacs occur in Nyctaginaceae and Aizoaceae, (9) endosperm of the Nuclear type, (10) Caryophyllad type of embryogeny in Phytolaccaceae, Molluginaceae, Portulacaceae, Caryophyllaceae, Asterad in Nyctaginaceae, Solanad in Aizoaceae, Chenopodiad in Basellaceae, Chenopodiaceae, Amaranthaceae, Didieraceae, and (11) generally exotestal arillate seed with a large curved embryo enclosing a massive perisperm.

16.2 Phytolaccaceae

The anther wall comprises the epidermis, fibrous endothecium, two or three ephemeral middle layers, and secretory tapetum of binucleate cells. The cells become multinucleate in *Phytolacca decandra* (Lewis 1905). Simultaneous cytokinesis follows meiotic divisions of microspore mother cells, and results in tetrahedral and isobilateral tetrads.

The pollen grains are tricolporate and three-celled at the dispersal stage in *Rivina humilis* (A.C. Joshi 1936a), and two-celled in *Phytolacca*.

The ovule is basal, ana-campylotropous, bitegmic and crassinucellate. The inner integument organizes the micropyle. An air-space is present between the two integuments, at the base of ovule. The cells of the nucellar epidermis, by periclinal divisions, produce a massive nucellar tissue. It is strongly curved, and the epidermal cells below the micropyle are stretched radially. This tissue becomes 12–14 cells thick in *Rivina* (A.C. Joshi 1936a). A funicular obturator and hypostase also differentiate.

The archesporial cell cuts off a parietal cell. The parietal tissue is massive. The megaspore mother cell undergoes meiotic divisions giving rise to a linear tetrad. The chalazal megaspore produces a Polygonum type of embryo sac. The synergids are elongate with a pointed apex, the polar nuclei fuse before fertilization, and the antipodals persist up to fertilization. In *Rivina humilis* (Kajale 1954a), the synergids are persistent, and the antipodal cells undergo secondary multiplication after fertilization.

The development of endosperm is of the Nuclear type. Wall formation commences at the late globular proembryo, and finally it becomes completely cellular. The enodsperm stores abundant starch.

The development of embryo follows Chenopodiad type in *R. humilis*, and Caryophyllad type in *Phytolacca dioica* (Kajale 1944a, 1954a). The suspensor is massive.

The seed is small and exotestal, and contains a large curved embryo enclosing the starchy perisperm. The scanty endosperm persists as one or two layers in the micropylar half of the seed. The seed is arillate in capsular genera, but the aril is absent in drupaceous genera. A vestigial aril occurs in *Rivina*. The outer epidermis of testa is composed of radially elongate, thick-walled cells with brown contents. The mesophyll and the inner epidermis become crushed. The tegmen is represented only by its inner epidermis since the outer collapses. The cells develop bar-like thickenings on the radial walls, and have brown contents.

AC Joshi (1936a), LB Kajale (1944a, 1954a), JF Lewis (1905), T Rocén (1927), O Rohweber (1970), P Satyanarayana and LL Narayana (1985), D Singh and CP Taneja (1985), MS Yakovlev (1983)

16.3 Gyrostemonaceae

The embryology has not been investigated.

16.4 Achatocarpaceae

The embryology has not been studied.

The anthers are bilocular. The ovule is basal and campylotropous. Seed exarillate, solitary, erect, lenticular, with strongly bent embryo surrounding the mealy endosperm, cotyledons linear. Fruit one-seeded berry.

J Hutchinson (1973), JC Willis (1973)

16.5 Nyctaginaceae

The anther wall comprises the epidermis, fibrous endothecium, two ephemeral middle layers, and a secretory tapetum of binucleate cells. The microspore mother cells undergo meiotic divisions, cytokinesis is simultaneous, and the tetrads tetrahedral and isobilateral.

The pollen grains are three- to four-colporate and two-celled at the dispersal stage. They are three-celled in *Oxybaphus micranthus* (Rocén 1927). Sterility of pollen grains occurs in *Boerhaavia repanda* (Bhargava 1932) and *Bougainvillea glabra* (D.C. Cooper 1931a).

The ovule is anatropous or campylotropous, and crassinucellar. It is bitegmic in *Bougainvillea*, *Mirabilis*, *Oxybaphus* and *Pisonia* with the inner integuments forming the micropyle, and unitegmic in *Ambronia* (Rocén 1927) and *Boerhaavia* (P. Maheshwari 1929, Bhargava 1932). The integumentary vascular bundle extends up to the micropyle in *Mirabilis*. Funicular obturator develops in *Pisonia aculeata* (Venkateswarlu 1947b). The single archesporial cell cuts off a parietal cell. The parietal tissue is massive. The nucellar epidermis produces a prominent nucellar beak in *Boerhaavia diffusa* (P. Maheshwari 1929) and *B. repanda* (Bhargava 1932). The megaspore mother cell, following meiotic divisions, produces a linear tetrad. In *Bougainvillea* (Das and Ramkrishnan 1974), a number of megaspore mother cells undergo meiotic divisions, and several linear tetrads are produced; most of them degenerate. The female gametophyte is of the Polygonum type. The synergids are ephemeral, and the three persistent antipodals multiply producing six to seven cells.

Several abnormalities in the formation of tetrads and female gametophyte have been reported. In *Oxybaphus nyctagineus* (Rocén 1927), a megaspore tetrad may contain a binucleate cell in the middle, and one uninucleate cell on either side. The binucleate cell never develops further. The suppression of one or more transverse walls in the linear tetrad in *Mirabilis jalapa* results in the formation of bi-, tri- or tetrasporic types of embryo sac (Hjelmqvist and Grazi 1965). Normally, monosporic development occurs. Sometimes, the partition walls are incomplete and the adjoining mature embryo sacs may fuse. Hjelmqvist and Grazi (1965) attributed these variations to the effect of temperature.

The entry of the pollen tube is porogamous. It finds its way between the two synergids leaving them intact (D.C. Cooper 1949). The pollen tube persists throughout the development of seed in *Oxybaphus*, and its haustorial role is suggested (D.C. Cooper 1949). An accessory pollen tube is reported in *Boerhaavia diffusa* (P. Maheshwari 1929). The persistence of both the synergids, and the haustorial role of the pollen tube are questionable.

Development of the endosperm is of the Nuclear type. After the formation of 64 free nuclei in *Oxybaphus* (D.C. Cooper 1949), the embryo becomes aggressive and the nuclear divisions in endosperm slow down. The endosperm becomes cellular in the micropylar region, but a larger part is absorbed in the free-nuclear state in *Boerhaavia* (Kajale 1938) and *Mirabilis jalapa* (Woodcock 1929).

The zygote divides transversely producing the apical and basal cells. The basal cell undergoes transverse divisions resulting in a filamentous six-celled proembryo. Simultaneously, a longitudinal division occurs in the apical cell followed by a vertical wall. Then the vertical divisions proceed in the second, third and fourth cell from the apex in *Boerhaavia* (Kajale 1936, 1938). Such a sequence of wall formation in the family Nyctaginaceae is totally different from that of the other related families of the order Centrospermae. The micropylar cells develop into a long and slender suspensor in *B. diffusa*, and a short and massive suspensor in *B. repanda* (Kajale 1938). Another peculiar feature is the formation of periblem from the dermatogen. It is not a sister layer of pleurome as in other angiosperms. The embryogeny conforms to the Asterad type. In *Oxybaphus* (see Crété 1963a), it conforms to Period I, Series A, Megarchetype III (Asterad type). The embryogeny in *Pisonia aculeata* (Venkateswarlu 1947b) is also of the Asterad type.

Synergid embryos have been observed in *B. repanda* (Kajale 1938).

The seed is small (except in *Pisonia*), with or without perisperm. The outer persistent epidermis of testa is composed of thick-walled cells with brown contents. These cells are not lignified. The mesophyll and inner epidermis of testa become crushed. The inner epidermis is thick-walled in *Mirabilis*. The inner epidermis of tegmen persists only in *Mirabilis* and *Pisonia*. Its cells show fine striations on the tangential walls. Otherwise, the tegmen becomes crushed along with the outer epidermis.

HR Bhargava (1932), DC Cooper (1931a, 1949), P Crété (1963a), VLD Das and V Ramkrishnan (1974), H Hjelmqvist and F Grazi (1965), LB Kajale (1936, 1937a, 1938), P Maheshwari (1929), P Maheshwari and BM Johri (1950b), T Rocén (1927), R Souèges (1938c), J Venkateswarlu (1947b), EF Woodcock (1929)

16.6 Molluginaceae

The anther wall comprises the epidermis, fibrous endothecium, an ephemeral middle layer, and secretory tapetum with bi- and trinucleate cells in *Limeum indicum*. Übisch granules occur in *Orygia decumbens* (H.S. Narayana and Lodha 1963), and nuclear fusions take place in the tapetal cells. Following meiotic divisions, cytokinesis is simultaneous, and the microspore mother cells produce tetrahedral, decussate and isobilateral tetrads.

The pollen grains are tricolporate; three-celled at the dispersal stage. Occasionally, they germinate in situ in *Mollugo nudicaulis* (Bhargava 1934).

The ovule is anatropous, bitegmic and crassinucellar with inner integument forming the micropyle. An air-space is present between the integuments, at the base of ovule, on one side. The nucellus is strongly curved in *Limeum indicum* (H.S. Narayana and Jain 1962). The cells of the nucellar epidermis divide to form a massive nucellar cap. A single hypodermal archesporial cell cuts off a parietal cell. Multicellular archesporium occurs in *Orygia*. The parietal tissue is two or three layers thick. The megaspore mother cell undergoes meiotic divisions resulting in a linear or T-shaped tetrad. The chalazal megaspore develops into a Polygonum type of embryo sac. The synergids are beaked, the polar nuclei fuse before fertilization, and the antipodals may be ephemeral. The development of a hypostase is a common feature. A funicular obturator develops in *Limeum*, whereas in *Orygia* the epidermis of the placenta develops multicellular filamentous outgrowths in the vicinity of micropyle.

Development of the endosperm is of the Nuclear type. Wall formation is initiated around the proembryo at the globular stage in *Limeum* and *Orygia decumbens* (H.S. Narayana and Lodha 1963), and finally the endosperm becomes completely cellular. The cells are uni- or binucleate, and store starch. In a mature seed the endosperm persists as a cap around the radicle.

The embryogeny conforms to the Caryophyllad type-Sagina variation in *Gisekia pharnaceoides* (A.C. Joshi and Rama Rao 1936), but in *Limeum indicum* H.S. Narayana and Jain (1962) report Caryophyllad type-Drosera variation. H.S. Narayana and Lodha (1963), however, suggest Solanad type-Linum variation in *Orygia decumbens*.

The reniform seed of *Orygia* contains a large curved embryo on the hilar side, with a massive mealy perisperm. The endosperm persists as one or two layers in the micropylar half of seed. Both the integuments contribute to the formation of seed-coat. The outer epidermis of testa is tanniniferous, and the inner is thin-walled. The outer epidermis of tegmen collapses due to the enlargement of the inner epidermis. The cells of the latter have a banded appearance due to bar-like thickenings on the radial walls (Fig. 16.6.1A, B).

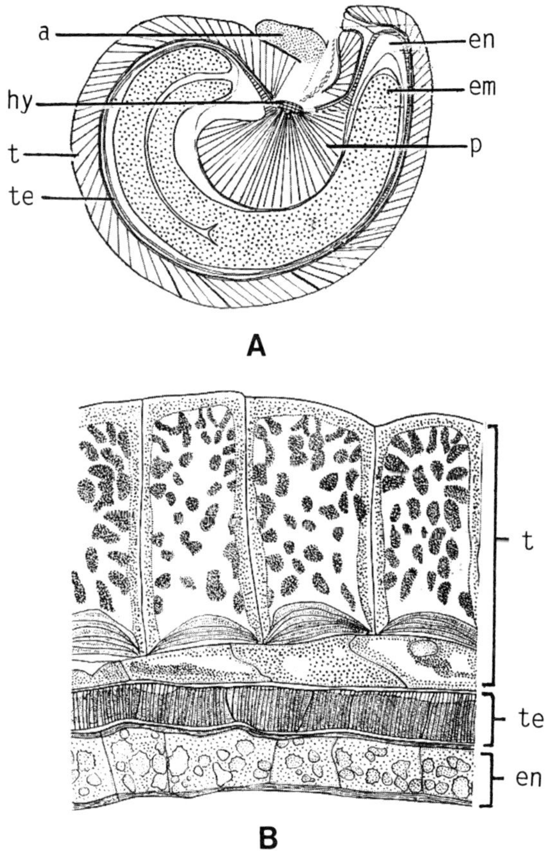

Fig. 16.6.1A, B. *Orygia decumbens.* **A** Longisection of mature camplyotropous seed. There is copious perisperm (*p*), scanty endosperm (*en*), curved embryo (*em*), and a short funicular aril (*a*); *hy* hypostase; *t* testa; *te* tegmen. **B** Transection of seed-coat to show two-layered testa and fibrous endotegmen. Outer epidermal cells of the testa are radially elongate, and lamellate thickenings occur on inner walls. Exotegmen has collapsed. (After H.S. Narayana 1962a)

The funicular aril surrounds the seed in *Mollugo oppositifolia*; it persists as a vestige in *Gisekia*, *M. nudicaulis* and *Orygia*. It is absent in *Limeum*.

HR Bhargava (1934), AC Joshi and V Rama Rao (1936), HS Narayana and K Jain (1962), HS Narayana and BC Lodha (1963), BS Siva Rao (1975)

16.7 Aizoaceae

The anther wall comprises an epidermis, fibrous endothecium, single ephemeral middle layer, and glandular tapetum with two- to four-nucleate cells in *Glinus lotoides* (H.S. Narayana and Lodha 1972). Simultaneous cytokinesis accompanies meiotic divisions in microspore mother cells. The tetrads are tetrahedral and decussate.

The pollen grains are tricolpate, and three-celled at the dispersal stage.

The ovule is ana-campylotropous, bitegmic and crassinucellar. The inner integument forms the micropyle. Both the integuments are two-layered. The apical cells form a collar-like structure in *Aptenia cordifolia* (N. Prakash 1967a) and *Delosperma cooperi* (Kapil and Prakash 1966). An air-space occurs between the integuments at the base. The nucellar epidermis is radially stretched in *Mesembryanthemum*, *Tetragonia* and *Trianthema*. The archesporium is uni- or multicellular. The archesporial cell cuts off a parietal cell, and the parietal tissue is well developed. The megaspore mother cell undergoes meiotic divisions to produce a linear tetrad. The chalazal megaspore develops into a Polygonum type of embryo sac. The synergids are hooked, polar nuclei fuse before fertilization, and antipodals are ephemeral. The latter may or may not organize into cells in *Delosperma*. Funicular obturator occurs in *Glinus lotoides* (H.S. Narayana and Lodha 1972), *Tetragonia tetragonioides* (N. Prakash 1967b), and *Trianthema argentina* (Cocucci 1961), and placental in *Aptenia cordifolia*. An operculum is formed in *Hypertelis bowkeriana* (Bregman and Bouman 1983).

Kapil and Prakash (1966) describe five types of embryo sacs in *Delosperma cooperi*: 14% Polygonum type, 18% Endymion type, and 68% tetrasporic – 47% Drusa type, 35% Penaea type, and 18% Adoxa type. Occasionally, an aposporic embryo sac (with reversed polarity) develops from the additional archesporial cell.

Development of the endosperm is of the Nuclear type. A weak chalazal haustorial caecum is reported in *Hereroa hesperantha* (N. Prakash 1967c). Wall formation is initiated at the globular proembryo. It is restricted to the micropylar region in *Aptenia cordifolia* and *Tetragonia tetragonioides* (N. Prakash 1967a, b).

The embryogeny conforms to Solanad type-Linum variation in *Glinus*. Solanad type also occurs in *Sesuvium portulacastrum*, *Trianthema decandra* and *T. monogyna*. Suspensor polyembryony occurs in *A. cordifolia*.

The exotestal and endotegmic seed contains a large curved embryo, well-developed perisperm, and scanty endosperm (H.S. Narayana 1962a). The testa is multiplicative in *Trianthema*. The outer epidermis of testa is thick-walled with brown contents, the inner one is unspecialized. The outer epidermis of tegmen is thick and persistent in *Mesembryanthemum.* The inner epidermis is uniformly thick. The aril surrounds the seed in *Trianthema* and *Sesuvium.* It is absent in *Aptenia cordifolia, Hereroa hesperantha* and *Tetragonia tetragonioides* (N. Prakash 1967a, b, c).

HR Bhargava (1935a), R Bregman and F Bouman (1983), AE Cocucci (1961), VR Dnyansagar and RS Malkhede (1963), RN Kapil and N Prakash (1966), HS Narayana (1962a), HS Narayana and BC Lodha (1972), F Pax (1889), N Prakash (1967a, b, c), W Schmid (1925)

16.8 Portulacaceae

The anther wall comprises the epidermis, fibrous endothecium, a single ephemeral middle layer, and secretory tapetum of binucleate cells. The tapetal cells become multinucleate, and the nuclei fuse in *Portulaca tuberosa* (Raghavan and Srinivasan 1941c). Granular cutinization occurs on the inner wall of tapetal cells. The archesporium is a single row of cells which function as microspore mother cells (Kajale 1942). They undergo meiotic divisions giving rise to tetrahedral and isobilateral tetrads. Cytokinesis is simultaneous.

The pollen grains are three- to four-colporate, and three-celled at the shedding stage.

The ovule is ana-campylotropous, bitegmic and crassinucellar. The inner integument forms the micropyle. At the base of the ovule an air-space occurs between the two integuments in *Portulaca.*

The hypodermal archesporial cell cuts off a parietal cell, the parietal tissue is three or four layers thick. The apical cells of the nucellus elongate radially. Following meiotic divisions, the megaspore mother cell produces a linear or T-shaped tetrad. The chalazal megaspore gives rise to the Polygonum type of embryo sac. The synergids are hooked, and the antipodals are small and ephemeral. Starch grains appear at the binucleate stage of the embryo sac, these have also been observed in the egg cell (D.C. Cooper 1940).

In *Portulaca* three to four hours elapse between pollination and fertilization. Through the micropyle the pollen tube enters the embryo sac between the two synergids, without damaging any of them. The tip of the pollen tube bifurcates, one branch passes to the egg and the other to the secondary nucleus (D.C. Cooper 1940). The fate of synergids, and branching of pollen tube need confirmation.

The development of endosperm is of the Nuclear type. It becomes cellular at the globular proembryo in *Portulaca tuberosa.*

The embryogeny conforms to Period I, Series A, Megarchetype III (Onagrad type) in *Claytonia,* and Period II, Series A, Megarchetype III (Caryophyllad type) in *Portulaca* (Souèges 1938b). The occurrence of three embryos in an embryo sac is reported in *Portulaca oleracea* (Kajale 1942). The largest of the three is zygotic, the origin of the other two is disputable. The mature embryo is curved.

The seed is small, reniform, albuminous, with a strophiole in *Claytonia* and *Talinum*, and has starchy perisperm. The seed is exotestal with the outer epidermis of cuboid cells with thickened and lignified outer wall. They may be finely pitted, papilliform or with short hair-like processes. The inner epidermis is a crystal-layer in *Portulaca*; otherwise it is crushed. The cells of the inner epidermis of tegmen have bar-like thickenings on the radial walls in *Portulaca*. The rest of the tissue is crushed (Corner 1976).

The seeds of *Portulaca crytopetala* and *Talinum patens* develop an operculum. In the seeds of *Anacampseros rufescens*, adapted to wind dispersal, the outer layer of the outer integument is papery and can be separated from the inner layer of the outer integument, forming a balloon (Bregman and Bouman 1983).

R Bregman and F Bouman (1983), R Carolin (1987), MT Cook (1903), DC Cooper (1935a, 1940), EJH Corner (1976), B Haccius (1954), LB Kajale (1942), AM Khan (1973, 1975), G Klebs (1881–1885), WH Murdy and MEB Carter (1987), TS Raghavan and AR Srinivasan (1941c), T Rocén (1927), R Souèges (1938b, 1945c)

16.9 Basellaceae

The anther wall comprises persistent epidermis, fibrous multiseriate endothecium, ephemeral middle layer, and glandular tapetum with multinucleate cells. The microspore mother cells undergo meiotic divisions, cytokinesis is simultaneous, and the tetrads tetrahedral and isobilateral.

The family is highly multipalynous, and the pollen grains are three-celled at the dispersal stage in *Basella rubra* (Maheswari Devi and Pullaiah 1975).

The ovule is ana- to campylotropous, bitegmic and crassinucellar with inner integument forming the micropyle. The inner integument overgrows the outer one. The cells of the nucellar epidermis divide and form a nucellar beak. A prominent air-space develops between the integuments in the basal region. The archesporial cell cuts off a parietal cell, the parietal tissue is well developed. The megaspore mother cell undergoes meiotic divisions forming a linear tetrad. The chalazal megaspore develops into Polygonum type of embryo sac, polar nuclei fuse before fertilization, antipodals are small and ephemeral, and starch grains are present in the mature embryo sac.

The development of endosperm is of the Nuclear type. Cytokinesis occurs at the globular proembryo in *Basella*.

The embryogeny conforms to Chenopodiad type in *Basella rubra* (Maheswari Devi and Pullaiah 1975).

The seed is small, exarillate, and contains perisperm. The outer epidermis of testa comprises short, radially elongate cells with thick outer walls. The tegmen is represented only by the inner epidermis. The cells have streak-like thickenings on the radial walls.

GL Hindmarsh (1965), H Maheswari Devi and T Pullaiah (1975)

16.10 Caryophyllaceae

The anther is bilocular, and the wall comprises the epidermis, fibrous endothecium, one or two ephemeral middle layers, and secretory tapetum of binucleate cells. Simultaneous cytokinesis accompanies meiosis in microspore mother cells, and tetrahedral and isobilateral tetrads are produced in *Stellaria media* (P.C. Joshi 1936, see also P.S. and L.L. Narayana 1986).

The pollen grains are three- or four-colporate, pancolpate, and pantoporate. They are three-celled at the dispersal stage. In *Polycarpea nebulosa* (Long 1965), some pollen grains precociously produce abortive pollen tubes within the anther. Normal pollen tubes develop on the stigma. A similar condition occurs in *Thylacospermum* (P.C. Joshi 1936).

The ovule is campylotropous, bitegmic and crassinucellar (Fig. 16.10.1A) with the micropyle formed by inner integument (S. Pal and Murty 1972). The cells of the nucellar epidermis undergo periclinal divisions. The hypodermal archesporium is uni- or multi-celled. A single cell cuts off a parietal cell. The

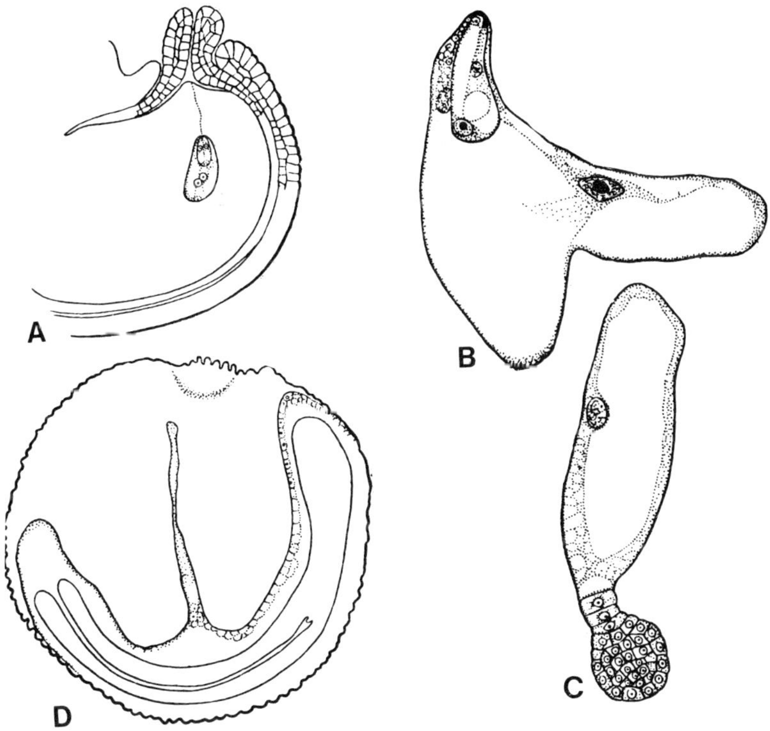

Fig. 16.10.1A–D. *Agrostemma githago.* **A** Longisection of ovule at eight-nucleate embryo sac. **B** Embryo sac with lateral extension containing secondary nucleus. **C** Proembryo with a large uninucleate suspensor. **D** Ovule at dicotyledonous embryo. Note the persistent lateral extension of the embryo sac. (After Rocén 1926)

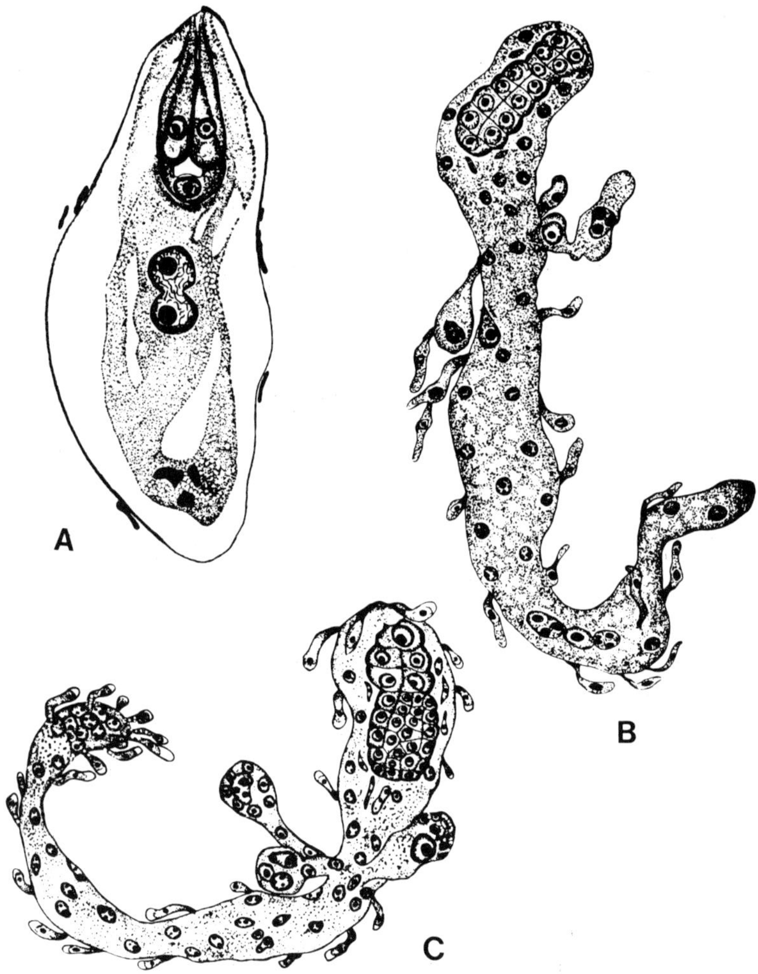

Fig. 16.10.2A, B. *Acanthophyllum kraschenninnikovii*, **B** *A. albidium*. **A** Mature embryo sac. **B**, **C** Nuclear endosperm with finger-like appendages at globular proembryo. (See Yakovlev 1983)

parietal tissue is single-layered in *Stellaria media* (P.C. Joshi 1936), and two- to three-layered in *S. paniculata* (S. Pal and Murty 1974). The megaspore mother cell undergoes meiotic divisions. The chalazal megaspore of a linear or T-shaped tetrad develops into a Polygonum type of embryo sac (Fig. 16.10.2A). The synergids are hooked, polar nuclei fuse before fertilization, and antipodals are ephemeral. The embryo sac may develop a lateral caecum (Fig. 16.10.1B, see also D).

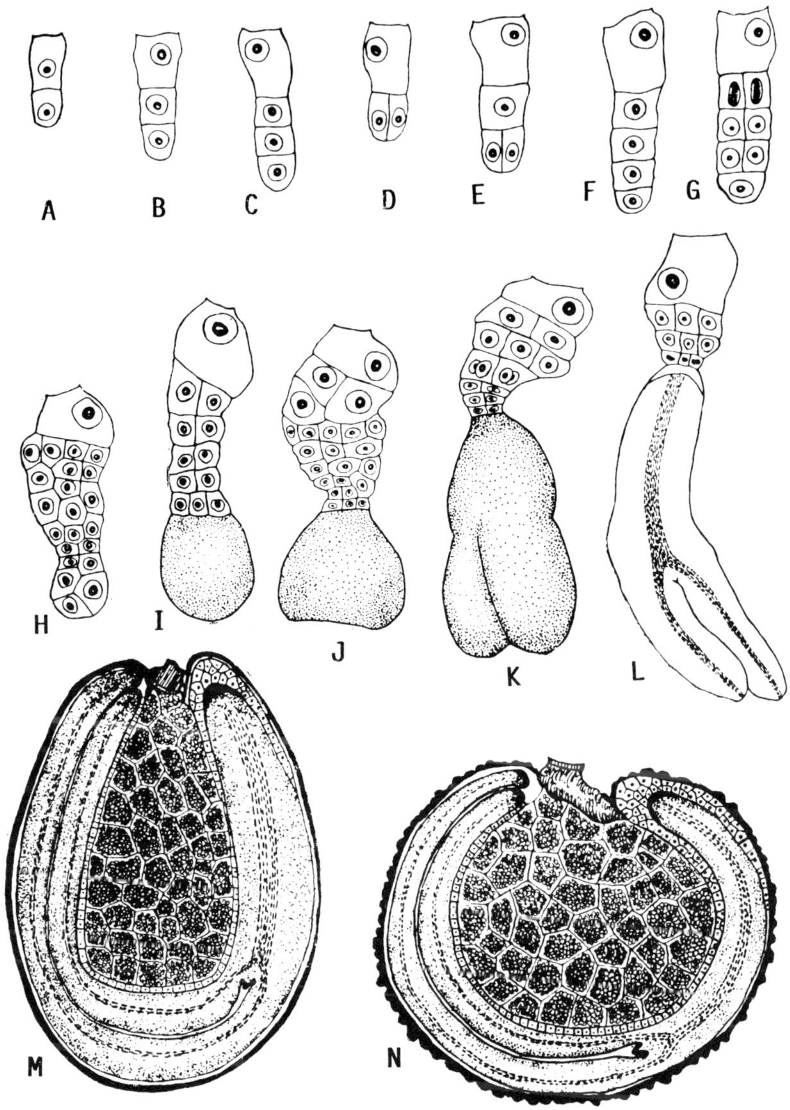

Fig. 16.10.3A–L. *Acanthophyllum cyrtostegium.* Development of embryo. Note the enlarged basal cell. **M** *A. paniculatum,* **N** *Gypsophila paniculata.* Longisection of mature seed. The endosperm persists only in the micropylar region, and the curved embryo encloses perisperm. (See Yakovlev 1983)

The development of endosperm is of the Nuclear type. In some taxa, the endosperm develops superficial extensions (Fig. 16.10.2B, C). Rocén (1927) further classified the development of endosperm into three types, according to the extent of cellular endosperm present in the seed: (1) Silene type – the endosperm occupies the entire seed, (2) Melandrium type – the endosperm occupies only the micropylar half of seed, and (3) Heliosperma type – a single layer of endosperm is present around the hypocotyl of embryo.

The embryogeny corresponds to the Caryophyllad type, and the basal cell of the conspicuous suspensor is vesicular and haustorial (Figs. 16.10.1C, 16.10.3A–L). The mature embryo is strongly curved around the copious mealy perisperm (Fig. 16.10.3M, N).

The seed is small, reniform, usually exarillate, and perispermous. In *Stegnosperma halimifolium* and *S. watsonii*, an aril develops from the chalazal region and completely surrounds the seeds (P.S. and L.L. Narayana 1986). The outer epidermis of testa comprises tabular cells with thick outer walls, the inner epidermis has tabular cells with thick inner walls, and the mesophyll is crushed. The tegmen is generally crushed, but the inner epidermis may persist.

JC Audran and M Batcho (1981), KM Buell (1952a, b), IP Chaudhury and HS Narayana (1978), RE Crang and MA Millay (1972), WD Creasy (1961), JL Guignard and JC Mestre (1967), PC Joshi (1936), K Kudlicka, M Kuras and B Rodkiewicz (1981), RW Long (1965), M Longevialle (1970, 1971), PS and LL Narayana (1986), W Newcomb and LC Fowke (1973, 1974), N Pal (1951d, 1952), S Pal (1974, 1975), S Pal and YS Murty (1972, 1974), MV Ramji (1975), T Rocén (1926, 1927)

16.11 Dysphaniaceae

The embryology has not been investigated.

16.12 Chenopodiaceae

The anther wall comprises an epidermis, fibrous endothecium and an ephemeral middle layer. The tapetum is mostly of the secretory type: *Arthrocnemum indicum* (Mahabale and Solanky 1954a), *Beta vulgaris* (Artschwager 1927), *Chenopodium ambrosioides* (Mahabale and Solanky 1954b), *Kochia scoparia* (Mahabale and Solanky 1953b) and *Suaeda maritima* (Maheswari Devi and Rama Rao 1976). The tapetal cells are uni-, bi- or four-nucleate and nuclear fusions are common.

In *A. indicum*, *C. ambrosioides*, *K. scoparia* and *S. fruticosa* (see Lakshmanan and Dulcy 1982), the cell walls break down and the protoplasts form a mass at the periphery.

In *Atriplex repens*, Lakshmanan and Dulcy (1982) report that in some loculi of a few anthers the tapetum is "amoeboid". The tapetal cells enlarge abnormally and, at the microspore stage, their inner walls break down and the protoplasts protrude into the loculi. The protoplast is densely cytoplasmic and forms tubular extensions around the microspores. According to Lakshmanan and Dulcy, the protoplasts are aggressive "haustorial" structures. If such extensions occur in the oppositely placed tapetal cells, these processes ramify in the locule around the microspores. They degenerate along with, or later than, the secretory tapetal cells.

The tapetum, in fact, is not of the amoeboid type, and such behaviour of tapetal protoplasts is of the intermediate type between the amoeboid and secretory. There is no question of the tapetal protoplasts being aggressive and haustorial. These protoplasts provide nutrition to the developing pollen grains.

Amoeboid type of tapetum is reported in *Chenopodium murale* (Mahabale and Solanky 1954c), and both secretory and amoeboid in male sterile *Beta vulgaris* (Artschwager 1947). The report of amoeboid tapetum in *Arthrocnemum indicum* by Mahabale and Solanky (1954a) seems to be erroneous.

All cases of aberrant behaviour of tapetum should be reexamined.

Sterile pollen grains occur in some loculi where the endothecium lacks fibrous thickenings, and the tapetum does not organize.

The microspore mother cells undergo meiotic divisions, and produce isobilateral and tetrahedral tetrads. Simultaneous cytokinesis occurs by furrowing.

The pollen grains are three-celled at the shedding stage. The occurrence of two vegetative nuclei in the pollen grains of double size is reported in *Chenopodium album* (Bhargava 1936), *Kochia scoparia* and *Suaeda fruticosa* (Mahabale and Solanky 1953a, b). Four-nucleate pollen grains with two vegetative and two male nuclei have been observed in *Arthrocnemum indicum* (Mahabale and Solanky 1954a), *Atriplex hymenelytra* (Billings 1934), *C. album* (Bhargava 1936) and *C. ambrosioides* (Mahabale and Solanky 1954b). The development of male gametophytes needs further study.

The ovule is campylotropous, bitegmic and crassinucellar. The inner integument forms the micropyle. A prominent air-space occurs between the two integuments in the chalazal region. The archesporium is one-, three- or four-celled in *Salicornia brachiata* and *S. fruticosa* (Maheswari Devi and Rama Rao 1982). The archesporial cell cuts off a parietal cell, the parietal tissue is not very conspicuous except in *Chenopodium ambrosioides* and *Salicornia brachiata* (Maheswari Devi and Rama Rao 1982). The nucellar epidermis divides, and produces a nucellar beak. Its cells elongate and become richly protoplasmic in the micropylar region. The megaspore mother cell undergoes meiotic divisions and produces a linear tetrad. The chalazal megaspore develops into a Polygonum type of embryo sac. The synergids have prominent hooks, and the three antipodal cells may be ephemeral or persist up to fertilization. In *Beta vulgaris* supernumerary antipodals are produced. Their number goes up to 15 in *C. amaranticolor*. The polar nuclei fuse before fertilization. Allium type of development of the embryo sac is reported in *Suaeda maritima* (Maheswari Devi and Rama Rao 1976). Twin embryo sacs occur in *Arthrocnemum*. A chalazal caecum develops in *Beta* and *Kochia*.

The pollen grains germinate in two to three hours on the stigma in *B. vulgaris*. Entry of the pollen tube is porogamous, and the richly cytoplasmic nucellar beak guides the growth of the pollen tubes (Artschwager and Starrett

1933). Double fertilization occurs, and the pollen tube persists up to six-celled proembryo.

The development of endosperm is of the Nuclear type. The endosperm becomes completely cellular at a young stage of the proembryo in *Arthrocnemum indicum* (Mahabale and Solanky 1954a). One or two layers of endosperm persist in the micropylar half of the seed.

The embryogeny conforms to the Chenopodiad type. The suspensor is massive except in *Arthrocnemum* and *Kochia* where it is uniseriate with a large basal haustorial cell. The mature embryo is curved or, generally, spirally-coiled. Polyembryony is reported in *Beta* and *Suaeda maritima* (Maheswari Devi and Rama Rao 1976).

The seed is small with starchy perisperm, and spirally coiled embryo. The integuments are reduced to exotesta and endotegmen. The cells of outer epidermis of the testa are somewhat radially elongated with their outer walls strongly thickened, and contain tannin. The inner epidermis of tegmen is persistent, and shows thickenings on the radial and inner walls.

E Artschwager (1927, 1947), E Artschwager and RC Starrett (1933), HR Bhargava (1936), FH Billings (1934), L Bruun and P Olesen (1989), GO Cooper (1935), MI Hakki (1971, 1973), B Jassem (1980), RL Jones (1974), KK Lakshmanan and EJ Dulcy (1982), TS Mahabale and IN Solanky (1953a, b, 1954a, b, c), H Maheswari Devi and MS Rama Rao (1976, 1982), DD Nautiyal and SC Gupta (1985), G Psaras (1984), R Souèges (1920a), ME Williams (1932), HJ Wilms (1980a, b, 1981a, b)

16.13 Amaranthaceae

The anther wall comprises an epidermis, fibrous endothecium, single ephemeral middle layer, and secretory tapetum of binucleate cells. The fibrous thickenings of endothecium are absent in *Achyranthes aspera* (Kajale 1937b). Übisch granules are studded on the inner walls of tapetal cells. In *Alternanthera paronychioides*, after the degeneration of tapetum, the Übisch granules come to lie appressed to the fibrous endothecium. The microspore mother cells undergo meiotic divisions and produce tetrahedral and isobilateral tetrads. Decussate tetrads are reported in *Psilostachys* (Bakshi 1952). Simultaneous cytokinesis occurs by furrowing.

The pantoporate pollen grains are three-celled at the shedding stage.

The basal ovule is orthotropous in *Psilostachys* (Bakshi 1952), campylotropous in *Celosia cristata* (Pal 1968) and amphitropous in *Amaranthus*. It is bitegmic and crassinucellar. The micropyle is formed by the inner integument, except in *Gomphrena* where both integuments are concerned. An air-space is present between the integuments in the basal region. A single cell of the multicellular archesporium cuts off a parietal cell producing parietal tissue. A nucellar cap is present. Following meiosis the megaspore mother cell produces a linear or T-shaped tetrad. The chalazal megaspore develops into a Polygonum type of embryo sac. The synergids are hooked, polar nuclei fuse before fertilization, and antipodals are ephemeral except in *Pupalia lappacea* (Kajale 1940, 1954b) where they increase up to 30–40 cells. In *Psilostachys* the embryo sac develops a chalazal caecum.

Double fertilization occurs as usual. The entry of accessory pollen tubes into the embryo sac is common in *Digera arvensis* (A.C. Joshi and Kajale 1937). This aspect needs further study.

The development of endosperm is of the Nuclear type. Wall formation commences in the micropylar part, and proceeds to the chalazal region. It becomes cellular at the globular proembryo. In *Alternanthera sessilis* the endosperm remains free-nuclear in the chalazal region.

The development of embryo follows the Chenopodiad type. In *Gomphrena celosioides* (Padhye 1962a), the embryogeny is of the Chenopodiad type, or Period I, Series C, Megarchetype V of Embryonic Group 3 of Souèges' system of classification.

The seeds are exotestal and endotegmic with funicular aril, and starchy perisperm. Both the integuments contribute to the formation of the seed-coat. The outer epidermis of testa is a mechanical layer of radially elongate cells with strongly thickened outer walls, and contains tannin. The inner epidermis is thin-walled. The inner epidermis of tegmen alone persists, and shows thickenings on the radial and inner walls.

In *Aerva tomentosa* (Sachar and Murgai 1958, 1959), the megaspore mother cell directly functions as uninucleate diploid embryo sac; meiosis does not occur (diplospory). Sometimes, mitotic division in the megaspore mother cell is followed by a wall and a "dyad" is formed. The micropylar dyad degenerates and the chalazal develops into a two-, four- and eight-nucleate embryo sac. The synergids, or even the egg apparatus, may degenerate. Due to the elongation of the embryo sac, the antipodals are shifted to a lateral position. Occasionally, twin embryo sacs develop, probably from two megaspore mother cells. The polar nuclei fuse to form the secondary nucleus which assumes the function of primary endosperm nucleus.

There is extreme rarity of male plants, and an almost total absence of pollen grains. This prevents fertilization, and pollen tubes have not been observed in the ovules. The Nuclear type of endosperm develops autonomously. The multinucleolate nuclei have an irregular outline. Cytokinesis occurs first at the micropylar end, and then proceeds to the chalaza.

The unfertilized unreduced egg develops parthenogenetically, and embryogeny is comparable to the Chenopodiad type. The suspensor may proliferate. The mature embryo is dicotyledonous.

Sachar and Murgai conclude that *Aerva tomentosa* is an apomict.

TS Bakshi (1952, 1954), AC Joshi (1936b), AC Joshi and LB Kajale (1937), LB Kajale (1935, 1937b, 1940, 1954b), R Mendoza (1973), MD Padhye (1962a), A Pal (1968), V Puri and B Singh (1935), RC Sachar and P Murgai (1958, 1959)

16.14 Didiereaceae

The embryology [see Schölch (1963) and Rauh and Schölch (1965)] has not been studied in sufficient detail, the following account is based on *Alluaudia dumosa*, *A. humbertii*, *A. montagnacii*, *A. procera* and *Didierea magascariensis*.

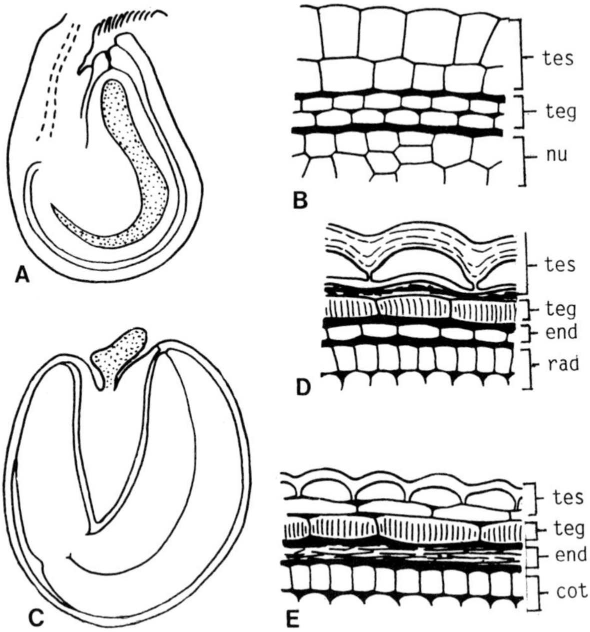

Fig. 16.14.1A–E. Didiereaceae, structure of seed. **A, B** *Alluaudia dumosa*; **A** Longisection of immature seed, micropyle is formed by inner integument; **B** Portion of seed-coat, there is a two-layered testa (*tes*) and two-layered tegmen (*teg*); *nu* nucellus. **C, D** *A. ascendens*; **C** Longisection of seed with curved embryo, and vestigial aril (*stippled*); **D** Portion of seed-coat from radicular region of **C** (endotesta and exotegmen crushed), exotesta with thick outer walls, endotegmen has thickenings, and there is a persistent layer of endosperm (*end*); *rad* radicular region of embryo. **E** *Didierea mirabilis*, portion of seed-coat in cotyledonary region, testa is two-layered, endotegmen shows thickenings, endosperm has degenerated, followed by a layer of cotyledon (*cot*). (After Perrot and Guérin 1903, see Corner 1976)

The embryological features of Didiereaceae are: secretory anther tapetum, simultaneous cytokinesis in pollen mother cells, bi- and tri-celled pollen, crassinucellate, ana- to campylotropous ovules with two-layered integuments, micropyle formed by the inner integument only, occasional presence of a nucellar cap formed by the division of nucellar epidermis, Polygonum type of embryo sac, Nuclear type of endosperm, Chenopodiad type of development of the embryo, exalbuminous (?), arillate seed with curved embryo (Fig. 16.14.1C), perisperm absent, and the structure of seed-coat is simple (Fig. 16.14.1A–E). The outer epidermal cells of testa develop thickenings, the inner may collapse (Fig. 16.14.1D) or persist (Fig. 16.14.1E). The outer epidermis of tegmen also collapses, and the inner develops bar-like thickenings (Fig. 16.14.1D, E). The endosperm is consumed; the presence of a single layer above the radicle is doubtful (see Corner 1976).

EJH Corner (1976), BM Johri (1967), E Perrot and P Guérin (1903), W Rauh and HF Schölch (1965), HF Schölch (1963)

16.15 Taxonomic Considerations

Engler and Prantl (1931) place all the 12 families (Melchior 1964) in Centrospermae. Hallier (1912) derives them from his Ranalian Podophylleae. Hutchinson (1973) distributes them over four orders: Nyctaginaceae in Thymelaeales; Didieraceae in Sapindales; Molluginaceae, Caryophyllaceae, Aizoaceae and Portulacaceae in Caryophyllales; Phytolaccaceae, Gyrostemonaceae, Chenopodiaceae, Amaranthaceae and Basellaceae in Chenopodiales.

There is a general uniformity of characters in the order.

The suborder Phytolaccineae comprises six families: Phytolaccaceae, Gyrostemonaceae, Achatocarpaceae, Nyctaginaceae, Molluginaceae and Aizoaceae.

Phytolaccaceae. Apart from the characters of the order, Phytolaccaceae have one basal ovule in each loculus of two-celled ovary, funicular obturator, persistent antipodals, Chenopodiad or Caryophyllad type of embryogeny, and schizocarpic fruit.

Nyctaginaceae. The position of Nyctaginaceae is rather controversial. The antipodals undergo secondary multiplication. The formation of mono-, bi- and tetrasporic embryo sacs in *Mirabilis jalapa* (Hjelmqvist and Grazi 1965) is similar to that in Aizoaceae. Kajale (1938) discusses the development of embryo in Nyctaginaceae (*Boerhaavia*) in relation to that in Caryophyllaceae, Amaranthaceae, Chenopodiaceae and Molluginaceae. According to him, there is a close similarity in the development of embryo in Centrospermales except in Nyctaginaceae. The proembryo is multicellular and uniseriate; four or five cells take part in the development of embryo proper. All the cells that form the embryo differentiate before any longitudinal division takes place in the proembryonal cells. The first longitudinal division occurs in the second, third or fourth cell from the apex, and then extends towards the apex. Contrary to this, in Nyctaginaceae the first vertical division occurs in the apical cell. Thus, the sequence in which longitudinal divisions appear in the proembryo is quite different from that in other families. Secondly, the periblem originates from the dermatogen, and is not a sister layer of plerome as in other angiosperms. All these developmental sequences need further investigation.

Venkateswarlu (1947c) supports the inclusion of Nyctaginaceae in Thymelaeales. The affinities between the two are: (1) secretory anther tapetum, (2) simultaneous cytokinesis by furrowing in pollen mother cells, (3) three-celled pollen grains, (4) presence of nucellar cap, (5) micropyle formed by inner integument, (6) supernumerary antipodals, and (7) Nuclear type of endosperm becoming cellular afterwards. Moreover, Nyctaginaceae differ from the rest of the families of Centrospermales in Asterad type of embryogeny in *Oxybaphus viscosus* (Souèges 1938c), *Boerhaavia diffusa*, and *B. repanda* (Kajale 1940). In Thymelaeales, Thymelaeaceae, Geissolomataceae and Penaeaceae have Asterad type of embryogeny. If embryogeny is the basis, Venkateswarlu (1947c) favours the placement of Nyctaginaceae with the members of Thymelaeales as also suggested by Hutchinson (1973). The characters cited by Venkateswarlu (1947c), except the embryogeny, also occur in Centrospermales.

Corner (1976) states that the seed structure of Nyctaginaceae is typically Centrospermalean, and rejects its placement in Thymelaeales. Takhtajan (1969, 1980) considers it as an advanced family with entomophilous mode of pollination and exarillate seed, and places it in Caryophyllales.

Ficoideae. Bentham and Hooker (1965c) place Ficoideae (Ficoidaceae) along with Cactaceae in Ficoidales, and split it into three tribes: Mesembryeae, Aizoideae and Mollugineae. Hutchinson (1973) erects two independent families, Molluginaceae and Aizoaceae (see also Melchior 1964).

Molluginaceae and Aizoaceae. Molluginaceae and Aizoaceae show marked similarity in the structure and development of anther, ovule, embryo sac, and endosperm, embryogeny of the Caryophyllad or Solanad type, and perispermous seed. A.C. Joshi and Rama Rao (1936) segregate Aizoaceae from Molluginaceae, on the basis of funicular aril, number of cell layers in outer integument, occurrence of periclinal divisions in nucellar epidermis, and presence of starch grains in the embryo sac. In addition, funicular aril, air-space between and at the base of integuments, and nucellar cap are poorly developed in Aizoaceae. Sometimes, the seed is exalbuminous (*Aptenia*). Aizoaceae also exhibit several types of development of the embryo sac. Schmid (1925) reports Adoxa type in *Mesembryanthemum pseudotruncatella*, but this needs confirmation. Kapil and Prakash (1966) report five types of embryo sacs (Sect. 16.7) in *Delosperma*.

N. Prakash (1967a) investigated the embryology of several genera of Aizoaceae, and concluded that there is no marked distinction between Molluginaceae and Aizoaceae, and Molluginaceae should be merged with Aizoaceae. This needs further study.

Portulacaceae and Basellaceae. The suborder Portulacineae comprises Portulacaceae and Basellaceae. Bentham and Hooker (1965c) treat Basellaceae as a subfamily Baselleae under Chenopodiaceae. Eichler (1878), Engler and Prantl (1931), Rendle (1959) and Lawrence (1951) raise it to a family rank under Centrospermales. Hutchinson (1973) recognizes Basellaceae as an independent family in Chenopodiales – an order evolved from Caryophyllales. Bessey (1915), Gunderson (1950), Takhtajan (1980) and Cronquist (1981) place it in Caryophyllales. Melchior (1964) includes Portulacaceae, Basellaceae, Chenopodiaceae and Caryophyllaceae in Centrospermales (Table 16.9.1). According to Takhtajan (1969, 1980), Basellaceae is related to Portulacaceae and differs from Chenopodiaceae. Portulacaceae and Basellaceae are in close alliance in most of the essential features which are in conformity with those of the order. However, Basellaceae do show some differences. The unilocular ovary with a single ovule on basal placenta, nucellar beak, Chenopodiad type of embryogeny, and drupaceous fruit do not match with those of Portulacaceae.

Chenopodiad type of embryogeny occurs in Chenopodiaceae. Chenopodiaceae show occasional presence of amoeboid tapetum, campylotropous ovule with an air space between the integuments, Polygonum as well as Allium type of embryo sac, and secondary multiplication of antipodals. These features clearly indicate the differences between Basellaceae and Chenopodiaceae. Maheswari Devi and Pullaiah (1975) favour the retention of Basellaceae in Centrospermales, as an independent family (Table 16.9.1).

Caryophyllaceae. Caryophyllaceae represent the suborder Caryophyllineae. Caryophyllaceae are characterized by free-central placentation. Eichler (1878), Wettstein (1935) and Rendle (1959) seek their ancestry in Phytolaccaceae (Table 16.9.1). On the other hand, Bessey (1915) supports their Ranalian ancestry. Phytolaccaceae and Caryophyllaceae both show glandular anther tapetum, three-celled pollen grains, ana-campylotropous ovule with a nucellar beak, Nuclear type of endosperm, and Caryophyllad type of embryogeny. According to Corner (1976), Caryophyllaceae has a simplified seed of Centrospermae. Embryologically, Caryophyllaceae are much closer to Phytolaccaceae than to Ranales. Ranalian families show variations in the number of integuments, and development of the embryo sac. Again, the campylotropous ovule with nucellar beak, ephemeral antipodals, Caryophyllad type of embryogeny and Centrospermalean seed do not occur in Ranales.

Stegnosperma. P.S. and L.L. Narayana (1986), who have studied the embryology of *S. halimifolium* and *S. watsonii*, suggest (without much justification) that this taxon should be raised to family rank, Stegnospermataceae, in the vicinity of Phytolaccaceae as a possible connecting link between the Phytolaccaceae and Caryophyllaceae. *Stegnosperma* is well-placed in the Caryophyllaceae, and we do not agree with P.S. and L.L. Narayana.

The suborder Chenopodiineae consists of Dysphaniaceae, Chenopodiaceae, Amaranthaceae and Didiereaceae. Chenopodiineae is supposed to have evolved from Caryophyllineae.

Chenopodiaceae, Amaranthaceae. Chenopodiaceae are characterized by fleshy habit, whereas Amaranthaceae are usually distinguished by the presence of scarious bracts. Embryologically, both Chenopodiaceae and Amaranthaceae have close affinities: secretory anther tapetum, bitegmic crassinucellar ovules with endostome forming the micropyle, presence of an air-space, massive nucellus (with nucellar beak or cap) persisting as perisperm, Polygonum type of embryo sac with persistent antipodals, Nuclear endosperm, and Chenopodiad embryogeny. The structure of seed is typically Centrospermalean. Their inclusion in the same order is justified. They also show affinities with Caryophyllaceae.

Didiereaceae. Didiereaceae have been included in the orders Sapindales, Centrospermales and Euphorbiales. The embryological features of Didiereaceae and Centrospermae are very similar: (1) secretory anther tapetum, (2) simultaneous cytokinesis in pollen mother cells, (3) trinucleate pollen, (4) ana- to campylotropous ovule, crassinucellate, (5) micropyle formed by inner integument, (6) occasional presence of nucellar cap formed from the divisions of nucellar epidermis, and (7) Chenopodiad type of embryogeny (Schölch

Table 16.9.1. Embryology of Phytolaccaceae, Aizoaceae, Portulacaceae, Basellaceae, Caryophyllaceae and Chenopodiaceae (Adapted after Maheswari Devi and Pullaiah 1975)

	Phytolaccaceae	Aizoaceae	Portulacaceae	Basellaceae	Caryophyllaceae	Chenopodiaceae
Anther	Tetrasporangiate	Tetrasporangiate	Tetrasporangiate	Tetrasporangiate	Bisporangiate	Tetrasporangiate
Wall	–	Basic type	Monocotyledonous	Monocotyledonous	Mono- or dicotyledonous	Monocotyledonous
Endothecium	Fibrous	Fibrous	Fibrous	Fibrous, multiseriate	Fibrous	Fibrous
Tapetum	Glandular, cells bi- or multinucleate	Glandular, cells bi- or four-nucleate	Glandular, cells bi- or multinucleate	Glandular, cells multinucleate	Glandular, cells binucleate	Glandular, cells uni- or binucleate; occasionally amoeboid
Tetrads	Tetrahedral, isobilateral	Tetrahedral, isobilateral, decussate	Tetrahedral, isobilateral	Tetrahedral, isobilateral	Tetrahedral, decussate	Tetrahedral, isobilateral, decussate
Pollen grain	Two- (*Phytolacca*) or three-celled (*Rivina*)	3-celled	3-celled	3-celled	3-celled	3-celled
Ovule	Anacampylotropous, bitegmic, crassinucellate	Anacampylotropous, bitegmic, crassinucellate, funicular obturator develops	Anacampylotropous, bitegmic, crassinucellate	Ana- or campylotropous, bitegmic, crassinucellate	Campylotropous, bitegmic, crassinucellate, nucellar beak conspicuous	Campylotropous, bitegmic, crassinucellate, nucellar beak conspicuous

Micropyle	Formed by inner integument	Inner integument	Inner integument	Inner integument	Inner integument	Inner integument
Air-space between the integuments at their base	Present	Present	Present	Present	Present	Present
Archesporium	2- or 3-celled	Multicelled	2- or 3-celled	2- or 3-celled	Uni- or multicelled	2- or 3-celled
Embryo sac	Polygonum type	Polygonum, Endymion, Drusa, Adoxa or Penaea	Polygonum	Polygonum	Polygonum	Polygonum; Allium type in *Sueda maritima*
Antipodal cells	Persistent, secondary multiplication in *Rivina*	Ephemeral	Ephemeral	Ephemeral	Ephemeral	Ephemeral, sometimes supernumerary due to divisions
Endosperm	Nuclear type	Nuclear, chalazal haustorium in *Hereroa hesparantha*	Nuclear	Nuclear	Nuclear, in some taxa superficial extensions develop	Nuclear
Embryogeny	Caryophyllad type, massive suspensor	Solanad, occasional suspensor polyembryony	Onagrad or Caryophyllad, occasional polyembryony	Chenopodiad	Caryophyllad, basal cell of suspensor haustorial	Chenopodiad, suspensor massive
Seed	Small, exotestal, perispermous	Exotestal, and endotegmic, perispermous	Small, exotestal, albuminous, perispermous	Small, exotestal, perispermous	Small, exotestal, perispermous	Small, exotestal and endotegmic, perispermous

1963). The structure of vegetative organs, pericarp, testa, and development of the embryo sac are also comparable (see B.M. Johri 1967). Schölch (1963) concludes that Didiereaceae should be placed in Centrospermales between Cactaceae and Portulacaceae, side by side with Nyctaginaceae. Wunderlich (1967) considers embryological, palynological and biochemical data of Didiereaceae, and favours their separation from Sapindaceae and placement in Centrospermae. Takhtajan (1980) places Didiereaceae in the Caryophyllales which includes a number of families assigned to Centrospermae by Melchior (1964).

17 Cactales

The family Cactaceae alone represents the order Cactales.

17.1 General Features

The embryological features are: (1) secretory anther tapetum with multinucleate cells, (2) simultaneous cytokinesis in pollen mother cells, (3) three-celled pollen grains, (4) bitegmic, crassinucellate ovule with an air-space between the integuments in the chalazal region, (5) micropyle formed by the swollen tip of the inner integument, (6) curved nucellus forming nucellar cap, (7) multicellular archesporium, (8) Polygonum type of embryo sac with ephemeral antipodals, (9) Nuclear type of endosperm, (10) occasional degeneration of zygotic embryo, (11) nucellar and suspensor polyembryony, and (12) perispermous, albuminous, arillate seed.

17.2 Cactaceae

The anther wall comprises the epidermis, fibrous endothecium, single middle layer, and tapetum in *Mammillaria carnea* (Y.D. Tiagi 1961). The epidermis is lignified and papillate in *Opuntia dillenii* (Y.D. Tiagi 1954). The tapetum is secretory with multinucleate cells; nuclear fusions are common. Meiotic divisions in microspore mother cells are followed by simultaneous cytokinesis. The tetrads are tetrahedral and decussate.

Pollen grains are three-colpate, pantoporate, and three-celled at the shedding stage.

The funicle is branched, and two or more ovules are borne on the same stalk (P. Maheshwari and Chopra 1955). The inner surface of the funicle is covered with papillate cells. The ovule is anatropous, hemianatropous, campylotropous or circinotropous. It is bitegmic and crassinucellate. The micropyle is formed by the swollen apex of the inner integument which protrudes and approaches the funicle. The cells of the nucellar epidermis produce a nucellar cap. The cells in the region of tip are radially elongate. In the chalazal region of ovule a prominent air-space occurs between the two integuments.

Y.D. Tiagi (1970a) has studied *M. tenuis*, *O. dillenii*, *O. ficusindica* and *Zygocactus truncatus*. A single cell from the hypodermal multicellular archesporium cuts off a parietal cell which gives rise to a parietal tissue. The megaspore mother cell undergoes meiotic divisions producing a triad, or linear or T-shaped tetrad. The chalazal megaspore develops into a Polygonum type of embryo sac. The synergids are hooked, and exhibit filiform apparatus in *M. tenuis* (Y.D. Tiagi 1957) and *O. dillenii* (P. Maheshwari and Chopra 1955). The antipodals are ephemeral; they may become binucleate. The embryo sac contains starch grains.

Double fertilization occurs, and persistent pollen tubes have been observed both in the micropyle and nucellus.

The development of endosperm is of the Nuclear type. It becomes cellular around the embryo in *Astrophytum*, *Thelocactus* and *Toumeya*, and persists as a cap in the mature seed (Engleman 1960). In *M. tenuis* (Y.D. Tiagi 1957) and *Opuntia* (P. Maheshwari and Chopra 1955), the endosperm remains free-nuclear throughout. In *Cereus* and *Ferocactus* (Kapil and Prakash 1969), the chalazal region of the nuclear endosperm contains dense cytoplasm, and the nuclei become hypertrophied. It forms a haustorium. Cytokinesis occurs at the dicotyledonous stage of embryo, and only a few cells are formed around the radicle.

The earlier divisions of zygote and its derivatives are transverse, resulting in a filamentous proembryo in *Astrophytum*, *Thelocactus*, *Toumeya* (Engleman 1960), and *Ripsalis* (Mauritzon 1934a). Kapil and Prakash (1969) report an oblique division of zygote in *Cereus jamacaru*, presence of a massive suspensor, and formation of additional embryos due to proliferation of suspensor cells. The mature embryo has well-developed cotyledons in primitive genera, but remains undifferentiated in advanced members.

The zygotic embryo generally degenerates, and additional embryos develop from the nucellus in several species of *Mammillaria* (Fig. 17.2.1A–D), *Opuntia* (P. Maheshwari and Chopra 1955, Naumova 1978), and *Pereskia* (Y.D. Tiagi 1970a). Generally, only one embryo reaches maturity. Sometimes, in *O. dillenii* (P. Maheshwari and Chopra 1955), due to fusion of embryonal masses, embryos with three and five cotyledons are formed. The seeds are polyembryonate in *O. elata* (Naumova 1978); the embryos are of nucellar origin. The embryogenesis in nucellar and sexual embryos is comparable, but the sexual embryos take 30 days to mature, while the nucellar ones only 15–20 days (Naumova 1978).

The seed is perispermous (perisperm absent in *Astrophytum myriostigma* – Engleman 1960), albuminous, and prominently exotestal.

The seeds of cactaceous taxa vary in size (approx 0.5 mm to 10 mm), shape, colour, epidermal structure, and position of funicle and micropyle (Barthlott and Voit 1979). The micropyle and hilum are sometimes separated by a zone of sclerotic cells of testa.

The hilum-micropylar zone, together with raphe and chalaza, form only a small area of seed structure. The embryo is oblong or globular, straight or

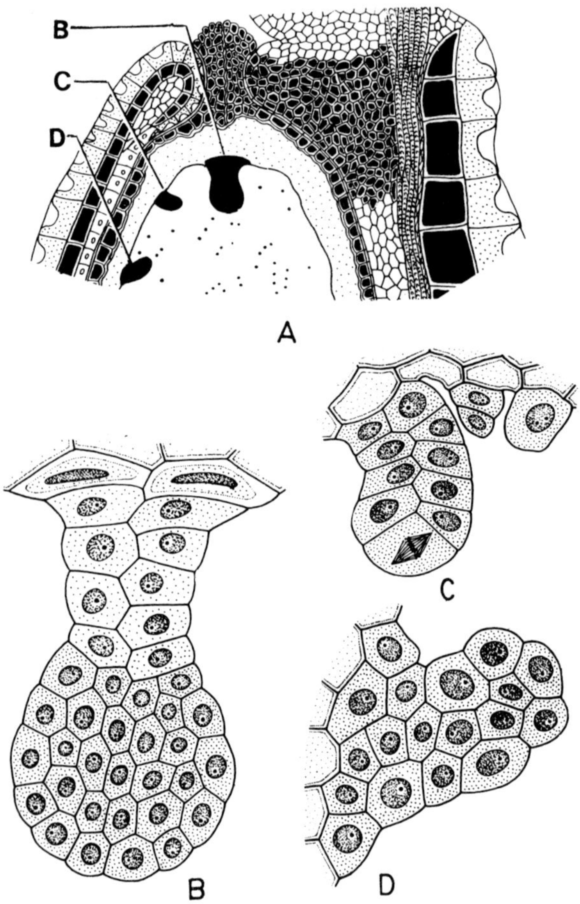

Fig. 17.2.1A–D. *Mammillaria tenuis*, nucellar polyembryo. **A** Longisection of micropylar portion of ovule to show embryo sac with three adventive embryos which have developed from the nucellus. **B–D** Nucellar proembryo marked *B*, *C* and *D* in **A**, note the suspensor in **B.** (After Y. D. Tiagi 1956, 1970a)

curved. It has well-developed hypocotyl and reduced cotyledons. Only in Pereskioideae and Opuntioideae are the cotyledons longer than the hypocotyl (Bregman and Bouman 1983).

In primitive subfamilies a well-developed perisperm is located between the cotyledons and hypocotyl, but in advanced genera the perisperm has disappeared and the function of storage has been taken over by the hypocotyl. A "hilum cup" is formed due to shrinkage of funicular cells, and aril or strophiole is formed due to the growth of funicular tissue. These cells form a collar around the hilum–micropylar zone, or develop into an amorphous mass. The strophiole aids in seed dispersal, and in water uptake prior to germination. Opercula are restricted to the subfamilies Cactoideae (Cereoideae) and Pereskioideae, and are absent in Opuntioideae.

The outer epidermis of testa is a mechanical layer with strongly thickened cells. The inner epidermis is thin-walled, and may persist. The inner epidermis of tegmen alone persists, and shows bar-like thickenings on radial walls. In

Opuntioideae the funicle, with its papillate outgrowths, becomes pulpy. The seeds of many cacti are strophiolate (Buxbaum 1955), and the swelling of funicle may contribute to its formation (Engleman 1960).

I Banerji and S Sen (1954), W Barthlott and G Voit (1979), R Bregman and F Bouman (1983), P Brough (1923), F Buxbaum (1944, 1955, 1958), RN Chopra (1957), EM Engleman (1960), RN Kapil and N Prakash (1969), P Maheshwari and RN Chopra (1954, 1955), J Mauritzon (1934a), TN Naumova (1978), M Neumann (1935), F Ritter (1980), YD Tiagi (1954, 1956, 1957, 1959, 1961, 1970a)

17.3 Taxonomic Considerations

Cactaceae. Bentham and Hooker (1965a) place Cactaceae in the order Ficoidales, between Passiflorales and Umbellales. Engler and Diels (1936) and Hutchinson (1973) group them with parietalean families. Y.D. Tiagi (1970a) negates the relationship with parietalean families, except for parietal type of placentation. Wettstein (1935) and Thorne (1963) associate them with Aizoaceae and Portulacaceae. Cronquist (1981) considers them to be allied to Caryophyllales; R. Dahlgren (1980b, 1981) places them in Caryophylliflorae (Caryophyllales). According to Hutchinson (1973), the position of Cactaceae (Cactales) is rather problematic.

P. Maheshwari (1950, 1964) emphasizes the relationship of Cactaceae with Aizoaceae and Portulacaceae. The embryological similarities are: (1) glandular anther tapetum with two- to four-nucleate cells, (2) simultaneous cytokinesis, (3) three-celled pollen grains, (4) bitegmic and crassinucellate ovules with curved nucellus, (5) occurrence of air-space between the two integuments in the chalazal region, (6) micropyle formed by the swollen tip of the inner integument, (7) presence of nucellar cap, (8) Polygonum type of embryo sac, (9) Nuclear type of endosperm, and (10) perisperm as storage tissue in the seed (Mauritzon 1934a, Neumann 1935, P. Maheshwari 1950, 1964, Banerji and Sen 1954, Kapil and Prakash 1969).

According to Martin (1946) and Kapil and Prakash (1969), the seed structure of Cactaceae is similar to that of Centrospermales. The radial elongation of nucellar epidermal cells is comparable to that in Aizoaceae. Engleman (1960) compares the funicular hump of Centrospermales with the strophiole of cacti. Wunderlich (1967) does not recognize the ordinal rank of Cactaceae, and she suggests their inclusion in Centrospermae.

P. Maheshwari (1950) cites several similarities between Portulacaceae and Cactaceae: (1) microspore mother cells in one or two (rarely more than two) rows in the pollen sac, (2) micropyle in close proximity to the funiculus, (3) T-shaped tetrad of megaspores, (4) long synergids, and (5) embryo sac constricted in the middle region (Neumann 1935).

P. Maheshwari (1950) rejects any affinity between Cactaceae and Passifloraceae. In the latter, periclinal divisions of nucellar epidermis are absent, the ovule is completely anatropous with a perfectly symmetrical nucellus, and the outer integument overgrows the inner one, and both form the micropyle. There is no parallel to these features in Cactaceae.

On the basis of primitive characters in the floral construction of Cactaceae, Y.D. Tiagi (1970a) mentions its relationship with Ranales or Ranalian ancestors. The floral anatomy, nodal type, and basic chromosome number are similar to those of Calycanthaceae. The similarities between Cactaceae and Calycanthaceae are: (1) secretory anther tapetum, (2) simultaneous cytokinesis of pollen mother cells, (3) multicellular female archesporium, (4) Polygonum type of embryo sac with ephemeral antipodals, and (5) nucellar polyembryony (S.L. Mathur 1968). On the other hand, the differences between the two are also well marked. Calycanthaceae show (1) two-celled, two- or three-sulcate pollen grains, (2) anatropous ovule with micropyle formed by both the integuments, (3) absence of parietal cell, (4) absence of double fertilization in some members, and (5) autonomous development of endosperm according to Cellular type (S.L. Mathur 1968). The seed is exalbuminous and endotestal, and does not show any similarity with that of Cactaceae.

Cactaceae form a bridge between Aizoaceae and Portulacaceae. Embryologically, the similarities are more pronounced with Aizoaceae. Buxbaum (1944) derives Cactaceae from Phytolaccaceae, and considers its parallel evolution with Aizoaceae.

18 Magnoliales

The Magnoliales comprise 22 families in six groups: (1) Magnoliaceae, Degeneriaceae, Himantandraceae, Winteraceae, Annonaceae, Eupomatiaceae, Myristicaceae, Canellaceae; (2) Schisandraceae, Illiciaceae; (3) Austrobaileyaceae, Trimeniaceae, Amborellaceae, Monimiaceae, Calycanthaceae, Gomortegaceae, Lauraceae, Hernandiaceae; (4) Tetracentraceae, Trochodendraceae; (5) Eupteleaceae; (6) Cercidiphyllaceae.

18.1 General Features

The Magnoliales are a large order, and exhibit structural variations: (1) secretory anther tapetum, occasionally amoeboid in Winteraceae and Monimiaceae, (2) simultaneous cytokinesis in pollen mother cells, successive in Myristicaceae, and both types in Annonaceae, (3) two-celled pollen grains, sterilization of sporogenous tissue in Annonaceae and Myristicaceae, (4) pollen grains shed in tetrads in Winteraceae, Monimiaceae, and rarely in Annonaceae, (5) anatropous, bitegmic and crassinucellate ovule, sometimes unitegmic in Monimiaceae (*Siparuna*), a middle integument also occurs in Annonaceae and Myristicaceae, (6) single-celled archesporium, multicellular in Annonaceae, Calycanthaceae, uni- or multicellular in Magnoliaceae, Monimiaceae and Lauraceae, (7) Polygonum type of embryo sac, Polygonum and Allium type in Schisandraceae and Monimiaceae, an embryo sac haustorium (chalazal) develops in Lauraceae and Trochodendraceae, (8) antipodals ephemeral, secondary multiplication in Monimiaceae and Lauraceae, (9) Cellular type of endosperm except in Myristicaceae and Canellaceae, Nuclear as well as Cellular in Lauraceae, ruminate condition in Degeneriaceae, Annonaceae, Myristicaceae and Canellaceae, (10) development of embryo irregular in Magnoliaceae and Winteraceae, Onagrad type in Magnoliaceae, Annonaceae, Schisandraceae and Lauraceae, Asterad in Schisandraceae, Illiciaceae, Monimiaceae and Lauraceae, Piperad in Lauraceae, Chenopodiad in Trochodendraceae, and Solanad type in Cercidiphyllaceae and Eupteleaceae, and (11) seed generally endotestal. Sarcotesta occurs in Magnoliaceae and Degeneriaceae, sclerotic exotesta in Winteraceae, Canellaceae and Schisandraceae, and woody mesotesta in Annonaceae.

18.2 Magnoliaceae

The anther wall comprises the epidermis, endothecium with reticulate thickenings, three to four ephemeral middle layers, and glandular tapetum with multinucleate cells. The endothecium is persistent, and irregularly biseriate. The tapetum is two- to three-layered in *Liriodendron* (Maneval 1914) and *Michelia* (Hayashi 1965). Übisch granules are studded on the inner wall of the tapetum. After meiotic divisions, the microspore mother cells produce tetrahedral, isobilateral, and T-shaped tetrads. Simultaneous cytokinesis takes place by furrowing.

The pollen grains are monosulcate (Hayashi 1960), and two-celled at the dispersal stage; three-celled in *Liriodendron tulipifera* (Kaeiser and Boyce 1962). Sometimes, they remain united in tetrads. During germination the pollen tube is not restricted to the sulcus (Hayashi 1965).

In *Talauma candollei*, the pollen grains are abnormal and contain a variable number of nuclei. In *Magnolia ovata* (Kapil and Bhandari 1964), the abnormal pollen grain has two generative cells and two vegetative nuclei.

The ovule is anatropous, bitegmic, and crassinucellar; both integuments form the micropyle. The outer integument is vascularized, and has stomata on its surface (Mohana Rao 1975a). Funicular obturator has been noted in *Liriodendron* (Kaeiser and Boyce 1962). A hypostase also differentiates. A single cell from the hypodermal multicellular archesporium cuts off a parietal cell; the parietal tissue is massive. The megaspore mother cell undergoes meiotic divisions, and produces a linear or T-shaped tetrad. The chalazal megaspore develops into a Polygonum type of embryo sac. The synergids and antipodals are ephemeral. The polar nuclei fuse before fertilization.

The endosperm is of the Cellular type. In *Magnolia obovata* (Kapil and Bhandari 1964), the chalazal chamber develops into a two- or three-celled haustorium.

The zygote divides by transverse, obliquely transverse, or vertical wall. Development of the embryo is either irregular, or conforms to the Onagrad type. The basal cell of the embryo divides to form a group of cells. In *Magnolia grandiflora* (Ly-Thi-Ba et al. 1970), the embryogeny corresponds to Period I, Megarchetype VI of the Souèges system of classification (Onagrad type).

The seed of *Magnolia* projects outside the fruit due to the elongation of funicule. The seed is massive with a small embryo, and oily endosperm. Both the cotyledons are well developed in a ripe fruit (Hayashi 1965). The seed is sarcotestal, and the testa is multiplicative with woody endotesta. The outer part of testa is fleshy, oily, and vascularized. The inner part, the endotesta, is made up of two to ten layers of radially elongate lignified cells with a single crystal in each cell.

NN Bhandari (1971), JE Canright (1952, 1953), PK Endress and LD Hufford (1989), A Guard (1943), Y Hayashi (1960, 1963b, 1964, 1965, 1966, 1984), M Kaeiser and SG Boyce (1962), RN Kapil and NN Bhandari (1964), Ly-Thi-Ba, JL Guignard and JC Mestre (1970), WE Maneval (1914), PR Mohana Rao (1975a), D Padmanabhan (1960a), HN Stoudt (1960), LB Thien (1974), K Yasui (1937)

18.3 Degeneriaceae

The anther wall comprises the epidermis, fibrous endothecium, three or four ephemeral middle layers, and glandular tapetum with binucleate cells. The fibrous thickenings extend to the connective region. The microspore mother cells undergo meiotic divisions, and tetragonal tetrads are produced. Cytokinesis is simultaneous. At the time of separation of microspores, a narrow groove develops on the distal-fold, and differentiates as a germinal furrow.

The pollen grains are monocolpate and two-celled, and shed singly. The generative cell is cut off towards the distal end in *Degeneria vitiensis* (Swamy 1949a).

The ovule is anatropous, bitegmic and crassinucellar, and the micropyle is formed by the inner integument. The funicular obturator of elongated cells is conspicuous. The raphal bundle extends beyond the chalaza (Fig. 18.3.1F, G). The single hypodermal archesporial cell divides into a parietal and a sporogenous cell. The parietal tissue is massive. The megaspore mother cell undergoes meiotic divisions, and produces a linear tetrad. The chalazal megaspore develops into a Polygonum type of embryo sac. The antipodals are ephemeral.

The endosperm is of the Cellular type. The mature endosperm is ruminate in *Degeneria vitiensis* (Periasamy 1962a). In later stages the endosperm grows vigorously, and pushes against the ingrowths of seed-coat (Fig. 18.3.1A–H). It contains oil as storage product.

The zygote divides transversely; vertical walls divide both the apical and basal cells producing two superposed tiers of two cells each (Fig. 18.3.2A–F). Further divisions are irregular (Fig. 18.3.2G–R). The mature embryo shows three or four cotyledons, a short suspensor, and a bulbous hypocotyl (Fig. 18.3.2S, T).

The albuminous, pachychalazal seed contains a small embryo.

The testa is multiplicative, and lacks fibres. The outer epidermis consists of thin-walled palisade-like cells with a thick cuticle. The mesophyll is thin-

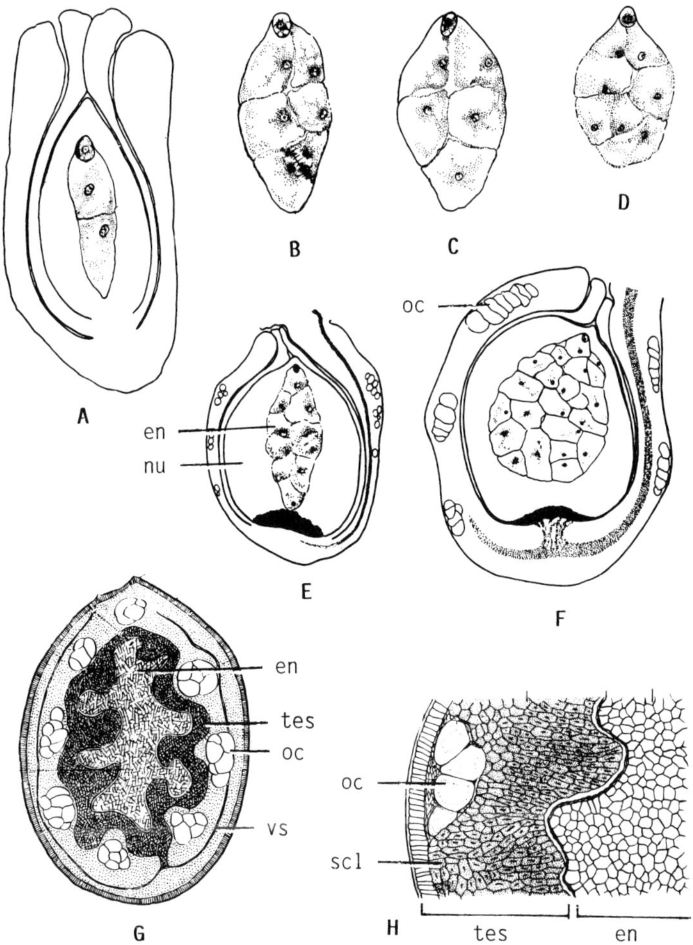

Fig. 18.3.1A–H. *Degeneria vitiensis*. **A** Bitegmic ovule, the embryo sac contains zygote and two-celled endosperm. **B–D** Early stages in the development of endosperm. **E–F** Later stages, oil-bearing cells (*oc*) differentiate in testa, the nucellar cells (*nu*) at chalaza degenerate (*black*); *en* endosperm. **G** Median longisection (diagrammatic) of mature seed; *tes* testa; *vs* vascular supply. **H** Portion from **G**; *scl* sclerenchyma. (After Swamy 1949a)

walled, and contains groups of oil cells. The endotesta is woody composed of thick-walled lignified fibrils. The inner cells undergo irregular divisions, and develop ingrowths. The tegmen gets crushed (Fig. 18.3.1G, H).

IW Bailey and AC Smith (1942), NN Bhandari (1970a, 1971), OA Dahl and JR Rowley (1965), BGL Swamy (1949a), K Periasamy (1962a)

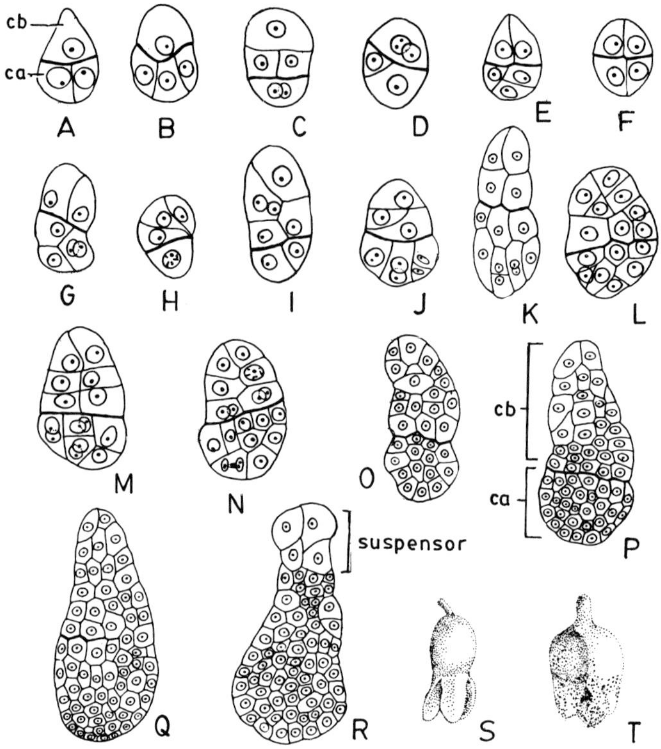

Fig. 18.3.2A–T. *Degeneria vitiensis*, embryogeny (*heavy line* indicates boundary between the daughter cells derived from basal (*cb*) and terminal (*ca*) cell of two-celled proembryo). **A** Three-celled proembryo, vertical division in *ca*. **B–K** Stages in the development of proembryo; there is no definite pattern. **L–Q** The derivatives of *ca* and *cb* form an ovoid embryo. **R** Differentiation of suspensor at the micropylar end. The demarcation between the derivatives of *ca* and *cb* could not be made out. **S, T** Embryos dissected from mature seeds. There are three cotyledons in each, the hypocotyl is swollen, and the suspensor is stalk-like. (After Swamy 1949a)

18.4 Himantandraceae

The development of endosperm, embryo and seed has not been investigated.

The four elongated microsporangia lie embedded in the tissue of laminar stamens. The anther wall is six layers thick below the epidermis, endothecium shows fibrous thickenings, tapetum is secretory with binucleate cells, and the anther dehisces by longitudinal slits. Successive cytokinesis follows meiotic divisions in pollen mother cells, the tetrads are tetrahedral, decussate, isobilateral, linear and T-shaped in *Galbulimima belgraveana* (N. Prakash et al. 1984).

The mature pollen grains are single, monosulcate, scabrate, atectate, spheroidal and two-celled. They do not contain starch.

The ovule has a funicular obturator, and is anatropous, bitegmic and crassinucellate with the inner integument forming the micropyle. The raphal

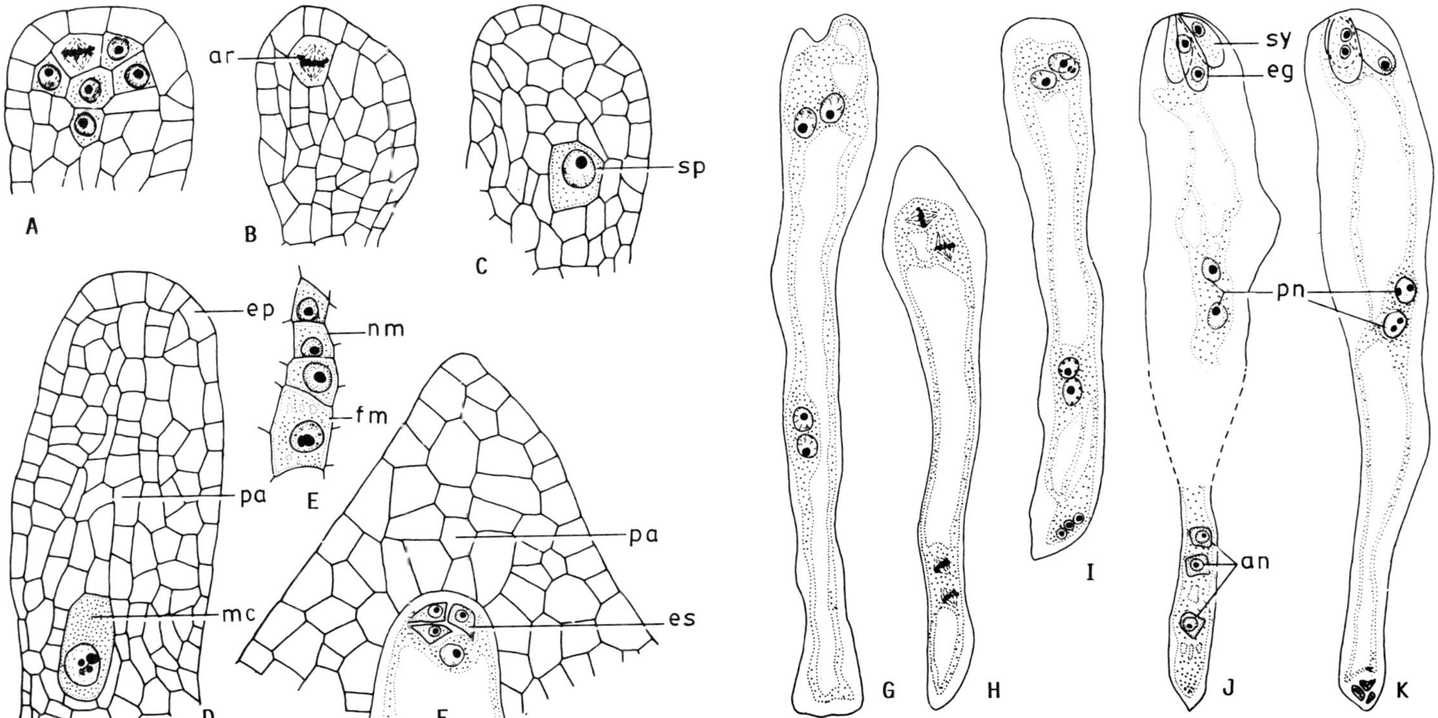

Fig. 18.4.1A–K. *Galbulimima belgraveana.* **A** Longisection of nucellus with a group of archesporial cells. **B** Archesporial cell (*ar*) in division. **C** Sporogenous cell (*sp*). **D** Megaspore mother cell (*mc*); *pa* parietal tissue; *ep* nucellar epidermis. **E** Linear tetrad; *fm* functional and *nm* nonfunctional megaspore. **F** Upper part of embryo sac (*es*) with egg apparatus and upper polar nucleus. **G** Four-nucleate embryo sac. **H** Nuclei in division. **I** Eight-nucleate embryo sac. **J** Mature embryo sac; *sy* synergid; *eg* egg; *pn* polar nuclei; *an* antipodal cells. **K** Later stage, the antipodals have degenerated. (After N. Prakash et al. 1984)

bundle terminates at the chalaza. The nucellus and the parietal tissue are massive so that the megaspore mother cell becomes deep-seated (Fig. 18.4.1A–D). Meiotic divisions in megaspore mother cell result in a linear tetrad (Fig. 18.4.1E), and the chalazal megaspore develops into a Polygonum type of embryo sac (Fig. 18.4.1F–K). Twin embryo sacs develop from two megaspore mother cells (N. Prakash et al. 1984).

The seeds have a small embryo with oily endosperm (Hutchinson 1973).

IW Bailey, CG Nast and AC Smith (1943), PK Endress (1977b), J Hutchinson (1973), N Prakash, DB Foreman and SJ Griffith (1984)

18.5 Winteraceae

The anther wall comprises the epidermis, fibrous endothecium, three or four middle layers, and tapetum. The epidermal cells contain phenol-like substances in *Zygogynum* (Swamy 1952). The tapetum is glandular in *Drimys winteri* (Bhandari and Venkataraman 1968), *Pseudowintera axillaris*, *Zygogynum* (Sampson 1963), and amoeboid in *Pseudowintera colorata* (Bhandari 1963a). The tapetal cells become polyploid. The fibrous endothecial thickenings extend to two or three layers of connective tissue (Fig. 18.5.1A–C). The meiotic divisions in microspore mother cells result in the formation of tetrahedral and decussate tetrads. Cytokinesis is simultaneous. In *Zygogynum* the divisions in microspores are synchronous; in other taxa the microspores in tetrads are at different stages of development.

The (modified) monocolpate (Hayashi 1965) pollen grains are shed as tetrads at the two-celled stage. The generative cell is cut off towards the proximal pole (Fig. 18.5.1A, C, D).

The ovule is laminar, anatropous, bitegmic, crassinucellar, and the micropyle is formed by the inner integument. In *Zygogynum* both the integuments are concerned. A hypodermal archesporial cell cuts off a parietal cell, the parietal tissue is massive. The cells of nucellar epidermis also divide and add to it. The megaspore mother cell undergoes meiotic divisions, and a linear or T-shaped tetrad is produced. The chalazal megaspore develops into a Polygonum type of embryo sac. The synergids exhibit filiform apparatus; the antipodals degenerate after fertilization.

Double fertilization occurs, and triple fusion precedes syngamy.

Development of the endosperm is of the Cellular type (Bhandari and Venkataraman, 1968). The cells are thin-walled, and store oil. It persists in the seed. According to Strasburger (1905), the development of the endosperm is of the Nuclear type in *Drimys winteri*, which needs confirmation.

The zygote divides transversely, and further divisions are irregular. The embryo is minute and organless (undifferentiated) in *Drimys*, and dicotyledonous in *Pseudowintera*.

The seed is exotestal with outer epidermis composed of lignified cells. The mesophyll is thin-walled. The tegmen is somewhat thick-walled in *Zygogynum* and *Belliolum*. In other members it does not persist. The cells of the seed-coat contain tannin.

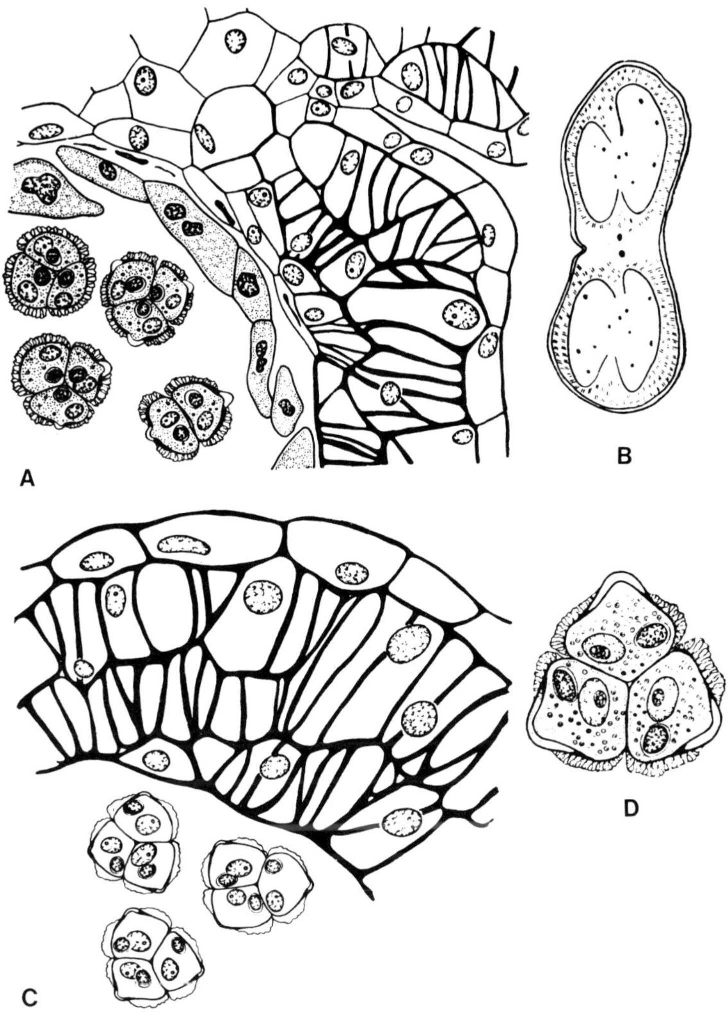

Fig. 18.5.1A–D. *Bubbia howeana*. **A** Portion of cross-section of anther. The epidermis is followed by two (or three) layers of fibrous endothecium and glandular tapetum. Two-celled pollen in tetrads. **B** Mature anther (cross-section), *stippled* portion represents fibrous endothecium. **C** Portion of mature anther with epidermis, fibrous endothecium, two-celled pollen in tetrads. **D** Mature bicelled pollen tetrad. (See Yakovlev 1981)

PR Bhagawathi (1938), NN Bhandari (1963a, 1971), NN Bhandari and R Venkataraman (1968), G Gottsberger, I Silberbauer-Gottsberger and F Ehrendorfer (1980), Y Hayashi (1965), FB Sampson (1963, 1970, 1974), E Strasburger (1905), BGL Swamy (1952), W Vink (1970, 1977), MS Yakovlev (1981)

18.6 Annonaceae

The anther wall comprises the epidermis, fibrous endothecium, three or four middle layers, and tapetum (Fig. 18.6.1A). The fibrous thickenings are absent

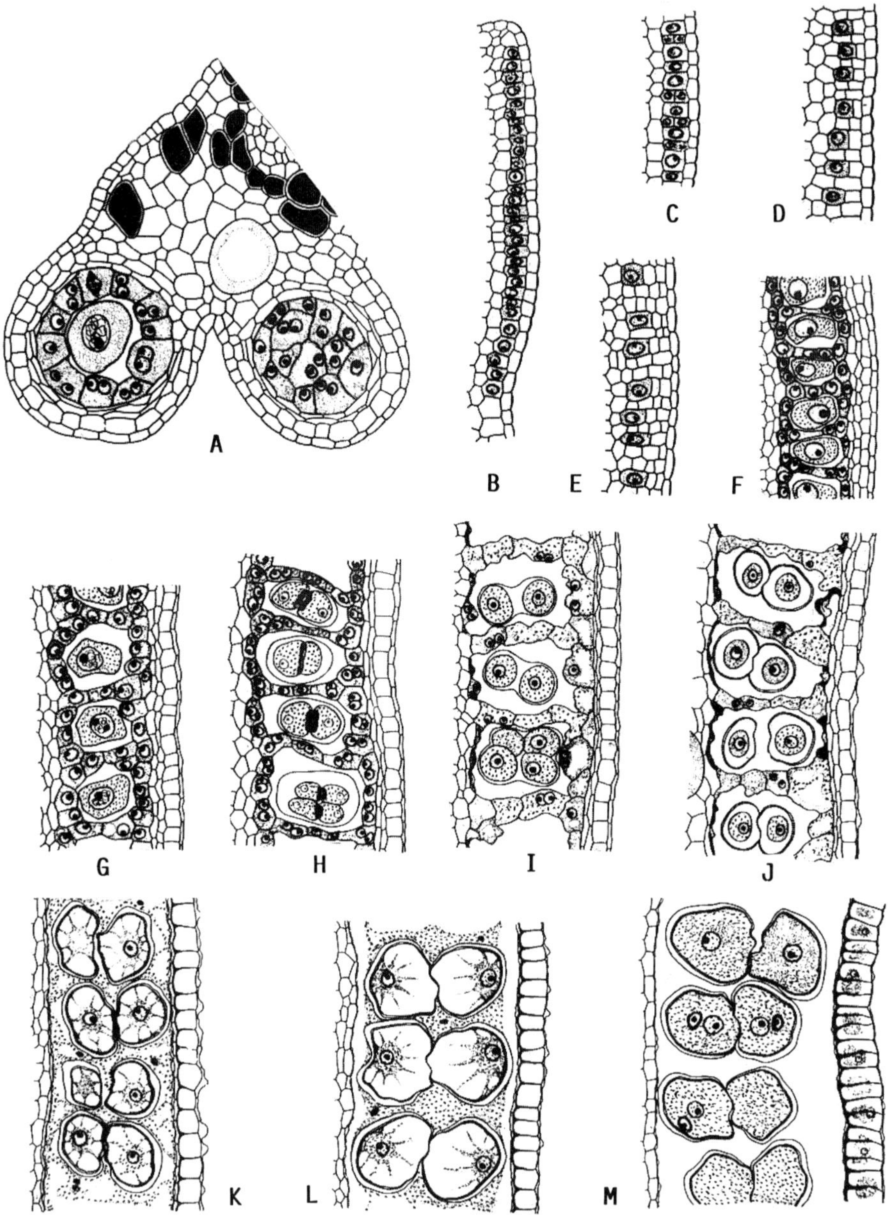

Fig. 18.6.1A–M. *Annona squamosa.* **A** Transection of half-anther through septum; one locule contains a sporogenous cell and the other shows only tapetal cells. **B–M** Longisection of portions of anther; **B** Single row of archesporial cells; **C** Alternate archesporial cells divide and differentiate as sterile septal initials; **D** Formation of primary parietal layer; **E** Three wall layers, and one- or two-layered sterile septa alternate with sporogenous cells; **F, G** The innermost parietal layer and septal layer form tapetum; **H, I** Formation of dyads and tetrads; **J** Microspores, the tapetum is still intact; **K, L** Uninucleate pollen grains embedded in tapetal periplasmodium; **M** Bicelled pollen grains, periplasmodium has been consumed. The epidermis becomes compressed (cf. **G–M**); endothecium shows fibrous thickenings in **M**. (After Periasamy and Kandasamy 1981)

in the endothecial cells of *Annona* (Juliano 1935a), which appears doubtful. The tapetum is glandular in *Artabotrys*, *Cananga* and *Miliusa*, and amoeboid in *Annona*, *Goniothalamus* and *Uvaria*. The tapetal cells are two- to four-nucleate, and due to fusion the nuclei become polyploid (Fig. 18.6.1A).

The male archesporium is multicellular in *Artabotrys* (Asana and Adatia 1947) and *Miliusa* (Periasamy and Swamy 1959), and it forms a row (Fig. 18.6.1B) in *Annona* (Juliano 1935a), *Asimina* (Locke 1936), *Cananga* (Periasamy and Swamy 1959), and *Mandora* and *Xylopia* (Lecompte 1896). Due to sterilization of sporogenous tissue (Fig. 18.6.1C–E), septa are produced so that individual or even groups of microspore mother cells become separated (Juliano 1935a, Periasamy and Swamy 1959). In *Annona* (Fig. 18.6.1F–J), *Asimina* and *Cananga*, the sterile septa behave as tapetum. The transverse septa are multilayered in *Goniothalamus*, and every spore mother cell has its own tapetum on all the sides (Parulekar 1970).

Periasamy and Thangavel (1988) point out that in *Xylopia nigricans*, in each lobe of the anther, up to 13 groups of single microspore mother cells differentiate separated by sterile septa. The cells adjoining the microspore mother cell organize into a single-layered glandular tapetum, and its cells remain uninucleate throughout. There may be 50–60 such microsporangia in each anther.

The tapetum is secretory and usually of parietal origin. Some of the sporogenous cells may also be involved. Periasamy and Kandasamy (1981) point out that in *Annona squamosa* the peripheral tapetum is of the secretory type but collapses. The cells of the transverse septa form a periplasmodium (Fig. 18.6.1K, L) at the tetrad stage. Probably, the peripheral tapetum also contributes to the periplasmodium. This aspect deserves further study.

The microspore mother cells undergo meiotic divisions, and cytokinesis is successive in *Annona* (Juliano 1935a, Sastri 1957), *Uvaria* (Sastri 1957), and *Xylopia* (Periasamy and Thangavel 1988), and simultaneous in *Annona* (Samuelsson 1914), *Asimina*, *Cananga*, *Polyalthia* and *Saccopetalum* (Sastri 1957). In *Cananga* and *Miliusa*, furrowing starts at the first meiotic division, but is completed only at the end of the second division. The tetrads are tetrahedral, isobilateral and decussate.

In *Annona* the pollen grains remain loosely attached in tetrads, in pairs in *Cananga*, and are free in *Artabotrys* and *Miliusa*. They are monocolpate and two-celled (Fig. 18.6.1M) at the shedding stage. The generative cell is cut off at the distal end.

Periasamy and Thangavel (1988) have tabulated the main features of the anther in Annonaceae – four, four to eight or numerous sporangia per anther; sporogenous tissue uniseriate with sterile partitions, biseriate or massive; cytokinesis successive or simultaneous; tapetum secretory, partly secretory and partly plasmodial, uni-, bi- or multinucleate cells; pollen grains single, dyads, loosely held tetrads, or tetrads enclosed in a common wall.

The ovule is anatropous, bitegmic or tritegmic, and crassinucellar. The micropyle is formed by both the integuments. The funicular vascular supply reaches up to the tip of the outer integument. In *Cananga*, *Mezzettia* and *Xylopia* a middle integument is present. In *C. odorata* (Periasamy and Swamy

1961), by its meristematic centripetal growth, the nucellus increases during post-fertilization development. An aril is common (Corner 1949), and a hypostase in *Artabotrys* and *Asimina.* One to multicelled hypodermal archesporium cuts off a parietal cell; the parietal tissue is massive. After meiotic divisions, the megaspore mother cell produces a linear tetrad. The chalazal megaspore develops into a Polygonum type of embryo sac. The antipodals are uninucleate. The inner region of the carpel wall, due to its meristematic activity, fills the locule with spongy tissue (Periasamy and Swamy 1961).

The endosperm is of the Cellular type. During early stages of development, the endosperm forms a uniseriate row of superposed cells, 20 in *Cananga odorata* and 30 in *Miliusa wightiana* (Periasamy and Swamy 1961). In later stages it becomes ruminated due to the folds of the seed-coat (Periasamy 1962a). In a mature seed it stores oil, and the cell walls become thick and pitted.

The embryogeny corresponds to the Onagrad type-Trifolium variation in *Polyalthia longifolia* (Sastri 1955a), and Onagrad type-Lotus variation in *Cananga odorata* (Periasamy and Swamy 1961). Rarely, two embryos develop concurrently in *Annona squamosa.*

The seed is massive, albuminous, arillate, perichalazal, and mesotestal. The testa is multiplicative. The outer epidermis of the testa is composed of radially elongate cells with stomata. The mesophyll differentiates into outer longitudinal fibres, and inner oblique or transverse fibres. The innermost layer of cells, together with fibres, in some parts, grows at a rapid rate and forms ingrowths. Ruminations in *Artabotrys odoratissimus* (Mohana Rao 1975b), *Miliusa wightiana* (Periasamy and Swamy 1961), and *Polyalthia longifolia* (Mohana Rao 1979) are plate-like and give a storied appearance in trans-longitudinal plane. The plates show irregular contour in *Cananga odorata.* The ruminations bifurcate and anastomose with the adjoining ones (Periasamy and Swamy 1961). The ruminations are conspicuously absent along the median and transmedian planes in *C. odorata*, and in transmedian plane in *Polyalthia.* Accordingly, the endosperm appears four- or two-lobed. The ruminations are formed by the outer, middle, or inner integument. In the majority of members, the outer integument develops ingrowths. In *Cananga odorata* (Periasamy and Swamy 1961), the middle and inner integument contribute to the formation of ruminations. However, in *Anaxagorea javanica* (Corner 1949) the inner integument alone forms the ruminations.

RD Adatia (1946), JJ Asana and RD Adatia (1947), JE Canright (1962), EJH Corner (1949), G Gottsberger (1989a, b), WB Herms (1907), JB Juliano (1935a), RK Lampton (1957), H Lecompte (1896), JF Locke (1936), PR Mohana Rao (1975b, 1979, 1982), EM Norman and D Clayton (1986), NK Parulekar (1970), K Periasamy (1962a), K Periasamy and MK Kandasamy (1981), K Periasamy and BGL Swamy (1959, 1961), K Periasamy and S Thangavel (1988), G Samuelsson (1914), RLN Sastri (1955a, 1957), A Thomas and B Lugardon (1972), H van der and F Bouman (1988), JW Walker (1971)

18.7 Eupomatiaceae

The embryology has not been fully investigated.

The anther is tetralocular in *Eupomatia laurina* (see Yakovlev 1981). The wall shows epidermis, fibrous endothecium, one middle layer, and collapsed tapetum at the binucleate pollen grain. The meiotic divisions in the microspore mother cells are followed by simultaneous cytokinesis. The tetrads are tetrahedral, isobilateral, decussate, occasionally linear. When the pollen grains are fully formed, the epidermal cells show conical projections. The endothecium

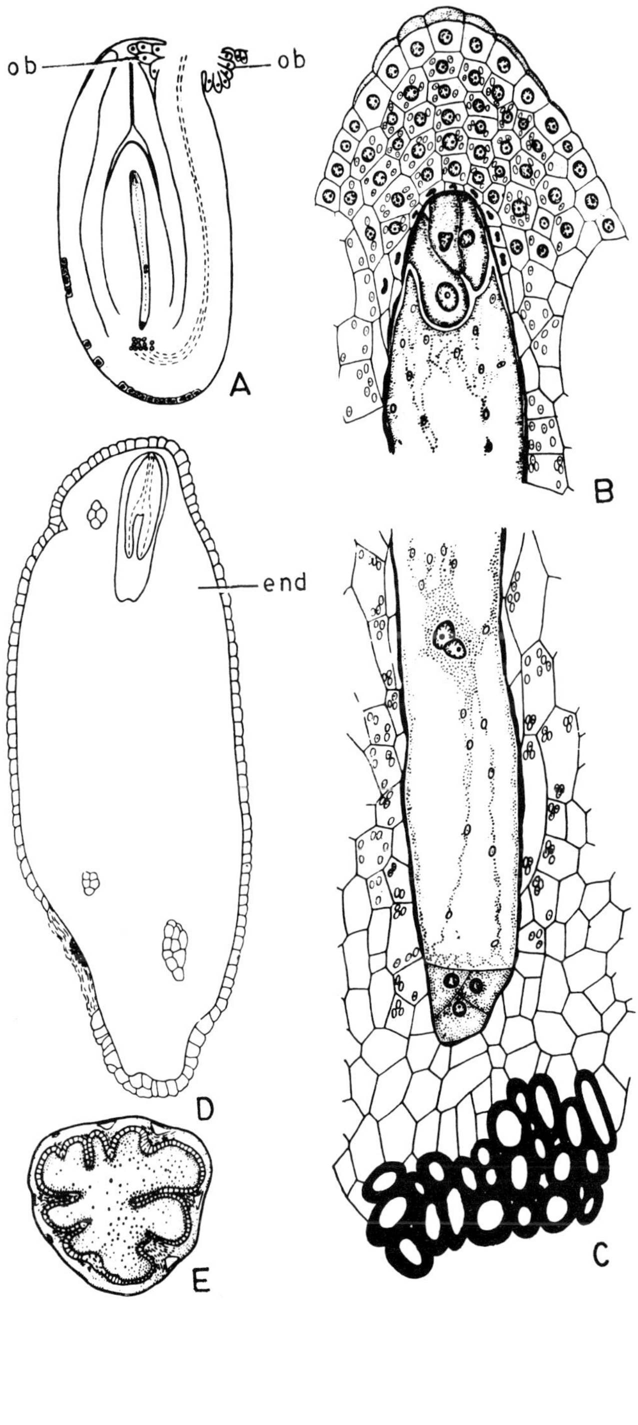

Fig. 18.7.1A–E. *Eupomatia laurina*. **A** Longisection of ovule at mature embryo sac, note funicular obturator (*ob*). **B** Micropylar part of ovule with embryo sac containing egg apparatus. **C** Chalazal portion of ovule; the embryo sac shows two polar nuclei, three antipodal cells, and starch. Note the well-developed hypostase. **D** Longisection of seed with embryo and invagination of testa into endosperm (*end*). (See Yakovlev 1981). **E** Cross-section of seed with ruminations in endosperm. (After Corner 1976)

becomes very prominent, and it alone persists at dehiscence. The two-celled pollen grains may germinate in situ.

There are two or three ovules in a carpel in lateral position. They are anatropous, bitegmic and crassinucellate with micropyle formed by the inner integument in *Eupomatia laurina* (Mohana Rao 1983). The outer integument is three or four cells thick, and the inner integument two or three cells thick. The hypostase and obturator are well developed at mature embryo sac (Fig. 18.7.1A).

The mature embryo sac is eight-nucleate (Fig. 18.7.1B, C). At the heart-shaped embryo the cellular endosperm is well developed. It is ruminated by the invasions of testa (Fig. 18.7.1D, E).

The embryo is small, and embedded in abundant oily endosperm.

The mesotestal seeds are pale-brownish, uneven, subangular, and coarsely (but shallowly) pitted. The raphe bundle remains unbranched and ends at the chalaza.

The young testa comprises six layers of cells. The outer epidermis consists of large flattened, tabular cells with thick cuticle, followed by four layers of fibres. The epidermis is crushed. The fibres elongate in longitudinal axis. In mature seed the outer epidermis of testa divides periclinally and produces a subepidermal layer of thin-walled parenchyma. At the point of rumination, the epidermal cells and fibres elongate transversely. The ruminations extend into the tegmen.

Of the three layers of tegmen, the outer epidermal cells elongate along longitudinal axis. The inner epidermal cells enlarge, and the middle layer becomes compressed between the two. In the mature seed only the inner epidermis persists.

The fruit is an urceolate, turbinate berry with one to three pale-brownish seeds in each locule.

EJH Corner (1976), AT Hotchkiss (1958), PR Mohana Rao (1983), MS Yakovlev (1981)

18.8 Myristicaceae

The anther wall comprises the epidermis, fibrous endothecium, two ephemeral middle layers, and glandular tapetum with uninucleate cells. Sterile septa divide the microsporangium into two or three compartments in *Myristica malabarica* (N.C. Nair 1972). The microspore mother cells undergo meiotic divisions, and produce isobilateral tetrads. Cytokinesis is successive.

The pollen grains are monocolpate, and two-celled at the dispersal stage.

The ovule is anatropous, bitegmic and crassinucellar. The inner integument forms the micropyle. Both the integuments are massive (Fig. 18.8.1A–E), and the outer epidermis of the outer integument contains stomata in *Myristica fragrans* and *M. malabarica* (N.C. Nair 1972). In both species, N.C. Nair reports a primordium of middle integument, between the inner and outer integuments, at mature embryo sac stage. It forms a collar only on antiraphe-side; Mohana Rao (1974b) does not mention its occurrence in *M. fragrans*.

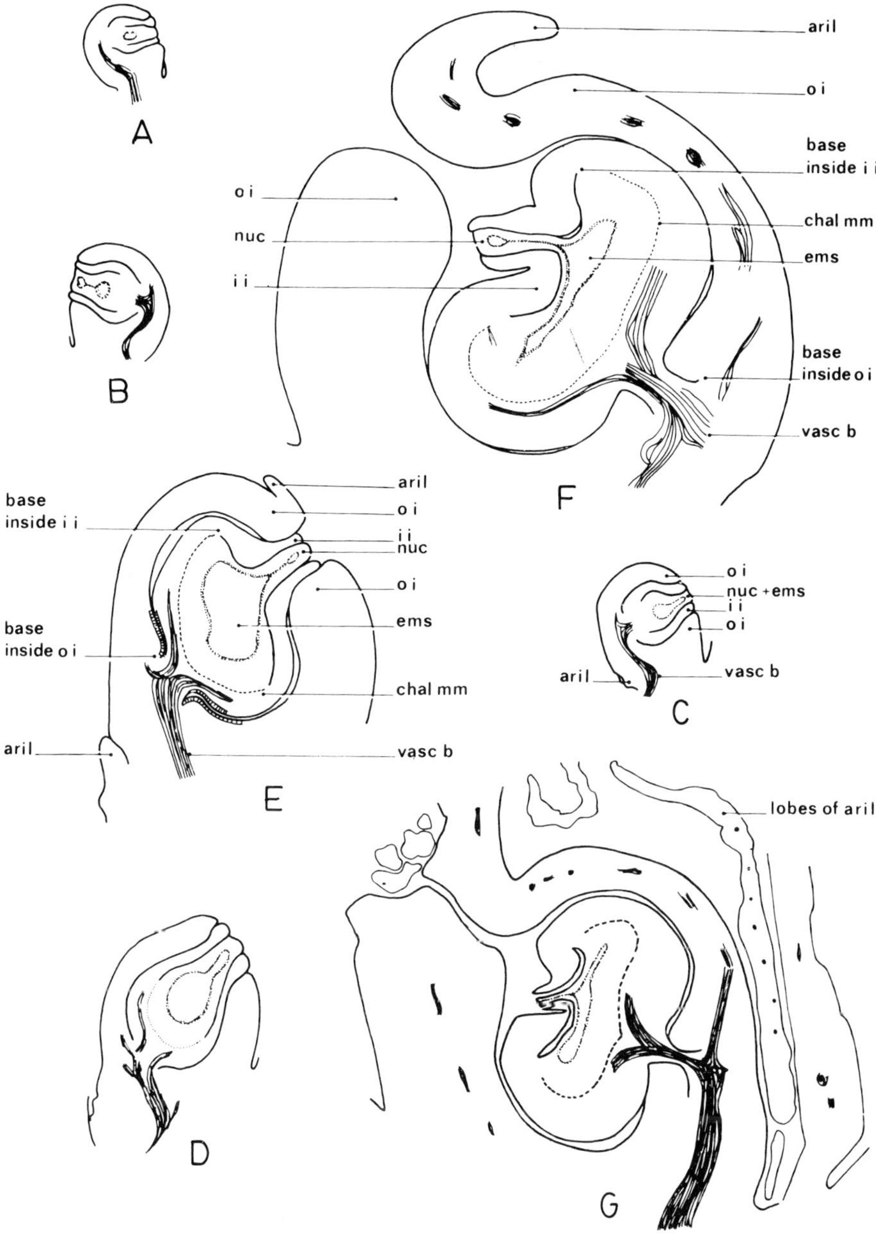

Fig. 18.8.1. **A** *Horsfieldia iryaghedi*, bitegmic, hemianatropous ovule with vascular supply terminating at chalaza, micropyle not formed. **B**, **C** *H. glabra*, later stage. The nucellar tissue in the micropylar region is prominent. **D–G** *H. sylvestris*; **D** Vascular supply in integuments; **E–G** Later stages. Note the aril and nucellar (*nuc*) cap. The inner base of inner integument (ii) has not grown as compared to outer integument (*oi*). The embryo sac (*ems*) is horseshoe-shaped. The demarcation of chalaza (*chal mm*), inner integument and nucellus is discernible; *vasc b* vascular bundle. (After van Heel 1982)

In *Myristica fragrans* (Voigt 1888, Periasamy 1961, N.C. Nair 1972), and in *Horsfieldia sylvestris* and *Knema laurina* (van Heel 1982), there is meristematic activity below the insertion of the inner integument. The newly formed tissue adds to the chalaza; the tip of the inner integument persists only at the

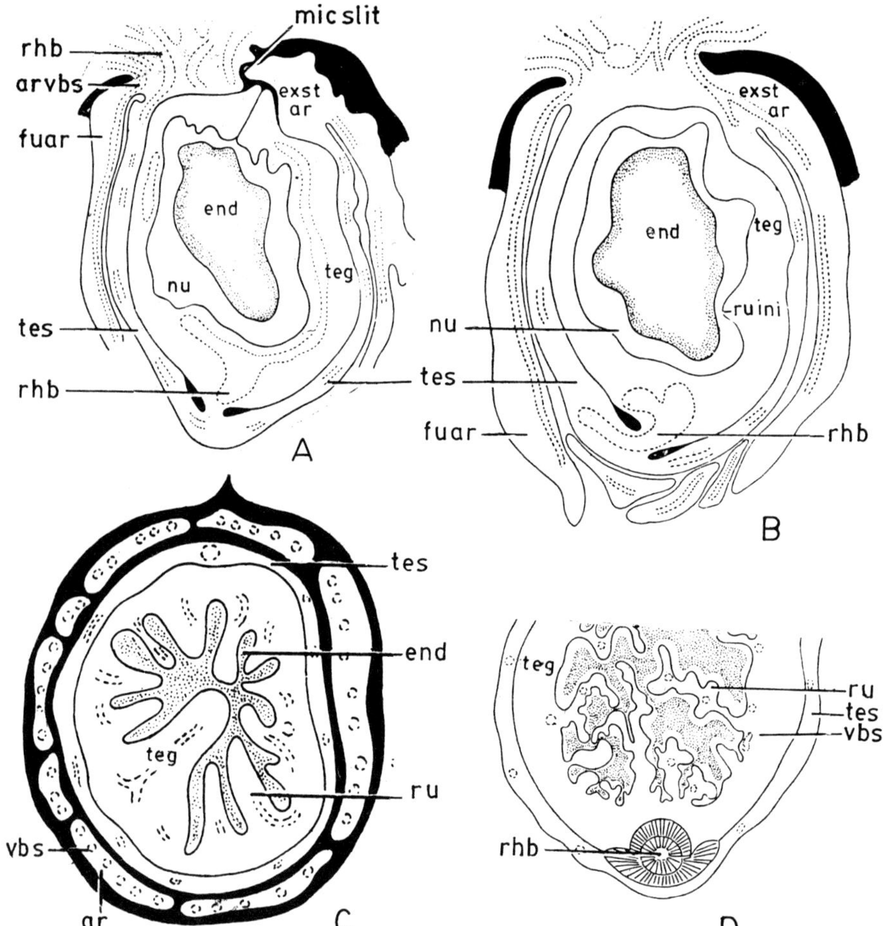

Fig. 18.8.2A–D. *Myristica fragrans.* **A** Median longisection of seed. Exostomal aril (*exst ar*), funicular aril (*fuar*), testa (*tes*) and tegmen (*teg*) are vascularized. Tegmen develops ruminations in the nucellus (*nu*); *arvbs* aril vascular bundles; *end* endosperm; *micslit* micropylar slit; *rhb* raphe bundle. **B** Longisection of seed, a little away from micropyle. Note entry of vascular strands at micropyle, and from funicle into exostome and funicular aril lobes; *ruini* rumination initials of tegmen. **C** Cross-section of fruit. The ruminations (*ru*) of tegmen in endosperm are vascularized, encircled by vascularized testa and aril (*ar*); *vbs* vascular strands. **D** Translongisection of chalazal region of seed. Palisade layers encircle raphe bundle. (After Mohana Rao 1974b)

micropylar region (Fig. 18.8.1D–G). The chalaza becomes massive and the embryo sac lies in it.

A single archesporial cell cuts off a parietal cell; the parietal tissue is massive. The megaspore mother cell undergoes meiotic divisions and produces a linear tetrad. The chalazal megaspore develops into a Polygonum type of embryo sac. According to van Heel (1982), the embryo sac undergoes enormous expansion. A postament also develops.

The entry of the pollen tube is porogamous. The pollen tube persists up to the young proembryo stage.

The development of endosperm is of the Nuclear type. Finally, it becomes cellular during later stages. It is thin-walled, ruminate, and stores starch and oil.

The mature embryo is small with divergent cotyledons, and a short radicle. Embryogeny has not been studied.

During the maturation of seed the chalazal tissue becomes massive and tanniferous. Tegmic ruminations occur at the chalaza and the vascular strand terminates at the base of the inner integument. According to N.C. Nair (1972), the boundary between nucellus and the inner integument remains vague in older ovules, and the ruminations are chalazal as well as lateral and originate from the inner integument. The inner basal portion of the inner integument shifts upward, the outer portion of base appears to remain close to the base of the outer integument. In *Horsfieldia* spp., the chalaza may be shortly columnar and lateral. The ruminations may be chalazal as well as lateral. In *Knema laurina* (van Heel 1982), the chalaza is tubular and ruminations develop from it as nine ridges around the endosperm.

Mohana Rao (1974b) and Roth and Lindorf (1974b) state that in *M. fragrans* the original condition of the two integuments does not change (Fig. 18.8.2A, B). Moreover, there is no pachychalazal growth and the inner seed-coat is formed by the inner integument. The cells of tegmen, on the inner side of the vascular bundle, develop ruminations which contain oil cells (Fig. 18.8.2C, D). Corner (1976) agrees with the interpretation of Mohana Rao (1974b) about the endotestal seed, normal chalaza, and tegmic ruminations.

The seed of Myristicaceae has exostomal and funicular aril (Fig. 18.8.2B). The cells of the outer epidermis of testa are thickly cuticularized. The mesophyll is composed of thin-walled cells. The inner epidermis consists of much-elongated tubular palisade cells with lignified pitted walls, and contain calcium oxalate crystals. The inner subepidermal layer is also a palisade layer, but is unlignified. Corner (1976) reports that the outer epidermis of tegmen is a single layer of longitudinally lignified fibres, compact or incomplete in places with thin-walled cells (*Horsfieldia irya*, *Myristica lowiana*), or it is a single layer of short sclerotic or tracheidal cells (*Knema*), or unspecialized (*M. fragrans*) at the chalaza with a counter-palisade of lignified cells recurved from the endotesta. The inner zone contains oil cells.

NN Bhandari (1971), EJH Corner (1976), WA van Heel (1982), AC Joshi (1943, 1946), PR Mohana Rao (1974b), NC Nair (1972), K Periasamy (1961), I Roth and H Lindorf (1974b), RLN Sastri (1955b), A Voigt (1888)

18.9 Canellaceae

The embryology of Canellaceae came to our attention through Parmeswaran's investigations (1961, 1962) on *Canella alba*, *Warburgia stuhlmannii* and *W. ugadensis*.

The anther wall comprises the epidermis, fibrous endothecium, two ephemeral middle layers, and glandular tapetum with two- to four-nucleate cells (Fig. 18.9.1A, B, F, G). Nuclear fusions may occur, and the cells become tetraploid. Übisch granules are present on the inner wall of tapetal cells. The microspore mother cells undergo meiotic divisions to produce tetrahedral tetrads (Fig. 18.9.1C, D). Cytokinesis is simultaneous.

The generative cell is produced towards the proximal pole. The pollen grains are monocolpate, and two-celled at the shedding stage (Fig. 18.9.1E).

The ovule is anatropous, bitegmic and crassinucellate with a zigzag micropyle (Fig. 18.9.1M). A single hypodermal archesporial cell cuts off a parietal cell. The parietal tissue is massive. The megaspore mother cell, by meiotic divisions, produces a linear tetrad. The chalazal megaspore develops into a Polygonum type of embryo sac (Fig. 18.9.1H–M). The polar nuclei fuse before fertilization. The antipodal cells persist up to fertilization.

The endosperm is oily, and ruminate in *Cinnamosma*. The mature embryo is small. The development of endosperm and embryo has not been studied.

Both the integuments form the seed-coat. The outer epidermis of the testa is composed of sclerosed cells. The mesophyll is thin-walled, and contains oil-cells. The testa is infiltrated with black contents. The tegmen is crushed.

NN Bhandari (1971), N Parmeswaran (1961, 1962), TK Wilson (1966)

18.10 Schisandraceae

The anther wall comprises an epidermis, fibrous endothecium, one or two ephemeral middle layers, and a glandular tapetum with multinucleate cells. Nuclear fusions occur in the tapetal cells. The epidermal cells become papillate, and contain tannin. Etherial oil cells are present in the connective. The microspore mother cells undergo meiotic divisions, and the tetrads are tetrahedral and decussate. Cytokinesis is simultaneous.

The pollen grains are heteropolar, hexacolpate, and shed at the two-celled stage (Hayashi 1960, Vijayaraghavan and Dhar 1975).

The ovule is ana- to campylotropous, bitegmic and crassinucellate (Hayashi 1963b). The micropyle is organized by both the integuments. The archesporium is multicelled in *Schisandra* and *Kadsura* (Hayashi 1963b). In *S. michaux* (Kapil and Jalan 1964), an archesporial cell cuts off a parietal cell; the parietal tissue is four layers thick. The megaspore mother cell undergoes meiotic divisions to produce dyads. The upper dyad fails to divide or degenerates, while the lower one divides in *S. michaux* (Kapil and Jalan 1964). The development of female gametophyte is of the Polygonum type; the antipodals are ephemeral. In *S. chinensis*, Swamy (1964a) describes monosporic Polygonum type, and bisporic Allium type embryo sacs. Polygonum type is

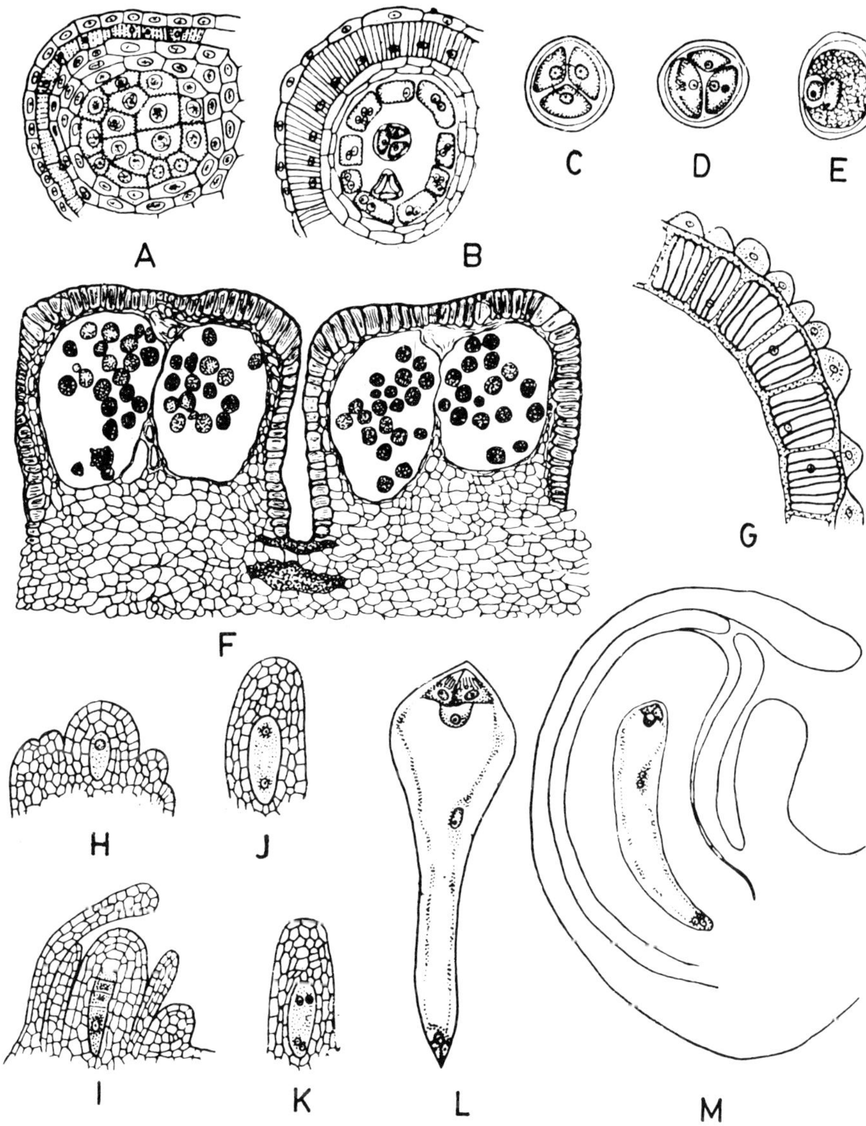

Fig. 18.9.1A–M. A–D, G, M *Canella alba*, **E, H–L** *Warburgia stuhlmannii*, **F** *Capsicodendron pimenteira*. **A** Cross-section of young anther lobe with microspore mother cells. **B** Fully differentiated wall, and tetrads. **C** Tetrad. **D** Tetrad with two-celled microspores. **E** Two-celled pollen grain. **F** Cross-section of staminal column with a pair of microsporangia in each anther lobe. **G** Portion of wall from mature anther. **H** Upper part of ovule with megaspore mother cell. **I** Linear tetrad. **J–L** Two-, four-nucleate and mature embryo sac. **M** Longisection of ovule at mature embryo sac. Note endostome and exostome. (**A–E** and **G–M** After Parmeswaran 1962, **F** after Wilson 1966)

exhibited by 50% ovules. The nucleus of the chalazal megaspore divides twice to form four nuclei lying at the micropylar pole of the embryo sac. They organize into an egg apparatus, and upper polar nucleus. In the Allium type, the two nuclei of chalazal dyad divide once, and the four nuclei again organize as the egg apparatus and upper polar nucleus. The antipodals and the lower polar nucleus are missing. Yoshida (1962) also describes similar development in *S. chinensis.*

The development of endosperm is of the Cellular type (Hayashi 1963b). The mature endosperm stores starch and oil.

The zygote divides by a transverse wall, and the apical cell plays a major role in the development of the embryo. The embryogeny corresponds to the Asterad or Onagrad type, and there is a massive suspensor.

The seed is small, campylotropous, albuminous and exotestal with a minute embryo. The cotyledons appear as two knobs of tissue until the time of seed dispersal (Hayashi 1963b). The outer epidermis of the vascularized testa consists of macrosclereids with hexagonal facets, and the hypodermal two to three layers of brachysclereids. The inner two or three layers are thin-walled with scattered sclerotic cells. The tegmen is crushed except at the micropyle. It is thick-walled.

The pericarp is thick and fleshy.

NN Bhandari (1970b, 1971), Y Hayashi (1960, 1963b), K-L Huynh (1976), S Jalan and RN Kapil (1964), RN Kapil and S Jalan (1964), BGL Swamy (1964a), MR Vijayaraghavan and U Dhar (1975, 1977), O Yoshida (1962)

18.11 Illiciaceae

The anther wall comprises an epidermis, fibrous endothecium, two or three ephemeral middle layers, and biseriate glandular tapetum of binucleate cells. The microspore mother cells undergo meiotic divisions producing tetrahedral tetrads; cytokinesis is simultaneous.

The pollen grains are tricolporate, and shed at the two-celled stage in *Illicium religiosum* (Hayashi 1960).

The ovule is anatropous, bitegmic and crassinucellate in *I. anisetum* (Hayashi 1963a). The micropyle is formed by exostome. It is thickened and lignified. The archesporial cell cuts off a parietal cell; the parietal tissue is massive. The megaspore mother cell undergoes meiotic divisions and produces a linear, or rarely T-shaped tetrad. The chalazal megaspore produces a Polygonum type of embryo sac. The three antipodals are ephemeral.

Development of the endosperm is of the Cellular type. The endosperm is copious and stores oil.

The embryogeny conforms to the Asterad type in *I. anisetum* (Hayashi 1963a). In a ripe seed the embryo shows two knob-like cotyledons.

The seed is albuminous and exotestal. The testa is multiplicative. The outer epidermis comprises microsclereids followed by thin-walled mesophyll. The inner two to three layers consist of sclerotic cells. The tegmen generally disappears.

NN Bhandari (1971), Y Hayashi (1960, 1963a)

18.12 Austrobaileyaceae

The stamens are laminar and massive with thecae borne on the adaxial surface (Fig. 18.12.1A). The anther wall (Fig. 18.12.1B) comprises an epidermis, well-differentiated fibrous endothecium, two to five middle layers, and irregularly biseriate secretory tapetum of parietal origin. The tapetal cells are binucleate in *Austrobaileya scandens* (Endress 1980a). During meiosis I of the microspore mother cells, a rudimentary cell plate is formed (Fig. 18.12.1C, D). Figure 18.12.1E shows meiosis II. Cytokinesis is simultaneous, and the tetrads are tetrahedral (Fig. 18.12.1F).

In *A. scandens* (Endress and Honegger 1980), the pollen grains are globose, anasulcate, tectate-perforate, regulose (reticulate) and columellate. The margin of sulcus is well marked; the endexine is considerably laminated at the border of aperture, and slightly laminated outside the aperture.

The pollen grain is two-celled at the dispersal stage (Fig. 18.12.1G).

The carpels are described as conduplicate in *A. scandens* by Bailey and Swamy (1949). However, according to Endress (1980a), they are ascidiate up to the stigmatic region, with a short plicate transition at the tip of style.

The ovules are anatropous, bitegmic and crassinucellate (Fig. 18.12.1H–K) with the inner or both integuments forming the micropyle. The raphe vascular strand extends into the outer integument along the antiraphe side. Tanniniferous cells occur on the ovular surface, the inner epidermis of the inner integument, chalazal region of vascular bundle, and base of nucellus. The nucellar epidermis undergoes periclinal divisions, and forms a "cap". The deeply seated megaspore mother cell (Fig. 18.12.1H), in *A. scandens* (Endress 1980a), undergoes meiotic divisions and a triad (possibly the upper dyad remains undivided) is formed (Fig. 18.12.1I). The development of the embryo sac is presumably of the Polygonum type; the antipodals are ephemeral (Fig. 18.12.1K).

The perichalazal, sarcotestal, albuminous seed is ruminate (Fig. 18.12.2A). The ruminations (Fig. 18.12.2B, C) are of the Annona type (Periasamy 1963). The testa is differentiated into outer thin-walled, and inner lignified tissue. The tegmen is crushed.

IW Bailey and BGL Swamy (1949), NN Bhandari (1971), PK Endress (1980a), PK Endress and R Honegger (1980), K Periasamy (1963)

18.13 Trimeniaceae

The embryology has not been studied in sufficient detail.

In *Trimenia papuana* the pollen sac is rudimentary and attached to the innermost tepal (Fig. 18.13.1A). The connective may extend beyond the anthers (Fig. 18.13.1B, C). In *T. neocaladonica* and *Piptocalyx moorei*, the anthers are much broader (Fig. 18.13.1D, E).

The anthers are tetrasporangiate (Fig. 18.13.1F–H), and the wall comprises persistent epidermis, fibrous endothecium (restricted only to the external part,

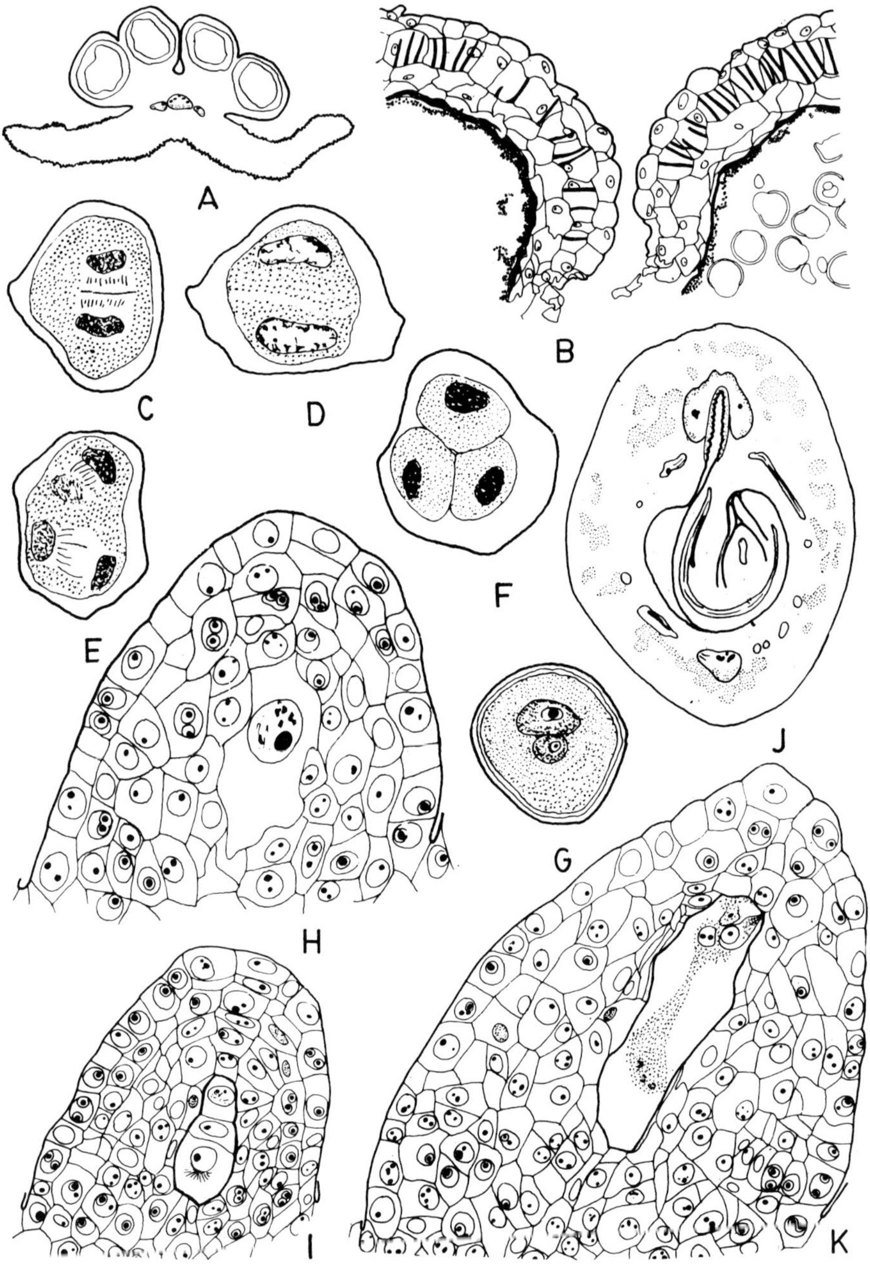

Fig. 18.12.1A–K. *Austrobaileya scandens.* **A** Transection, laminar stamen with two bisporangiate thecae on adaxial surface. **B** Part of anther at anthesis. **C–F** Pollen mother cell in meiosis I and II. **G** Two-celled pollen grain. **H** Longisection of nucellus at megaspore mother cell. **I** Linear triad. **J** Transection of ovary, ovule at mature embryo sac. **K** Nucellus with mature embryo sac. (After Endress 1980a)

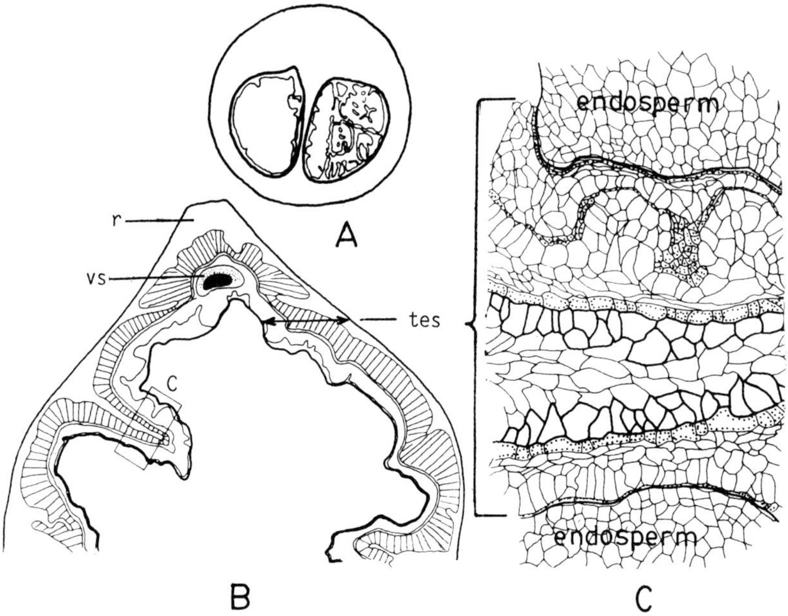

Fig. 18.12.2A–C. *Austrobaileya scandens.* **A** Transection of fruitlet with two ruminate seeds, the right seed cut superficially. **B** Part of transection of seed to show ruminations; lignified tissue *hatched*, followed by tanniniferous tissue marked with a *thin line*. **C** *Inset* C from **B** (lignified cells with *thick walls*, tanniniferous cells *dotted*). (After Endress 1980a)

absent in the region of the connective), and one or two ephemeral middle layers. The secretory tapetum shows binucleate cells in *Piptocalyx moorei* and *Trimenia papuana* (Endress and Sampson 1983). The microspore mother cells undergo meiosis, cytokinesis is successive, and the tetrads isobilateral and decussate.

The pollen is biaperturate with the centres of apertures located at the two pointed ends of each grain. In *T. papuana* the pollen is inaperturate, globose, spherical and polyporate (Sampson and Endress 1984). Both types have rugulate structure. The exine is tectate-columellate. In *T. papuana* and *T. weinmanniifolia*, the innermost tectum, columella and foot-layer have a partly granular form. The endexine is lamellate in non-apertural regions. The pollen is shed at the two-celled stage.

The surface of ovary is marked by longitudinal shallow furrows indicating the position of anthers in the bud. The stigma is hairy. A strand from the ventral bundle supplies the ovule, and extends up to the micropyle. A hypostase differentiates.

The ovule is anatropous and bitegmic (Fig. 18.13.2A–G). The outer integument has a short and discrete tip which forms the micropyle (Fig. 18.13.2A). The outer integument is six to seven layers thick, and the inner three to four layers. The outer epidermis of outer integument has radially elongate cells, and it develops into a stony layer in the seed. The inner epidermis of the inner

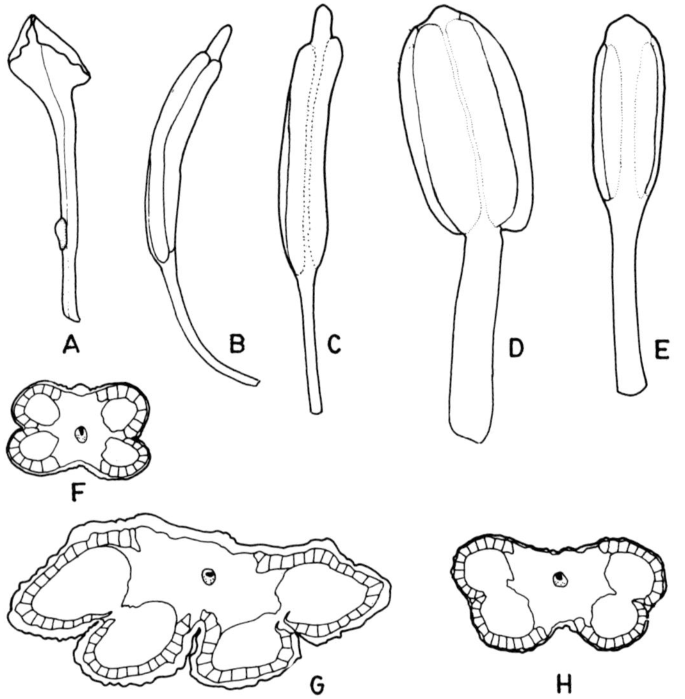

Fig. 18.13.1A–H. Stamens. **A**–**C**, **F** *Trimenia papuana* **D**, **G** *T. neocaledonica*, **E**, **H** *Piptocalyx moorei*. **A** Innermost tepal with rudimentary pollen sac (adaxial view) from mature bud. **B** Stamen at anthesis, lateral view (right side – adaxial). **C**–**E** Stamens at anthesis (abaxial view). Note the prolonged connective in **B** and **C**. In **D** and **E** the anthers are much broader and extrosely dehiscent. **F**–**H** Transections of tetralocular anthers shortly before dehiscence (adaxial side above), endothecium *hatched*. (After Endress and Sampson 1983)

integument is tanniniferous, and it forms a massive cap over the micropyle in *T. neocaladonica* and *P. moorei* (Endress and Sampson 1983).

At the time of initiation of integument in the ovule, the archesporium is not distinct. But, later, multicellular archesporium differentiates (Fig. 18.13.2E, F). There is no information about the formation of a parietal cell. After meiosis several embryo sacs grow simultaneously but only one reaches maturity. Megasporogenesis has not been described.

The embryo sac is long and narrow, and occupies the entire length of nucellus (Fig. 18.13.2G). The egg apparatus and polar nuclei lie at the apex, and the antipodals probably at the base of the embryo sac.

The development of endosperm and embryo has not been worked out.

PK Endress and FB Sampson (1983), FB Sampson and PK Endress (1984)

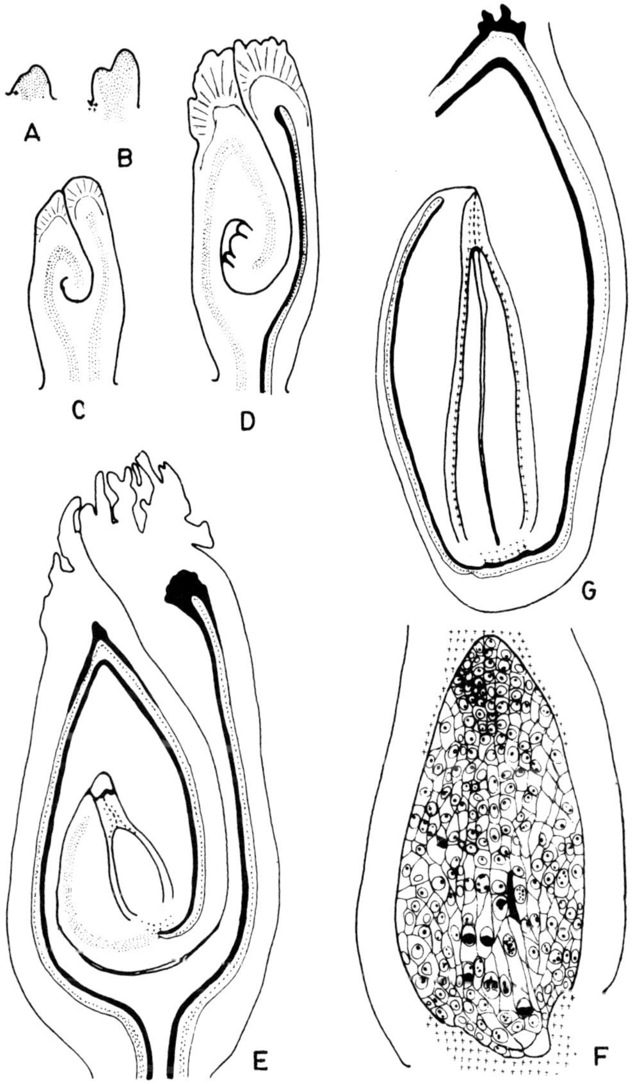

Fig. 18.13.2A–G. *Trimenia papuana*, development of carpel; median longisections (left side adaxial). *Dotted areas* in meristematic tissue, represent procambium, and phloem in differentiated vascular bundles. *Black areas* xylem in differentiated vascular bundles. *Plus signs* tanniniferous tissue. *Black lines in stigmatic region* cell rows. **A** Origin of ventral "cross-zone". **B** "Chair-like" stage. **C** Early epidermal periclinal divisions in future stigmatic region (cell rows of epidermal origin indicated); inception of inner integument. **D** Rotation of ovule almost completed, ventral vascular bundle still procambial. **E** Ovule with meiocytes, integumentary vascular bundle still procambial. **F** Nucellus with numerous meiocytes (from **E**). **G** Ovule at anthesis, elongated nucellus, and long tubular embryo sac. (After Endress and Sampson 1983)

18.14 Amborellaceae

Information on embryology is very scanty. The family includes a single genus, *Amborella* (Willis 1973).

The anthers are bilocular and dehiscence occurs by slits. The microspore tetrads are tetrahedral, and the pollen grains are large.

Each carpel contains a single anatropous ovule (orthotropous according to Willis 1973) which is bitegmic and crassinucellate. The inner integument is three layers thick, and the outer one five to seven layers.

The seed has a small and straight embryo with copious endosperm. The mature embryo displays only initial stages of cotyledonary primordia.

IW Bailey and BGL Swamy (1948), J Hutchinson (1973), J Perkins and E Gilg (1911), JC Willis (1973)

18.15 Monimiaceae

In some genera the flowers are highly specialized (Endress 1980b). An extreme case is that of *Tambourissa*. The male and female flowers, on the same inflorescence, are globular and fig-like (Fig. 18.15.1A, B). The carpels are situated around the floral cup (Fig. 18.15.1C, D). The secretory transmitting tissue covering the carpels is continued into the short channel at the mouth of the floral cup (Fig. 18.15.1D, E). This constitutes the "hyperstigma". All the genera with hyperstigma are shown in Fig. 18.15.1F–J. Other taxa lack hyperstigma.

In *Hortonia angustifolia* (Endress 1980c), the fertile stamens have two basal appendages, and their inner epidermis is nectariferous.

The anther wall comprises an epidermis, fibrous endothecium, two or three ephemeral middle layers, and a glandular tapetum of binucleate cells in *Laurelia novae-zelandiae* (Sampson 1969c) and *Tambourissa religiosa* (Endress 1980b). In *H. angustifolia* the tapetum is irregularly one- or two-layered. The tapetum is periplasmodial in *Antherosperma* (Sastri 1963). Meiotic divisions in microspore mother cells result in the formation of tetrahedral tetrads. Cytokinesis is simultaneous. The centripetal plates do not develop until the end of meiosis II.

The pollen grains are in permanent tetrads in *Hedycarya* sp. (Money et al. 1950, Sampson 1969a) and *Laurelia*. The generative cell is cut off towards the distal face (Sampson 1969a). They are inaperturate, two- to three-sulcate and shed at the two- or three-celled stage.

The development of carpel in *Hortonia* is shown in Fig. 18.15.2A–D, and a longisection of ovule at mature embryo sac in E.

The ovule is anatropous and crassinucellate, unitegmic in *Siparuna* (Heilborn 1931), bitegmic in *Hortonia* (Fig. 18.15.2F, G), *Laurellia* (Sampson 1969c), and *Tambourissa*. The micropyle is organized by both the integuments. The cells of the nucellar epidermis divide periclinally to produce a nucellar cap. The female archesporium is uni- to multicelled. An archesporial cell cuts off

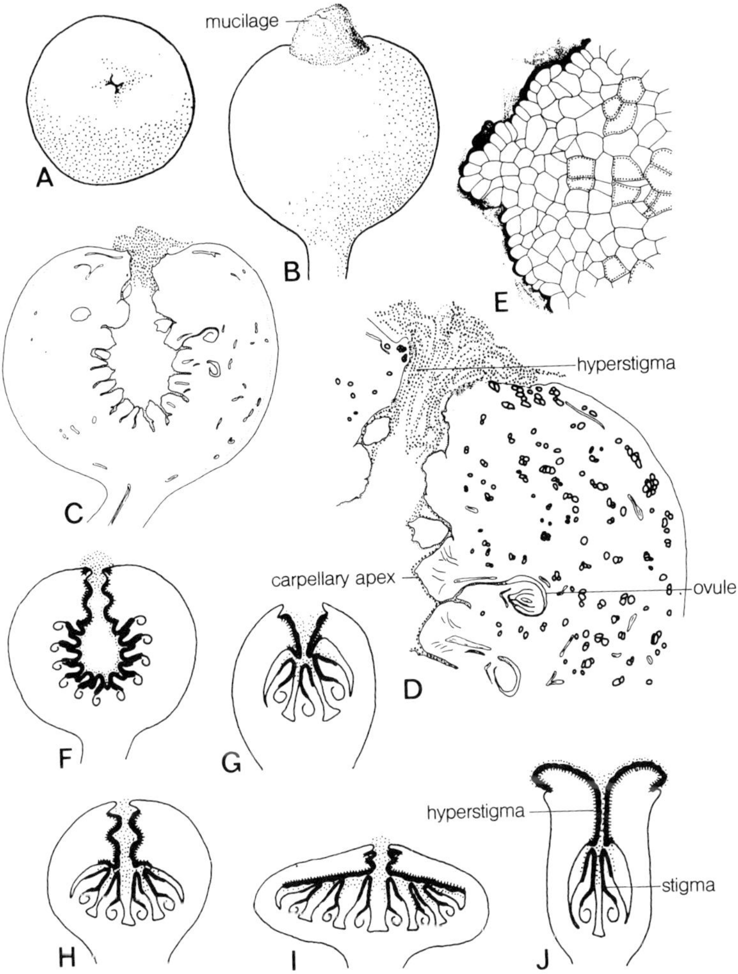

Fig. 18.15.1A–J. **A–F** *Tambourissa religiosa*, **G** *Kibara*, **H** *Wilkiea macrophylla*, **I** *W. huegeliana*, **J** *Hennecartia*. **A, B** Female flower, before (**A**) and at anthesis (**B**). **A** As seen from above, mucilage removed from pore. **B** Lateral view, with mucilage. **C** Entire female flower in longisection. **D** Upper portion (*right*) of **C**. Floral channel with hyperstigma (mucilage *dotted*), and upper carpels. Note groups of stone cells. **E** Carpellary apex with secretory epidermis from **D**, mucilage *dotted*. **F–J** Schematic longisections of female flowers. *Dark lines studded with short stubs* represent hyperstigma (secretory), *plain dark lines* represent stigma and ventral slit, mucilage *dotted*. (After Endress 1980c)

a parietal cell; the parietal tissue is massive. The megaspore mother cell (Fig. 18.15.2F) undergoes meiotic divisions and produces a linear tetrad. The chalazal megaspore (Fig. 18.15.2G) gives rise to a Polygonum type of embryo sac in *Hortonia*, *Mollinedia* (Peter 1920) and *Laurelia*. A peculiar feature has been observed in *Siparuna*. The chalazal megaspore forms a long, tortuous

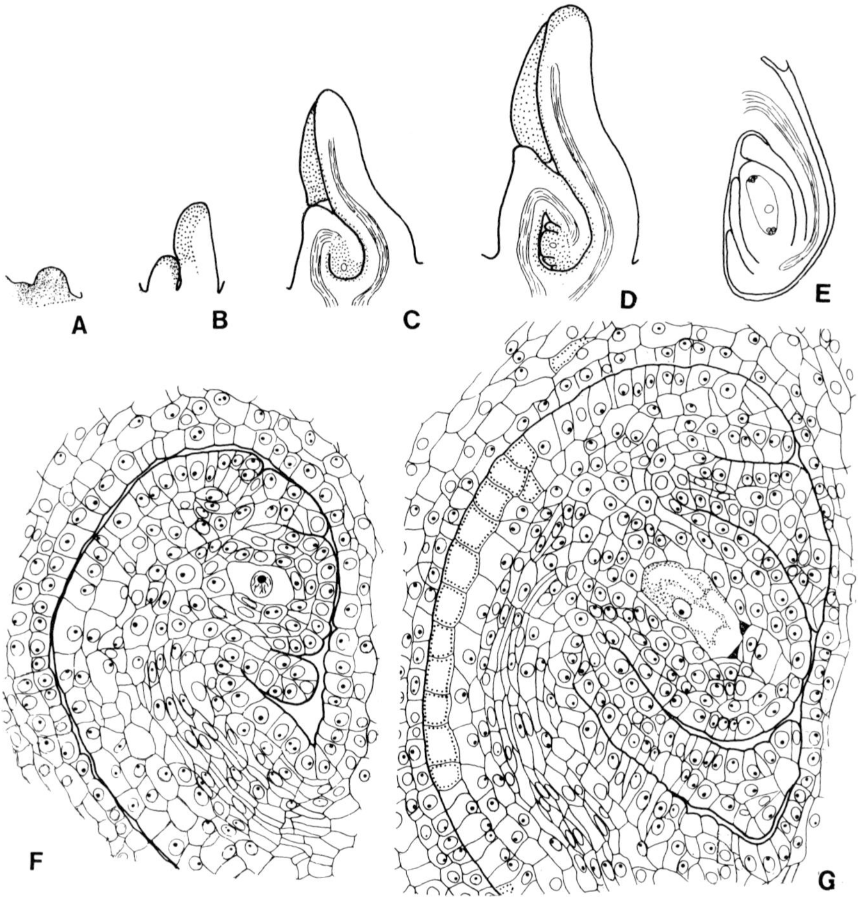

Fig. 18.15.2A–G. *Hortonia angustifolia.* **A, B** Initiation of ovular primordium. **C, D** Longisection of carpels to show the curvature of ovule, the stigma has ventral slits. **E** Longisection of bitegmic, crassinucellate ovule at mature embryo sac (seven-nucleate, polars fused). **F, G** Longisection of ovule at megaspore mother cell (**F**) and functional megaspore (**G**). Tannin-bearing cells marked with *bold dots* in **G**. (After Endress 1980b)

tube with vesicular tip. The tip bursts in another cell, and discharges its contents. This cell forms a four- to five-nucleate unorganized female gametophyte. These observations require confirmation, and a further investigation is called for. In *Peumus* the upper of the two dyad cells degenerates, and the lower is functional giving rise to an Allium type of female gametophyte. An increase in the number of antipodal cells (5–20) has been reported in *Peumus boldus* (see Davis 1966). A hypostase is present.

At anthesis, in *Tambourissa religiosa*, the hyperstigma secretes mucilage (Fig. 18.15.1B–D) so that there is a continuous transmitting medium for the pollen tube from the mouth of the floral cup (where pollen is deposited) up to the micropyle. Syrphids and drosophilids have been observed as pollinators of *T. peltata* (see Endress 1980b). The pollinators deposit their eggs in the male and female flowers by piercing the wall of the floral cup. The adult pollinator flies out of the male flower carrying pollen which is deposited in the mucilage at the mouth of hyperstigma. It would be interesting to know if there are symbiotic relationships with insect parasites serving as pollinators.

The development of endosperm is of the Cellular type. It is massive, and stores oil.

The embryogeny corresponds to the Asterad type.

The seed is albuminous, endotestal with a small straight embryo; the cotyledons may be appressed or divergent. The outer two or three layers of the testa comprise thin-walled cells, and oil cells. The inner epidermis shows lignified tracheidal cells. The tegmen is thin-walled, and may not persist.

NN Bhandari (1971), GL Davis (1966), PK Endress (1979, 1980b, c), O Heilborn (1931), LL Money, IW Bailey and BGL Swamy (1950), J Peter (1920), FB Sampson (1969a, b, c, 1975, 1977), RLN Sastri (1963)

18.16 Calycanthaceae

The anther wall comprises an epidermis, fibrous endothecium, one or two ephemeral middle layers, and secretory tapetum in *Calycanthus fertilis* (S.L. Mathur 1968). Following meiosis, the microspore mother cells produce tetrahedral and decussate tetrads. Cytokinesis is simultaneous.

The pollen grains are two- to three-sulcate, and shed at the three-celled stage. Occasionally, they remain united in tetrads. Their germination in situ has also been observed (S.L. Mathur 1968).

The ovule is anatropous, bitegmic and crassinucellate. The micropyle is organized by both the integuments. The exostome and endostome are lined with microscopic hairs. The nucellar cells divide to produce a nucellar cap. The archesporium is multicellular. These cells function as megaspore mother cells, undergo meiotic divisions, and produce linear tetrads. The chalazal megaspore develops into a Polygonum type of embryo sac (Peter 1920). Supernumerary embryo sacs and a hypostase develop. The three antipodals are ephemeral; one of the polar nuclei also degenerates. Syngamy and triple fusion (the polars may fuse and form secondary nucleus) do not occur in *Calycanthus fertilis* (Davis 1966, S.L. Mathur 1968). According to Peter (1920), Schürhoff (1923) and Brofferia (1930), *Calycanthus* and *Chimonanthus* are apomicts. While Peter and Brofferia suggest nucellar polyembryony as the source of supernumerary embryos, Schürhoff (1923) considers the possibility of somatic parthenogenesis (see Schnarf 1931).

The development of endosperm is autonomous and of the Cellular type.

The development of a sexual embryo is shown in Figure 18.16.1A–E and its structure in H–K; nucellar embryos are common (Fig. 18.16.1F, G).

The seed is endotestal, exalbuminous and enclosed in a hard pericarp. The outer epidermis of testa comprises thin-walled cells with brown contents. It is

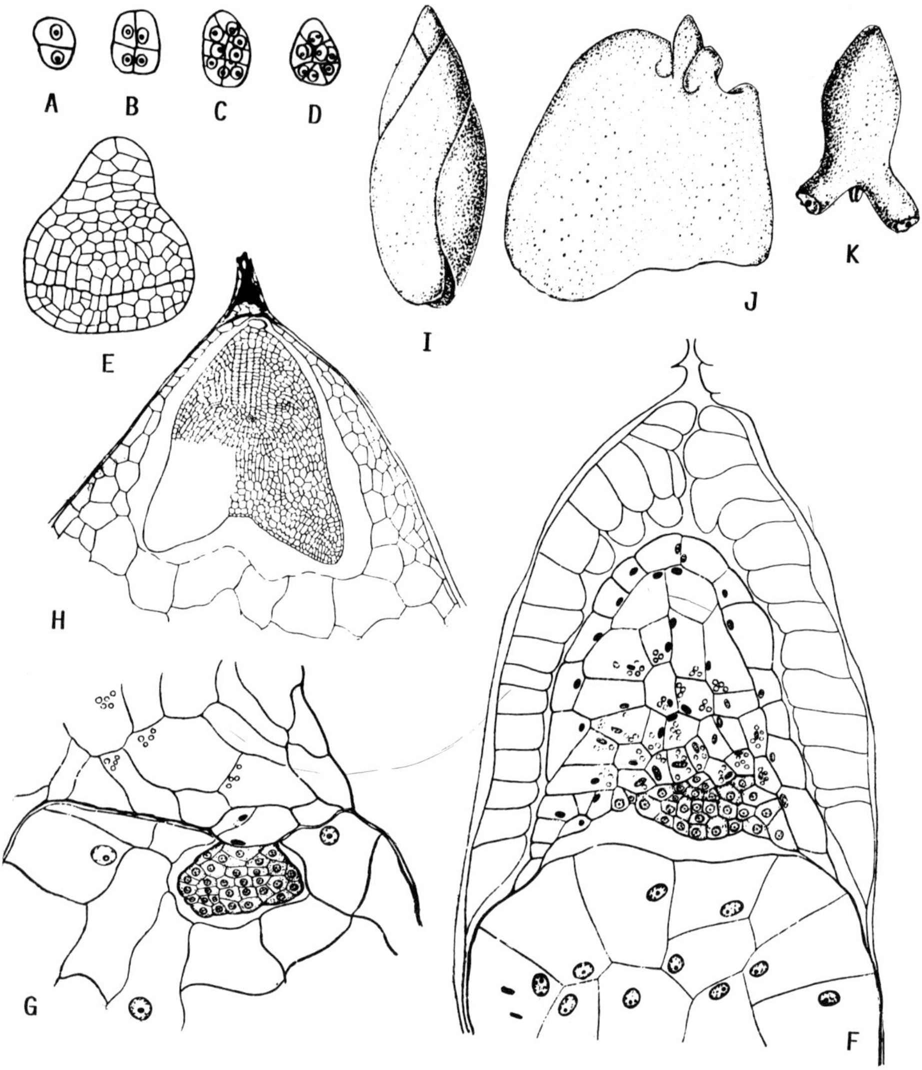

Fig. 18.16.1A–E. *Chimonanthus fragrans*, Embryogeny. **F–H** *C. occidentalis*; **F** Nucellar cap with meristematic zone adjacent to cellular endosperm; **G** Portion of nucellus with multicellular proembryo surrounded by cellular endosperm; **H** Heart-shaped sexual embryo and cellular endosperm, note remnants of pollen tube. **I–K** *C. praecox*; **I** Mature convolute embryo; **J** Embryo with expanded cotyledons; **K** Hypocotyl and shoot-apex, cotyledons removed. (**A** –**E** After Ly-Thi-Ba 1962, for **F–K** see Yakovlev 1981)

slightly lignified in *Chimonanthus*. The inner epidermis is formed of tangentially elongate, lignified cells with reticulate thickenings. The tegmen is crushed.

NN Bhandari (1971), J Brofferia (1930), GL Davis (1966), Ly-Thi-Ba (1962), SL Mathur (1968), J Peter (1920), K Schnarf (1931), PN Schürhoff (1923), YD Tiagi (1963)

18.17 Gomortegaceae

The embryology has not been investigated.

18.18 Lauraceae

The anther wall comprises a persistent epidermis, fibrous endothecium, two middle layers, and amoeboid tapetum. The tapetum is glandular in *Cassytha* (Sastri 1962). The tapetal cells may be bi-, tetra- or multi-nucleate. The microspore mother cells undergo meiotic divisions to produce tetrahedral, isobilateral, T-shaped and linear tetrads. Cytokinesis is successive.

The pollen grains are acolpate and two-celled at the shedding stage.

The ovule is pendulous, anatropous, bitegmic and crassinucellate. Generally, the micropyle is produced by both the integuments. Only the outer integument is concerned in *Cinnamomum* and *Litsea* (Sastri 1958a), and the inner in *Sassafras* (Coy 1928). In *Cassytha* the integuments do not grow beyond the nucellus so that the micropyle is broad, and the overarching funiculus comes in direct contact with the nucellus. The archesporium is one-celled in *Laurus* and *Umbellularia*, and multicelled in *Cassytha* and *Cinnamomum*. The archesporial cells cut off parietal cells, and a massive parietal tissue is produced. The epidermal cells of the nucellus also divide and add to it. Several megaspore mother cells function. They undergo meiotic divisions, and linear tetrads are produced. The chalazal megaspore gives rise to a Polygonum type of embryo sac. In *Cassytha* (Sastri 1956), in a single ovule as many as six embryo sacs develop simultaneously. They show tubular haustorial processes in the direction of micropyle, protruding beyond the nucellus and entering the overarching funicule. The antipodals are ephemeral, but in *Laurus nobilis* (Bambacioni-Mezzetti 1935) they undergo multiplication. In *Cassytha* they do not organize into cells.

In *Persea americana* (Schroeder 1952, Tomer and Gottreich 1978), the ovules are frequently sterile either due to the absence of formation of embryo sacs, or their early degeneration. Twin embryo sacs may originate from two archesporial cells.

The endosperm is of the Nuclear type. Later, it becomes cellular at the globular proembryo in *Cinnamomum* (Sastri 1958a); in *Cassytha* (Sastri 1962) it is ab initio Cellular.

The embryogeny corresponds to the Piperad type in *Sassafras* (Coy 1928), Asterad type in *Persea americana* (Schroeder 1952), and Onagrad type-Trifolium variation in *Cinnamomum iners* and *Litsea sebifera* (Sastri 1958a). The division of zygote is transverse in *Cassytha* (Sastri 1962) and *Laurus nobilis*; further development is irregular.

The seed is exalbuminous, and endotestal. The embryo is large with thick, massive cotyledons, and a very short radicle. The testa alone forms the seed-coat. The outer zone of testa comprises thin-walled cells, and the outer epidermis contains tannin. The inner epidermis consists of longitudinally and transversely elongate narrow tracheidal cells with annular or spiral thickenings. The endocarp is stony.

V Bambacioni-Mezzetti (1935, 1938), E Battaglia (1947a), NN Bhandari (1971), JK Choudhury and JN Mitra (1953), GV Coy (1928), PK Endress (1972), RO Gardner (1974), PR Mohana Rao (1980, 1986), F Pax (1891), RLN Sastri (1952, 1956, 1958a, 1962, 1963), CA Schroeder (1952), M Sedgley (1976, 1979), G Täckholm and E Soderberg (1918), E Tomer and S Gazit (1979), E Tomer and M Gottreich (1975, 1978), MS Yakovlev (1981)

18.19 Hernandiaceae

The embryology has not been studied in sufficient detail (Table 18.24.2).

The flowers are unisexual. The anthers are bilocular with glandular appendages. The tapetum is of the amoeboid type. The pollen mother cells

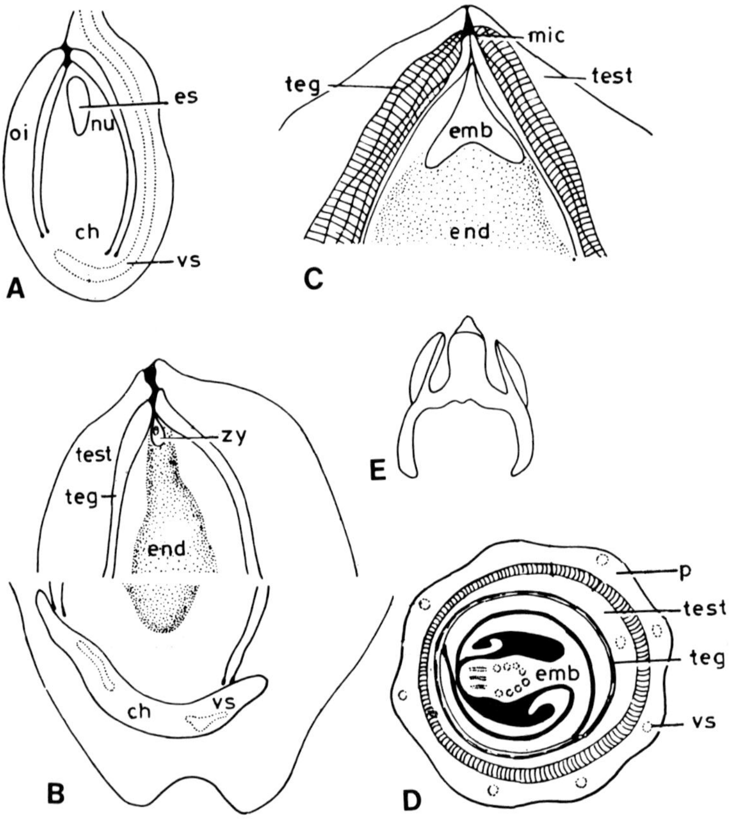

Fig. 18.19.1A–E. *Gyrocarpus americanus.* **A** Longisection of ovule; *es* embryo sac; *nu* nucellus; *oi* outer integument; *ch* chalaza; *vs* vascular supply. **B** Ovule after fertilization; *zy* zygote; *end* endosperm; *test* testa; *teg* tegmen. **C** Translongisection of upper part of immature seed; *emb* embryo; *mic* micropyle. **D** Transection of fruit; *p* pericarp. **E** Mature embryo with leafy and convolute cotyledons. (After Mohana Rao 1986)

undergo meiosis and successive cytokinesis. The pollen grains are two-celled at shedding, and the exine shows spinescent outgrowths (see Yakovlev 1981).

The carpel is two-trace as in Lauraceae. The vascular bundle enters the funiculus, reaches up to the base of nucellus and branches dichotomously in the chalaza (Fig. 18.19.3 C, E). The raphal bundle is reported to enter the outer integument in *Hernandia peltata.*

The single pendulous ovule in the inferior ovary is at first apotropous and then becomes anatropous. It is bitegmic and crassinucellar with micropyle formed by both integuments (Fig. 18.19.1A–C). The inner integument is three- to five-layered at the base, and more massive in the micropylar region. The apical region contains large starch grains. The outer integument is massive, and multiplicative (Fig. 18.19.3A, B).

The distally lobed integuments in *Hernandia peltata* (van Heel 1971b) are absent in *Gyrocarpus americanus* (Mohana Rao 1986).

The embryo sac of *Gyrocarpus* is eight-nucleate. The endosperm is of the Cellular type. The embryogeny (Fig. 18.19.2A–J) is of the Onagrad type [Period I, Series A, Megarchetype VI (B.S.M. Dutt 1978b)].

In *Hernandia*, the exalbuminous testal pachychalazal seed, enclosed in a woody endocarp (Corner 1976, Hutchinson 1973), has a straight embryo with large cotyledons which are plano-convex, folded or lobate and more or less connate in the ruminations of the perisperm (Fig. 18.19.2K, L cf. Fig. 18.19.1D, E). The folded perisperm is massive. This is an example of labyrinth seed (van

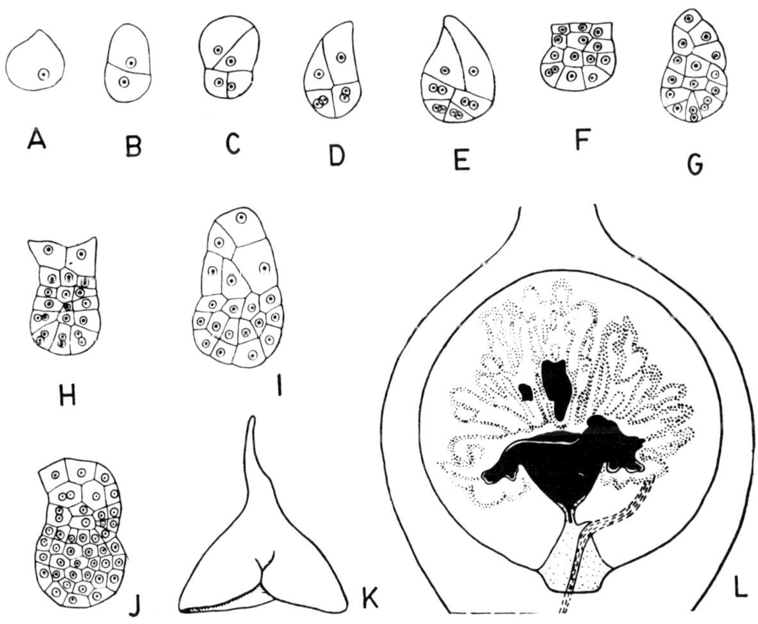

Fig. 18.19.2A–J. *Gyrocarpus americanus*, embryogeny. **K, L** *Hernandia peltata*; **K** Mature embryo; **L** Longisection of mature fruit, the embryo (*black*) shows two large folded (or lobed) cotyledons and ruminate perisperm. (**A–J** After B.S.M. Dutt 1978b, **K, L** After van Heel 1971a)

Heel 1971a). The chalazal region of the ovule develops enormously to form the body of the seed. The wall of this region becomes ruminate from profuse lobings of the enlarging nucellus and the cotyledons eventually bulge into these lobings to substitute the nucellus (Corner 1976).

In *Gyrocarpus* the testa is thick and undifferentiated, the tegmen becomes crushed (Fig. 18.19.3A, B). The pachychalaza and ruminate outgrowths are absent (Mohana Rao 1986).

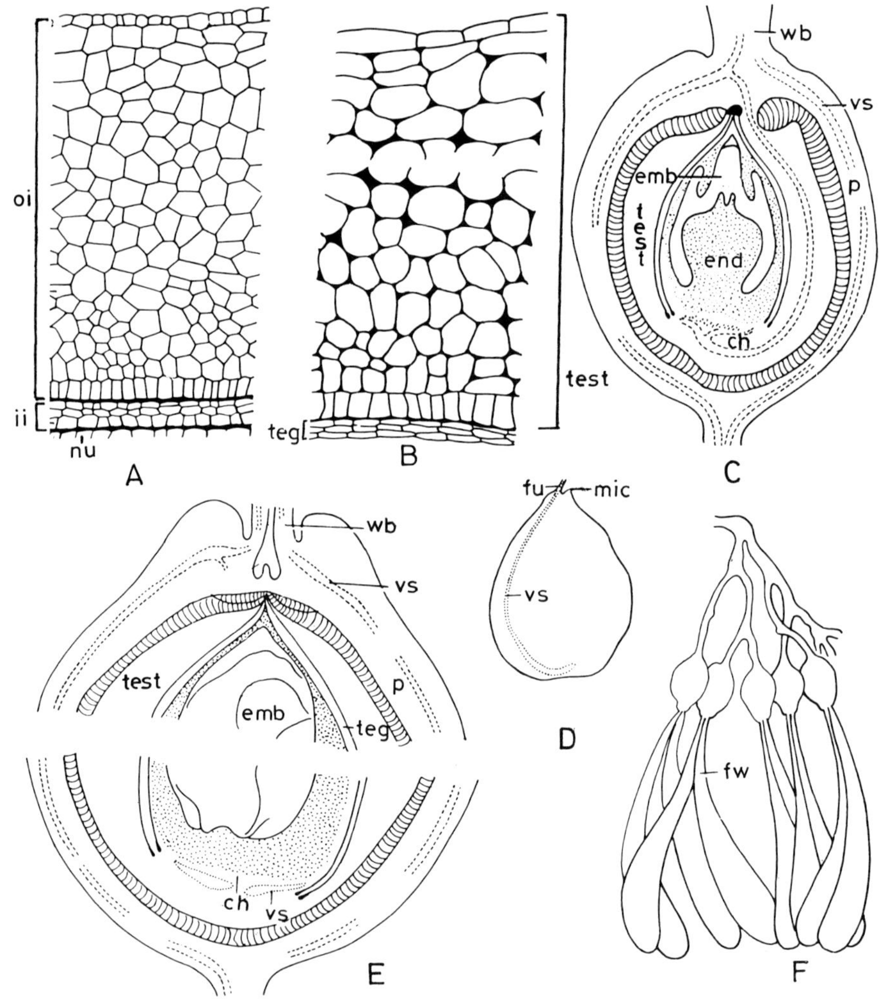

Fig. 18.19.3A–F. *Gyrocarpus americanus.* **A** Outer (*oi*) and inner (*ii*) integument from immature seed; *nu* nucellus. **B** Testa (*test*) and tegmen (*teg*) from older seed. **C** Longisection of younger fruit; *wb* wing base; *vs* vascular supply; *p* pericarp; *emb* embryo; *end* endosperm; *ch* chalaza. **D** Seed (dissected from fruit); *fu* funicle; *mic* micropyle. **E** Longisection of older fruit, seed cut in transmedian longisection. **F** Mature winged fruits; *fw* fruit wing. (After Mohana Rao 1986)

The seed closely adheres to the inner epidermis of the pericarp, and can be easily dissected out from a ripe and undried fruit (Figs. 18.19.1D, E, 18.19.3C, D).

In *Gyrocarpus americanus* (Mohana Rao 1986), the ovoid fruit has eight longitudinal ridges and two spatulate wings which extend from the distal end (Fig. 18.19.3F). The wing epidermis has unicellular hairs and stomata. The wings help in the dispersal of dry fruit.

BSM Dutt (1978b, 1979a, 1980), EJH Corner (1976), PK Endress (1972), WA van Heel (1970a, 1971a, b, c), J Hutchinson (1973), K Kubitzki (1969), PR Mohana Rao (1986), RLN Sastri (1958a, 1963), CF Shutts (1960), AL Takhtajan (1988), MS Yakovlev (1981)

18.20 Tetracentraceae

The information on the embryology is very scanty.

The anther is tetralocular with lateral microsporangia which open by slits (Hutchinson 1973). The endothecium with fibrous thickenings differentiates only on the outer side of anther lobe. The mature pollen grain is bicelled.

The ovule is anatropous, bitegmic and crassinucellate in *Tetracentron* sp.

The albuminous seed has a minute embryo with primordial cotyledons. The endosperm is oily.

JL Brewbaker (1967), J Hutchinson (1973), CG Nast and IW Bailey (1945)

18.21 Trochodendraceae

The anther wall comprises an epidermis, fibrous endothecium, two or three ephemeral middle layers, and biseriate glandular tapetum with two-nucleate cells. The epidermal cells develop protuberances and cuticular fibrillar thickenings along their outer surface in *Trochodendron aralioides* (Yoffe 1962). The endothecium shows broad bands of fibrous thickenings. The microspore mother cells undergo meiotic divisions and produce tetrahedral tetrads. Cytokinesis is simultaneous.

The pollen grains are tricolpate, and shed at the two-celled stage.

The ovule is anatropous, bitegmic, and crassinucellate. The micropyle is produced by the inner integument. The chalazal end of the ovule develops into a long tapering projection. The single hypodermal archesporial cell cuts off a parietal cell. The parietal tissue is three- to four-layered. Following meiotic divisions, the megaspore mother cell produces a linear or T-shaped tetrad. The chalazal megaspore develops into a Polygonum type of embryo sac. The antipodals are uninucleate, and persistent.

The development of endosperm (Yoffe 1965) is of the Cellular type. The endosperm is thin-walled, and oily.

The zygote divides transversely followed by another transverse division in both the cells producing a four-celled linear proembryo. The terminal cell forms the octant, and the remaining cells produce the suspensor.

The embryogeny appears to be of the Chenopodiad type.

The seed is small, endotestal and exotegmic with a very small embryo. The outer epidermis of the testa is thin-walled, whereas the inner comprises cuticularized cells with thick inner and radial walls. The outer epidermis of tegmen shows small tangentially elongate thick-walled cells; the inner one is thin-walled.

The seeds of *Trochodendron aralioides* (Mohana Rao 1981) are linear and ellipsoid with micropylar and chalazal projections. The testa is two layers thick and the inner layer forms the characteristic crystal layer. The tegmen is three layers thick, and the outer epidermis is a conspicuous fibrous layer. The raphe bundle terminates below the hypostase. The fruit is a follicle. The mature pericarp is 30 layers thick, and the ground tissue is parenchymatous.

NN Bhandari (1971), H Harms (1897a), PR Mohana Rao (1981), CG Nast and IW Bailey (1945), MD Yoffe (1962, 1965)

18.22 Eupteleaceae

The anther wall comprises the epidermis, fibrous endothecium, two or three ephemeral middle layers, and secretory tapetum with one- to four-nucleate cells in *Euptelea polyandra* (Endress 1969a). Simultaneous cytokinesis accompanies meiotic divisions in microspore mother cells, and the tetrads are tetrahedral.

The pollen grains are two-celled at the dispersal stage.

The ovule is pendulous, anatropous, bitegmic and crassinucellate with micropyle formed by the inner integument. The archesporial cell cuts off a parietal cell. Due to repeated divisions of the latter and of nucellar epidermal cells, the parietal tissue is very conspicuous. The megaspore mother cell undergoes meiotic divisions producing a linear tetrad. The chalazal megaspore develops into a Polygonum type of embryo sac. Two months elapse between pollination and fertilization.

The development of endosperm is of the Cellular type.

The embryogeny is irregular. The hypocotyl is very conspicuous in the mature dicotyledonous embryo.

The endotestal seed is small with a minute embryo. The outer epidermal cells of the testa are thin-walled with reddish-brown contents. The mesophyll and inner epidermal cells are sclerotic. The tegmen comprises thin-walled cells.

NN Bhandari (1971), PK Endress (1969a, b), R Lemesle (1946), CG Nast and IW Bailey (1946)

18.23 Cercidiphyllaceae

Swamy and Bailey (1949) have made a detailed study of *Cercidiphyllum japonicum.*

The anther tapetum is secretory with binucleate cells. Simultaneous cytokinesis follows meiotic divisions in microspore mother cells forming tetrahedral and decussate tetrads.

The pollen grains are tricolpate. Their furrows are very broad and ill-defined, and are generally described as immersed. They are shed at the two-celled stage.

The ovule is anatropous, bitegmic and crassinucellate. Both the integuments form the micropyle. The hypodermal archesporial cell cuts off a parietal cell producing five or six layers of parietal tissue. The nucellar epidermis undergoes periclinal divisions. The megaspore mother cell undergoes meiotic divisions and produces a linear tetrad. The chalazal megaspore develops into a Polygonum type of embryo sac. The egg apparatus is prominent, and the antipodals are small. The fertilized embryo sac elongates conspicuously.

The development of endosperm is of the Cellular type (Fig. 18.23.1A–E). A large micropylar and a small chalazal chamber are produced following the

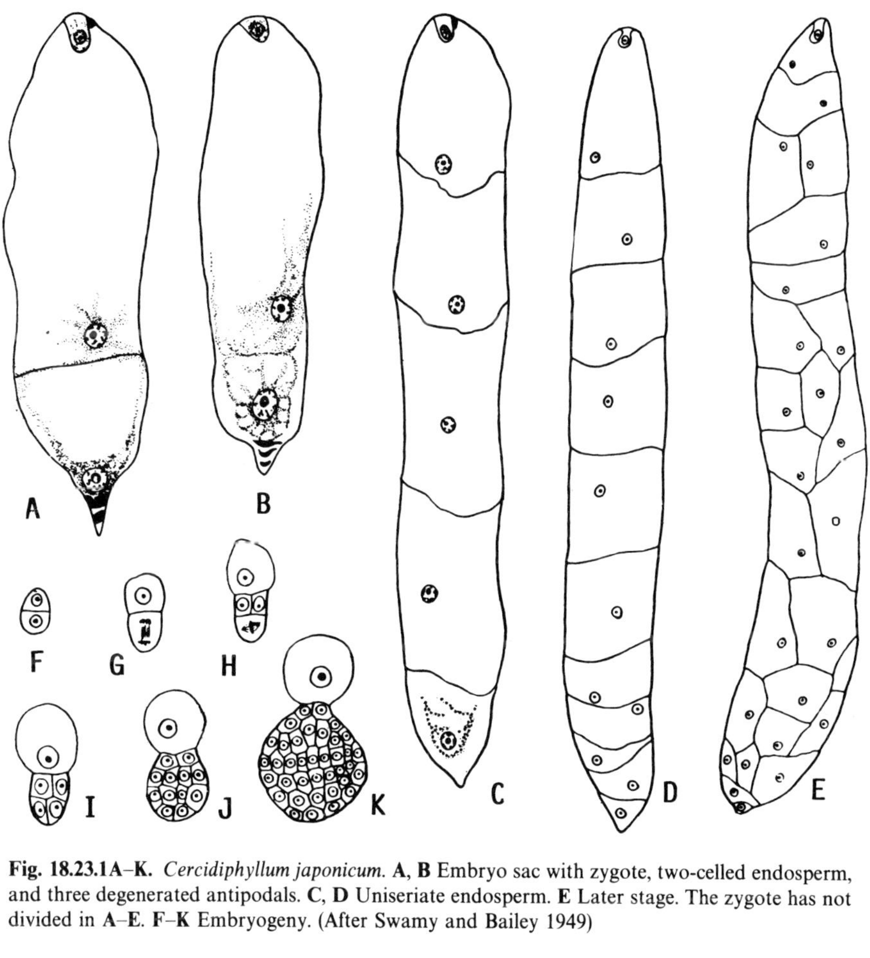

Fig. 18.23.1A–K. *Cercidiphyllum japonicum.* **A, B** Embryo sac with zygote, two-celled endosperm, and three degenerated antipodals. **C, D** Uniseriate endosperm. **E** Later stage. The zygote has not divided in **A–E**. **F–K** Embryogeny. (After Swamy and Bailey 1949)

transverse division of the primary endosperm nucleus. Both the chambers undergo transverse divisions, and the endosperm gives a uniseriate appearance; vertical and oblique walls develop afterwards.

The zygote divides transversely forming apical and basal cell. The apical cell develops into a quadrant; the basal cell does not divide further and becomes enlarged (Fig. 18.23.1 F–I). J and K are proembryos. The embryogeny corresponds to the Caryophyllad type. The mature embryo is large with a well-developed hypocotyl.

The seed is small, winged and albuminous. The chalazal wing has a curved vascular bundle. The testa is thin-walled, and the tegmen is crushed.

NN Bhandari (1971), L Diels (1936), H Harms (1918), K Subramanyam (1962), BGL Swamy and IW Bailey (1949)

18.24 Taxonomic Considerations

Bhandari (1970a, b, 1971) has comprehensively reviewed the embryology of Magnoliales, and discusses the relationships (Table 18.24.1).

Melchior (1964) includes 22 families (of Magnoliales) in six groups, while Cronquist (1981) reckons only ten families. Takhtajan (1980) distributes 20 families in three orders: Magnoliales (8), Laurales (10) and Illiciales (2); R. Dahlgren (1980b, 1981) in four orders: Magnoliales (6), Laurales (8) and Illiciales (2) in Magnoliiflorae, and Trochodendrales (2) in Rosiflorae (see also Cronquist 1975a).

Previously, several genera of doubtful affinity had been included in Magnoliaceae. Now they have been assigned to separate families: *Drimys* to Winteraceae, *Illicium* to Illiciaceae, *Euptelea* to Eupteleaceae, *Cercidiphyllum* to Cercidiphyllaceae, and *Schisandra* and *Kadsura* to Schisandraceae (see also Hutchinson 1973). According to Bailey and Smith (1942), and Swamy (1949a), Magnoliaceae are closely allied to Degeneriaceae and Himantandraceae (Table 18.24.1). Hutchinson places *Degeneria* in Winteraceae. Now it has been removed from Winteraceae, and Bailey and Smith (1942) recognize it as an independent family, Degeneriaceae (see also Melchior 1964).

Degeneriaceae. Magnoliaceae and Degeneriaceae (Table 18.24.1) have several common features indicating their alliance: secretory anther tapetum, simultaneous cytokinesis in pollen mother cells, monocolpate two-celled pollen grains with generative cell produced at distal end, anatropous, bitegmic, crassinucellate ovules, Polygonum type of embryo sac with ephemeral antipodals, and funicular obturator. The Cellular type of endosperm develops a two-celled chalazal haustorium in Magnoliaceae; it is ruminate in Degeneriaceae. The seed structure is more or less alike. A sarcotesta develops in Magnoliaceae and Degeneriaceae. The follicular fruit opens by a dorsal suture in Magnoliaceae, and a ventral suture in Degeneriaceae.

Himantandraceae. According to Bailey et al. (1943), Himantandraceae, Magnoliaceae and Degeneriaceae are closely related. *Eupomatia*, *Galbulimima*,

Magnoliaceae and Annonaceae form a coherent group (Endress 1980b). *Galbulimima* has not been fully investigated.

Himantandraceae, Magnoliaceae, Degeneriaceae and Annonaceae resemble each other in several features (Table 18.24.1): tetrasporangiate, abaxial anthers with a massive wall, secretory tapetum of binucleate cells, two-celled monosulcate pollen, anatropous, bitegmic, crassinucellate ovule with massive parietal tissue, and Polygonum type of embryo sac.

The differences are less conspicuous; cytokinesis in pollen mother cells is simultaneous in Annonaceae, Degeneriaceae and Magnoliaceae, and successive in Himantandraceae and Annonaceae. The organization of microspore tetrads shows much variation (N. Prakash et al. 1984).

The embryological characters confirm a close relationship between Himantandraceae, Annonaceae and Degeneriaceae (Table 18.24.1).

Winteraceae. Bentham and Hooker (1965a) include *Drimys* in the tribe Winterae of Magnoliaceae. Hutchinson (1973) recognizes Winteraceae as a separate family which has some distinctive characters: extensive fibrous endothecium in anther wall, pollen grains in permanent tetrads, monoporate with conspicuous to minute reticulations, generative cell cut off towards the proximal pole, phenolic compounds in the outer epidermis of outer integument, embryo sac of the Polygonum type, endosperm of the Cellular type, embryogeny irregular, and exotesta sclerotic. The family shows Magnolian (Table 18.24.1) affinity, and is related to Illiciaceae and Degeneriaceae (Willis 1973).

Annonaceae. Bentham and Hooker (1965a) place Annonaceae with Magnoliaceae in Ranales and Hutchinson (1973) with Eupomatiaceae in Annonales.

Annonaceae show certain affinities with Magnoliaceae and Degeneriaceae: transversely septate anthers in some genera (similar to Myristicaceae), simultaneous or successive cytokinesis in pollen mother cells, tetragonal tetrads, monocolpate pollen grain, anatropous, bitegmic, crassinucellar ovule with vascularized outer integument. The embryo sac is of the Polygonum type with ephemeral antipodals, as also in Magnoliaceae. Cellular type of endosperm with ruminations also occurs in Degeneriaceae. Onagrad type of embryogeny, and follicular fruit is reported in Magnoliaceae. Its woody mesotesta shows affinity to Illiciaceae. Annonaceae definitely belongs to the Magnolian complex (Table 18.24.1).

Eupomatiaceae. Eupomatiaceae have been previously included in Annonaceae, but now they are treated as a separate family (Hutchinson 1973). In both families the ovule is anatropous, bitegmic and crassinucellate, with oily ruminate endosperm. The fibrous testa forming ruminations is common to both families. The seeds are mesotestal, but in Annonaceae there is a well-developed perichalaza, which is absent in Eupomatiaceae. Eupomatiaceae and Annonaceae probably arose from the same ancestral stock but, in the course of evolution, they deviated considerably (Mohana Rao 1983).

Table 18.24.1. Embryology of Magnoliaceae (Sect. 18.2), Degeneriaceae (18.3), Himantandraceae (18.4), Winteraceae (18.5), Annonaceae (18.6), Schisandraceae (18.10) and Illiciaceae (18.11). (See Hayashi 1960, Padmanabhan 1960a, Kapil 1962, Bhandari 1963a, Kapil and Bhandari 1964, Kapil and Jalan 1964, Bhandari and Venkataraman 1968, N. Prakash et al. 1984)

	Magnoliaceae	Degeneriaceae	Himantandraceae	Winteraceae
Anther tapetum	1–3-layered, secretory, cells multinucleate	1-layered, secretory, cells 2-nucleate	1- or 2-layered, secretory, cells 2-nucleate	1-layered, secretory, cells become polyploid; amoeboid in some taxa
Cytokinesis	Simultaneous	Simultaneous	Successive	Simultaneous
Tetrad	Isobilateral, decussate	Tetragonal	Tetrahedral, isobilateral, decussate, linear, T-shaped	Tetrahedral, decussate
Pollen	Monosulcate, 2-celled	Monocolpate, 2-celled, generative cell cut off towards distal end	Monosulcate, 2-celled	Monocolpate, shed in tetrads, 2-celled, generative cell cut off towards proximal pole
Ovule	Anatropous, bitegmic, crassinucellate, micropyle formed by both integuments	Funicular obturator, anatropous, bitegmic, crassinucellate, micropyle formed by inner integument	Funicular obturator, anatropous, bitegmic, crassinucellate, micropyle formed by inner integument	Laminar, anatropous, bitegmic, crassinucellate, micropyle formed by inner or both integuments
Embryo sac	Polygonum type, synergids and antipodals ephemeral	Polygonum, synergids and antipodals ephemeral	Polygonum	Polygonum, antipodals degenerate after fertilization
Endosperm	Cellular type, in *Magnolia obovata* the chalazal chamber develops into a 2- or 3-celled haustorium	Cellular, ruminate in *Degeneria vitiensis*	–	Cellular
Embryo	Onagrad type or irregular, suspensor massive	–	–	Dicotyledonous with small suspensor in *Pseudowintera*, undifferentiated and organless in *Drimys*

Annonaceae	Schisandraceae	Illiciaceae
1-layered, secretory, cells 2–4-nucleate; amoeboid in some taxa	1-layered, secretory cells multinucleate	2-layered, secretory, cells 2-nucleate
Simultaneous or successive	Simultaneous	Simultaneous
Tetrahedral, isobilateral, decussate	Tetrahedral, decussate	Tetrahedral
Anasulcate, tectate, atectate, tetrads in a common wall, loose tetrads, in pairs, or free, 2-celled, generative cell cut off at distal end	Heteropolar, hexacolpate, 2-celled	Tricolporate, 2-celled
Anatropous, bi- or tritegmic, crassinucellate, micropyle formed by inner integument	Ana- to campylotropous, bitegmic, crassinucellate, micropyle formed by both integuments	Anatropous, bitegmic, crassinucellate, micropyle formed by outer integument
Polygonum	Polygonum, antipodals ephemeral; in *Schisandra chinensis* 50% embryo sacs of Polygonum type and others Allium type, the lower polar and antipodals are not formed	Polygonum, antipodal cells ephemeral
Cellular, lobed due to rumination	Cellular	Cellular
Onagrad	Asterad or Onagrad, suspensor short and inconspicuous, mature embryo shows only primordia of cotyledons	Asterad, mature embryo shows only primordia of cotyledons

Table 18.24.1. (*Contd.*)

	Magnoliaceae	Degeneriaceae	Himantandraceae	Winteraceae
Seed	Sarcotestal, testa multiplicative woody endotesta, albuminous, embryo small	Pachychalazal, albuminous, embryo small, suspensor short, hypocotyl bulbous, 3 or 4 cotyledons	Albuminous, embryo small	Exotestal, albuminous, embryo small

Myristicaceae. Hutchinson (1973) places Myristicaceae in Laurales, and Canellaceae in Magnoliales. Parmeswaran (1962) records their close affinities, and states that they have diverged from a common ancestor. Myristicaceae and Canellaceae have simultaneous cytokinesis of pollen mother cells, two-celled monocolpate pollen grains, generative cell cut off towards the proximal pole, anatropous, bitegmic, crassinucellar ovule, Polygonum type of embryo sac, ruminate endosperm, and a small embryo. The presence of oil cells and trilacunar node are additional features. A.C. Joshi (1946) relates Myristicaceae to Annonaceae. Both have trilacunar nodes, trimerous flowers, five-layered anther wall with secretory tapetum, fibrous endothecium, single row of microspore mother cells, sterile septa in anther, single-furrowed two-celled pollen grains with reticulate exine, anatropous, bitegmic, crassinucellate ovule with post-fertilization chalazal growth, Polygonum type of embryo sac with ephemeral antipodal cells, Onagrad type of embryogeny, ruminate endosperm, fleshy aril, and follicular fruit. The middle integument in Myristicaceae is rudimentary (N.C. Nair 1972). Both have complicated seed structure; the mesotesta is woody. The exotegmen is sclerotic in Myristicaceae, and in Annonaceae it forms a sclerotic endostome.

Myristicaceae differ from Lauraceae (Table 18.24.2), and need not be transferred to Laurales. The occasional occurrence of amoeboid tapetum, successive cytokinesis in pollen mother cells, acolpate pollen grains, embryo sac haustoria, secondary multiplication of antipodals, occasional Cellular type of endosperm, and exalbuminous seed of Lauraceae have no parallel in Myristicaceae. The embryogeny may also reflect their relationships. The seed structure differs in both. N.C. Nair (1972) negates the relationship of Myristicaceae with Magnoliaceae. He states that as the family has several distinctive features, it should be raised to an independent order, Myristicales.

Canellaceae. The taxonomic position of Canellaceae is rather debatable. Hallier (1912), Wettstein (1935), Melchior (1964) and Hutchinson (1973) associate them with Magnoliales, whereas Engler and Diels (1936) and Rendle (1959) place them in Parietales. Embryologically, they closely resemble Myristicaceae. Their exotestal seed shows alliance with Winteraceae. But the syncarpous ovary with parietal placentation does not go well with Magnolian families.

Annonaceae	Schisandraceae	Illiciaceae
Mesotestal, arillate, pachychalazal, albuminous	Exotestal, albuminous, embryo small	Exotestal, albuminous, embryo small

Schisandraceae and Illiciaceae. Schisandraceae and Illiciaceae are grouped together in the Magnoliales (Melchior 1964), and Illiciales (R. Dahlgren 1980a, b, Takhtajan 1980). Bentham and Hooker (1965a), Engler and Prantl (1931) and Rendle (1952) include the tribe Schisandrae in Magnoliaceae. *Illicium* has been placed in the tribe Winterae of Magnoliaceae by Bentham and Hooker (1965a). Hutchinson (1973) treats both as separate families, Schisandraceae and Illiciaceae, in Magnoliales. Kapil and Jalan (1964) have evaluated their embryological features, and conclude that they are not closely related. Schisandraceae have hexacolpate pollen grains, Polygonum, sometimes modified bisporic type of embryo sac, Asterad or Onagrad type of embryogeny, and a berry type of fruit with succulent pericarp, in contrast to tricolpate pollen grains, Polygonum type of embryo sac, Asterad type of embryogeny, and follicetum with sclerotic pericarp of Illiciaceae. According to Corner (1976), *Schisandra* is primitive to *Illicium* due to a thick testa, palisade-like endotesta, and absence of lignified endocarp, whereas it is advanced due to campylotropous small-sized seed. The exotestal structure of the seed allies *Illicium* with Schisandraceae and Winteraceae, not with Magnoliaceae. (see Table 18.24.1)

The third group of the order Magnoliales comprises eight families: Austrobaileyaceae, Trimeniaceae, Amborellaceae, Monimiaceae, Calycanthaceae, Gomortegaceae, Lauraceae and Hernandiaceae. All these families, except Amborellaceae and Calycanthaceae, belong to Laurales, according to Hutchinson (1973). Calycanthaceae belong to Rosales.

Formerly, several genera were included in Monimiaceae. Now they have been assigned to separate families: *Austrobaileya* to Austrobaileyaceae, *Trimenia* and *Piptocalyx* to Trimeniaceae, *Amborella* to Amborellaceae, *Calycanthus* to Calycanthaceae, and *Gomortega* to Gomortegaceae.

Sufficient embryological data are not available to determine the systematic position of the above families, except Austrobaileyaceae, Monimiaceae, Calycanthaceae and Lauraceae. These four families show close affinities.

Austrobaileyaceae. Bentham and Hooker (1965a) associate Austrobaileyaceae with Monimiaceae in series Microembryae, whereas Engler and Prantl (1889–1897) place them in Ranales along with Magnoliaceae, Myristicaceae and Monimiaceae. Bailey and Swamy (1949) also relate them to Monimiaceae. Cronquist (1968, 1981) treats them as an isolated taxon of Magnoliales.

Table 18.24.2. Embryology of Hernandiaceae (Sect. 18.19), Lauraceae (18.18), *Cassytha* (18.18), Myristicaceae (18.8), Monimiaceae (18.15), Calycanthaceae (18.16) and Lactoridaceae (20.5). (See Bhandari 1971, Bouman 1971a, Dutt 1978b, Endress 1979, 1980a, b, Mathur 1968, Mohana Rao 1980, 1986, Sastri 1956)

	Hernandiaceae	Lauraceae	*Cassytha*	Myristicaceae
Anther	Bisporangiate with glandular appendages, tapetum amoeboid, cytokinesis successive, pollen 2-celled	Bi-tetrasporangiate, wall 5-layered, tapetum amoeboid, cytokinesis successive, tetrads tetrahedral, isobilateral, T-shaped, linear, pollen 2- celled	Bisporangiate, tapetum secretory, cells 2-, 4- or multinucleate	Tetrasporangiate, in *Myristica* each microsporangium becomes divided into 2 or 3 compartments by sterile septa, wall 5-layered, tapetum secretory, cells uninucleate, cytokinesis successive, tetrads isobilateral, pollen 2-celled
Ovule	Anatropous, bitegmic, outer vascularized, crassinucellate; exo- and endostome; normal or pachychalaza; chalaza vascularized; nucellus causes ruminations in chalaza	Anatropous, bitegmic, crassinucellate; endostome; peri- or pachychalaza; periphery of chalaza vascularized; nucellus absorbed or persists in ruminations	Anatropous, bitegmic, crassinucellate; micropyle occluded by nucellus, funiculus in direct contact with nucellus	Anatropous, funicular aril conspicuous and vascularized, bitegmic, crassinucellate, postament, endostome, pachychalaza, postchalazal branches extend into tegmen and ruminations
Embryo sac	Not studied	Polygonum type, antipodals usually ephemeral but multiply in *Laurus nobilis*	Polygonum, supernumerary embryo sacs common, antipodal nuclei do not organize into cells, tips of embryo sacs produce haustorial tubular extensions in the direction of "micropyle", protrude beyond the nucellus and enter the funicle	Polygonum, in *Horsfieldia* the embryo sac undergoes much expansion
Endosperm	Cellular type	Nuclear, becomes cellular at globular proembryo	Cellular, abundant, reserve food starchy	Nuclear, becomes cellular, ruminate, oily or starchy

Monimiaceae	Calycanthaceae	Lactoridaceae
Tetrasporangiate, wall 4- to 6-layered, tapetum secretory or amoeboid, cells binucleate, cytokinesis simultaneous, tetrads tetrahedral, pollen free or remain attached in tetrads, 2- or 3-celled	Tetrasporangiate, wall 4–5-layered, tapetum secretory, cytokinesis simultaneous, tetrads tetrahedral, decussate, pollen 3-celled, occasionally remain attached in tetrads	Pollen grains united in tetrads
Anatropous, uni- or bitegmic, crassinucellate; hypostase; exostome, nucellar beak persists at endostome; nucellar cap; perichalaza; raphal bundle extends up to two-thirds of length of seed	Anatropous, bitegmic crassinucellate, nucellar cap; hypostase; both exo- and endostome lined with microscopic hairs; raphal bundle expands into shallow cupular chalaza	Anatropous, bitegmic, tenuinucellate (?), endostome
Polygonum and Allium type, antipodals 5 20	Polygonum, supernumerary embryo sacs dcvclop, onc polar may degenerate, antipodals ephemeral	Not studied
Cellular, massive, stores oil	Autonomous, Cellular type	Nuclear, later becomes cellular, copious and oily food reserve

Table 18.24.2. (*Contd.*)

	Hernandiaceae	Lauraceae	*Cassytha*	Myristicaceae
Embryo	Onagrad type, straight with large cotyledons, more or less connate in the ruminations of perisperm	Piperad, Asterad or Onagrad, large, thick massive cotyledons, radicle short	Zygote divides transversely, further development irregular, small with well-developed cotyledons	Small, cotyledons divergent, radicle short
Seed	Exalbuminous, testal, pachychalazal, large subglobose	Exalbuminous, endotestal	Moderately large, albuminous, exo- and endotesta and two outer layers of mesophyll persist	Albuminous, arillate (exostomal and funicular), seed-coat formed by both testa and tegmen
Fruit	Basically achene, sometimes 2–4-winged samara	Drupe or berry, persistent perianth surrounds base, pericarp with oil cells, endocarp stony	Drupe, enclosed by inflated calyx tube, pericarp with outer and inner epidermis and inner hypodermis of palisade cells	Fleshy to subwoody, dehiscent drupe

Hutchinson (1973), Takhtajan (1973, 1980), Thorne (1976) and R. Dahlgren (1977a, 1980b, 1981) associate Austrobaileyaceae with Monimiaceae in Laurales.

Austrobaileyaceae are characterized by spiral flowers, laminar, massive stamens with thecae on adaxial surface (similar to Magnoliales), anasulcate, distally aperturate, pollen grains with irregularly reticulate exine, two-celled pollen grains at the dispersal stage, several ascidiate closed carpels with transversely, strongly bilobed stigma (weakly bilobed in Annonaceae–*Goniothalamus*).

The resemblances with the Annonaceae are: floral arrangement, staminal structure, and sarcotestal, albuminous, reticulate, perichalazal seed. The ruminations in the endosperm are of the Annona type. Austrobaileyaceae can be segregated from the Annonaceae on the basis of spiral perianth, imbricate tepals, anasulcate pollen with reticulate-regulate exine, and ascidiate carpels (Endress 1980a).

On the contrary, infrequent presence of amoeboid tapetum, successive cytokinesis, monocolpate pollen grains, several ovules per carpel, embryo sac with haustoria, and secondary multiplication of antipodals of Lauraceae are not met with in Austrobaileyaceae.

The similarity of features with Monimiaceae is more pronounced. Secretory anther tapetum, simultaneous cytokinesis in pollen mother cells, structure of ovule and seed, and the type of embryo sac exhibit their close affinity.

Monimiaceae	Calycanthaceae	Lactoridaceae
Asterad	Sexual embryo uncommon, nucellar polyembryony common, cotyledons convolute	Minute, organless embryo
Albuminous, endotestal, embryo small and straight, cotyledons adpressed or divergent, oil cells in outer layers of testa	Exalbuminous, endotestal	Albuminous, testa and tegmen collapse, meso- and endocarp provide protection
Achenes or drupe, enclosed by hypanthium	Pericarp hard, encloses seed	Follicle, beaked

However, occasional occurrence of amoeboid anther tapetum, unitegmic ovule, Allium type of embryo sac, and supernumerary antipodals segregate Monimiaceae from Austrobaileyaceae. The most important feature, which separates these two families, is the type of pollen grains. They are monocolpate, aperturate and globose, and are shed singly in Austrobaileyaceae, whereas the pollens are inaperturate, two- to three-sulcate in Monimiaceae, and are often dispersed in tetrads.

Endress (1980a) has discussed the resemblances with Magnoliales, and differences with Annonaceae. He concludes that Austrobaileyaceae point to a closer relationship with Magnolian, and not Lauralian families, and they should not be included in the Annonaceae. He emphasizes that "the pollen resembles, more than in almost all other Ranales (and angiosperms), *Clavatipollenites*, the earliest known angiosperm pollen fossils of the lower Cretaceous (Endress and Honegger 1980)". Therefore, Austrobaileyaceae are one of the most prominent "living fossils among the angiosperms".

Further information about the development of endosperm and embryo of *Austrobaileya* may prove helpful in taxonomic assignment.

Trimeniaceae. Endress (1987b) points out that the Trimeniaceae (Sect. 18.13) seem to be the closest relatives of Chloranthaceae (Sects. 20.4, 20.6). The stamens are the visually attractive organs in the entomophilous members of both families. The pollen is polyporate in some members of both families

(*Trimenia* and *Sarcandra*), and reticulate. The drupe-like fruits are berries (except *Hedyosmum*), since the hard layer is produced by testa and not by the ovary wall. Another relatively closely related relict family is the Amborellaceae (Sect. 18.14). As in Chloranthaceae, the ovule is pendant and orthotropous. The pollen is inaperturate as in some Trimeniaceae. The Chloranthaceae also share some features with the Piperaceae (Sect. 20.3) and, therefore, have frequently been placed in the Piperales (Cronquist 1981). A combination of unusual characters points to a possible phylogenetic relationship with the Hamamelidaceae (Sect. 26.3) via Trochodendraceae (Sect. 18.21). The Chloranthaceae seem to have a key position and probably represent a link between Laurales (Sect. 18.18) and Piperales within the Magnoliidae, as well as between Magnoliidae and Hamamelididae in the context of the primitive angiosperms. This issue can be further considered when we have adequate embryological data on the related families.

Lauraceae, Monimiaceae. Lauraceae are related to Monimiaceae (Table 18.24.2): secretory or amoeboid anther tapetum, simultaneous cytokinesis of pollen mother cells, bitegmic, crassinucellar ovule with a massive parietal tissue, Polygonum type of embryo sac with secondary increase in antipodals, Cellular type of endosperm, Asterad type of embryogeny, and tracheidal endotesta. But, occasionally, successive cytokinesis, Nuclear or Cellular type of endosperm, and Piperad or Onagrad type of embryogeny occur in Lauraceae. The embryo sac haustorium in Lauraceae resembles that of Trochodendraceae. Monimiaceae are segregated due to the occurrence of pollen grains in permanent tetrads, nucellar cap, unitegmic ovule in some members, and occasional occurrence of Allium type of embryo sac.

Cassytha. The systematic position of the genus *Cassytha* (Sastri 1962) has been much debated (Bhandari 1971). The structure and development of anther, pollen, ovule, and seed-coat with endocarp closely resemble that of Lauraceae (Table 18.24.2). However, Bhandari erects a separate family, Cassythaceae, due to several dissimilarities reported in *Cassytha*: (1) broad nucellar tip, wide micropyle formed by both integuments, (2) multiple "haustorial" embryo sacs which become extramicropylar and invade the funiculus, (3) ephemeral (three) antipodal nuclei which do not organize into cells, and (4) Cellular type of endosperm.

Hortonia, endemic to Ceylon (Sri Lanka), is considered as especially primitive in Monimiaceae. It has been raised to the rank of a subfamily, Hortonioideae (Money et al. 1950), or even of a family, Hortoniaceae (Smith 1972). Endress (1980b) states that *Hortonia* has an intermediate position between Atherospermataceae and Monimiaceae, showing greater affinity with the latter. Further consideration should await a detailed study of reproductive aspects.

Endress (1980b) concludes that "the evolutionary step from the floral type of *Hortonia* to the type of *Tambourissa* represents an extraordinary morphological grade comparable to the step from gymnospermy to angiospermy".

Calycanthaceae. Morphologically, Calycanthaceae resemble Magnoliaceae and Annonaceae. Due to opposite arrangement of leaves, cup-like receptacle, and exalbuminous seed with a large embryo, Hutchinson (1973) relates them to Rosaceae. Their floral anatomy, nodal type, and basic chromosome number are similar to Cactaceae (Y.D. Tiagi 1970a). The relationship with Cactaceae has already been discussed (see Cactales, Sect. 17.3). Embryologically, Calycanthaceae are allied to Monimiaceae (Table 18.24.2) in the development of anther, ovule, embryo sac, and seed-coat. The formation of more than one embryo sac per ovule, and the Cellular type of endosperm also occur in Lauraceae [*Cassytha filiformis* (Sastri 1963)]. But, the absence of amoeboid anther tapetum, ephemeral antipodals, lack of fertilization due to male sterility in some members and, consequently, autonomous development of endosperm, and nucellar polyembryony are distinct features of Calycanthaceae. They have no relationship with the Rosales, or Myrtales, but are allied to Magnoliaceae, Annonaceae and Eupomatiaceae.

Hernandiaceae. Hernandiaceae are closely allied to Lauraceae (Table 18.24.2). The presence of two-trace carpel, amoeboid anther tapetum, multicellular female archesporium, Cellular type of endosperm (as in *Cassytha*), and the presence of helical thickenings in the seed are some of the similarities. The embryogeny resembles that of *Cinnamomum* and *Litsea* (Sastri 1958a). The study of wood anatomy (Shutts 1960) also supports their alliance.

Pax (1891) established the family Hernandiaceae and included the genera *Gyrocarpus*, *Sparattanthelium*, *Illigera* and *Hernandia*. Mohana Rao (1986) has discussed whether all the four genera should be included in the Hernandiaceae, or *Hernandia* and *Illigera* in Hernandiaceae, and *Gyrocarpus* and *Sparattanthelium* in Gyrocarpaceae (see Takhtajan 1969). In his later classification, Takhtajan (1980) does not recognize the family Gyrocarpaceae. The relationships of the Hernandiaceae have been pointed out with Lauraceae, Gomortegaceae, Myristicaceae, Calycanthaceae, Lactoridaceae, Monimiaceae and Cassythaceae; except Myristicaceae, all the others are included in the order Laurales by Takhtajan (1980). The chief embryological characters are listed in Table 18.24.2 (Gomortegaceae have been omitted because the embryology has not been studied).

This table clearly brings out that the Hernandiaceae are quite distinct from Myristicaceae, Calycanthaceae, Lactoridaceae and Monimiaceae, but are close to Lauraceae (*Cassytha* is included in Lauraceae by Melchior 1964). Further resemblances with Lauraceae are: non-aperturate, spheroidal and tenuexinous pollen grains, presence of unicellular hairs, rubiaceous type stomata, needle-type crystals in leaf and cortex, medium-length vessel members with simple perforations, alternate intervascular pitting, three-cell wide homogeneous rays, paratracheal parenchyma, septate fibres of medium length with simple pits, p-type of sieve element plastids, and unilacunar nodes (see Mohana Rao 1986).

Of the four families – Tetracentraceae, Trochodendraceae, Eupteleaceae and Cercidiphyllaceae – of Magnoliales (Melchior 1964), Trochodendraceae and Cercidiphyllaceae are retained in Magnoliales by Hutchinson (1973). Previously, *Euptelea* had been included in the Trochodendraceae (Hutchinson

1959, 1973); it has been separated as an independent family, Eupteleaceae (see Willis 1973).

Eupteleaceae. On embryological grounds, Endress (1969a, b) relates Eupteleaceae to *Cercidiphyllum*, and retains them in Magnoliales. In conformity with the order Magnoliales, Eupteleaceae show: (1) secretory anther tapetum, (2) simultaneous cytokinesis of pollen mother cells, (3) tricolpate, two-celled pollen grains, (4) anatropous, bitegmic, crassinucellar ovule, (5) Polygonum type of embryo sac, and (6) Cellular type of endosperm. The endotestal seed corresponds to that of Trochodendraceae.

Cercidiphyllaceae. The family Cercidiphyllaceae has been included in Hamamelidaceae by Baillon (1871), in Magnoliaceae by Maximowicz (1872), Bessey (1915) and Lemsle (1946). It is treated as an independent family, Cercidiphyllaceae, by Harms (1918), Takhtajan (1966, 1980), Cronquist (1968, 1981) and Thorne (1968). Swamy and Bailey (1949) include it in Ranales, as it is intimately related to Trochodendraceae and Eupteleaceae.

The pollen grains of *Tetracentron*, *Trochodendron*, some members of Hamamelidaceae, and *Cercidiphyllum* are tricolpate, but, in the last-named genus, the furrows are considerably broader and poorly defined, and the exine is also different. In *Cercidiphyllum* the nucellar epidermal cells divide periclinally, and the embryo sac elongates conspicuously after fertilization. The early development of endosperm is rather typical, and reminds one of *Sarracenia purpurea* (Shreve 1906). The seed is small with a large embryo, and the chalazal wing has a vascular bundle. But, this simplified seed is not primitive (Corner 1976).

According to Bhandari (1971), Trochodendraceae and Cercidiphyllaceae are closely related. Both have anatropous, bitegmic, crassinucellate ovule with a tapering chalazal projection, a single-celled hypodermal female archesporium, massive parietal tissue, Polygonum type of embryo sac, Cellular type of endosperm, and tapering wing of the seed.

19 Ranunculales

The order Ranunculales comprises two suborders: Ranunculineae and Nymphaeineae with seven families: Ranunculineae – Ranunculaceae, Berberidaceae, Sargentodoxaceae, Lardizabalaceae, Menispermaceae; Nymphaeineae – Nymphaeaceae and Ceratophyllaceae.

19.1 General Features

The order Ranunculales exhibits the following features: (1) secretory anther tapetum, amoeboid in Ceratophyllaceae, and of both types in Ranunculaceae and Nymphaeaceae, (2) simultaneous cytokinesis in pollen mother cells, successive in Nymphaeineae, and of both types in Berberidaceae, (3) variations in the number of integuments in Ranunculaceae and Menispermaceae, ovules undifferentiated (ategmic), unitegmic, or bitegmic; usually bitegmic, only unitegmic in Ceratophyllaceae; anatropous, crassinucellate, crassi- or tenuinucellate in Ranunculaceae and Nymphaeaceae, orthotropous in Lardizabalaceae, and hemi- to amphitropous in Menispermaceae, (4) Polygonum type of embryo sac; mono-, bi- or tetrasporic in Ranunculaceae, and mono- or tetrasporic in Berberidaceae, (5) antipodals quite prominent, except in Lardizabalaceae, (6) Nuclear type of endosperm, Cellular in Lardizabalaceae and Ceratophyllaceae, and Nuclear, Cellular and Helobial in Nymphaeaceae, (7) embryogeny of Onagrad type in Menispermaceae, Onagrad, Solanad and irregular in Ranunculaceae, Onagrad and Solanad in Berberidaceae, and Asterad type in suborder Nymphaeineae, and (8) seed mostly exotestal.

19.2 Ranunculaceae

The anther wall comprises the epidermis, and fibrous endothecium; *Anemone* and *Aquilegia* lack the thickenings (Bhandari 1968). There are two or three ephemeral middle layers, and secretory tapetum except in *Actaea spicata* (Jalan 1963) where it is amoeboid. Both types of tapetum are reported in different anthers of the same flower in *Anemonopsis macrophylla* (Tobe 1974b). The development of wall layers is of the Basic type (Davis 1966) in *Nigella damascena* (Vijayaraghavan and Marwah 1969). Both the middle layers persist in *Aquilegia*, and only one in *Ceratocephalus falcatus* (Bhandari and Asnani 1968). Übisch granules are present on the tangential and radial walls of tapetal cells in *Helleborus foetidus* and *Nigella* (Bhandari et al. 1976). The tapetal cells are bi-, tri- or multinucleate. The nuclei fuse and become polyploid.

Turala-Szybowska (1984) analyzed the mitotic behaviour of the tapetal cell in *Ranunculus fluitans* during microsporogenesis. Mitotic divisions and vacuolation in tapetal cells start in the central part of the internal tapetum, and the tapetal layer situated in the apical part of the anther locule. In these regions the tapetum may be more than one layer thick. During the premeiotic phase of microspore mother cells, the tapetal cells remain uninucleate (2n) (Fig. 19.2.1A). While the mother cells undergo meiotic divisions, the tapetal cells become binucleate (2n) either by mitosis or by disturbed mitosis (Fig. 19.2.1B). Due to inhibited prophase the nuclei of the binucleate tapetal cells become tetraploid, or due to mitosis the cells become four-nucleate (2n), or due to disturbed mitosis the cells become three-nucleate (two diploid nuclei and one tetraploid nucleus). A further inhibited prophase in tetraploid nuclei results in eight- or 16-ploid nuclei. In addition, some of the four-nucleate diploid cells undergo inhibited prophase to form four-nucleate tetraploid cells. At uninucleate pollen grain stage, 7% of tapetal cells were 2n and binucleate, 19% 4n binucleate, 51% 8n binucleate, and 19% 16n binucleate. Other tapetal cells were 1% 2n four-nucleate, 2% 2n four-nucleate, and 1% with two nuclei 2n and one 4n (Fig. 19.2.1C). Thus, polyploidization occurs due to endomitosis, disturbed mitosis, or inhibited prophase.

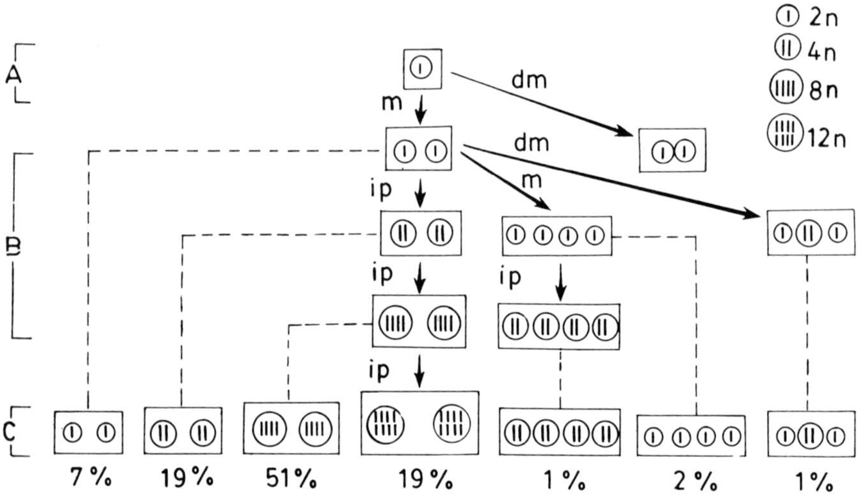

Fig. 19.2.1A–C. *Ranunculus fluitans* Lam. (2n = 16), main pathways in the development of tapetal cells. **A** At pre-meiotic stage of microspore mother cells. Mitosis (*m*) and disturbed mitosis (*dm*) occur in diploid nuclei of tapetal cells. **B** At meiotic divisions I and II in microspore mother cells. Mitosis in tapetal cells occurs without cytokinesis (2n), or mitosis may get inhibited (*ip*) at prophase. Then the nuclei of these cells remain 4n or 8n. When disturbed mitosis occurs, the cell contains two diploid nuclei and one tetraploid nucleus. **C** At primary pollen grain stage. The mitosis in tapetal cells may get inhibited in octaploid nuclei so that the cell becomes 12-ploid. The percentage of nuclei with different ploidy-level is also shown. (After Turala-Syzbowska 1984)

Simultaneous cytokinesis follows meiotic divisions of microspore mother cells, and the tetrads are tetrahedral and decussate.

The tricolporate pollen grains have smooth exine, and are two-celled at the dispersal stage.

The ovules are anatropous, unitegmic in *Anemone*, *Ceratocephalus*, *Clematis*, *Helleborus* and *Ranunculus*, bitegmic in *Actaea*, *Adonis*, *Caltha* and *Thalictrum*, and undifferentiated sterile ovules in *Adonis*, *Anemone* and *Clematis*; tenuinucellate in *Clematis gouriana* and *R. sceleratus*, crassinucellate in *Adonis*, *Aquilegia*, *Caltha*, *Ceratocephalus*, *Ranunculus*, *Thalictrum* and *Trollius*, both tenui- and crassinucellate in *Nigella damascena* and *Adonis annua*, and occasionally pseudocrassinucellate in *A. annua*, *H. foetidus* and *N. damascena* (see Vijayaraghavan 1970a). They are unitegmic and pseudocrassinucellar in *Hepatica nobilis* var. *japonica* (Tobe et al. 1975), and *Trautvetteria japonica* (Tobe 1973). Bitegmic crassinucellar ovules are reported in *Pulsatilla aurea*, and unitegmic tenuinucellar in *P. chinense* (Sun and Wang 1984).

The archesporium is uni- or multicelled. The archesporial cell functions directly as megaspore mother cell, or cuts off a parietal cell as in *Caltha palustris* (Kapil and Jalan 1962). The parietal tissue is two or three layers thick in *Actaea*, *Ceratocephalus* and *Trollius*. The cells of the nucellar epidermis also contribute to the formation of parietal tissue. A nucellar "cap" is prominent in *Ceratocephalus falcatus*.

The megaspore mother cell undergoes meiotic divisions, and produces a linear or T-shaped tetrad. The chalazal megaspore develops into a Polygonum type of embryo sac in *Actaea* (Jalan 1963), *Anemone* (Bhandari 1968), *Anemonella*, *Aquilegia*, *Batrachium*, *Caltha*, *Ceratocephalus*, *Delphinium*, *Helleborus*, *Myosurus*, *Ranunculus* and *Thalictrum* (see Schnarf 1929), *Ceratocephalus orthoceros* (Kordyum 1959), *Cimicifuga acerina* (Tobe 1974a), *Clematis gouriana* (*Vijayaraghavan* 1962), *Hepatica* (Tobe et al. 1975), *Nigella* (Vijayaraghavan and Marwah 1969), *Pulsatilla chinense* (Sun and Wang 1984), *Thalictrum javanicum* (Vijayaraghavan 1970a) and *Trollius europaeus* (Bhandari and Kapil 1964). The embryo sac is of the Allium type in fertile and sterile ovules of *Adonis* (Bhandari 1962, 1966), *Ceratocephalus falcatus* (Bhandari and Asnani 1968) and *Trollius* (Bhandari and Kapil 1964). The occurrence of mono- and bisporic types of embryo sac development, in different species of the same genus, has been noted in *Ceratocephalus* and *Trollius*. Picci (1964) reported tetrasporic-Pyrethrum type in *Thalictrum angustifolium* var. *flavum*. Tetrasporic type occurs in the sterile ovules of *Anemone obtusifolia* (Bhandari 1968), and *Clematis gouriana* (Vijayaraghavan 1962). Such a tendency is also reported in *Adonis annua* (Bhandari 1962). The variation in the development of embryo sacs is to such an extent that, in the same species, mono-, bi- and tetrasporic types occur, e.g. in *Anemone vitifolia* (Bhandari, 1965).

In *Pulsatilla chinense* (Sun and Wang 1984), sometimes the embryo sac shows two egg apparatus and six antipodals; presumably due to the dissolution of intervening wall of the twin embryo sacs.

Fish (1970) reports several variations in development of the embryo sac of *Clematis* spp. In fertile ovules, in all species, the development is generally of the Polygonum type, and in sterile ovules of tetrasporic type. In the sterile ovules of *C. orientalis* he observed that the four megaspore nuclei lying in the centre of coenomegaspore fuse in pairs. These two divide once mitotically. The secondary four-nucleate embryo sacs show the egg, two polar nuclei, and one antipodal, or the egg, one polar nucleus, and two antipodals. If another division occurs, a secondary eight-nucleate embryo sac develops. In one case he noticed the egg apparatus at the chalazal end of the embryo sac. The embryo sacs in the sterile ovules finally degenerate soon after the fertilization of fertile ovules. The sterile ovules of *C. vesicolor* exhibit normal eight-nucleate embryo sacs. However, in sterile ovules of *Hepatica nobilis* var. *japonica* (Tobe et al. 1975) only the archesporial cell differentiates and further development is arrested.

The embryo sac develops a caecum at its chalazal end in *Nigella damascena*, and the antipodals are left behind in situ (Vijayaraghavan and Marwah 1969). In Ranunculaceae the antipodals are noteworthy. They are ephemeral in *Adonis aestivalis* (Bhandari 1963b). They enlarge, undergo endomitosis, and become polyploid in *Caltha palustris* (Kapil and Jalan 1962), *Hepatica* (Tobe et al. 1975), and *Thalictrum* (Picci 1969). They are octaploid in *Helleborus niger*, 32-ploid in *Anemone hepatica*, and 64- or 128-ploid in *Aconitum* (Tschermak-Woess 1956).

Osterwalder (1898) reported large antipodal cells with polyploid nuclei in the post-fertilized embryo sac (containing young globular proembryo and free-nuclear endosperm) of *Aconitum napellus*. Zhukova and Sokolovskaya (1977) also studied the antipodals (in the same species), before fertilization. Bohdanowicz and Turala-Szybowska (1985, 1987) examined the antipodal cells in *A. vulparia*. In mature embryo sac the antipodals are very rich in organelles, have giant endopolyploid nuclei with lobed outline especially at the chalazal side. The antipodals resemble transfer cells with marked wall ingrowths which are particularly well-developed in the chalazal part. During the development of free-nuclear endosperm, the antipodal cells enlarge considerably, the wall ingrowths increase in size and number, the cytoplasm becomes dense with weak vacuolation, and they are at their most active metabolic state for transfer and secretory activity.

Syngamy precedes triple fusion in *Helleborus*; in *Adonis* one sperm fuses either with the upper or lower polar nucleus prior to polar fusion. Whether this is the normal condition, it requires further study.

The development of endosperm is of the Nuclear type. Enucleate cytoplasmic vesicles are present in *Thalictrum japonicum* (Vijayaraghavan and Bhandari 1970). Wall formation is initiated either from the micropylar or chalazal region, and is centripetal. The initiation of walls occurs at 256-nucleate stage in *Adonis* (Bhandari 1966), and ca. 1000-nucleate in *Anemone* (Bhandari 1968) at the globular stage of proembryo. Kordyum (1957) reported a chalazal endosperm haustorium in *Nigella sativa*.

The embryogeny also exhibits variation. It is irregular in *Adonis* (Bhandari 1966), Onagrad type in *Actaea spicata* (Jalan 1963), *Anemone* (Bhandari 1968), *Caltha palustris* (Kapil and Jalan 1962), *Ceratocephalus calcaratus* (Fig. 19.2.2A–Q), *Thalictrum japonicum* (Vijayaraghavan and Bhandari 1970), *Ranunculus sceleratus* (Souèges 1914) and *R. tripartitus* (Patel and Cook 1972), and Solanad type in *Helleborus foetidus* (Vijayaraghavan 1970a) and *Nigella damascena* (Vijayaraghavan and Marwah 1969). Tobe et al. (1975) report Caryophyllad or Solanad type in *Hepatica nobilis* var. *japonica*.

Somatic apospory of Hieracium type has been reported in *Ranunculus* spp. (Håfliger 1943, Rousi 1956). In *R. auricomus* (Rozanova 1932, Håfliger 1943, Rutishauser 1954), the egg develops pseudogamously, but triple fusion is normal. Antipodal embryos occur in *Nigella arvensis* (Derschau 1918); this needs confirmation.

In *Eranthis hiemalis* (Fig. 19.2.3), at the shedding stage the mature seed has a 0.1-mm long pear-shaped undifferentiated embryo with a conspicuous suspensor. The cotyledonary primordia and the radicle develop in the soil. The process of intraseminal growth and differentiation continues for about eight months, and the mature embryo attains a length of 2 mm. It comprises 90% cotyledons, and 10% hypocotyl and radicle. In contrast to other angiosperms, the cotyledons are closely adpressed to each other in the upper half region (called "cotyledonary sheath" by the author); the lower half of the cotyledons is free.

The seed is small, albuminous, ex-arillate, with a minute embryo. The chalaza is swollen in *Caltha*, and a swollen raphe-elaiosome is present in

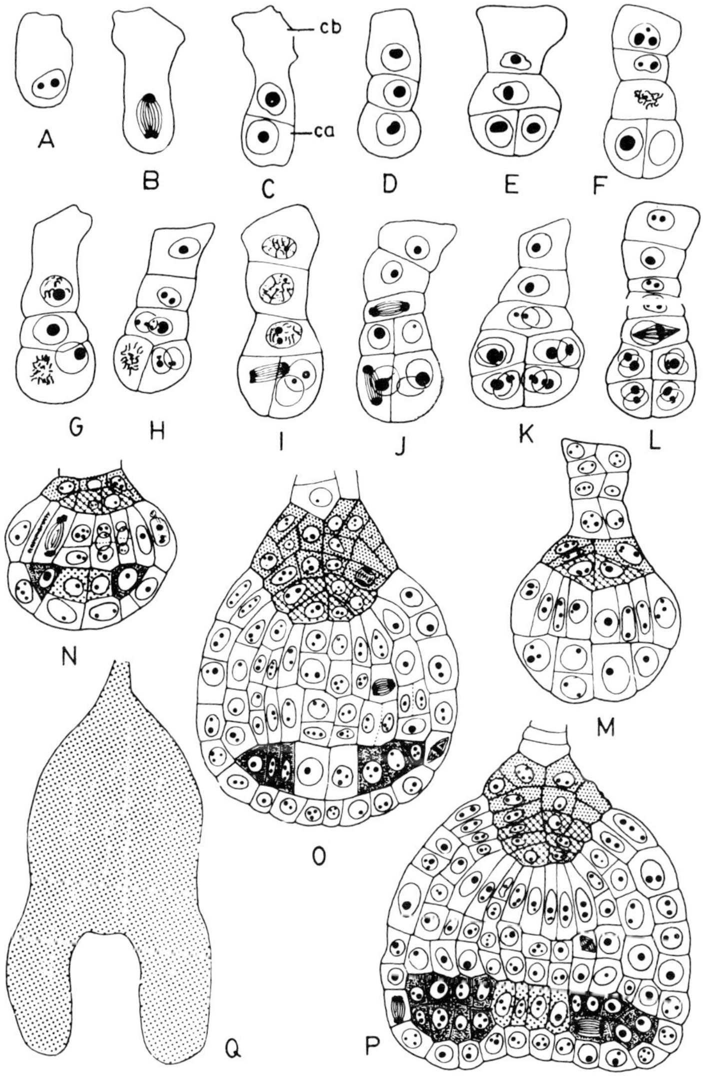

Fig. 19.2.2A–Q. *Ceratocephalus calcaratus*, development of embryo. **A–C** Transverse division in zygote to form the cells *ca* and *cb*. **D**, **E** Three- and four-celled proembryo. **F–J** Development of quadrant. **K–L** Proembryo at octant stage. **M–O** Globular proembryo. Note the delimitation of dermatogen, cotyledonary and shoot apex initials. **P** Longisection of early heart-shaped embryo. **Q** Longisection of mature dicotyledonous embryo (diagrammatic). (After Bhandari and Asnani 1968)

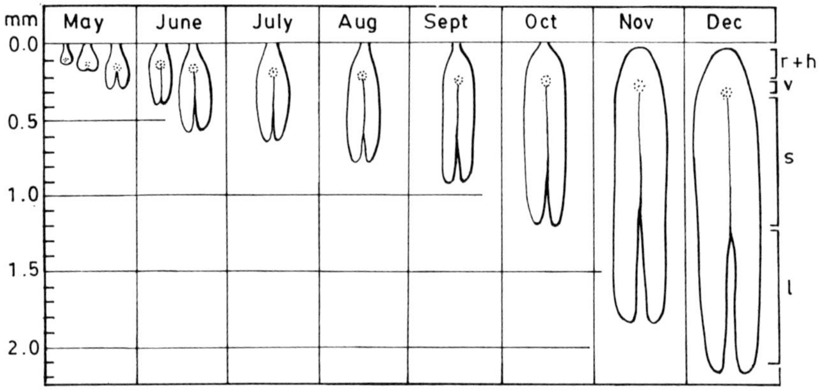

Fig. 19.2.3. *Eranthis hiemalis*, diagrammatic representation of the maturation of embryo; *r* + *h* radicle + hypocotyl; *v* shoot apex; *s* sheath; *l* lamina of cotyledons. In mature seed (0.1 mm long – stage A) the embryo is organless (undifferentiated). The formation of cotyledonary primordia and radicle takes place in the soil (stages B, C). Then the embryo grows in the region of cotyledonary sheath (*s*), and not in the hypocotyl (*h*) region. Until December there is no further change in the outer form, but the embryo grows larger and procambial tissue develops. The mature embryo represents 90% of cotyledons, and only 10% of hypocotyl with radicle. Germination begins in December or early January. (After Haccius 1963)

Helleborus. The seed-coat is formed by both the integuments in bitegmic ovules. The outer epidermis of the outer integument is hard in *Caltha*, and sclerotic in *Aquilegia*. The inner integument has fine bar-like thickenings on the radial walls in *Trollius*. The testa is crushed in *Anemone* and *Clematis*. The tegmen is more or less crushed, but in *Helleborus* the outer epidermis is thick-walled, and the inner shows fine thickenings on radial walls.

In several respects *Circaeaster* differs from other members of Ranunculaceae and is, therefore, described separately.

The anther is bilocular. The anther wall comprises the epidermis, fibrous endothecium, an ephemeral middle layer and secretory tapetum with uninucleate cells in *C. agrestis* (Junell 1930). After meiosis in microspore mother cells, cytokinesis is simultaneous, and the tetrads are tetrahedral. The mature pollen is two-celled.

The ovule is orthotropous, unitegmic and tenuinucellate. The integument is two to three cell layers thick. The archesporial cell functions as megaspore mother cell. The megaspore tetrad is linear or T-shaped, and the chalazal megaspore develops into a Polygonum type of embryo sac. The polars fuse before fertilization. The antipodals are ephemeral, or may persist up to a few divisions in the endosperm.

The path of the pollen tube is chalazogamous (Junell 1930) or mesogamous. The pollen tube penetrates the integument near the antipodals, enters the embryo sac, elongates up to the synergids, ruptures and pours its contents into one of them. This synergid degenerates, whereas the other synergid remains healthy for some time. The remnants of pollen tube persist. The entry of pollen tube and fertilization should be reinvestigated.

The development of the endosperm is of the Cellular type. Division of the primary endosperm nucleus is followed by a vertical wall and then there is an asymmetric transverse division separating two small cells at the chalazal end. The two upper larger cells undergo another transverse division.

The zygote divides by a transverse wall and the linear proembryo shows five cells. The development of embryo probably conforms to the Chenopodiad type.

The greater part of seed is occupied by oily and proteinaceous endosperm. The large, terate, straight embryo shows only primordial cotyledons. Its differentiation takes place during germination. The seed-coat is crushed so that the endosperm remains in contact with the pericarp (Corner 1976).

H Babis (1966, 1973, 1974, 1976a, b, 1978, 1979a, b, c), NN Bhandari (1962, 1963b, 1965, 1966, 1968), NN Bhandari and S Asnani (1968), NN Bhandari and P Chitralekha (1989), NN Bhandari and RN Kapil (1964), NN Bhandari and Ram Kishori (1973), NN Bhandari, Ram Kishori and S Natesh (1976), J Bohdanowicz and K Turala-Szybowska (1985, 1987), F Bouman (1978), F Bouman and JIM Calis (1977), M Brouland (1975), M Brouland and C Guerven (1978), M Brouland, C Guerven and C Le Cog (1976), M Charzynska (1973), EJH Corner (1976), JM Coulter (1898), GL Davis (1966), MV Derschau (1918), RK Fish (1970), M Fougêre-Rifot (1975, 1981), Å Gustafsson (1946, 1947a, 1947b), B Haccius (1963), E Håfliger (1943), M Hesse (1978), T Hiroshi (1973, 1974), T Hiroshi, A Toshiko and I Osamu (1975), R Izmailow (1965, 1973), S Jalan (1963), S Jalan and NN Bhandari (1963), S Junell (1930), RN Kapil and S Jalan (1962), KY Kavathekar and A Pillai (1981), EL Kordyum (1957, 1959), M Kumazawa (1972), O Langlet (1928), C Le Cog, M Brouland and C Guerven (1978), Ly-Thi-Ba (1971, 1974), CR Metcalfe (1936), Y Miyaji (1927), Mu-Xi-jin (1983, 1984), Mu-Xi-jin and Wang Fu-Xiong (Wang Fu-hsiung) (1985b), A Osterwalder (1898), E Pacini, M Cresti and G Sarfatti (1972), TK Patel and CDK Cook (1972), V Picci (1964, 1969), E Pogan and H Wcislo (1981), M Rifot (1973), PN Robeva-Davidova and MA Fileva-Mikhajlova (1979), A Rousi (1956), MA Rozanova (1932), A Rutishauser (1954), RLN Sastri (1969b), J Sâto (1972), K Schnarf (1929), B(Bahadur) Singh (1936), R Souèges (1974), AC Sun and FH Wang (1984), M Tamura (1972), M Tamura and I Mizumoto (1972, 1974), H Tobe (1973, 1974a, 1974b, 1981), H Tobe, T Adachi and O Yoshida (1975), H Tobe and RC Keating (1985), Z Trela-Sawicka (1974, 1977), K Turala-Szybowska (1981, 1984), K Turala-Szybowska and J Wolańska (1990), E Tschermak-Woess (1956), M Ünal (1978), MR Vijayaraghavan (1962, 1970a), MR Vijayaraghavan and NN Bhandari (1970), MR Vijayaraghavan and KN Marwah (1968, 1969), CLF Woodcock and PR Bell (1968), M Yoffe (1969), GYA Zhukova and RB Sokolovskaya (1977)

19.3 Berberidaceae

The anther wall comprises the epidermis, fibrous endothecium, two or three ephemeral middle layers, and secretory tapetum. The thickenings of the endothecium are U-shaped in *Berberis umbellata* (Sastri 1969a). The nuclei of the tapetal cells divide mitotically, and as many as eight nuclei occur in each cell. Nuclear fusions take place and they become polyploid (B.M. Johri 1935c). The microspore mother cells undergo meiotic divisions, and give rise to tetrahedral and isobilateral tetrads. Cytokinesis takes place by furrowing.

The mature pollen grain is three-colpate, and shed at the two-celled stage. The generative cell is ovoid.

The ovule is anatropous, bitegmic and crassinucellate. Integumentary vascular bundles occur in *Caulophyllum thalictroides*, and funicular obturator and hypostase in *Berberis aristana*. The outer integument takes part in the formation of micropyle; both integuments in *Nandina domestica* (Ehdaie and

Russell 1984), and the micropylar passage is zig-zag. The archesporium is multicellular, but only one cell functions. It cuts off a parietal cell in some species of *Mahonia* (see Schnarf 1931). The parietal cell is not cut off in *Jeffersonia* (Andrews 1895), and *Podophyllum* (Lubliner 1925). In *Berberis nepalensis* it may or may not be cut off (B.M. Johri 1935c). The nucellar epidermal cells divide to produce a massive tissue. The megaspore mother cell undergoes meiotic divisions, and produces a linear tetrad. The chalazal megaspore develops into a Polygonum type of embryo sac. Modified Peperomia type of embryo sac is reported in *Caulophyllum*, but this needs confirmation. The synergids are ephemeral; Mauritzon (1936a) mentions haustorial synergids in *Berberis* and *Mahonia* which appears to be incorrect. The antipodals are very conspicuous and persistent. The polar nuclei fuse before fertilization. The nucellus, hypostase and mature embryo sac contain starch grains in *N. domestica*.

The development of endosperm is of the Nuclear type. The embryogeny conforms to the Onagrad type–Myosurus variation; Solanad type in *Berberis aristata*. The embryo is tricotyledonous in *Podophyllum peltatum*. Polyembryony occurs in *B. aristata*.

The seed is albuminous and exotestal. The embryo is well-developed with a long radicle. The testa alone forms the seed-coat. The outer epidermis comprises thick-walled, lignified, pitted, longitudinally elongate fibres, whereas the inner epidermis consists of small, thick-walled cells. The outer layers of mesophyll are slightly thickened and aerenchymatous, the inner one is crushed. A vestigial aril occurs rarely (Corner 1976).

FM Andrews (1895), EJH Corner (1976), M Ehdaie and SD Russell (1984), BM Johri (1935c), OK Lubliner (1925), Ly-Thi-Ba (1976), J Mauritzon (1936a), JW Nowicke and JJ Skvarla (1981), RLN Sastri (1969a), K Schnarf (1931), YF Shen (1954), K Subramanyam (1962), SD Swanson and SH Sohmer (1976)

19.4 Sargentodoxaceae

The embryology of the family Sargentodoxaceae has not been investigated.

19.5 Lardizabalaceae

The anther wall comprises the epidermis, fibrous endothecium, two or three ephemeral middle layers, and secretory tapetum. It is of the amoeboid type in *Stauntonia hexaphylla* (Yoshida and Nakajima 1978). The tapetal cells are two- to four-nucleate; the nuclei fuse to form polyploid masses. The microspore mother cells undergo meiotic divisions, and form tetrahedral (Fig. 19.5.1A) and decussate tetrads. Cytokinesis is simultaneous.

Pollen grains are three-colporate, and shed at the two-celled stage.

The young ovule is orthotropous, bitegmic and crassinucellate. It becomes campylotropous in *Akebia* spp. (Yoshida and Michikawa 1973), and anatropous in *Decaisnea* (Swamy 1953c) and *Stauntonia hexaphylla* (Yoshida and

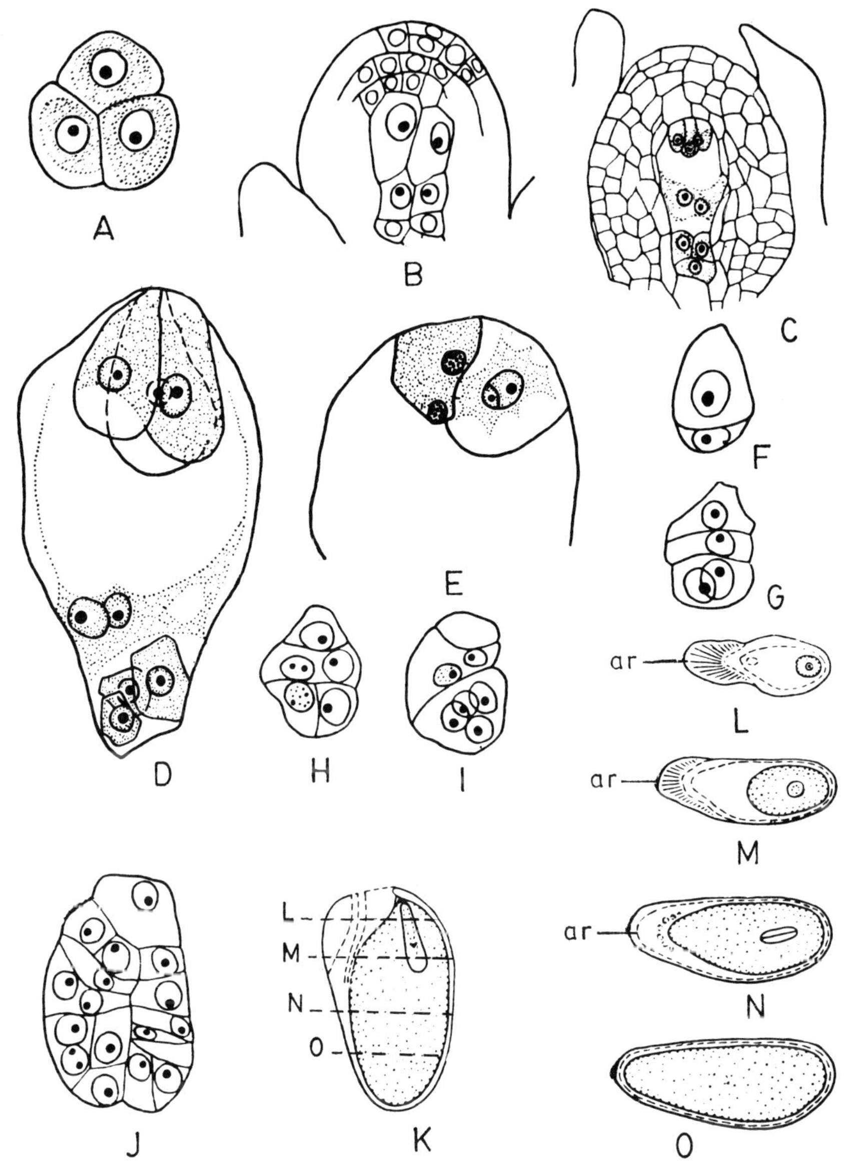

Fig. 19.5.1A–O. **A**, **B**, **F–J** *Stauntonia hexaphylla*, **C** *Akebia quinata*, **D**, **E** *A. pentaphylla*, **K–O** *A. lobata*. **A** Tetrahedral tetrad of microspores. **B** Sporogenous cells in nucellus. **C**, **D** Mature embryo sac. **E** Micropylar portion of embryo sac to show syngamy. **F–J** Embryogenesis. **K** Longisection of seed. **L–O** Transections of **K** at respective levels; *ar* aril. (**A**, **B**, **F–J** After Yoshida and Nakajima 1978, **C–E** after Yoshida and Michikawa 1973, **K–O** after Corner 1976)

Nakajima 1978). The micropyle is formed by the inner integument only. The hypodermal archesporial cell cuts off a parietal cell. Multiple archesporium occurs in *S. hexaphylla*, but only one cell functions. The parietal cell produces a well-developed parietal tissue. The cells of the nucellar epidermis also divide forming two to three layers (Fig. 19.5.1B, C). The megaspore mother cell undergoes meiotic divisions, produces a T-shaped, or, rarely, linear tetrad in *Holboellia latifolia* (S.P. Bhatnagar 1965b). In *Stauntonia hexaphylla*, during the formation of a linear tetrad, the micropylar dyad divides earlier than the chalazal, but in all cases the chalazal megaspore develops into a Polygonum type of embryo sac (Fig. 19.5.1C, D). The antipodals are small and ephemeral in *Akebia pentaphylla*, *A. trifoliata* (Yoshida and Michikawa 1973), *Holboellia latifolia* (S.P. Bhatnagar 1965b), and *S. hexaphylla* (Yoshida and Nakajima 1978), but persist in *Decaisnea insignis* (Swamy 1953c).

The polar nuclei fuse before fertilization, and the secondary nucleus lies near the antipodals in *Decaisnea*. During double fertilization in *Akebia trifoliata* (Yoshida and Michikawa 1973) one sperm nucleus fuses first with one polar nucleus and then triple fusion occurs. Syngamy in *A. pentaphylla* is shown in Fig. 19.5.1E.

The endosperm is of the Cellular type. Both micropylar and chalazal chamber contribute to the formation of endosperm. It contains oil as a storage product.

The embryogeny conforms to the Asterad type (Fig. 19.5.1F–J).

The seed is albuminous with a vestigial aril in *Akebia* (Fig. 19.5.1K–O). The outer epidermis of testa comprises narrow cells with thick inner and radial walls. The mesophyll cells show brown, unlignified walls. In *Akebia* the mesophyll is composed of longitudinal fibres. The tegmen and inner epidermis of the testa do not show any remarkable feature.

SP Bhatnagar (1965b), EJH Corner (1976), BGL Swamy (1953c), O Yoshida and A Michikawa (1973), O Yoshida and Y Nakajima (1978)

19.6 Menispermaceae

The anther wall comprises persistent epidermis, fibrous endothecium, ephemeral middle layer, and secretory tapetum with uninucleate cells. The tapetum is irregularly biseriate. Cytokinesis is simultaneous, and the tetrads are tetrahedral in *Cocculus triloba* (Yoshida and Yamada 1973).

The tricolpate pollen grains are two-celled at the dispersal stage.

The ovule is hemianatropous or amphi- to campylotropous, bitegmic (*Cocculus* and *Tiliacora*), or unitegmic (*Cissampelos*, *Cocculus*, *Stephania* and *Tinospora*), and crassinucellar. The micropyle is formed by the inner integument. The outer integument is smaller in *Cocculus villosus* (A.C. Joshi 1937a). The single hypodermal archesporial cell cuts off a parietal cell, the parietal tissue is well-developed. The cells of the nucellar epidermis also divide, and form a cap two to four cells thick. The megaspore mother cell undergoes meiotic divisions, and produces a linear or T-shaped tetrad. The chalazal megaspore develops into a Polygonum type of embryo sac. The synergids are egg-like in *Tinospora*. The polar nuclei fuse before fertilization. The three antipodals are large, and they persist up to or even after fertilization in *Cocculus* and *Tinospora* (A.C. Joshi 1937a, 1939a). The number of antipodals increases in *Tinospora*.

The development of the endosperm is of the Nuclear type. Wall formation is initiated (at the globular proembryo) in the micropylar region, and gradually progresses up to the chalaza. In *Tiliacora racemosa* and *Tinospora cordifolia* (Sastri 1964), the infoldings of the outer integument lead to ruminations (Annona type – Periasamy 1962a), whereas in *Cissampelos* and *Cocculus* the infoldings of fibrous endocarp cause a wavy surface of the endosperm (weakly ruminate) (Sastri 1964). The endosperm is oily.

The embryogeny conforms to the Onagrad type. The zygote divides six to eight weeks after fertilization in *Tinospora cordifolia* (Sastri 1964). In *Cocculus triloba* (Yoshida and Yamada 1973), the growth of embryo is said to be vertical, horizontal, globular, and irregular. This interpretation is not quite correct even though the earlier stages of embryogeny show considerable variation.

The exarillate seed is albuminous with a large, curved embryo. It is characterized by an epiphysis. The seed-coat is unspecialized, and generally gets crushed. The outer epidermis of testa persists in some species. It comprises tabular cells with thin, lignified walls. The thick-walled endocarp gives protection in *Cocculus triloba* (Yoshida and Yamada 1973).

AC Joshi (1937a, 1939a), AC Joshi and BVR Rao (1934, 1935), PM Mathew (1958), K Periasamy (1962a), RLN Sastri (1954a, b, 1964), B Singh (1934), NK Tiwary (1934), O Yoshida and H Yamada (1973)

19.7 Nymphaeaceae

The anther wall comprises the epidermis, fibrous endothecium, two to five middle layers, and tapetum. The endothecial thickenings are well-developed in *Brasenia*, *Nelumbo*, *Nymphaea* and *Victoria* (P. Khanna 1965a, 1967), and are absent in *Cabomba* (Padmanabhan and Ramji 1966). They are masked by tannin in *Nelumbo nucifera* (S.C. Gupta and Ahluwalia 1979). The two middle layers persist in *Nelumbo*. The tapetum is reported to be amoeboid in *Castalia* and *Nuphar* (see Schnarf 1931), and secretory with uni- to multinucleate cells in other members. This aspect needs further study. In *Nelumbo* and *Euryale* (P. Khanna 1964a), the cells are multi-nucleate. "A large number of Übisch granules are studded on a prominent tapetal membrane on the inner tangential surface of tapetal cells at the uninucleate microspore stage" (S.C. Gupta and Ahluwalia 1979). The microspore mother cells undergo meiotic divisions and form tetrahedral and isobilateral tetrads. Cytokinesis is successive in *Cabomba*, and simultaneous in *Castalia*, *Nelumbo* and *Nuphar* (see Schnarf 1931, S.C. Gupta and Ahluwalia 1979), and *Brasenia* (P. Khanna 1965a). Compound pollen grains are reported in *Nymphaea* and *Victoria* (P. Khanna 1967).

Pollen grains are shed at the three-celled stage in *Euryale*, *Nymphaea* and *Victoria* (P. Khanna 1964a, 1967) and at the two-celled stage in *Cabomba*

(Padmanabhan and Ramji 1966). In *Nelumbo*, the pollen grains are three-colpate, although in all other members they are monocolpate (see P.K.K. Nair 1970).

The pollen grains in *Barclaya longifolia* (Schneider 1978) are spherical, bean-shaped or psilate to scabrate with a perforate tectum. They are inaperturate.

Sometimes, in *Nelumbo* the pollen tubes enter an embryo sac in which the synergids have degenerated before its entry. This needs confirmation.

The ovule is anatropous (initially orthotropous in *Nelumbo*), bitegmic, and crassinucellate. It is orthotropous (permanently) in *B. longifolia*. The integument forms a beak, or an operculum develops. The nucellar tissue lies below the operculum. The micropyle is formed by the inner integument (Fig. 19.7.1A). In *Cabomba caroliniana* (Ramji and Padmanabhan 1965), the nucellar tissue grows in height due to the presence of a basal meristem. A glandular, aerenchymatous, arillate outgrowth from the funicle has been observed in *Euryale*, *Nymphaea* and *Victoria*. S.C. Gupta and Ahluwalia (1977, 1979) report occasional occurrence of a middle integument in *Nelumbo nucifera*. The funicle is massive with air-spaces in this species, and produces secretory hairs which function as an obturator. A single hypodermal archesporial cell cuts off a parietal cell. The parietal tissue is three to four layers in *Victoria*, two to six layers in *Nelumbo* (S.C. Gupta and Ahluwalia 1979), and only a single layer in *Nymphaea* (P. Khanna 1967). The megaspore mother cell undergoes meiotic divisions producing a linear or T-shaped tetrad. The chalazal megaspore develops into a Polygonum type of embryo sac (Fig. 19.7.1B, C). The antipodals are ephemeral.

In *Barclaya longifolia* (Schneider 1978), the embryo sac has an elongated tubular extension which has been interpreted as "haustorium-like". This assumption is incorrect and the extension should be considered as an embryo sac caecum. Similar extension of the embryo sac is also reported in *Nuphar* (Chifflot 1902), *Nymphaea* (Conrad 1905) and *Euryale* (P. Khanna 1964a). The caecum enlarges in the basal nucellar region, and in a mature embryo sac contains three antipodals. The basal nucellar tissue grows in files of radiating rows, and after fertilization forms perisperm. It is packed with packets of starch grains.

In *Ondinea purpurea* (Schneider 1983), anthesis occurs over a three-day period. First-day flowers are protogynous and characterized by colourful reflexed perianth and staminal parts. The pollen-receptive stigmas of first-day flowers secrete a fluid that fills the small stigmatic cup. As potential pollinators are attracted and attempt to land on the distally exposed gynoecium, they contact the stigmatic fluid which loosens pollen from the insect's body and pollination is achieved. Second- and third-day flowers are functionally staminate.

During double fertilization, syngamy and triple fusion occur simultaneously.

According to P. Khanna (1965a), the antipodals multiply and become multinucleate (Fig. 19.7.1C), and haustorial in *Nelumbo nucifera* during post-fertilization stages. However, according to S.C. Gupta and Ahluwalia (1979),

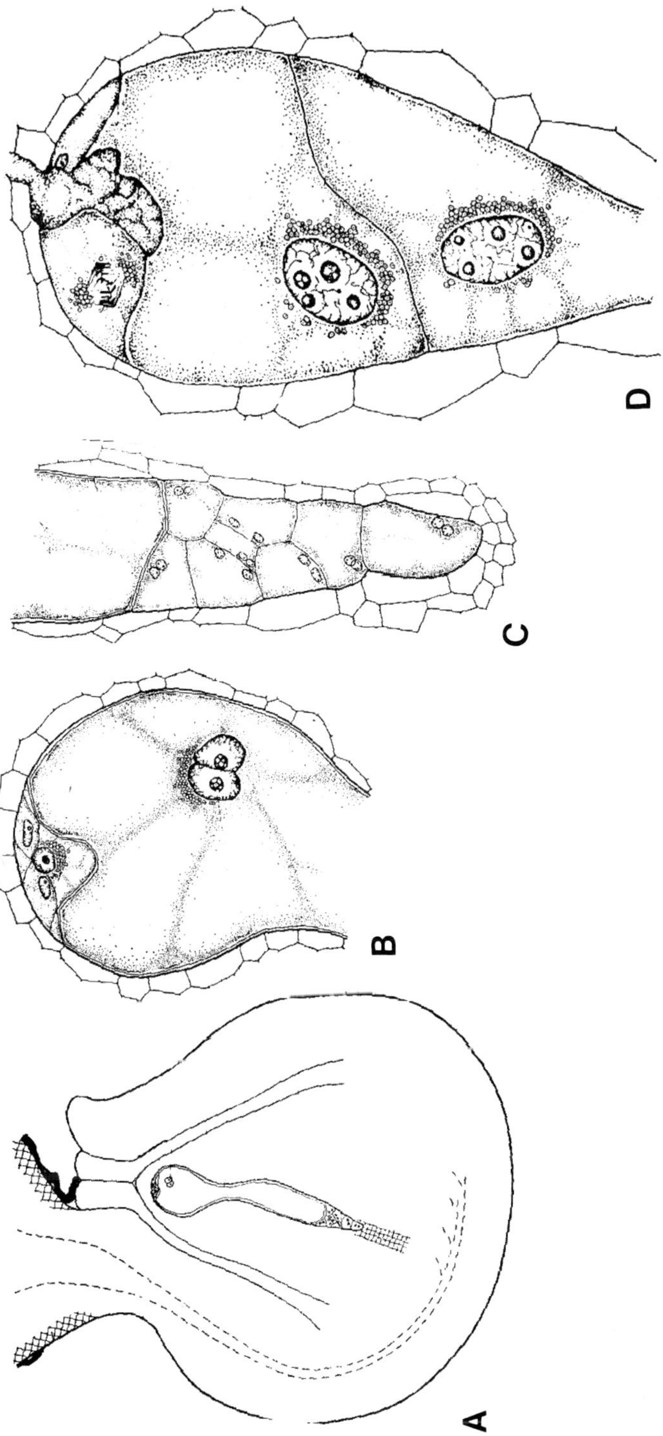

Fig. 19.7.1A–D. *Nelumbo nucifera.* **A** Longisection of ovule at pollination. **B**, **C** Micropylar (**B**) and chalazal (**C**) part of embryo sac, the antipodals are binucleate. **D** Upper part of embryo sac at two-celled endosperm, the zygote nucleus is in division, note the synergid and remnants of pollen tube. (After Batygina et al. 1983)

there is no multiplication of antipodals in this species. This feature needs further investigation.

The endosperm is of the Cellular type in *Castalia ampla*, *C. pubescens*, *Nelumbo nucifera* (Fig. 19.7.1D), *Nuphar luteuma* (see Schnarf 1929), *Ondinea purpurea* (Schneider and Ford 1978) and *Victoria cruziana* (P. Khanna 1967), and Helobial type in *Nymphaea stellata* (P. Khanna 1967). A tube-like structure develops from the primary chalazal chamber in *Castalia* and *Nymphaea* (M.T. Cook 1906).

In *Euryale* and *Nelumbo* (P. Khanna 1965a), the endosperm is Nuclear. Wall formation commences from the micropylar end.

The embryogeny corresponds to the Asterad type in *Castalia odorata*, *Euryale ferox* and *Nymphaea advena*. S.C. Gupta and Ahuja (1967) report that in *Nelumbo* two distinct meristematic zones representing cotyledonary initials differentiate in the apical region of the proembryo, on either side of the plumule. The suspensor is absent. The cotyledons show meristematic activity at the tips, become massive, and usually one of them grows rapidly. In mature seed it covers the arrested cotyledon at the distal end.

The embryo is large in *Nelumbo* (Fig. 19.7.2A–M), as compared to other members of the family. The growth of plumule is precocious, and in mature seed the shoot apex develops four leaves with primordium of the fifth. The first two leaves do not bear sheaths, but they are present in the latter two leaves and the primordium of the fifth. The growth of the radicle is suppressed, and adventitious roots develop at the nodal region (S.C. Gupta and Ahuja 1967).

The seed is exotestal with abundant starch in perisperm. It is exalbuminous in *Nelumbo* (Fig. 19.7.3) and arillate in *Euryale*, *Nymphaea* and *Victoria*. The

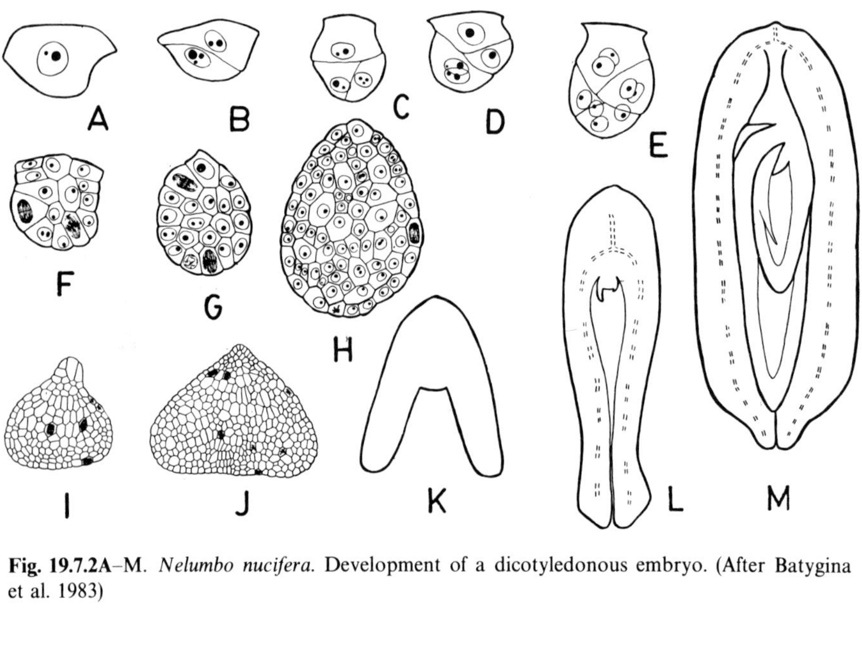

Fig. 19.7.2A–M. *Nelumbo nucifera*. Development of a dicotyledonous embryo. (After Batygina et al. 1983)

testa alone forms the seed-coat. The outer epidermis comprises thick-walled, lignified, columnar cells. The mesophyll and the inner epidermis are unspecialized.

The seed of *Ondinea purpurea* (Schneider and Ford 1978) is small, hard, brown, arillate and ellipsoidal with slightly protruding and flattened micropyle-hilum end. The surface is marked by longitudinal parallel ridges, converging at the micropylar and chalazal end (Fig. 19.7.4A). The raphe is a conspicuous ridge (Fig. 19.7.4B). The sclereid coat is covered by a thin cuticle and each cell has highly thickened radial and outer tangential walls. At the micropylar region the cells of tegmen differentiate into macrosclereids to form an operculum (Fig. 19.7.4C) which helps in rupturing the testa during germination. The endosperm is one or two layers thick, but the rest of the seed is filled with perisperm gorged with starch grains. The operculate seed has a minute embryo, and the plumular leaves are enclosed by two cotyledons (Fig. 19.7.4C). The embryo bears two large separate cotyledons, concave to conform to the shape of epicotyl. The first and second plumular leaves lie decussate to the cotyledons (Fig. 19.7.5A–C).

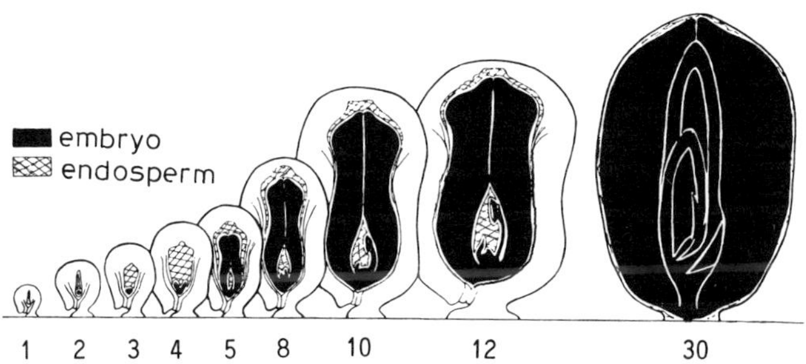

Fig. 19.7.3. *Nelumbo nucifera.* Development of ovule to seed, longisections. The *numbers at the base line* indicate days after pollination. (After Batygina et al. 1983)

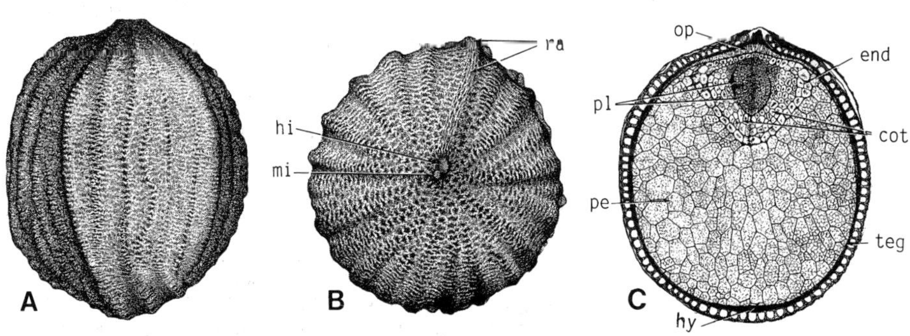

Fig. 19.7.4A–C. *Ondinea purpurea.* **A** Seed, lateral view, aril removed, micropyle-hilum-end flattened. **B** End-view; *ra* raphe; *hi* hilum; *mi* micropyle. **C** Near-median longisection, aril removed; *op* operculum; *end* endosperm; *pl* plumular leaf; *cot* cotyledons; *pe* starchy perisperm; *teg* tegmen; *hy* hypostase. (After Schneider and Ford 1978)

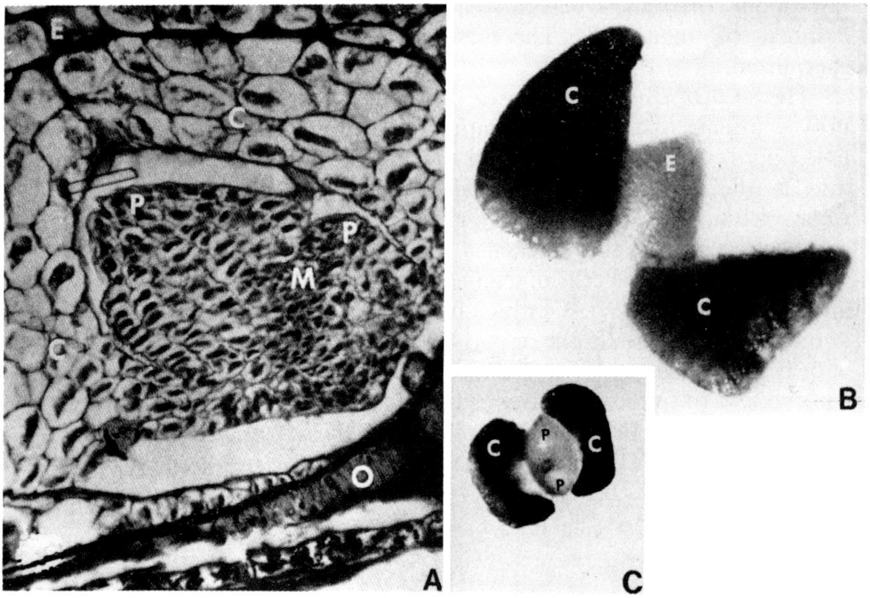

Fig. 19.7.5A–C. *Ondinea purpurea.* **A** Near-median longisection through embryo; *E* epicotyl; *C* cotyledons; *P* plumular leaves; *M* apical meristem; *O* operculum. **B** Dissected mature embryo with two separate cotyledons (*C*) pulled away from epicotyl (*E*). **C** Excised embryo (wholemount), distal view to show concave cotyledons (*C*) which conform to the shape of epicotyl; *P* first and second plumular leaves decussate to cotyledons. (After Schneider and Ford 1978)

The exotestal seed of *Barclaya longifolia* (Schneider 1978) is exarillate and bears large radiating spines except in the micropylar region (spiny testa). The spine is covered with cuticular papillae. The tip of the spine bends into a hook-like structure and becomes a means of dispersal. The cells of operculum, in *B. longifolia*, form macrosclereids and those in the chalaza become sclerified forming a hypostase.

Upon germination, the cotyledons always remain in the seed forcing the operculum. The operculum ruptures the surrounding testa and the hypocotyl-radicle axis and epicotyl emerge from the seed (Conrad 1905). The cotyledons provide reserve food from the perisperm to the developing sporophyte.

TB Batygina, GE Kolesova and VE Vasiljeva (1983), JBJ Chifflot (1902), HS Conrad (1905), MT Cook (1906), SC Gupta and R Ahluwalia (1977, 1979), SC Gupta and R Ahuja (1967), C den Hartog (1970), P Khanna (1964a, 1965a, 1967), KI Meier (1960), PKK Nair (1970), D Padmanabhan (1970a), D Padmanabhan and MV Ramji (1966), MV Ramji and D Padmanabhan (1965), K Schnarf (1929, 1931), EL Schneider (1978, 1983), EL Schneider and J Buchanan (1980), EL Schneider and EG Ford (1978), KT Sundri, M Radhakrishnan and LL Narayana (1982), JW Walker (1976, 1978), R Wheeler and KA Lye (1975)

19.8 Ceratophyllaceae

The embryology has not been studied in detail. The innermost layer of anther wall develops into amoeboid tapetum. Due to the absence of endothecial thickenings, the dehiscence of anther is irregular. The microspore mother cells undergo meiotic divisions, and cytokinesis is successive (Sastri 1955c).

The pollen grains are two-celled at the dispersal stage, in *Ceratophyllum demersum* (Sastri 1955c).

The ovule is anatropous, unitegmic, and crassinucellar. The archesporial cell cuts off a parietal cell. The parietal tissue is well developed. The megaspore mother cell undergoes meiotic divisions giving rise to a linear or T-shaped tetrad. The chalazal megaspore develops into a Polygonum type of embryo sac. The polar nuclei fuse before fertilization. The antipodals are small.

The development of endosperm is of the Cellular type. The division of the primary endosperm nucleus results in the formation of a micropylar and chalazal chamber. The micropylar chamber undergoes repeated divisions and comprises small endosperm cells. The chalazal chamber divides only once.

The embryogeny conforms to the Asterad type-Penaea variation. The mature embryo is well developed, and green with a conspicuous plumule.

The exalbuminous seed has a seed-coat of one to two layers of thin-walled cells. The endocarp gives protection.

RLN Sastri (1955c), K Schnarf (1929, 1931), E Strasburger (1902), K Subramanyam (1962)

19.9 Taxonomic Considerations

Bentham and Hooker (1965a) place Ranunculaceae, Berberidaceae, Menispermaceae and Nymphaeaceae under Ranales; Lardizabalaceae are treated as a tribe under Berberidaceae. Hutchinson's Ranales (1973) comprises Paeoniaceae, Helleboraceae, Ranunculaceae, Nymphaeaceae, Podophyllaceae, Ceratophyllaceae and Cabombaceae. Sargentodoxaceae, Lardizabalaceae, Menispermaceae, Nandinaceae, Circaeasteraceae and Berberidaceae are included in Berberidales.

Paeonia, once included in Ranunculaceae, has been raised to an independent family, Paeoniaceae (see Y.D. Tiagi 1970b). Embryological features (Sect. 19.2) also support its independent status (Sect. 22.3). Hutchinson (1973) raised the tribe Helleboraceae to family rank. Vijayaraghavan (1970a) investigated *Helleborus foetidus* and other multiovular genera. He is not in favour of the separation of Helleboraceae from Ranunculaceae on embryological grounds.

Ranunculaceae. Ranunculaceae show a wide range of variation in floral structure, as well as embryology. Variations occur in the type of anther tapetum, number of integuments, extent of development of nucellus, types of embryo sac, and embryogeny. The anther tapetum is secretory or amoeboid;

the tricolporate pollen grains are two- or three-celled. The anatropous ovule may be bitegmic, unitegmic or undifferentiated (ategmic), and crassi- or tenuinucellar. The development of embryo sac is mono-, bi- or tetrasporic. Even in one and the same species, *Anemone vitifolia* (Bhandari 1965), all the three types occur. The antipodals are exceptionally large, and persistent with polyploid nuclei. The ploidy level in some species is very high. The embryogeny is either of the Onagrad or Solanad type, or irregular. The albuminous, exarillate seed has a straight, minute embryo.

Glaucidium. A monotypic genus, *Glaucidium*, is usually included in Ranunculaceae (Engler and Gilg 1924, Melchior 1964, Cronquist 1981). Melchior places it in the subfamily Hydrastoideae of Ranunculaceae. Hutchinson (1973) includes *Glaucidium* in Helleboraceae – a family separated from Ranunculaceae. Langlet (1928) and Miyaji (1927) suggest that this genus should be assigned to Berberidaceae. The relationship with *Paeonia* has also been pointed out (see Tobe 1981). Tamura (1972) emphasizes that *Glaucidium* should be raised to a unigeneric family. R. Dahlgren (1975b) and Takhtajan (1980) accept the family rank of Glaucidiaceae.

Several workers have pointed out that on the basis of morphology, anatomy, cytotaxonomy and chemotaxonomy, *Glaucidium* should be removed from Ranunculaceae (see Tobe 1981).

We do not have enough information on the embryology of Hydrastidaceae, and the family rank of Paeoniaceae is already well-established. The development of embryo in *Paeonia* is so unusual (Sect. 22.3) and different from that of *Glaucidium* that any relationship with *Paeonia* is out of question.

Tobe (1981) has studied the embryology of *G. palmatum*. The embryology of *Glaucidium*, Ranunculaceae and Berberidaceae is compared in Table 19.9.1 to show the similarities and differences between them. Tobe (1981) points out that the embryology of *Glaucidium* is so distinctive that it confirms the family rank of Glaucidiaceae. The similarities between *Glaucidium* and Ranunculaceae are far more significant than the differences. Therefore, we do not agree with Tobe's conclusions, and wish to emphasize that *Glaucidium* is correctly placed in Ranunculaceae.

Circaeaster. The genus *Circaeaster* (Melchior 1964) is included in Ranunculaceae. However, embryologically it shows significant differences. In *Circaeaster* the ovule is orthotropous and unitegmic. The embryo sac is monosporic, Polygonum type. Fertilization is chalazo-, or mesogamous, and endosperm is of the Cellular type. The mature embryo has only primordial cotyledons and differentiation occurs during germination. The seed-coat is crushed and pericarp covers the endosperm.

In *Eranthis hiemalis*, *Ranunculus ficaria* and some other taxa of Ranunculaceae the embryo lacks organs at the shedding of seeds. The growth is intraseminal (Natesh and Rau 1984).

According to Hutchinson (1973), the placement of *Circaeaster* with *Kingdonia* in Ranunculaceae is not justified. He considers it as a very reduced

Table 19.9.1. Embryology of *Glaucidium*, Ranunculaceae and Berberidaceae (Adapted after Tobe 1981)

	Glaucidium	Ranunculaceae	Berberidaceae
Ovule	Numerous, anatropous, bitegmic	Numerous, anatropous, bitegmic, sometimes unitegmic	Numerous, anatropous, hemianatropous in *Podophyllum* and *Ranzania*, bitegmic
Nucellus	Nucellar cap 15–20 layers thick, pseudo-crassinucellar	2 or 3 layers, tenui-, crassi- or pseudocrassinucellar	6 to 7 layers, pseudo- or crassinucellar
Integument	Outer integument vascularized up to the tip, inner integument of dermal origin, outer subdermal, inner integument 3–5 layers thick, outer 7–10 layers	Unvascularized, inner integument of dermal origin, outer subdermal, inner integument 2–4 layers thick, outer 4–6 layers	Unvascularized, inner integument 2–3 layers thick, outer 4–8 layers
Micropyle	Formed by inner integument	Inner integument	Both integuments
Archesporium	Multicelled, 3–4 cells function, parietal cell not formed	Uni- or multicelled, only one cell functions, parietal cell may or may not be formed	Multicelled, only one cell functions, parietal cell may or may not be formed
Embryo sac	Polygonum type, one, rarely up to 3 embryo sacs	Polygonum type, only one	Polygonum type, Peperomia type in *Caulophyllum*; only one
Antipodal cells	Small, ephemeral or persistent	Large, commonly multinucleate, persistent	Large with hypertrophied nuclei, persistent
Endosperm	Nuclear type	Nuclear	Nuclear
Embryo	Unclassified, comparable to Onagrad	Onagrad or Solanad	Onagrad

relative of Podophyllaceae or Berberidaceae. There is substantial embryological evidence to raise *Circaeaster* to a family rank, Circaeasteraceae, near Ranunculaceae and Berberidaceae.

Berberidaceae. Berberidaceae show close alliance with Ranunculaceae (Table 19.9.1). Both have secretory anther tapetum, tricolporate two-celled pollen grains, anatropous, bitegmic and crassinucellar ovule, mono- or tetrasporic embryo sac with conspicuous antipodals, Nuclear endosperm, and Onagrad and Solanad type of embryogeny. The seed structure is also comparable to some extent. The similarities between the two are very pronounced, and B.M. Johri (1935c) states that on the basis of embryology Berberidaceae can be included in Ranunculaceae. However, there are certain features, such as occasional occurrence of successive cytokinesis in the pollen mother cells, presence of integumentary vascular bundle, massive nucellus, absence of bisporic embryo sac, and well-developed embryo with a long radicle, which separate it from Ranunculaceae.

Nandina. *Nandina* has been traditionally classified within the Berberidaceae (Cronquist 1981). However, *Nandina* shows aberrant characters in vegetative structure, floral anatomy and cytology which justify its segregation from Berberidaceae and its raising to family rank, Nandinaceae (see Hutchinson 1973, Metcalfe and Chalk 1950, Nowicke and Skvarla 1981, Shen 1954). Embryology does not support such a segregation. In *Nandina domestica* (Ehdaie and Russell 1984), the hypostase is well developed. The micropyle is zigzag. In *Berberis* and *Mahonia* the synergids persist and are haustorial (Mauritzon 1936a), in *B. nepalensis* ephemeral (B.M. Johri 1935c). In *Nandina* the synergids degenerate about the time of fertilization. *Nandina* does not differ significantly from *Berberis* and *Mahonia*, and there is no justification to raise it to an independent family, Nandinaceae.

Lardizabalaceae. Lardizabalaceae have been treated as a tribe of Berberidaceae by Bentham and Hooker (1965a). They can be distinguished from Berberidaceae on the basis of ana- or orthotropous ovule with a nucellar cap. The micropyle is formed by the inner integument alone. The archesporium is single-celled, parietal tissue is well developed, and the antipodals are ephemeral. Only the monosporic type of embryo sac development occurs. Lardizabalaceae alone have Cellular type of endosperm in the suborder Ranunculineae. The family deserves an independent status (Table 19.9.2).

Menispermaceae. Menispermaceae are a multipalynous family with hemianatropous to amphitropous, uni- or bitegmic, crassinucellar ovule. Their embryological features, such as secretory anther tapetum, simultaneous divisions in pollen mother cells, Polygonum type of embryo sac, Nuclear endosperm, and Onagrad type of embryogeny, also occur in Ranunculaceae. The ruminate nature of endosperm is their special feature.

On the whole, the four families Ranunculaceae, Berberidaceae, Lardizabalaceae and Menispermaceae (Table 19.9.2) of the suborder Ranunculineae

Table 19.9.2. Embryology of Berberidaceae, Lardizabalaceae and Menispermaceae (Adapted after Sastri 1969a)

	Berberidaceae	Lardizabalaceae	Menispermaceae
Anther wall	5- or 6-layered	5–6-layered	4–5-layered
Tapetum	Secretory, cells 4–8-nucleate	Secretory, cells 2–4-nucleate; amoeboid in *Stauntonia hexaphylla*	Secretory, cells 1- or 2-nucleate
Cytokinesis	Simultaneous	Simultaneous	Simultaneous
Pollen	2-celled, 3 colpate	2-celled, 3-colporate	2-celled, 3-colpate
Ovule	Anatropous, bitegmic, crassinucellate	Ortho-, campylo- or anatropous, bitegmic, crassinucellate	Hemiana-, amphi- or campylotropous, uni- or bitegmic, crassinucellate
Parietal tissue	Massive, develops from nucellar epidermis	Well developed	Well developed
Archesporium	Uni- or multicellular, one cell functions	Uni- or multicellular, one cell functions	Uni- or multicellular, one cell functions
Megaspore tetrad	Linear	T-shaped or linear	T-shaped or linear
Embryo sac	Polygonum type	Polygonum	Polygonum
Polar nuclei	Fuse before fertilization	Fuse before fertilization	Fuse before fertilization
Antipodal cells	3, large, persistent	3, small, ephemeral, sometimes persistent	3, small, ephemeral
Endosperm	Nuclear type	Cellular	Nuclear, ruminate
Embryo	Onagrad type	Asterad	Onagrad
Seed	Exotestal, albuminous, rarely arillate	Exotestal, albuminous, vestigial aril in *Akebia*	Seed-coat generally crushed, or endotegmic, exarillate, albuminous

exhibit close affinities. Cronquist (1968, 1981) and R. Dahlgren (1975a, 1980b, 1981) include them in the order Ranunculales.

Nymphaeaceae. Nymphaeaceae and Ceratophyllaceae are vessel-less aquatics, and constitute the suborder Nymphaeineae. The family Nymphaeaceae comprises three subfamilies: Nymphaeoideae, Cabomboideae and Nelumbonoideae (Melchior 1964).

Cabomba. *Cabomba* differs from other members of the family in the (1) absence of endothecial thickenings in anther wall, (2) two-nucleate tapetal cells, (3) successive cytokinesis in pollen mother cells, (4) two-celled pollen grains, and (5) nucellar growth due to a basal meristem. Its inclusion in the subfamily Cabomboideae is justified. R. Dahlgren (1977a, 1980b) treats Cabomboideae as an independent family, Cabombaceae, in the superorder Nymphaeiflorae.

Nelumbo. The systematic position of *Nelumbo* is controversial. Assuming that *Nelumbo* shows monocotyly, Cook (1906) and Seaton (1908) place it in Monocotyledons, while Arber (1920), Lawrence (1951) and Hutchinson (1973) retain it in Dicotyledons (reticulate venation of leaf; tricolpate pollen).

Li (1955) and Takhtajan (1959, 1980) propose the erection of the order Nelumbonales, which has been supported by Snigirevskaya (1964) and P. Khanna (1965a). Takhtajan emphasizes features such as tricolpate pollen grains, morphology of the gynoecium, absence of suspensor, absence of idioblasts, and the basic chromosome number. P. Khanna (1965a) draws attention to additional features: persistence of middle layers of the anther wall, secondary multiplication of antipodals and their mutlinucleate condition (proliferation of antipodals is refuted by S.C. Gupta and Ahluwalia née Ahuja – 1979), Nuclear endosperm, exalbuminous seed with a large embryo, and absence of suspensor, perisperm, and hypostase.

On the basis of phytochemical data (absence of ellagic acid and presence of benzylisoquinoline alkaloids), and tricolpate pollen grains, R. Dahlgren (1975a, 1977a, 1980b) erects the order Nelumbonales with the lone family Nelumbonaceae, in superorder Magnoliiflorae.

S.C. Gupta and Ahuja (1967) and S.C. Gupta and Ahluwalia née Ahuja (1977) emphasize the dicotyledonous nature of the embryo in *Nelumbo* (see also Batygina et al. 1983). They also refer to the occurrence of a middle integument, as in *Cananga odorata* (Periasamy and Swamy 1961), and to the open carpel with papillate ovarian canal comparable to that in *Degeneria* (Swamy 1949a). The floral structure strongly supports Ranalian affinity. The ovarian canal, stigmatic hairs, obturator, middle integument, and ovarian hump distinguish the genus *Nelumbo* (S.C. Gupta and Ahluwalia née Ahuja 1977).

The precocious development of embryo forming the shoot apex with four to five leaves and sheaths, two massive cotyledons, and suppression of radicle in mature seed are noteworthy (S.C. Gupta and Ahuja 1977). These investigators favour the position of *Nelumbo* in Nymphaeaceae, in the Ranalian complex, and refute its segregation until other members are fully investigated.

Ondinea. Structural similarities between *Ondinea* and *Nymphaea* together with similar pollination and seed dispersal syndromes lend support to den Hartog's (1970) placement of *Ondinea* in Nymphaeaceae. Further support for the inclusion of *Ondinea* in the Nymphaeaceae s. str. comes from comparative studies of stem vascularization (see Schneider 1983).

Huber (1977), in his proposed system of Monocotyledons, erects a superorder Nymphaeiflorae with Nymphaeales – an order with some dicotyledonous characters – and transfers Nymphaeaceae from Dicotyledons to Monocotyledons.

Cronquist (1968, 1981) and Takhtajan (1969, 1980) consider the Monocotyledons to be derived from Dicotyledons. The aquatic or semiaquatic plants like proto-Nymphaeales must be a transitional series to proto-Alismatales. R. Dahlgren et al. (1985) emphasize that Nymphaeales do show some features of alismatiflorous monocotyledons as laticifers, stipules, extrose anthers, laminar-dispersed placentation and Helobial type of endosperm (rare in Nymphaeaceae). *Cabomba* has liliiflorean floral diagram. The presence of two cotyledons, S-type sieve-element plastids, and perispermous seed do suggest that Nymphaeales are not the starting point of Alismatales. The seeds of Nymphaeales have scanty endosperm and a well-developed perisperm as in Piperales. The absence of perisperm and endosperm in the seeds of Alismatiflorae distinguish them from Nymphaeales.

The reports that Nymphaeaceae s.l. are monocotyledonous, or ancestors to monocotyledons, based on the presence of a single bilobed cotyledon, have also been debated. The studies of Schneider (1983) and Batygina et al. (1983) on the development of embryo have clearly demonstrated that *the embryo has two distinct cotyledons, and not a single bilobed cotyledon.*

Therefore, the assignment of Nymphaeaceae to moncots is not justified at all. Embryological evidences put forward by S.C. Gupta and Ahluwalia née Ahuja (1977) and Batygina et al. (1983) support the dicotyledonous affinities of Nymphaeaceae. Melchior's inclusion of this family in the Ranalian group is in order.

Ceratophyllaceae. Ceratophyllaceae are closely allied to Nymphaeaceae. The similarities are: (1) amoeboid anther tapetum, (2) successive cytokinesis in pollen mother cells, (3) anatropous, crassinucellate ovule with single-celled archesporium, (4) Polygonum type of embryo sac, (5) Cellular type of endosperm, and (6) Asterad type of embryogeny with a straight mature embryo (Sastri 1955c, Subramanyam 1962).

According to Sundri et al. (1982), the Ceratophyllaceae are closely related to *Cabomba* in abaxial microsporangia (Eames 1961), absence of benzylisoquinoline alkaloids, presence of tannins, S-type plastids in sieve tubes, voluminous starchy perisperm (R. Dahlgren 1980b, Gibbs 1974), and the same number of chromosomes, i.e. $2n = 24$ (Darlington and Wylie 1955). Ceratophyllaceae are, therefore, related to Nymphaeaceae in general, and *Cabomba* in particular, rather than to *Nelumbo*.

The presence of unitegmic ovules, exalbuminous seed, and thin-walled testa segregate it from Nymphaeaceae. The former two features indicate its more advanced character. Ceratophyllaceae have been derived from a nymphaeaceous ancestry (Cronquist 1968, 1981, R. Dahlgren 1981).

20 Piperales

The order Piperales comprises Saururaceae, Piperaceae, Chloranthaceae and Lactoridaceae.

20.1 General Features

The general features of the order are: (1) secretory anther tapetum, (2) simultaneous cytokinesis of pollen mother cells, (3) monosulcate pollen grains, two-celled at the dispersal stage, (4) orthotropous, bitegmic, crassinucellate ovule; unitegmic in *Peperomia*, tenuinucellate in *Houttuynia*, (5) single-celled archesporium with parietal tissue, (6) monosporic type of embryo sac; tetrasporic Fritillaria type in Piperaceae, and Peperomia type in *Peperomia*, (7) a large number of antipodals in Piperaceae and *Chloranthus*; ephemeral in *Sarcandra* and Saururaceae, (8) Cellular endosperm; Nuclear type in *Piperaceae*, (9) embryogeny of Piperad type in Piperaceae, and Onagrad in Chloranthaceae (*Sarcandra*), and (10) tegmic seed with a minute embryo and perisperm; perisperm is totally absent in Chloranthaceae.

20.2 Saururaceae

The anther wall comprises the epidermis, fibrous endothecium, single middle layer, and secretory tapetum with binucleate cells. The microspore mother cells undergo meiotic divisions, and the tetrads are tetrahedral, isobilateral and decussate. Cytokinesis is simultaneous.

The pollen grains are monosulcate with a smooth exine. They are two-celled at the dispersal stage.

Houttuynia cordata (Takahashi 1986) is a monotypic genus represented by only one species. Failure of cytokinesis after meiosis results in the formation of coenocytic microspores which vary widely in shape and size. The tapetal cells do not show any demonstrable abnormality, but the microspores degenerate after callose dissolution. Later, the loculus contains numerous coenocytic microspores which consist only of ektexine. This species is completely, or nearly completely, male sterile.

The ovule is orthotropous, bitegmic and crassinucellate; tenuinucellate in *Houttuynia cordata* (Y.S. Murty 1960). The outer integument is two-layered and gets crushed in mature seed. The inner integument is three-layered in *Houttuynia* and *Saururus* (Y.S. Murty 1960), and four-layered in *Anemopsis* (Quibell 1941). A zigzag micropyle, formed by both the integuments, occurs in

Houttuynia. The single hypodermal archesporial cell functions directly in *Houttuynia* (Y.S. Murty 1960); a parietal cell is produced in *Anemopsis* and *Saururus* and the parietal tissue is two layers thick. The megaspore mother cell undergoes meiotic divisions. The chalazal megaspore of a linear or T-shaped tetrad, or a triad, develops into a Polygonum type of embryo sac. The antipodals are ephemeral, and the polar nuclei fuse before fertilization (Fig. 20.2.1A–D).

The development of endosperm is of the Cellular type (Fig. 20.2.1E–G) in *Saururus* (Johnson 1900). The chalazal chamber develops into a haustorium, and the micropylar produces the endosperm proper. Yoshida (1961) reports Helobial type of development in *S. loureiri.* The development of endosperm requires further study.

The embryogeny is of the Asterad type (Fig. 20.2.1H-Q).

The endospermic seed (Fig. 20.2.1G) is small, and has abundant perisperm (Y.S. Murty 1959a). The testa gets crushed. The tegmen forms the seed-coat. Its outer and inner epidermis becomes thick-walled.

P Crété (1957c), DS Johnson (1900), YS Murty (1959a, 1960), M Takahashi (1986), CH Quibell (1941), MVS Raju (1961), O Yoshida (1961)

20.3 Piperaceae

Of the four microsporangia in the anther in *Piper* (Kanta 1962), the development of one, two or three sporangia may be suppressed. Even when both thecae appear normal externally, one may be completely sterile.

The anther wall comprises persistent epidermis, fibrous endothecium, one or two ephemeral middle layers, and secretory tapetum of binucleate cells. The microspore mother cells undergo meiotic divisions; cytokinesis is simultaneous.

The tetrads are isobilateral, decussate and tetrahedral, and usually the microspores separate. Occasionally, in *Peperomia pellucida* and *P. sandersii* (Y.S. Murty 1959a), for some time the four microspores remain attached together in a linear row, or in some other configuration by small connections. After separation such microspores show one or two small protuberances. [This arrangement recalls the condition in *Halophila* (Kausik and Rao 1942)]. Pollen grains of about double the normal size have been observed occasionally in *P. fraseri.*

Pollen grains are monocolpate, and shed at the two-celled stage. In *Peperomia* they are non-sulcate, and show warty protuberances.

The ovule is orthotropous and crassinucellate. It is bitegmic in *Heckeria* (see Schnarf 1931) and *Piper* (Kanta 1962), and unitegmic in *Peperomia* (Y.S. Murty 1959a). The micropyle is produced by both the integuments in *Heckeria*, but only the inner integument in *Piper.*

The archesporial cell cuts off a parietal cell, and four- to five-layered parietal tissue is formed in *Heckeria* (P. Maheshwari and Gangulee 1942). The megaspore mother cell undergoes meiotic divisions and a coenomegaspore

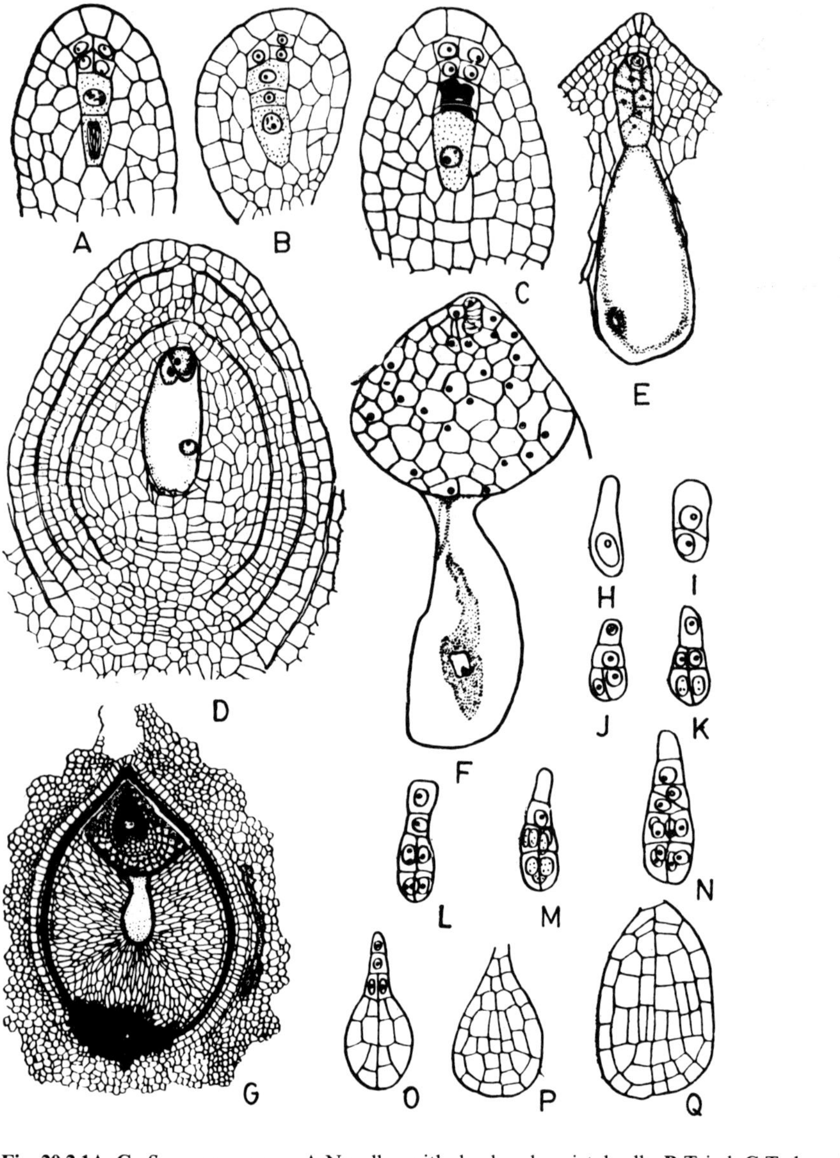

Fig. 20.2.1A–G. *Saururus cernuus*; **A** Nucellus with dyad and parietal cells; **B** Triad; **C** T-shaped tetrad, chalazal megaspore functional; **D** Ovule with mature embryo sac, micropyle zigzag; **E**, **F** Cellular endosperm with chalazal haustorium, zygote in **E** and proembryo in **F**; **G** Longisection of mature seed, embryo (black) surrounded by endosperm (remnants of the chalazal haustorium persist). Note the perisperm and hypostase. **H–Q** *Houttuynia cordata*, embryogenesis, (**A–C**, **F** After M.V.S. Raju 1961, **D**, **E**, **G** after Johnson 1900, **H–Q** after Crété 1957c)

with four nuclei is produced. Cytokinesis does not occur, and the arrangement of nuclei is 1 + 3.

Further development is of the Fritillaria type in *Heckeria* (P. Maheshwari and Gangulee 1942, Swamy 1945b), *Piper* (Swamy 1944b, Kanta 1962), *P. medium* var. *ceanothifolium* (Maugini 1953), and *P. gaudichandianum* (Y.S. Murty 1959a). In *Piper nigrum* (Kanta 1962), the triploid antipodal cells multiply to form a massive tissue of a hundred or more cells. In *Peperomia* (Fisher 1914), the megaspore mother cell undergoes meiotic divisions and produces a four-nucleate coenomegaspore, the arrangement of the nuclei is tetrahedral. Rudimentary walls appear in between the nuclei, but later they disappear. Further development is of 16-nucleate Peperomia type (Fisher 1914, Johnson 1914, Y.S. Murty 1959b). In several species of *Peperomia* the egg is associated with a single synergid (Fig. 20.3.1A).

A high degree of sterility of ovules, and embryolessness is reported in *Piper nigrum* (Kanta 1961, 1962). It may be due to the failure of fertilization, or excess of deposition of calcium oxalate crystals in embryo sacs.

A forked pollen tube is reported in *Peperomia* (Fagerlind 1939a). One of its branches passes to the egg, the other to the secondary nucleus. This needs confirmation. Figure 20.3.1A–C shows syngamy.

The development of endosperm is of the Cellular type in *Heckeria* and *Peperomia*, (Fig. 20.3.1C–F), and Nuclear type in *Piper*. Later, it becomes cellular.

The embryogeny generally conforms to the Piperad type (Fig. 20.3.1D, E). The division of the zygote is followed by a vertical, sometimes transverse or oblique wall (Kanta 1962). The embryo of *Peperomia* (Johnson 1914) is an undifferentiated globular mass, showing 8–10 cells in *P. hispidula*. The histogens, except for epidermal initials or epidermal cells, are usually unrecognizable in such embryos. The embryo in *Macropiper* (Balfour 1957) shows a high degree of differentiation since not only epidermis and ground tissue are present, but also a provascular cone and cotyledonary lobes. In *Piper nigrum* (Kanta 1961, 1962), older embryos show two well-differentiated cotyledons. The embryo has prominent ridges and furrows and appears hexagonal when seen from the top.

The seed is endotegmic, and contains a small, starchy endosperm, and copious perisperm with starch and oil as storage products (Fig. 20.3.1G). The embryo is microscopic. The testa gets almost crushed. The tegmen is distinct only in the micropylar region; it is crushed into a brown layer around the seed.

In *Peperomia* (Y.S. Murty 1959a), the outer layer of integument is crushed, the inner layer becomes brown and thickened, and its inner wall develops numerous infoldings (few or absent at the base of seed) which change the contour of the outermost layer of nucellus. The remnants of the outer layer and the thick brown inner layer form the protective covering of the mature seed.

In *Piper nigrum* (Kanta 1962), the innermost layer of the inner integument has densely cytoplasmic radially elongate cells. Near the micropyle they give out finger-like projections which meet similar processes, directed outward from the nucellar epidermis so that the two fit tightly against each other.

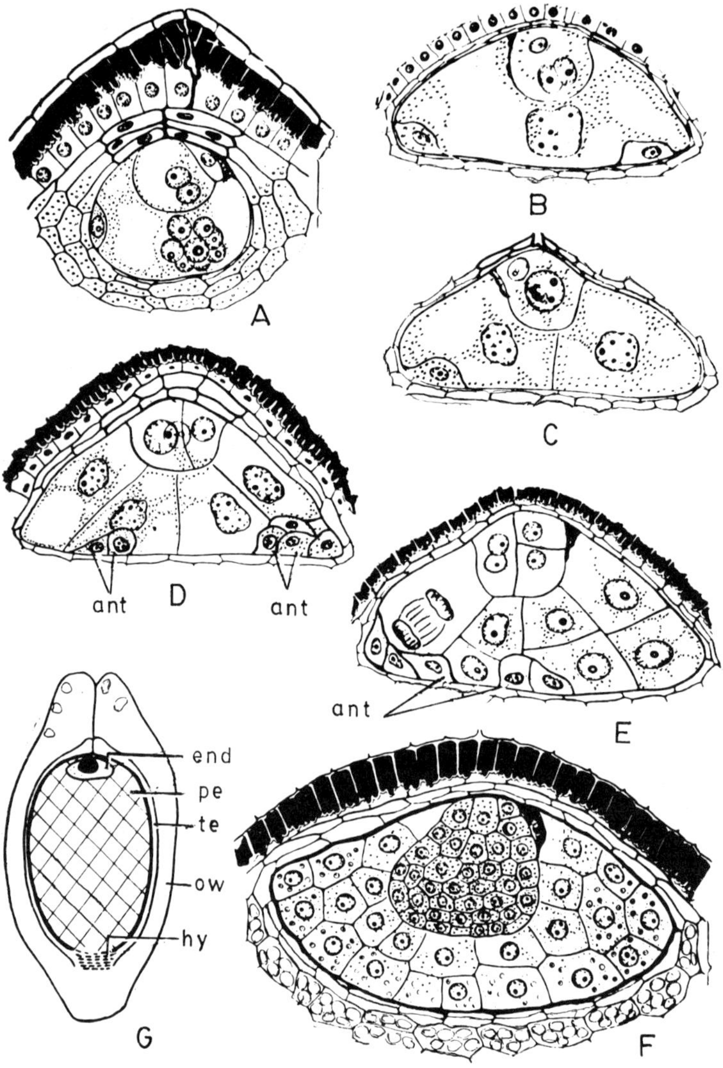

Fig. 20.3.1A. *Peperomia pellucida*, fertilized embryo sac, male and female nuclei in egg cell, one synergid, central cell with a group of polar nuclei, and one lateral antipodal. **B–E** *P. blanda*; **B** Syngamy, primary endosperm nucleus, and two antipodals; **C–E** Progressive stages of development of proembryo and cellular endosperm; *ant* antipodals. **F**, **G** *P. eburnea*; **F** Immature seed with embryo and endosperm; **G** Mature fruit. Embryo *black*; *end* endosperm; *pe* perisperm; *te* testa; *ow* ovary wall; *hy* hypostase. (See Yakovlev 1981)

During the maturation of seed, the outer integument and outer layers of the inner integument collapse. The inner epidermis of tegmen becomes thick-walled and filled with tannin. A few layers of cells just below the perisperm, at the base of ovule, also become thick-walled and tanniniferous. The epidermis of pericarp and several layers below it become thick-walled. Thus, the embryo and endosperm are well protected.

E Balfour (1957), F Fagerlind (1939a), GC Fisher (1914), DS Johnson (1914), AC Joshi (1944), K Kanta (1961, 1962), SB Kausik and PVK Rao (1942), P Maheshwari and H Gangulee (1942), E Maugini (1953), YS Murty (1959a, b), ZI Nikiticheva (1981), K Schnarf (1931), KS Semple (1974), BGL Swamy (1944b, 1945b)

20.4 Chloranthaceae

The anther wall comprises the epidermis, fibrous endothecium, two or three middle layers, and secretory tapetum of binucleate cells. The epidermal cells contain tannin, and a "spiny" cuticle covers the surface in *Sarcandra irvingbaileyi* (Vijayaraghavan 1964). The microspore mother cells undergo meiotic divisions producing isobilateral and tetrahedral tetrads. Cytokinesis is simultaneous.

The pollen grains are inaperturate, and two-celled at the shedding stage.

The ovule is orthotropous, bitegmic and crassinucellar. The micropyle is formed by the inner integument initially, but it is overgrown by the outer integument. A rudimentary third integument develops in *Sarcandra* (Vijayaraghavan 1964).

The hypodermal archesporial cell cuts off a parietal cell. The parietal tissue is well developed. The megaspore mother cell undergoes meiotic divisions and gives rise to a linear tetrad. The chalazal megaspore develops into a Polygonum type of embryo sac. One of the synergids degenerates before fertilization. The antipodals are small; supernumerary antipodals are formed (by their multiplication) in *Chloranthus japonicus* (Yoshida 1957). They are ephemeral in *Sarcandra* (Yoshida 1957, 1959, 1960). A hypostase differentiates.

Development of the endosperm is of the Cellular type (Fig. 20.4.1A, B). It is copious, persistent, and stores oil.

The embryogeny conforms to the Onagrad type in *Chloranthus* and *Sarcandra* (Fig. 20.4.1C–M). In the ripe seed the embryo is at the globular stage in *Chloranthus* and *Hedyosmum nutans* (Edwards 1920), but dicotyledonous in *Sarcandra* (Fig. 20.4.1N, O).

The seed is albuminous. The outer epidermis and mesophyll of testa comprise thin-walled cells, and scattered sclerotic cells. The inner epidermis comprises radially elongate cells with a fine cellulosic reticulum and crystals on their walls. The outer epidermis of tegmen is thin-walled; the inner is a compact layer of longitudinally elongate cells with firm brown walls. The tegmen may get crushed. In *Hedyosmum* the seed-coat is unspecialized.

J Edwards (1920), PK Endress (1971, 1987a, b), BGL Swamy (1953b, d), MR Vijayaraghavan (1964), O Yoshida (1957, 1959, 1960)

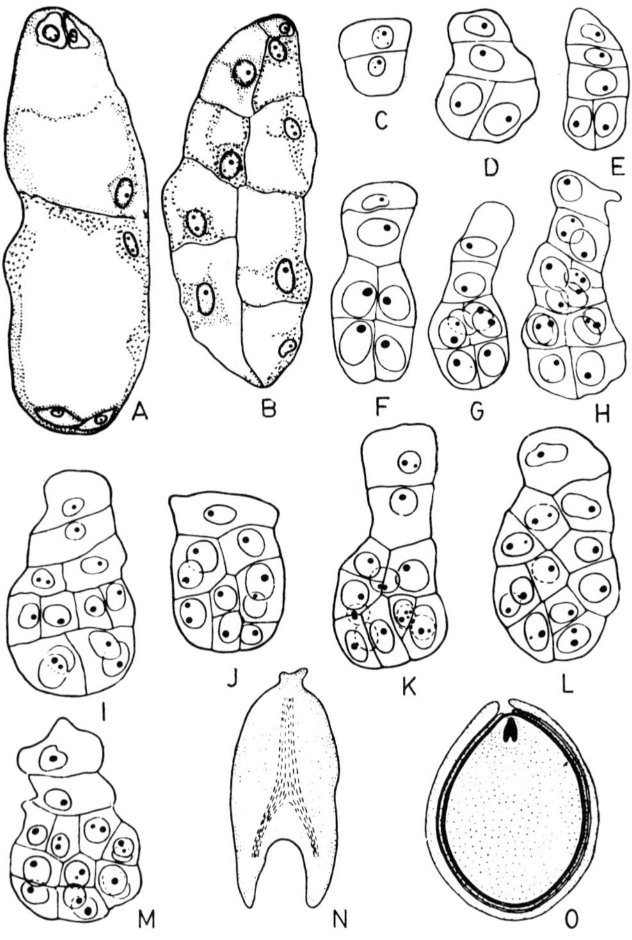

Fig. 20.4.1A–O. *Sarcandra irvingbaileyi.* **A** Embryo sac with zygote, persistent synergid, two-celled endosperm, and two antipodal cells. **B** Zygote and eight-celled endosperm. **C–M** Development of proembryo. **N** Dicotyledonous embryo. **O** Longisection of seed with dicotyledonous embryo and endosperm (diagrammatic). (After Vijayaraghavan 1964)

20.5 Lactoridaceae

The embryological data are very meagre (see Table 18.24.2). The pollen grains remain attached in tetrads. The anatropous bitegmic ovule is tenuinucellar with a cap-like nucellus. The inner integument forms the micropyle. The development of endosperm is of the Nuclear type (Cronquist 1981). The mature seed has copious oily endosperm, and a minute undifferentiated embryo.

F Bouman (1971a), S Carlquist (1964), EJH Corner (1976), A Cronquist (1981), MS Yakovlev (1981)

20.6 Taxonomic Considerations

Bentham and Hooker (1965c) include Piperaceae and Chloranthaceae in Piperales. Engler and Diels (1936) and Hutchinson (1973) include Saururaceae also.

Saururaceae, Piperaceae. Saururaceae and Piperaceae have trilacunar or multilacunar nodes, secretory anther tapetum, simultaneous cytokinesis in pollen mother cells, monosulcate, two-celled pollen grains, orthotropous, bitegmic, crassinucellate ovules, and seed with a minute embryo and abundant perisperm. It appears that the two families are closely allied. However, in Saururaceae the ovule may be tenuinucellate as in *Houttuynia*, both the integuments organize the micropyle, and the archesporium is single-celled. The development of embryo sac is of the Polygonum type with ephemeral antipodals. The Cellular endosperm has a chalazal haustorium. The division of zygote is transverse. In contrast, in Piperaceae the endostome alone forms the micropyle. The female archesporium is multicellular. The development of the embryo sac is of the Fritillaria type, and the number of antipodals is more than 100. The endosperm is of the Nuclear type. The division of zygote is vertical, and the embryogeny is of the Piperad type.

In the family Piperaceae, the position of the genus *Peperomia* is rather debatable. In contrast to the other members of the family, *Peperomia* shows non-sulcate pollen grains with warty protuberances, unitegmic ovules with single-celled archesporium, Peperomia type of embryo sac, Cellular endosperm, and perisperm with starch and oil. These are rather marked differences. Its assignment to a separate family is appropriate.

Chloranthaceae. Chloranthaceae differs from the other two families in having unilacunar nodes (Money et al. 1950). Brongniart (see Swamy 1953b) assigns Chloranthaceae to Santalineae, and Endlicher to Ceratophyllaceae (see Vijayaraghavan 1964). Bessey (1915) places them in Ranales. Vijayaraghavan (1964) investigated the embryology of *Sarcandra irvingbaileyi*.

Chloranthaceae show much more affinity with Saururaceae than Piperaceae in the structure and development of anther, ovule, embryo sac, and endosperm. The division of zygote is also transverse as in Saururaceae. But it can be distinguished due to monosulcate pollen grains with reticulate exine, Cellular type of endosperm without any haustorium, total absence of perisperm, well-developed persistent endosperm, and Onagrad type of embryogeny. Moreover, in Saururaceae and Piperaceae the seeds are tegmic, whereas in Chloranthaceae the seeds have a well-developed seed-coat with exotestal and exotegmic fibres. The inner epidermis of the testa is a crystal layer, and that of the tegmen a compact palisade layer.

Vijayaraghavan (1964) compares the embryological features of Chloranthaceae with those of Santalineae and Ceratophyllaceae. He negates any relationship with both taxa, and points out some affinities with Ranales. The shoot apex of *Sarcandra* is similar to the other vesselless Ranalian genera, especially *Trochodendron* (Ramji 1961). The occurrence of the third rudimentary integument in *Sarcandra* also suggests Ranalian affinity.

The ancestry of Piperales is sought in Magnoliales. The presence of etherial oil cells and uniaperturate pollen shows Magnolian affinities. The position of the families Piperaceae, Chloranthaceae and Saururaceae in the order Piperales, assigned by Hutchinson (1973) and Cronquist (1981) is appropriate. Takhtajan (1980) includes Chloranthaceae in Laurales, and R. Dahlgren (1981) in Magnoliales; the other two families in Piperales.

21 Aristolochiales

The order Aristolochiales comprises three families: Aristolochiaceae, Rafflesiaceae and Hydnoraceae.

21.1 General Features

The order Aristolochiales is characterized by (1) secretory anther tapetum, (2) simultaneous or successive cytokinesis in pollen mother cells, (3) two-celled pollen grains, (4) anatropous, bitegmic, and crassinucellate ovule, (5) single-celled archesporium, massive parietal tissue, (6) Polygonum type of embryo sac, (7) prominent hypostase, (8) Cellular type of endosperm; ruminate in Aristolochiaceae, (9) albuminous seed with a minute embryo, and (10) exo- or endotegmic seeds.

21.2 Aristolochiaceae

The anther wall comprises persistent epidermis, fibrous endothecium, two or three middle layers, and a secretory tapetum of binucleate cells. In *Aristolochia* (B.M. Johri and Bhatnagar 1955a), at places the endothecium and tapetum become biseriate. Übisch granules are present on the inner wall of tapetal cells. The microspore mother cells undergo meiotic divisions. Cytokinesis is simultaneous in *A. europaeum*, *A. sipho* (Täckholm and Söderberg 1918), *A. bracteata* (B.M. Johri and Bhatnagar 1955a) and *Bragantia wallichi* (N.C. Nair and Narayanan 1961), and successive in *Aristolochia clematitis* (Samuelsson 1914), *A. elegans* (B.M. Johri and Bhatnagar 1955a) and *A. fimbriata* (Täckholm and Söderberg 1918). The tetrads are linear, T-shaped, isobilateral, tetrahedral and decussate. Polyspory occurs.

Pollen grains are dispersed at the two-celled stage. They are inaperturate, monosulcate, or three-colpoidate.

The ovule is anatropous, bitegmic, and crassinucellate. It is circinotropous in *Bragantia* (N.C. Nair and Narayanan 1961). The micropyle is formed by endostome. A nucellar cap is produced. A single hypodermal archesporial cell

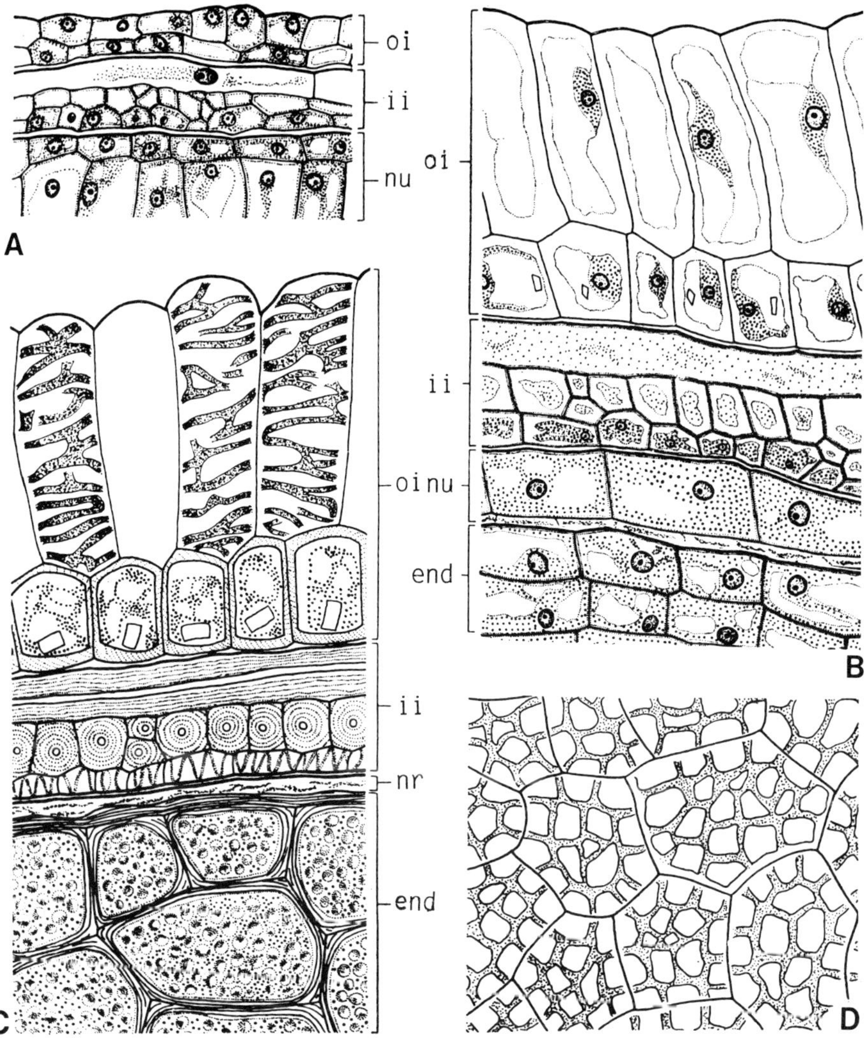

Fig. 21.2.1A–D. *Aristolochia bracteata.* Development of seed-coat, transections. **A** Outer (*oi*) and inner (*ii*) integument, and part of nucellus (*nu*). **B** Later stage, the cells of the exotesta have elongated radially. Cellular endosperm (*end*) is also shown. **C** Reticulate thickenings in exotesta, and thick-walled endotesta with crystals. The exo- and mesotegmen become fibrous, the sclerenchymatous fibres are at right angles to each other. The endotegmen is also fibrous. The endosperm is thick-walled because of deposition of hemicellulose. *nr* nucellus. **D** Funicular cells with reticulate thickenings. (After B.M. Johri and Bhatnagar 1955a)

cuts off a parietal cell. The latter gives rise to a massive parietal tissue. The megaspore mother cell undergoes meiotic divisions, and a linear or T-shaped tetrad is produced. The chalazal megaspore develops into a Polygonum type of embryo sac. A hypostase is prominent in *Bragantia.* The antipodals persist after fertilization and may exhibit haustorial activity. The synergids are ephemeral.

Development of the endosperm is of the Cellular type; ruminations of Annona type occur in *Apama* (Periasamy 1962a). The endosperm is thick-walled, and stores oil and starch.

The embryogeny is probably of the Solanad type. Synergid polyembryony occurs in *A. bracteata* (B.M. Johri and Bhatnagar 1955a).

The development of the seed-coat is represented in Figure 21.2.1A–C. The outer epidermis of testa comprises thick-walled, lignified, radially elongate cells. It is followed by thick-walled cells with crystals. The two outer layers of tegmen consist of crossed lignified fibres. The inner epidermis develops reticulate thickenings, and contains tannin.

The seed is triangular, albuminous with a small embryo, and the testa has an irregular surface. Both the integuments are fused with the raphe for most of their length. The funicular cells develop reticulate thickenings (Fig. 21.2.1D).

BM Johri and SP Bhatnagar (1955a), A Leeman (1927), NC Nair and KR Narayanan (1961), K Periasamy (1962a), G Samuelsson (1914), G Täckholm and E Söderberg (1918), RL Wyatt (1955)

21.3 Rafflesiaceae

The anther wall comprises an epidermis, fibrous endothecium, two middle layers and secretory tapetum in *Cytinus hypocistis* (Guzowska 1964). The microspore mother cells undergo meiotic divisions, cytokinesis is simultaneous, and the tetrads are isobilateral and decussate.

The pollen grains are three- or four-colpate, or two- or three-porate (poroidate).

The ovules are not fully developed in open flowers, and development continues after pollination. The ovule is ana- or orthotropous (*Cytinus*), bi- or unitegmic (*Rafflesia*), and tenuinucellate. *Pilostyles* has a bitegmic ovule. In *Cytinus* the outer integument is a small ring-like swelling at the chalazal end of ovule (see Bouman and Meijer 1987). In *Rafflesia* the ovule should be interpreted as unitegmic because the chalazal swelling (outer integument) is not very prominent (Fig. 21.3.1A–C). The nucellus is single-layered. The archesporial cell functions directly as megaspore mother cell. Development of the embryo sac is of the Polygonum type. The antipodals are well developed.

The development of the endosperm is of the Nuclear type. Later, it becomes cellular and stores fatty food reserve.

The division of zygote and its daughter cells is transverse, and the embryogeny probably conforms to the Caryophyllad or Solanad type (Fig. 21.3.1D–Z). In a mature seed, the unorganized (Fig. 21.3.2) embryo is only four-celled in *Mitrastemon*, and has ten cells in six tiers in *Cytinus* (see Davis 1966). In *Pilostyles*, the embryo consists of eight cells; two cells may be comparable to suspensor, and the embryo proper comprises three tiers of two cells each (see Kuijt 1969).

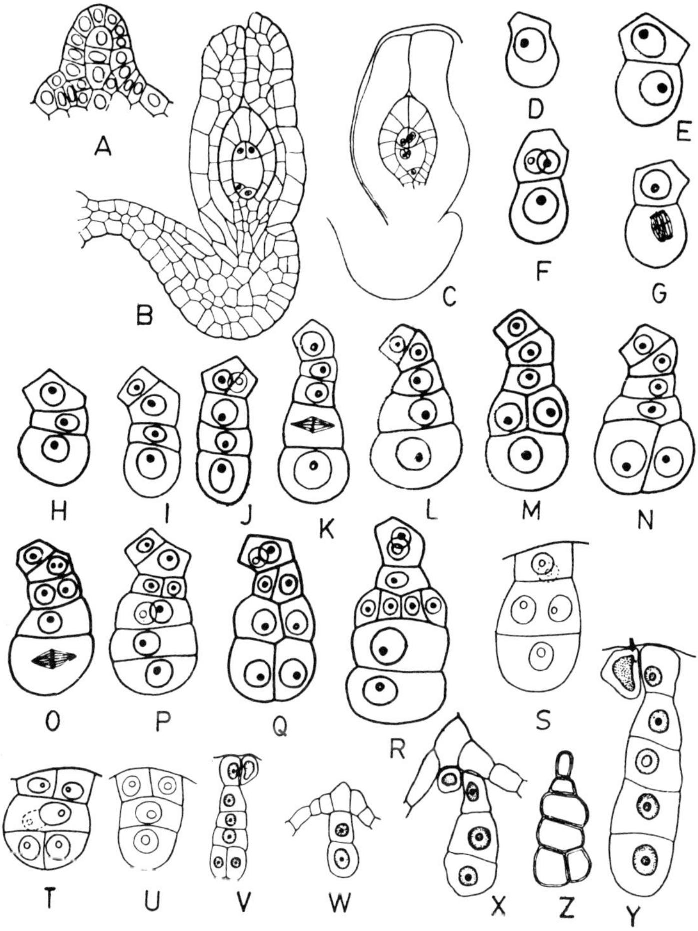

Fig. 21.3.1. A–C *Rafflesia patma*, longisection of ovules, mature embryo sac in **C**. **D–R** *Cytinus ruber*. **S–U** *R. patma*, **V** *R. hasseltii*, **W–Z** *Pilostyles ingae*, early embryogeny. (**A–C**, **S–V** After Ernst and Schmid 1913, for **D–R** see Yakovlev 1981, **W–Z** after Endriss 1902)

The minute seed is albuminous with only a few-celled embryo (Fig. 21.3.2). The seed-coat (outer epidermis of tegmen) is sclerotic with pitted cells. In *Rafflesia* the sclerenchymatous outermost layer of tegmen forms the seed-coat. The chalazal swelling also becomes sclerenchymatous. In *Cytinus*, the chalazal swelling is fleshy and may serve as an elaisome. In *Pilostyles*, the seeds have a single compact layer of sclereids surrounding the endosperm. The sclereids have very thick walls with a complex system of pits (see Kuijt 1969, Corner 1976).

The four tribes – Apodantheae, Rafflesieae, Mitrastemoneae, Cytineae – of Rafflesiaceae show striking differences in ovular structure, reflected in the

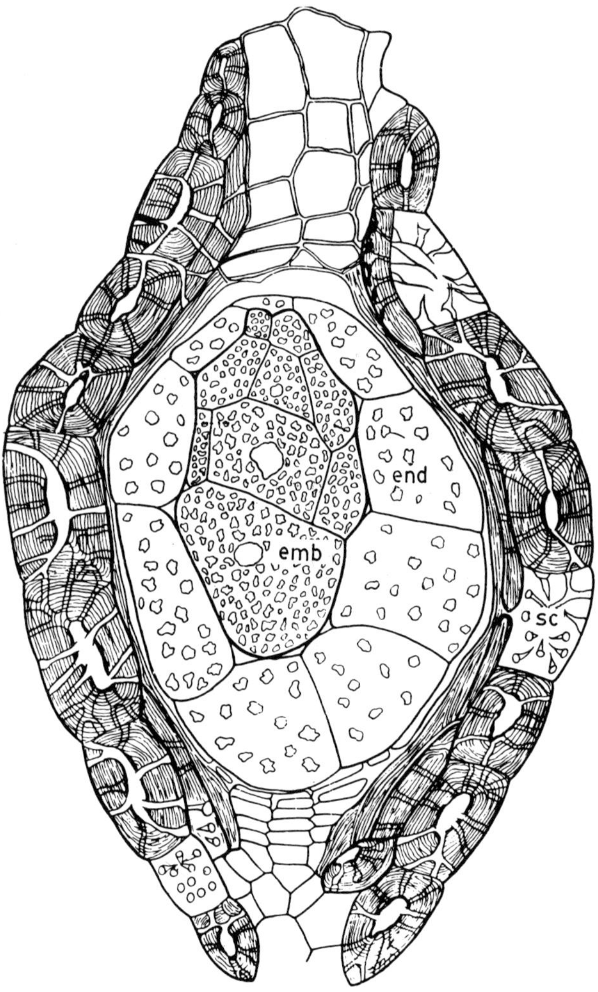

Fig. 21.3.2. *Cytinus ruber*. Longisection of seed with endosperm (*end*) and undifferentiated embryo (*emb*). The testa comprises sclerotic cells (*sc*) with strongly pitted walls. (See Yakovlev 1981)

mature seeds. These are always exotegmic but show differences in the type of cell wall thickenings (Bouman and Meijer 1987).

F Bouman and W Meijer (1987), EJH Corner (1976), GL Davis (1966), W Endriss (1902), A Ernst and E Schmid (1913), I Guzowska (1964), J Kuijt (1969), K Schnarf (1931), K Watanabe (1936), MS Yakovlev (1981)

21.4 Hydnoraceae

The development of anther, pollen grains, and endosperm has not been investigated.

The ovule is orthotropous, unitegmic, and tenuinucellate. The archesporial cell functions as megaspore mother cell. Development of the embryo sac is of the Allium type in *Prosopanche bertoniensis* (Chodat 1916), and tetrasporic Adoxa type in *Hydnora africana* (Dastur 1922). The polar nuclei fuse before fertilization, and the antipodals are ephemeral.

The embryogeny in *Hydnora africana* conforms to the Solanad type, but this needs confirmation (see Davis 1966).

The seed is minute with an inconspicuous embryo, copious endosperm (Hutchinson 1973), and a single layer of perisperm. The outer epidermis of seed is thin-walled, and the inner epidermis has thickened inner walls, except in *Hydnora*.

R Chodat (1916), AE Cocucci (1975, 1976), RH Dastur (1922), GL Davis (1966), J Hutchinson (1973), K Schnarf (1931)

21.5 Taxonomic Considerations

Hallier (1912) considers the origin of Aristolochiales from Lardizabalaceae (Ranales), and Wettstein (1935) from myrtaceous and annonaceous ancestors. Hutchinson (1973) derives it from herbaceous members of Berberidales (Ranales), and includes Aristolochiaceae, Hydnoraceae, Cytinaceae and Nepenthaceae in Aristolochiales. Bessey (1915) assigns Aristolochiaceae to Myrtales.

Cronquist (1968, 1981) places the order Aristolochiales in the subclass Magnoliidae, and R. Dahlgren (1980b, 1981) treats it in the superorder Magnoliiflorae. The presence of etherial oil cells, the uniaperturate pollen, and seeds with copious endosperm in Aristolochiaceae clearly indicate their Magnolian ancestry.

Aristolochiaceae. Aristolochiaceae show close affinities with Annonaceae: (1) secretory anther tapetum, (2) simultaneous or successive cytokinesis, (3) anatropous, bitegmic, and crassinucellate ovule with a massive parietal tissue, (4) Polygonum type of embryo sac, (5) Cellular endosperm which becomes ruminate due to infoldings of the seed-coat, Annona type in *Apama siliquosa* (Periasamy 1962a), and (6) albuminous seed.

Wettstein's view that they have annonaceous ancestors is generally accepted.

Rafflesiaceae, Hydnoraceae. Rafflesiaceae are allied to Hydnoraceae. Both have ana- or orthotropous (*Cytinus*), unitegmic (*Rafflesia*) and tenuinucellate ovules; the single archesporial cell functions directly as megaspore mother cell. The minute, albuminous seed has an undifferentiated embryo, and thin-walled or pulpy exotesta (absent in *Rafflesia*). Both families differ from Aristolochiaceae in triaperturate pollen grains, tenuinucellate ovule, bi- or tetrasporic embryo sac in Hydnoraceae, Nuclear endosperm in Rafflesiaceae, and absence of ruminations. The endotegmic seed of *Rafflesia* negates its affinities with that

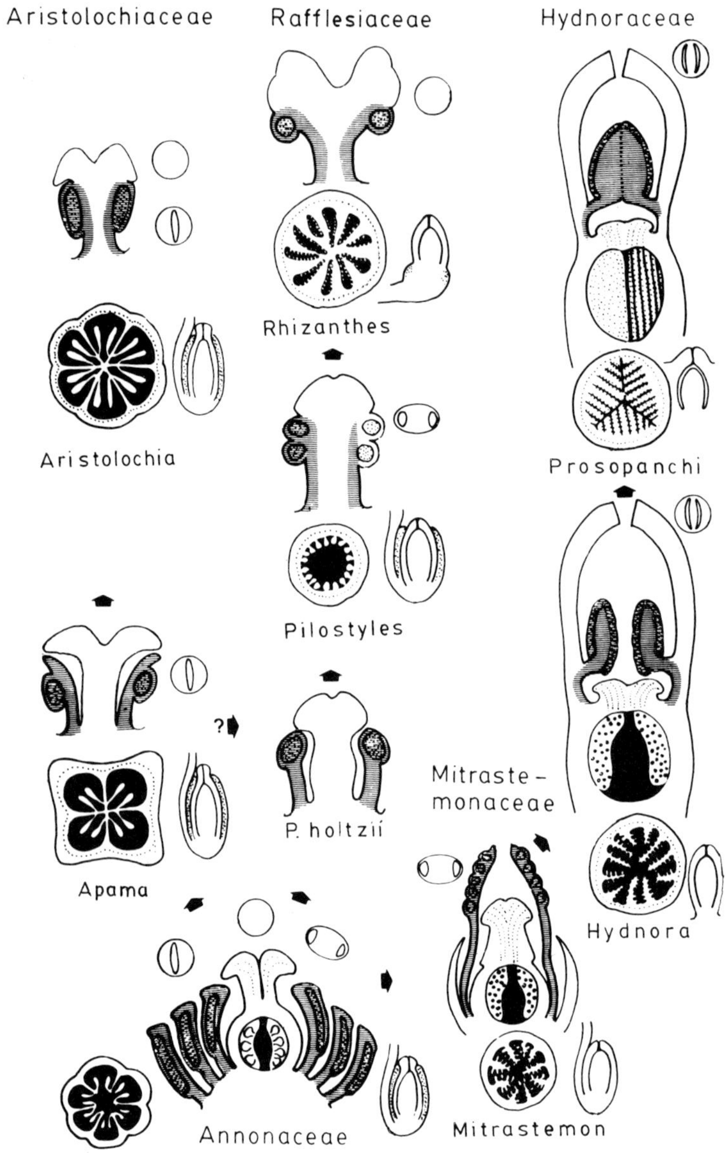

Fig. 21.5.1. Phylogenetic relationships between Annonaceae (embryo sac Polygonum type, endosperm Cellular type), Mitrastemonaceae (embryo sac Polygonum type, endosperm Cellular type), Aristolochiaceae (embryo sac Polygonum type, endosperm Cellular type), Rafflesiaceae (embryo sac Polygonum type, endosperm Cellular type), and Hydnoraceae (embryo sac Allium type in *Prosopanche* and Adoxa type in *Hydnora*, endosperm Cellular type). Each family is represented by a longisection of flower, or only the androecium (*shaded*), with style and stigma, transection of ovary, longisection of ovule, and the pollen type. (After Cocucci 1976, see also Cocucci 1983)

of Aristolochiaceae, while the exotegmic seed of *Apodanthus* shows relationship with Piperaceae and Saururaceae (Corner 1976).

Hydnoraceae can be singled out from the other families due to bisporic Allium, tetrasporic Adoxa type of embryo sac development, and perispermous seed.

R. Dahlgren (1980b, 1981) and Cronquist (1981) include Rafflesiaceae and Hydnoraceae in Rafflesiales (see also Kuijt 1969).

Hydnoraceae (Sect. 21.4) are related to Rafflesiaceae (21.3) and Aristolochiaceae (21.2). The floral and embryological characters indicate that the annonaceous stock may be considered to be the basis for three different evolutionary lines. One leads to Aristolochiaceae, the second to Rafflesiaceae, and the third to Hydnoraceae with Mitrastemonaceae (21.3) in an intermediate position (Fig. 21.5.1). We agree with Cocucci's (1983) interpretation.

22 Guttiferales

The order Guttiferales comprises sixteen families, distributed over four suborders: suborder Dilleniineae – Dilleniaceae, Paeoniaceae, Crossosomataceae, Eucryphiaceae, Medusagynaceae and Actinidiaceae; Ochnineae – Ochnaceae, Dioncophyllaceae, Strasburgeriaceae and Dipterocarpaceae; Theineae – Theaceae, Caryocaraceae, Marcgraviaceae, Quiinaceae and Guttiferae (Clusiaceae); Ancistrocladineae—Ancistrocladaceae.

The embryology of the families Eucryphiaceae and Medusagynaceae of Dilleniineae, Strasburgeriaceae of Ochnineae, Quiinaceae of Theineae, and Ancistrocladaceae of Ancistrocladineae has not been investigated. The development of the embryo of Ochnaceae and Marcgraviaceae either has not been fully worked out.

22.1 General Features

The order is characterized by (1) secretory anther tapetum, (2) simultaneous cytokinesis in pollen mother cells, (3) two-celled pollen grains at the dispersal stage; in Marcgraviaceae they germinate in situ, (4) anatropous, bitegmic ovules, unitegmic in Actinidiaceae, fusion of integuments at places in Marcgraviaceae, and appressed integuments in Dilleniaceae, (5) Polygonum type of embryo sac, Allium type in Theaceae, (6) Nuclear endosperm, Cellular type in Actinidiaceae and Marcgraviaceae, (7) zygotic mantle, and coenocytic aril in a member of Dilleniaceae, peculiar embryogeny in Paeoniaceae, Onagrad type in Dilleniaceae and Crossosomataceae, Solanad type in Actinidiaceae and Theaceae, and Asterad type in Dipterocarpaceae, and (8) pachychalazal seed in Ochnaceae, mesotestal in Theaceae, endotestal in Dilleniaceae, and exotestal in other members.

22.2 Dilleniaceae

The anther is tetrasporangiate. The anther wall comprises four layers in *Acrotrema* (Swamy and Periasamy 1955) and *Dillenia indica* (Sastri 1958b), and five layers in *Hibbertia* (Sastri 1958b). The epidermis is tannin-filled in *H. stricta* and *D. indica* (Sastri 1958b), and is a persistent and dominant layer in *Acrotrema*. The persistent epidermis in *Wormia burbidgei* (Nagaraja Rao 1961) shows fibrous thickenings. The endothecium develops fibrous thickenings in *H. stricta* (Sastri 1958b), but in *Acrotrema* and *H. billardieri* only a few cells acquire thickenings. They are totally absent in *Dillenia*. The one or two middle layers are ephemeral. The tapetum is secretory, and of binucleate cells in *Acrotrema*, *Dillenia* and *Hibbertia*. Amoeboid tapetum is reported in *Wormia suffruticosa* (Paetow 1931), which needs confirmation. Simultaneous cytokinesis by furrowing follows meiotic divisions in the microspore mother cells. The tetrads are tetrahedral, isobilateral, linear and rhomboidal.

Pollen grains are two-celled at the shedding stage; they are three colporate. A large protein crystal occurs in the pollen of *Wormia suffruticosa* (Paetow 1931), and disappears during its maturation.

The ovule is anatropous, bitegmic and crassinucellate. A small space is present between the closely appressed integuments in the micropylar region. The micropyle is zigzag, or closed by the exostome.

In *D. indica* the inner integument gets crushed soon after fertilization, while in *H. stricta* the outer epidermis of the inner integument becomes radially elongate and develops spiral thickenings in *Acrotrema*, *Hibbertia* and *Wormia*. Due to further curvature, after fertilization, the ovule becomes amphitropous (Sastri 1958b). A hypodermal archesporial cell cuts off a parietal cell which gives rise to a two- to three-layered parietal tissue. A nucellar cap has been noted. The megaspore mother cell produces a linear or T-shaped tetrad, and the chalazal megaspore gives rise to a Polygonum type of embryo sac. The synergids are elongated and hooked in *Hibbertia*, and pear-shaped in *Dillenia* (Nagaraja Rao 1955a). The antipodals are ephemeral.

The endosperm is of the Nuclear type, and becomes completely cellular due to centripetal wall formation at the globular stage of proembryo in *Hibbertia* (Sastri 1958b).

A unique feature (see Vijayaraghavan et al. 1990) is the presence of a zygotic mantle in the embryo sac (Fig. 22.2.1A). It is pronounced in *Acrotrema* and *Wormia*, but absent in *Dillenia*. The inner surface of the zygotic membrane becomes gelatinous and lamellated, and the outer surface swells and grows out in the form of a coralloid protuberance between the adjoining endosperm cells (Fig. 22.2.1B). In the mature seed most of the space is occupied by the endosperm, and the embryo is very small.

The embryogeny conforms to the Onagrad type.

The seed of *Acrotrema* is arillate and the aril develops from the funicle (Fig. 22.2.2A, B). During post-fertilization stages, the aril covers the seed partially (Fig. 22.2.2C, D). The embryo sac becomes curved due to the growth of raphe (Fig. 22.2.2C). To begin with, the cells of the aril are uninucleate

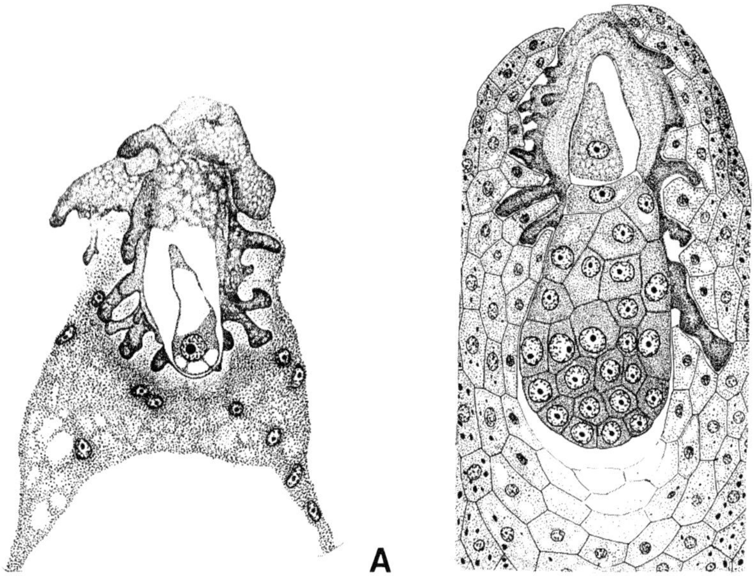

Fig. 22.2.1A–B. *Acrotrema arnottianum.* **A** Zygote, mantle with appendages, and nuclear endosperm. **B** Proembryo with one-celled suspensor, mantle, and cellular endosperm. (After Swamy and Periasamy 1955)

and tubular, then they elongate and become coenocytic (Fig. 22.2.2E–I). In *Hibbertia* the aril is rudimentary.

The albuminous seed is endotestal. At fertilization, the outer integument is two-layered, and the inner one three-layered. The outer epidermis of testa is tanniniferous, and becomes compressed in the mature seed. The cells of the inner epidermis of testa undergo sclerification, and show branched pits. The cells of the outer epidermis of tegmen elongate and become spirally thickened, but collapse in the mature seed. The inner epidermis of tegmen alone persists and becomes tanniniferous. The middle layer of tegmen, and nucellus also collapse.

JD Elizabeth and KK Lakshmanan (1984), JE Dulcy and KK Lakshmanan (1981, 1984), A Nagaraja Rao (1955a, 1957a, 1961), W Paetow (1931), RLN Sastri (1958b), BGL Swamy and K Periasamy (1955), MR Vijayaraghavan, Sudesh and Nidhi Gautham (1990)

22.3 Paeoniaceae

The anther wall comprises an epidermis, two to three layers of endothecium, one or two ephemeral middle layers, and secretory tapetum of binucleate cells. The tapetum becomes biseriate. The microspore mother cells undergo meiotic

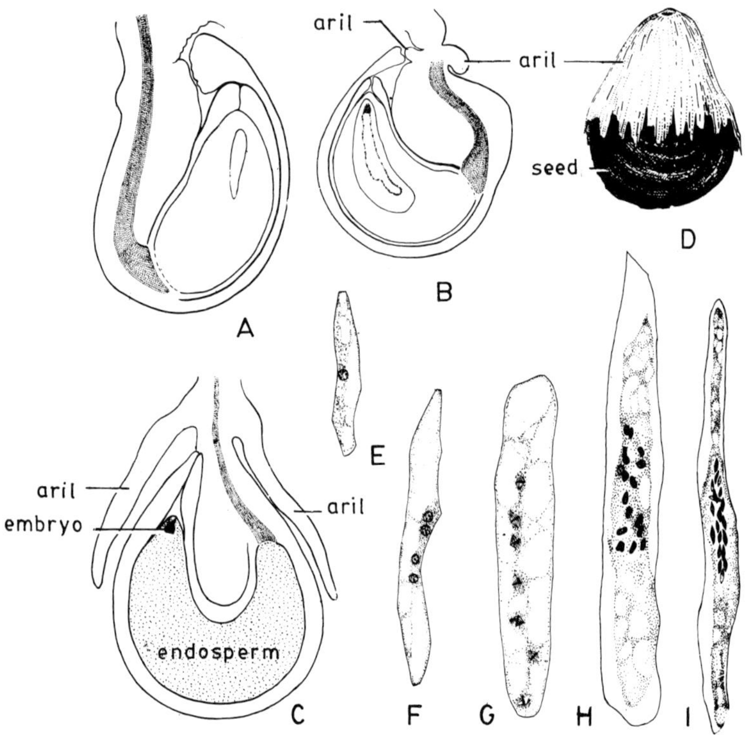

Fig. 22.2.2A–I. *Acrotrema arnottianum.* **A** Longisection of anatropous, bitegmic, arillate ovule. **B** At globular proembryo. The aril is funicular, and the endosperm of the Nuclear type. **C** Immature seed at heart-shaped embryo. **D** The aril covers the upper half of mature seed. **E** Uninucleate cell of aril. **F–I** Elongated coenocytic cells of aril. (After Swamy and Periasamy 1955)

divisions, cytokinesis is simultaneous, and the tetrads are tetrahedral and decussate.

The pollen grains are two-celled at the shedding stage. They are por(oid)ate, prolate and spheroidal. The sexine is thinner than nexine, and is reticulate.

The ovule is anatropous, bitegmic and crassinucellate. The outer integument is massive, and forms the micropyle. A multicellular archesporium differentiates in the hypodermal region in *Paeonia anomala* (Murgai 1962). The formation of a parietal layer is doubtful. The megaspore mother cell undergoes meiotic divisions and linear or T-shaped tetrad is produced. The chalazal megaspore produces a Polygonum type of embryo sac. The polar nuclei fuse in the centre of the embryo sac. The antipodals are large, persistent and multinucleolate in *P. delavayi* (Murgai 1962). They may undergo divisions and produce accessory cells. Formation of twin embryo sacs is reported in *P. tenuifolia* (Murgai 1962). Yakovlev and Yoffe (see Y.D. Tiagi 1970b) observed the concurrent formation of several embryo sacs, and their fusion. A prominent hypostase is present.

The nucellus is absorbed by the embryo sac even before pollination.

During double fertilization one of the synergids perishes (Y.D. Tiagi 1970b). According to Yakovlev and Yoffe (1957), both the synergids remain intact, which is doubtful.

Development of the endosperm is of the Nuclear type. The free nuclei in the chalazal region fuse, and form irregular polyploid masses. Wall formation is centripetal, and commences from the micropylar end at an early stage of proembryo. In *P. officinalis* (Murgai 1962), the free-nuclear chalazal part of the endosperm develops into a haustorium, when the proembryo has a coenocytic suspensor. The haustorium attains its maximum activity and becomes com-

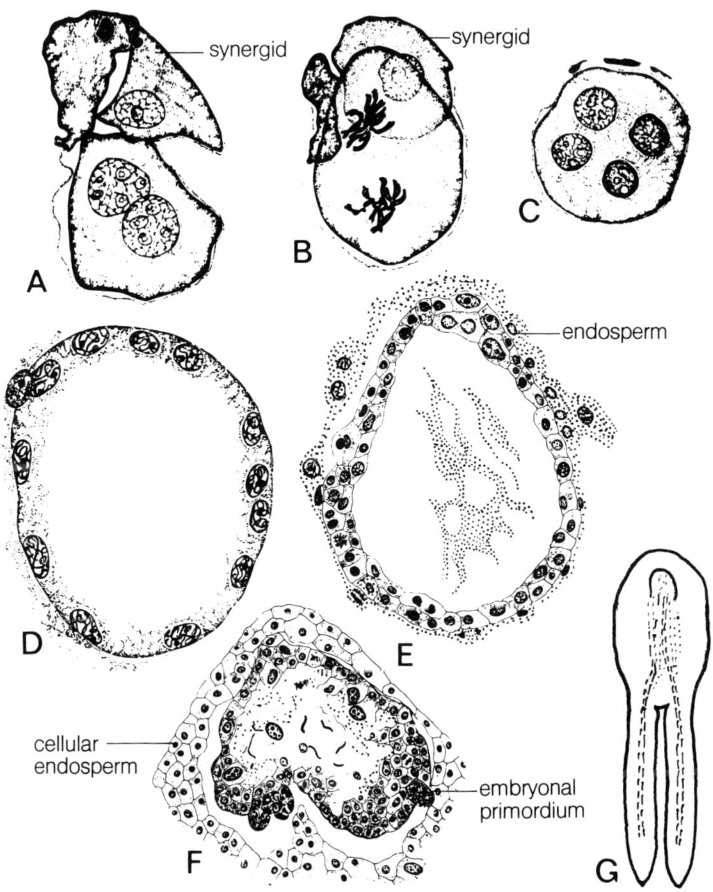

Fig. 22.3.1A–G. *Paeonia lactiflora.* **A** Two-nucleate, proembryo. **B** Both nuclei in division. **C** Four-nucleate proembryo. **D** Multinucleate proembryo with peripheral arrangement of nuclei. The dark nucleus in the upper region (*left*) is an endosperm (free-nuclear) nucleus. **E** Wall formation has taken place in the proembryo, the endosperm is still free-nuclear. **F** The proembryo undergoes active cell division, and embryonal primordia differentiate. Note the cellular endosperm. **G** Dicotyledonous embryo. (After Yakovlev 1969)

pletely cellular at the early globular stage of the proembryo. It degenerates at an advanced stage of globular proembryo. The endosperm is persistent, and stores fat and aleurone grains.

The embryogeny of *Paeonia* has become a controversial issue. The findings of Yakovlev and Yoffe (1957, 1961) have been supported by Cave et al. (1961), Matthiessen (1962) and Moscov (1964).

The nucleus of the zygote undergoes repeated free-nuclear divisions to form a coenocyte and the nuclei become distributed around a large central vacuole (Fig. 22.3.1A–D). Centripetal wall formation occurs producing a proembryonal mass (Fig. 22.3.1E). Several peripheral cells develop meristematic centres, and produce embryonal primordia (Fig. 22.3.1F). Only one of these primordia matures into a dicotyledonous embryo (Fig. 22.3.1G). Such a development is unparalleled in angiosperms, and reminds one of gymnosperms.

Contrary to the above mode, Murgai (1959, 1962) reported that the division of zygote is followed by a wall resulting in an apical and a basal cell. The apical cell may undergo one longitudinal division; the basal cell becomes coenocytic with a vacuole in the centre. It becomes cellular followed by proembryonal meristematic centres. Finally, only one of these matures into an embryo. Murgai's observations require confirmation.

The mature embryo is small and dicotyledonous surrounded by copious endosperm.

The testa is massive, and forms the seed-coat. Its outer epidermis and hypodermis are composed of thick-walled, tangentially elongate cells. The mesophyll is thin-walled, and contains starch. Its inner layers and tegmen get crushed. The inner epidermis of testa is thin-walled. A slightly vascularized aril develops at the base (Camp and Hubbard 1963).

WH Camp and MM Hubbard (1963), K Carniel (1967), MS Cave, HJ Arnott and SA Cook (1961), M Kumazawa (1935), A Matthiessen (1962), IV Moscov (1964), Mu-Xi-Jin and Wang Fu (1985a), P Murgai (1959, 1962), YD Tiagi (1970b), JL Walters (1962), WC Worsdell (1908), R Wunderlich (1966), MS Yakovlev (1951, 1959, 1961, 1965, 1967, 1969), MS Yakovlev and MD Yoffe (1957, 1961)

22.4 Crossosomataceae

A major contribution to the embryology of this family is by Kapil and Vani (1963) on *Crossosoma californicum*.

The anther wall comprises an epidermis, fibrous endothecium, two or three middle layers, and secretory tapetum. The endothecium and tapetum become irregularly two- to three-layered. The tapetal cells are two- to four-nucleate; the nuclei fuse and become polyploid. Simultaneous cytokinesis follows meiotic divisions of microspore mother cells, and tetrahedral and decussate tetrads are produced.

The pollen grains are tricolpate, and two-celled at the shedding stage.

The ovule is amphitropous, bitegmic and crassinucellar. The micropyle is organized by both the integuments. A hypostase differentiates in the chalazal region.

The hypodermal archesporium is multicellular. A massive parietal tissue is produced. The functional megaspore mother cell undergoes meiotic divisions and produces a linear tetrad. The chalazal megaspore develops into a Polygonum type of embryo sac. The antipodals divide and form five or six cells.

The entry of pollen tube is porogamous, and triple fusion precedes syngamy.

The development of the endosperm is of the Nuclear type. Enucleate cytoplasmic vesicles, and aggregation of nuclei in the chalazal region has been noted. Wall formation commences from the micropylar region at the globular proembryo, and proceeds to the chalazal region.

The embryogeny corresponds to the Onagrad type. The mature embryo is large and curved.

The seed is arillate, and thinly albuminous. The testa is highly sclerified. The inner epidermis of the inner integument develops fibrous thickenings. A fimbriate aril, surrounding the seed, develops from the funicle.

RN Kapil (1970a), RN Kapil and RS Vani (1963)

22.5 Eucryphiaceae

The embryology has not been studied.

22.6 Medusagynaceae

The embryological data are very meagre.

The anther opens longitudinally. The microspore tetrads are tetrahedral. The pollen grains are two-, three- or four-porate.

The anatropous ovule has a long funicle in *Medusagyne oppositifolia* (Engler and Melchior 1925).

The seeds are winged (Hutchinson 1973).

A Engler and K Melchior (1925), J Hutchinson (1973)

22.7 Actinidiaceae

The anther is tetrasporangiate, and its wall consists of the epidermis, fibrous endothecium, two or three ephemeral middle layers, and secretory tapetum of multinucleate cells. Due to fusions the tapetal nuclei become polyploid. Raphides are present in the anther wall and connective. Simultaneous cytokinesis, following meiotic divisions of microspore mother cells, results in tetrahedral tetrads.

The pollen grains are two-celled at the dispersal stage.

The ovule is anatropous, unitegmic and tenuinucellar. The nucellus is thin and ephemeral, and calcium oxalate crystals occur in the ovule. A persistent conspicuous endothelium and a well-developed hypostase also develop. Of the two or three hypodermal archesporial cells, one functions directly as megaspore mother cell. It undergoes meiotic divisions and gives rise to a linear tetrad. The chalazal megaspore develops into a Polygonum type of embryo sac in *Actinidia* (Vijayaraghavan 1965). The synergids and antipodals are ephemeral. Nagaraja Rao (1953a) reported inverted polarity in the embryo sac of *Saurauia nepalensis*.

Development of the endosperm is of the Cellular type.

The embryogeny follows the Solanad type. Additional embryos are produced due to the proliferation of suspensor cells in *A. chinensis* (Crété 1944).

The seed is arillate, exalbuminous, and exotestal. The inner walls of the cells are highly thickened and lignified; the upper part of the cell contains tannin and a prominent vacuole. The endothelium appears as a compressed layer.

DI Brundell (1975), P Crété (1944), WC Dickison (1972), CF Harvey and LG Fraser (1988), ME Hopping (1976), ME Hopping and EM Jerram (1979), HL Li (1952), A Nagaraja Rao (1953a), R Schmid (1978a, b), R Souèges (1943b), Ph van Tieghem (1899), MR Vijayaraghavan (1965, 1970b), J White (1990)

22.8 Ochnaceae

The embryogeny has not been investigated.

The anther wall comprises the epidermis, fibrous endothecium, an ephemeral middle layer, and secretory tapetum of binucleate cells in *Ochna atropurpurea* and *O. kirkii* (L.L. Narayana 1975). The microspore mother cells undergo meiotic divisions, and produce tetrahedral tetrads. Cytokinesis is simultaneous.

The pollen grains are three-colpate and two-celled at the shedding stage.

The ovule is anatropous, bitegmic, and tenuinucellate. The integuments are free only at the micropyle. A thin-walled prominent hypostase develops. The hypodermal archesporial cell develops directly into the megaspore mother cell. An additional archesporial cell has been observed in the inner integument in *Ochna atropurpurea* (R.B. Ghosh 1964). Following meiotic divisions, the megaspore mother cell produces a linear or T-shaped tetrad. The chalazal megaspore develops into a Polygonum type of embryo sac (Fig. 22.8.1A–C, E). The formation of a "false" endothelium from the nucellar epidermis has been reported in *O. squarrosa* (Chikkannaiah 1954). The antipodals are large with hypertrophied nuclei.

Somatic apospory of the Hieracium type occurs in *O. serrulata* (Fig. 22.8.1 D, F–L); F, G, K, L show one, I two and J three aposporous "embryo sacs". Generally, somatic cells develop into an eight-nucleate embryo sac, but in *Ochna* the embryo sac shows fewer nuclei.

The development of the endosperm is of the Nuclear type; wall formation commences from the micropylar end. The endosperm is soon absorbed by the developing embryo.

The formation of a synergid embryo has been observed in *O. serrulata* (Chiarugi and Francini 1930). In some species integumentary embryos may also develop (Fig. 22.8.2A, B).

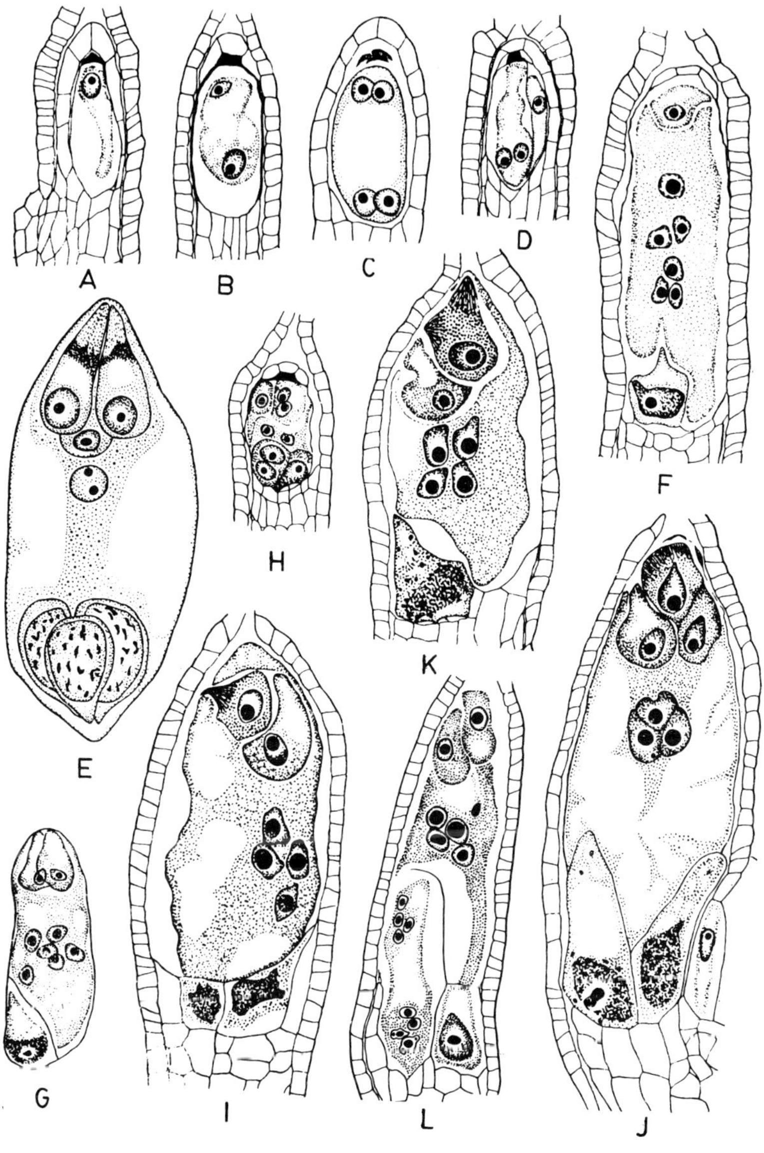

Fig. 22.8.1A–L. A, B, D, F–L *Ochna serrulata,* **C, E** *O. squarrosa.* **A** Chalazal functional megaspore. **B, C** Bi- and tetranucleate embryo sac. **D** Binucleate embryo sac and an aposporous embryo sac initial. **E** Mature embryo sac with hypertrophied antipodal nuclei. **F–H** Abnormal embryo sacs. **I** Zygote, persistent healthy synergid, and four-nucleate endosperm. Note one or two aposporous embryo sacs in **F, G** and **I–L. J** Embryo sac with zygote and two egg-like synergids, and primary endosperm nucleus (?). One cell of "endothelium" has differentiated like a functional megaspore. **K** Zygote with hypertrophied synergid, and four-nucleate endosperm. **L** Three embryo sacs enclosed within a common endothelium. One uninucleate, other unorganized eight-nucleate (two lower nuclei tri-nucleolate) and one embryo sac with zygote, synergid and four-nucleate endosperm. (**A, B, D, F–L** After Chiarugi and Francini 1930, **C, E** after Chikkanaiah 1954)

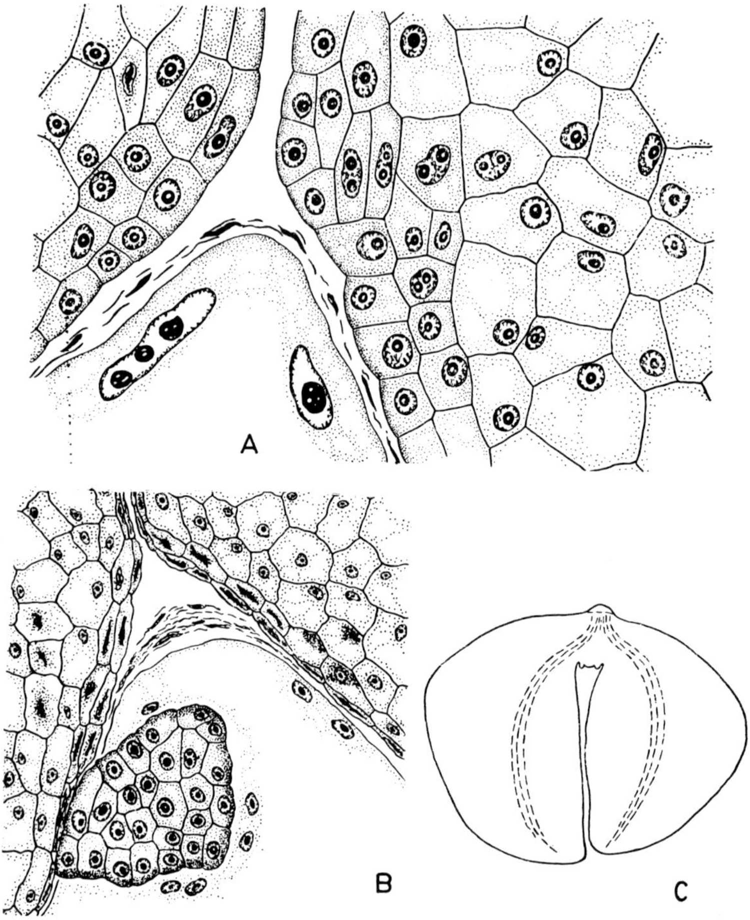

Fig. 22.8.2A–C. A, C *Ochna multiflora*, **B** *O. atropurpurea*. **A** Micropylar part of embryo sac with free endosperm nuclei; the inner integumentary cells are richly cytoplasmic. **B** Development of integumentary embryo, lying inside the embryo sac surrounded by free endosperm nuclei **C** Mature embryo. (See Yakovlev 1983)

The mature embryo is straight with thick cotyledons, and a short radicle (Fig. 22.8.2C).

An exalbuminous pachychalazal seed is enclosed in the sclerotic endocarp of a drupe. The massive seed-coat of 14–18 cell-layers is highly aerenchymatous, thin-walled and vascularized. It gets crushed during further development, except for the outer epidermis which is brownish and membraneous.

A Chiarugi and E Francini (1930), PS Chikkannaiah (1954), EJH Corner (1976), RB Ghosh (1964), LL Narayana (1975)

22.9 Dioncophyllaceae

The embryology has not been studied in sufficient detail.

The anthers are tetralocular, introse and open longitudinally. The pollen grains are three- to four-colporate.

The ovule is anatropous, bitegmic, crassinucellate, with the inner integument longer than the outer (Fig. 22.9.1A–C). The micropyle is formed by the inner integument, which has a beak-shaped process at the apex. The outer integument forms a cover (Fig. 22.9.1D, E).

The archesporial cell cuts off a parietal cell, the parietal tissue is two-layered. In *Triphyophyllum peltatum* (Schmid 1964), the megaspore mother cell undergoes meiotic divisions, and the chalazal megaspore of a linear tetrad develops into Polygonum type of embryo sac (Fig. 22.9.1F–H).

The large circular winged seed has a large discoid-obconic dicotyledonous embryo (Fig. 22.9.1I–K) and copious starchy endosperm in *Triphyophyllum* (Cronquist 1981).

A Cronquist (1981), R Schmid (1964)

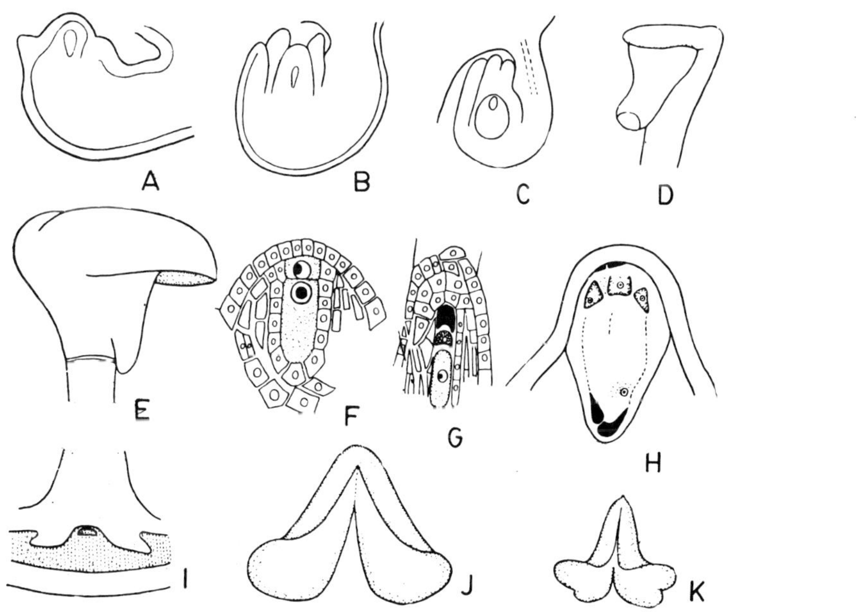

Fig. 22.9.1A–K. *Triphyophyllum peltatum.* **A** Longisection of ovule with initials of integuments and archesporial cell. **B** The integuments have developed further. **C** Micropyle formed by inner integument. **D, E** Outer integument forms a cover. **F** Nucellus with megaspore mother cell and parietal cell. **G** Linear tetrad with chalazal megaspore functional. **H** Embryo sac with egg apparatus and secondary nucleus, antipodals have degenerated. **I** Portion of embryo with endosperm (diagrammatic). **J, K** Dicotyledonous embryo with thin, funnel-shaped cotyledons. The margins of cotyledons have bifurcated in **K**. (After Schmid 1964)

22.10 Strasburgeriaceae

The embryology has not been studied.

22.11 Dipterocarpaceae

The anther comprises cutinized epidermis, endothecium, three ephemeral middle layers, and glandular tapetum of binucleate cells. The endothecium is devoid of fibrous thickenings in *Shorea talura, S. robusta, Vateria indica* and *Hopea racophloea* (Nagaraja Rao 1953c, 1955b, c, 1956, 1962). The dehiscence of anther takes place due to disorganization of cells between the two loculi. The inner walls of the tapetal cells are studded with globular bodies. The microspore mother cells undergo meiotic divisions and produce tetrahedral, isobilateral and decussate tetrads. Quadripartition of the tetrads takes place by centripetal furrowing.

The pollen grains are three-colpate, and shed at the two-celled stage (A. Kaur et al. 1986).

The ovule is anatropous, bitegmic, and crassinucellate. The inner integument forms the micropyle (Fig. 22.11.1A), and is fairly broad at the micropylar end; the outer integument contains a vascular strand.

A hypodermal archesporial cell cuts off a parietal cell. The parietal tissue is well developed. The megaspore mother cell undergoes meiotic divisions, and produces a linear or T-shaped tetrad. The chalazal megaspore develops into a Polygonum type of embryo sac. The antipodals persist up to fertilization. The polar nuclei fuse in the centre of the embryo sac.

The entry of pollen tube is porogamous, and one of the synergids is destroyed.

The development of endosperm is of the Nuclear type (Fig. 22.11.1A). The free-nuclear endosperm, in the lower half of the embryo sac, is aggressive and absorbs nourishment from the adjoining nucellar and integumentary tissue. During later stages, the endosperm consumes the tissue of seed-coat, irregularly leading to a lobed appearance (Fig. 22.11.1B–I). Some of the endosperm nuclei lie in these lobes (Fig. 22.11.1D–H). The chalazal region of ovule shows meristematic activity (Fig. 22.11.1G–I). Further studies may reveal some more interesting features.

Wall formation is initiated in the micropylar region at an early stage of proembryo (Fig. 22.11.1A), and proceeds towards the chalazal region. In *Vateria* nucleated endosperm nodules are formed. The lower part of the embryo sac remains free-nuclear for a long time. Finally, it becomes completely cellular. The mature endosperm is fatty.

The zygote divides transversely forming an apical and a basal cell. Vertical divisions occur in both cells resulting in a globular tetrad. Further divisions are irregular. The embryogeny is of the Solanad type, Chenopodiad in *Shorea macroptera* (A. Kaur et al. 1986). The mature embryo is well-differentiated, and broad with unequal fleshy cotyledons enclosing the radicle. Polyembryony is reported in *S. robusta* (R.B. Ghosh and Shahi 1957).

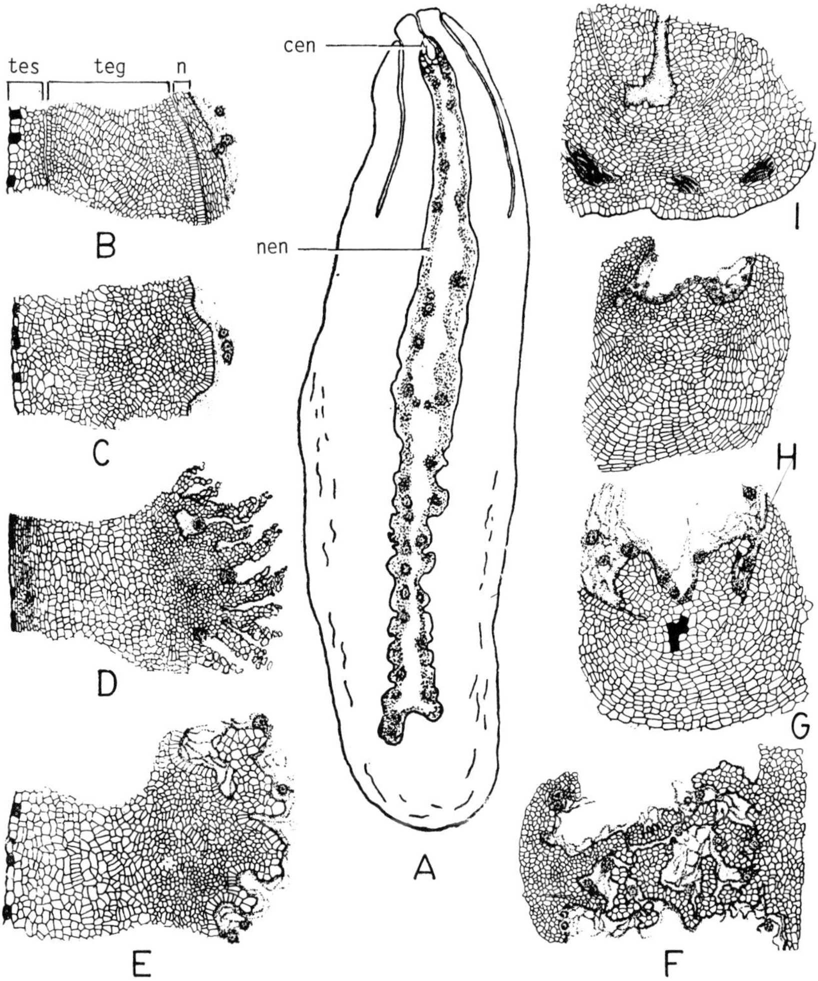

Fig. 22.11.1A–I. *Vateria indica.* **A** Longisection of bitegmic ovule, the two integuments are separate only in the upper region. The embryo sac has globular proembryo and cellular endosperm (*cen*) around it, nuclear endosperm (*nen*) in the rest of the embryo sac. The nuclear endosperm invaginates into the integument. **B** Portion of seed-coat to show testa (*tes*), tegmen (*teg*), nucellus (*n*), and free-nuclear endosperm. **C** Portion of seed-coat in the fused region of testa and tegmen, the nucellus has collapsed, endosperm nuclear. **D**, **E** Invaginations of nuclear endosperm into seed-coat, endosperm nuclei lie in the lobes of seed-coat. **F** Surface view of seed to show the invaginations of endosperm in seed-coat. **G**–**I** Basal region of ovule; stages of invagination of endosperm in the chalazal tissue. (After Nagaraja Rao 1955d)

The seed is exalbuminous, and exotegmic. Otherwise, the seed-coat is unspecialized and appears thin and papery. The exotegmen is sclerotic. The woody capsule protects the seeds.

RB Ghosh and R Shahi (1957), A Kaur, K Jong, VE Sands and E Soepadmo (1986), G Maury (1980), A Nagaraja Rao (1953c, 1955b, c, d, 1956, 1962), K Swarupanandan (1986)

22.12 Theaceae

The anther wall comprises an epidermis, fibrous endothecium, two ephemeral middle layers, and secretory tapetum of binucleate cells. Crystals (druses) are present in the connective region in *Ploiarium alternifolium* (N. Prakash and Lau 1976). Simultaneous cytokinesis follows meiotic divisions of microspore mother cells, and results in the formation of tetrahedral, isobilateral and decussate tetrads.

The pollen grains are three-colpate, and shed at the two-celled stage. They are tricolporate, subporolate, reticulate and heterobrochate in *Camellia sasanqua* (Mathew 1978).

The ovule is anatropous, bitegmic, and tenuinucellate. The inner integument alone forms the micropyle. The vascular supply reaches up to the tip of the outer integument in *Camellia sinensis* (Kapil and Sethi 1963). A hypostase has been observed.

A hypodermal archesporial cell (or two- to three-celled archesporium) develops into the megaspore mother cell. The latter undergoes meiotic divisions, and dyads are produced. In *Thea* (Mikatadze 1975), the micropylar dyad may develop up to the binucleate embryo sac. The chalazal dyad functions and the development of female gametophyte is of the Allium type in *Camellia* (Kapil and Sethi 1963, Mathew 1978) and *Thea* (Fagerlind 1939b, Mikatadze 1975). The synergids are ephemeral and exhibit hooks or hood-like expansions in *C. sasanqua* (Mathew 1978). The antipodals may persist up to or even after fertilization.

In *Ploiarium alternifolium* (N. Prakash and Lau 1976) the megaspore tetrad is either linear or T-shaped. Sometimes, the upper dyad cell degenerates before its division. The chalazal megaspore is functional and gives rise to a Polygonum type of embryo sac. The antipodals do not organize into cells and degenerate before the fusion of polar nuclei or soon after it.

Fertilization is porogamous, and the remnants of pollen tube persist after fertilization.

The development of endosperm is of the Nuclear type, later it becomes cellular. The mature seed is exalbuminous.

Development of the embryo corresponds to the Solanad type. The cotyledons show backward growth on either side of the radicle (Sethi 1965). Adventitious embryos originate from the placental cells, as well as the inner integument (Cavara 1899). Cavara also mentions the formation of twin embryos from the same egg apparatus. Supernumerary embryos and twin seedlings have been observed by Venkataramani (1950) and Wu (1960). Placental origin of embryos is doubtful; development of integumentary embryos need further study.

In a mature seed of *Ploiarium* the embryo is straight, narrow, cylindrical and dicotyledonous with a short uniseriate suspensor.

The seed is mesotestal. The testa alone forms the seed-coat. The cells of the outer epidermis are thin-walled with lignified inner walls. The cells of outer zone, except the outer epidermis, undergo sclerification and develop into a hard shell. The inner zone is slightly thickened, and contains tannin. The tegmen gets crushed.

In *P. alternifolium* (Corner 1976), the cells of the second layer of seed-coat are lignified, undulate and oblong, but whether they belong to endotesta or exotegmen is uncertain.

SB Bawa (1970a), F Cavara (1899), EJH Corner (1976), F Fagerlind (1939b), CE Kobuski (1948), RN Kapil and SB Sethi (1963), CJ Mathew (1978), TA Mikatadze (1975), N Prakash and YY Lau (1976), SB Sethi (1965), KS Venkataramani (1950), HK Wu (1960)

22.13 Caryocaraceae

The embryology has not been fully studied.
The anther is tetralocular and opens longitudinally.
The ovule is anatropous and bitegmic.
The endosperm, in mature seed, is scanty with two or three layers of thin-walled cells near the chalaza. The cells contain fats.

The embryo of *Caryocar glabrum* (Pilger 1925) has a strongly thickened hypocotyl which stores proteins and fats. The embryo of *Anthodiscus* is thin, long and spirally coiled. The cotyledons are small and hooked.

The seed-coat is formed by thin-walled aerenchymatous testa in *Caryocar*. The mesotesta and tegmen collapse; the mesocarp and endocarp give protection.

EJH Corner (1976), R Pilger (1925)

22.14 Marcgraviaceae

The embryogeny has not been investigated.

The major contribution to the embryology of Marcgraviaceae is that of Swamy (1948b) on *Marcgravia cuyuniensis* and *M. purpurea.*

The epidermis of the anther wall is persistent, and tanniniferous. The endothecium shows fibrous thickenings, and is followed by two or three ephemeral middle layers. The tapetum is secretory, and consists of uninucleate cells. Cytokinesis is simultaneous.

The pollen grains are three-colpate, and germinate in situ.

The ovule is anatropous, bitegmic, and tenuinucellar. The integuments are two layers thick, and the inner integument forms the micropyle. The outer epidermis of the outer integument contains phenol-like substances. The integuments show signs of fusion at varying levels from the base. The nucellus is highly reduced (Fig. 22.14.1A–C, G).

The hypodermal archesporial cell functions directly as megaspore mother cell. Following meiotic divisions it produces a linear or T-shaped tetrad. The chalazal megaspore develops into a Polygonum type of embryo sac. The

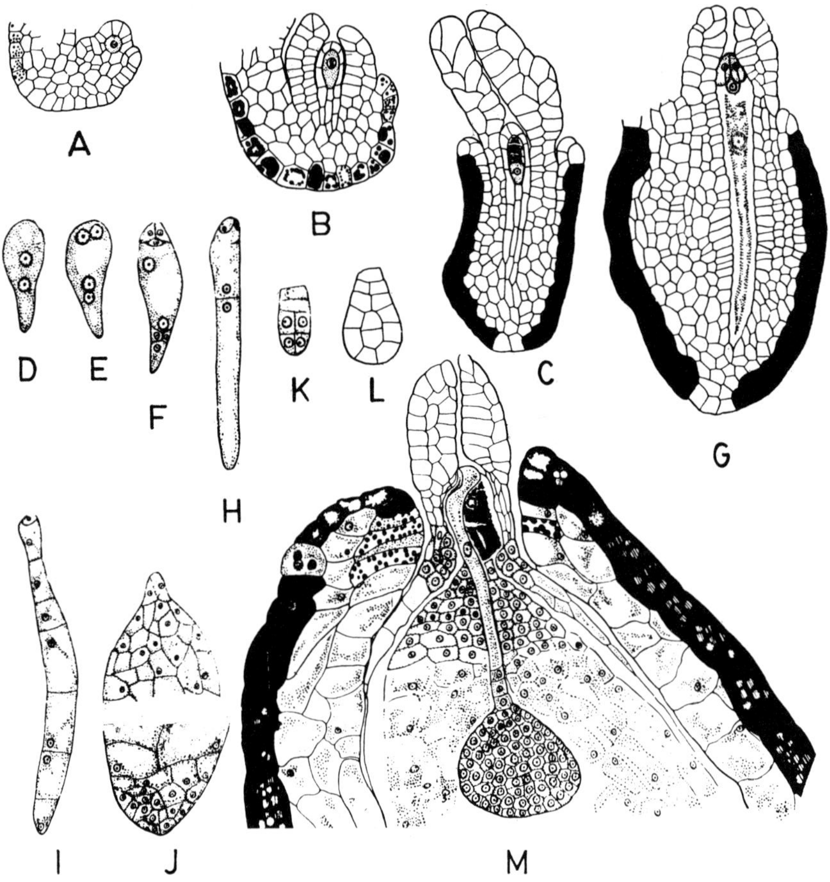

Fig. 22.14.1A–M. A–C, G, M *Marcgravia purpurea*, **D–F, H–L** *M. umbellata*. **A, B** Longisections of ovule at hypodermal archesporial and megaspore mother cell. **C** Longisection of bitegmic ovule at linear tetrad. Note the endostome. **D–F** Two-, four- and eight-nucleate embryo sac. **G** Longisection of ovule at mature embryo sac, the antipodals have degenerated. **H** Zygote and two-celled endosperm. **I** Zygote and uniseriate endosperm. **J** Cellular endosperm. **K–L** Proembryo. **M** Longisection of micropylar portion of bitegmic ovule at globular proembryo with a long suspensor, endosperm cellular. (**A–C, G, M** After Swamy 1948b, **D–F, H–L** after Mauritzon 1939a)

antipodals are ephemeral. The antipodal end of the gametophyte extends in the chalaza. An incipient endothelium develops in this region (Fig. 22.14.1A–G).

The endosperm is of the Cellular type. The first few divisions are transverse resulting in an eight-celled uniseriate tissue. Longitudinal divisions occur later, and the cell at the extreme tip of cellular endosperm, towards the micropyle, develops into a non-aggressive haustorium (Fig. 22.14.1H, I, J, M).

Figure 22.14.1K, L shows early proembryos. The terminal cell of the suspensor develops into a tube, and the tip of the tube comes in contact with the endosperm haustorium (Fig. 22.14.1M).

The seed is arillate, and slightly albuminous. The outer epidermis contains orange-brown deposits, and the rest of the seed-coat collapses.

J Mauritzon (1939a), BGL Swamy (1948b)

22.15 Quiinaceae

The embryology has not been investigated.

22.16 Guttiferae (Clusiaceae)

The anther wall comprises the epidermis containing darkly stained bodies, fibrous endothecium, two ephemeral middle layers, and glandular tapetum of binucleate cells. The microspore mother cells undergo meiotic divisions giving rise to tetrahedral, isobilateral, and decussate tetrads. Cytokinesis takes place by centripetal cleavage furrows.

The pollen grains are tricolpate, and shed at the two-celled stage in *Hypericum* (Nagaraja Rao 1957b).

The ovule is anatropous, bitegmic, and tenuinucellate. The micropyle is organized by both the integuments. The hypodermal archesporial cell functions directly as megaspore mother cell (Fig. 22.16.1A–C). Following meiotic divisions, it gives rise to a linear tetrad. The chalazal megaspore develops into a Polygonum type of embryo sac (Fig. 22.16.1D–F) in *Garcinia mangostina* (Sprecher 1919), *G. livistonia* (V. Puri 1939b) and *Hypericum* (Schnarf 1914, Nagaraja Rao 1957b).

Sometimes, two embryo sacs develop concurrently (Fig. 22.16.1E), due to the presence of two megaspore mother cells (V. Puri 1939b, Nagaraja Rao 1957b).

The antipodals are ephemeral; their multiplication is reported in *Hypericum gentianoides* and *H. punctatum* (see Davis 1966).

The development of endosperm (Fig. 22.16.1G, R) is of the Nuclear type (Stenar 1938a, Swamy 1946a), and not Helobial type as reported by Palm (1922). A chalazal "cyst" is formed at eight- to 16-nucleate stage of endosperm (Fig. 22.16.1H), but it gradually merges with the general endosperm. Wall formation is initiated only when the cotyledons differentiate in the embryo.

Autonomous development of endosperm has been observed in *H. virginicum* (Myers 1964). A partition wall develops following division of the primary endosperm nucleus. It suggests Helobial type. The development of endosperm requires further study.

The embryogeny conforms to the Solanad type. In *Hypericum* (Nagaraja Rao 1957b), it conforms to Megarchetype IV and Period II of Souèges system

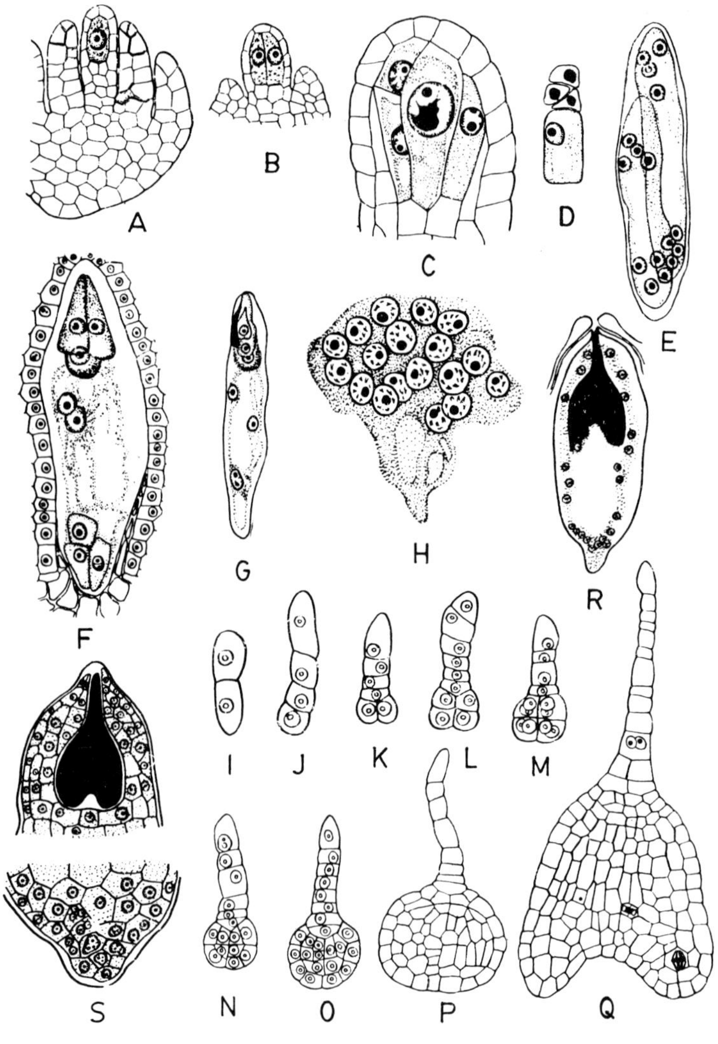

Fig. 22.16.1A–S. A, B, D–G, I–O, R, S *Hypericum patulum,* **C** *Garcinia livingstonii,* **H, P, Q** *H. mysorense.* **A** Bitegmic ovule with megaspore mother cell. **B** Two megaspore mother cells. **C** Two megaspore mother cells and a dyad. **D** Tetrad of megaspores. **E** Twin embryo sacs, each with eight nuclei. **F** Mature embryo sac, and endothelium. **G** Embryo sac with two-celled proembryo, remnants of pollen tube, and nuclear endosperm. **H** Aggregation of endosperm nuclei in the chalazal region of the embryo sac. **I–Q** Development of embryo. **R** Early dicotyledonous embryo (*black*) and free-nuclear endosperm. **S** Endosperm cellular. (**A, B, D–S** After Nagaraja Rao 1957b, **C** after V. Puri 1939b)

of classification (Fig. 22.16.1I–P). Later stages of embryo are shown in Figure 22.16.1Q–S.

Polyembryony, either from synergid by apogamy, or by the proliferation of suspensor cells, is reported in *H. tetrapterum* (Bugnicourt 1969). Myers (1964) described apomictic embryo sacs from the aposporic nucellar cells in *H. virginicum*. Parthenocarpic fruits have been observed in *Garcinia mangostina* (Horn 1940). The fruit contains seed-like structures which may or may not contain the adventive embryo.

The seed is exalbuminous, and arillate with a large embryo in Clusieae. It is exotestal in Clusieae, mesotestal in Callophylleae, and the testa is unspecialized in Garcinieae. In *Hypericum* the seed is exarillate, thinly albuminous, and exotegmic. The testa is represented by the outer epidermis alone. The cells have slightly thickened radial and inner walls, and contain tannin. The exotegmen is thick-walled with stellate-undulate facets. The mesophyll, and inner epidermis are unspecialized.

M Bugnicourt (1969, 1971a, b), KVO Dahlgren (1923), GL Davis (1966), DA Govindappa (1956a), CL Horn (1940), O Myers Jr (1964), A Nagaraja Rao (1957b), B Palm (1922), V Puri (1939b), K Schnarf (1914), YN Seetharam (1989), R Souèges (1925a), A Sprecher (1919), H Stenar (1938a), BGL Swamy (1946a)

22.17 Ancistrocladaceae

The embryology has not been studied in detail.

The anther is tetralocular, introse and opens longitudinally in *Ancistrocladus tectorius* (Keng 1967). The anther wall comprises the epidermis, fibrous endothecium with very large cells, and two ephemeral middle layers (Keng 1967).

The solitary ovule has a short funicle. It is hemianatropous and bitegmic. The outer integument is fairly thick, and the inner integument has only two to three layers.

The development of the endosperm is of the Cellular type. It is starchy and ruminate (Netolitzky 1926).

The mature embryo is relatively small and straight. The testa intrudes between the folds of the cotyledons which enclose the radicle (Hutchinson 1973).

EJH Corner (1976), J Hutchinson (1973), H Keng (1967), F Netolitzky (1926)

22.18 Taxonomic Considerations

The suborder Dilleniineae includes the families Dilleniaceae, Paeoniaceae, Crossosomataceae, Eucryphiaceae, Medusagynaceae and Actinidiaceae.

Bentham and Hooker (1965a), Hallier (1912) and Bessey (1915) include Dilleniaceae in Ranales. The genus *Paeonia* has been placed in Ranunculaceae. Hutchinson (1973) places Dilleniaceae in Dilleniales, and Paeoniaceae in Ranales. Bentham and Hooker treat Crossosomataceae as a member of Dilleniaceae. Engler and Prantl (1897) assign the genus *Crossosoma* to a separate family, Crossosomataceae, in Rosales. Bessey (1915), Wettstein (1935),

Gundersen (1950), Lawrence (1951), Benson (1962), Hutchinson (1973) and Takhtajan (1980) support this view. Lemesle (1948, 1955) proposed the transfer of Crossosomataceae to an intermediate position between Paeoniaceae and Ranunculaceae. Cronquist (1981) places Dilleniaceae and Paeoniaceae in the order Dilleniales, and Crossosomataceae in Rosales.

All the three families – Dilleniaceae, Paeoniaceae and Crossosomataceae – have several common features, such as development of pollen grains, bitegmic, crassinucellate ovule with the micropyle formed by both the integuments, Polygonum type of embryo sac, Nuclear endosperm, and albuminous arillate seeds (Table 22.18.1).

Dilleniaceae. Dilleniaceae can be distinguished from the others due to the absence of fibrous thickenings in endothecium (anther wall), and differentiation of hypostase and endothelium. In the lower region the integuments are closely appressed to each other, but the micropyle is quite wide (see Sastri 1958b). They also have a nucellar cap, a single-celled archesporium, and ephemeral antipodals. The occurrence of the zygotic mantle and coenocytic aril in *Acrotrema* are noteworthy features. The embryogeny is probably of the Onagrad type, and the mature embryo is small.

Paeoniaceae. Paeoniaceae and Dilleniaceae have centrifugal stamens, scalariform vessels, abundant wood rays, a massive bitegmic ovule with the outer integument longer than the inner one and exotestal seed. Corner (1976) favours its assignment close to Dilleniaceae. On the basis of wood anatomy, Keefe and Moseley (1978) support its inclusion in Dilleniales.

On the basis of vascular anatomy (Worsdell 1908), wood anatomy (Kumazawa 1935), palynology (Wodehouse 1936), cytology (Gregory 1941), and floral anatomy (Eames 1951) it has been suggested that *Paeonia* should be assigned to a separate family, Paeoniaceae.

R. Dahlgren (1981) includes Paeoniaceae in Paeoniales, close to Dilleniaceae in the Dilleniales. Takhtajan (1980) agrees with both these orders except that he also includes Crossosomataceae in the Dilleniales. Cronquist (1981) includes both Dilleniaceae and Paeoniaceae in Dilleniales. Y.D. Tiagi (1970b) emphasizes that while *Paeonia* differs from Ranunculaceae, it resembles Dilleniaceae and Crossosomataceae (Table 22.18.1). Willis (1973) states that *Paeonia* is related to Ranunculaceae. Hutchinson (1973) refutes any relationship of *Paeonia* with the Dilleniaceae. He considers it to be much more related to Helleboraceae (separated from Ranunculaceae) than to the Magnoliaceae.

The justification for an independent family, the Paeoniaceae, is provided by Subramanyam (1960) and B.M. Johri (1963b). The embryological features in which *Paeonia* differs from other members of the Ranunculaceae (Table 22.3.1) are: bitegmic ovule, massive outer integument longer than the inner, presence of hypostase, long embryo sacs (sometimes multiple gametophytes), persistent antipodals and their multiplication, chalazal endosperm haustorium, a very unusual mode of development of the embryo, and seed-coat derived from the testa alone.

A comparison of the embryology of *Paeonia*, Ranunculaceae, Dilleniaceae and Crossosomataceae (Table 22.18.1) indicates that Paeoniaceae are much more closely related to Ranunculaceae, than to the other two families. Moreover, Paeoniaceae are correctly placed under the Ranales (Hutchinson 1973), and there is no need to raise them to an ordinal rank, Paeoniales. Also, there is no justification to include the Paeoniaceae in the order Dilleniales.

Crossosomataceae. Crossosomataceae can be distinguished from Dilleniaceae (Table 22.18.1) on the basis of campylotropous ovule with multicellular archesporium, massive parietal tissue, hypostase, secondary multiplication of antipodals, and a large curved embryo (Kapil and Vani 1963). Nuclear endosperm shows enucleate vesicles, and aggregation of nuclei in the chalazal region. The seed has a fimbriate aril in *Crossosoma*. The embryogeny and seed structure are comparable to that in Dilleniaceae. The separation of *Crossosoma* from Dilleniaceae to a separate family Crossosomataceae is justified. Kapil and Vani (1963) recognize its separate family rank, and negate its relationship to Rosales or Ranales. The members of the order Ranales exhibit variations in the number of integuments, nature of ovule, types of embryo sacs, large, persistent polyploid antipodals, and occasionally irregular embryogeny. None of the above-mentioned three families show marked similarities with Ranales (Ranunculales of Cronquist 1981).

Actinidiaceae. Actinidiaceae have been treated previously as a component family of Dilleniaceae (see Lawrence 1951). Hutchinson (1973) considers them as a member of Theales. The genus *Saurauia* is treated as a separate unigeneric family (Saurauiaceae) in Theales by Hutchinson. Vijayaraghavan (1965, 1970b) points out that Actinidiaceae show marked differences from Dilleniaceae (Tables 22.18.1, 22.18.2). They have unitegmic, tenuinucellate ovules on axile placenta, raphides in ovary wall, persistent endothelium, hypostase, Cellular type of endosperm, absence of zygotic mantle, and Solanad type of embryogeny. These features support their separation from Dilleniaceae. He further considers the affinities with Marcgraviaceae, and suggests transfer to Theales. The ovules of *Marcgravia* show fusion of integuments at varying levels from base upward. Mauritzon (1939a) considers it as an approach to unitegmic condition in *Actinidia*. Both Actinidiaceae and Marcgraviaceae (Table 22.18.2) have secretory anther tapetum, ephemeral nucellus, Polygonum type of embryo sac, ephemeral antipodals, and Cellular type of endosperm. Centrifugal stamens, hypostase, and Solanad type of embryogeny occur in Theales. Therefore, the assignment of Actinidiaceae to Theales, as proposed by Vijayaraghavan (1970b), is appropriate.

Takhtajan (1980) and R. Dahlgren (1975a, 1981) include Actinidiaceae in Ericales. Poricidal dehiscence of anthers, unitegmic, tenuinucellate ovule with endothelium, Cellular type of endosperm, Solanad type of embryogeny, and presence of iridoids are some of the features allied to Ericales. But the remaining features enumerated by P. Maheshwari (1950) and Palser (1961) for Ericaceae are absent in Actinidiaceae.

Table 22.18.1. Embryology of *Paeonia*, Ranunculaceae, Dilleniaceae and Crossosomataceae

	Paeonia	Ranunculaceae	Dilleniaceae	Crossosomataceae
Stamen	Spirally arranged, centrifugal	Spirally arranged, centripetal	Numerous, not spirally arranged	Numerous, not spirally arranged
Anther tapetum	Irregularly 1- or 2-layered, glandular, cells binucleate	1-layered, glandular, cells uni- to multinucleate; sometimes amoeboid	1-layered, glandular, cells binucleate	Irregularly 2–3-layered, glandular, cells 2–4-nucleate
Pollen	2-celled	2-celled	2-celled	2-celled
Ovule	Anatropous, bitegmic, outer integument vascularized, crassinucellate, parietal tissue formed only by divisions in nucellar epidermis, nucellus absorbed by embryo sac even before pollination, hypostase present	Anatropous, uni-, bi- or ategmic, integuments non-vascularized, tenui- or crassinucellate, rarely nucellar epidermis divides and contributes to parietal tissue, hypostase rarely present	Ana- to amphitropous, bitegmic, integuments non-vascularized, crassinucellate, 2–3-layered parietal tissue, nucellar cap is formed	Amphitropous, bitegmic, integuments non-vascularized, crassinucellate, parietal tissue massive, hypostase present
Archesporium	Multicelled, several megaspore mother cells function	Uni- or multicelled, one cell functions	Unicelled	Multicelled, only one megaspore mother cell functions
Embryo sac	Polygonum type, twin to multiple embryo sacs may develop	Polygonum, rarely Allium type; tetrasporic Pyrethrum type in *Thalictrum angustifolium*	Polygonum type	Polygonum

Antipodal cells	Large and persistent, multinucleate, secondary multiplication occurs	Persistent, ephemeral in *Adonis*, each with one or more nuclei, become polyploid	Ephemeral	Divide and forms 5 or 6 cells
Endosperm	Nuclear type, chalazal haustorium develops	Nuclear, chalazal haustorium present (*Nigella*)	Nuclear, chalazal haustorium absent	Nuclear, enucleate cytoplasmic vesicles occur, chalazal haustorium absent
Embryogeny	The zygote undergoes free-nuclear divisions followed by centripetal wall formation and differentiation of several peripheral embryonal primordia, only one primordium develops into a dicotyledonous embryo	Onagrad type, rarely Solanad or irregular	Onagrad; a "mantle" develops around the zygote	Onagrad
Seed	Arillate, albuminous, testa massive and epidermis and hypodermis are thick-walled, mesophyll, inner epidermis and tegmen become crushed	Ex-arillate, albuminous in bitegmic ovules, testa formed chiefly by outer integument, in achenes seed-coat is crushed	Arillate, albuminous, exotestal	Arillate, thinly albuminous, exotestal
Fruit	Follicle	Achene	Dehiscent or baccate	Bivalved capsule

Table 22.18.2. Embryology of Actinidiaceae, Theaceae and Marcgraviaceae (See Mauritzon 1939a, Vijayaraghavan 1965)

	Actinidiaceae	Theaceae	Marcgraviaceae
Anther tapetum	Secretory, cells binucleate	Secretory, cells binucleate	Secretory, cells binucleate
Ovules	Numerous, anatropous, unitegmic, tenuinucellar, hypostase present	4 or 5, anatropous, bitegmic, tenuinucellar, micropyle formed by inner integument; hypostase present	Numerous, anatropous, bitegmic[a], tenuinucellar, micropyle formed by inner integument, hypostase absent
Embryo sac	Polygonum type	Allium	Polygonum
Antipodals	Degenerate before fertilization	Degenerate after fertilization	Antipodal nuclei (cells not formed) degenerate before fertilization
Endosperm	Cellular type	Nuclear	Cellular, the uppermost cell becomes conspicuous, penetrates into the micropyle and behaves as a haustorium
Embryogeny	Solanad type	Solanad	—
Polyembryony	Occurs	Occurs	—
Seed-coat	Exotestal, outer epidermis persists, cells filled with tannin, endothelium compressed	Mesotestal, outer integument persists, cells filled with tannin	Exotestal, outer epidermis of outer integument persists, cells filled with tannin

[a] According to Mauritzon (1939a), the two integuments of *Marcgravia* fuse to varying length from the base upward. He points out that this condition is an intermediate step in the derivation of unitegmic condition in *Actinidia*. Although the integuments arise distinctly and in succession, during later stages their identity at the chalazal half of the ovule becomes obliterated. This is partly due to addition of cells which increase their thickness, and partly to mutual pressure. The cells of the outermost layers of the outer integument become filled with orange-brownish contents (perhaps some phenolic compounds) at the time the embryo sac is ready for fertilization. The endothelium is restricted only to the chalazal half of the embryo sac.

The presence of unitegmic and tenuinucellate ovules, endothelium towards the chalazal end of the embryo sac, Polygonum type of embryo sac, small and ephemeral antipodal cells, and Cellular type of endosperm in *Saurauia* indicate its close affinity to Actinidiaceae. Its removal from Actinidiaceae to a separate family (Saurauiaceae) by Hutchinson (1973) does not seem proper on embryological grounds.

Of the ten families belonging to the last three suborders, embryological data are available for only five families: Ochnaceae, Dipterocarpaceae, Theaceae, Marcgraviaceae and Guttiferae. Bentham and Hooker (1965a) place Ochnaceae in Geraniales, and Dipterocarpaceae in Guttiferales along with Guttiferae and Hypericaceae. Hutchinson (1973) groups Ochnaceae and Dipterocarpaceae in Ochnales, Theaceae, Actinidiaceae, Saurauiaceae and Marcgraviaceae in Theales, and Hypericaceae and Clusiaceae in Guttiferales.

Schnarf (1931) states that the Ochnaceae, Marcgraviaceae, Theaceae and Guttiferae are embryologically similar. Cronquist (1981) places the first three families in his Theales. The secretory anther tapetum, simultaneous cytokinesis in pollen mother cells, two-celled tricolpate pollen grains, anatropous, bitegmic and tenuinucellate ovules, absence of parietal cell, Polygonum type of embryo sac, Nuclear type of endosperm, and exalbuminous seed are some of the common features. The endosperm is of the Cellular type, and the seed is albuminous in Marcgraviaceae. Apospory and synergid polyembryony occur in Ochnaceae and Guttiferae.

Ochnaceae. The distinctive features of Ochnaceae are: prominent hypostase, false endothelium (from nucellus), hypertrophied antipodals, and membraneous seed-coat. It also differs from Geraniales in embryological features, such as presence of a well-developed parietal tissue, hypertrophied antipodals, and exalbuminous seed. Corner (1976) prefers to treat the similarity of seed structure of Ochnaceae to that of Simarubaceae (Rutales). The study of its embryogeny would be helpful in determining the phyletic position. At present, its assignment to Guttiferales, rather than to Geraniales, appears to be more satisfactory.

Dipterocarpaceae. Dipterocarpaceae show several notable features. The fibrous thickenings of endothecium are absent in several members, outer integument is vascularized, a meristematic zone is present at the base of nucellus, and the embryo sac is highly aggressive in the chalazal region (in *Vateria*). The Nuclear endosperm contains uni- or multinucleate vesicles. The embryogeny is probably of the Asterad type, or irregular, and the seeds are exotegmic.

Theaceae and Marcgraviaceae. The affinities of Theaceae and Marcgraviaceae (Table 22.18.2) have already been discussed (see paragraph preceding Ochnaceae). The presence of a vascular strand in the outer integument of ovule, Allium type of embryo sac, Solanad type of embryogeny, and sclerotic mesotesta in seed-coat segregate Theaceae. The signs of fusion of integuments, Cellular type of endosperm, and arillate albuminous seed distinguish the family Marcgraviaceae.

Ploiarium. The embryology of *P. alternifolium* has been studied by N. Prakash and Lau (1976). This taxon shares several features with *Camellia* (Kapil and Sethi 1963, Sethi 1965), but there are some differences too. In *Ploiarium* the embryo sac is monosporic, Polygonum type, the antipodals do not organize into cells, hypostase is absent, the shape of ovule and structure of embryo are markedly different, and whether the seed is endotestal or exotegmic is not certain. In *Camellia* the embryo sac is bisporic, Allium type, antipodals organize as cells, hypostase is present, and seed is mesotestal. We do not have any information on the embryology of *Bonnetia.* According to Hutchinson (1973), *Bonnetia* and *Ploiarium* should both be included in Bonnetiaceae; and N. Prakash and Lau support the transfer of *Ploiarium* from Theaceae to Bonnetiaceae. Melchior (1964) assigns *Bonnetia* to Theaceae. The differences in the embryology of *Camellia* and *Ploiarium* do warrant their separation and transfer of *Ploiarium* to Bonnetiaceae. An embryological study of *Bonnetia* may provide further support.

Hypericaceae. Wettstein (1935), Engler and Diels (1936), and Melchior (1964) include Hypericaceae in Guttiferae.

There is a controversy about the development of endosperm in *Hypericum.* Schnarf (1914) describes it as of the Nuclear type, whereas Palm (1922) mentions Helobial type. According to Swamy (1946a), the endosperm is of the Nuclear type in *H. mysorense*; a coenocytic cyst develops in the chalazal region at the 16-nucleate stage of endosperm. It persists for a long time, and then degenerates. Nagaraja Rao (1957b) states that the coenocytic cyst in the Nuclear type of endosperm merges with the general cytoplasm in *H. patulum* and *H. mysorense.* Govindappa (1956a) made similar observations in *H. japonicum.* Myers (1964) reports the fusion of polar nuclei to form a secondary nucleus, prior to opening of the flower. Fertilization may not occur, and the endosperm develops autonomously. The division of secondary nucleus is followed by wall formation in *H. virginicum.* He also mentions the occurrence of aposporic embryo sacs from nucellar cells. The development of endosperm in *Hypericum* needs further investigation.

Guttiferae. Vestal (1937) and Hutchinson (1973) consider that the family Guttiferae seems to have been derived from Theaceae, or from stocks ancestral to them. The exotegmic seed-coat segregates Guttiferae from Theaceae where the seed-coat is mesotestal. In Guttiferae the exotegmen has characteristic large, lignified, tabular, stellate cells. Mesotestal condition occurs only in the subfamily Calophylloideae (Guttiferae).

23 Sarraceniales

The Sarraceniales, an order of insectivorous plants, are considered by most taxonomists to comprise a natural taxon. It has three families: Sarraceniaceae, Nepenthaceae and Droseraceae.

The development of anther, pollen, endosperm, embryo and seed structure has not been investigated in Nepenthaceae. The embryogeny in Sarraceniaceae needs detailed study.

23.1 General Features

The characteristic features of the order Sarraceniales are: (1) secretory anther tapetum of binucleate cells, (2) simultaneous divisions in pollen mother cells, (3) colporate, two-celled pollen grains at the dispersal stage; united in tetrads in Nepenthaceae and Droseraceae, (4) anatropous, bitegmic, tennuinucellar ovule; unitegmic in Sarraceniaceae, (5) Polygonum type of embryo sac, conspicuous antipodals, and a nucellar jacket in Droseraceae, (6) Nuclear type of endosperm with chalazal haustorium; Cellular type in Sarraceniaceae, and (7) endospermic seed, winged in Sarraceniaceae.

23.2 Sarraceniaceae

Shreve (1906) has given a detailed account of embryology of *Sarracenia purpurea* (see also Nichols 1908, Livingston 1950).

The anther wall comprises persistent epidermis, fibrous endothecium, two or three ephemeral middle layers, and secretory tapetum of binucleate cells in *Sarracenia purpurea* (Shreve 1906). The microspore mother cells undergo meiotic divisions. Cytokinesis is simultaneous, and the tetrads are tetrahedral.

The pollen grains have eight meridional grooves in *Sarracenia*, and are two-celled at the dispersal stage.

The ovule is anatropous, unitegmic, and tenuinucellar. The archesporial cell functions directly as megaspore mother cell. It undergoes meiotic divisions, and a linear or T-shaped tetrad is produced. The chalazal megaspore develops into a Polygonum type of embryo sac (Fig. 23.2.1A–F). Sometimes twin gametophytes may develop (Fig. 23.2.1G). A hypostase differentiates. An endothelium develops at the two-nucleate embryo sac. The polar nuclei fuse at the time of fertilization, and the three antipodals are pushed to one side in a pouch during downward elongation of the embryo sac.

The development of the endosperm is of the Cellular type (Fig. 23.2.1H–J). The primary endosperm nucleus undergoes a transverse division. Due to repeated transverse divisions, a row of eight cells is produced. Longitudinal divisions follow later on. The endosperm stores aleurone grains.

A transverse division of zygote occurs after the formation of about 150 endosperm cells. The terminal cell of a uniseriate row of five to seven cells of the proembryo develops into the embryo proper. The suspensor is strongly curved (Fig. 23.2.1J–L).

The seed is endospermic. The outer epidermis of the testa has thick outer walls with conspicuous pores in lateral walls. The raphe develops a wing on one side.

M Livingston (1950), M Nichols (1908), F Shreve (1906)

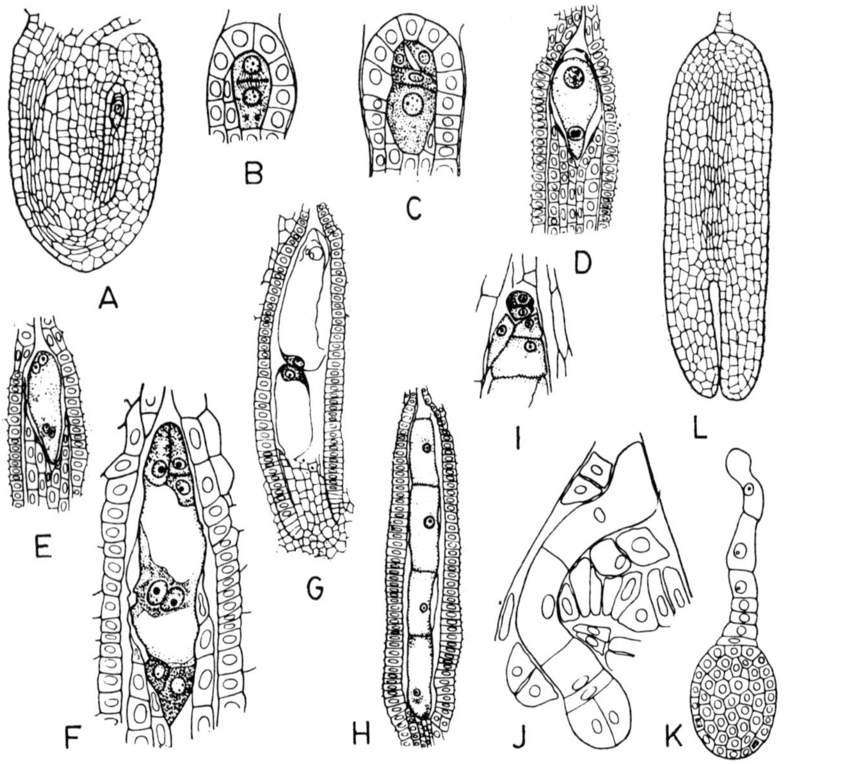

Fig. 23.2.1A–L. *Sarracenia purpurea.* **A** Longisection of young ovule. **B** Megaspore mother cell at telophase I. **C** T-shaped tetrad. **D**, **E** Two- and four-nucleate embryo sac. **F** Mature embryo sac and endothelium. **G** Twin embryo sacs lying one upon the other. **H** Four-celled endosperm, and prominent endothelium. **I** Micropylar part of embryo sac with two-celled proembryo and a few endosperm cells. **J** Early globular proembryo with suspensor. **K** Later stage. **L** Mature embryo. (After Shreve 1906)

23.3 Nepenthaceae

The development of anther, pollen, endosperm and embryo has not been investigated.

The pollen grains remain attached in permanent tetrads.

The ovule is anatropous, bitegmic, and crassinucellate in *Nepenthes* (Stern 1917). The archesporial cell cuts off a parietal cell; the parietal tissue is single-layered. The megaspore mother cell undergoes meiotic divisions. The chalazal megaspore of a linear tetrad develops into a Polygonum type of embryo sac (see Davis 1966).

GL Davis (1966), R Kühl (1933), K Stern (1917), K Subramanyam, LL Narayana and KT Sundari (1985)

23.4 Droseraceae

The anther wall comprises the epidermis, fibrous endothecium, an ephemeral middle layer, and secretory tapetum. The epidermis is persistent and irregular. The tapetal cells are binucleate; nuclear fusions may occur producing tetraploid and octaploid nuclear masses. Venkatasubban (1950) described occasional intrusion of tapetal cells into the pollen sac dividing it into chambers, and the formation of false periplasmodium in *Drosera indica* and *D. peltata.* The microspore mother cells undergo meiotic divisions resulting in tetrahedral and isobilateral tetrads. During simultaneous cytokinesis, definite walls are laid down to organize the microspores. Compound pollen grains are formed due to the failure of their separation in *D. burmanni* (Narasimhachar 1949).

The generative cell develops in the distal region. Occasionally, in *D. longifolia*, *D. rotundifolia* and their hybrids (Rosenberg 1909), the vegetative and generative cells are quite similar to each other. The pollen grains are three-colpate, pantoporate (e.g. *Drosophyllum*), and spinulate. They are two-celled at the shedding stage. Sometimes, they germinate in situ. The male gametophyte deserves further study.

The ovule is anatropous, bitegmic, and crassinucellate. The inner integument forms the micropyle (Fig. 23.4.1A, E, H). The hypodermal archesporial cell cuts off a parietal cell in *D. burmanni* (Narasimhachar 1949, Patankar 1956). There are one to three archesporial cells. A single layer of parietal cells develops. In *Drosera rotundifolia* (Pace 1912), *D. indica* (Narasimhachar 1951b) and *D. peltata* (Venkatasubban 1950), the archesporial cell directly functions as the megaspore mother cell. The latter divides meiotically producing a T-shaped tetrad in *Aldrovanda vesiculosa.* The chalazal megaspore develops into a Polygonum type of embryo sac (Fig. 23.4.1A–F). The antipodals are conspicuous in *D. peltata* and *D. burmanni.* The cells of the nucellus are radially elongated, and form a jacket around the embryo sac (false endothelium) in *D. indica* (Narasimhachar 1951b).

During double fertilization triple fusion precedes syngamy.

Development of the endosperm is of the Nuclear type (Fig. 23.4.1G, H), and free nuclei aggregate in the chalazal region. Centripetal wall formation is initiated in the micropylar region, after the formation of 32–64 free nuclei. The chalazal part remains free-nuclear for a fairly long time, and acts as a haustorium. Ultimately, it becomes cellular. The persistent endosperm contains plenty of starchy food reserve.

A transverse division in the zygote results in an apical, and a basal cell. The apical cell develops into the embryo proper, and also contributes to the formation of the suspensor. The basal cell may or may not divide, and add to the suspensor in *D. burmanni* and *D. indica.* The embryogeny (Fig. 23.4.2A–I) may be of the Caryophyllad type. In *Dionaea muscipula* the terminal cell of the two-celled proembryo divides vertically, indicating the Asterad or Onagrad type of development. The embryogeny of *Drosera* (see Crété 1963a) conforms to Period II, Megarchetype IV, Group XI (Caryophyllad type). The mature embryo is microscopic.

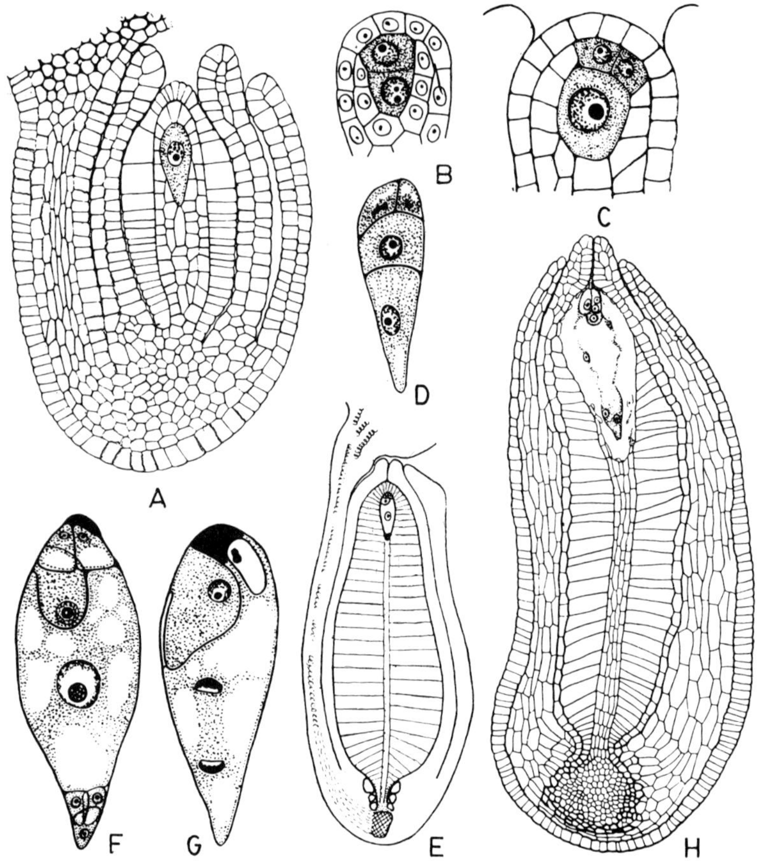

Fig. 23.4.1A–G. *Aldrovanda vesiculosa.* **A** Longisection of anatropous, bitegmic, crassinucellate ovule at megaspore mother cell; **B, C** Megaspore mother cell with parietal cell/s; **D** T-shaped tetrad; **E** Longisection of ovule at mature embryo sac (diagrammatic). Note the conspicuous nucellus in **E** and **H**; **F** Embryo sac from **E**; **G** Embryo sac with zygote, remnants of synergid and pollen tube, and division of primary endosperm nucleus. **H** *Dionaea muscipula*, longisection of ovule with well-developed hypostase, also in **E.** The embryo sac shows zygote, synergids, and free-nuclear endosperm. One synergid has two nuclei. There are no remnants of pollen tube. (See Batygina and Yakovlev 1985)

The outer epidermis of testa is thin-walled with radially elongate cells (see Fig. 23.4.2J). The inner epidermis of tegmen is persistent only in *Drosera intermedia* and *D. rotundifolia*. The inner and radial walls of cells are lignified.

FD Boesewinkel (1989), P Crété (1963a), SG Narasimhachar (1949, 1951b), L Pace (1912), TB Patankar (1956), JE Planchon (1848), O Rosenberg (1909), R Souèges (1936b), H Takahashi (1988), K Venkatasubban (1950)

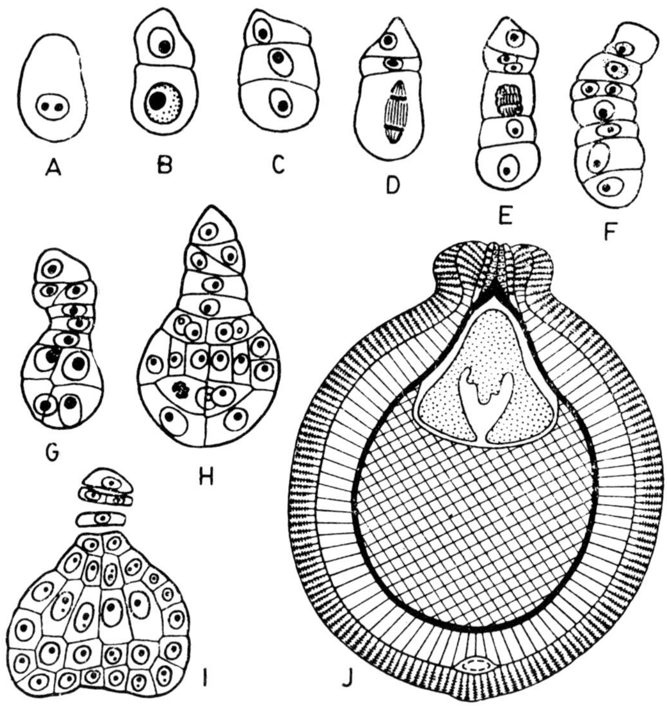

Fig. 23.4.2A–J. A–I *Drosera burmanni*; embryogenesis of Caryophyllad type. **J** *Aldrovanda vesiculosa*, longisection of endospermic seed (diagrammatic). Outer epidermis of testa composed of palisade-like cells with transverse thickenings on radial walls followed by mesophyll; tegmen thickened (*bold black line*). Embryo *dotted*, nuclear endosperm *cross-hatched*. (**A–I** After Patankar 1956, for **J** see Batygina and Yakovlev 1985)

23.5 Taxonomic Considerations

Wettstein (1935) places Droseraceae in Parietales due to parietal placentation, while other taxonomists assign all the three families to Saxifragaceae on morphological basis. Engler and Diels (1936) and Melchior (1964) group these families – Sarraceniaceae, Nepenthaceae and Droseraceae – in Sarraceniales. Hutchinson (1973) transfers Nepenthaceae to Aristolochiales, and retains Sarraceniaceae and Droseraceae in Sarraceniales, near Saxifragaceae. R. Dahlgren (1975a, 1981) erects three orders: Sarraceniales (Sarraceniaceae) in superorder Corniflorae; Theales (Nepenthaceae) and Droserales (Droseraceae) in superorder Theiflorae.

Nepenthaceae, Droseraceae and Sarraceniaceae. Scanty data are available on the embryology of Nepenthaceae. It has a bitegmic, crassinucellate ovule with a single layer of parietal cells, and Polygonum type of embryo sac. Nepenthaceae and Droseraceae both show pollen grains in permanent tetrads. The formation of a parietal layer is mentioned in *Drosera burmanni* (Patankar 1956), otherwise the ovules are tenuinucellate. Its affinities to Aristolochiales cannot be determined due to lack of sufficient information.

The Droseraceae show occasional development of false periplasmodium, compound pollen grains, bitegmic ovules with single-layered parietal tissue and a nucellar jacket, Nuclear type of endosperm with chalazal haustorium, a minute embryo, and sometimes fibrous endotegmen.

The Sarraceniaceae and Droseraceae are always grouped together. But Bentham and Hooker (1965a) treat the family Sarraceniaceae separately, and place it between Nymphaeaceae and Papaveraceae. Takhtajan (1980) treats Sarraceniaceae as the sole family in the order Sarraceniales, and places this order near Ranunculales. Thorne (1968) places it in Sarraceniiflorae between Theiflorae and Annoniflorae, and R. Dahlgren (1977a, 1981) includes the order Sarraceniales in superorder Corniflorae.

According to de Buhr (1975), there is insufficient evidence to support the alliance between Sarraceniaceae, Droseraceae and Nepenthaceae except their insectivorous habit. Cronquist (1981) places these three families in Nepenthales. Morphologically, Sarraceniaceae are akin to Theales, particularly Theineae, and their affinities are with Theales (Cronquist 1968, Thorne 1968, see de Buhr 1975). The embryological similarities between Sarraceniaceae and members of Theineae are: (1) secretory anther tapetum with two-nucleate cells, (2) simultaneous cytokinesis, two-celled pollen grains, (3) anatropous, bi- or unitegmic tenuinucellate ovule, (4) differentiation of endothelium and hypostase, (5) archesporial cell functions directly as megaspore mother cell, (6) Polygonum type of embryo sac, and (7) tanniniferous seed as in Actinidiaceae, Marcgraviaceae, Theaceae and Aquifoliaceae (see de Buhr 1975).

Following Thorne's (1968) classification, de Buhr (1975) proposes that the family Sarraceniaceae should be treated as a suborder Sarraceniineae near Theineae in Theales.

Pace (1912) transfers the genus *Parnassia* of Saxifragaceae to Droseraceae, on account of the structure of ovule and embryo sac. V.K. Sharma (1968b) compares the embryology of *Parnassia* with Saxifragaceae and Droseraceae (see Rosales, Sect. 26). The bitegmic, tenuinucellar ovule has a single-celled archesporium which functions as megaspore mother cell, and T-shaped tetrads, Polygonum type of embryo sac, persistent antipodals, and Nuclear type of endosperm are common features between *Parnassia* (V.K. Sharma 1968b) and Droseraceae. However, the embryogeny of *P. palustris* (Lebégue 1953) conforms to Period I, Megarchetype III, Group I (Asterad type). It is very similar to that in *Claytonia perfoliata* (Portulacaceae) (see Rau 1962). The embryogeny of *Drosera* conforms to Period II, Megarchetype IV, Group XI (Caryophyllad type), and is far removed from that of *Parnassia*. V.K. Sharma (1968b) suggests that *Parnassia* forms a link between Saxifragaceae and Droseraceae (see Table 26.21.1).

Nepenthaceae and Sarraceniaceae show exotestal seeds similar to those of Saxifragaceae. Droseraceae (*Drosophyllum*) has fibrous exotegmen (Corner 1976).

24 Papaverales

The order Papaverales comprises four suborders: Papavarineae represented by Papaveraceae; Capparineae by Capparaceae, Cruciferae and Tovariaceae; Resedineae by Resedaceae; Morigineae by Moringaceae.

24.1 General Features

The order Papaverales shows: (1) secretory anther tapetum, (2) simultaneous cytokinesis in pollen mother cells, (3) two-celled pollen grains, (4) bitegmic, crassinucellate ovule, (5) Polygonum type of embryo sac with ephemeral antipodals except in Papaveraceae and *Maerua* (Capparaceae), (6) Nuclear type of endosperm, and (7) Chenopodiad and Caryophyllad type of embryogeny in Papaveraceae; Onagrad in Capparaceae, Cruciferae and Resedaceae; Onagrad, Caryophyllad and Solanad in Capparaceae; Asterad in Moringaceae. In all the families, except Moringaceae, the testa is thin-walled and the inner integument (tegmen) serves the function of protection. In Moringaceae the mesotesta is fibrous, and tegmen is membraneous. On the whole, the order appears to be homogeneous except for embryogeny.

24.2 Papaveraceae

The anther wall comprises the epidermis, fibrous endothecium, three ephemeral middle layers, and secretory tapetum with multinucleate cells. The number of middle layers varies: two in *Eschscholzia californica* (Sachar and Mohan Ram 1958), only one in *Fumaria* (H.B. Saksena 1954). The nuclei of tapetal cells fuse to form irregular polyploid masses. The tapetum, at times, becomes two- to three-layered (Sachar 1953). The microspore mother cells undergo meiotic divisions resulting in tetrahedral and isobilateral tetrads. Cytokinesis is simultaneous.

The family is highly multipalynous. The pollen grains are three-colpate, colporate, pantoporate, and pantocolpate. They are two-celled at the dispersal stage.

The transformation of anthers into carpellary tissue bearing naked ovule and stigmatic surfaces, and differentiation of pollen sacs in the ovary wall in *Argemone mexicana* are occasional features (Sachar 1953).

The ovule is anatropous, bitegmic, and crassinucellar. Both integuments take part in the formation of the micropyle. The ovule is campylotropous in *Fumaria*. Stomata occur on the integuments in *Argemone mexicana* (Sachar 1955). In *Eschscholzia* (Sachar and Mohan Ram 1958), the inner integument projects beyond the outer, and forms the micropyle. A single cell from uni- or multicellular archesporium cuts off a parietal cell. The megaspore mother cell undergoes meiotic divisions, and produces a linear or T-shaped tetrad. The chalazal megaspore develops into a Polygonum type of embryo sac. The antipodals become polyploid and hypertrophied, and are persistent (A.C. Joshi 1933). In *Papaver rhoeas*, they become 64- or 128-ploid (Hasitschka-Jenschke 1956). They undergo secondary multiplication in *Dendromecon* (Berg 1967).

In *P. somniferum* (Bhandari and Bhargava 1983), at the mature embryo sac stage, the antipodals show transfer cell wall ingrowths from the basal wall into the cytoplasm. The ingrowths become conspicuous after fertilization. The degenerated antipodals show scanty cytoplasm dispersed between the labyrinth of wall ingrowths. A similar condition is reported in *Chelidonium majus* and *E. californica* (Fougère-Rifot 1979).

Double fertilization occurs, and branching of pollen tubes is common in *Papaver* and *Argemone* (Sachar 1953).

The development of the endosperm is of the Nuclear type. The nuclei fuse in the chalazal region to form irregular masses in *Eschscholzia*. The rest of the endosperm becomes cellular at the young globular proembryo. It stores food in the form of oil. In *Argemone*, the entire endosperm becomes cellular due to the activity of the peripheral layer. Cytokinesis occurs in *Fumaria* after the formation of several hundred endosperm nuclei, at the globular proembryo.

The division of zygote is transverse, rarely oblique. Further development is irregular in *Argemone*. The embryogeny conforms to Period I, Series C, Megarchetype V (Chenopodiad type) in *Papaver rhoeas* (Souèges 1926b, c, 1928). It conforms to Period II, Series C, Megarchetype V (Caryophyllad type) in *Eschscholtzia* (Souèges 1949a), and to Period II, Series B, Megarchetype VI, Group XIII or XIV (Caryophyllad type) in *Fumaria* (Souèges 1941e, see Crété 1963a).

Solanad type is reported in *Platystemon*, and Caryophyllad in *Hypecoum*. The basal cell/s become hypertrophied and polyploid (see Yakovlev 1981).

The origin of an additional embryo in *Argemone mexicana* is controversial. Sachar (1953, 1956a) states that it must be due to the development of a persistent synergid. Souèges (1949b) and Crété (1956a), however, consider that it must be due to the proliferation of suspensor cells. This aspect requires further study.

A notable feature in *Eschscholtzia californica* (Sachar and Mohan Ram 1958) is the occurrence of four, rarely five, cotyledons in the embryo.

The seed is albuminous with a small chlorophyllous embryo. The exotesta is composed of elongate cells with their outer walls thickened. The inner layer consists of radially elongate thin-walled cells with fibrillar cellulose reticulum, and contain crystals or brown pigment. The exotegmen comprises narrow longitudinal fibres, and the endotegmen is a tracheidal layer.

A rudimentary aril develops in *Sanguinaria canadensis* (Nagaraja Rao and Shamanna 1963), and an exostomal-funicular aril in *Dendromecon*. Some cells in the hilum-region develop into an elaiosome in *Dicentra* (Berg 1969). The elaiosome is very conspicuous in *Corydalis* and *Chelidonium* (Fig. 24.2.1A–C); the seed surface of *Argemone* is shown in Fig. 24.2.1D.

RY Berg (1967, 1969), NN Bhandari and M Bhargava (1983), NN Bhandari, M Bhargava and A Sachdeva (1985), PK Bose and I Banerji (1933), P Crété (1956a, 1963a), ÅE Dahl (1989), M

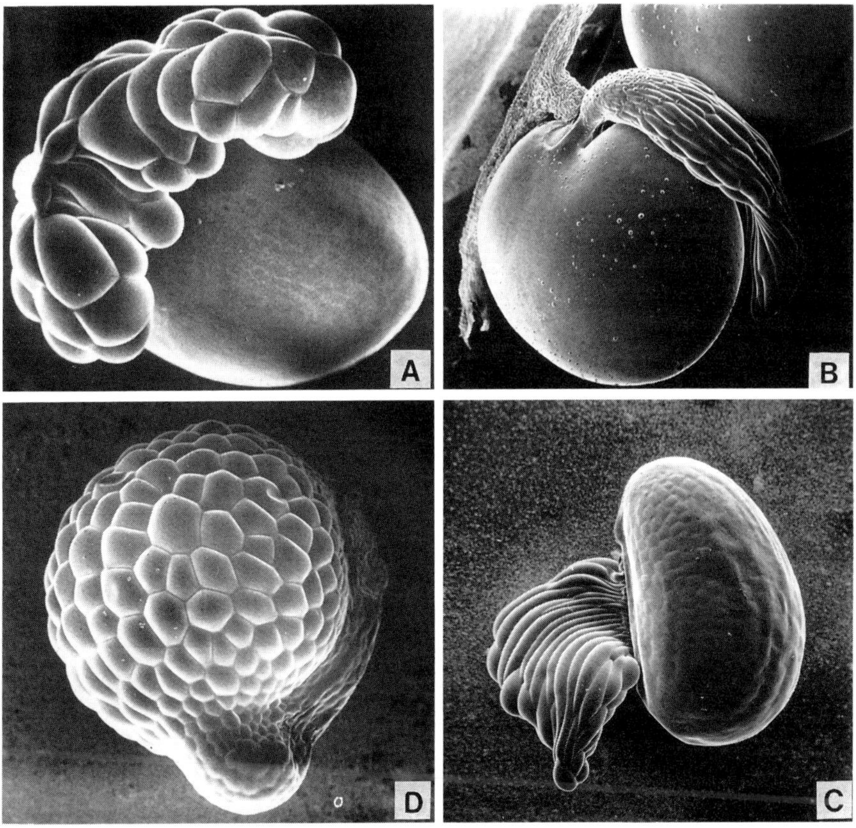

Fig. 24.2.1A–D. (SEM micrographs) **A**, **B** *Corydalis cava*, **C** *Chelidonium majus*, **D** *Argemone platiceras*. **A–C** Seeds with elaiosome. **D** Seed. (Courtesy Dr F. Bouman, Amsterdam)

Fougère-Rifot (1979), V Gupta and LC Lamba (1985), G Hasitschka-Jenschke (1956), JA Jernstedt and C Clark (1979), AC Joshi (1933), RN Kapil and A Nanchahal (1989), A Nagaraja Rao and S Shamanna (1963), RC Sachar (1953, 1955, 1956a), RC Sachar and HY Mohan Ram (1958), HB Saksena (1954), EL Schneider and DM Nichols (1984), RP Singh and AM Khan (1974), R Souèges (1926b, c, e, 1928, 1941e, f, 1949a, b), MS Yakovlev (1981)

24.3 Capparaceae

The anther wall comprises an epidermis, fibrous endothecium, two or three ephemeral middle layers, and secretory tapetum of multinucleate cells. In *Cadaba* (H.S. Narayana 1965) even the epidermal cells develop fibrous thickenings. The tapetum becomes biseriate in *Capparis*. Simultaneous cytokinesis follows meiotic divisions in microspore mother cells, resulting in tetrahedral, decussate and isobilateral (*Cleome monophylla* – A.V.N. Rao 1967) tetrads.

The pollen grains are two- to three-colporate. They are two-celled at the dispersal stage in *Capparis* (H.S. Narayana 1962c, Hanumantha Rao and Prakasa Rao 1977), *Cadaba* and *Crataeva* (H.S. Narayana 1965), and three-celled in *Capparis* and *Cleome* (Tasneem and Kusuma Kumari 1970). Binucleate microspore mother cells and germination of pollen grains in situ have been observed in *Capparis decidua* (H.S. Narayana 1962c).

The ovule is campylotropous, bitegmic, and crassinucellar with a zigzag micropyle. It is anatropous in *Crataeva*. The outermost layer of the inner integument develops into a stony layer in *Isomeris arborea* (P. Maheshwari and Khan 1953). Rudimentary stomata develop along the funicular side of the ovules of *I. arborea* (Sachar 1956b). A hypostase differentiates, and tracheidal cells occur in the nucellus.

A hypodermal cell of uni- or multicellular archesporium produces a parietal cell. The latter forms a few layers of parietal tissue. The megaspore mother cell undergoes meiotic divisions, produces a linear or T-shaped tetrad (T.S. Raghavan and Venkatasubban 1941a), or a linear triad – the upper one is a dyad cell (P. Maheshwari and Khan 1953). The chalazal megaspore develops into a Polygonum type of embryo sac. The synergids are ephemeral in *Isomeris* (P. Maheshwari and Sachar 1954a). The antipodals are usually ephemeral but in *Maerua arenaria* (V.S. Rao 1936a) they are large and occupy nearly half the space of the embryo sac. Twin embryo sacs occur in *Capparis aphylla* (Vaidya 1961a).

The development of the endosperm is of the Nuclear type. It becomes cellular at the late globular or heart-shaped stage of proembryo. P. Maheshwari and Khan (1953) report, in *Isomeris arborea*, nucleated nodules in the middle of the embryo sac, and cambium-like strips in cellular endosperm. The nodules also become cellular.

According to Billings, the embryo develops from one of the multinucleate endosperm nodules. P. Maheshwari and Sachar (1954a), and Sachar (1956b), refute Billings's assumption (1937) of endospermic origin of the embryo in *I. arborea*. The abnormal position of the embryo is partly due to the extension of the micropylar end of the embryo sac leaving the embryo behind. It *does not* arise from the endosperm. The zygote produces a long, uniseriate, filamentous proembryo (Sachar 1956b). The suspensor becomes massive, haustorial, and coiled (Fig. 24.3.1A–F′). Its cells are vesicular and multinucleate (P. Maheshwari and Khan 1953, P. Maheshwari and Sachar 1954a). The embryo proper develops from the two or three uninucleate cells at the tip of suspensor (Fig. 24.3.1D, F).

The embryogeny corresponds to the Caryophyllad type in *Capparis* and *Cadaba*, Solanad in *Crataeva* (H.S. Narayana 1962c, 1965), and Onagrad in *Cleome chelidonii* (T.S. Raghavan 1937) and *Gynandropsis pentaphylla* (Phouphas 1952). Nucellar polyembryony is reported in *Capparis frondosa* (Mauritzon 1934b).

The seed is campylotropous, and exalbuminous with a folded embryo. The testa comprises thin-walled cells with or without stomata. The tegmen develops into a protective tissue. The exotegmen comprises palisade cells with thickened walls or longitudinal fibres followed by lignified endotegmen.

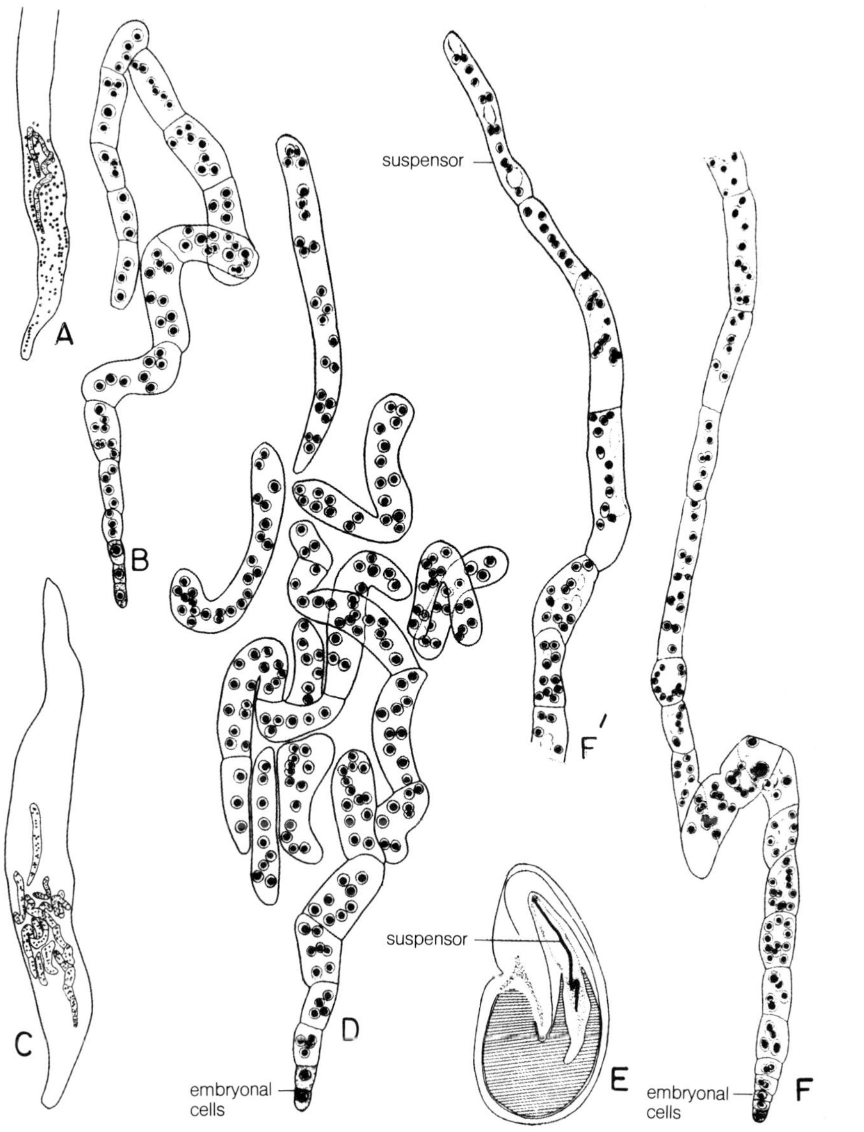

Fig. 24.3.1A–F. *Isomeris arborea.* **A** Lower part of embryo sac (wholemount) with proembryo and free-nuclear endosperm. **B** Proembryo from **A**. **C** Embryo sac (wholemount). Note the proembryo with a long suspensor; endosperm omitted. **D** Proembryo with a highly coiled, uniseriate, suspensor with coenocytic cells, from **C**, **E** Older ovule, the proembryo has a long suspensor. **F**, **F′** Proembryo with suspensor from **E**. The uninucleate cells at the tip **(B, D, F′)** are progenitors of the embryo proper. (After Sachar 1956b)

V Arunalakshmi (1989), FH Billings (1937), B Hanumantha Rao and PS Prakasa Rao (1977), P. Maheshwari and R Khan (1953), P. Maheshwari and RC Sachar (1954a, 1955), J Mauritzon (1934b), HS Narayana (1962c, 1965, 1970a), MY Orr (1921), C Phouphas (1952), TS Raghavan (1937), TS Raghavan and KR Venkatasubban (1941a), AVN Rao (1967), PS Rao and B Hanumantha Rao (1975, 1976, 1977), VS Rao (1936a, c, 1938), RC Sachar (1956b), F Tasneem and P Kusuma Kumari (1970), NK Tiwary (1936), PB Vaidya (1961a)

24.4 Cruciferae

The anther wall comprises the epidermis, fibrous endothecium, one to three middle layers, and glandular tapetum of binucleate cells. Following meiotic divisions, the microspore mother cells produce tetrahedral, isobilateral and decussate tetrads. Cytokinesis is simultaneous.

The pollen grains are three-colpate and shed at the three-celled stage, except in *Cardamine* and *Chenopodifolia* where they are two-celled. Degeneration of anther, arrested growth of pollen grains, and their improper differentiation have been noted in *Erysimum cheiranthoides* (Belyayeva 1968).

The ovule is ana- to campylotropous (*Brassica juncea* – Sulbha 1957), and bitegmic with a zigzag micropyle formed by both the integuments. It is crassinucellar in *B. campestris* (Ahuja and Bhaduri 1956), and tenuinucellar in *Malcolmia africana*. The hypodermal archesporium is one- to many-celled, but only one cell develops further. It cuts off a parietal cell in *B. juncea*, whereas it functions directly as megaspore mother cell in *Arabidopsis thalianum* and *Erophila verna*. The megaspore mother cell undergoes meiotic divisions and produces a linear or T-shaped tetrad. The chalazal megaspore develops into a Polygonum type of embryo sac. The three antipodals are ephemeral, and may become multinucleate (Sulbha 1957). Hypostase develops in *Matthiola tricuspidata* (Renzoni-Cela 1969). An endothelium, and twin embryo sacs are reported in *B. juncea*.

In *Brassica campestris* (S.P. Bhatnagar and Kaur 1984), the style is hollow, whereas in other Cruciferae the style is solid: *Brassica rapa*, *Capsella bursa-pastoris*, *Diplotaxis tenuifolia* and *Raphanus sativus*. It contains a uniform canal which is lined by a single row of conspicuous thick-walled cells (canal cells). They constitute the transmitting tissue for the growth of pollen tube. The canal cells are larger than the cortical cells, and bulge into the canal. These vacuolated cells are rich in metabolites, and their cytoplasm contains proteins and starch.

During pollination and pollen tube growth, the canal cells shrivel, lose contents, walls break down, and the collapsed cells form a loose mass of cells around the canal.

Development of the endosperm is of the Nuclear type. The endosperm forms prominent nodular, nuclear aggregations in *Capsella bursa-pastoris* (P. Maheshwari and Sachar 1954b). At the initial stage each nodule contains only two to three nuclei, and they are confined to the chalazal region. In later stages they become five- to six-nucleate, and occur in the lower half of the embryo sac. Enucleate cytoplasmic nodules are reported in *Brassica campestris*

(Ahuja and Bhaduri 1956, Rathore and Singh 1968). Nuclear fusions occur in the endosperm of *B. juncea* (Il'ina 1962). The tissue becomes cellular.

The embryogeny corresponds to the Onagrad type-Capsella variation (R.C. Misra 1962, Rathore and Singh 1968). In *Capsella* it conforms to Period I, Series A, Megarchetype IV. The suspensor is thin and filamentous, and the uppermost cell is vesicular. The embryogeny in *Lunaria annua* (see Rodionova 1983) differs from the Capsella variation in the occurrence of a linear tetrad of embryo in addition to the T-shaped, four-celled proembryo. There is no distinct hypophysis cell, and the differentiation of histogens is delayed.

The development of embryo of *Brassica napus* (Tykarska 1976) conforms to Onagrad type, or Megarchetype IV of Souèges:

$$ca = pvt + pco + phy + icc$$

$$cb = iec + co + s$$

Tykarska (1979) describes the development in three phases. Phase I includes growth of the apical cell (*ca*) up to its division into two layers of octants (*l* and *l'*). Phase II comprises the increase of spherical proembryo and change of its symmetry from radial to bilateral. Phase III concerns the growth of embryo from heart-shaped up to the end of embryogenesis. In each phase the growth is very vigorous in the beginning, but later declines.

According to Ramji (1975), the bilateral symmetry is determined much earlier, at the formation of an intercept at the quadrant and octant stage during phase I, and not during phase II as claimed by Tykarska.

Polyembryony is reported in several genera. In *Arabis halleri* and *A. hirsuta* (Mianny 1972), the supernumerary embryo develops from synergid, either by apogamy or by fertilization. In *A. lyalli* (Lebègue 1948b), it develops from an unfertilized synergid. Reduced pseudogamy, i.e. the formation of an embryo without fertilization from a reduced egg, has been observed in *Brassica campestris* × *B. oleracea* var. *gummifera* (Noguchi 1928). *A. holboellii* (Böcher 1947, 1951) exhibits aneuspory of the Taraxacum type: aneuspory results due to irregularities in meiosis during megaspore formation. In this taxon the lower cell of an unreduced dyad develops into an eight-nucleate embryo sac. The same species also shows unreduced pseudogamy. The sperm degenerates inside or outside the embryo sac, but may provide stimulus to the egg to develop pseudogamously.

The seed is endotestal, and exalbuminous. The outer epidermis of testa comprises cells with reticulate thickenings on their radial walls. The mesophyll either collapses or is collenchymatous. The inner epidermis consists of cells with U-shaped thickenings on the inner and radial walls. The outer epidermis of tegmen is pigmented; the inner one consists of palisade cells.

Spooner (1984) studied the sexual population of *Dentaria laciniata*. Following meiosis, the microspore mother cells undergo simultaneous cytokinesis. The tetrads are tetrahedral, and the pollen grains three-celled. The archesporial cell functions directly as megaspore mother cell, a triad is formed, and the embryo sac is of the Polygonum type. The endosperm is of the Nuclear type (later becomes cellular), and the embryogeny is of the Onagrad type-Capsella variation. The seeds germinate readily in nature. Similar development occurs in *D. diphylla* up to embryo sac formation but it does not organize and degenerates. The reproductive events and other data indicate that the eastern North American species of *Dentaria* may form a sexual polyploid complex with some sexual populations and some sterile ones.

YR Ahuja and PN Bhaduri (1956), LY Belyayeva (1968), R Bergfeld, T Kühnl and P Schopfer (1980), SP Bhatnagar and I Kaur (1984), TW Böcher (1947, 1951), F Bouman (1975), R Czapik (1974), J Eigner (1973), S Ellerström and L Zagorcheva (1977), D Gerlach-Cryse (1970), I Grant, WD Bevers-dorf and RL Peterson (1986), MW Green (1978), Å Håkansson (1956), MI Hakki (1974), GM Il'ina (1962), J Kallarackal and SP Bhatnagar (1978), M Kroh and A Munting (1967), M Kuras (1978), LC Lamba (1975), LC Lamba and BB Arora (1981a, b), Lebègue (1948b), T Mackiewicz (1973), P Maheshwari and RC Sachar (1954b), SJ Mayer, AA Snow and ML Stanton (1986), N Mianny (1972), RC Misra (1962, 1966), Y Noguchi (1928), G Norton and JF Harris (1975), PM Pechan (1988), K Prasad (1977a, b, 1979), MV Ramji (1975), RKS Rathore and RP Singh (1968), GC Renzoni-Cela (1969), GB Rodionova (1972, 1983), P Schulz and WA Jensen (1986), C Simoncioli (1974), LD Smith and DH Scarisbrick (1990), W Soerodikoesomo (1970), DM Spooner (1984), AD Stead, IN Roberts and HG Dickinson (1979), K Sulbha (1957), MJ Sumner and L van Caeseele (1989), T Tykarska (1976, 1979), L van Caeseele, JT Mills, M Sumner and R Gillespie (1981), JG Vaughan (1968), JG Vaughan and JM Whitehouse (1971), MS Yakovlev and GK Alimova (1976), MD Yoffe (1952)

24.5 Tovariaceae

The development of anther and embryo has not been investigated.

The anther tapetum is of the secretory type. The microspore mother cells undergo meiotic divisions, and cytokinesis is simultaneous.

The pollen grains are two-celled at the dispersal stage.

In *Tovaria pendula* (Mauritzon 1934b), the ovule is campylotropous, bitegmic, and crassinucellar with a zigzag micropyle. The archesporial cell cuts off a parietal cell which subsequently divides repeatedly. The megaspore mother cell undergoes meiotic divisions. The chalazal megaspore of a linear tetrad develops into a Polygonum type of embryo sac. The antipodals are ephemeral.

The development of endosperm is of the Nuclear type. Later it becomes cellular.

The seed is small, and exarillate. The embryo is bent, contains oil, and is surrounded by copious endosperm. The testa is thin-walled. The outer epidermis of tegmen, in the region of endostome and raphe, comprises palisade cells with reticulate thickenings on the walls.

J Hutchinson (1973), J Mauritzon (1934b), P Schürhoff (1926)

24.6 Resedaceae

The development of the anther wall has not been investigated.

The microspore mother cells undergo meiotic divisions, and tetrahedral and isobilateral tetrads are produced. Cytokinesis is simultaneous.

The pollen grains are two-celled at the dispersal stage.

The ovule is ana- to campylotropous, bitegmic, crassi- or tenuinucellar with a single vascular strand in the integument. The archesporium is multicellular, and each cell cuts off a parietal cell. Several megaspore mother cells undergo meiotic divisions in *Reseda odorata* (Gori 1957). The chalazal megaspore of linear tetrads develops into Polygonum type of embryo sac. The polar nuclei fuse before fertilization, and antipodals are ephemeral. A hypostase develops.

The development of the endosperm is of the Nuclear type. Later it becomes cellular.

The embryogeny corresponds to the Onagrad type. The suspensor is uniseriate.

The seed is reniform, exotegmic, and albuminous with a curved embryo. The outer epidermis of testa consists of papillate cells with thick outer walls, and the inner epidermis is a crystal layer in *Oligomeris linifolia* (D. Singh and Gupta 1968). In *Reseda odorata* (D. Singh and Gupta 1968), in the mature seed the testa becomes thin and papery. The outer epidermis of tegmen forms the mechanical layer of thick-walled, lignified, palisade-like cells. In *Reseda* it forms ridges and furrows. The inner epidermis has dark contents.

P Crété (1955a), C Gori (1957), D Singh and S Gupta (1968), MS Yakovlev (1983)

24.7 Moringaceae

The anther wall comprises an epidermis, fibrous endothecium, a single ephemeral middle layer, and a glandular tapetum of binucleate cells (H.S. Narayana 1962b). The development of the anther wall follows the "Basic" type (Davis 1966) in *Moringa concanensis* (B.S.M. Dutt 1978a). The endothecial thickenings extend to four or five layers in the sterile theca. Übisch granules are studded on the inner walls of tapetal cells. The microspore mother cells undergo meiotic divisions, and produce tetrahedral, isobilateral and decussate tetrads. Cytokinesis is simultaneous, and quadripartitioning occurs by furrowing.

The pollen grains are three-colpate, and shed at the two-celled stage. B.S.M. Dutt and Prakasa Rao (1977) report the presence of pollen grains in the stylar canal of *M. concanensis*.

The ovule is anatropous, bitegmic, and crassinucellate. The integuments are vascularized; the hypostase is persistent. A single hypodermal archesporial cell cuts off a parietal cell which gives rise to three to four layers of parietal tissue. Following meiotic divisions, the megaspore mother cell produces a T-shaped or linear tetrad, sometimes a triad (V. Puri 1941). The development of the female gametophyte is of the Polygonum type. The antipodals are ephemeral.

The occurrence of twin nucelli, twin embryo sacs, egg-like synergids, and supernumerary nuclei in the embryo sac has been reported in *Moringa oleifera* (V. Puri 1934).

The endosperm is of the Nuclear type, and the nuclei are crowded at the micropylar end. Wall formation commences from the micropylar end at the late globular stage of proembryo, and the entire endosperm becomes cellular.

The development of the embryo corresponds to the Asterad type in *M. oleifera* (H.S. Narayana 1962b). The suspensor is absent. However, B.S.M. Dutt (1978c) states that the suspensor is persistent and disintegrates only in mature seed. The embryogeny conforms to Myosurus-variation, Onagrad type in *M. concanensis* (Period I, Megarchetype IV). The embryo consists of thick cotyledons, and the radicle is short. The development of the embryo requires further study.

The seed is mesotestal, exalbuminous, and the testa alone forms the seed-coat. It is multiplicative. The outer and inner epidermis of testa are thin-walled. The thin-walled mesophyll, containing chlorophyll and starch, develops into a thick-walled, pitted, strongly lignified compact tissue. At an early stage, the meristematic activity of hypodermal cells, at three corners, gives rise to wings which become vascularized. The tegmen persists as a thin membraneous layer.

RM Datta and JN Mitra (1947), GL Davis (1966), BSM Dutt (1978a, c, 1979b), BSM Dutt and PS Prakasa Rao (1977), HS Narayana (1962b, 1970b), K Periasamy and Indira (1986), V Puri (1934, 1941), FL Rutgers (1923)

24.8 Taxonomic Considerations

Bentham and Hooker (1965a) place Cruciferae, Capparidaceae, Resedaceae and Papaveraceae in Parietales. Hutchinson (1973) includes Papaveraceae and Fumariaceae in Rhoeodales; Capparaceae and Moringaceae in Capparales; Cruciferae (Brassicaceae) in Cruciferales (Brassicales), and Resedaceae in Resedales. Engler and Diels (1936) include all the families in Rhoeodales. Cronquist (1968, 1981) treats the orders Papaverales and Capparales separately as they differ in the nature of secretory cells, chemical substances, and orientation of stamens. Palynological and serological data also support this separation. He assigns Papaveraceae under Magnoliidae, and Capparales under Dilleniidae.

Papaveraceae. Papaveraceae are characterized by hypertrophied persistent antipodals which undergo secondary multiplication, Chenopodiad and Caryophyllad type of development of embryo, thick-walled exotesta, and endotestal crystal layer. The exotegmen is fibrous with tracheidal endotegmen.

Fumaria. Melchior (1964) places the genus *Fumaria* in Papaveraceae. Hutchinson (1973) and Cronquist (1981) recognize it as a separate family, Fumariaceae, and place it in Rhoeodales along with Papaveraceae. *Fumaria* differs from Papaveraceae in the absence of latex; zygomorphic flowers with spurs, two lateral tripartite stamens, anther wall with a single middle layer, pantoporate pollen grains, campylotropous ovule, cytokinesis in the nuclear endosperm after the formation of several hundred free nuclei (H.B. Saksena

1954), and embryogeny conforms to Period II, Megarchetype VI, Group XIII and, at times, Group XIV (Souèges 1941d, e) – Caryophyllad type. H.B. Saksena (1954) justifies the erection of the family Fumariaceae.

Capparaceae. H.S. Narayana (1970a) states that Capparales are different from Papaverales. Capparales (Capparaceae, Tovariaceae, Moringaceae and Resedaceae) have three-colpate spheroidal pollen with sexine as thick as nexine, while Papaverales (Papaveraceae and Fumariaceae) are characterized by two- to nine-colpate, rupate, and rugate pollen, or pollen provided with irregular apertures. Also, Capparales are different from Papaverales in ephemeral antipodals, and large polar nuclei. According to H.S. Narayana, Tovariaceae and Capparaceae have probably originated from a common ancestor. He suggests the erection of the orders Papaverales and Capparales.

Capparaceae resemble most to Resedaceae. The campylotropous ovule with parietal tissue and hypostase, Polygonum type of embryo sac with ephemeral antipodals (rarely hypertrophied as in *Maeruva*), Nuclear type of endosperm, Onagrad type of embryogeny, curved embryo, albuminous seed, and thin-walled testa and fibrous tegmen are common features. In addition, Capparaceae show cytoplasmic endosperm nodules as in Cruciferae, and Solanad type of embryogeny. The long, massive suspensor in *Isomeris* is an unusual feature.

Resedaceae. According to D. Singh and Gupta (1967), the endotegmen in the seed of Resedaceae has been misinterpreted as the third integument. The seed-structure of Resedaceae is comparable to violaceous seed. Papaveraceae, Capparaceae and Resedaceae have exotegmic seed. As in the seeds of Capparaceae, the exotegmen is one- to many-layered. D. Singh and Gupta consider Capparaceae as the "Basic" type from which seed-structure of some other families can be derived (see also Netolitzky 1926).

Cruciferae show close alliance with Capparaceae. The occurrence of chlorophyllous embryos in both families indicate a close affinity (Yakovlev and Zhukova 1975). The exalbuminous seed with characteristic endotesta, and fruit of Cruciferae indicate greater specialization.

Tovariaceae. Scanty information is available on Tovariaceae, but the development of ovule and endosperm is in general conformity with that of the order.

Moringaceae. De Jussieu (1789) and Hallier (1912) place *Moringa* in Leguminoseae. Lindley recognizes it as a separate family, Moringaceae, and assigns to the order Violales. Hutchinson (1973) associates Moringaceae with Capparaceae in the order Capparales. V. Puri (1941) considers it to be related to Cruciferae and Capparaceae. R.M. Datta and Mitra (1947) support Lindley, and relate Moringaceae to Violaceae and Tovariaceae, on the basis of zygomorphy of flower, perigyny, and parietal placentation. H.S. Narayana (1970b) compares the embryology of Moringaceae with that of Violaceae, Leguminoseae and Polygalaceae, and negates any relationship with these families. The presence of myrosin cells in Moringaceae is similar to that in Capparaceae and

Resedaceae. The structure of anther, pollen, ovule, embryo sac, and endosperm is similar to that in Capparaceae and Cruciferae. A hypostase, and exalbuminous seeds also occur in Cruciferae. However, the occurrence of a short gynophore, anatropous ovule, Asterad type of embryogeny, thin-walled testa containing chlorophyll, lignified mesotesta, and membraneous tegmen are their characteristic features. Consequently, Moringaceae may have diverged from the primitive Capparaceae. The mesotestal seed-coat indicates ancestry to Theales.

25 Batales

The order Batales is represented by the family Bataceae. Its embryology has not been studied.

26 Rosales

The order Rosales comprises 19 families, distributed over four suborders: Hamamelidineae – Platanaceae, Hamamelidaceae, Myrothamnaceae; Saxifragineae – Crassulaceae, Cephalotaceae, Saxifragaceae, Brunelliaceae, Cunoniaceae, Davidsoniaceae, Pittosporaceae, Byblidaceae, Roridulaceae, Bruniaceae; Rosineae – Rosaceae, Neuradaceae, Chrysobalanaceae; Leguminosineae – Connaraceae, Leguminosae, Krameriaceae.

26.1 General Features

The order exhibits a great deal of variation, but some common features are: (1) secretory anther tapetum, both secretory and amoeboid in Saxifragaceae, (2) simultaneous cytokinesis in pollen mother cells, (3) two-celled pollen grains, two- to three-celled in Papilionaceae, and three-celled in Pittosporaceae, (4) sterile transverse septa delimit groups of archesporial cells (in the pollen sac), each group is encircled by its own tapetum, and pollinia are produced in Mimosaceae; pollen embryo sacs observed in Saxifragaceae, (5) ovule ortho-, ana-, hemiana- or campylotropous, crassinucellar, tenuinucellar in Pittosporaceae and Byblidaceae, and both types in Saxifragaceae and Caesalpiniaceae, usually bitegmic, unitegmic in Brunelliaceae, and uni- and bitegmic in Saxifragaceae and Rosaceae; integumentary vascular strand present in Mimosaceae and Caesalpiniaceae, (6) unicellular archesporium, multicellular in Crassulaceae and Cunoniaceae, and uni- and multicellular in Papilionaceae, (7) Polygonum type of embryo sac, Allium type in Crassulaceae, Rosaceae, Saxifragaceae and Papilionaceae, and Adoxa type in Saxifragaceae, (8) Nuclear type of endosperm in Cunoniaceae, Pittosporaceae, Rosaceae, Connaraceae and Leguminosae, Cellular in Crassulaceae, Nuclear and Cellular in Hamamelidaceae, and Nuclear, Cellular and Helobial types in Saxifragaceae; endosperm haustoria develop in Crassulaceae, Rosaceae, Byblidaceae and Leguminoseae, (9) embryogeny follows Solanad type in Hamamelidaceae, Caryophyllad in Crassulaceae, Solanad or Caryophyllad in Saxifragaceae, Asterad in Pittosporaceae, Caryophyllad and Asterad in Rosaceae, Onagrad in Mimosaceae, Asterad in Caesalpiniaceae, and Asterad, Onagrad, Solanad and Caryophyllad in Papilionaceae, (10) exotestal seed, mesotestal in Rosaceae, and (11) apospory in Rosaceae.

26.2 Platanaceae

The embryology has not been fully investigated.

The anther tapetum is glandular with two- to four-nucleate cells in *Platanus acerifolia.* Following meiosis in microspore mother cells, cytokinesis is simultaneous (Bretzler 1924). The pollen grains are three-colpate and two-celled.

The ovule in *Platanus* is orthotropous, bitegmic and crassinucellate. The micropyle is formed by both the integuments, and the vascular bundle terminates at the chalaza. The archesporial cell cuts off a parietal cell; the parietal tissue is well developed. The development of the embryo sac is of the Polygonum type (Bretzler 1924).

The development of the endosperm is of the Cellular type.

The seeds are very small with a thin endosperm, and linear embryo with flat cotyledons. The seed-coat is almost crushed.

E Bretzler (1924), J Brouwer (1924), PK Endress (1977a), PR Mohana Rao (1974a), T Nicoloff (1911), K Schnarf (1931), MS Yakovlev (1981)

26.3 Hamamelidaceae

The anther wall comprises an epidermis, fibrous endothecium, two or three middle layers, and secretory tapetum. The tapetum becomes two- to three-layered in *Exbucklandia populnea* (Kaul and Kapil 1974). The tapetal cells are multinucleate, their nuclei fuse and become polyploid. Orbicules are studded on the inner wall of tapetal cells. Simultaneous cytokinesis follows meiotic divisions in microspore mother cells. The tetrads are tetrahedral, isobilateral and decussate.

The pollen grains are tri-, tetra- and pentacolpate. They are two-celled at the dispersal stage.

The ovule is anatropous, bitegmic and crassinucellate; both integuments form the micropyle. The outer integument elongates, and serves as a partial obturator. The cells of the nucellar epidermis undergo periclinal divisions. The hypostase is thick-walled. The sporogenous tissue is massive in *Corylopsis willmotiae* (Endress 1977a). The archesporial cell cuts off a parietal cell. The parietal tissue is eight to ten layers thick in *Parrotiopsis jaquemontiana* (Kapil and Kaul 1972). The megaspore mother cell undergoes meiotic divisions, and produces a linear or T-shaped tetrad. The second megaspore (from the micropylar end) develops into a binucleate embryo sac, or the lower two develop up to the two-nucleate stage. However, the chalazal megaspore develops into a Polygonum type of embryo sac. One of the synergids may persist up to the octant stage of proembryo. The polar nuclei fuse before fertilization. The antipodals are generally persistent.

During double fertilization, triple fusion precedes syngamy. The pollen tube undergoes overwintering in *Hamamelis* (Endress 1977a).

The development of the endosperm is of the Cellular type in *Parrotiopsis jaquemontiana* (Kaul 1969, Kapil and Kaul 1972), and Nuclear type in *Exbucklandia populnea* (Kaul and Kapil 1974) and *Hamamelis virginiana* (Mathew 1980). In *Exbucklandia*, the endosperm becomes cellular after a few nuclear divisions. It is persistent and stores fats and proteins.

The development of the embryo conforms to the Solanad type in *Parrotiopsis*, and Chenopodiad type-Hamamelis variation in *H. virginiana*. The mature embryo has two large leafy cotyledons in *Exbucklandia*.

The seeds are winged; the wing develops from the free-end of the outer integument. Its cells contain tannin. The testa is massive and sclerotic. The tegmen is thin-walled and tanniniferous (Mohana Rao 1974a).

PK Endress (1977a), FF Flint (1957a, b, 1959), LD Hufford and PK Endress (1989), RN Kapil and U Kaul (1972), U Kaul (1969), U Kaul and RN Kapil (1974), CJ Mathew (1980), AP Melikian (1973), PR Mohana Rao (1974a), F Niedenzu (1891), E Shaw and RD Gibbs (1961), KY Tong (1930), RF Thorne (1973)

26.4 Myrothamnaceae

The development of the embryo has not been fully investigated.

The anther wall comprises the epidermis, endothecium, three or four middle layers, and secretory tapetum of binucleate cells. Cytokinesis is simultaneous, and the tetrads are tetrahedral (Jager-Zürn 1966).

The binucleate pollen grains remain in permanent tetrads in *Myrothamnus flabellifolia* and *M. moschata* (Jager-Zürn 1966).

The ovule is anatropous, bitegmic, and crassinucellate. The inner integument forms the micropyle in *M. flabellifolia*. The outer integument overgrows the inner one in *M. moschata* (Jager-Zürn 1966). The archesporial cell cuts off a parietal cell. The parietal tissue is three to four layers thick. The megaspore mother cell undergoes meiotic divisions and the chalazal dyad develops into a bisporic Allium type of embryo sac. The antipodals are ephemeral.

The development of the endosperm is of the Nuclear type. The embryogeny (Jager-Zürn 1966) appears to be of the Crucifer type.

I Jäger-Zürn (1966), E Shaw and RD Gibbs (1961)

26.5 Crassulaceae

The anther wall comprises the epidermis, fibrous endothecium, two middle layers, and secretory tapetum with uninucleate cells. The microspore mother cells undergo meiotic divisions, and produce tetrahedral and isobilateral tetrads. Cytokinesis is simultaneous, and furrows develop centripetally.

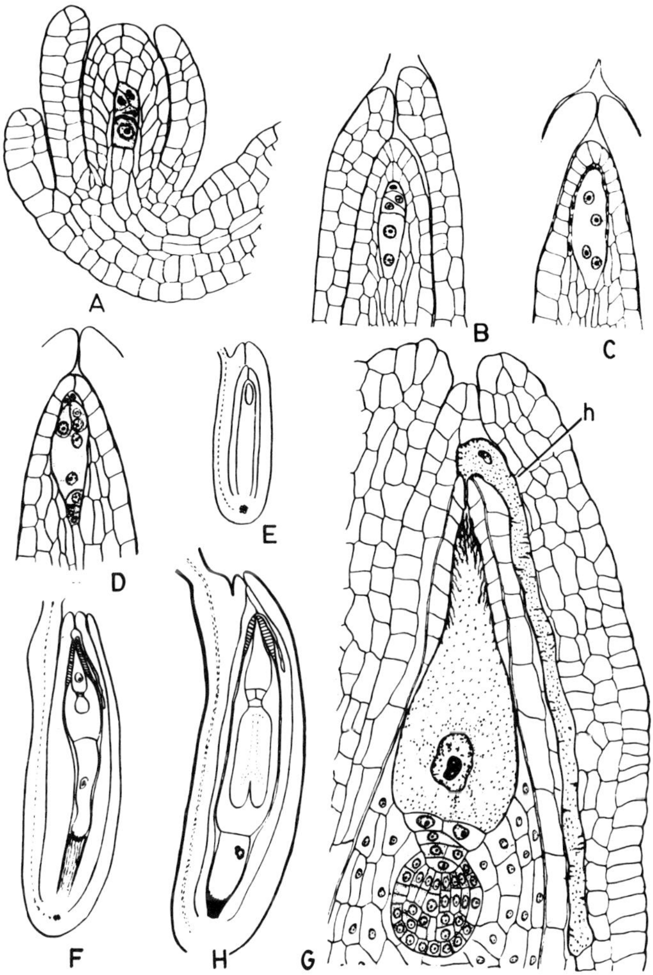

Fig. 26.5.1A. *Kalanchoe faustii*, bitegmic ovule with functional chalazal megaspore, the upper three megaspores have degenerated. **B–H** *Sedum hispanicum* (**E**, **G** and **H** diagrammatic); **B–D** Two- and four-nucleate, and mature embryo sac; **E** Anatropous, bitegmic ovule; **F** Outline figure for **G**; **G** Micropylar region of ovule. The globular proembryo has a highly expanded haustorial uppermost cell of suspensor. Note the wall-ingrowths (transfer cell) in the upper portion. The suspensor haustorium (*h*) invades the inner integument. Cellular endosperm surrounds the proembryo; **H** Mature seed. The lower cell in **F** and **H** is the chalazal endosperm haustorium. (See Batygina and Yakovlev 1985)

The pollen grains are three- or four-colporate and two-celled at the shedding stage.

The ovule is anatropous, bitegmic, and crassinucellate (Fig. 26.5.1A, E–H). The archesporial cell cuts off a parietal cell. Following meiotic divisions, the megaspore mother cell produces linear, isobilateral, T-shaped or even oblique T-shaped tetrad (Fig. 26.5.1A, B). A peculiar feature is the development of megaspore haustoria in *Sedum sempervivoides* (Mauritzon 1933a), and *S. chrysanthum* (Fig. 26.5.2A–D). The multiple archesporium produces several megaspore mother cells, each gives rise to a tetrad. Each megaspore develops a tubular haustorial process which grows upward towards the micropyle. Several such megaspore haustoria intertwine, and form a tangle in the upper part of nucellus. Some of them become extranucellar, and lie at the base of the micropyle. Development of the embryo sac is of the Polygonum type (Fig. 26.5.1A–D). Allium type of development is reported in *Sedum fabaria*, *S. populifolium* and *S. populifolium* var. *notarjanni* (Mauritzon 1933a). The synergids show a cap-like structure at the apex, or from their apex arise three or four slender tubular processes in *Sedum*. The synergids develop haustoria in *S. sempervivoides* (Mauritzon 1933a). The polars fuse before fertilization. There are three small antipodals which develop haustoria in *S. fabaria* (Mauritzon 1933a). The embryo sac grows considerably towards the chalaza, becomes

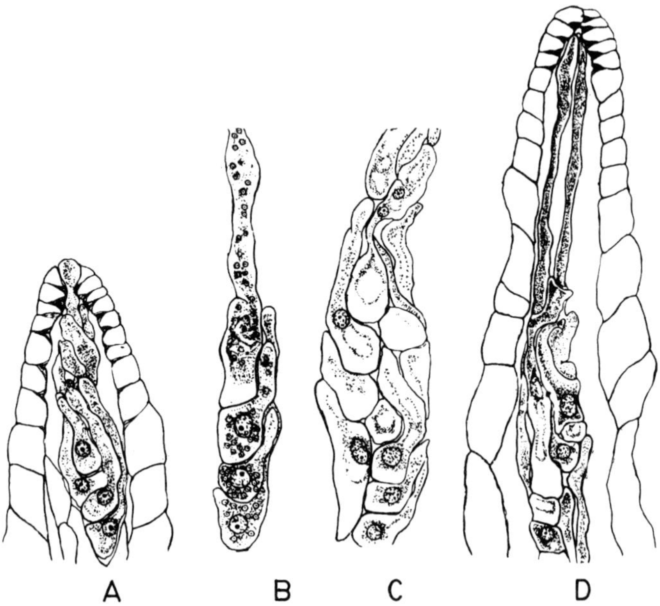

Fig. 26.5.2A–D. *Sedum chrysanthum*. **A** Upper part of nucellus to show a haustorium from each megaspore of tetrad. There are additional haustoria (at the apex) from other megaspores in the same nucellus. **B** Megaspores and haustoria contain starch grains. **C, D** Groups of megaspore tetrads with haustoria. A binucleate embryo sac has developed in **C**. (After Subramanyam 1967)

haustorial, and accumulates a large quantity of starch. These observations need confirmation.

The entry of pollen tube is porogamous, and double fertilization occurs.

The development of the endosperm is of the Cellular type (Fig. 26.5.1F–H). The division of primary endosperm nucleus is followed by a transverse wall producing a larger micropylar and a smaller chalazal chamber. The chalazal

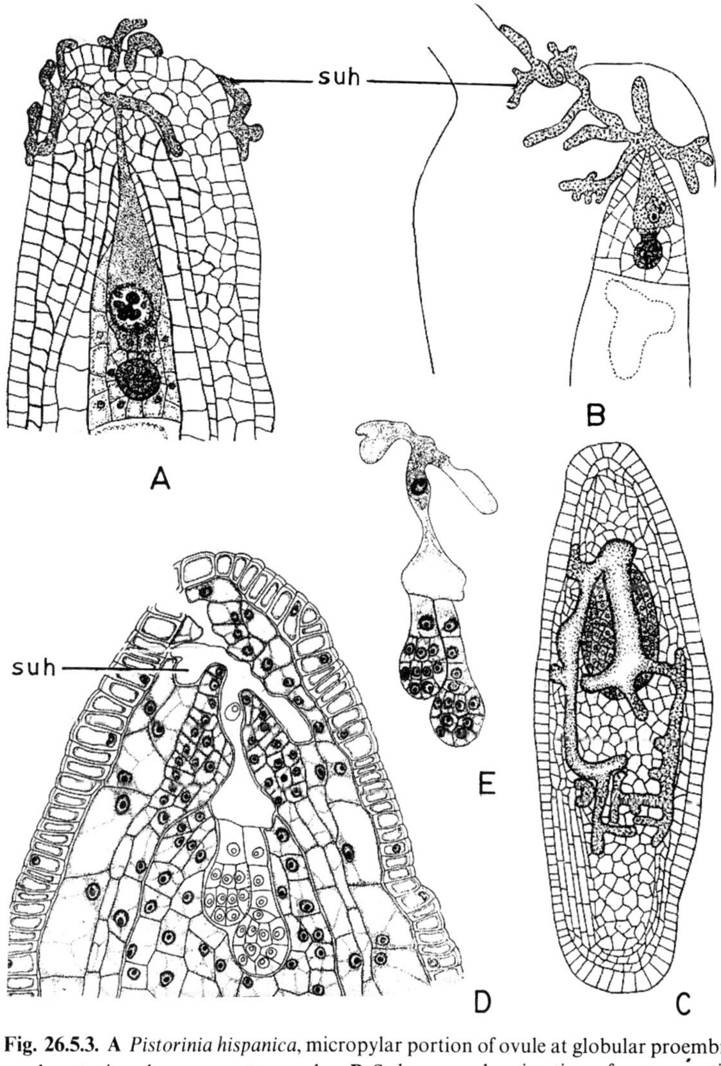

Fig. 26.5.3. A *Pistorinia hispanica*, micropylar portion of ovule at globular proembryo, the suspensor haustorium becomes extra-ovular. **B** *Sedum acre*, longisection of upper portion of seed, the suspensor haustorium (*suh*) branches and ramifies in the seed. **C** *Aeonium bentejui*, longisection (paradermal) of ovule at globular proembryo, the branched suspensor haustorium invades the integument. (After Mauritzon 1933a). **D**, **E** *S. spectabile*; **D** Micropylar part of seed showing twin embryos and cellular endosperm, suspensor haustorium (*suh*) invades the seed-coat; **E** Twin embryos with basal cell as suspensor haustorium, from **D**. (After Vignon-Fétré 1966)

chamber develops into a uninucleate haustorium. Five subtypes of development are recognized, according to the mode of wall formation in the micropylar chamber: Acre, Spathulata, Rupestre, Crassula, and Tatra subtype (Mauritzon 1933a).

The embryogeny conforms to the Caryophyllad type-Sedum variation, or Period II, Series A, Megarchetype IV (Souèges 1936d, Crètè 1946c, Subramanyam 1963). Sometimes, twin proembryos with a common basal cell may develop (Fig. 26.5.3D, E). An interesting feature in embryogeny is the haustorial behaviour of the basal cell. It becomes hypertrophied and vesicular, and develops fine haustorial processes penetrating the seed-coat (Figs. 26.5.1F–H, 26.5.3A–E). The presence of an antipodal embryo has been noted in *Sedum fabaria* (Mauritzon 1933a); this needs confirmation.

The testa alone forms the seed-coat. Its outer epidermis comprises radially elongate cells with thickened outer walls and brownish contents. The outer two layers are parenchymatous.

P Crètè (1946c), J Mauritzon (1930, 1933a, 1939b), R Souèges (1925c, 1927, 1936d, e), K Subramanyam (1962, 1963, 1967, 1970a), J Vignon-Fétré (1966)

26.6 Cephalotaceae

The embryology has not been investigated.

26.7 Saxifragaceae

The anther wall comprises the epidermis, fibrous endothecium, one or two ephemeral middle layers, and secretory tapetum with two- to four-nucleate cells in *Heuchera micrantha* (Vijayaraghavan and Ratnaparkhi 1977). The epidermis may be tanniniferous, or cutinized. The microspore mother cells undergo meiotic divisions and form tetrahedral and decussate tetrads, isobilateral in *Vahlia* (*Bistella*) *oldenlandioides* (Raghavan and Srinivasan 1942) and *Bauera capitata* (N. Prakash and McAllister 1977). Cytokinesis is simultaneous.

The three-colporate pollen grains are two- or three-celled at the dispersal stage, and are dimorphic. They are two-colpate in *Bauera capitata*, and gorged with starch grains.

Some noteworthy features have been observed in *Heuchera micrantha* (Vijayaraghavan and Ratnaparkhi 1977). Idioblasts and clustered crystals occur in the connective tissue. It also shows fibrous thickenings. The tapetal cells become four-nucleate. In addition, occurrence of "pollen embryo sacs" is unique. The uninucleate microspores, at times, enlarge, undergo three successive divisions, and give rise to eight-nucleate embryo sac-like structures. The organization simulates Polygonum, Allium or Adoxa type of female gametophyte. Further divisions of nuclei occur in some cases. The nuclear arrangements simulate Penaea and Plumbago variations, but the cells are not organized. The formation of pollen embryo sacs is associated with the degeneration of tapetal cells.

The ovules are anatropous, bitegmic, and crassinucellate in most species of the subfamilies Penthoroideae, Saxifragoideae (*Bistella* is tenuinucellate; *Peltiphyllum* and *Saxifraga pallida* are unitegmic), and Baueroideae; bitegmic and tenuinucellate in Brexioideae and Parnassioideae; unitegmic and tenuinucellate in Kirengshomoideae and Hydrangeoideae (Rau 1970). The micropyle is organized by both the integuments.

In *Ribes* spp. (Mauritzon 1936a, see also Batygina and Yakovlev 1985), the archesporium consists of one or two cells. It cuts off a parietal cell which develops a small parietal tissue. The sporogenous cell functions directly as megaspore mother cell. The megaspore tetrads are linear, and occasionally T-shaped. The chalazal megaspore develops into a Polygonum type of embryo sac. The polar nuclei fuse before or at the time of fertilization, and the antipodals are ephemeral in *Saxifraga* and *Bistella*. However, the antipodals are prominent in *Parnassia nubicola* (V.K. Sharma 1968b), and their number increases up to five in *Parnassia* and *S. diversifolia* (Saxena 1964a, b), and seven in *Bauera capitata* (N. Prakash and McAllister 1977). At times, the development of the embryo sac is bisporic in *S. pallida* (Rau and Sharma 1967). In *Bistella oldenlandioides*, the embryo sac grows laterally, and an endothelium with binucleate cells develops. Twin embryo sacs are common in *Bergenia ciliata* (Saxena 1969b), *S. diversifolia* (Saxena 1963, 1964a), and *S. verginensis* (Chapman 1933).

The position of the secondary nucleus is variable. In the species with Helobial type of endosperm development it lies near the antipodals, otherwise in the upper half of the embryo sac.

Rodgersia asculifolia, *Tellima grandiflora* and *Tolmiea menziesii* (Ratnaparkhi 1972, Vijayaraghavan and Ratnaparkhi 1975) show Polygonum as well as Allium type of development. The embryo sac becomes extramicropylar in *Philadelphus tomentosus* (Vijayaraghavan and Ratnaparkhi 1975).

A good deal of variation occurs in the mode of endosperm formation. It may be Cellular, Helobial, or Nuclear. Both Cellular and Helobial types occur in *Boykinia*, *Mitella*, *Ribes* and *Tiarella* (Rau 1970), Helobial type in *Chrysosplenium* and *Saxifraga* (Rau and Sharma 1967, Kaplan 1976), and Cellular type in *Astilbe*, *Bergenia* (Saxena 1969a, b), *Kirengeshoma*, *Penthorum* and *Bistella*. Nuclear type is recorded in *Bauera*, *Brexia*, *Fendlera*, *Francoa* and *Parnassia*. Wall formation is initiated at the globular proembryo in *Parnassia nubicula*, and a single layer persists in seed. In *Bistella*, four uninucleate cells of the chalazal endosperm become richly cytoplasmic, suggesting their haustorial role. In *Mitella pentandra*, a non-aggressive cellular haustorium is formed from the chalazal chamber.

The embryogeny corresponds to the Caryophyllad type or Period II, Megarchetype IV (Rau and Sharma 1967). *Parnassia palustris* belongs to the Asterad type, or Period I, Megarchetype III (Lebègue 1953). Synergid polyembryony occurs in *Bergenia delavayi* (Lebègue 1949). *Penthorum* and *Tellima* also show the presence of additional embryos.

The seed is small, and albuminous with a minute embryo. The testa is represented by radially elongate or thick-walled papillate cells. The tegmen gets crushed, and persists only in the micropylar region. The inner epidermis of tegmen is thick-walled and persistent in *Heuchera*. The SEM studies of seed coats of four species of *Saxifraga* (Kaplan 1976) show differences in the number and distribution of humps and papillae, and the fine structure of cuticle.

TB Batygina and MS Yakovlev (1985), CR Bensel and BF Palser (1975), M Chapman (1933), A Engler (1894), K Kaplan (1976), K Kaplan and M Strohschneider (1984), MS Hils, WC Dickison, TW Lucansky and WL Stern (1954), A Lebègue (1949, 1953), C Le Cog (1967), Vya Mandryk (1972), Vya Mandryk and MA Andrusyak (1969), J Mauritzon (1933a, 1936a), L Pace (1912), BP Pandey and NP Saxena (1977), N Prakash and EJ McAllister (1977), TS Raghavan and VK Srinivasan (1942), S Ratnaparkhi (1972), MA Rau (1970), MA Rau and VK Sharma (1967), NP Saxena (1963, 1964a, b, 1969a, b, 1970, 1971, 1973, 1976), VK Sharma (1968b), R Souèges (1936a), K Subramanyam (1962), MR Vijayaraghavan and S Ratnaparkhi (1975, 1977, 1979), IL Wiggins (1959)

26.8 Brunelliaceae

The embryology has not been studied.

26.9 Cunoniaceae

The embryogeny has not been investigated.

The anther wall comprises the epidermis containing tannin, fibrous endothecium, a single middle layer, and secretory tapetum with binucleate cells in *Weinmannia fraxinea* (Govil and Saxena 1976). Simultaneous cytokinesis follows meiosis in microspore mother cells, and the tetrads are tetrahedral.

The pollen grains are two-celled at the dispersal stage.

The ovule is anatropous, bitegmic, and crassinucellar with a zigzag micropyle. The archesporium is multicellular, and each cell cuts off a parietal cell. The parietal tissue is well developed. The cells of the nucellar epidermis also divide and increase the parietal tissue. Kennedy and Prakash (1981) report a nucellar cap and hypostase in *Callicoma serratifolia.*

A single megaspore mother cell functions, and undergoes meiotic divisions. The chalazal megaspore of a linear or T-shaped tetrad develops into a Polygonum type of embryo sac. Epichalazal megaspore functions in *Schizomeria* and, occasionally, in *Weinmannia*. The polar nuclei fuse before fertilization, and the secondary nucleus lies at the chalazal end of embryo sac. The antipodals are ephemeral. A hypostase differentiates in *Weinmannia.*

The development of the endosperm is of the Nuclear type. Later, it becomes cellular. It is persistent.

The seed is exotegmic. The outer epidermis of the testa comprises tangentially elongate cells; the cells of the inner epidermis are vacuolate. The outer epidermis of tegmen forms the sclerotic layer with radially elongate cells; the radial walls are strongly lignified. The inner epidermal cells are tangentially elongate, and contain tannin.

KVO Dahlgren (1939a), CM Govil and NP Saxena (1976), MJ Kennedy and N Prakash (1981), J Mauritzon (1939b)

26.10 Davidsoniaceae

The embryology has not been worked out.

26.11 Pittosporaceae

The anther wall comprises the epidermis, fibrous endothecium, one or two ephemeral middle layers, and secretory tapetum (Fig. 26.11.1A). The tapetum is irregularly bi- and tri-seriate in *Billardiera cymosa*, *B. lehmamiana*, *B. latifolia* (L.L. Narayana and Sundri 1983) and *Pittosporum heterophyllum* (L.L. Narayana and Sundri 1978). The tapetal cells are binucleate in *Hymenosporum flavum* and *Pittosporum floribundum*, and vesicular and multinucleate (Fig. 26.11.1B) in *Bursaria spinosa* (Sheela and Narayana 1966, L.L. Narayana and Sundri 1978). In *Marianthus bignonaceus* (Sundri and Narayana 1977), they are multinucleate and become polyploid due to nuclear fusions. Simultaneous cytokinesis in microspore mother cells is followed by furrowing, and tetrahedral and decussate (*Sollya heterophylla* – L.L. Narayana and Sundri 1979) tetrads are produced; rhomboidal and isobilateral tetrads are recorded in *P. heterophyllum*.

The three-colporate pollen grains are three-celled at the dispersal stage (Fig. 26.11.1C); two-celled in *Billardiera verifolia* (L.L. Narayana and Sundri 1983). At times, the pollen grains germinate in situ in *Marianthus*. Several abnormalities, such as double pollen grains, occurrence of two to six nuclei in a pollen grain, and both the nuclei of pollen grain of similar size, are reported in *Marianthus* (Sundri and Narayana 1977).

The orientation of ovules in the ovary of *Pittosporum floribundum* shows variation. These are hypotropous in the micropylar half and epitropous in the chalazal half.

The ovule is anacampylotropous, unitegmic, and tenuinucellar. The hypodermal archesporial cell functions directly as megaspore mother cell, undergoes meiotic divisions, and produces a triad, or linear, or T-shaped tetrad. The chalazal megaspore develops into a Polygonum type of embryo sac. The polar nuclei fuse prior to fertilization, and the antipodals are ephemeral (Fig. 26.11.1D–I).

The development of the endosperm is of the Nuclear type (Fig. 26.11.1J). Five- to six-nucleate nodules, invested by a cytoplasmic membrane, occur in *Bursaria spinosa* (L.L. Narayana and Sundri 1978). Later, the endosperm becomes cellular (Fig. 26.11.1K) and packed with food reserve.

The embryogeny (Fig. 26.11.1L–R) conforms to the Geum-variation of Asterad type (L.L. Narayana and Sundri 1978, 1979). The mature embryo is very small with two to five cotyledons. Suspensor polyembryony has been noted in *Bursaria spinosa* (Fig. 26.11.1S) and *Pittosporum* spp. (L.L. Narayana and Sundri 1978, 1979).

The seed-coat comprises three to 12 layers of aerenchymatous cells. The outer epidermis consists of compact and thick-walled cells.

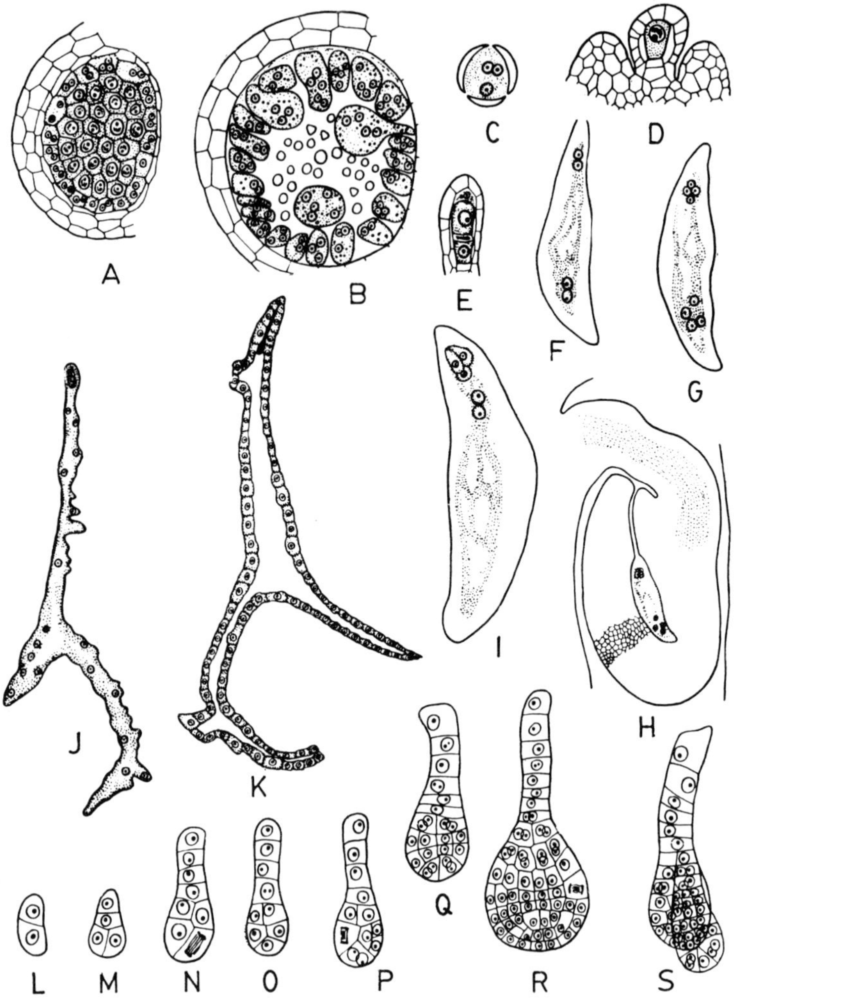

Fig. 26.11.1A–S. *Bursaria spinosa.* **A** Cross-section of anther lobe at microspore mother cell. **B** Later stage, tapetal cells multinucleate and protoplasts project inward. **C** Three-celled pollen. **D** Hypodermal megaspore mother cell. **E** Linear tetrad. **F**, **G** Four- and eight-nucleate embryo sac. **H** Longisection of unitegmic ovule at eight-nucleate embryo sac. **I** Embryo sac with egg-apparatus and two polar nuclei. **J** Nuclear endosperm at globular proembryo. **K** Cellular endosperm at globular proembryo. **L–R** Embryogeny. **S** Twin proembryos, additional embryo from proliferation of suspensor. (After L.L. Narayana and Sundri 1978)

The fruit is a loculicidal capsule with seven or eight layers thick wall. The outer epidermis is followed by two or three layers of thin-walled cells and four to six layers of lignified cells.

G Bremer (1915), LL Narayana and KT Sundri (1977, 1978, 1979, 1981, 1983, 1984), R Sheela and LL Narayana (1966), KT Sundri and LL Narayana (1977)

26.12 Byblidaceae

The development of anther, pollen, female gametophyte, embryo, and seed-coat has not been investigated.

The differentiation of endothecium only at the apex of anther has been noted. The outer walls of cells are thickened.

The ovule is anatropous, bitegmic, and tenuinucellate. The inner integument forms the micropyle. The archesporial cell functions directly as megaspore mother cell. An endothelium differentiates.

The development of endosperm is probably of the Cellular type, and multicellular terminal haustoria are formed in *Byblis gigantea* (Lang 1901).

The seed is small, and albuminous. The cells of the outer epidermis of testa are tangentially elongate, and have thick and pitted walls.

FX Lang (1901)

26.13 Roridulaceae

The anther is tetrasporangiate. The development of anther wall is of the Basic type in *Roridula gorgonias* (Vani-Hardev 1972). Fibrous thickenings are absent in the endothecium. The secretory tapetum is irregularly biseriate, and nuclear divisions and fusions occur (Fig. 26.13.1A). Simultaneous cytokinesis follows meiotic divisions in microspore mother cells. The tetrads are tetrahedral.

The pollen grains are tricolporoidate, and tenuinexinous with reticulate sexine and striate nexine. They are two-nucleate at the dispersal stage (Fig. 26.13.1B, C).

The ovule is anatropous, unitegmic and tenuinucellate (Fig. 26.13.1J). The archesporial cell develops directly into the megaspore mother cell. The latter undergoes meiotic divisions, and the chalazal megaspore of the linear tetrad develops into a Polygonum type of embryo sac (Fig. 26.13.1D–K). The synergids are elongated with tapering ends. The polar nuclei fuse before fertilization. The three antipodals persist up to fertilization. A hypostase differentiates.

Fertilization is porogamous, and the pollen tube persists for some time (Fig. 26.13.1K, L).

The development of endosperm is of the Cellular type. The division of primary endosperm nucleus results in a micropylar and a chalazal chamber (Fig. 26.13.1L). A vertical division occurs in the micropylar chamber followed by further divisions so that it becomes completely cellular (Fig. 26.13.1M, N). Vani-Hardev (1972) states that the derivatives of the micropylar chamber "encroach into the micropyle and behave like a haustorium".

The first and a few subsequent divisions in the chalazal chamber are transverse. Except the lowermost cell, all the other derivatives of the chalazal chamber divide vertically. The penultimate zone of chalazal endosperm undergoes active cell divisions forming a distinct tissue of small, elongated cells with dense cytoplasm and prominent nuclei.

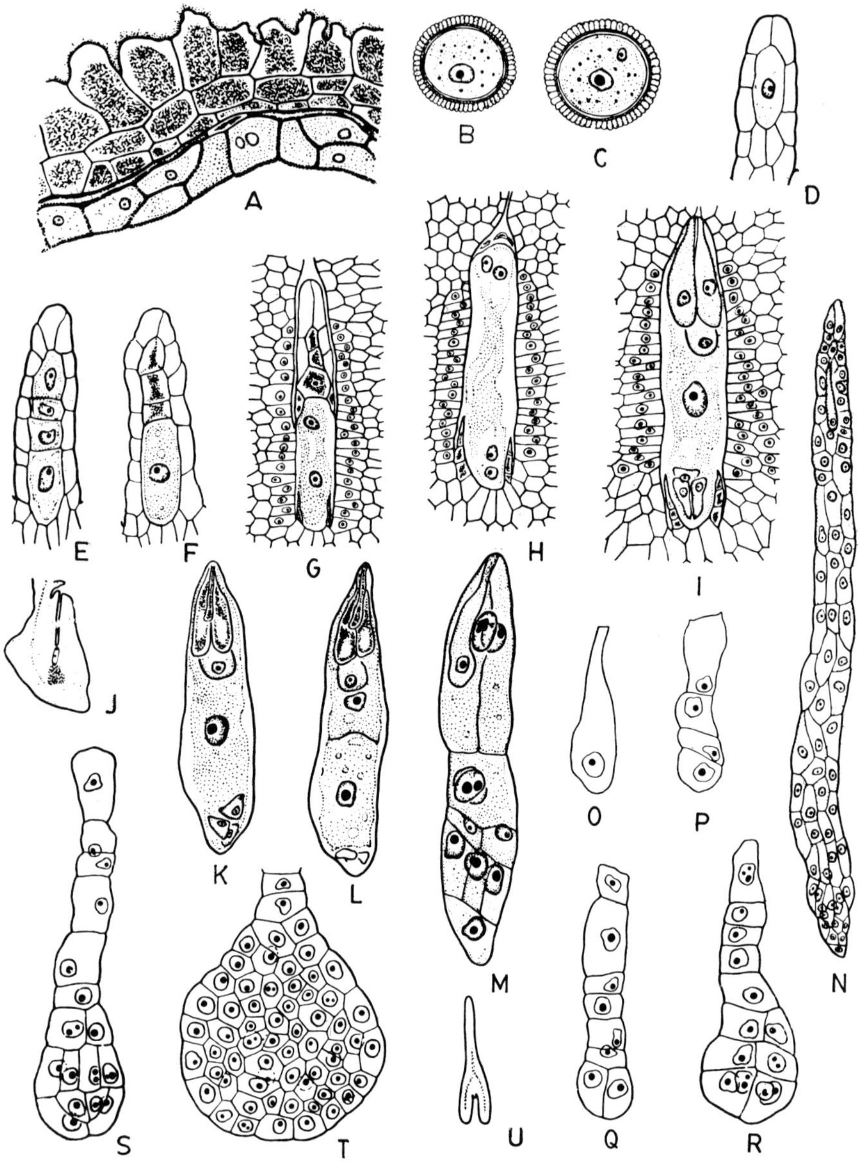

Fig. 26.13.1A–U. *Roridula gorgonias.* **A** Portion of anther wall at uninucleate pollen grain, tapetum irregularly biseriate. **B**, **C** Uninucleate and bi-celled pollen grain. **D** Hypodermal megaspore mother cell. **E** Linear tetrad. **F** Chalazal megaspore functional. **G**, **H** Bi- and four-nucleate embryo sac, note endothelium in **G–I**. **I**, **K** Mature embryo sac. **J** Longisection of ovule, note the hypostase. **L** Embryo sac with zygote, remnants of pollen tube and antipodals, and two-celled endosperm. **M** Zygote with multicellular endosperm. **N** Later stage. **O–T** Embryogeny. **U** Early dicotyledonous embryo. (**A–T** After Vani-Hardev 1972, **U** after Solereder 1908 – see Batygina and Yakovlev 1985)

The embryogeny conforms to the Solanad type (Fig. 26.13.1O–T). The seeds are small and albuminous with a small embryo (Fig. 26.13.1U), the endosperm is tanniniferous. The cells of the testa become thick-walled. The epidermal cells enlarge, and their tangential and radial walls become thick.

EJH Corner (1976), J Hutchinson (1973), R Marloth (1903), Vani-Hardev (1972)

26.14 Bruniaceae

The development of anther, pollen grains, endosperm, embryo, and seed-coat has not been investigated.

The ovule is anatropous, unitegmic (due to fusion of two primordia), and crassinucellar with a long micropyle. The mode of formation of parietal tissue is uncertain; it is represented by a single layer (Saxton 1910). The megaspore mother cell undergoes meiotic divisions, and produces a linear tetrad. The chalazal megaspore develops into a Polygonum type of embryo sac. The polar nuclei fuse before fertilization. The antipodal cells are ephemeral. Starch grains are present in the mature embryo sac.

EJH Corner (1976), J Mauritzon (1939b), WI Saxton (1910)

26.15 Rosaceae

The anther wall comprises the epidermis, fibrous endothecium, two ephemeral middle layers, and secretory tapetum of multinucleate cells in *Exochorda racemosa* (Johnston 1961). Simultaneous cytokinesis follows meiotic divisions in microspore mother cells, resulting in tetrahedral and, rarely, decussate tetrads. Johnston reported stickiness of chromosomes at anaphase I in *Exochorda*.

The pollen grains are tricolporate, and two-celled at the dispersal stage.

The ovule is ana- to hemianatropous, uni- or bitegmic, and crassinucellate. Unitegmic condition is reported in *Adenostoma*, *Alchemilla*, *Aphanes*, *Aruncus*, *Geum*, *Holodiscus*, *Neviusia*, *Potentilla*, *Rhodotypos*, *Rubus*, *Sibiraea* and *Ulmaria* (see Davis 1966). In bitegmic ovules both the integuments, or only the inner integument, take part in the formation of micropyle. The micropyle is not organized in *Prinsepia utilis* (M. Sharma and Narayana 1971). In *Sanguisorba minor* (Nordberg 1967), a single integument is formed by the accretion of the two. The two sides fuse, and surround the nucellus. In such cases the micropyle is absent. The development of postament and hypostase is reported in *Rubus idaeus* (Topham 1970). An obturator develops in *Eriobotrya japonica* (H.S. Narayana and Sharma 1976), *Malus sieboldii* (Hjelmqvist 1957b) and *Prinsepia utilis* (M. Sharma and Narayana 1971).

The hypodermal archesporium is multicellular. The archesporial cells cut off parietal cells, and the parietal tissue is well developed. The cells of the nucellar epidermis divide, and add to the parietal tissue. The megaspore mother cells undergo meiotic divisions. Generally, only one megaspore mother cell develops into a linear or T-shaped tetrad, but dyads, triads or tetrads, functional megaspores, two-, four- and eight-nucleate embryo sac are reported in the same ovule (H.S. Narayana and Sharma 1976). Similar features have been noted in *Rubus spectabilis* (Virdi and Eaton 1969) and *Sanguisorba* (Nordberg 1967). The chalazal megaspore develops into a Polygonum type of embryo sac. At times, the second megaspore from the micropylar end develops in *Prunus* and *Cerasus* (Baciu 1972). The antipodals are ephemeral, but in *Alchemilla* spp. they are persistent, and multiply in *A. orbiculata*. The polar nuclei fuse prior to fertilization.

In *Alchemilla*, *Aphanes*, *Cotoneaster*, *Rubus* and *Sorbus* the megaspore mother cell degenerates. One or more secondary megaspore mother cells differentiate, undergo meiosis, and produce embryo sacs in the normal way.

Allium type of development is reported by H.S. Narayana and Sharma (1976). They also point out several abnormalities in the pattern of embryo sac, arrangement of nuclei, presence of plural embryo sacs, and multiple embryo sac containing variable nuclei in "odd-ploid" members of *Malus* and *Rubus*.

The carpelloid stamens of flowers of *Rubus trivials* (Shealy and Herr 1973) show anatropous, unitegmic, and crassinucellar ovules. The development of embryo sacs in such ovules is of the Polygonum type.

The development of endosperm is of the Nuclear type. In *Prunus* and *Rubus* the chalazal part of the embryo sac develops a long haustorial tube and remains free-nuclear. In *Sanguisorba* a single layer of cells is produced, the rest of the endosperm remains free-nuclear.

The embryogeny conforms to the Asterad type in *Rubus idaeus* (Topham 1970), and Caryophyllad with some deviation from Sedum variation in *R. caesium* and *R. idaeus* (Mandryk and Petrus 1971). Synergid polyembryony, either by fertilization or by apogamy, is reported in *Geum rivale* (Juguet 1966a). Adventive polyembryony (apogamous embryos) occurs in *Potentilla* (Popova and Kolesnikov 1972).

A tendency to apospory is predominant in the family. Somatic apospory of Hieracium type occurs in *Malus hupehensis* (Dermen 1936), *M. sieboldii* (Hjelmqvist 1957b) and *Sanguisorba minor* (Nordberg 1967). In *M. sieboldii* the development of the original megaspore mother cell is suppressed, and some of the cells in the vicinity develop into mature embryo sacs. The pollen tube enters the embryo sac, but the secondary nucleus alone is fertilized; pseudogamy occurs.

Normal fertilization of apomictic eggs is reported in *Malus* and *Rubus* spp. (see Battaglia 1963). H.S. Narayana and Sharma (1976) refer to the tendency of somatic apospory in diploid and polyploid species of *M. hupehensis* and *M. sieboldii*.

Diplospory is strongly exhibited in *Potentilla arenaria* (Czapik 1962), *Rubus caesius* and *R. idaeus* (Mandryk and Petrus 1971). *R. caesius* is an obligate apomict. The unreduced eggs develop due to pseudogamy.

In *Alchemilla arvensis* (Böös 1924), the megaspore mother cell enters meiosis but the nucleus soon degenerates. The adjoining cells, potentially sporogenous, undergo mitotic divisions producing a few parietal cells and two to six axial cells which may divide meiotically or mitotically. The reduced embryo sac must be fertilized before the development of embryo. The unreduced embryo sacs function without fertilization.

In some species of *Potentilla* (Rutishauser 1943), there is a multicelled archesporium. Embryo sacs may develop from these cells by mitotic divisions as in *P. vera* (generative apospory), or from the chalazal cells of the nucellus (somatic apospory) as in *P. canescens*, *P. praecox* and *P. argentea* (see also P. Maheshwari 1950).

Parthenogenetic development of reduced or unreduced eggs occurs in *Alchemilla*, *Crataegus*, *Malus* (Gustafsson 1946), and *Aphanes* (Hjelmqvist 1959).

In *Malus* (Einset 1945, Olden 1953) and *Rubus* spp. (Crane 1940a, b, Einset 1951), the gametophytes develop as usual. Pollination is essential for embryogenesis. Fertilization does not occur even if the sperms enter the embryo sac, and the development of reduced egg is pseudogamous (reduced pseudogamy).

Nordberg (1967) states that in *Sanguisorba* both tetraploid and hexaploid hybrids show a tendency to form unreduced embryo sac. If the egg cell differentiates, it does not develop into an embryo. Then the apomictic embryo formation follows. The unreduced egg cells develop parthenogenetically, or after fertilization. Only the hybrids produce embryos by facultative apospory.

Campbell et al. (1985) studied ovule and embryo sac development in tetraploid (n = 34) individuals of *Amelanchier laevis*. Nomarski differential interference contrast microscopy of cleared whole-ovules, and microtome-stained sections indicated failure of meiosis in ovules. The megasporocyte or its derivatives degenerate, and one to six nearby cells develop into aposporous initials. One of these produces a Polygonum type of embryo sac. The egg forms a proembryo parthenogenetically, but seed maturation requires pollination. This apospory and pseudogamy conforms to the general pattern common in other apomictic genera of Maloideae (Rosaceae).

Unreduced embryo sac may arise from unreduced megaspore mother cell which undergoes mitotic divisions (diplospory). Apospory involves a nucellar cell close to archesporium. Only in *Malus* and *Crataegus* do apomicts produce endosperm autonomously without pollination (Olden 1953). In spite of autonomous development of embryo from egg, seeds may not complete development if endosperm formation is not stimulated by fertilization of polar nuclei.

Apomixis is rare in diploids. In Maloideae all apomictic taxa are either triploid or tetraploid.

Generally, the seed has a sclerotic mesotesta in woody members of the family. In herbaceous members the testa is thin-walled. The outer epidermis is a thin-walled palisade layer with reticulate thickenings. In several members the outer wall of epidermis is thick, and becomes mucilaginous. The inner epidermis of the outer integument, and that of the inner integument, are thin-walled. In unitegmic members the seed-coat is unspecialized.

S Asker (1970, 1977), E Baciu (1971, 1972), I Banerji and KL Chaudhari (1944), E Battaglia (1963), NP Berezenko (1971), G Böös (1924), SC Campbell, CW Greene, BF Neubauer and JM Higgins

(1985), TJ Costa and KAD Mackenzie (1990), MB Crane (1940a, b), R Czapik (1962, 1981, 1985), GM Darrow and GF Waldo (1933), GL Davis (1966), H Dermen (1936, 1949), GJ Dowrick (1961), J Einset (1945, 1948, 1951), Å Gustafsson (1930, 1946), H Hjelmqvist (1957b, 1959, 1962), GW Johnston (1961), M Juguet (1966a), M Labrecque and D Barabe (1984), A Lijefors (1953), P Maheshwari (1950), Vyu Mandryk and YY Petrus (1971), HS Narayana and MK Sharma (1976, 1979), G Nordberg (1967), EJ Olden (1947, 1953), DF Petrov (1939), E Pimienta and VS Polito (1983), LK Popova and SM Kolesnikov (1972), K Ptak (1990), KM Purohit and G Panigrahi (1983), H Rutishauser (1943), M Sharma and HS Narayana (1971), HE Shealy Jr and JM Herr Jr (1973), UP Singh (1959), GL Smith (1963), PB Topham (1970), BV Virdi and GW Eaton (1969)

26.16 Neuradaceae

The embryology has not been investigated.

26.17 Chrysobalanaceae

The embryogeny has not been studied in detail.

The anthers are tetrasporangiate. The anther wall comprises five layers: persistent epidermis, fibrous endothecium, two middle layers, and glandular tapetum with binucleate cells in *Chrysobalanus icaco* (Tobe and Raven 1984b). The anther dehisces by longitudinal slits. Meiosis in microspore mother cells is followed by simultaneous cytokinesis. The tetrads are tetrahedral, occasionally decussate.

The pollen grains are two-celled at the shedding time.

The ovary bears two ovules ascending from the base. The ovule is anatropous, bitegmic and tenuinucellar. The vascular supply shows post-chalazal branches (Netolitzky 1926). The outer integument is more than five layers thick, the inner three to five layers. The outer integument becomes highly tanniniferous and develops into a thick protective structure. The micropyle is formed by both the integuments, and is zigzag in *Licania michauxii* (Tobe and Raven 1984b). A hypostase develops in *C. icaco*, but is absent in *Licania*. The endothelium is a common feature. The archesporium comprises three to five cells. The megaspore mother cells undergo meiotic divisions. Usually there are two to three functional megaspores per nucellus, but only one develops into a modified Polygonum type of embryo sac.

The nucleus of the chalazal megaspore divides mitotically twice to form a four-nucleate embryo sac. The two chalazal nuclei soon unite and move towards the central region of the embryo sac. A mitotic division occurs in the two micropylar nuclei to form four nuclei, three organize into an egg apparatus, while the fourth functions as the upper polar nucleus. The latter unites with the nucleus formed by the fusion of two chalazal nuclei. There are no nuclei at the chalazal end, and, therefore, the antipodal cells are not formed. The mature embryo sac usually contains the egg apparatus and one central nucleus.

Fertilization is porogamous and the development of endosperm is of the Nuclear type. The endosperm remains scanty throughout seed development.

The fusion of two chalazal nuclei, and then with the upper polar nucleus requires confirmation, and development of the endosperm deserves detailed study.

The zygote develops into a globular proembryo with a short suspensor. It later differentiates into a dicotyledonous embryo. The cotyledons are rather large, unfolded and fleshy (Willis 1973).

The seed is ovoid in *Chrysobalanus icaco*, and slightly curved-terete in *Licania apetala*. The seed-coat is smooth. It is mesotestal. Thirty to 50 cells thick tissue (mesotesta) intervenes between the endotesta and exotesta. It is composed principally of undulated "vessels". In *L. michauxii* the mesotesta is much thinner and three to four cells thick. In *L. rigida* Vaughan (1970) reports single-layered endosperm, and five- to six-layered testa that contains vascular bundles. The tegmen collapses.

GT Prance (1972), F Netolitzky (1926), H Tobe and P Raven (1984b), JG Vaughan (1970), JC Willis (1973)

26.18 Connaraceae

The development of anther, pollen, and embryo has not been worked out.

The ovule is hemianatropous with pronounced preraphe (with free integuments between the hilum and the micropyle on the raphe side), bitegmic, and crassinucellar. The micropyle is formed by the outer integument, and funicle is absent (Corner 1976). The single archesporial cell cuts off a parietal cell, and the parietal tissue is two cells thick. The megaspore mother cell undergoes meiotic divisions; the micropylar dyad may not divide, resulting in a triad. The chalazal megaspore of a linear triad develops into a Polygonum type of embryo sac in *Connarus lucçus* (Mauritzon 1939b). The polar nuclei fuse before fertilization, and the secondary nucleus lies near the antipodals.

Development of the endosperm is of the Nuclear type, and it becomes cellular after the formation of numerous free nuclei. The embryo has thick, straight cotyledons, but thin in albuminous seeds (Corner 1976).

Corner (1976) summarizes the main features of connaraceous seed: preraphe form which separates the micropyle widely from the hilum, black testa with firm but not lignified exotestal palisade, fibrous exotegmen, preraphe vascular bundle, chalazal aril, large embryo, and often bean-like smell of the crushed tissue.

EJH Corner (1976), J Mauritzon (1939b)

26.19 Leguminosae (Fabaceae)

The family Leguminosae comprises three subfamilies: Mimosoideae, Caesalpinioideae and Faboideae. The subfamilies are described independently.

Mimosoideae (Mimosaceae)

The anther wall comprises the epidermis, fibrous endothecium, two ephemeral middle layers, and secretory tapetum of uninucleate cells. An intermediate form between the secretory and amoeboid tapetum is reported in *Adenanthera pavonia* (Dnyansagar 1954a). During meiosis in microspore mother cells, the tapetal cells lose their contact with each other and the protoplasts wander between the microspore mother cells and get absorbed. This is doubtful and needs careful reinvestigation.

The microspore mother cells undergo meiotic divisions, and cytokinesis is simultaneous. Simple pollen grains are formed only in a few plants, e.g. *Neptunia* (B. Singh and Shivapuri 1935), *Leucaena* (Dnyansagar 1949) *Prosopis* and *Desmanthus* (Dnyansagar 1957).

In *Dichrostachys cinerea* (Dnyansagar 1954b), occasionally, a transverse septum of tapetal cells is formed in each anther lobe so that its sporogenous mass becomes partitioned into two compartments. In *Parkia biglandulosa* (Dnyansagar 1954b), several sterile transverse septa appear and, eventually, each compartment contains only one pollinium nourished by its own tapetum. In *Mimosa pudica* (Narasimhachar 1951a), the tetrads form pollinia, but in *M. hamata* (Dnyansagar 1951a) single tetrads as well as twin tetrads form pollinia. In *Albizzia lebbek* (P. Maheshwari 1931), *Acacia baileyana* (Newman 1934), and *Calliandra* (Dnyansagar 1958) all the microspores (in a sporangium) together form a single pollinium.

The pollen grains in a polyad (twin tetrad) are unicolpate, trizonocolpate, or trizonocolporate with a smooth surface. They are shed together in masses of eight or 16 pollen grains (Fig. 26.19.1A) in *Acacia dealbata* (Poddubnaya-Arnoldi 1976), *A. farnasiana* (Narasimhachar 1948) and *A. leucophloea* (Ugemuge 1982). Compound pollen grains also occur in *Mimosa hamata*. The

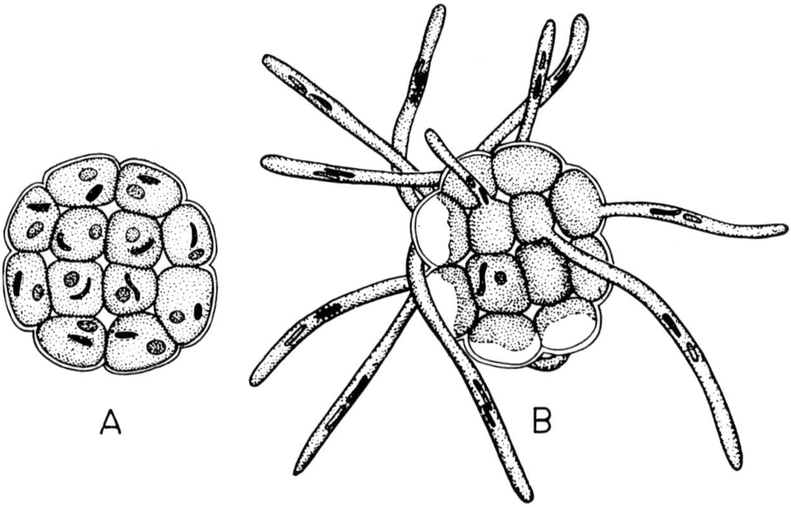

Fig. 26.19.1A, B. *Acacia dealbata.* **A** Pollinium with two-celled pollen grains. **B** Germinated pollen grains. (After Poddubnaya-Arnoldi 1954, see 1964)

pollen grains are two-celled at the shedding stage; three-celled in *Prosopis juliflora* (Deshpande and Gomkale 1982).

The peripheral pollen grains in a pollinium in *A. dealbata* (Fig. 26.19.1B) germinate earlier than the central ones.

In *Albizzia lebbek* (P. Maheshwari 1931), degeneration occurs in anthers only after the formation of microspores. The sterility is partly due to the failure in the mechanism of pollination.

The ovules are ana-, campylo-, or amphitropous, bitegmic, and crassinucellate. The micropyle is usually formed by the outer integument. In *Acacia baileyana* (Newman 1934) it is produced by both the integuments, and in *Parkia biglandulosa* only the endostome is formed. The vascular strand develops in the outer integument up to the tip of micropyle in *Leucaena glauca* (Dnyansagar 1954e). A nucellar cap is produced in *Acacia auriculaeformis* (Dnyansagar 1958).

The archesporium is uni- or multicelled. An archesporial cell functions directly as megaspore mother cell in *Leucaena glauca* (Dnyansagar 1949), or it cuts off a parietal cell. The parietal tissue is well developed in *Desmanthus virgatus* (Gaekwad and Phatak 1952). The megaspore mother cell undergoes meiotic divisions. The chalazal megaspore of a linear or T-shaped tetrad develops into a Polygonum type of embryo sac. The polar nuclei fuse before or at the time of fertilization. The antipodals are persistent.

The development of endosperm is of the Nuclear type. Wall formation is initiated at the micropylar end, and proceeds to the chalaza. In *Mimosa hamata* (Dnyansagar 1951a), wall formation occurs at the four-celled proembryo, whereas the endosperm remains free-nuclear up to the dicotyledonous embryo in *Neptunia triquetra* and *Prosopis spicigera* (Dnyansagar 1954c). The chalazal part remains free-nuclear for a fairly long time, assumes a vesicular form, and becomes haustorial. In *Prosopis juliflora* (B.M. Johri and Garg 1959), the haustorium shows finger-like processes at the chalazal tip. In older seeds it shrinks and collapses, and only traces persist.

The embryogeny conforms to the Onagrad type-Trifolium variation, or Period I, Series B, Megarchetype VI (Dnyansagar 1954e). In several members the embryo lacks a suspensor (Lersten 1983): *Acacia baileyana* (Newman 1934), *Adenanthera pavonia* (Dnyansagar 1958), *Albizzia lebbek* (Pillai et al. 1974), *Mimosa hamata* (Dnyansagar 1951a, 1954c), and *Prosopis spicigera* (Dnyansagar 1954a, 1957).

The seed contains a large, straight embryo with a short radicle and thick cotyledons. A funicular aril is present in some members, such as *Pithecolobium*. The seed-coat is similar to that in Papilionaceae.

PA Buss Jr and NR Lersten (1975), MA Caccavari (1986), B Dell (1980), PK Deshpande and KD Gomkale (1982), VR Dnyansagar (1949, 1951a, b, 1952, 1954a, b, c, d, e, 1955, 1955–56, 1957, 1958, 1970), S Feuer, CJ Niezgoda and LI Nevling (1985), LK Gaekwad and VG Phatak (1952), BM Johri and S Garg (1959), J Kenrick and RB Knox (1979, 1982), DA Larston and CW Lewis (1962), NR Lersten (1971, 1982, 1983), P Maheshwari (1931), NC Nair and S Kahate (1961), SG Narasimhachar (1948, 1951a), IV Newman (1934), SK Pillai, K Ramasita and D Ishwar (1974), VA Poddubnaya-Arnoldi (1976), B Roy (1933), VG Serrato, P Modenesi, G Roti-Michelozzi and L Bevilacqua (1986), B Singh and TN Shivapuri (1935), K Tybirk (1989), NR Ugemuge (1982)

Caesalpinioideae (Caesalpiniaceae)

The anther wall comprises the epidermis, fibrous endothecium, three ephemeral middle layers, and glandular tapetum of two- to multinucleate cells (Venkatesh 1956a). The tapetal cells are tetranucleate in *Cassia didymobotrya* (M.L. Sethi 1930), and *C. tora* (R.M. Datta 1934). The tapetum is periplasmodial in *Cassia absus*, *C. kleinii* and *C. mimosoides* (Venkatesh 1956a). The microspore mother cells undergo meiosis, cytokinesis is simultaneous, and tetrahedral, decussate, isobilateral and T-shaped tetrads are produced.

The pollen grains are three-(four)-zonoaperturate and smooth. They are two-celled at the dispersal stage.

The ovule is anatropous, bitegmic, and crassinucellate with a zigzag micropyle. The outer integument is thicker than the inner one, and the integumentary vascular strand extends up to the tip of micropyle (Rau 1950a, d).

The archesporium is uni- or multicelled, parietal cell is cut off, and the parietal tissue is well developed. A nucellar cap is also a common feature (see Dnyansagar 1970).

The megaspore mother cell undergoes meiotic divisions. The chalazal or epichalazal megaspore of a linear or T-shaped tetrad develops into a Polygonum type of embryo sac. In *Cassia nictitans* (Rembert 1969a), the micropylar dyad fails to undergo meiosis II, and the chalazal megaspore of the lower dyad functions. The polar nuclei fuse prior to fertilization, and the three antipodals persist, except in *Cassia mimosoides*, and *Parkinsonia*. Occasionally, double linear tetrads occur in *Gleditsia triacanthos* (Rembert 1969a).

The development of the endosperm is of the Nuclear type. Wall formation is initiated at the micropylar end, and extends to the chalaza. The chalazal part remains free-nuclear, and develops into a vesicular haustorium. In some species of *Cassia* (Anantaswamy Rau 1951a, Pantulu 1951), the haustorium is tubular with a vesicular tip. In *C. mimosoides* and *C. occidentalis* (Rau 1951a), one more free-nuclear bulbous portion lies below the cellular endosperm. In *C. sophera* (B.M. Johri and Garg 1959), the haustorial nuclei become spindle-shaped. The haustorium collapses in older seeds.

The embryogeny conforms to the Asterad type in *Bauhinia acuminata* (Gomkale and Deshpande 1979) and *Cassia occidentalis* (Pantulu 1951). The suspensor is poorly developed or is absent (see Lersten 1983). Guignard (1881) refers to a small suspensor in *Caesalpinia mimosoides*, *Cassia corymbosa* and *Cercis siliquastrum*, and its absence in *Ceratonia siliqua*, *Gymnocladus dioica*, *Parkinsonia aculeata* and *Poinciana gilliesii*. A short suspensor of cells larger than those of the proembryo, but not sharply delimited from it, develops in *Caesalpinia mimosoides*, *Cassia corymbosa* and *Cercis siliquastrum* (Guignard 1881).

In *Cassia occidentalis*, the suspensor is uniseriate, filamentous, of about five short cells (Pantulu 1951). Deshpande and Thakkar (see Suri and Deshpande 1981) misinterpret the suspensor in *C. mimosoides* as endosperm haustoria.

Nucellar polyembryony is of common occurrence in *B. acuminata* (Gomkale and Deshpande 1979).

The seed-coat is more or less similar to that of Mimosaceae. The linea lucida is absent, and the inner epidermis of testa develops into hour-glass cells. The funicle is thickened, and arilloid.

L Bevilacqua, P Profumo, P Gastaldo and P Barella (1988), PA Buss Jr and NR Lersten (1975), EJH Corner (1951, 1976), RM Datta (1934), PK Deshpande and SJ Thakkar (1982), VR Dnyansagar (1955–56, 1970), KD Gomkale (1980), KD Gomkale and PK Deshpande (1979), L Guignard (1881), BM Johri and S Garg (1959), NR Lersten (1971, 1983), SK Mukherjee (1952), PK Mukherjee and OS Thakur (1959), JV Pantulu (1945, 1951), AK Paul (1937), M Anantaswamy Rau (1950a, d, 1951a), DH Rembert Jr (1969a), ML Sethi (1930), RK Suri and PK Deshpande (1981), M Takahashi (1989), NR Ugemuge (1982), CS Venkatesh (1956a), J Venkateswarlu and V Seshavataram (1971)

Faboideae (Papilionaceae)

The anther wall comprises the epidermis, fibrous endothecium, two ephemeral middle layers, and secretory tapetum of uninucleate cells. The sterile sporogenous tissue divides the loculus into several chambers in *Desmodium glutinosum*, *D. illinense* (Buss et al. 1969), and *Phaseolus aconitifolius* (Deshpande and Bhasin 1974). Shukla (1954) reported intermediate behaviour of tapetal cells in *Lens esculenta.* By the time uninucleate pollen grains are formed, the walls of the tapetal cells collapse and the protoplasts form a peripheral layer around the microspores. The periplasmodium disappears at the binucleate stage of pollen. Calcium oxalate crystals have been observed in *Desmodium* (Buss et al. 1969). The microspore mother cells undergo meiotic divisions followed by simultaneous cytokinesis by furrowing. The tetrads are tetrahedral, decussate (Misra and Sahu 1970), and isobilateral (Mercy et al. 1974).

The pollen grains are spheroidal, and trizonocolpate in *Indigofera enneaphylla* (Untawale and Deshpande 1968). They are two- or three-celled at the dispersal stage. Occasionally, they germinate in situ in *Cicer arietinum* and *C. soonagaricum* (Mercy et al. 1974). They are polysiphonous in *Lupinus.*

The ovule is anatropous, hemianatropous, amphitropous or campylotropous, bitegmic, and crassinucellate with a zigzag micropyle.

An obturator develops in *Amorpha glabra* (Kostrikova 1972), and an endothelium differentiates in *Alfalfa* (Reeves 1930b), *Erythrina indica* (Chandravadana 1963), *Tephrosia* (Narang and Chandra 1977), and *Vicia faba* (Mitchell 1975).

The hypodermal archesporium is uni- or multicelled. A single archesporial cell cuts off a parietal cell, and the parietal tissue is one- to two-layered. The megaspore mother cell undergoes meiotic divisions, and the tetrads are linear, T-shaped, oblique-linear and ⊥-shaped (Rembert 1969b). The chalazal megaspore develops into a Polygonum type of embryo sac. Sometimes, epichalazal megaspore functions as in *Trifolium repens* (Martin 1914) and *Vicia faba* (Mitchell 1975). The polar nuclei fuse before fertilization, and the antipodals are ephemeral. Allium type of development is reported in *Laburnum anagyroides* (Rembert 1966) and *Pongamia*, and Oenothera type in *Vigna unguiculata* (Ojeaga and Samyaolu 1970).

Several abnormal megagametophytes have been reported in *Sesbania aculeata* and *C. aegyptica* (Salgare 1974a, b): three-nucleate, six-nucleate, nine-nucleate, and 15-nucleate with eight nuclei at the micropylar end and seven at the chalazal end. The origin of abnormality, and the frequency of occurrence, have not received attention.

A nucellar cap is present in *Milletia ovalifolia* (N. Pal 1960). In *Indigofera enneaphylla* (Deshpande and Untawale 1971) also a cap-like structure develops; at the late globular proembryo the nucellar epidermis becomes glandular. A reinvestigation is suggested. The nucellar epidermal layer persists in the micropylar region of seed in *Desmodium gangeticum* (Pantulu 1942), *Phaseolus vulgaris* (Weinstein 1926) and *P. aconitifolius* (Deshpande and Bhasin 1974). In *Zornia diphylla* (Deshpande and Bhasin 1976), the tip of embryo sac extends into the micropyle, and comes in direct contact with the outer integument. The nucellar cells in contact with the inner integument elongate and give endothelium-like appearance. The chalazal end of the nucellus is extensively developed in *Phaseolus* spp. (Seshavatharam 1978). The chalazal tissue comprises polygonal, thick-walled, refractive cells in the seeds of *Cicer arietinum* (Behl and Tiagi 1977a).

In *Vigna catjang* the organization of a tapetum-like layer, successively by the nucellus, inner integument, endosperm and, finally, even by the epidermal layer of the hypophysis region of embryo is noteworthy (Anantaswamy Rau, 1951b). These features require confirmation.

Trifolium pratense (Hindmarsh 1964) gives an extremely low yield. Nearly nine per cent of the microspore tetrads are abnormal and uninucleate; sterile pollen grains are produced. The sterility of ovule, formation of twin gametophytes, arrested growth of embryo sacs, and degeneration of egg and synergid are some of the causes of low yield. In addition, the nucellus forms a plug and prevents the entry of pollen tube through the micropyle.

In *T. repens* (Rembert 1977), the young gametophyte grows vigorously, and its expansion is checked by the adjoining nucellus. This causes damage to the gametophyte which results in reduced seed-set. Further information is necessary to confirm this observation.

In *Indigofera enneaphylla* (Deshpande and Untawale 1971), the pollen tube enters the embryo sac between the egg and one synergid without destroying the latter, and remnants of the pollen tube persist. It is rather unlikely that the pollen tube does not destroy one of the synergids. Cook (1924b) states that in *Crotalaria sagittalis* the pollen tube is "parasitic" and destroys the contents of the embryo sac. Perhaps Cook based his observations on fertilized embryo sacs which had degenerated in the normal course. There is no record of parasitic behaviour of pollen tube in angiosperms.

Occasionally, single fertilization occurs in *Arachis hypogea* (B.W. Smith 1956), which needs confirmation.

Sterility in *Crotalaria juncea* (Samal 1936) is due to the failure of growth of pollen tube, so that fertilization does not occur.

The development of the endosperm is of the Nuclear type. Endosperm nodules develop in *Lupinus albus*, *L. luteus* and *L. augustifolius* (Jaranowski 1962). Wall formation is totally absent in *Stylosanthes mucronata* (Anantaswamy Rau 1953). Only one or two layers of cells are formed around the embryo in *Clitoria ternata* (Rau 1951e) and *Lathyrus* spp. (Terzijski and Dumanova 1974).

In *Arachis hypogea* (Prakash 1960a) and *Indigofera enneaphylla* (Deshpande and Untawale 1971), wall formation in endosperm commences from the periphery to the centre, while in most members it commences from the micropylar region. Wall formation is initiated at the globular proembryo but, in some cases, it may be initiated only after the formation of cotyledonary primordia in the embryo.

In most taxa the endosperm becomes cellular in the micropylar region, and in the chalazal region it remains free nuclear for a long time. The latter functions as a haustorium and encroaches upon the chalazal tissue (Anantaswamy Rau 1951f, 1953, B.M. Johri and Garg 1959, Sehgal and Gandhi 1985). The free nuclei in the chalazal region become hypertrophied, and undergo amitotic and endomitotic divisions and nuclear fusions (Terzijski and Dumanova 1974). Amitotic divisions in most cases have turned out to be aberrant mitotic figures. The haustorium may be broad, or narrow and tubular. The haustorium becomes cellular in *Cyamopsis psoralioides* and *Desmodium pulchellum* (Anantaswamy Rau 1953). A pad of thick-walled cells, similar to postament, checks the encroachment of haustorium in the chalazal region.

The development of the endosperm in *Trifolium repens* × *T. uniflorum* (Chen and Gibson 1971), and *T. semipilosum* × *T. repens* (White and Williams 1976) is abnormal. In the former the endosperm collapses during early stages of development, and the growth of hybrid embryo is arrested. According to Chen and Gibson, this must be due to genetic imbalance of the endosperm itself, or unfavourable interaction of the endosperm with the maternal tissue.

Deshpande and Thakkar (see Suri and Deshpande 1981) state: " . . . in *Cassia mimosoides* . . . the cells of the endosperm situated above the suspensor not only elongate but show a tendency to divide mitotically". Such elongated multinucleate cells, according to them, take up haustorial function; so far this is the only record of micropylar endosperm haustorium for the entire Leguminosae. Further, "The endosperm cells, in the micropylar region above the suspensor, in *C. absus*, elongate to form a cap-like structure." It is obvious that in both cases the authors misinterpret suspensor cells as "endosperm haustoria" and "endosperm cap". They must have examined non-median sections which did not have the suspensor connected to the embryo sac wall. Such an error could have been avoided if they had also examined dissections and transections.

Variations occur in the development of embryo (see Rau 1954, Seshavatharam 1969). The embryogeny conforms to the Onagrad, Asterad, Caryophyllad and Solanad type.

In Papilionaceae (Lersten 1983), the suspensor shows interesting features and there is much diversity in the form and structure. It may be massive, pear-shaped, haustorial, extremely elongated, thread-like, or rudimentary. It is large with inflated cells toward the base in *Sophora flavescens* (Nägl 1962). It has bulging cells in *Cyamopsis psoralioides* (Rau 1954), and *Glottidium vescarium*

(N. Prakash and Herr 1979). The suspensor is biseriate and filamentous in *Atylosia barbata* (Crété 1958c), and massive in *Hedysarum coronarium* (Guignard 1881). In *Vicia* the haustorial suspensor protrudes into the micropyle.

In the tribe Loteae the suspensor is a short, biseriate filament but the ploidy levels increase in the polytenic suspensor nuclei (Freed and Grant 1976). Extremely inflated suspensor cells develop in *Anthyllis tetraphylla* (Guignard 1881) and *A. vulneraria* (Souèges 1951a). In the tribe Viscicae the suspensor is most distinctive and uniform, and consists of four cells in each tier, each containing several nuclei at various ploidy-levels (Nägl 1962): *Lathyrus* (Guignard 1881), *Lens* (Crété 1948a) and *Pisum* (Cooper 1938, Yoffe 1957). A long, slender, biseriate filamentous suspensor develops in *Cicer arietinum* (Hegelmaier 1880).

In *Crotalaria juncea* (Samal 1936), the suspensor is much larger and massive and like a carrot tap-root with epidermal cells protruding from the proximal half. The suspensor cells contain chloroplasts and the tubular peripheral cells have a haustorial role (Anantaswamy Rau 1950b).

In *Phaseolus coccineus* (Yeung and Clutter 1978), at the globular proembryo the suspensor is broad, massive and multiseriate and shows conspicuous wall-ingrowths. The nuclei in the basal region of the cells are hypertrophied.

According to Terzijski and Dumanova (1974), "the deleterious effects of the quantitative reduction of the endosperm are compensated by the differentiation of a well-developed suspensor, composed of four hypertrophied coenocytic structures in *Lathyrus*".

Cleavage polyembryony is reported in *Crotalaria sagittalis* (Cook 1924b), and nucellar polyembryony in *Millettia ovalifolia* (N. Pal 1960). Dhawan and Bhojwani (1985) observed twin embryos in the mature seeds of *Leucaena leucocephala*, 0.7% in CVK-28 and 0.2% in CVK-8. Both the embryos were viable; their origin was not determined.

In *Melilotus alba* (Hubam) × *M. messanensis* (Jaranowski 1961), the diploid zygotic embryo, 6 days after pollination, showed budding resulting in an additional embryo from the massive globular proembryo. When a cross with changed partners was made, that is *M. messanensis* × *M. alba* (Hubam), sometimes haploid-diploid twins developed. It is assumed that the diploid zygotic embryo and the haploid synergid embryo would have started developing at the same time. But, at this stage, 140 h after pollination, the zygotic embryo takes the lead and grows faster than the synergid embryo. Two dissimilar embryos resulted from a cross between *M. messanensis* and *M. polonica* × *M. alba*; keeping the male parent *M. polonica* and using *M. alba* (Hubam) as female, two haploid and one diploid proembryo developed.

The embryos produced with *M. polonica* × *M. alba* as male parent had an elongated suspensor, and a higher potential for the development of additional embryos. In Figure 26.19.5B, C, there are haploid-diploid twins but with a different structure of the suspensor.

While the embryology of *Arachis hypogea* is normal, in *A. hagenbeckii* (Bharathi and Murty 1984) the embryo aborts at the octant stage due to absence of endosperm and hyperplastic growth of endothelium. In *A. glabrata* (Bharathi and Murty 1984), some of the embryos reach the heart-shaped stage, but viable seeds are not formed. The endosperm is of the Nuclear type with a tubular coenocytic chalazal haustorium. The embryological differences between wild and cultivated species account for the differences in their fruit characters only to a limited extent. The failure of seed-set in *A. hagenbeckii* is due to malfunction of endosperm.

The testa is generally multiplicative. The outer epidermis of testa comprises a palisade layer of thick-walled malpighian cells with linea lucida and hexagonal facets. The outer wall may become mucilaginous. The outer hypodermis is differentiated as hour-glass cells, or osteosclereids. It is followed by the aerenchymatous mesophyll. The rest of the tissue becomes compressed. The tegmen is unspecialized, and it gets crushed (Corner 1951). In *Canavalia* spp. the osteosclereids form two or three hypodermal layers (Ambegaokar 1976).

The structure of epidermal macrosclereids and hypodermal osteosclereids form the basis for classification of seeds (see Chowdhury and Buth 1970, Ambegaokar 1976, Behl and Tiagi 1976, 1977a, b).

S (Shikha) Agarwal (1984), KB Ambegaokar (1976), M Anantaswamy Rau (1950b, c, 1951b, c, d, e, f, 1953, 1954, 1955), AI Atabekova and CH Ling (1962), HM Behl and B Tiagi (1976, 1977a, b, 1978), NL Benikova (1952), RS Benavente, H Skorupska, RG Palmer and RC Shoemaker (1989), LR Bevilacqua, F Fossati and G Dondero (1987), M Bharathi, PB Kirti, VR Murty and NGP Rao (1982), M Bharathi and VR Murty (1984), JA Biddle (1979), CL Briggs (1989), RA Brink and DC Cooper (1936), MM Brown (1917), PA Buss Jr and NR Lersten (1975), PA Buss Jr, FD Gallen and NR Lersten (1969), BG Cameron and N Prakash (1990), J Cebrat (1967), P Chandravadana (1963, 1965), C Chen and PB Gibson (1971), S Chopra, S Agarwal and RP Singh (1984), KA Chowdhury and GM Buth (1970), MT Cook (1924b), DC Cooper (1938), EJH Corner (1951), P Crété (1948a, 1958c), B Dell (1980), PK Deshpande and RK Bhasin (1974, 1976), PK Deshpande and AG Untawale (1971), V Dhawan and SS Bhojwani (1985), MW Folsom and DD Cass (1989), A Ganesan, SS Shah and MS Swaminathan (1957), VR Dnyansagar (1970), RR Dute, CM Peterson and AE Kushing (1989), HM Farley and AH Hutchinson (1941), HJ Freed and WF Grant (1976), GP and RA George and JM Herr Jr (1979), L Guignard (1881), TM Halward and HT Stalker (1987), WA van Heel (1974), F Hegelmaier (1880), GJ Hindmarsh (1964), M Hossaert and M Valevo (1988), J Jaranowski (1961, 1962), R Jayalakshmi and KK Lakshmanan (1990), SS Jha (1985), SS Jha and AK Pandey (1989), BM Johri and S Garg (1959), AC Joshi (1938b), MK Kandasamy and M Vivekanandan (1984), AT Kapuskar (1964), JC Kennel and HT Horner (1985), JR King (1947), LN Kostrikova (1972), S Krupko (1973), NR Lersten (1982, 1983), KH Makde (1971), JN Martin (1914), ST Mercy, SN Kakkar and TM Varghese (1974), RC Misra and RC Sahu (1970), JP Mitchell (1975), DR Murray (1987), UR Murty, NGP Rao, PB Kirit and M Bharathi (1981), W Nägl (1962, 1974), AK Narang and N Chandra (1977), OO Ojeaga and MO Samyaolu (1970), N Pal (1960), KK Pandey (1955), AK Pandey, SS Jha and A Jha (1990), JV Pantulu (1942), HE Pattee and SC Mohapatra (1987), N Prakash and YY Chan (1976), N Prakash and JM Herr Jr (1979), S Prakash (1960a), KA Pramanik and K Thothathri (1986), NS Rangaswamy and L Nandkumar (1985), HL Reed (1924), RG Reeves (1930a, b), DH Rembert Jr (1966, 1969b, 1971, 1977), SA Salgare (1974a, b), KK Samal (1936), N Sangduen, GL Kreitner and EL Sorensen (1983a, b), JH Schaffner (1901), E Schnepf and W Nägl (1970), CB Sehgal and V Gandhi (1985), V Seshavatharam (1969, 1978, 1982), N Seth and MR Vijayaraghavan (1990), TC Shukla (1954), BW Smith (1956), DL Smith (1971), R Souèges (1949c, 1951a), R Souèges and P Crété (1952), D Subramanian (1983), D Terziyski (1971), VR Tilton, LW Wilcox and RC Palmer (1984), D Terzijski and A Dumanova (1974), NR Ugemuge and MD Padhye (1989), AG Untawale and PK Deshpande (1968), AJ Weinstein (1926), DWR White and E Williams (1976), B Wojciechowska (1971), EC Yeung (1980), EC Yeung and MJ Cavey (1988), EC Yeung and ME Clutter (1978), MD Yoffe (1957)

26.20 Krameriaceae

The embryology has not been investigated in sufficient detail. Simpson and Skvarla (1981) studied the pollen morphology and ultrastructure, and Verkerke (1986b) ovule ontogeny and seed-coat development of *Krameria*.

The anthers are tetrasporangiate, and dehiscence occurs by terminal pores. The isopolar, spheroidal, three-(four)-colporate to triporate pollen grains do not contain an appreciable amount of starch, and are rarely collected by bees (Simpson and Skvarla 1981).

The ovular primordia are trizonate, the inner and outer integuments are of dermal origin. The ovule curves anatropically, at the same time it rotates

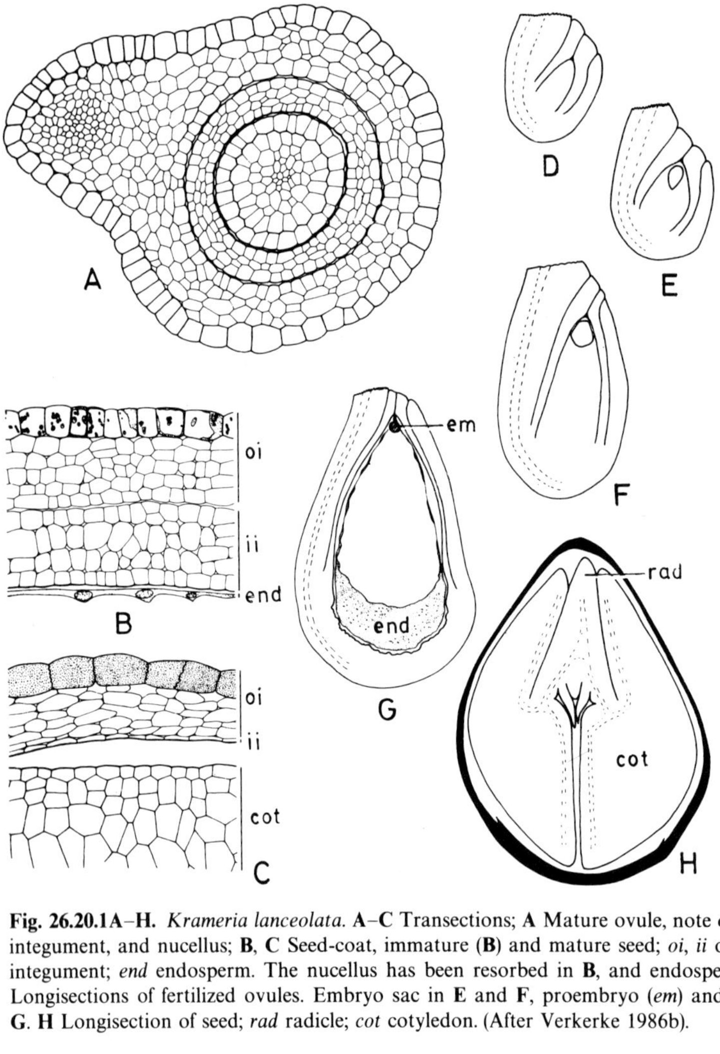

Fig. 26.20.1A–H. *Krameria lanceolata*. **A–C** Transections; **A** Mature ovule, note outer and inner integument, and nucellus; **B**, **C** Seed-coat, immature (**B**) and mature seed; *oi*, *ii* outer and inner integument; *end* endosperm. The nucellus has been resorbed in **B**, and endosperm in **C**. **D–G** Longisections of fertilized ovules. Embryo sac in **E** and **F**, proembryo (*em*) and endosperm in **G**. **H** Longisection of seed; *rad* radicle; *cot* cotyledon. (After Verkerke 1986b).

outward over 90 degrees. Eventually, the superimposed curvature and rotation lead to an anatropous, epitropous ovule with a curved funicle. The curved funicle and the raphe have anomocytic stomata.

The pyriform ovule has a narrow and elongated nucellus. The inner integument is three to four cells thick, and the outer three to five cells (Fig. 26.20.1A). The micropyle is formed entirely by the inner integument and the endostome is very narrow. The raphal vascular strand extends into the outer integument. A hypostase-like inner layer of flattened strongly tanniniferous cells extends over the whole nucellus-chalaza connection.

The hypodermal archesporial cell cuts off a parietal cell which undergoes a few divisions. Following meiosis, the megaspore mother cell produces a linear, sometimes T-shaped, tetrad and the chalazal megaspore produces the embryo sac. Further details have not been worked out.

The structure of the testa and tegmen is shown in Figure 26.20.1B, C, and an enlargement of post-fertilized ovule in D–G. The embryo sac is at first situated apically in the nucellus, then (at the globular proembryo) it enlarges a great deal and consumes the nucellus as well as the vast chalazal region (Fig. 26.20.1 cf. E–G).

The endosperm is of the Nuclear type, and most of the nuclei aggregate in the chalazal region of the embryo sac which is lined by hypertrophied cells. Wall formation does not occur in the endosperm, and it is consumed by the growing embryo.

The embryogeny has not been investigated. The mature seed (Fig. 26.20.1H) contains a large embryo with two plano-convex cotyledons. At the upper end each cotyledon extends backward and the two auricles enclose the radicle leaving only its apex exposed. In a cross section there would be four auricles (?), a condition reported earlier in *Floerkea* by P. Maheshwari and Johri (1956).

During seed development, the inner integument thickens by periclinal divisions (Fig. 26.20.1B). The tegmen is mostly absorbed in the ripe seed, and only the innermost layer and cuticle persist. The outer integument also thickens, and the tanniniferous exotesta with slightly thickened walls and a few layers of parenchymatous cells persist (Fig. 26.20.1C). The mesophyll is traversed by numerous vascular bundles (Corner 1976).

The exalbuminous seed, without any especial development of seed-coat, is enclosed in a barbed fruit.

TB Batygina and MS Yakovlev (1985), EJH Corner (1976), J Hutchinson (1973), P Maheshwari and Johri (1956), BB Simpson and JJ Skvarla (1981), W Verkerke (1986b)

26.21 Taxonomic Considerations

The suborder Hamamelidineae comprises Platanaceae, Hamamelidaceae and Myrothamnaceae. Wettstein (1935), Takhtajan (1966, 1980) and Cronquist (1968, 1981) treat these families in Hamamelidales. Hutchinson (1973) groups them in his Hamamelidales alongwith Bruniaceae and Tetracentraceae. According to Hallier (1903, 1912), Tippo (1938), Tang (1943), Takhtajan (1966, 1969, 1980) and Cronquist (1968, 1981), Hamamelidaceae (with inferior or semi-inferior ovary) are derived from Magnoliales. Bentham and Hooker (1965a), Baillon (1871, 1874), Bessey (1915), Harms (1930), Melchior (1964), Endress (1967), Jay (1968) and Hutchinson (1973) presume that Hamamelidaceae arose from a Protorosalian stock with affinities with the living Rosaceae, Saxifragaceae and Cunoniaceae. According to Tippo (1938), Hamamelidaceae gave rise to Casuarinaceae, Fagales and Urticales.

Platanaceae, Myrothamnaceae. The embryology of Platanaceae and Myrothamnaceae has not been adequately worked out. Platanaceae show alliance with Hamamelidaceae: three-colpate pollen grains, bitegmic, crassinucellate ovule, raphe bundle ending at chalaza, formation of micropyle by both integuments, oil and aleurone as reserve food in endosperm, and a narrow embryo with flat cotyledons. The simplified, much-reduced seed of Platanaceae is probably derived from hamamelidaceous seed (Mohana Rao 1974a). Anatomical and cytological data support their close relationship, as also does a chemotaxonomic study (Jay 1968, 1969).

Hamamelidaceae. Hamamelidaceae have secretory anther tapetum, simultaneous cytokinesis in pollen mother cells, anatropous, bitegmic, crassinucellar ovule with a massive parietal tissue, Polygonum type of embryo sac, and Cellular type of endosperm in some members as in Magnoliaceae. The development of an endosperm haustorium is absent in the family, and Solanad type of embryogeny occurs in Eupteleaceae of the magnoliaceous families. Embryologically, Myrothamnaceae differ from Platanaceae and Hamamelidaceae, to some extent, but show similarity with Monimiaceae, another magnoliaceous family: two-celled pollen grains in permanent tetrads, anatropous, crassinucellate ovule, and bisporic Allium type of embryo sac. Hamamelidaceae, Myrothamnaceae and Platanaceae, the phytochemically allied families forming a natural group (Shaw and Gibbs 1961), are related to Magnoliales. Hamamelidaceae also show alliance with Saxifragaceae. Both have massive parietal tissue, Polygonum type of embryo sac, Nuclear type of endosperm, and Solanad type of embryogeny.

The suborder Saxifragineae includes Crassulaceae, Cephalotaceae, Saxifragaceae, Bruneliaceae, Cunoniaceae, Davidsoniaceae, Pittosporaceae, Byblidaceae, Roridulaceae and Bruniaceae.

Crassulaceae. The family Crassulaceae has been placed in the order Rosales, adjacent to Saxifragaceae, by Bentham and Hooker (1965a), Bessey (1915), Wettstein (1935), Engler and Diels (1936), and Rendle (1952). Hutchinson (1973) includes the families Crassulaceae and Saxifragaceae in Saxifragales.

Subramanyam (1962) compares the embryological features of Crassulaceae and Saxifragaceae, and points out their close similarities: multilayered anther wall, simultaneous cytokinesis in pollen mother cells, three to polycolpate, two-celled pollen grains, indefinite number of crassinucellate, usually bitegmic ovules, uni- or multicelled archesporium, Polygonum type of embryo sac, Cellular type of endosperm with chalazal haustorium, and Caryophyllad type of embryogeny. But secretory anther tapetum, apocarpous superior ovary with crassinucellate, bitegmic ovule, total absence of endothelium, development of haustoria from megaspores, synergids and antipodals, and haustorial behaviour of the basal cell of two-celled proembryo are features characteristic of Crassulaceae.

Crassula aquatica has a mode of life somewhat similar to Podostemaceae. According to Subramanyam (1962), its reduced endosperm is a transitional stage to complete suppression of endosperm formation in Podostemaceae. The Crassulaceae have affinities with Hydrostachyaceae (Sect. 27.2) and Podostemaceae (Sect. 28.2).

Saxifragaceae. Saxifragaceae show diversity of features as regards anther tapetum, integuments, parietal tissue, and uni- or multi-celled archesporium. The occasional presence of Allium type of embryo sac and endothelium, varying position of primary endosperm nucleus, Nuclear, Cellular or Helobial type of endosperm development, and Solanad or Caryophyllad type of embryogeny segregate it from the other members of Rosales.

Occurrence of pollen embryo sacs in *Heuchera micrantha* is rather interesting.

Pace (1912) relates *Parnassia* to Droseraceae due to structural similarity of ovule and embryo sac (see Sarraceniales, Sect. 23). Hallier (1901) includes *Parnassia* in Sarraceniales; in 1908 he proposed to place the genus in a separate family near Ranunculaceae, Nymphaeaceae, Droseraceae and Sarraceniaceae. Hutchinson (1959, 1973) also assigns *Parnassia* to a separate family, Parnassiaceae. V.K. Sharma (1968b) compares the embryological features of *Parnassia* with Saxifragaceae, Droseraceae and Hypericaceae (Table 26.21.1). He proposes that *Parnassia* be assigned to a separate family, Parnassiaceae, which forms a link between Saxifragaceae and Droseraceae, and shows similarities with Hypericaceae. *Parnassia* differs from Saxifragaceae: tenuinucellate, bitegmic ovule, large persistent antipodals, Nuclear type of endosperm, and Asterad type of embryogeny. The embryogeny in *Parnassia* belongs to Period I (Asterad type) in Souèges system of classification, whereas in the other three families it corresponds to Period II (Caryophyllad type).

Similarly, *Vahlia* (syn. *Bistella*, subfamily Vahlioideae) also differs in certain respects from the family Saxifragaceae: tenuinucellate ovule, presence of endothelium with binucleate cells, chalazal growth of embryo sac, and richly cytoplasmic chalazal endosperm cells which probably have a haustorial function (Saxena 1973). Anatomical data also support its segregation from Saxifragaceae to an independent family, Vahliaceae.

Hutchinson (1973) assigns the genera *Parnassia* and *Bistella* to separate families, Parnassiaceae and Vahliaceae. In fact, Saxifragaceae are not a uniform family from a systematic and embryological point of view, as stated by Melchior (1964).

Table 26.21.1. Embryology of Guttiferales (Sect. 22.16), Droseraceae (23.4), Saxifragaceae (26.7), *Parnassia*, *Bauera* and Cunoniaceae (26.9) (See V.K. Sharma 1968b, N. Prakash and McAllister 1977)

	Guttiferales	Droseraceae	Saxifragaceae	*Parnassia*	*Bauera*	Cunoniaceae
Anther tapetum	Secretory, cells 2-nucleate	Secretory, cells 2-nucleate; sometimes pseudoplasmodial	Secretory, cells 2–4-nucleate	Secretory, cells 1- or 2-nucleate	Secretory, cells uninucleate	Secretory, cells binucleate
Cytokinesis	Simultaneous	Simultaneous	Simultaneous	Simultaneous	Simultaneous	Simultaneous
Tetrads	Tetrahedral, isobilateral, decussate	Tetrahedral, isobilateral	Tetrahedral, isobilateral, decussate	Tetrahedral	Tetrahedral, decussate	Tetrahedral
Pollen	Free, tricolpate, 2-celled	Free or united in tetrads, generative cell cut off distally, tricolpate, 2-celled	Free, tricolporate or bicolpate, 2- or 3-celled, pollen embryo sacs reported in *Heuchera micrantha*	Free, tricolporate, 2-celled	Free, bicolpate, 2-celled	Free, 2-celled
Ovule	Anatropous, bitegmic, tenuinucellate, micropyle formed by both integuments	Anatropous, bitegmic, tenui- or crassinucellate, micropyle formed by inner integument	Anatropous, uni- or bitegmic, tenui- or crassinucellate, micropyle formed by both integuments in bitegmic ovules	Anatropous, bitegmic, tenuinucellate, micropyle formed by inner integument	Anatropous, bitegmic, crassinucellate, micropyle zigzag and formed by both integuments	Ana- to hemianatropous, bitegmic crassinucellate, micropyle zigzag and formed by both integuments, hypostase present
Tetrad	Linear	T-shaped	Linear or T-shaped	Linear or T-shaped, triads (upper dyad and two megaspores) common	Linear, twin tetrads develop	Linear or T-shaped

Embryo sac	Polygonum type, antipodals, ephemeral, supernumerary in *Hypericum*	Polygonum, antipodals conspicuous, nucellar cells elongate radially around the embryo sac	Polygonum, sometimes bisporic, antipodals ephemeral or persistent, supernumerary in *Saxifraga*	Polygonum, antipodals persistent, supernumerary	Polygonum, rarely supernumerary antipodals occur, ephemeral	Polygonum antipodals ephemeral
Endosperm	Nuclear type, a chalazal cyst is formed, becomes cellular	Nuclear, nuclei aggregate in chalazal region which functions as a haustorium, entire endosperm becomes cellular	Cellular, Helobial or Nuclear, the entire endosperm becomes cellular in Helobial or Nuclear type	Nuclear, becomes cellular	Nuclear, later stages not examined	Nuclear, becomes cellular
Embryo	Solanad or Caryophyllad type	Caryophyllad, Asterad (?) or Onagrad (?)	Caryophyllad or Asterad	Asterad	—	—
Polyembryony	Apogamous from synergid, proliferation of suspensor cells, aposporous from nucellar cells	Absent	Synergid polyembryony	Absent	Absent	Absent
Seed-coat	Exotegmic, mesotestal or unspecialized	Not multiplicative, in some taxa testa larger than tegmen with an air-space between them	Not multiplicative, testa of radially-elongate or thick-walled cells, in *Heuchera* thick-walled endotegmen persists	Multiplicative, exotestal (?)	Multiplicative	Exotegmic
Seed	Arillate, exalbuminous, embryo large in Clusieae	Small exarillate, albuminous, embryo microscopic, cotyledons short	Small, exarillate, albuminous, embryo minute	Exarillate, exalbuminous, embryo large	Exarillate, albuminous, embryo large	Exarillate, albuminous, embryo small

Bauera. N. Prakash and McAlister (1977) investigated the embryology of *Bauera capitata.* Melchior (1964) includes the genus *Bauera* in Saxifragaceae, but Cronquist (1968), Thorne (1968) and Takhtajan (1969) place it in Cunoniaceae. Airy Shaw (1973) and Hutchinson (1973) raise it to a monogeneric family, Baueraceae. N. Prakash and McAlister (1977) compare the embryological features of *Bauera,* Cunoniaceae and Saxifragaceae (Table 26.21.1). Even though there are several similarities within the three, Saxifragaceae differ in (1) occasional amoeboid tapetum, (2) uni- or bitegmic ovules, (3) single-celled archesporium, (4) sometimes persistent antipodals, (5) absence of starch in the embryo sac, and (6) absence of tannin in endosperm. Dickison (1975a) supports the placement of *Bauera* in or near the Saxifragaceae, on the basis of floral morphology and anatomy.

Bauera shows close resemblance to Cunoniaceae in the type of pollen grains, endosperm, and seed structure.

After assessing the available embryological features, N. Prakash and McAlister (1977) propose an independent family, Baueraceae, between Saxifragaceae and Cunoniaceae, as also suggested by Airy Shaw (1973) and Hutchinson (1973). We do not consider that there is enough justification to assign *Bauera* to an independent family, Baueraceae. Its inclusion in the family Cunoniaceae is fully justified (see also Takhtajan 1980, Batygina and Yakovlev 1985).

Cunoniaceae. Cunoniaceae are assigned to the order Cunoniales (Hutchinson 1973). They have been treated in Saxifragaceae under different status as series, tribe, or sub-order (see Govil and Saxena 1976). They show close alliance with Saxifragaceae in embryological features (Table 26.21.1): anatropous, bitegmic and crassinucellar ovule with multicellular archesporium, well-developed parietal tissue, Polygonum type of embryo sac, ephemeral antipodals, and Nuclear type of endosperm. The knowledge about their embryogeny may add some more facts. The study of floral anatomy (Dickison 1975b) supports their position near Saxifragaceae.

Pittosporaceae. Pittosporaceae and Byblidaceae are placed in Pittosporales by Hutchinson (1973). Pittosporaceae show secretory anther tapetum, simultaneous cytokinesis in pollen mother cells, unitegmic tenuinucellate ovule, single-celled female archesporium, Polygonum type of embryo sac with ephemeral antipodals, and Nuclear type of endosperm. All these features are met with in Saxifragaceae as well as Umbelliferae. According to L.L. Narayana and Sundri (1978), the Pittosporaceae show a close alliance with Saxifragaceae in floral anatomy, vegetative anatomy, palynology, and chemotaxonomy. R. Dahlgren (1977b) places the order Pittosporales in his non-iridoid sympetalous superorder Araliiflorae (Umbelliflorae). Pittosporaceae have sympetalous corolla. The embryological features such as pollen grains, ovule, embryo sac and endosperm are similar to those of Umbelliferae.

The tenuinucellar, unitegmic ovule occurs, occasionally, in Saxifragaceae, but the three-celled pollen grains at the dispersal stage, and two to five cotyledons of mature embryo, have no similarity in Rosales.

The bitegmic, tenuinucellar ovules with endothelium, and the probably Cellular type of endosperm with haustoria in Byblidaceae, are resemblances with *Bistella* (Saxifragaceae), and differences with Pittosporaceae.

Byblidaceae, Roridulaceae. Planchon (1848) indicated the affinities of *Roridula* with Ochnaceae, and Engler (1907) places it in Ochnaceae. Bentham and Hooker (1965a), Eichler (1875), Engler and Prantl (1894), Netolitzky (1926), Solereder (1908) and Wettstein (1911) treat it under Droseraceae along with *Byblis*. Diels (1930a), however, excludes *Roridula* from Droseraceae. Hutchinson (1973), Riley (1963) and Metcalfe and Chalk (1950) treat it under Byblidaceae. Hutchinson does not recognize their independent family status and considers Roridulaceae as a synonym of Byblidaceae (see also Willis 1973). Due to its non-insectivorous habit, Marloth (1925) transfers *Roridula* from the family Byblidaceae to Roridulaceae. Diels (1930b), Erdtman (1952), Good (1957), Melchior (1964), Takhtajan (1980) and Wettstein (1935) agree with this assignment.

Bitegmic ovule with false endothelium developing from the nucellus, occurrence of gonial or somatic apospory, and single-layered testa in Ochnaceae are differences with Roridulaceae. The floral structure and wood anatomy also differ, and the two families do not show any alliance.

Though the "tentacles" of Roridulaceae are always compared with those of Droseraceae, they do not trap insects. They secrete a resinous balsam-like substance, and not digestive enzymes as in Droseraceae. In addition, some of the features, such as pollen in tetrads, bitegmic crassi- or tenuinucellar ovule with false endothelium, Caryophyllad type of embryogeny, and hard seed-coat formed by both the integuments in Droseraceae, do not match with those of Roridulaceae. The floral structure, anatomy, acid physiology of tentacles, and stem anatomy are also different in both (Vani-Hardev 1972).

Vani-Hardev (1972) compares the morphological and embryological features of *Roridula* and *Byblis* (Table 26.21.2). *Byblis* is a perennial carnivorous herb with morphologically and physiologically different tentacles from those of *Roridula*. The wood anatomy is also different. *Roridula* has reflexed anthers with poricidal dehiscence (introrse anthers with apical slits in *Byblis*), tricarpellary, syncarpous, trilocular ovary with pendulous ovules, ovary wall without tentacles (bicarpellary, syncarpous, bilocular ovary with erect ovule, and ovary wall studded with tentacles in *Byblis*), chalazal megaspore develops into a Polygonum type of embryo sac (micropylar megaspore functional in *Byblis*), and Cellular type of endosperm without prominent haustoria (the haustoria are terminal and prominent in *Byblis*).

The segregation of *Roridula* from Byblidaceae (a non-iridoid family) and its elevation to an independent family, Roridulaceae (see Marloth 1925), are justified on embryological basis.

Chemotaxonomic data show the presence of small quantities of iridoid in the leaves of *Roridula* (Jensen et al. 1975) which supports its placement in or near Ericales. Hallier (1912) suggests its relationship with Clethraceae. The embryological similarity between Roridulaceae and Ericales (particularly Clethraceae) are: (1) absence of fibrous endothecium in anther wall, (2)

Table 26.21.2. Floral morphology and embryology of Roridulaceae and Byblidaceae (See Lang 1901, Vani-Hardev 1972)

	Roridulaceae (*Roridula*)	Byblidaceae (*Byblis*)
Inflorescence	Long, slender, racemes	Solitary, axillary
Stamens	5, free, reflexed in bud, anthers more or less cylindrical, connectives form massive nectaries, dehiscence poricidal, endothecium lacks fibrous thickenings	5, slightly fused with petals at the base, anthers conical, connective non-nectariferous, anthers introse dehiscing by small apical pores, endothecium weakly fibrous
Pollen	2-celled	2-celled
Gynoecium	Tricarpellary syncarpous, trilocular ovary, wall devoid of tentacles, ovules pendulous	Bicarpellary syncarpous, bilocular ovary, wall studded with tentacles, ovules erect
Ovule	Anatropous, unitegmic, tenuinucellate	Anatropous, bitegmic, tenuinucellate, inner integument forms the micropyle, endothelium differentiates
Embryo sac	Chalazal megaspore produces Polygonum type of embryo sac	Micropylar megaspore produces Polygonum type of embryo sac
Endosperm	Cellular type, without prominent haustoria	Cellular (?), prominent multicelled micropylar and chalazal haustoria
Embryo	Solanad type	—
Seed	Endospermic, exotestal	Endospermic, exotestal

glandular tapetum with binucleate cells, (3) two-celled pollen shed singly as in Clethraceae, (4) tricarpellary gynoecium as in Clethraceae, (5) unitegmic, tenuinucellate ovule with endothelium, (6) single-celled archesporium which develops into megaspore mother cell, (7) Polygonum type of embryo sac, and synergids with long tapering tips, (8) fusion of polar nuclei before fertilization, (9) Cellular type of endosperm which becomes haustorial (haustoria weakly developed in Clethraceae), (10) Solanad type of embryogeny, and (11) loculicidal capsule.

The placement of Roridulaceae near or in the Ericales is supported by embryological and chemotaxonomic data.

Bruniaceae. The unitegmic, crassinucellate ovule, and Polygonum type of embryo sac with ephemeral antipodals in Bruniaceae can be compared with those of *Saxifraga pallida* (Rau and Sharma 1967).

Rosaceae. Rosaceae, Neuradaceae and Chrysobalanaceae comprise the suborder Rosineae. No information is available on the Neuradaceae. Bentham and Hooker (1965a), Engler and Diels (1936), and Hutchinson (1973) place Rosaceae in Rosales, near Leguminoseae. The occurrence of carpelloid stamens, and spirally arranged free carpels on elongated torus in some genera indicate their direct Ranalian origin. The accretion of integuments in *Sanguisorba minor* leading to unitegmic condition suggests reduction.

The nature of anther, pollen, crassinucellar ovule with multicellular archesporium, Polygonum and Allium type of embryo sac, Nuclear type of endosperm, and Caryophyllad type of embryogeny in Rosaceae are resemblances with Saxifragaceae. The development of chalazal haustorium from the Nuclear type of endosperm, and its absence in seed shows affinity with Leguminoseae. Asterad type of development of the embryo occurs in *Parnassia* and Caesalpiniaceae. Apospory is prevalent in Rosaceae. Mesotestal seed occurs in woody members, and is comparable to that of Hamamelidaceae, Theaceae and Myrtaceae.

Chrysobalanaceae. Traditionally, the family Chrysobalanaceae is assigned to the order Rosales (Melchior 1964, Hutchinson 1973, Takhtajan 1980, Cronquist 1981). However, wood anatomy (Metcalfe and Chalk 1950) differs significantly from that of Rosales. Palynologically also they are different (Erdtman 1952).

Tobe and Raven (1984a) point out that Chrysobalanaceae show affinity with Theaceae, Ochnaceae and even Clusiaceae (Guttiferae).

Chrysobalanaceae and Theaceae resemble each other in several embryological features: (1) anatropous, bitegmic, tenuinucellate ovule with hypostase, (2) multicelled female archesporium, (3) Nuclear type of endosperm, (4) mesotestal exalbuminous seed, and (5) the tegmen becomes crushed. The Theaceae have Allium type of embryo sac with persistent antipodals, and polyembryony occurs.

With Ochnaceae also Chrysobalanaceae show some resemblances: (1) anatropous, bitegmic, tenuinucellar ovule with hypostase, (2) Nuclear type of endosperm which later becomes cellular but is absorbed by the developing embryo, and (3) exalbuminous seed. But the dissimilarities are more significant. In Ochnaceae there is a (1) false endothelium which differentiates from the nucellus, (2) Polygonum type of embryo sac with large antipodals containing hypertrophied nuclei, (3) apospory of Hieracium type, and (4) Polyembryony.

Like Chrysobalanaceae, Clusiaceae also have anatropous, bitegmic, tenuinucellate ovule, Nuclear type of endosperm, and mesotestal exalbuminous seed (Clusieae). However, Clusiaceae have one-celled female archesporium, and Polygonum type of embryo sac with well-developed antipodals. Polyembryony and apomictic embryo sacs have also been reported.

R. Dahlgren (1983a, b) and R. Dahlgren and Thorne (1984) suggest a relationship between Chrysobalanaceae and Myrtales. The resemblances are: (1) glandular anther tapetum, (2) pollen of Chrysobalanaceae resembles that of Aizateaceae, (3) micropyle formed by both integuments, endothelium differentiation in *Axinandra* (Crypteroniaceae), (4) antipodals ephemeral or absent, (5) modified Polygonum type of embryo sac, (6) endosperm of Nuclear type, and (7) exalbuminous seed. But the thicker inner integument and tenuinucellate ovule in Chrysobalanaceae do not match with Myrtales.

According to Wettstein (1933) and Prance (1972), Chrysobalanaceae stand between Rosaceae and Fabaceae (Papilionaceae). The differences with Fabaceae are more prominent: The anther tapetal cells are binucleate in Chrysobalanaceae, multinucleate in Fabaceae. The ovule is tenuinucellate with short-lived nucellus, crassinucellate with persistent nucellar cap in Fabaceae. The mesotestal seed of Chrysobalanaceae differs from the exotestal seed of Fabaceae.

Chrysobalanaceae share many embryological features with Rosaceae (Table 26.21.3): (1) persistent anther epidermis, (2) fibrous endothecium, (3) two-celled pollen grains, (4) multicelled female archesporium, (5) thick outer and inner integument, (6) Polygonum type of embryo sac, and (7) mesotestal seed with fibrous testa.

There are also important differences: Chrysobalanaceae have tenuinucellar ovule with endothelium, and scanty nucellus which disintegrates at an early stage. In Rosaceae the nucellus is massive, endothelium is absent, and apospory is common.

The embryology of Chrysobalanaceae, Rosaceae, Theaceae and Mrytales is compared in Table 26.21.3.

There is no doubt that there are some similarities in the embryological features of Chrysobalanaceae, and Ochnaceae, Clusiaceae, Fabaceae and Myrtales but these similarities do not warrant any close relationship. The affinities with Theaceae can be discussed more meaningfully when further embryological data on this taxon become available, as also pointed out by Tobe and Raven (1984a). Wettstein's (1933) suggestion that Chrysobalanaceae stand between Fabaceae and Rosaceae is more appropriate, and the assignment of Chrysobalanaceae to Rosales (Melchior 1964) is justified.

The suborder Leguminosineae comprises Connaraceae, Leguminosae and Krameriaceae.

Connaraceae. The occurrence of bitegmic, crassinucellar ovule, Polygonum type of embryo sac, and Nuclear type of endosperm is common to Cunoniaceae and Leguminoseae. Exotegmic seeds occur in Cunoniaceae and Connaraceae.

Bentham and Hooker (1965a), Engler and Prantl (1931), Rendle (1959) and Lawrence (1951) consider Leguminoseae as a family, while Hutchinson (1959, 1973) assigns to them an ordinal rank. The family comprises three subfamilies: Mimosoideae (Mimosaceae), Caesalpinioideae (Caesalpiniaceae) and Faboideae (Papilionaceae).

These three subfamilies show similar features among themselves and also in conformity with those of the order: secretory anther tapetum, two-celled pollen grains, bitegmic, crassinucellate ovules, Polygonum type of embryo sac, Nuclear type of endosperm with chalazal haustorium, and characteristic seed-coat. However, they differ from one another in several respects, and their separation to independent families has been proposed by several taxonomists.

Mimosaceae. Mimosaceae are in closer alliance with Caesalpiniaceae than with Papilionaceae. They are considered as primitive. The formation of sterile septa in anther locule, pollinia of single tetrads, twin tetrads, and all the microspores in a sporangium, shedding of pollen grains singly or in masses, vascular strand reaching up to the micropyle in the integument, bitegmic tenui- or crassinucellar ovule, persistent antipodals, Onagrad type of embryogeny, and arillate seeds in some members segregate Mimosaceae from the others. Dnyansagar (1970) proposed the division of Mimosaceae into two tribes, Prosopoideae and Mimoseae, according to single and compound pollen grains, respectively.

Caesalpiniaceae. Caesalpiniaceae show occasional occurrence of anther tapetum forming periplasmodium, bitegmic crassinucellar ovule with an integumentary vascular strand reaching up to the micropyle, Polygonum type of embryo sac with persistent antipodals, and Asterad type of embryogeny. Both families, Mimosaceae and Caesalpiniaceae, are homogeneous.

Papilionaceae. In Papilionaceae, occasional occurrence of division of the sporogenous tissue by sterile septa as in Mimosaceae, and the anther tapetum forming periplasmodium (as in some members of Caesalpiniaceae) are noteworthy. The glandular nucellar epidermis in *Indigofera*, and the organization of a tapetum-like layer successively by the nucellus, inner integument, endosperm and, finally, even the epidermal layer of the hypophysis region of embryo in *Vigna* (Anantaswamy Rau 1951b) are worth mentioning.

Development of the endothelium, ephemeral antipodals, abnormal megagametophytes, occasional occurrence of Allium type of embryo sac, single fertilization in some members (rather doubtful), and absence of integumentary vascular strand differentiate Papilionaceae from the Mimosaceae and Caesalpiniaceae.

Table 26.21.3. Embryology of Chrysobalanaceae, Rosaceae, Fabaceae, Theales and Myrtales (Adapted after Tobe and Raven 1984b)

	Chrysobalanaceae	Rosaceae	Fabaceae (Papilionaceae)	Theales	Myrtales
Anther	Tetrasporangiate, wall 4–5 layers thick, tapetum secretory of binucleate cells, cytokinesis simultaneous	Bisporangiate, wall 4 layers, tapetum secretory of multinucleate cells, cytokinesis simultaneous	Bisporangiate, wall 4–5 layers, tapetum secretory of uninucleate cells, cytokinesis simultaneous	Bisporangiate, wall 4–5 layers, tapetum secretory of binucleate cells, cytokinesis simultaneous	Tetrasporangiate, wall 4–5 layers, tapetum secretory of binucleate cells, cytokinesis simultaneous
Ovule	Anatropous, bitegmic, tenuinucellar, outer integument 5 layers thick, inner 3–5 layers, micropyle formed by both integuments, endothelium and hypostase present, archesporium multicellular	Ana- or hemianatropous, uni- or bitegmic, crassinucellate, outer integument 3–14 layers, inner 2–6 layers, micropyle formed by both integuments, endothelium absent, hypostase present, archesporium multicellular	Ana-, hemiana- or amphitropous, bitegmic, crassinucellate, outer integument 2–10 layers, inner 2–3 layers, micropyle formed by both integuments, endothelium rare, hypostase absent, archesporium uni- or multicellular	Anatropous, bitegmic, tenuinucellar, outer integument 3–10 layers, inner 3–11 layers, micropyle formed by inner integument, endothelium absent, hypostase present, archesporium 1–3-celled	Ana- or campylotropous, bitegmic (unitegmic in *Eugenia*), crassinucellar, outer integument 2–4 or 6–8 layers, inner 2–4 layers, micropyle formed by both integuments, endothelium absent, hypostase present, archesporium 1-celled

Embryo sac	Polygonum type, at the 4-nucleate stage the two chalazal nuclei fuse and the diploid nucleus does not divide	Polygonum or Allium	Polygonum, Oenothera or Allium	Allium	Polygonum
Antipodals	Not formed	3, persistent and multiply	3, ephemeral	3, persist after fertilization	3, ephemeral
Endosperm	Nuclear type	Nuclear	Nuclear	Nuclear	Nuclear
Embryo	—	Asterad or Caryophyllad type	Onagrad, Asterad, Caryophyllad or Solanad	Solanad	Solanad or Onagrad
Seed	Exalbuminous, mesotestal, tegmen crushed	Albuminous, mesotestal, tegmen unspecialized	Exalbuminous, exotestal, tegmen crushed	Exalbuminous or albuminous, mesotestal, tegmen crushed	Exalbuminous, exotestal, tegmen crushed
Polyembryony	—	Synergid polyembryony	Cleavage and nucellar	Placental and inner integumental	Nucellar
Apospory	—	Hieracium type, pseudogamy, diplospory	—	—	Nucellar

The embryogeny varies a great deal: Asterad, Onagrad, Caryophyllad and Solanad type. Similarly, variations occur in the structure of suspensor.

Lersten (1983) has reviewed the structure of suspensor in the Leguminosae. The Mimosoideae and Caesalpinioideae have a rudimentary or a small suspensor in a few taxa; mostly the embryo is without a suspensor. The Papilionoideae have many suspensorless taxa, others are with small suspensor, and still others with very conspicuous suspensor. The scheme of tribal delimitation and relationships for the Papilionoideae is fully supported by the distribution and structure of suspensor. The enormous spherical, many-celled suspensor is unique, as is the variable biseriate filamentous to columnar suspensor in *Lupinus*. There seems to be no correlation between the amount of endosperm and final suspensor size.

Dnyansagar (1970) proposes to split the family Papilionaceae, as suggested also by Souèges and Crété (1952), and Souèges (1956).

In contrast to the variability in the development of embryo, Papilionaceae show marked similarity in the structure of the seed-coat. In an exotestal, arillate seed, the epidermal palisade layer with linea lucida is followed by a layer of hypodermal hour-glass cells in the testa. A similar structure of seed, with some variation, occurs in the other two families.

Krameriaceae. Hutchinson (1973), Takhtajan (1980) and Cronquist (1981) include Krameriaceae in the order Polygalales along with several common families.

Simpson and Skvarla (1981) compare the pollen morphology of Krameriaceae, Leguminosae, Malpighiaceae, Trigoniaceae, Vochysiaceae, Tremandraceae and Polygalaceae, and conclude that the pollen studies do not provide convincing evidence for affinities with any family.

Verkerke (1986b, see also Corner 1976) points out that ovule ontogeny and seed-coat development of *Krameria* are very different from those of Leguminosae; there are though some exceptions. The thickened inner integument with an inner cuticle, and the absence of a vestigial endotesta separate Krameriaceae also from Polygalaceae. The endotegmic seeds of Krameriaceae suggest affinity with Malpighiaceae. The ovule ontogeny and seed-coat development of Krameriaceae differ much less from those in Malpighiaceae. Some members of the Malpighiaceae have pyriform, anatropous, bitegmic ovule with a curved funicle. There are other resemblances too. For the present, it may be emphasized that the Krameriaceae have no affinity with either Leguminosae or Polygalaceae, and the affinity with Malpighiaceae may be further confirmed following detailed studies of malpighiaceous seed and the reduction trends they exhibit.

Rosales. According to Bessey's system of classification (1915), Rosales are placed near Ranales. Hutchinson (1973) considers that the Saxifragales, a herbaceous group, are closely connected with Ranales and slightly advanced over them. Subramanyam (1962) compares the embryological features of both orders. The structure of anther, pollen, embryo sac, and small antipodals (except Ranunculaceae and Berberidaceae) are similar in both. Cellular type of endosperm occurs, except in some Magnoliaceae (sensu stricto), Lauraceae, Ranunculaceae and Berberidaceae. The peculiar development of megaspore tube (haustorium) in two species of Crassulaceae is also reported in one member of Lauraceae (Bambacioni-Mezzetti 1935). Subramanyam also considers such features as multiple archesporium, presence of starch grains in mature embryo sac, and enlargement of basal cell of two-celled embryo in Crassulaceae. He concludes that there are several similar features in Crassulaceae, Saxifragaceae, and Ranalian families (Magnoliales and Ranales of Melchior 1964), and suggests the possible origin of Rosales from Ranales, as assumed also by Bessey and Hutchinson.

27 Hydrostachyales

The order Hydrostachyales comprises a single family, Hydrostachyaceae.

The development of anther wall and seed-coat has not been investigated.

27.1 General Features

The order Hydrostachyales exhibits the following embryological features: (1) secretory anther tapetum, (2) simultaneous cytokinesis, (3) non-aperturate pollen grains formed in permanent tetrads, (4) anatropous, unitegmic, tenuinucellar ovule on parietal placenta, (5) a narrow elongated nucellus with a tiny gametophyte located at its apex, (6) Polygonum type of embryo sac, (7) starch grains in mature embryo sac, (8) Cellular type endosperm (Stachys type), and (9) embryogeny of Onagrad type-Capsella variation.

27.2 Hydrostachyaceae

The anther wall shows secretory anther tapetum with binucleate and diploid cells (Jäger-Zürn 1965). The cytokinesis in microspore mother cells is simultaneous.

The non-aperturate mature pollen grains are shed in permanent tetrads.

The ovule is anatropous, unitegmic and tenuinucellar. The archesporial cell functions directly as megaspore mother cell. The latter undergoes meiotic divisions producing a linear tetrad. The chalazal megaspore develops into a Polygonum type of embryo sac; polar nuclei fuse before fertilization. The mature embryo sac contains starch grains. An endothelium also differentiates.

The development of endosperm is of the Cellular type, Stachys type of Schnarf (1917). The division of primary endosperm nucleus is followed by a transverse wall. The micropylar chamber divides transversely forming a linear row of three cells. The uppermost cell develops into an aggressive micropylar haustorium, the central cell gives rise to cellular endosperm, and the chalazal cell does not form a haustorium.

The embryogeny conforms to the Onagrad type-Capsella variation, and there is no distinct suspensor haustorium.

G Erdtman (1945, 1952), I Jäger-Zürn (1965), J Mauritzon (1933b, 1939b), W Rauh and I Jäger-Zürn (1966), K Schnarf (1917, 1931), K Subramanyam (1962)

27.3 Taxonomic Considerations

Hydrostachyaceae. Hallier (1901, 1903, 1905, 1908) assigns the genus *Hydrostachys* to Podostemaceae; Engler and Prantl (1931) places Hydrostachyaceae in Rosales along with Podostemaceae, Crassulaceae, Saxifragaceae, etc. Hutchinson (1973) assigns it to the order Podostemales along with Podostemaceae. According to Melchior (1964), Hydrostachyales occupy an intermediate position between Rosales and Podostemales. Cronquist (1968) and Takhtajan (1969, 1980) place Hydrostachyaceae in Scrophulariales. Subsequently, Cronquist (1981) includes them in Callitrichales. R. Dahlgren (1977a, 1980b, 1981) erected the order Hydrostachyales near Scrophulariales.

Subramanyam (1962) compares the embryological features of Crassulaceae with those of Podostemaceae and Hydrostachyaceae. They all have a large number of small ovules, and a narrow elongated nucellus with tiny embryo sac located at the apex. Hydrostachyaceae and Crassulaceae are closely allied and have Polygonum type of embryo sac which contains starch grains, and the endosperm is of the Cellular type.

Jäger-Zürn (1965) discusses the morphology and embryology of Hydrostachyaceae and comments on their systematic position. In contrast to Podostemaceae, Hydrostachyaceae exhibit simultaneous type of cytokinesis in pollen mother cells (both types in the former), pollen in permanent tetrads (single grains or dyads in the former), unitegmic ovules on parietal placenta (bitegmic ovules on central placenta in the former), long chalaza, absence of pseudo-embryo sac, Polygonum type of embryo sac with endothelium and starch grains in mature gametophyte, Cellular (Stachys type – transitory) type of endosperm with micropylar haustorium, Onagrad type of embryogeny, and absence of suspensor haustorium.

Jäger-Zürn (1965) suggests the inclusion of Hydrostachyaceae in tubiflorous families, Scrophulariales, on the basis of similarity in the development of gametophytes, endosperm, and embryo.

28 Podostemales

28.1 General Features

The order Podostemales is represented by the family Podostemaceae which shows several important features: (1) binucleate glandular tapetum, (2) pollen grains in pairs in some members, (3) anatropous, bitegmic, tenuinucellate ovules, (4) formation of pseudo-embryo sac, (5) Apinagia, Dicraea, and Podostemum type (bisporic) of development of female gametophyte, (6) absence of triple fusion and, consequently, absence of endosperm, (7) Solanad type of embryogeny, (8) development of haustoria from the binucleate, basal cell of suspensor, and (9) tegmic seed.

28.2 Podostemaceae

The anther wall comprises the epidermis, fibrous endothecium, an ephemeral middle layer, and glandular tapetum with binucleate cells. They are uninucleate in *Griffithella* (Razi 1949). The microspore mother cells undergo meiotic divisions, and produce tetrahedral and isobilateral tetrads. Cytokinesis is simultaneous in *Indotristicha* (Chopra and Mukkada 1966) and *Terniola* (Mukkada 1969), and successive in *Dicraea stylosa* (Mukkada 1962). In *Dicraea* the microspores separate in pairs to form double pollen grains.

Pollen grains are three-colpate, and shed at two-celled stage individually or in pairs in *Angolaea, Lophogyne, Myopsis* (Went 1910, 1912), *Dicraea* (Fig. 28.2.1A) *Griffithella* and *Lawia* (Razi 1949), and *Podostemon* (Magnus 1913).

In *Indotristicha ramosissima* (Mukkada and Chopra 1973), the ovary contains abundant starch grains; the cells of the inner epidermis of the ovary wall are binucleate and simulate tapetum (Fig. 28.2.1B, C).

The ovule is anatropous, bitegmic and tenuinucellate. The outer integument develops before the inner one, and organizes the micropyle. The pseudo-embryo sacs (in the nucellar region) have a nutritive role in the absence of endosperm. These develop before fertilization in all the genera of Podostemoideae, and after fertilization in Tristichoideae (Arekal and Nagendran 1977b). The nucellar cells located below the embryo sac undergo longitudinal stretching, their walls break down, the protoplasts unite, and a pseudo-embryo sac is formed (Fig. 28.2.1D, E). The inner integument encloses this unique structure. Arekal and Nagendran (1975a) designate it as "nucellar plasmodium".

The hypodermal archesporial cell functions directly as megaspore mother cell. The latter undergoes the first meiotic division giving rise to micropylar and chalazal dyads.

The development of the female gametophyte (Fig. 28.2.1D, E) is rather peculiar. Battaglia (1971) reviewed the different types and, according to him, only two types occur: Apinagia type and Dicraea type, and the third, Podostemum type, is doubtful. However, Arekal and Nagendran (1975b) confirm the occurrence of Podostemum type in 30% ovules of *Hydrobryopsis sessilis.*

1. Apinagia type: Battaglia (1971) subdivided this type into A, B, C and D forms. The C and D forms are mere variations.

Apinagia A form:	five-nucleate, four-celled,
Apinagia B form:	four-nucleate, four-celled,
Apinagia C form:	five-nucleate, five-celled, and
Apinagia D form:	six-nucleate, four-, five- or six-celled.

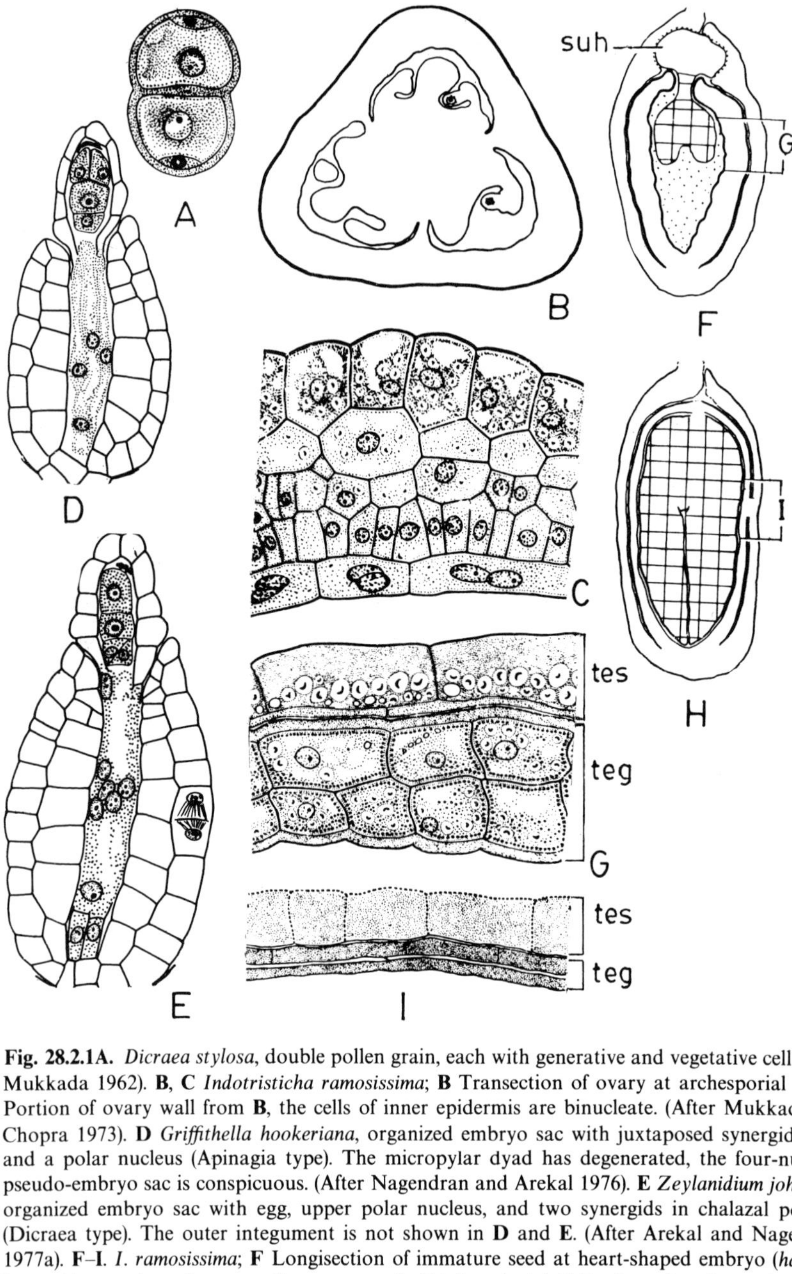

Fig. 28.2.1A. *Dicraea stylosa*, double pollen grain, each with generative and vegetative cell. (After Mukkada 1962). **B, C** *Indotristicha ramosissima*; **B** Transection of ovary at archesporial cell; **C** Portion of ovary wall from **B**, the cells of inner epidermis are binucleate. (After Mukkada and Chopra 1973). **D** *Griffithella hookeriana*, organized embryo sac with juxtaposed synergids, egg, and a polar nucleus (Apinagia type). The micropylar dyad has degenerated, the four-nucleate pseudo-embryo sac is conspicuous. (After Nagendran and Arekal 1976). **E** *Zeylanidium johnsonii*, organized embryo sac with egg, upper polar nucleus, and two synergids in chalazal position (Dicraea type). The outer integument is not shown in **D** and **E**. (After Arekal and Nagendran 1977a). **F–I**. *I. ramosissima*; **F** Longisection of immature seed at heart-shaped embryo (*hatched*) located in pseudo-embryo sac, the suspensor (*suh*) is haustorial; **G** Portion of seed-coat, *inset G* in **F**, note the starch grains in testa (*tes*) and tegmen (*teg*), and thickenings in tegmen; **H** Longisection of mature seed, the embryo (*hatched*) occupies the entire pseudo-embryo sac; **I** Portion of seed-coat, *inset I* in **H**, the testa (*tes*) is mucilagenous, and tegmen (*teg*) thick-walled. (After Mukkada and Chopra 1973)

In Apinagia A form the nucleus of the micropylar dyad may divide, but it degenerates. The chalazal dyad undergoes second meiotic division. Of the two nuclei, the chalazal megaspore nucleus degenerates. The upper spore nucleus divides twice, and all the four nuclei organize into two juxtaposed synergids followed by a large egg cell, and a small lower cell (Fig. 28.2.1D). The last-named cell is very often termed as an antipodal, or upper polar nucleus, or polar cell. As it degenerates after fertilization, and as there is no nuclear fusion with the male cell, Battaglia (1971) prefers to call it an antipodal.

Apinagia A form occurs in *Anastrophea abyssinica* (Jäger-Zürn 1967), *A. divertens*, *A. imthurnii* (Went 1909, 1910), *A. versteegiana* (Went 1910, 1926), *Cladopus nymani* (Went 1926), *Indotristicha ramosissima* (Chopra and Mukkada 1966), *Inversodicraea minutiflora* (Jäger-Zürn 1967), *Mourera fluviatilis* (Went 1909, 1910), *Oserya perpusilla* (Went 1910, 1926), *Terniola zeylanica* (Magnus 1913, Razi 1949, 1955, Mukkada 1969), *Tristicha trifaria* (Went 1910, 1926, Jäger-Zürn 1967), and *Zeylanidium olivaceum* (Arekal and Nagendran 1977a) (cf. Battaglia 1971).

In Apinagia B form there is a precocious degeneration and disappearance of the chalazal megaspore nucleus. The embryo sac becomes four-nucleate, four-celled; the arrangement of cells is similar to that in A form. This type has been observed in *Dalzellia zeylanica*, *Indotristicha ramosissima* (Arekal and Nagendran 1977b), *Farmeria indica* (Arekal and Nagendran 1975a), *Hydrobryum griffithii* (Nagendran et al. 1976), *Griffithella hookeriana* (Razi 1949, 1955), *Podostemon ceratophyllum* (Hammond 1937), *Polypleurum filifolium* (Nagendran et al. 1977), and *Zeylanidium olivaceum* (Razi 1955) (cf. Battaglia 1971). Nagendran and Arekal (1976) reinvestigated the development of female gametophyte in *G. hookeriana* and confirmed its classification. Nagendran (1974) considered the Apinagia type as monosporic, as all the nuclei are derived from only one megaspore nucleus.

In Apinagia C form, noted by Razi (1949, 1955) in *Lawia* (*Terniola*) *zeylanica*, the chalazal megaspore nucleus is cut off as a small antipodal cell. The embryo sac is five-nucleate, five-celled.

The Apinagia D form occurs as an abnormal feature (Razi 1949). The chalazal megaspore nucleus undergoes second somatogenic division producing a six-nucleate and four-, five- or six-celled stage. The lower chalazal cell, with derivatives of the chalazal megaspore nucleus, forms a three-nucleate antipodal cell (Razi 1949).

After examining seven species of Podostemaceae conforming to Apinagia type of embryo sac, Arekal and Nagendran (1977b) confirm that the chalazal megaspore nucleus never persists in the organized embryo sac, as reported by Chopra and Mukkada (1966), Jäger-Zürn (1967) and Mukkada (1969). Moreover, it does not divide to produce a five-, or six-celled embryo sac, as reported by Razi (1949, 1955). According to Arekal and Nagendran, the C and D forms of the Apinagia type (Battaglia 1971) do not occur in Podostemaceae.

2. Dicraea type: This was proposed by Magnus (1913) in *Dicraea elongata*. The first meiotic division in the megaspore mother cell results in the formation of two dyads. The micropylar dyad generally degenerates. The nucleus of the chalazal dyad divides. According to Magnus (1913), this division is followed by a wall. However, Mukkada (1962) confirmed that no such wall is laid down in *D. stylosa.* The development is of the bisporic four-nucleate type. The two nuclei of the chalazal dyad undergo simultaneous mitotic division. The four nuclei are arranged as an egg cell in the centre, superposed by a synergid at the micropylar end, and two juxtaposed antipodal cells (Fig. 28.2.1E). The synergid and egg are sister cells, and the polar nucleus is absent. Dicraea type occurs in *D. elongata* (Magnus 1913, Mukkada 1964), *D. stylosa* (Mukkada 1962, 1964) and *Hybrobryopsis sessilis* (Arekal and Nagendran 1975b).

Arekal and Nagendran (1977a) noted Dicraea type in *Zeylanidium johnsonii.* Nagendran (1974) designated it Polypleurum type due to the change in the name of the genus *Dicraea* (Hall 1971). The development of the embryo sac in *Polypleurum dichotomum* and *P. munnarense* (Nagendran et al. 1977) is according to the Dicraea type, as described by Mukkada (1962, 1964) and Razi (1966). However, Nagendran et al. differ with the interpretation of Magnus (1913), Mukkada (1962), Razi (1966) and Battaglia (1971). According to these authors, Dicraea type of embryo sac consists of one synergid, one egg – both considered sister cells, and two antipodal cells. In Podostemaceae, there is a trend in the elimination of nuclei at the chalazal end.

Arekal and Nagendran (1975b) refute the formation of antipodal cell/s in this family. This is essentially true except when the degenerated chalazal nucleus of the two-nucleate embryo sac persists up to the stage of organized gametophyte.

Arekal and Nagendran give a new interpretation to the Dicraea type of embryo sac, and consider the organized gametophyte "an inverted image of the Podostemum type". The mature gametophyte invariably comprises four cells – two synergids, egg cell, and one polar cell. If the Dicraea type is regarded as a case of permanent reverted polarity (inverted image), the position of the egg cell remains the same in all types of embryo sacs of Podostemaceae. The interpretation that a polar cell tops the egg and the two synergids are below the egg, completely changes the conventional structure of the egg apparatus and disposition of the upper polar nucleus. For the present, the interpretation of Arekal and Nagendran should be accepted with reservation.

3. Podostemum type: The micropylar dyad degenerates as in the Dicraea type. The development is of the bisporic, four-nucleate type. The nucleus of the chalazal dyad undergoes two consecutive divisions, and the resulting four nuclei organize as two juxtaposed synergids at the micropylar pole followed by an egg cell and a polar cell at the chalazal end. Arekal and Nagendran (1975b) report this type in 30% of the ovules of *Hydrobryopsis sessilis.*

4. Willisia type: Arekal and Nagendran (1976) investigated *Willisia selaginoides*, and suggested an additional type, "Willisia". The development of the embryo sac is bisporic and four-nucleate; the chalazal dyad is functional. The components of the embryo sac are disposed in decussate or isobilateral manner. The organized embryo sac shows two large juxtaposed synergids confined to the micropylar half of the ovoid gametophyte, while the egg and polar cell, also juxtaposed, occupy the chalazal half of gametophyte. Thus, according to these investigators, four types of embryo sacs occur in Podostemaceae – Apinagia (monosporic ?), Podostemum, Polypleurum and Willisia (all bisporic). We do not see enough justification for adding the Willisia type, although there is some variation in the organization of the embryo sac.

Nagendran (1974) has raised a fundamental issue. "Is the embryo sac of Podostemaceae bisporic?". Of the three main types of gametophytes in the family, Polypleurum and Podostemum are "truly bisporic" since both the

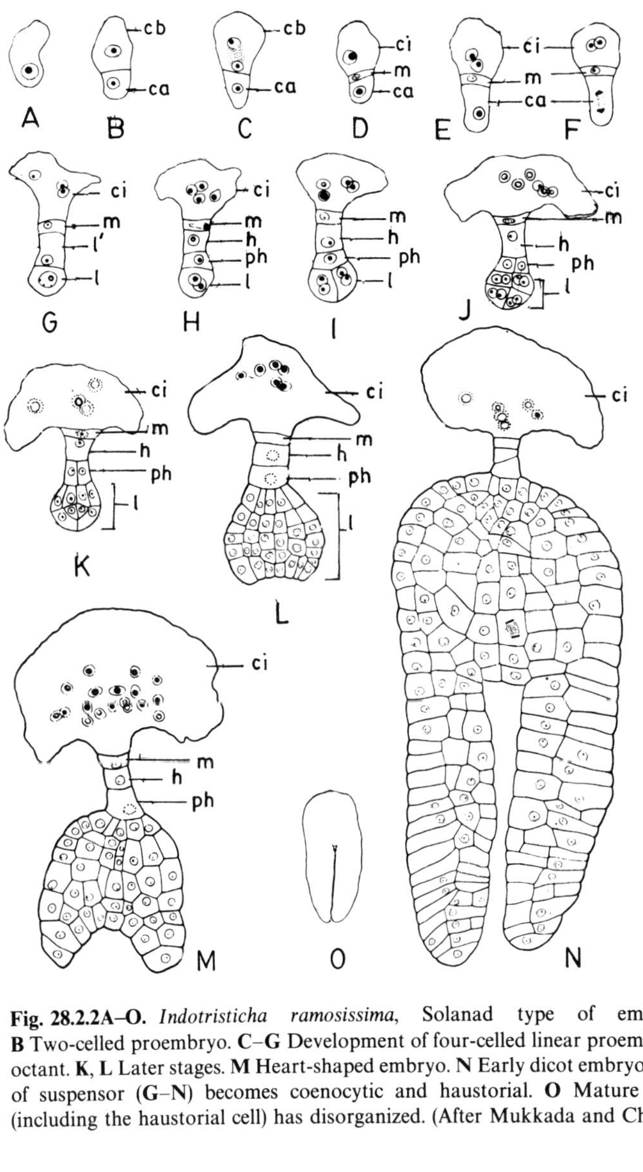

Fig. 28.2.2A–O. *Indotristicha ramosissima*, Solanad type of embryogeny. **A** Zygote. **B** Two-celled proembryo. **C–G** Development of four-celled linear proembryo. **H–J** Formation of octant. **K, L** Later stages. **M** Heart-shaped embryo. **N** Early dicot embryo. The upper-most cell (*ci*) of suspensor (**G–N**) becomes coenocytic and haustorial. **O** Mature embryo, the suspensor (including the haustorial cell) has disorganized. (After Mukkada and Chopra 1973)

megaspore nuclei of the lower dyad cell produce the mature four-nucleate embryo sac. The third type – Apinagia – is interpreted as monosporic because only the micropylar nucleus of the lower dyad cell produces all the four nuclei present in the organized embryo sac. "Whether the chalazal megaspore nucleus persists in the mature embryo sac or not, it is logical and reasonable to conclude that the Apinagia type of embryo sac development be treated as monosporic and tetranucleate". The fact that the chalazal megaspore nucleus persists (may even be ephemeral) in the embryo sac is adequate justification to designate the development as bisporic and not monosporic. Another similar example is *Butomopsis lanceolata* (B.M. Johri 1935e), and Roper (1952) gave a more or less similar reasoning for considering the embryo sac in Alismaceae as monosporic. This has already been discussed at great length by B.M. Johri (1963a).

The entry of pollen tube is porogamous, and syngamy is normal. The second male gamete may not discharge from the pollen tube (Chopra and Mukkada 1966) or, if it is discharged, may not reach and fuse with the polar nucleus. The polar nucleus degenerates, consequently there is no triple fusion and endosperm formation. In the absence of endosperm, the pseudo-embryo

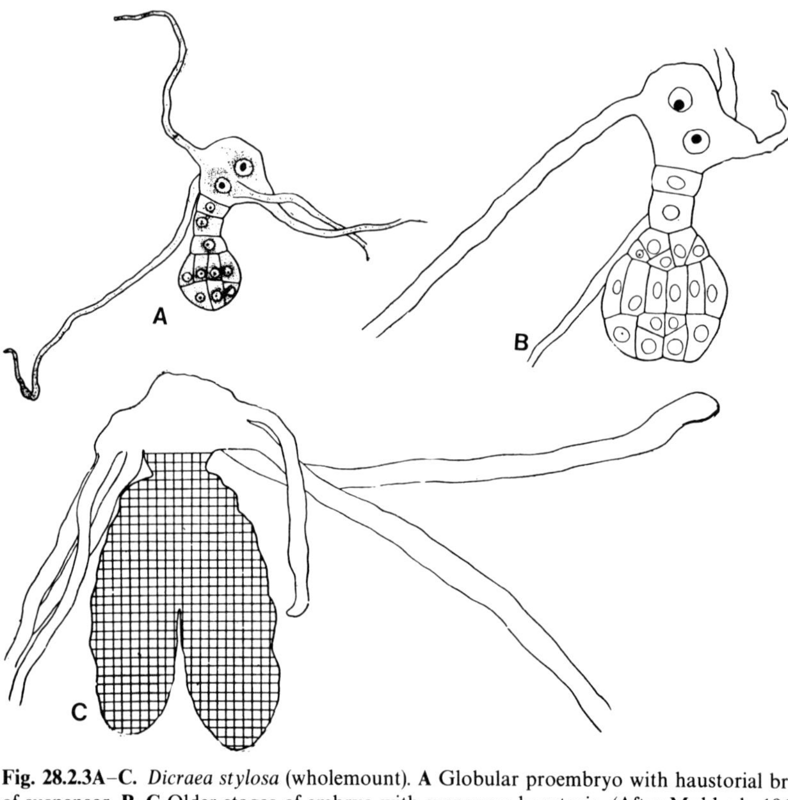

Fig. 28.2.3A–C. *Dicraea stylosa* (wholemount). **A** Globular proembryo with haustorial branches of suspensor. **B, C** Older stages of embryo with suspensor haustoria. (After Mukkada 1962)

sac plays the nutritive role and the developing embryo comes to lie in it (Fig. 28.2.1F, H; cf D, E).

The embryogeny conforms to the Solanad type (Fig. 28.2.2A–N). The uppermost cell of the suspensor enlarges, becomes coenocytic and haustorial (Fig. 28.2.2B–N). It develops thin-walled haustorial branches which penetrate the integuments (Fig. 28.2.3A–C).

The seeds are minute, and show a tegmic seed-coat (Fig. 28.2.1G, I). The cells of the outer epidermis of testa have thick pectic walls and a narrow lumen. They become mucilaginous. The cells of the tegmen are thickened and lignified, and only the thickenings of the walls persist in mature seed (Fig. 28.2.1I).

GD Arekal and CR Nagendran (1975a, b, 1976, 1977a, b), E Battaglia (1971), RN Chopra and AJ Mukkada (1966), JB Hall (1971), BL Hammond (1937), I Jäger-Zürn (1967), BM Johri (1935e, 1963a), RN Kapil (1970b), W Magnus (1913), J Mauritzon (1933b), AJ Mukkada (1962, 1964, 1969), AJ Mukkada and RN Chopra (1973), CR Nagendran (1974), CR Nagendran and GD Arekal (1976, 1979), CR Nagendran, VV Anand and GD Arekal (1980), CR Nagendran, GD Arekal and K Subramanyam (1977), CR Nagendran, K Subramanyam and GD Arekal (1976), BA Razi (1949, 1955, 1966), RB Roper (1952), K Subramanyam (1962), FAFC Went (1909, 1910, 1912, 1926)

28.3 Taxonomic Considerations

Podostemaceae. Podostemaceae are placed with Crassulaceae, Saxifragaceae, and Hydrostachyaceae in Rosales by Engler and Prantl (1931). Hutchinson (1926, 1973) groups them with Hydrostachyaceae in the order Podostemales. P. Maheshwari (1945), on the basis of embryological findings, concludes that it is "almost certain that Podostemaceae are much reduced apetalous derivatives of the Crassulaceae". Subramanyam (1962) compares the embryological features of Crassulaceae with those of Podostemaceae and Hydrostachyaceae, and emphasizes that the differences between these families are probably due to their special mode of life.

Podostemaceae show alliance with Crassulaceae in secretory anther tapetum, two-celled pollen grains, bitegmic ovules, embryo sac located in the upper part of nucellus, presence of starch in pollen and embryo sac, haustorial basal cell of suspensor, and "dwarf" type of seed with a capsular fruit. However, Podostemaceae show several unique features. The pollen grains are shed in pairs in some members. The Apinagia, Dicraea, and Podostemum types of embryo sac exhibit reduction in the development of gametophyte. The pseudo-embryo sac, or "nucellar plasmodium", serves a nutritive role, and is not met with elsewhere. The absence of triple fusion and degeneration of the polar nucleus lead to a total absence of endosperm. The embryogeny corresponds to the Solanad type. The haustorial basal cell of suspensor is non-septate. The thin haustorial branches penetrate the integuments which are rich in starch. The seeds are distinctly tegmic, and remind one of Piperaceae and Saururaceae (Corner 1976). In Crassulaceae, the above-mentioned features are absent. The seed of Crassulaceae is testal, the tegmen gets crushed.

Podostemaceae, likewise, also differ from Hydrostachyaceae (see Hydrostachyales, Sect. 27).

The embryological features justify the erection of an independent order, Podostemales, with a single family, Podostemaceae.

29 Geraniales

The order Geraniales comprises three suborders and nine families: Limnanthineae – Limnanthaceae; Geraniineae – Oxalidaceae, Geraniaceae, Tropaeolaceae, Zygophyllaceae, Linaceae and Erythroxylaceae; Euphorbiineae – Euphorbiaceae and Daphniphyllaceae.

29.1 General Features

The order Geraniales is characterized by the (1) occurrence of secretory anther tapetum (amoeboid in *Erodium* and *Geranium*), (2) simultaneous cytokinesis in pollen mother cells, (3) three-colporate pollen grains, (4) anatropous, bitegmic ovules, integumentary vascular bundle in Euphorbiaceae, (5) development of the embryo sac mostly of the Polygonum type, Allium type in Oxalidaceae and Euphorbiaceae, and tetrasporic in Limnanthaceae and Euphorbiaceae, (6) development of the endosperm of the Nuclear type, chalazal haustorium present in Linaceae, the endosperm is ruminate in Oxalidaceae and *Guaicum* (Zygophyllaceae), (7) embryogeny conforms to the Asterad type in Oxalidaceae and Geraniaceae, and Solanad type in Zygophyllaceae, Linaceae and Erythroxylaceae, and (8) outer epidermis of the inner integument is generally fibrous, or the seeds are exotegmic.

29.2 Limnanthaceae

The anther wall comprises the epidermis, fibrous endothecium, two ephemeral middle layers, and secretory tapetum with binucleate cells. Following meiotic divisions in microspore mother cells, tetrahedral and decussate tetrads are produced. Cytokinesis occurs by furrowing (P. Maheshwari and B.M. Johri 1956, N. Mathur 1956).

The pollen grains are tetracolporate, and two-celled at the dispersal stage in *Floerkea proserpinacoides* and *Limnanthes douglasii*.

The ovule is anatropous (Fig. 29.2.1A, B), bitegmic, and tenuinucellar with integumentary vascular supply reaching up to the tip of integument. The archesporium is multicellular, only one cell functions as megaspore mother cell. It undergoes meiotic division resulting in a larger nucleus which remains in the centre, and a smaller nucleus which moves to the chalazal end. The latter usually degenerates as in *Limnanthes*, or may rarely divide as in *Floerkea*. The central nucleus undergoes second meiotic division and again produces two unequal nuclei. The larger one remains in the centre of the embryo sac, while the other migrates downward and often divides.

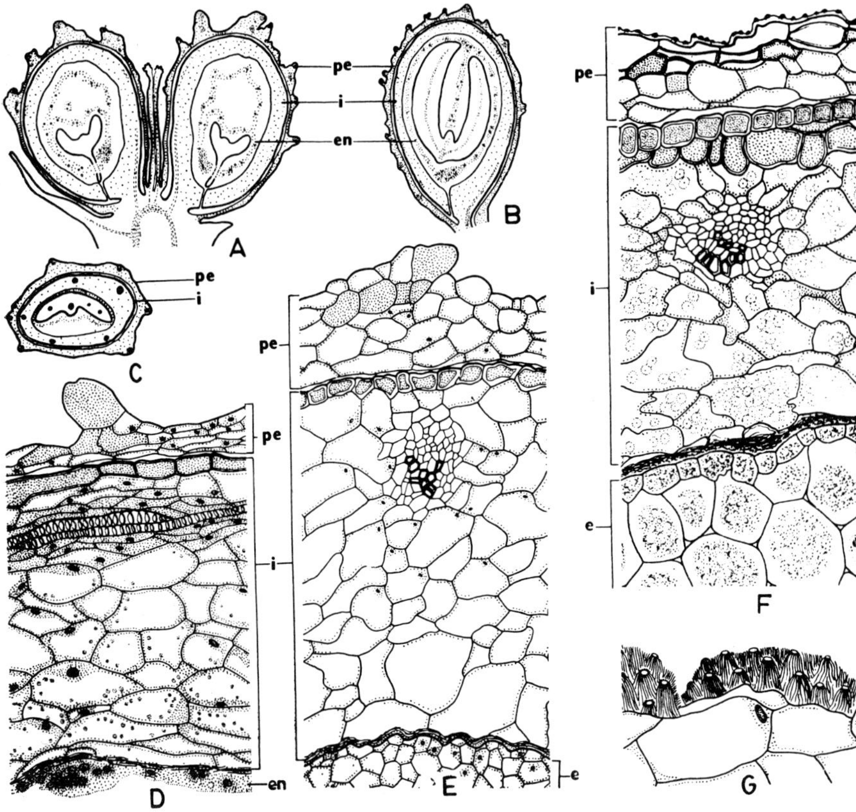

Fig. 29.2.1A–G. *Floerkea proserpinacoides.* **A** Longisection of fruits at early dicotyledonous embryo to show pericarp (*pe*), testa (*i*) and nuclear endosperm (*en*). Note the persistent gynobasic style. **B** Longisection of older fruit. **C** Transection of fruit. **D** Portion of ovary wall, vascularized testa (*i*) and nuclear endosperm from seed approximately at the same age as in **B**. The cells of the testa contain starch. **E** Pericarp and testa from **C**, and peripheral cells of embryo (*e*). **F** Pericarp, testa (*i*), degenerated remnants of endosperm, and portion of a cotyledon (*e*), from **B**. **G** Thickenings on the outer tangential walls of epidermal cells of pericarp. (After P. Maheshwari and Johri 1956)

The upper nucleus divides mitotically, but the daughter nuclei do not become separated by a vacuole. These nuclei undergo one more mitotic division, and the four daughter nuclei organize into egg apparatus and upper polar nucleus. Sometimes, the chalazal end has two nuclei which may be the daughter nuclei of the chalazal nucleus formed after Meiosis I or, if it degenerates, these nuclei may have originated from the lower nucleus formed after Meiosis II. One of these functions as lower polar nucleus, while the other may degenerate.

The mature embryo sac contains only the egg apparatus and two polar nuclei, the antipodal cells are absent. The synergids lack a basal vacuole in *Floerkea.*

The development of the embryo sac in *Floerkea* and *Limnanthes* is tetrasporic and biphasic. Occasionally, six- or seven-nucleate gametophytes have been observed in both taxa.

Double fertilization occurs, and one of the synergids persists up to the globular proembryo.

Development of the endosperm is of the Nuclear type, and it remains nuclear throughout (Fig. 29.2.1A–D). During its development the endosperm digests the cells of the inner integument and forms a haustorial pouch at the micropylar end, towards the funicular side. Nearly 30 to 35 nuclei aggregate in it in *Floerkea*.

The zygote divides transversely. The terminal cell develops into the embryo proper, while the basal cell produces a long suspensor. The mature embryo comprises two large cotyledons, a small radicle, and stem tip with a pair of leaves. The cotyledons show backward extensions (auricles) with vascular supply; they bifurcate on either side of the radicle.

The seed is exalbuminous. The vascularized testa consists of several layers of thin-walled cells (Fig. 29.2.1D–F) which become lignified in the micropylar and chalazal region. The outer tangential walls of the epidermal cells of pericarp develop pyramidal thickenings (Fig. 29.2.1G).

F Fagerlind (1939b), BM Johri (1970a, b), BM Johri and P Maheshwari (1951), P Maheshwari and BM Johri (1956), CT Mason (1950), N Mathur (1956), A Russell (1919), H Stenar (1925a)

29.3 Oxalidaceae

The anther wall comprises the epidermis, fibrous endothecium, one or two middle layers, three or four in *Averrhoa* and *Biophytum* (Thathachar 1942a), and secretory tapetum. The tapetal cells are binucleate, except in *Averrhoa bilimbi*, *A. carambola* and *Biophytum sensitivum* (Thathachar 1942a). The microspore mother cells undergo meiotic divisions and form tetrahedral tetrads. Simultaneous cytokinesis occurs by furrowing.

The pollen grains are three-colporate, and shed at the two-celled stage.

The ovule is anatropous, bitegmic, and tenuinucellate. It is crassinucellate in *A. bilimbi* and *A. carambola* (Thathachar 1942a, Govil 1978, Boesewinkel 1985b). The micropyle is organized by both the integuments. An endothelium differentiates in the inner epidermis of the inner integument. The archesporium is uni- or multicellular. A single cell develops directly into megaspore mother cell. The latter undergoes reduction divisions, and produces a linear or T-shaped tetrad. The chalazal megaspore develops into a Polygonum type of embryo sac in *Averrhoa* (Thathachar 1942a, Govil 1978), *Biophytum* (Govindappa and Boriah 1956) and *Oxalis latifolia* (L.L. Narayana 1959). According to Herr and Maryhruce (1968), in *Oxalis corniculata* the development is mostly of the Allium type, Polygonum type occurs less frequently. The synergids are pyriform, and antipodals are ephemeral.

The entry of pollen tube is porogamous, and double fertilization occurs.

Development of the endosperm is of the Nuclear type, it becomes cellular after the formation of about 32 free nuclei. The mature seed is albuminous with oil as storage product. It is subruminate in *Biophytum* and *Oxalis*.

The embryogeny conforms to the Asterad type (L.L. Narayana 1962). The development of the embryo of *Oxalis* is assigned to Period I, Series A, Megarchetype II (Asterad type).

The outer epidermis of testa comprises a short palisade layer, while the inner one contains crystals (Boesewinkel 1985b). The mesophyll is thin-walled. The outer epidermis of tegmen is composed of longitudinal fibres, and the hypodermis of tracheidal fibres. The tegmen develops folds causing rumination in the endosperm in *Biophytum* and *Oxalis*. The rest of the tissue is crushed.

FD Boesewinkel (1985b), CM Govil (1978), CM Govil and Maninder Kaur (1989), DA Govindappa and G Boriah (1956), CJ Guth and SG Weller (1986), JM Herr Jr (1972), JM Herr Jr and D Maryhruce (1968), VL Kumari and LL Narayana (1980), LL Narayana (1959, 1962, 1970a), W Noll (1935), Souèges (1939d), T Thathachar (1942a), J Venkateswarlu (1936b)

29.4 Geraniaceae

The anther wall comprises the epidermis, fibrous endothecium, a middle layer, and secretory tapetum with multinucleate cells in *Monsonia* (H.S. Narayana and Arora 1963), *Erodium stephanianum*, *Geranium* and *Pelargonium hortatum* (Kumar 1976). Amoeboid tapetum is reported in *Erodium corsicum* (Neogodi 1937), *Geranium robertianum* and *G. pyrenaicum* (Schürhoff 1924). The formation of polyploid nuclei due to nuclear fusions occurs in *Geranium grevellianum*, *G. nepalenses* and *G. ocellatum*. Meiotic divisions in microspore mother cells result in tetrahedral and decussate tetrads. Simultaneous cytokinesis occurs by furrowing.

The pollen grains are tricolpate and shed at the two-celled stage in *M. senegalensis* (H.S. Narayana and Arora 1963), *Erodium* and *Monsonia* (Kumar 1976), and at the three-celled stage in *E. corsicum* (Neogodi 1937), *E. cicutarium*, *P. zonale* (see Davis 1966), and *Geranium* and *P. hortatum* (Kumar 1976).

The ovule is ana- to campylotropous, bitegmic, and crassinucellate. The outer integument organizes the micropyle. An endothelium is absent. The archesporial cell cuts off a parietal cell, and the latter gives rise to a four- to six-layered parietal tissue. A nucellar cap, obturator and hypostase are reported in *Monsonia senegalensis* (H.S. Narayana and Arora 1963). The hypostase is suberized. Following meiotic divisions, the megaspore mother cell produces a linear tetrad. The chalazal megaspore develops into a Polygonum type of embryo sac. The antipodals are ephemeral. Multicellular archesporium, twin megaspore mother cells, and T-shaped tetrads are reported in *G. nepalense* and *G. wallichianum* (Kumar 1976). There is a high degree of abortion in the developmental stages of the embryo sac in *Pelargonium hortatum* (Kumar 1976).

Double fertilization occurs, and the entry of pollen tube is porogamous. The remnants of the pollen tube persist for some time after fertilization.

The development of the endosperm is of the Nuclear type, later it becomes cellular and is mostly consumed by the developing embryo.

The embryogeny conforms to the Asterad type in *Erodium circutarium* (Souèges 1923b), *E. stephanianum* and *G. wallichianum* (Kumar 1976), and Onagrad type in *G. molle* (Souèges 1923c).

The seed is mainly exotegmic and campylotropous. The outer epidermis of testa comprises round, thin-walled cells. The inner epidermis shows thickenings on the outer walls, exhibits a light line, and contains crystals. The outer epidermis of tegmen is a compact layer of thick-walled, lignified cells. The inner one shows thin-walled, compressed cells (Boesewinkel and Been 1979).

FD Boesewinkel (1988), FD Boesewinkel and W Been (1979), GL Davis (1966), A Kumar (1976), HS Narayana and PK Arora (1963), LL Narayana (1970b), G Neogodi (1937), PN Schürhoff (1924), R Souèges (1923b, c), AH Tsai, PM Harney and RL Peterson (1973)

29.5 Tropaeolaceae

The anther wall comprises the epidermis, endothecium, one or two ephemeral middle layers, and secretory tapetum (L.L. Narayana 1970c). Meiotic divisions in microspore mother cells result in tetrahedral tetrads. Simultaneous cytokinesis is followed by ephemeral cell-plate formation (Bolenbach 1928).

The pollen grains are tricolpate, and shed at the three-celled stage (LL. Narayana 1970c).

The ovule is anatropous, bitegmic, and tenuinucellar with the micropyle formed by the inner integument. The archesporial cell directly functions as megaspore mother cell. The latter undergoes meiotic divisions, and produces a linear or T-shaped tetrad. The chalazal megaspore develops into a Polygonum type of embryo sac. The synergids are pyriform, and the antipodals ephemeral. Polar nuclei fuse before fertilization.

The entry of pollen tube is porogamous. During its entry it does not destroy any of the synergids (Walker 1947). This is doubtful.

The development of the endosperm is of the Nuclear type, and it remains nuclear throughout (Walker 1947). However, L.L. Narayana (1970c) mentions that it becomes cellular.

In *Tropaeolum majus* (Walker 1947), the zygote divides transversely, followed by another transverse division in each cell. The apical cell of the four-celled linear proembryo gives rise to the embryo proper, the subapical cell forms the suspensor, and the two remaining cells give rise to suspensor haustoria. At first vertical and then transverse divisions in the four-celled proembryo results in a 14–16-celled stage.

The behaviour of the derivatives of the two uppermost cells of the proembryo is very unusual. They divide repeatedly to produce the basal mass of cells which give rise to an extensive suspensor haustorial system.

The peripheral cells of the basal cell mass, close to the suspensor, on the opposite side of the funiculus, become meristematic and a root-like haustorium

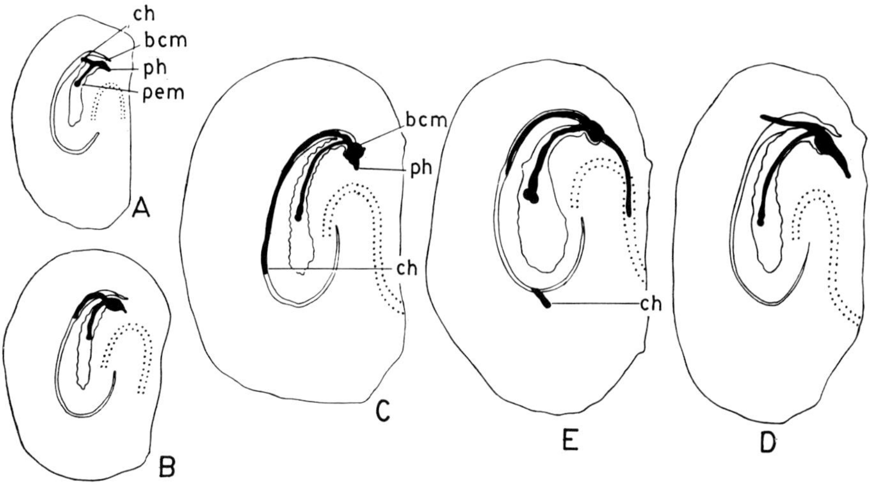

Fig. 29.5.1A–E. *Tropaeolum majus.* **A** Longisection of ovary to show proembryo (*pem*) with suspensor, and origin of carpellary (*ch*) and placental (*ph*) haustorial branches, and basal cell mass (*bcm*). **B** Later stage. **C** Longisection of ovary to show globular proembryo (*black*) with a long suspensor. Note the carpellary (*ch*) and placental (*ph*) haustorial branch. **D** Advanced stage. **E** Longisection of ovary to show heart-shaped embryo with collar and suspensor, the carpellary haustorial branch (*ch*) extends around the ovule, opposite to the funiculus, and enters the carpel wall. The placental branch enters the nucellus and grows close to the vascular bundle. The basal cell mass shows peripheral projections. (After R.I. Walker 1947)

is initiated which penetrates the integument near the micropyle, grows around the ovule in the carpellary cavity, and into the carpel wall. The haustorium penetrates the carpel wall opposite to the micropyle (Fig. 29.5.1A–E).

Shortly after the formation of carpellary suspensor haustorium, another protuberance arises from the peripheral cells of the basal cell mass, close to the funiculus (Fig. 29.5.1A–C). The apical cells are smaller and appear to be meristematic, while the remaining portion consists of large vacuolate cells. This is the placental suspensor haustorium which grows through the integuments and funiculus, reaches up to the vascular bundle of placenta, and grows along its curvature (Fig. 29.5.1D, E).

Besides the carpellary and placental suspensor haustoria, the peripheral cells of the basal mass produce finger-like haustorial cells. The nuclei of these cells are located at the base, whereas the upper portion is vacuolate.

The proembryo and the suspensor grow concurrently with the suspensor haustoria and, in due course, the globular proembryo develops into a heart-shaped embryo with epicotyl surrounded by nuclear endosperm. As the heart-shaped embryo grows to dicotyledonous form, the cells located at the junction of embryo and suspensor proliferate, and elongate to form a collar. These cells are also haustorial in function.

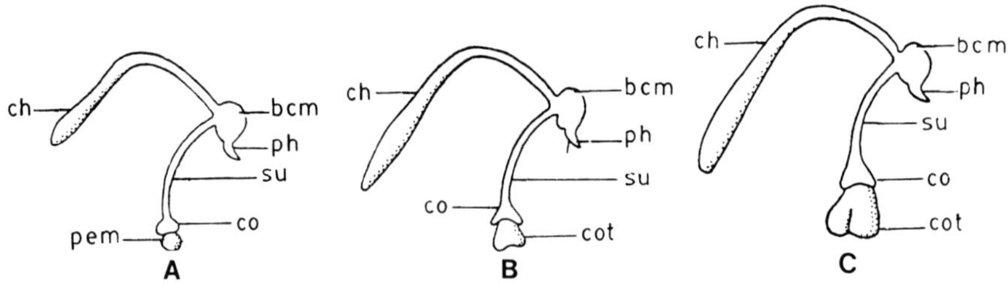

Fig. 29.5.2A–C. *Tropaeolum majus*, semi-diagrammatic drawings of dissected globular (**A**), heart-shaped (**B**), and cotyledonary embryo (**C**); *bcm* basal cell mass; *ch* carpellary haustorium; *co* collar; *cot* cotyledon; *pem* proembryo; *ph* placental haustorium; *su* suspensor. (After Bhalla et al. 1979)

According to Bhalla et al. (1982), embryo development is inhibited until the suspensor has fully grown, and ramifications become established in the placentum and fruit wall. Figure 25.5.2A–C represents the disposition of the haustorial system at globular, heart-shaped and dicotyledonous embryo.

The mature embryo is dicotyledonous with an epicotyl, hypocotyl, and remnants of suspensor haustorium in the integument.

Bhalla et al. (1979) point out that the basal cells of the suspensor in the majority of species display wall labyrinths and resemble transfer cells. The wall-ingrowths in the micropylar region of basal cells indicate that their main function is absorption of nutrients from integuments and other tissues.

After 5 days of pollination the embryo is very small but the suspensor is well developed with two distinct haustoria (Bhalla et al. 1982). At the globular proembryo the suspensor is 2 mm long, while the carpellary haustorium traverses about 3 mm around the ovule, protrudes outside and its tip becomes embedded in the inner fruit wall. At the heart-shaped embryo the suspensor proper is 4 mm long, the rosette of apical cells grows in width, and the carpellary haustorium is 6 mm long.

The seed is pachychalazal and exalbuminous with a straight embryo. Tiwari et al. (1977) point out that at the mature embryo sac the chalaza shows massive growth. A meristematic inwardly curved cup-shaped strip develops in the chalaza above the vascular strand. Anti- and periclinal divisions lead to diffuse growth. The cells derived from the meristematic tissue enlarge and grow along the micropylar-chalazal axis. The endosperm develops all along the periphery of the embryo sac. The embryo sac with the embryo and its well-developed suspensor haustorial system, and the endosperm consume the adjoining nucellar, integumentary, and chalazal tissue. The embryo proper is deep-seated in the chalaza. The suspensor haustoria and the embryo consume the remaining integumentary covering. Finally, the integumentary tissue shrivels completely.

In a mature seed, the chalazal tissue gives protection to the embryo, and can be differentiated into somewhat periclinally elongated cells of the epidermis which do not show any thickening, eight to ten layers of parenchymatous tissue with vascular strands, and three to four layers of tanniniferous tissue. The meristematic zone and the innermost parenchymatous zone are resorbed (Tiwari et al. 1977).

PL Bhalla and CP Malik (1982), PL Bhalla, MB Singh and CP Malik (1979, 1982), A Bolenbach (1928), M Chadefaud (1974), EJH Corner (1976), W Nägl and S Kuhner (1976), LL Narayana (1970c), SC Tiwari, F Bouman and RN Kapil (1977), RI Walker (1947)

29.6 Zygophyllaceae

The anther wall comprises the epidermis, fibrous endothecium, two or three middle layers, and secretory tapetum with binucleate cells. In *Peganum harmala* (Kapil and Ahluwalia 1963), the epidermis is heavily cutinized. The middle layers persist in *Balanites roxburghii* (N.C. Nair and Jain 1956), *Guaiacum officinale* (Masand 1970) and *Seetzenia orientalis* (H.S. Narayana and Prakasa Rao 1963). The outer middle layer may develop fibrous thickenings. The tapetum becomes multinucleate, and irregularly biseriate in *Peganum* and *Seetzenia*. The nuclei become polyploid due to nuclear fusions. Simultaneous cytokinesis follows meiotic divisions in microspore mother cells, and the tetrads are tetrahedral and decussate.

The pollen grains are three-colporate, and shed at the two-celled stage in *Peganum*, *Seetzenia* and *Zygophyllum* (Masand 1963b), and three-celled in *Fagonia* (N.C. Nair and Gupta 1961). In *Peganum*, some of the pollen grains germinate in situ.

The ovule is anatropous, hemianatropous or campylotropous, bitegmic and crassinucellate. The micropyle is organized by the inner integument in *Fagonia*, *Tribulus* and *Zygophyllum*, and by both the integuments in *Guaiacum*, *Peganum* and *Seetzenia*. An endothelium is very prominent, except in *Peganum* where it is absent. The nucellus is well developed in *Peganum*, in contrast to the other members. A hypostase develops.

The archesporium is multicellular, and each cell cuts off a parietal cell. Supernumerary megaspore mother cells, dyads, triads and tetrads have been observed in *Peganum*. Only one embryo sac of the Polygonum type develops. The synergids are pyriform, and the antipodals ephemeral.

The development of the endosperm is of the Nuclear type, and wall formation commences from the micropylar end at the young proembryo stage. In *Fagonia* and *Seetzenia* the chalazal endosperm nuclei become amoeboid and multinucleolate. In *S. orientalis* (H.S. Narayana and Prakasa Rao 1963), the caecum-like lowermost cell of the endosperm at the chalazal end is large and multinucleate. Testal ingrowths in *Guaiacum* (Masand 1970) bring about ruminations in endosperm.

The embryogeny conforms to the Solanad type. Synergid polyembryony occurs in *Peganum* (Kapil and Ahluwalia 1963); proliferation of suspensor cells is reported in *Zygophyllum* (Masand 1963b). The embryogeny of *Tribulus* and *Zygophyllum* corresponds to Period II, Series A, Megarchetype II (Caryophyllad type), and that of *Peganum* to Period I, Series C, Megarchetype IV (Solanad type).

The outer epidermis of testa is slightly thick-walled and lignified in *Peganum* (Corner 1976). It gets crushed in other members. The mesophyll is thin-walled. The inner epidermis comprises small, lignified cells with crystals in *Tribulus*; the cells are radially elongate and large and do not contain crystals in *Peganum*. The inner epidermis of tegmen alone persists, and serves as the mechanical layer. It is composed of narrow, longitudinal, lignified rectangular cells.

EJH Corner (1976), BM Johri (1963b), RN Kapil and K Ahluwalia (1963), P Masand (1963b, 1970), HS Narayana and CG Prakasa Rao (1963), NC Nair and I Gupta (1961), NC Nair and RK Jain (1956), MA Rau (1962), R Souèges (1953)

29.7 Linaceae

The anther wall comprises the epidermis, fibrous endothecium, one or two middle layers, and secretory tapetum with two- to multinucleate cells. A stomium has been observed in *Ochthocosmus africanus* (D. Rao and Narayana 1965). Polyploid nuclei produced by nuclear fusions occur in the tapetum of *Linum flavum*. In *O. africanus*, *Ixonanthes grandiflora* and *I. icosandra* (D. Rao and Narayana 1965), granular thickenings occur on the inner surface of tapetum. The microspore mother cells undergo meiotic divisions, and produce tetrahedral tetrads. Simultaneous cytokinesis occurs by furrowing.

The three-colpate pollen grains are shed at the two-celled stage in *O. africanus* and *Roucheria griffithiana* (D. Rao and Narayana 1965), and three-celled in *Durandea pentagyna*, *Hugonia mystax*, *Linum grandiflorum*, *L. rubrum* and *Reinwardtia trigyna* (L.L. Narayana 1964a).

The ovule is anatropous, bitegmic and crassinucellate; tenuinucellate in *L. mysorens* (Dorasami and Gopinath 1945). A tendency towards tenuinucellate condition has been observed in *Radiola linoides* (Mauritzon 1934d), and *L. perenne* (L.L. Narayana 1970d). Both conditions occur in *Anisadenia*. In *Octhocosmus* and *Ixonanthes* the nucellus is massive. An endothelium, obturator and hypostase are present.

In *Humiria balsamifera* (Boesewinkel 1985a), the ovule is trizonate and both the integuments are of dermal derivation. They are initially two- to three-layered. The micropyle is organized by both integuments.

A hypodermal archesporial cell cuts off a parietal cell. The parietal tissue is one to four layers thick. The megaspore mother cell undergoes meiotic divisions, and produces a linear tetrad, linear or T-shaped in *H. balsamifera*. The chalazal megaspore develops into a Polygonum type of embryo sac. The antipodals are ephemeral. Twin embryo sacs are present in *L. perenne* (see L.L. Narayana 1970d).

The development of endosperm is of the Nuclear type; Helobial in *Linum flavum*, *L. perenne* and *L. usitatissimum* var. *hungaricum* (Schürhoff 1924) which is doubtful. Erdelská (1967) reports the formation of a pectic wall at the time of division of the primary endosperm nucleus. It has no connection with the nuclear divisions. The nuclei pass through the perforation of the wall from

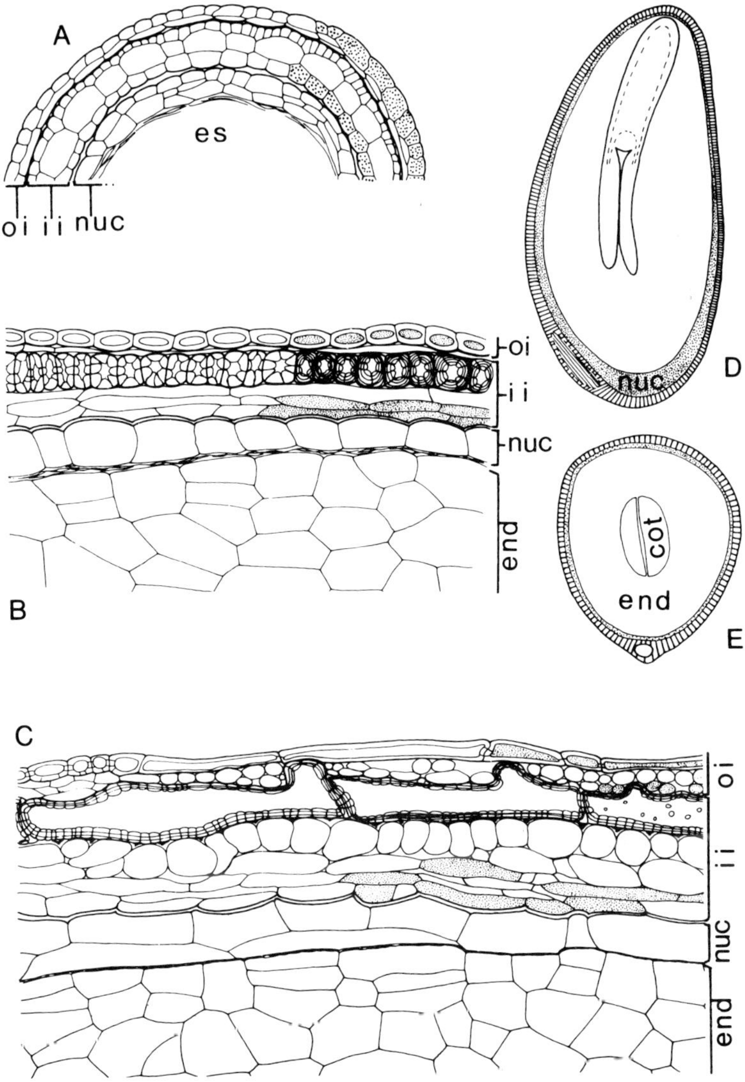

Fig. 29.7.1A–E. *Humiria balsamifera.* **A** Cross-section of a portion of mature ovule to show the disposition of outer (*oi*) and inner integument (*ii*), and nucellus (*nuc*) surrounding the embryo sac (*es*). The cells of the middle layer of inner integument enlarge. Its inner epidermis and the outer epidermis of the outer integument contain tannin. **B** Cross-section of a portion of mature seed-coat, testa two-layered and tegmen and nucellus multiplicative. The cells of the middle layer of tegmen do not divide but become stretched; *end* endosperm. **C** Longisection of a portion of mature seed-coat. Exotegmen thick-walled, the original middle layer of tegmen is longitudinally stretched. **D**, **E** Longi- and transection of mature seed; *cot* cotyledon. (After Boesewinkel 1985a)

micropylar region to chalazal region and, hence, the development is of the Nuclear and not Helobial type (in *Linum austriacum*).

In *Humiria balsamifera* the development of endosperm is of the Nuclear type. Later, it becomes cellular and the cells contain oil.

A chalazal haustorium is present in *Anisadenia* and *Linum* (L.L. Narayana 1970d), and *Reinwardtia* (L.L. Narayana 1964a).

The embryogeny conforms to the Solanad type. The development in *Linum* corresponds to Period I, Series C, Megarchetype III (Asterad type), and in *Radiola linoides* to Period I, Series C, Megarchetype IV (Chenopodiad type) (see Crété 1963a). Haploid-diploid twins are reported in *Linum usitatissimum* (Kappert 1933).

The cells of the outer epidermis of testa are thick-walled, often mucilaginous in *Linum*. The mesophyll is sclerotic in *Hugonia*. In *L. usitatissimum* the outer epidermis of tegmen comprises longitudinal, thick-walled, pitted fibres. The inner epidermis, endothelium, develops into rectangular cells with pitted walls and contains tannin (Boesewinkel 1980a).

The seed is exotestal and exotegmic. The outer integument remains two- to three-layered throughout, while the inner is multiplicative and contains tannin (Fig. 29.7.1A–C). The outer epidermis of the outer integument (exotesta) consists of thick-walled cells with lamellate and lignified pitted walls. These cells contain tannin and are stretched in longitudinal direction of the seed (Fig. 29.7.1B, C). The cells of the original middle layer of the inner integument do not divide, remain thin-walled, and gradually become stretched perpendicular to the longitudinal axis of seed. The cells of the outer epidermis of the inner integument develop into fibrous exotegmen of thick-walled, lignified pitted cells, stretched in longitudinal direction of the seed (Fig. 29.7.1A–C).

The mature seed of *Humiria* contains a straight embryo with rather short plano-convex cotyledons (Fig. 29.7.1D, E). One to three layers of nucellus persist in the mature seed and are covered by a thick cuticle.

G Bentham (1862b), FD Boesewinkel (1980a, 1984a, 1985a), P Crété (1963a), U Dhar and MR Vijayaraghavan (1979), LS Dorasami and DM Gopinath (1945), R Dulberger (1989), O Erdelská (1967), S Ghosh and KR Shivanna (1984), DM Gopinath (1942), H Kappert (1933), J Mauritzon (1934d), LL Narayana (1964a, 1970d), LL Narayana and D Rao (1978), D Rao and LL Narayana (1965), PN Schürhoff (1924), DL Secor and SD Russell (1988), R Souèges (1937a)

29.8 Erythroxylaceae

The anther wall comprises the epidermis, fibrous endothecium, three ephemeral middle layers, and secretory tapetum with binucleate cells. The microspore mother cells divide meiotically, and produce tetrahedral and decussate tetrads. Simultaneous cytokinesis occurs by furrowing.

The pollen grains are tricolporate, and shed at the three-celled stage (L.L. Narayana 1964b). The pollen grains have a uniform pattern of ornamentation of exine (stenopalynous).

The conspicuously elongated ovule is anatropous, bitegmic and crassinucellar in *Erythroxylum coca* (Boesewinkel and Geenen 1980). Both the integuments are of dermal origin. The inner integument is multiplicative. The outer integument is three layers thick, and remains shorter than the inner one. The inner integument alone forms the micropyle (L.L. Narayana 1970e). The endothelium extends up to the base of nucellus. The archesporial cell cuts off a parietal cell, the parietal tissue is two- or three-layered. The megaspore mother cell undergoes meiotic divisions, and produces a linear tetrad. The chalazal megaspore develops into a Polygonum type of embryo sac. The synergids are beaked, and the antipodals ephemeral. The mature embryo sac lies in direct contact with the endothelium (L.L. Narayana 1960b).

The endosperm is of the Nuclear type. Wall formation is initiated at the globular proembryo. Eventually, it becomes cellular and stores starch.

The embryogeny conforms to the Solanad type. The mature embryo is straight with thick, flat cotyledons.

The outer epidermis of testa is thin-walled. The outer epidermis of tegmen comprises thick-walled, lignified, pitted fibres with sclerotic cells at the endostome.

FD Boesewinkel and J Geenen (1980), J Mauritzon (1934c), LL Narayana (1960b, 1964b, 1970e), D Rao (1968a), PN Schürhoff (1924)

29.9 Euphorbiaceae

The anther wall comprises the epidermis, fibrous endothecium, one or two ephemeral middle layers, and secretory tapetum. The fibrous thickenings are reported to be absent in the endothecium in *Chrozophora rottleri* (G. Sharma 1956), but Kapil (1956b) reports the thickenings in *C. obliqua* and *C. parviflora*. The tapetal cells are binucleate; multinucleate in *Sebastiania chamaelea* (N.C. Nair and Maitreyi 1962). The tapetum is amoeboid in *Acalypha indica* (B.M. Johri and Kapil 1953). The microspore mother cells undergo meiotic divisions. Cytokinesis is simultaneous, and the tetrads are tetrahedral, isobilateral and decussate. Polyspory is reported in *Euphorbia corollata* (Lyon 1898). Degeneration of tapetal cells and pollen mother cells occurs in *Casava* (P.N. Rao and Sarveswara Rao 1976), and the pollen grains are sterile in *Euphorbia dulcis* (Kapil 1961).

Pollen grains are generally triporate. They are non-aperturate, quadriporate, hexaporate, or 10- to 12-porate. The exine is smooth or reticulate. They are two- or three-celled at the dispersal stage.

The ovule is anatropous, hemianatropous, or orthotropous (*Melanthesa*), bitegmic and crassinucellate. Both the integuments are vascularized in *Dalechampia*, only the inner in *Baliospermum*, *Croton*, *Jatropha*, *Ricinus* and others. The vascular supply terminates at the base of the inner integument in *Euphorbia geniculata* (S.P. Singh 1959), whereas it terminates at the chalaza in *Manihot esculentum* (P.N. Rao and Sarveswara Rao 1976). The vascular supply enters the nucellus in *Acalypha indica* (B.M. Johri and Kapil 1953). The obturator is generally placental in origin; of stylar origin in *Breynia* (Venkateswarlu et al. 1974). It is absent in *Cicca* and *Phyllanthus*. In many taxa the nucellar beak is long, slender and prominent (Fig. 3.4.6A). The beak protrudes beyond the micropyle, and a proper micropyle is not formed. Otherwise, the micropyle is produced by both the integuments, or the outer integument only. A hypostase develops during post-fertilization stages.

The archesporium is uni- or multicelled. The archesporial cell cuts off a parietal cell, and the parietal tissue is massive. The nucellar cap is very prominent in *Tragia involucrata* var. *angustifolia* (N.C. Nair and Abraham 1980). The megaspore mother cell undergoes meiosis. The chalazal megaspore of a linear or T-shaped tetrad develops into a Polygonum type of embryo sac, in the majority of the species. Monosporic five-nucleate gametophyte occurs in *Codiaeum veriegatum* (Galimberti 1963). Three nuclei at the micropylar end form the egg apparatus, and the two polar nuclei form a secondary nucleus.

Sanchez's (1938) report of Adoxa type of embryo sac in *Euphorbia heterophylla* has been corrected by P. Maheshwari (1942), and the gametophyte is of the Polygonum type. Dessiatoff (1911) refers to 16-nucleate gametophyte in *E. virgata* which is again of the Polygonum type, as pointed out by Modilewski (1911). In *E. pulcherrima*, Donati (1913) mentions a 16-nucleate embryo sac but K.D. Sharma (1955) considers the development to be of the Adoxa or Polygonum type.

The antipodals are ephemeral in *Hevea* (Nagaraja Rao 1964a), *Phyllanthus niruri* and *E. thymifolia*, and persistent in *Putranjiva roxburghii* and *Treuia nudiflora* (Banerji and Dutt 1944).

There is much variation in the development of the embryo sac in Euphorbiaceae. Bisporic, Allium type occurs in *E. amygadaloides*, *E. characias* (D'Amato 1939), *E. lagascae* (D'Amato 1946a) and *E. mauritanica* (Ventura 1933). Penaea type is reported in *E. procera*, *E. palustris* (Modilewski 1909a, 1910, 1911), *Acalypha australis* (Tateishi 1927), *A. tricolour* (Swamy and Balakrishnan 1946), and *A. malabarica* (Mukherjee 1958). "Acalypha indica" type – a modified Penaea type (two polar and two lateral groups of two cells each, and eight nuclei fuse in the centre) – occurs in *A. indica* (P. Maheshwari and Johri 1940, 1941, B.M. Johri and Kapil 1953), *A. fallax* (Banerji 1949) and *A. ciliata* (Kajale and Murthy 1954). Peperomia type is reported in *A. lanceolata* (Thathachar 1952). *Euphorbia dulcis* (Kapil 1961) shows Fritillaria type of development. Drusa type has been observed in *Mallotus japonicus* (Ventura 1933), and Adoxa type in *E. pulcherrima*. Chrysanthemum parthenium type (16-nucleate) and C. cineraraefolium type (12-nucleate) are reported in *E. epithymoides* (Cesca 1969). *E. dulcis* is an apomict.

The development of the endosperm is of the Nuclear type. It is autonomous in *E. dulcis*. The primary endosperm nucleus is three- to ten-ploid; during early stages nuclear divisions are synchronous. Enucleate cytoplasmic nodules are present in the micropylar region of the embryo sac in *Sebastiana chamaelea* (N.C. Nair and Maitreyi 1962) and *E. helioscoria* (R.P. Singh 1969). A coenocytic chalazal haustorium develops in *Croton klotzschianus*, *C. sparsiflorus* (Venkateswarlu and P.N. Rao 1963), *C. bonplandianum* (R.P. Singh and Chopra 1970) and *C. oblongifolius* (A. Pal 1974). The endosperm finally becomes cellular. Wall formation is initiated at the 32-nucleate stage in

Acalypha lanceolata (Thathachar 1952), 64-nucleate in *A. indica* (B.M. Johri and Kapil 1953) and 256-nucleate in *A. australis* (Tateishi 1927). The laying down of walls commences at the micropylar region, and proceeds to the chalaza, or is centripetal. The endosperm stores starch, fat globules and proteins.

The embryogeny generally conforms to the Onagrad type (Mukherjee 1961a, 1964, 1965, 1968, N.C. Nair and Abraham 1963, 1980, N.C. Nair and Maitreyi 1962, Mukherjee and Padhye 1964, Dnyansagar and Tiwari 1966, Samant 1973, Prakasa Rao and Rajeswari Devi 1974, Sarveswara Rao 1978). In addition, four other major types of development of embryo have been noted: Asterad, Chenopodiad, Solanad and Piperad. The occurrence of Piperad type in *Euphorbia rothiana* is refuted by Venkateswarlu and Rao (1967) who consider the development to be of Onagrad type. Other reports of the Piperad type should be reinvestigated.

Nucellar polyembryony occurs in *E. dulcis* (Kapil 1961).

The seed is albuminous with a large, almost straight embryo with a short radicle, and thin flat cotyledons. A caruncle is characteristic of Crotonoideae. In *Acalypha indica* (B.M. Johri and Kapil 1953), lateral flowers develop into carunculate seeds, whereas the terminal flower lacks caruncle. It develops from the outer integument, at the micropylar region, after fertilization. The caruncle is absent in Phyllanthoideae.

The seed is exotegmic. The testa is thin-walled, pulpy or dry. Its inner epidermis is a thin-walled layer of palisade cells. The outer epidermis of tegmen is a mechanical palisade layer of malpighian cells in Crotonoideae, and sclerotic in Phyllanthoideae. The testa is corky in *Sapium* spp.

PN Bajpai (1965), I Banerji (1949), I Banerji and MK Dutt (1944), J Bor and F Bouman (1974), J Bor and RN Kapil (1975, 1976), A Bruni and B Tosí (1978), G Cesca (1969), S Chopra (1970), S Chopra and RP Singh (1969), F D'Amato (1939, 1946a), N Dessiatoff (1911), VR Dnyansagar and HN Tiwari (1966), G Donati (1913), MK Dutt (1943), ML Frean (1979), Z Galimberti (1963), P Gori (1977), JS Greenwood and JD Bewley (1982), MI Hakki (1985), I Hurusawa (1954), BM Johri and RN Kapil (1953), LB Kajale (1954c), LB Kajale and KSN Murthy (1954), LB Kajale and GV Rao (1943), RN Kapil (1956a, b, 1960, 1961), RN Kapil and AK Bhatnagar (1972), SS Kelkar (1960), M Landes (1946), FN Lyon (1898), P Maheshwari (1942, 1945), P Maheshwari and OR Chowdhury (1937), P Maheshwari and BM Johri (1940, 1941), J Modilewski (1909a, 1910, 1911), LA Morrow (1979), PK Mukherjee (1957, 1958, 1961a, c, 1962, 1964, 1965, 1968), PK Mukherjee and MD Padhye (1964), PK Mukherjee and KN Santhianathan (1985), M Murgia, HJ Wilms, M Cresti and G Cesca (1986), A Nagaraja Rao (1964a), NC Nair and V Abraham (1963, 1980), NC Nair and M Maitreyi (1962), JW Nowicke and JJ Skvarla (1984), A Pal (1974), F Pax (1896a), VG Phatak and SD Savant (1955), PS Prakasa Rao and KR Rajeswari Devi (1974), PN Rao (1962, 1970, 1976), KR Rajeswari Devi and PS Prakasa Rao (1975), MVS Raju and A Nagaraja Rao (1953), SS Raman (1954), PN Rao and D Sarveswara Rao (1976), PS Rao, DK Prakasa and KR Rajeswari Devi (1975, 1977), EUB and CS Reddi (1985), L and M Rossignol and R Haicour (1987), K Samant (1973), ST Sanchez (1938), D Sarveswara Rao (1978), AM Schwarzenbach (1972), G Sharma (1956), KD Sharma (1955), RP Singh (1954, 1956, 1962, 1965, 1968, 1969, 1970a, b), RP Singh and S Chopra (1970), RP Singh and JL Jain (1965), SP Singh (1959), RK Srivastava and GP Agarwal (1953), BGL Swamy and BP Balakrishna (1946), S Tateishi (1927), T Thathachar (1952, 1953), T Thathachar and A Nagaraja Rao (1956), J Venkateswarlu and PN Rao (1963, 1967, 1975b), J Venkateswarlu, PN Rao and D Sarveswara Rao (1974), M Ventura (1933, 1934)

29.10 Daphniphyllaceae

The embryogeny has not been investigated.

In *Daphniphyllum himalayense* (A.K. Bhatnagar and Garg 1977), the anthers are fused and the connective slightly projects beyond the anthers (Fig. 29.10.1A–D). In transections the anthers show four to eight microsporangia (Fig. 29.10.1E–H). The anther wall comprises the epidermis, endothecium with thickened radial and inner tangential walls, four or five middle layers, and two- to three-layered glandular tapetum of multinucleate cells (Fig. 29.10.1I, J). Following meiosis I and II, there is simultaneous cytokinesis

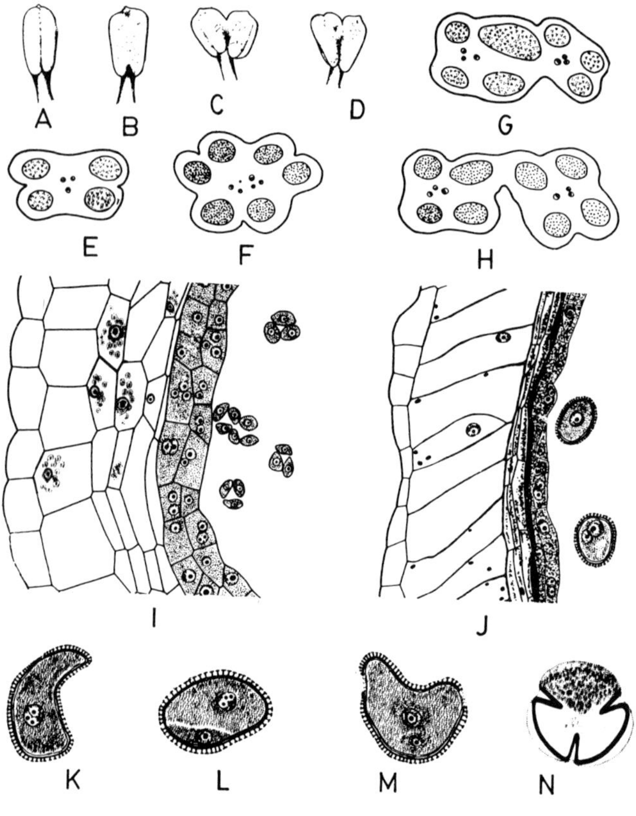

Fig. 29.10.1A–N. *Daphniphyllum himalayense* (**E–H** Semidiagrammatic). **A, B** Dorsal (**A**) and ventral (**B**) view of a tetrasporangiate anther with slightly prolonged connective. **C, D** Dorsal and ventral views of fused stamens. **E** Cross-section of **A** with four microsporangia, the connective has three vascular traces. **F–H** Transections of fused anthers with six, seven and eight microsporangia. The connective has two separate groups of vascular traces. **I** Part of transection of young anther at uninucleate pollen, there are four middle layers, and biseriate secretory tapetum with two- or three-nucleate cells. **J** Portion of mature anther at two-celled pollen grains, endothecial cells radially elongated, middle layers and tapetum have collapsed. **K** Uninucleate, intectate microspore, section passes through one aperture. **L, M** Two-celled pollen grain. **N** Palynogram shows three colpi, apocolpium, and reticulate pattern of exine. (After A.K. Bhatnagar and Garg 1977)

and mostly tetrahedral, occasionally isobilateral and decussate tetrads are formed. The mature pollen grains are two-celled (Fig. 29.10.1K–M). They are three-zonocolporoidate with a thick reticulate exine (Fig. 29.10.1N).

The ovule is anatropous, bitegmic and crassinucellate with a small obturator. The micropyle is closed by the exostome (Corner 1976). The archesporium is single-celled; formation of a parietal cell is uncertain. The development of the embryo sac follows the Polygonum type in *Daphniphyllum macropodum* var. *humile* (Sâto 1972). The antipodals are ephemeral.

The development of the endosperm is of the Cellular type (see Schnarf 1931).

The seed is albuminous and exarillate. The endosperm is thin-walled and oily, and the mature embryo is minute and straight. The testa and tegmen are thin-walled, and get crushed (Mohana Rao 1974a).

The fruit is a drupe, and endocarp is lignified with two layers of cross fibres; mesocarp has scattered stone-cells (Corner 1976).

AK Bhatnagar and M Garg (1977), EJH Corner (1976), PR Mohana Rao (1974a), Y Sâto (1972), K Schnarf (1931), M Ventura (1930)

29.11 Taxonomic Considerations

Limnanthaceae. Limnanthaceae are a monotypic family of the suborder Limnanthineae. Engler and Prantl (1897) place it in the order Sapindales, and Hutchinson (1973) in Geraniales. Lindley (1833, quoted in Takhtajan 1966) created the order Floerkeales. P. Maheshwari and Johri (1956) evaluate the embryological features of Limnanthaceae, and conclude that Limnanthaceae differ both from Sapindales and Geraniales (Table 29.11.1).

The Limnanthaceae comprise only herbaceous members with perfect flowers, and stamens in two whorls. They show: (1) persistent middle layers in anther wall, (2) two- to four-colporate pollen grains, (3) gynobasic style, (4) unitegmic, tenuinucellate, basal-parietal ovules, (5) an unusual type of tetrasporic embryo sac, (6) straight embryo with large, upwardly extended cotyledons (bifurcate in *Floerkea*), (7) pachychalazal, unitegmic seed without a characteristic exotegmen, and (8) schizocarpic fruit.

The Sapindales are shrubs or trees with compound leaves, and unisexual flowers with a single whorl of stamens. The seed is exo- or mesotestal, and the embryo is generally crumpled with curved cotyledons.

The occasional occurrence of amoeboid anther tapetum, three-colporate pollen grains, bitegmic crassinucellate ovules, mostly monosporic embryo sac, and nucellar polyembryony in some members of Sapindales and Geraniales differentiate them from Limnanthaceae.

P. Maheshwari and Johri (1956) propose that Limnanthaceae should be raised to an ordinal rank, Limnanthales, with *Limnanthes* as the type genus.

Oxalidaceae, Geraniaceae, Tropaeolaceae. Oxalidaceae, Geraniaceae and Tropaeolaceae have been treated as sub-families of Geraniaceae by Bentham

Table 29.11.1. Embryology of Limnanthaceae, Geraniales and Sapindales (see P. Maheshwari and Johri 1956)

	Limnanthaceae	Geraniales	Sapindales
Anther tapetum	Glandular, cells binucleate	Glandular, amoeboid in some taxa, cells bi- or multinucleate	Glandular, amoeboid in some taxa, cells 1–4-nucleate
Pollen	2–4-colporate, 2-celled	Usually 3-colporate, 2- or 3-celled	Usually 3-colporate, 2-celled
Ovule	Anatropous, unitegmic tenuinucellate, integument vascularized	Anatropous, bitegmic, crassinucellate (Linaceae, Oxalidaceae), integument vascularized in Euphorbiaceae	Anatropous, bitegmic, crassinucellate, tenuinucellate in Balsaminaceae, integument rarely vascularized
Embryo sac	Tetrasporic, but development differs from that in tetrasporic members of Euphorbiaceae and Malpighiaceae	Mostly monosporic, tetrasporic in Malpighiaceae, mono-, bi- or tetrasporic in Euphorbiaceae	Mostly monosporic, Allium type in *Hydrocera triflora* (Balsaminaceae)
Endosperm	Nuclear type, in *Floerkea* a coenocytic pouch is formed adjacent to the funicular-side	Nuclear	Nuclear, a large multinucleate micropylar haustorium is formed in *Hydrocera triflora*
Embryo	Straight, cotyledons large, there is an upward extension (auricle) and bifurcation of cotyledons in *Floerkea*, nucellar polyembryony absent	Straight, nucellar polyembryony occurs in Euphorbiaceae	Curved, cotyledons crumpled, nucellar polyembryony occurs in Anacardiaceae
Seed	Exalbuminous	Exalbuminous in Geraniaceae, Tropaeolaceae, Balsaminaceae, albuminous in Oxalidaceae	Exalbuminous in Sapindaceae, Sabiaceae and Anacardiaceae, albuminous in Melianthaceae
Fruit	Schizocarpic, 1-seeded achene (nutlet), indehiscent, style persists until a late stage	Capsular, dehiscent, style often persists in fruit	Berry, capsule, or drupe

and Hooker (1965a). Engler and Prantl (1931), Hutchinson (1973), Takhtajan (1966, 1980) and Cronquist (1968, 1980b, 1981) recognize their family rank and include them in Geraniales.

All the three families have similar embryological features indicating their close affinities (L.L. Narayana 1970a, b, c). However, each also has some distinctive features.

In Oxalidaceae the ovules are tenuinucellate with endothelium (crassinucellate in *Averrhoa*), occasional Allium type of embryo sac development, subruminate endosperm in *Biophytum* and *Oxalis*, and endospermic seeds. The seed structure is comparable with that of Caricaceae.

Geraniaceae show occasional presence of amoeboid anther tapetum, crassinucellate ovules, absence of endothelium, and presence of nucellar cap, parietal tissue, obturator, suberized hypostase, and exalbuminous seed.

Oxalidaceae, Geraniaceae and Tropaeolaceae exhibit extreme multipalyny (L.L. Narayana 1970a, b, c). Oltmann (1971) concluded that palynologically they are not similar. Chemotaxonomic studies (Hegnauer 1973) also support this view. Chadefaud (1974), on the basis of floral morphology of Tropaeolaceae, suggests their placement in Capparales. R. Dahlgren (1980b, 1981) includes them in a separate order, Tropaeolales (position uncertain), next to Geraniales. Tropaeolaceae have tenuinucellar ovule, and aggressive suspensor haustoria. Tiwari et al. (1977) point out that Tropaeolaceae differ from the families of Geraniales in ovular characters, and their inclusion in Geraniales is not justified. The outer integument is of subdermal origin (dermal in others), and the seeds are pachychalazal. The massive chalaza forms the seed-coat; crystalliferous and sclerotic cells are absent in the inner epidermis of the outer integument, whereas the characteristic exotegmen does not develop into a fibrous layer. In view of these facts, Dahlgren's separation of Tropaeolaceae from Geraniales deserves attention.

Embryologically, these three families (L.L. Narayana 1970a, b, c) resemble Balsaminaceae (Sapindales, Sect. 31).

Zygophyllaceae. Most phylogenists retain the family Zygophyllaceae in Geraniales. The genus *Balanites* is placed in Simaroubaceae by Bentham and Hooker (1965a), as a subfamily of Zygophyllaceae by Engler and Prantl (1931), as a subfamily Balanitoideae by Melchior (1964), and in a monogeneric family Balanitaceae in Malpighiales by Hutchinson (1973). The other genus *Peganum* is included in Rutaceae by Hooker (1885), and in Zygophyllaceae by Hutchinson (1973). Souèges (1953) and Rau (1962) suggest the creation of a separate family Peganaceae, to be placed near Linaceae.

Peganum differs from Rutaceae: three-colporate pollen grains, occurrence of supernumerary dyads and tetrads of megaspores, and absence of nucellar cap and nucellar polyembryony. In contrast, it resembles Zygophyllaceae in essential features (Table 29.11.2). The occurrence of biseriate tapetum, massive nucellus, and the absence of endothelium in *Peganum* are features similar to those of *Seetzenia* (Zygophyllaceae). However, (1) heavily cutinized epidermis of anther wall, (2) occasional germination of pollen in situ, (3) a persistent synergid and synergid polyembryony, and (4) presence of sclereids in pericarp

Table 29.11.2. Embryology of *Peganum*, Zygophyllaceae (Sect. 29.6), *Balanites* and Simaroubaceae (30.4). (See N.C. Nair and Jain 1956, Kapil and Ahluwalia 1963, Masand 1970, Corner 1976)

	Peganum	Zygophyllaceae	*Balanites*	Simaroubaceae
Anther tapetum	Irregularly biseriate, secretory, cells multinucleate	Secretory, cells binucleate	Secretory, cells multinucleate	Secretory, cells bi- or multinucleate
Cytokinesis	Simultaneous	Simultaneous	Simultaneous	Simultaneous
Tetrads	Tetrahedral, isobilateral, decussate	Tetrahedral, decussate	Tetrahedral, decussate	Tetrahedral, isobilateral, decussate
Pollen	Tricolporate, 2-celled	Tricolporate, 3-celled	Tricolpate, 2-celled	Triporate, 2-celled
Ovule	Funicular small ligulate aril, anatropous, bitegmic, crassinucellate, micropyle formed by both integuments, zigzag, feebly developed hypostase	Ana-, hemiana- or campylotropous, bitegmic, crassinucellate, micropyle formed by both or inner integument, endothelium differentiates, hypostase develops	Campylotropous, bitegmic, crassinucellate, endothelium conspicuous, cells 1–3-nucleate, micropyle formed by both integuments, zigzag, there is a space in the chalazal region between the nucellus and inner integument, and the inner and outer integument	Ana-, hemiana- or campylotropous, uni- or bitegmic, crassinucellate, integuments separated by a space in chalazal region, micropyle formed by both integuments, zigzag, obturator, nucellar cap and hypostase develop
Tetrad	Multiple dyad, triad and tetrad	Linear	Linear, T-shaped	Linear, T-shaped

Embryo sac	Polygonum type, antipodals uninucleate, persist up to 2-nucleate endosperm	Polygonum, antipodal ephemeral	Polygonum	Polygonum, antipodals ephemeral, twin embryo sacs may develop
Endosperm	Nuclear type, becomes cellular, thick-walled, reserve food oil and starch	Nuclear, becomes cellular	Nuclear	Nuclear, becomes cellular; chalazal haustroium in *Suriana*
Embryo	Solanad type, suspensor conspicuous	Solanad or Caryophyllad	—	Onagrad
Polyembryony	Synergid polyembryony	Suspensor polyembryony in *Zygophyllum*	Absent	In *Brucea* an additional embryo may develop
Seed coat	Not multiplicative, exotestal palisade, less distinct thin-walled endotestal palisade, endotegmen lignified fibriform, cells of seed-coat contain oil globules and starch	Not multiplicative	Testa thick, some cells contain crystals, others sclerotic and pitted, vascularized	Testa multiplicative, testa and tegmen become more or less crushed
Seed	Small, subconic, angled, brown, arillate (?), albuminous, a few nucellar cells persist at the micropylar end	Small to medium size, exarillate, albuminous or exalbuminous	Regulose, yellow, exalbuminous, embryo with thick cotyledons	Small to fairly large, exarillate, enclosed in woody endocarp, albuminous or exalbuminous, perisperm reported in *Brucea*, embryo straight or curved, cotyledons thick

support its assignment to a separate tribe, Peganoideae under Zygophyllaceae. The embryogeny of *Peganum harmala* (Souèges 1953, Kapil and Ahluwalia 1963) is similar to that of *Radiola* (Linaceae – see Rau 1962), and differs from that of Rutaceae and other members of Zygophyllaceae.

B.M. Johri (1963b, see also Nair and Jain 1956) summarizes the embryological features of *Balanites*. It shows secretory anther tapetum, Polygonum type of embryo sac, and Nuclear type of endosperm similar to that in Zygophyllaceae and Simaroubaceae (Table 29.11.2). The syncarpous gynoecium, two- or three-celled pollen grains at the dispersal stage, bitegmic, crassinucellate, campylotropous ovule – anatropous in Zygophyllaceae – and conspicuous endothelium occur in both *Balanites* and Zygophyllaceae. In Simaroubaceae the gynoecium is apo- or syncarpous, mature pollen is two-celled, the ovule may be ana-, hemiana-, campylo- or amphitropous, and uni- or bitegmic with a massive nucellus and a nucellar cap. The endothelium is absent, and embryogeny is of the Onagrad type. Therefore, *Balanites* is more closely related to Zygophyllaceae than to Simaroubaceae. The knowledge about its embryogeny may be helpful in determining its phyletic relationship. Masand (1970) discusses the systematic position of *Peganum* and *Balanites*, and suggests their assignment to separate subfamilies, Balanitoideae and Peganoideae, in Zygophyllaceae, under Geraniales. We agree with this arrangement.

Linaceae, Erythroxylaceae. Bentham and Hooker (1965a), Bessey (1915), Engler and Prantl (1931), Melchior (1964), and Takhtajan (1966, 1980) place Linaceae under Geraniales. Bentham and Hooker (1965a) treat Erythroxylaceae as a tribe Erythroxylae under Linaceae, while Engler and Prantl (1931), Melchior (1964), and Takhtajan (1966, 1980) treat them as an independent family under Geraniales. Hutchinson (1973) places both Linaceae and Erythroxylaceae in Malpighiales.

Both the families have essentially a similar type of development of anther, pollen, ovule, embryo sac, endosperm and embryo (L.L. Narayana 1970d, e). Linaceae are multipalynous, whereas Erythroxylaceae are uni-palynous. There is an occasional occurrence of Helobial endosperm, chalazal endosperm haustorium, haploid-diploid twin embryos; obturator and hypostase are absent in Erythroxylaceae. This may justify its separation from Linaceae. Otherwise, in both families, the testa is thin-walled as in Oxalidaceae, but lacks the crystal layer. The characteristic exotegmen occurs in both; the endothelium contains tannin.

Humiria. Melchior (1964) places *Humiria* in Linaceae, subfamily Humirioideae (Humiriaceae). Takhtajan (1980) and R. Dahlgren (1980b, 1981) treat Humiriaceae as a separate family in the order Geraniales which also includes Linaceae, Hutchinson (1973) in Malpighiales and Cronquist (1981) in Linales.

Comparable to Linaceae, in *Humiria* (Boesewinkel 1985a) the ovule is anatropous, bitegmic and crassinucellate, the embryo sac of the Polygonum type, and endosperm of the Nuclear type. In the seed-coat the middle layer of the inner integument remains thin-walled and gradually becomes stretched perpendicular to the longitudinal axis of seed and resembles the "cross cells" of *Linum*. In some respects, *H. balsamifera* differs from other members of Linaceae: there is no endothelium, obturator, or hypostase. The seed is exotestal and exotegmic. A few layers of nucellus persist in the mature seed. Mucilaginous wall formation of exotesta, as in *Linum*, is absent.

The embryology of *H. balsamifera* resembles that of other members of Linaceae, and the differences are not so significant as to warrant the erection of a separate family, Humiriaceae. The placement of *Humiria* in a subfamily Humirioideae, of Linaceae, is quite appropriate.

The suborder Euphorbiineae comprises Euphorbiaceae and Daphniphyllaceae.

Euphorbiaceae. Bentham and Hooker (1965c) place Euphorbiaceae in Unisexuales, Hallier (1912) in Passionales, Wettstein (1935) in Tricoccae, and Hutchinson (1973) and Cronquist (1968, 1981) in Euphorbiales. Bentham and Hooker (1965c) treat Buxaceae as a tribe Buxeae under Euphorbiaceae. Embryologically, it is different from Euphorbiaceae (Celastrales, Sect. 33).

Baillon (1874) includes Dichapetalaceae as a tribe Dichapetalae in Euphorbiaceae. Engler and Diels (1936) place them in Geraniales, and Melchior (1964) in Thymelaeaceae. As embryological information is not available, their relationships cannot be determined.

Wunderlich (1967) calls attention to the similarities between Euphorbiaceae and Thymaeflorae (carotenoid pollen, epitropous ovule with obturator, supernumerary antipodals, and straight embryo).

In fact, embryologically, Euphorbiaceae are a very heterogeneous family. It shows secretory or amoeboid tapetum, two- or three-celled pollen grains, anatropous, hemianatropous or orthotropous, bitegmic and crassinucellate ovules, placental obturator, nucellar beak, hypostase, massive parietal tissue, several types of embryo sac development, Nuclear type of endosperm, different types of development of embryo, and exotegmic seeds.

Thatachar (1953) proposes the relationship of Euphorbiaceae with Malpighiaceae. P.N. Rao (1970) observed apparent similarity between the two, but points out that Polygonum type of embryo sac occurs in the majority of genera, whereas bi- and tetrasporic types occur only in *Acalypha*, *Euphorbia* and *Mallotus*. Failure of fertilization, and frequent polyembryony occur commonly in Malpighiaceae.

Radcliffe-Smith (1987) emphasizes that the Euphorbiaceae differ from (a) Buxaceae which have apotropous ovules with a dorsal raphe, (b) Aextoxicaceae with apotropous, anatropous ovules borne in only one of the two loculi, and seeds with ruminate endosperm, (c) Daphniphyllaceae with a minute apical embryo in copious endosperm, (d) Pandaceae with ortho- or anatropous ovules which lack obturator, (e) Stilaginaceae with small drupaceous fruits, often flattened and somewhat obliquely oriented endocarp, (f) Icacinaceae where the ovules have a dorsal raphe, (g) Hymenocardiaceae with large extrose anthers with a terminal gland which folds back after dehiscence, and bi- or triporate pollen, (h) Bischofiaceae with spherical pollen grains, and fruit a small globose fleshy drupe, and (i) Androstachydaceae with inaperturate pollen grains. The author does not consider phylogenetic relationships.

Daphniphyllaceae. The genus *Daphniphyllum* was once placed under the tribe Phyllantheae of Euphorbiaceae (Hooker 1885), under subfamily Phyllanthoideae of Euphorbiaceae (Wettstein 1935), and as a tribe Daphniphylleae in the subfamily Phyllanthoideae (Rendle 1959). Engler and Prantl (1931), Rosenthal (1931) and Scholz (1964) assign it to a separate family, Daphniphyllaceae, close to Euphorbiaceae. Croziat (1941) finds its alliance with Eucommiaceae, and Thorne (1968) classifies it in Pittosporales. Hallier (1912), Hutchinson (1973), R. Dahlgren (1977a) and Takhtajan (1980) include it in the order Hamamelidales. Airy Shaw (1973) agrees with this assignment, but relates it also with Magnoliales. Hurusawa (1954) and Cronquist (1981) erect the order Daphniphyllales for the monogeneric family Daphniphyllaceae.

The heteropolar, three-colporoidate and intectate pollen grains of *Daphniphyllum himalayense* (A.K. Bhatnagar and Garg 1977) have no resemblance with those of Euphorbiaceae. The absence of an obturator and a Cellular type of endosperm in the family do not justify its alliance with Euphorbiaceae. The occurrence of apomixis (see Schnarf 1931) and precocious degeneration of antipodals (Sâto 1972) seem to be similar features between the two. The wood anatomy is also distinct from that in Euphorbiaceae (Metcalfe and Chalk 1950).

In contrast, it shows resemblance with Hamamelidaceae in floral structure, endothecial thickenings on the radial and inner tangential walls, multinucleate tapetal cells, similar pattern of two-celled pollen grains (A.K. Bhatnagar and Garg 1977), and the structure of wood (Metcalfe and Chalk 1950).

Both Daphniphyllaceae and Hamamelidaceae have anatropous, bitegmic, and crassinucellate ovules, Polygonum type of embryo sac (Sâto 1972), Cellular type of endosperm, large seeds, and straight embryo (Mohana Rao 1974a). In Daphniphyllaceae, however, the entire seed-coat collapses during development, and iridoid substances are present.

30 Rutales

The order Rutales includes 12 families distributed in three suborders. Rutineae comprises seven families – Rutaceae, Cneoraceae, Simaroubaceae, Picrodendraceae, Burseraceae, Meliaceae and Akaniaceae; Malpighiineae three families – Malpighiaceae, Trigoniaceae and Vochysiaceae; and Polygalineae two families – Tremandraceae and Polygalaceae.

30.1 General Features

The order Rutales exhibits: (1) secretory anther tapetum, (2) simultaneous cytokinesis in pollen mother cells, (3) bitegmic, crassinucellate ovule, massive parietal tissue and nucellar cap, (4) hypostase, (5) Polygonum type of embryo sac with ephemeral antipodals, Polygonum and Allium type in Rutaceae, and Allium and Penaea type in Malpighiaceae, (6) Nuclear type of endosperm, (7) Asterad type of embryogeny in Rutaceae (sometimes Onagrad or Solanad), Cneoraceae and Polygalaceae, and Onagrad type in Simaroubaceae and Meliaceae, (8) nucellar polyembryony in Rutaceae, Meliaceae and Malpighiaceae, and (9) endospermic seeds, except in Malpighiaceae.

30.2 Rutaceae

The anther wall comprises persistent epidermis, fibrous endothecium, two or three ephemeral middle layers, and glandular tapetum with multinucleate cells. Nuclear fusions occur in the tapetal cells in *Aegle marmelos*. The microspore mother cells undergo meiotic divisions, simultaneous cytokinesis occurs by furrowing, and the tetrads are tetrahedral, decussate and isobilateral.

Pollen grains are shed at the two-celled stage. They are two- to eight-colporate with reticulate exine.

The ovule is anatropous, bitegmic and crassinucellate. The zigzag micropyle is formed by both the integuments. A third integument develops in *Coleonema album* (Desai 1962a). A nucellar cap develops in *Chloroxylon* and *Glycosmis* (L.L. Narayana 1963a). The hypodermal archesporium is multicellular. A single cell cuts off a parietal cell, the parietal tissue is well developed. The megaspore mother cell undergoes meiotic divisions, and produces a linear or T-shaped tetrad. The chalazal megaspore develops into a Polygonum type of embryo sac. The antipodals are ephemeral. A hypostase differentiates. Large-scale degeneration of gametophytes occurs in *Aegle marmelos* (B.M. Johri and Ahuja 1956). Sâto (1971) reports Polygonum type of embryo sac in *Aucuba japonica*. But, in some cases, the development of female gametophyte is peculiar. The two megaspores of a tetrad, either micropylar or chalazal, develop into four-nucleate embryo sacs. The separating wall, between the two embryo sacs, disintegrates and an eight-nucleate embryo sac is produced. This suggests bisporic development and needs confirmation.

In *Aegle marmelos* (B.M. Johri and Ahuja 1956); triple fusion occurs normally. The egg apparatus degenerates, syngamy has not been observed.

The development of endosperm is of the Nuclear type. Centripetal wall formation is initiated at the micropylar end, after the formation of about 256 free nuclei. The chalazal end of the embryo sac grows into a coiled aggressive haustorium (B.M. Johri and Ahuja 1957), or several tubular projections develop in the chalazal region in *Coleonema album*, *Empleurum serrulatum*, *Ptelea trifoliata* and *Triphasia aurantiola* (Mauritzon 1935a).

The embryogeny conforms to the Onagrad type in *Ruta graveolens* (Johansen 1950), Solanad in *Citrus trifoliata*, and Asterad in *Ravenia spectabilis* (R.B. Ghosh 1969, 1975).

Nucellar polyembryony occurs in the majority of genera. As many as six embryos have been noted in *Xanthoxylum* (Desai 1962b). However, only one matures. In *Citrus* 13 viable embryos occur in a seed; fertilization is essential

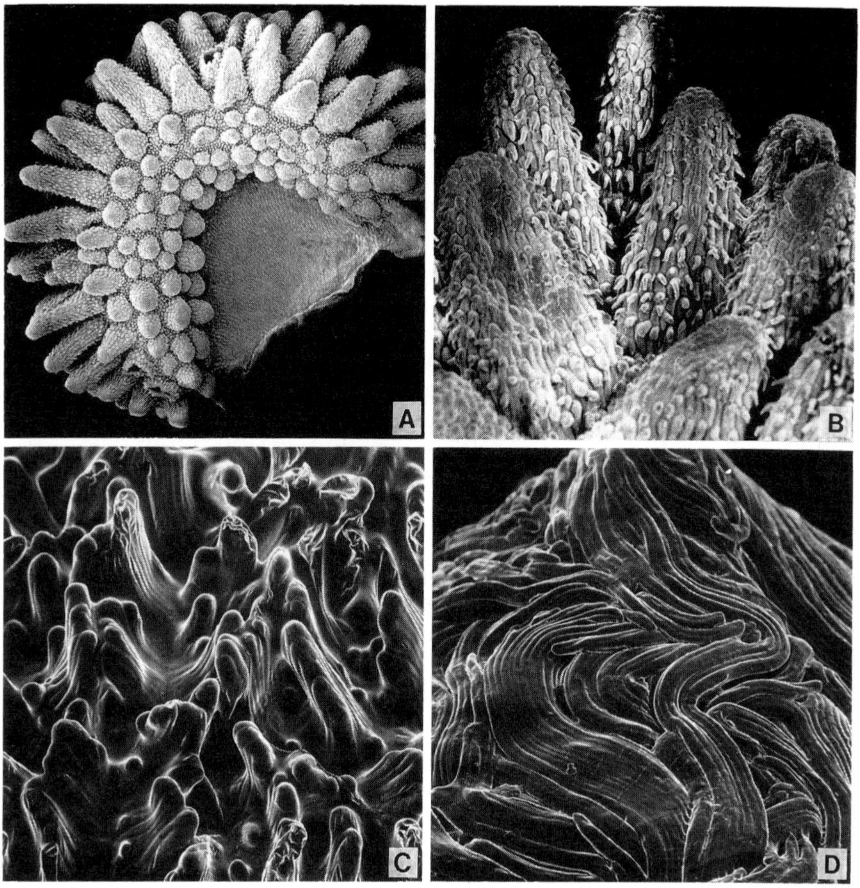

Fig. 30.2.1A–D (SEM) **A**, **B** *Thamnosma africanum*, **C**, **D** *Murraya paniculata*. **A** Seed; **B** Portion of seed-coat. **C** Young seed-coat with trichomata; **D** Same, with adpressed trichomata. (Courtesy Dr F. Bouman, Amsterdam)

for the development of these embryos (H.J. Webber and Batchelor 1943). At times, more than two embryo sacs develop in the same ovule, resulting in two zygotic embryos (Bacchi 1943).

The outer epidermis of testa is fibrous. The mesophyll is either sclerotic, or with scattered stone cells. The cells of tegmen are tracheidal with spiral and annular thickenings. The seed and seed-coat are shown in Figure 30.2.1A–D.

O Bacchi (1943), I Banerji (1954), FD Boesewinkel (1977, 1978), DK Bruck and DB Walker (1985), RS Chakravarthy (1935, 1936), C Chandra and G Shankar (1964), S Desai (1961, 1962a, b), HB Frost (1938), K Furusato (1951, 1953), K Furusato, Y Ohta and K Ishibashi (1957), RB Ghosh (1969, 1975, 1983a, b), HT Horner Jr and NR Lersten (1971), L Iglesias, H Lima and JP Simon (1974), M Iwamasa, V Isamu and N Masao (1970), DA Johansen (1950), BM Johri and MR Ahuja (1956, 1957), B Longo (1908), J Mauritzon (1935a), LL Narayana (1963a), V Narmatha Bai and KK Lakshmanan (1982), AP Pieringer and GJ Edwards (1967), Y Sâto (1971), R Souèges (1926d), DR Thakur and BS Bajwa (1971), MD Vijaya (1986), A Wakana and S Uemoto (1988), JM Webber (1940), HJ Webber and LD Batchelor (1943)

30.3 Cneoraceae

The development of anther and pollen has not been studied. We have some information on *Cneorum tricoccum* (Schürhoff 1924, Souèges 1955, Rau 1962). Boesewinkel (1984b) has made a detailed study of the ovule and seed-coat of this taxon.

The pollen grain is two-celled at dispersal.

The anatropous ovule is bitegmic and crassinucellate (Boesewinkel 1984b). The ovular primordium is trizonate. Both the integuments are two- to three-layered (Fig. 30.3.1A–C), the inner one is longer than the outer and forms the micropyle. The primary parietal cell produces a massive parietal tissue, and there is a distinct nucellar cap of about four cells high.

An obturator is present, but is hardly distinguishable in a mature seed from the outer integument on the funicular side. Gradually, the ovule changes from anatropy to campylotropy due to the intercalary growth of the raphal region, appreciable growth in the length of the nucellus and the integuments on the antiraphal side, and the relatively small ovarian locule and adpression of the seed against the endocarp.

The chalazal megaspore develops into a Polygonum type of embryo sac. The polar nuclei fuse before fertilization, and the secondary nucleus comes to lie near the three antipodals. The endosperm is of the Nuclear type but later becomes cellular due to centripetal wall formation (Fig. 30.3.1C–F). The embryogeny conforms to the Asterad type, or Period I, Group II, Megarchetype II (Souèges 1955, Rau 1962). The mature embryo is strongly curved due to the change in the ovule from anatropy to campylotropy.

During the maturation of seed (Fig. 30.3.1C–F), the nucellus and inner integument are crushed and resorbed. The inner layer of the outer integument undergoes division so that it becomes four- to five-layered. The outer tangential walls of the exotesta become thick, the cells of the meso- and endotesta show conspicuous intercellular spaces. A distinct nucellar cuticle and nucellar remains are present close to the seed-coat. The pericarp has taken over the protective function of the seed-coat. The inner layers of the pericarp become multilayered and form the sclerotic endocarp. The mesocarp and the endocarp become lignified. The seeds are uncinately conduplicate with fleshy endosperm and uncinate embryo (Hutchinson 1973). The cotyledons are situated in the transmedian plane. The endosperm and embryo are rich in lipids, and have scanty starch.

FD Boesewinkel (1984b), J Hutchinson (1973), MA Rau (1962), PN Schürhoff (1924), R Souèges (1955)

30.4 Simaroubaceae

The anther wall comprises the epidermis, fibrous endothecium, two or three ephemeral middle layers, and secretory tapetum with bi- or multinucleate cells.

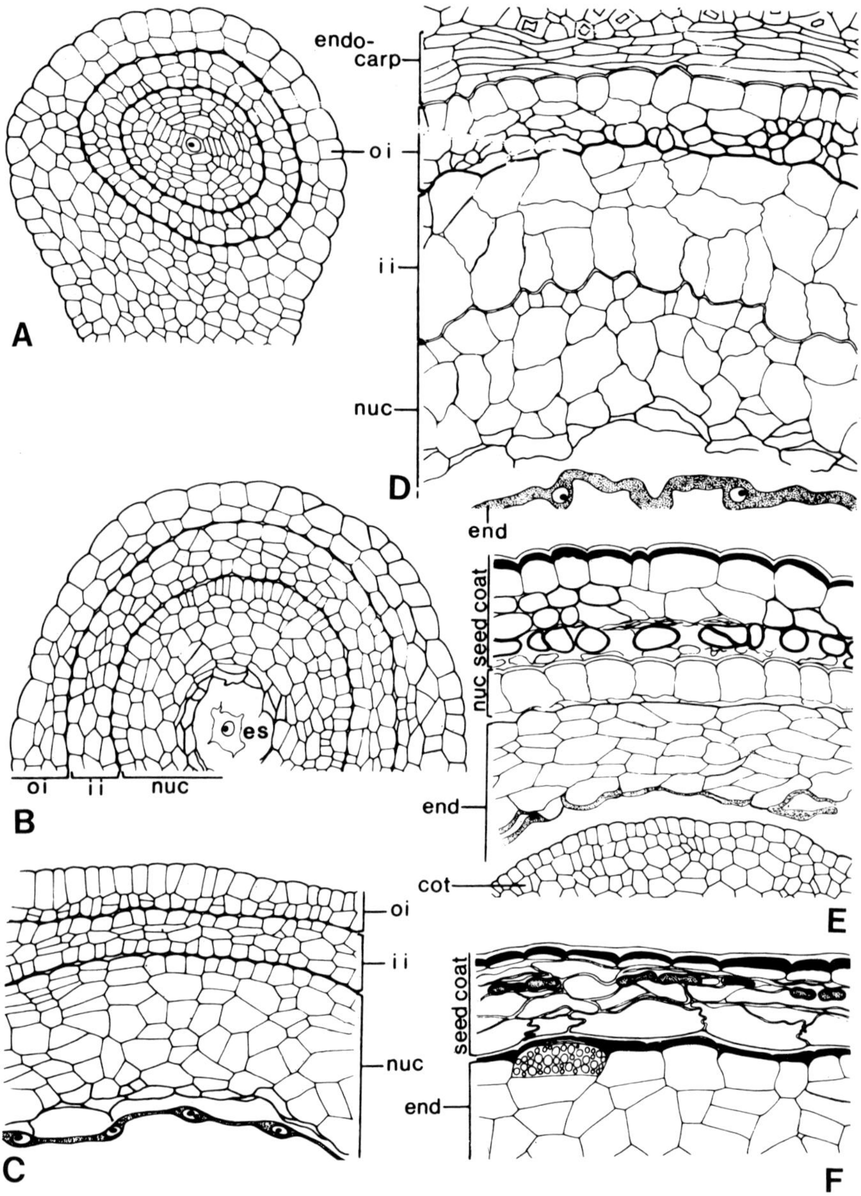

Fig. 30.3.1A–F. *Cneorum tricoccum.* **A, B** Transection of ovules; *oi*, *ii* outer and inner integument; *nuc* nucellus; *es* embryo sac. **C–F** Transection of seed-coat (endocarp is shown in **D**), nucellus prominent in **C** and **D**, mostly crushed in **E**, and completely in **F**. Endosperm (*end*) nuclear in **C**, **D**, and cellular in **E**, **F**; *cot* cotyledon. (After Boesewinkel 1984b)

Due to nuclear fusions the tapetal cells become polyploid in *Samadera indica* (N.C. Nair and Joseph 1957) and *Brucea amarissima* (N.C. Nair and Sukumaran 1960). The microspore mother cells undergo meiotic divisions, cytokinesis is simultaneous, and the tetrads are tetrahedral, decussate and isobilateral.

The triporate pollen grains have smooth exine, and are two-celled at the dispersal stage.

The ovules are hemianatropous in *Ailanthus excelsa* (L.L. Narayana 1957) and *Brucea*, anatropous in *Samadera* (N.C. Nair and Joseph 1957), and campylotropous in *Balanites roxburghii* (L.L. Narayana 1957) and *Suriana maritima* (Rau 1940). They are bitegmic, unitegmic in *Suriana*, and crassinucellate. The vascular strand extends beyond the chalaza in *Ailanthus excelsa* (L.L. Narayana 1957). The integuments are separated by a space in the chalazal region. The micropyle is zigzag. A hypostase is present in *Brucea*, *Samadera* and *Suriana*. A nucellar cap develops except in *Balanites*. An obturator occurs in *Brucea*, *Irvingia* and *Picrodendron* (Wiger 1935). In *Samadera* the nucellar cells become two- to 14-nucleate in the chalazal region, and the nuclei fuse and exhibit polyploid condition (N.C. Nair and Joseph 1957).

The archesporium is uni- or multicelled, only one cell is functional. It cuts off a parietal cell, the parietal tissue is massive except in *Balanites*. An endothelium develops in *Balanites* (L.L. Narayana 1957). The megaspore mother cell undergoes meiotic divisions and produces a linear tetrad, T-shaped tetrads in *Ailanthus* and *Brucea*. The chalazal megaspore develops into a Polygonum type of embryo sac. The antipodals are ephemeral. Twin embryo sacs occur in *Suriana*.

Development of the endosperm is of the Nuclear type. At the chalazal end of the embryo sac, an unbranched haustorium develops in *Suriana maritima* (Rau 1940). Later, the endosperm becomes cellular.

The embryogeny conforms to the Onagrad type. An additional embryo is produced in *Brucea*.

The mature seed contains a straight or curved embryo with a short radicle, and thick cotyledons. The outer epidermis of testa is thin-walled. Its mesophyll contains idioblasts in *Simarouba*, and the inner epidermis is slightly lignified with fine spirals in *Brucea*, *Picrasma* and *Simarouba*. The tegmen becomes crushed. The nucellus persists for a long time. The seed is enclosed in a woody endocarp.

In *Suriana* the outer epidermis of seed-coat is thick-walled, and the inner few layers are thin-walled.

NC Nair and RK Jain (1956), NC Nair and TC Joseph (1957), NC Nair and NP Sukumaran (1960), LL Narayana (1957), D Rao (1970a), MA Rau (1940), J Wiger (1935)

30.5 Picrodendraceae

The embryology has not been investigated.

30.6 Burseraceae

The embryogeny has not been investigated.

The anther wall comprises the epidermis, fibrous endothecium, two ephemeral middle layers, and secretory tapetum of binucleate cells. The epidermal cells are papillose in *Boswellia* (L.L. Narayana 1959, 1960c). Degeneration of anthers is reported in *Santiria, Balsamodendron* (Shukla 1954) and *Bursera* (L.L. Narayana 1960b). After meiotic divisions, the microspore mother cells produce tetrahedral tetrads. Cytokinesis is simultaneous, and quadripartition occurs by furrowing.

The three-colporate pollen grains are shed at the two-celled stage. In *Boswellia* and *Garuga* (L.L. Narayana 1960c), they are three-celled at the shedding stage.

The ovule is hemianatropous, bitegmic and crassinucellate. The inner integument alone organizes the micropyle. Unitegmic condition occurs in *Commiphora, Canarium asperum, C. oleosum* and *Santiria rubiginosa* (Wiger 1935). Shukla (1954) has reported two integuments in *Commiphora* (*Balsamodendron*).

The hypodermal archesporial cell cuts off a parietal cell. A massive parietal tissue is produced. A nucellar cap develops due to periclinal divisions of the nucellar epidermal cells. Following meiotic divisions, the megaspore mother cell gives rise to a linear tetrad. The functional chalazal megaspore develops into a Polygonum type of embryo sac. More than one megaspore mother cell develops up to the tetrad stage, but only one survives in *Balsamodendron* (Shukla 1954). The antipodals are ephemeral.

The pollen tube is monosiphonous in *Garuga* (R.B. Ghosh 1970). Its entry is porogamous, and double fertilization occurs.

The development of endosperm is of the Nuclear type, later it becomes cellular. Rarely, the cells become multinucleate in *Garuga* (R.B. Ghosh 1970). In *Bursera* (Srivastava 1964) a few free nuclei are distributed around the vacuole. The endosperm stores oil.

The mature embryo is large with folded or palmately lobed cotyledons, and a short radicle.

The outer epidermis of testa consists of thin-walled tabular cells. The inner epidermis persists as a pellicle of small, lignified, tracheidal cells with fine spiral and annular thickenings. The tegmen becomes crushed.

RB Ghosh (1970), LL Narayana (1959, 1960c), D Rao (1970b), RD Shukla (1954), GN Srivastava (1964), J Wiger (1935)

30.7 Meliaceae

The anther wall comprises the epidermis, fibrous endothecium, two or three ephemeral middle layers, and secretory tapetum with multinucleate cells. The epidermis is persistent and papillate in *Azadirachta indica* (N.C. Nair and Kanta 1961), *Cipadessa baccifera* (L.L. Narayana 1958), *Melia azedarach*

(N.C. Nair 1959b), *Naregamia alata* (N.C. Nair 1959a) and *Walsura piscidia* (R.B. Ghosh 1966b). The nuclei of tapetal cells fuse and polyploidy occurs. In *Cipadessa* star-shaped crystals are present in the middle layers. The microspore mother cells undergo meiotic divisions, and produce tetrahedral, decussate and isobilateral tetrads; cytokinesis is simultaneous. Bernardo and Ramirez (1959) reported several irregularities in microsporogenesis in "lanzones" (a variety of *Lansium domesticum*) leading to sterility, and formation of polyspores. The "duku" and "langsat" varieties of *L. domesticum* (N. Prakash et al. 1977) show extensive degeneration of microspore mother cells, and formation of viable pollen is totally absent. However, the endothecium develops characteristic fibrous thickenings.

The tetraporate pollen grains are two- or three-celled at the shedding stage. Polyspory has been observed in *Azadirachta indica*, *Cipadessa baccifera* (L.L. Narayana 1958), *Melia azedarach* and *Naregamia alata* (N.C. Nair 1959a, b, N.C. Nair and Kanta 1961).

The ovule is anatropous, bitegmic and crassinucellate. The inner integument produces the micropyle, except in *Dysoxylum* (Paetow 1931) where the outer integument is also concerned. The nucellar epidermal cells divide to produce a prominent nucellar cap. Formation of an obturator is reported in *Cipadessa*, *Lansium domesticum* (N. Prakash et al. 1977), *Melia*, *Narembergia* and *Sandoricum* (Juliano 1934).

The archesporium is uni- to multi-celled, only one cell functions. It cuts off a parietal cell. The parietal tissue is four to seven cells thick. The nucellar tissue comprises eight to ten cell layers in *Aphanamixis polystachya* (R.B. Ghosh 1966a), it is even more extensive in *Walsura piscidia* (R.B. Ghosh 1966b). The megaspore mother cell divides meiotically, and produces a linear, or sometimes T-shaped tetrad. The chalazal megaspore develops into a Polygonum type of embryo sac. The synergids may simulate the egg. The antipodals are ephemeral. In *Melia azedarach*, at times, they become multinucleate. The polar nuclei fuse before fertilization. The degeneration of gametophytes is common in *Melia azedarach* (N.C. Nair 1959b). The development of multiple embryo sacs has been noted in *Aphanamixis* (R.B. Ghosh 1966a), *Azadirachta* (N.C. Nair and Kanta 1961), *Dysoxylum* (Wiger 1935), *Lansium* (Fig. 30.7.1A, B) and *Melia* (N.C. Nair 1959b).

Double fertilization occurs, and the pollen tube persists for some time after fertilization. All varieties of *L. domesticum* (Bernardo et al. 1961, N. Prakash et al. 1977) develop apomictically.

The development of the endosperm is of the Nuclear type. The nuclei aggregate in the chalazal region of the embryo sac. Multinucleate endosperm nodules have been reported in *Azadirachta indica* (N.C. Nair and Kanta 1961). The endosperm becomes cellular at the heart-shaped stage of proembryo, except in *Dysoxylum* (Wiger 1935) where it remains free-nuclear in the chalazal region. It stores oil.

The development of endosperm is autonomous in the "duku" and "langsat" varieties (N. Prakash et al. 1977).

The embryogeny conforms to the Crucifer type in *Naregamia alata* (N.C. Nair 1959a). It is assigned to Period I, Megarchetype III, Group V

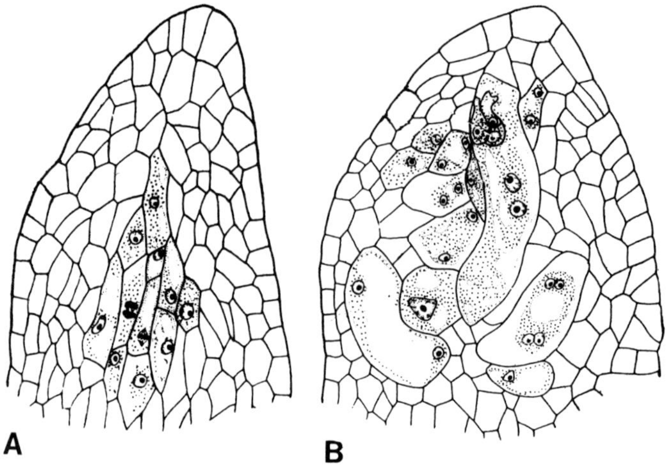

Fig. 30.7.1A, B. *Lansium domesticum*, longisections of nucelli showing a number of sporogenous cells (**A**), and a sexual and five uni-, five bi- and one four-nucleate aposporous embryo sac (**B**). (After N. Prakash et al. 1977)

(Chenopodiad) in *Azadirachta indica* (Garudamma 1956), and Period I, Megarchetype III (Asterad) in *Cipadessa baccifera* (L.L. Narayana 1958). N.C. Nair and Kanta (1961), however, report that the sequence of development is variable in *A. indica*. Synergid polyembryony occurs in *A. indica* (Garudamma 1956, N.C. Nair and Kanta 1961) and *Melia azedarach* (N.C. Nair 1959b). In *Aphanamixis polystachya* (R.B. Ghosh 1962a), nucellar polyembryony has been noted in 25–30 per cent of ovules. It is common in *L. domesticum* (N. Prakash et al. 1977). The egg cell may develop parthenogenetically up to two-celled proembryo, but degenerates. The nucellar embryo reaches maturity. The seeds are polyembryonate in the "langsat" variety, whereas in "duku" a single embryo matures in 70% of the seeds. An unusual feature is the formation of additional embryos by budding of the cotyledons of nucellar embryos.

The seed is albuminous and exotegmic, and shows much variation in different genera. The massive seeds are arillate, the aril arises from the funicle and exostome as in *Aglaia*, *Aphanamixis*, *Dysoxylon*, *Lansium* and *Naregamia*. In *Lansium* the succulent aril covers the entire seed in spite of the degeneration of seed within. The seeds are winged in *Cedrella* and *Cedrelopsis*.

In many genera the seed is pachychalazal. The outer epidermis of testa comprises thick-walled, cuticularized cells with stomata. They contain tannin in *Melia*. The mesophyll is thin-walled, and the inner epidermis contains crystals. The outer epidermis of tegmen consists of longitudinal, lignified, pitted fibres, and the inner contains tannin.

In *Trichilia grandifolia* (Boesewinkel 1981), the seed is exalbuminous, exarillate, and pachychalazal with a massive raphe bundle. The chalaza and outer integument form the sarcotesta with slightly thickened cells and intercellular spaces. The inner epidermis of testa contains crystals. The tegmen gets crushed, except at the micropyle where it becomes lignified.

According to Corner (1976), the expansion of the chalaza is associated with an extensive vascular supply. These vascular bundles are concerned in the transport of nutrients to the large embryo.

"The pachychalaza, the lack of endosperm, the strongly multiplicative integuments, the very large seed, the sarcotesta, the pre-raphe bundles, and perhaps also the loss of an aril, indicate a derived condition of the seed of *Trichilia grandiflora*" (Boesewinkel 1981).

FA Bernardo, CC Jessena Jr and DA Ramirez (1961), FA Bernardo and DA Ramirez (1959), FD Boesewinkel (1981), EJH Corner (1976), GK Garudamma (1956, 1957), SG Gavade (1963), RB Ghosh (1962a, 1966a, b), JB Juliano (1934), NC Nair (1958, 1959a, b, 1970a), NC Nair and K Kanta (1961), LL Narayana (1958), W Paetow (1931), N Prakash, AL Lim and R Manurung (1977), J Wiger (1935)

30.8 Akaniaceae

The embryology has not been investigated.

30.9 Malpighiaceae

The anther wall (Fig. 30.9.1A, B) comprises persistent epidermis, fibrous endothecium, two ephemeral middle layers, and secretory biseriate tapetum with two- to four-nucleate cells. The tapetum forms periplasmodium in *Stigmaphyllon periplocaefolium* (Munshi and Farooq 1969). The microspore mother cells undergo meiotic divisions, cytokinesis occurs by furrowing, and the tetrads (Fig. 30.9.1B–D) are rhomboidal, decussate, isobilateral and tetrahedral in *Malpighia coccigera* (Siddiqui 1968).

The pollen grains are two-celled at the dispersal stage (Fig. 30.9.1E). They degenerate prior to dehiscence of anthers in *S. periplocaefolium* (Munshi and Farooq 1969).

The ovule is semiana- or anatropous, bitegmic and crassinucellar with micropyle showing endostome, or exostome, or both (Fig. 30.9.1L). The nucellus projects beyond the micropyle in *Hiptage* and *Stigmatophyllum.* The archesporium is multicellular (Fig. 30.9.1F). Each hypodermal archesporial cell cuts off a parietal cell, and the parietal tissue is massive. A single megaspore mother cell functions, undergoes meiotic divisions, cytokinesis is absent, and a coenomegaspore is produced (Fig. 30.9.1G, H). The development of the embryo sac is of the Penaea type in *M. urens, M. glabra* (Fig. 30.9.1I–K) and *S. periplocaefolium.* The secondary nucleus (4n) is produced by the union of four nuclei, one from each quartet. Cortini (1958) reported *Allium* type, Penaea type, and several modifications of Penaea type in *Malpighia fulcata*, and also "Malpighia" type in this species. Allium type of development (Fig. 30.9.1M–S) occurs in *Galphimia gracilis* (Stenar 1937a), *M. glauca* (Subba Rao 1939) and *Thryallis glauca* (B. Singh 1959). The antipodals are ephemeral and polars fuse before fertilization.

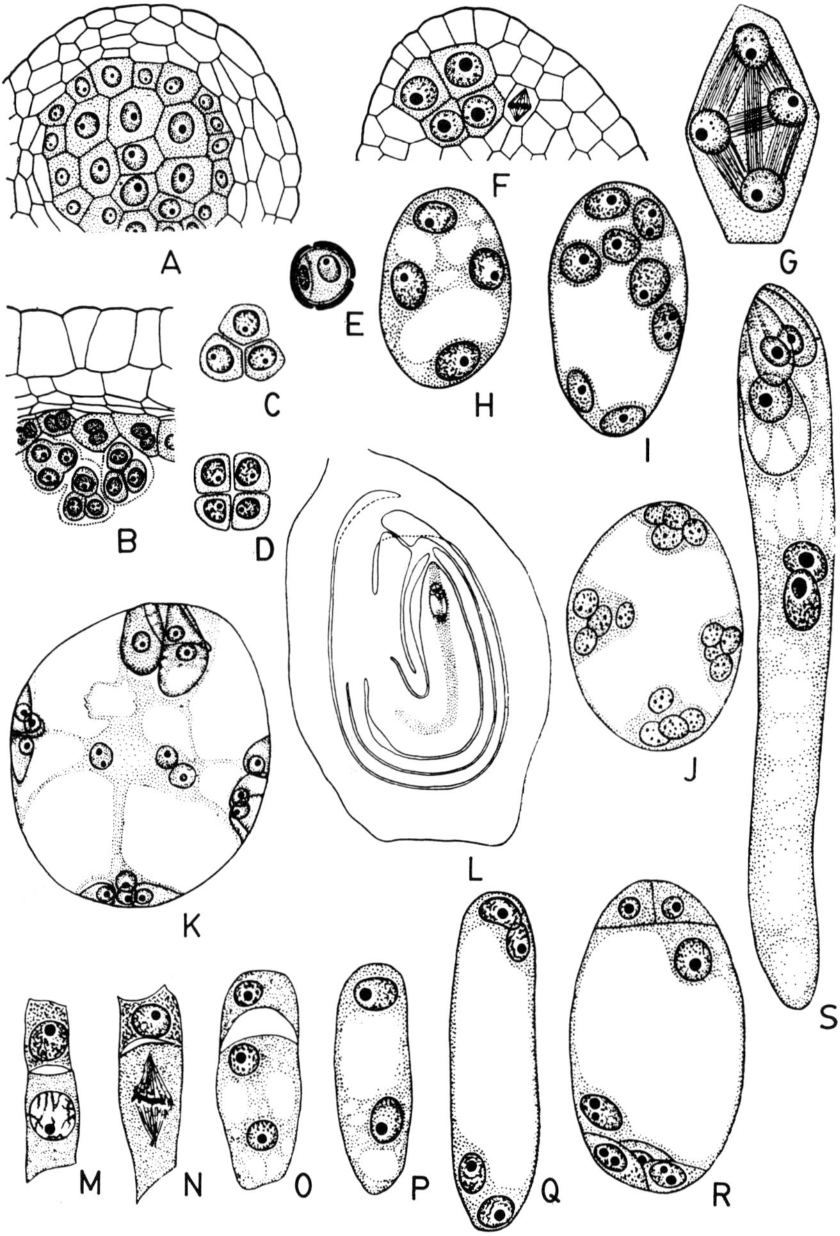

Fig. 30.9.1A–S. A, **F**, **G**, **K** *Malpighia glabra*, **B–E**, **L** *Thryallis glauca*, **H–J** *M. urens*, **M–S** *Galphimia gracilis.* **A** Cross-section of anther lobe at microspore mother cells. **B** Tetrads, tapetal cells binucleate. **C**, **D** Tetrahedral and isobilateral tetrad. **E** Two-celled pollen grain. **F** Nucellus with hypodermal archesporium. **G** Meiosis II. **H** Embryo sac with cross-wise arrangement of megaspore nuclei. **I** Embryo sac with eight nuclei. **J** 16 nuclei arranged in four quartets. **K** Embryo sac, Penaea type. **L** Longisection anatropous, bitegmic ovule at mature embryo sac. **M** Dyad. **N** Chalazal dyad functional. **O**, **P** Binucleate embryo sac. Micropylar dyad persists in **O**. **Q** Four-nucleate embryo sac. **R** Embryo sac, Allium type. **S** Mature embryo sac. (**A**, **F**, **G**, **K** After B. Singh 1961a, **B–E**, **L** after B. Singh 1959, **H–J**, **M–S** after Stenar 1937a)

Pollination and fertilization do not take place as the pollen degenerates, stigma is nonreceptive, and embryo sacs abort before the opening of flowers. Subba Rao (1940b) describes failure of fertilization in *Banisteria laurifolia*, *Hiptage madablota* and *Stigmatophyllum aristatum*.

The development of the endosperm is of the Nuclear type. Small, nucleated nodules develop in the endosperm during early stages in *Thryallis glauca* (B. Singh 1959). The nuclei aggregate in the chalazal region to form a haustorium in *Thryallis glauca* (Fig. 30.9.2A) and *M. glabra* (B. Singh 1961a, b). The chalazal haustorium remains free-nuclear in *M. glabra*, whereas it becomes

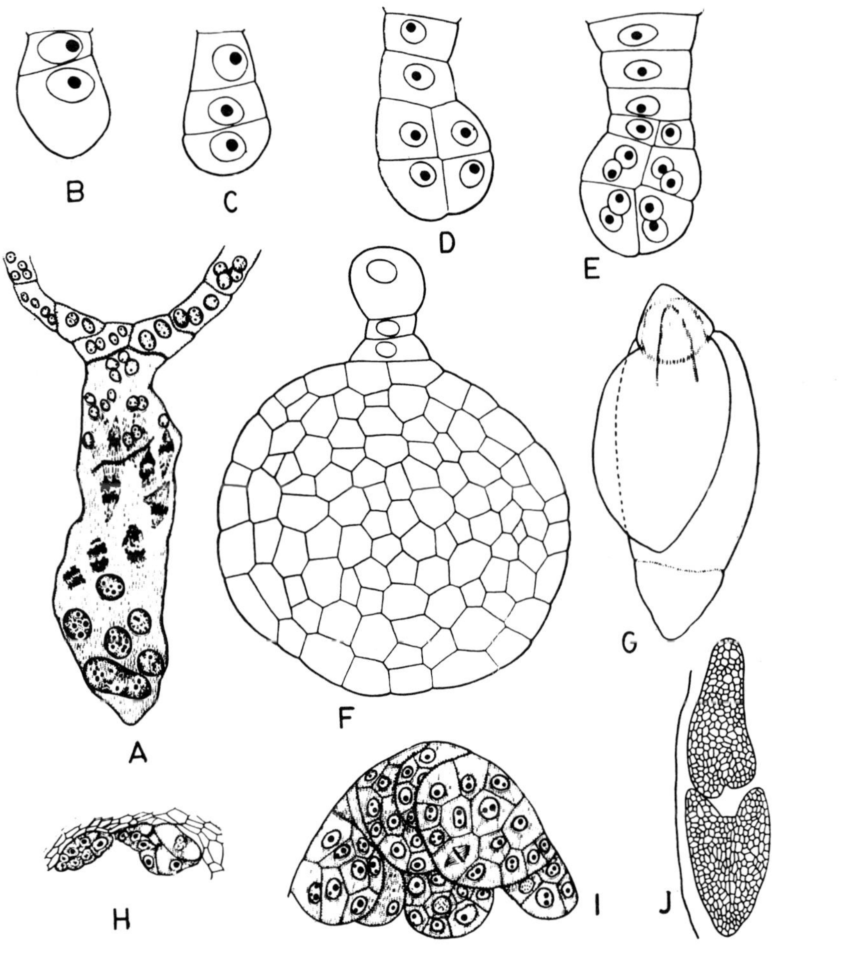

Fig. 30.9.2A. *Thryallis glauca*, chalazal portion of embryo sac to show endosperm haustorium with polyploid nuclei, the endosperm cells have two to four nuclei (some polyploid). **B–G** *Malpighia glabra*; **B–F** Embryogenesis of Solanad type; **G** The cotyledons are thick and unequal. **H–J** *Hiptage madablata*, nucellar polyembryony. (**A** After B. Singh 1959, **B–G** after B. Singh 1961a, b, **H–J** after Subba Rao 1940b)

cellular in *Thryallis*. Due to failure of fertilization in several members, the secondary nucleus forms endosperm autonomously.

The failure of fertilization is a common feature in Malpighiaceae; *M. glabra* is an exception and the embryo develops from the zygote. The embryogeny (Fig. 30.9.2B–F) conforms to the Solanad type (see Davis 1966). Nucellar polyembryony is a common feature (Fig. 30.9.2H–J).

The seed is exalbuminous with a straight or curved embryo with thick cotyledons (Fig. 30.9.2G). The testa is subaerenchymatous; the tegmen forms the main protective part. The outer epidermis of tegmen is fibrous, and the inner sclerotic. The inner epidermis is firmly adherent to the nucellus.

C Cortini (1958), GL Davis (1966), AN Fotidar (1939), M Munshi and M Farooq (1969), SG Narasimhachar (1938), PN Schürhoff (1924), SA Siddiqui (1968), B Singh (1959, 1961a, b), H Stenar (1937a), AM Subba Rao (1937, 1939, 1940b, 1941), W Verkerke (1984), W Verkerke and F Bouman (1980), H Wirz (1910)

30.10 Trigoniaceae

The embryology has not been fully studied. Boesewinkel (1987) gives a good account of the ovule and seed of Trigoniaceae.

The ovular primordium is trizonate, and gradually an anatropous, bitegmic, tenuinucellar ovule is formed in *Trigonia cipoensis*. The outer integument is suppressed on the funicular side. The zigzag micropyle is formed by both the integuments. The dermal cells of the funicle, facing the micropyle, are somewhat papilliferous and may function as an obturator.

The hypodermal sporogenous cell functions directly as the megaspore mother cell. The report of a parietal layer in *Trigonia parviflora* by Mauritzon (1936b) is erroneous. The eight-nucleate embryo sac contains starch, and probably conforms to the Polygonum type (Boesewinkel 1987). The scanty nucellus collapses so that the inner integument surrounds the embryo sac.

After fertilization the ovule enlarges. The endosperm is of the Nuclear type and later becomes cellular. In the mature seed, the walls of the endosperm cells are thickened and pectinaceous. These cells contain one large or several globules as in Vochysiaceae. The endosperm has more fats than starch, as reserve food. Between the inner integument and endosperm develops a conspicuous cuticle which reacts strongly with Sudan IV (Fig. 30.10.1B).

The seed-coat anatomy of *T. cipoensis*, *T. rugosa* and *T. villosa* is quite similar. The middle layers of the inner integument are completely crushed (Fig. 30.10.1A, B). The cells of the fibrous exotegmen enlarge and develop tracheidal lignified wall thickenings. The inner and middle layers of the outer integument divide so that it becomes thicker (Fig. 30.10.1B). All or some of the epidermal cells grow out into unicellular hairs, up to 5 mm long, and become thick-walled and lignified over their entire length (Fig. 30.10.1A–C). In the mature seeds of *T. rugosa* and *T. cipoensis*, the outer integument is nearly completely crushed so that the bases of the hairs come to lie against the exotegmen (Fig. 30.10.1C). The outer walls of testa cells become strongly thickened.

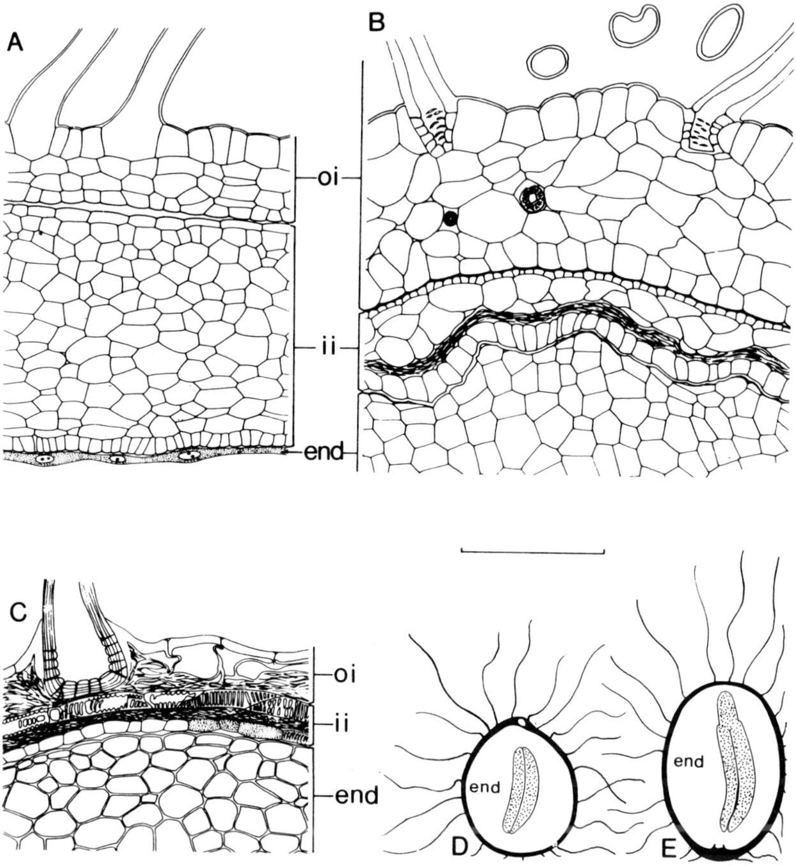

Fig. 30.10.1A–E. **A, B, D, E** *Trigonia cipoensis*, **C** *T. rugosa*. **A** Longisection of seed-coat, note free-nuclear endosperm; *oi*, *ii* outer and inner integument; *end* endosperm. **B** Transection, the epidermal cells of testa elongate into unicellular thick-walled hairs with strongly pitted wall in basal region. The mesophyll of the tegmen has collapsed. **C** Longisection, mature seed-coat. **D** Transection, hairy seed. Cotyledons *stippled*. **E** Longisection. (After Boesewinkel 1987)

The mature seeds of *T. cipoensis*, *T. rugosa* and *T. villosa* are covered with white cottony hairs. The embryo is linear with flat cotyledons (Fig. 30.10.1D, E). The albuminous seeds are adapted to anemo- or hydrochorous dispersal.

FD Boesewinkel (1980a, b, c, 1985a, 1987), EJH Corner (1976), J Mauritzon (1936b), W Verkerke (1985)

30.11 Vochysiaceae

The embryology of Vochysiaceae has not been fully investigated. The ovule and seed structure have been described by Boesewinkel and Venturelli (1987, see also Mauritzon 1936b, Davis 1966).

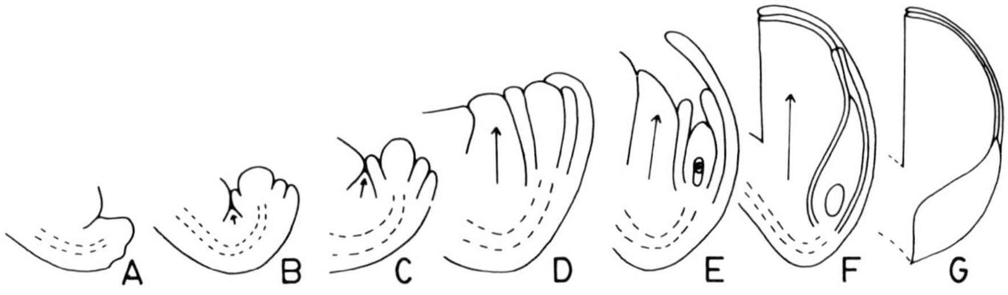

Fig. 30.11.1A–G. *Qualea cordata.* **A–F** Development of ovule. **G** Seed. Outline diagrams, *arrow* points to wing. (After Boesewinkel and Venturelli 1987)

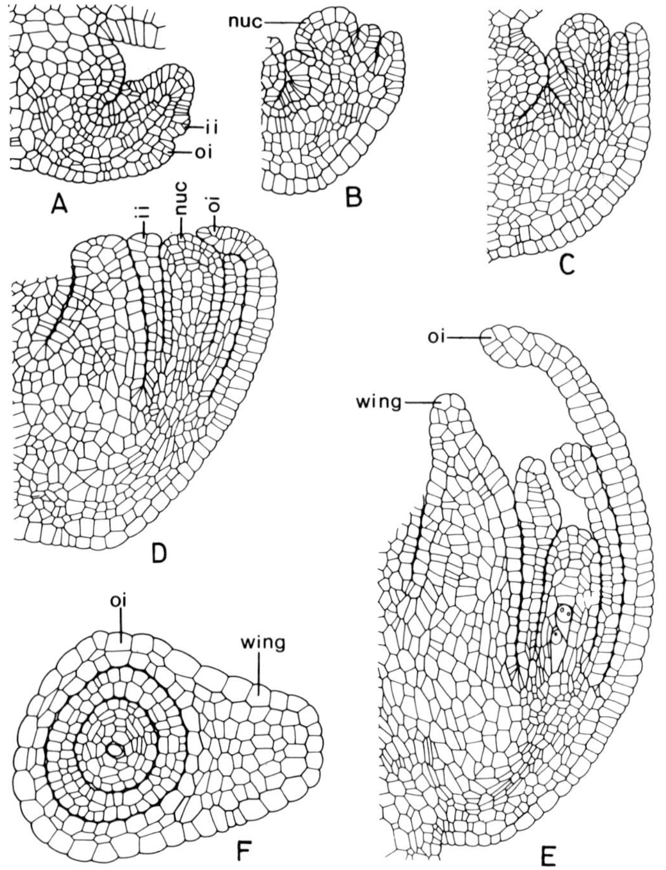

Fig. 30.11.2A–F. *Qualea cordata,* development of ovule, Longi- (**A–E**) and transection (**F**); *oi, ii* outer and inner integument; *nuc* nucellus. (After Boesewinkel and Venturelli 1987)

The development of the ovule is shown in Fig. 30.11.1A–F. The ovular primordium is trizonate. The differentiation and structure of inner and outer integuments is represented in Fig. 30.11.2A–F. The ovule is bitegmic, hemi-anatropous and crassinucellate with a massive and elongated nucellus (Fig. 30.11.1F). The nucellus is completely resorbed but the basal part (postament) persists in the mature seed in *Qualea cordata*. The inner integument is three to four layers thick, and the outer six to seven layers. The micropyle is formed by both the integuments, and the exostome extends beyond the endostome (Fig. 30.11.1E–G). The funicle is short and elongated. Between the tip of the outer integument and the funicle develops a flat tissue which gives rise to the wing in the seed (Fig. 30.11.1B–G).

The archesporial cell has not been observed. The megaspore mother cell (sometimes two cells) is situated deep in the nucellus (Fig. 30.11.2E). The development of the embryo sac, endosperm and embryo has not been studied.

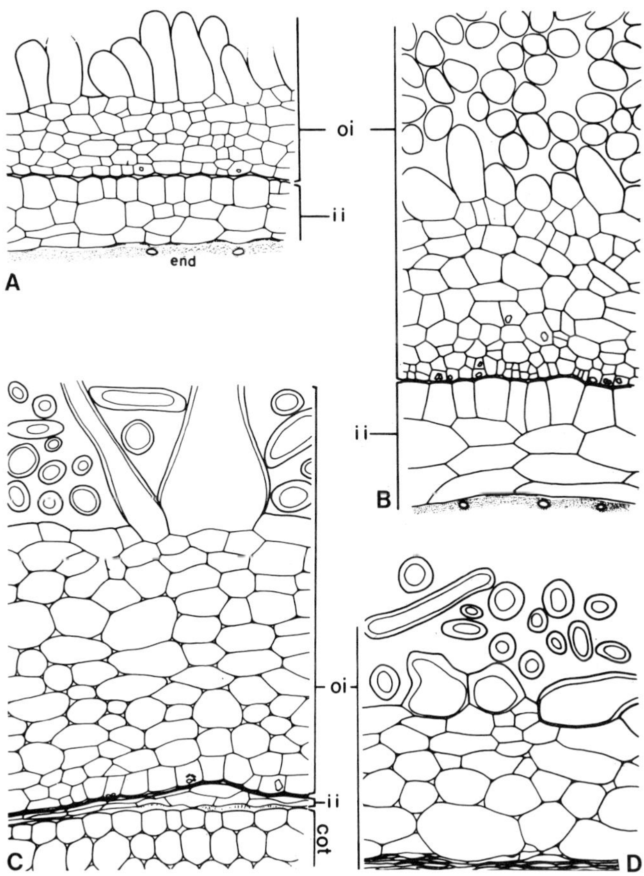

Fig. 30.11.3A–D. *Vochysia pygmaea*, development of seed-coat; *oi*, *ii* outer and inner integument; *end* endosperm; *cot* cotyledon. (After Boesewinkel and Venturelli 1987)

The endosperm is of the Nuclear type (Fig. 30.11.3B), wall formation was not observed, the seed is exalbuminous. The mature embryo completely fills the seed cavity and the cotyledons are convolute in *Callisthene fasiculata* and *C. major* (Boesewinkel and Venturelli 1987), often plicate or flat (Corner 1976). In *Vochysia pygmaea* (Boesewinkel and Venturelli 1987), the straight embryo has two large spirally folded cotyledons which are rich in lipids.

The development of the seed-coat is variable. In *V. pygmaea*, the parenchymatous cells of the inner integument enlarge, become vacuolate, and are gradually consumed by the endosperm (Fig. 30.11.3A–D). The outer integument increases in thickness. Shortly after fertilization, some of the radially elongate epidermal cells grow out into unicellular hairs (Fig. 30.11.3A). Subsequently, the hairs elongate and become thick-walled (Fig. 30.11.3C, D). The cells of the middle layers enlarge and intercellular air-spaces appear. The inner layers get crushed and the mature seed-coat comprises only four to seven layers of the outer integument (Fig. 30.11.3D).

In *Qualea cordata* and *Q. densiflora* (Boesewinkel and Venturelli 1987), the inner integument remains two- to three-layered. The endotegmen is tanniniferous and forms the inner layer of the seed-coat. The outer layers elongate, become fibrous, and thick-walled and pitted. The outer integument remains two-layered. The radial and inner walls of the inner layer become thickened, and small spaces appear between endotesta and exotegmen. The exotesta remains thin-walled. The seed-coat in *Callisthene* (Boesewinkel and Venturelli 1987) resembles that of *Qualea*.

Among the anemochorous seeds of the Vochysiaceae three types of wings can be distinguished. In *Vochysia*, the seed is covered with hairs and the dorsal unilateral (antiraphal) wing is compounded of matted hairs. The hairs partaking in the formation of the wing remain thin-walled. These hairs grow into the empty part of the locule, and therefore the wing simulates a cast of the locule (Fig. 30.11.4A–D).

In *Qualea* (Boesewinkel and Venturelli 1987), the wing is initiated on the funicle-raphal side (Fig. 30.11.2E, F). In hemianatropous ovules there is some

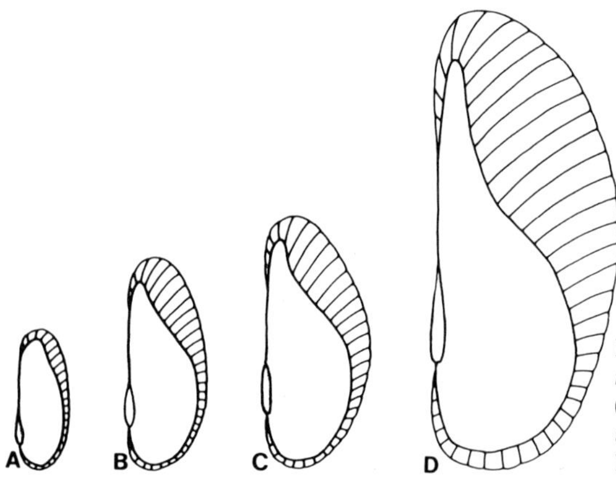

Fig. 30.11.4A–D. *Vochysia pygmaea*, development of wing (*hatched*), longisection of immature and mature seed. (After Boesewinkel and Venturelli 1987)

space for the development of outer integument on the funicular side. The seed wing is a lateral outgrowth of the outer and middle layer of the outer integument, and is inserted median-ventrally. This type of wing formation is not known in any other plant family. In *Callisthene* also the wing originates from the outer integument, but it is inserted transmedian-circumferentially.

FD Boesewinkel and M Venturelli (1987), EJH Corner (1976), GL Davis (1966), J Mauritzon (1936b), FA Stafleu (1952)

30.12 Tremandraceae

The development and structure of anther, pollen grain, embryo sac, endosperm, and embryo has not been worked out.

The ovule is anatropous, bitegmic and crassinucellate. In *Tetratheca* the outer integument is four- and the inner 15-layered, the inner epidermis forms an endothelium. Both the integuments form the micropyle. A parietal cell is cut off by the archesporial cell, and probably it produces a parietal layer in *Tetratheca* sp. The embryo sac shows three antipodals (Mauritzon 1936b).

Seeds small, glabrous or hairy with a conspicuous chalazal appendage (aril?), copious endosperm, and microscopic embryo. Seed-coat thin-walled, outer layers tangentially elongate, and inner layer strongly sclerotic.

JA Biddle and DC Christophel (1978), EJH Corner (1976), GL Davis (1966), J Mauritzon (1936b), JC Willis (1973)

30.13 Polygalaceae

The anther wall comprises the epidermis, fibrous endothecium, one or two ephemeral middle layers, and glandular tapetum with binucleate cells. At maturity only the epidermis and endothecium persist. The microspore mother cells divide meiotically, cytokinesis is simultaneous, and the tetrads are tetrahedral.

The three-colporate pollen grains are three-celled in *Polygala* and *Salomonia*, otherwise two celled. In *Salomonia* (Nagaraja Rao 1964b), they mostly germinate in situ.

The ovule is trizonate, anatropous, bitegmic and crassinucellate. The outer integument is two- to multilayered and the inner two-layered. Only the outer integument forms the micropyle in *Salomonia*, otherwise both the integuments are concerned and the micropyle is zigzag. In *Salomonia* the nucellus extends up to the tip of the inner integument.

Periclinal divisions occur in the dermal layer of the nucellus giving rise to a nucellar cap. Much of the parental tissue is resorbed by the growing embryo sac.

A hypodermal archesporial cell cuts off a parietal cell which produces eight to ten layers of parietal tissue. The megaspore mother cell undergoes meiotic divisions, produces mostly a linear, sometimes a T-shaped tetrad, and the chalazal megaspore develops into a Polygonum type of embryo sac. A pseudomonosporic, tetrasporic type of development (only one nucleus of the four-nucleate coenomegaspore functions) is reported in *Polygala polygama* (Reeves 1957). This needs confirmation. The antipodals are usually ephemeral, but they are persistent in *P. erioptera* (Mukherjee 1961b), and multiply up to 12 in *P. sibrica* (Srinivasachar 1942). A hypostase is present.

The entry of pollen tube is porogamous, and double fertilization occurs.

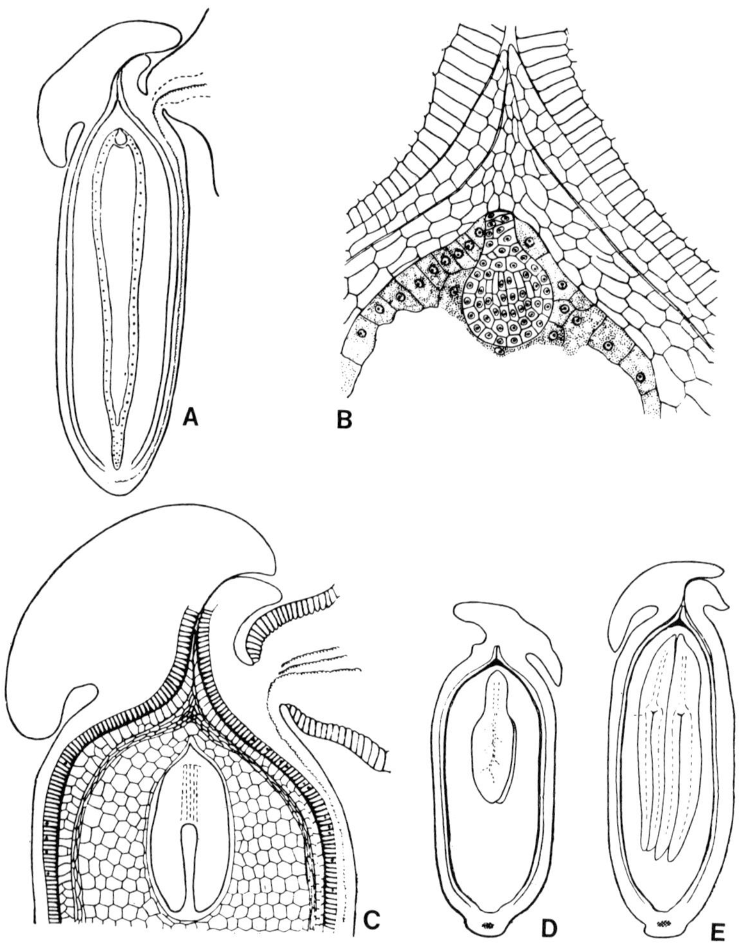

Fig. 30.13.1A–E. *Polygala major*. **A** Longisection of seed at globular proembryo, endosperm nuclear. Note the aril (strophiole) in **A**, **C–E**. **B** Later stage, endosperm partly cellular. **C** Longisection of micropylar region at dicotyledonous embryo. **D** Longisection of seed at dicotyledonous embryo; cellular endosperm not shown in **D** and **E**. **E** Twin embryos. (See Batygina and Yakovlev 1985)

The endosperm is of the Nuclear type. Centripetal wall formation (Fig. 30.13.1A–C) is initiated from the micropylar end at the late globular stage of the proembryo. The mature seed contains only a few layers of endosperm. In *Xanthophyllum fragrans* (Verkerke 1984), copious endosperm envelops foliaceous cotyledons. Both endosperm and cotyledons have fatty food reserve. In contrast, in *X. arnottianum* (Verkerke 1984) the endosperm is scanty and free-nuclear, and does not become cellular. The embryo is rich in starch grains, but not in fatty substances.

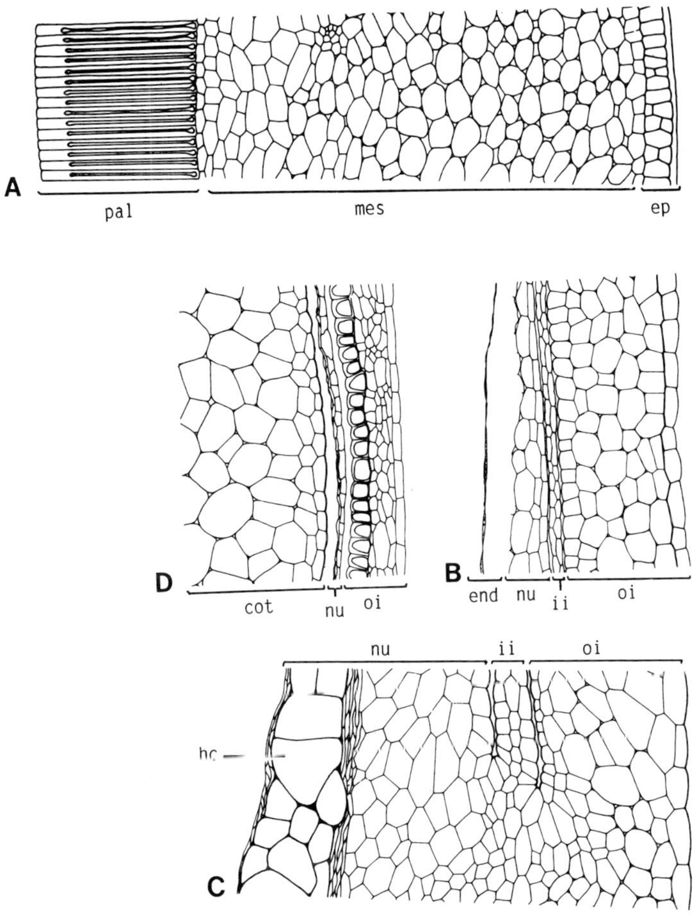

Fig. 30.13.2. A *Xanthophyllum fragrans*, cross-section of mature seed-coat. The epidermis (*ep*) of the testa is followed by many-layered mesophyll (*mes*) and palisade layer (*pal*) of the endotesta, tegmen collapses. **B–D** *X. arnottianum*, longisections; **B** Seed-coat at free-nuclear endosperm (*end*), many-layered outer integument (*oi*) and two-layered inner integument (*ii*); *nu* nucellus; **C** Section at lower end of seed through junction of integument and nucellus; *hc* hypertrophied cells; **D** Mature seed-coat, endotesta shows U-shaped thickenings, remnants of nucellus persist, tegmen collapses, and endosperm is consumed; *cot* cotyledon. (After Verkerke 1984)

The embryogeny corresponds to the Asterad type, and sometimes twin embryos may develop (Fig. 30.13.1E).

The seed is endotestal, and arillate in *Polygala* (Figs. 30.13.1A, C–E, 30.13.3A). The aril covers the seed, and may be pulpy or horny. The testa is multiplicative and forms the seed-coat. The outer epidermis comprises sub-sclerotic cells. The mesophyll is parenchymatous with thick or thin walls. The inner epidermis consists of a compact palisade of thick-walled cells (Fig. 30.13.2A) with crystals. The thick wall is faintly laminated and prominently pitted.

The structure of seed-coat in *Xanthophyllum arnottianum* is somewhat different (Fig. 30.13.2B–D). The testa is four- to five-layered and thin-walled. The cells of the inner epidermis do not elongate to form a true palisade layer. The tegmen collapses. The remnants (two- to three-layers) of nucellus persist. In the chalazal region some of the perisperm cells become hypertrophied (Fig. 30.13.2C). The seed is exalbuminous.

Two features of the testa are very prominent and of great taxonomic value: length of the palisade cell, and the presence of a parenchymatous middle layer (see Verkerke and Bouman 1980).

In *Polygala chamaebuxus*, *P. myrtifolia*, *P. vayredae* and *P. vulgaris* (Fig. 30.13.3A, B), the seed is covered with many one-celled striate acicular hairs (Fig. 30.13.3B) which are up to 500 μm long and 8 μm wide. The hairs are typically surrounded by their radially arranged neighbouring cells (Verkerke and Bouman 1980). In *P. klotzschii* the acicular hair is curled and thick-walled. In *Bredemeyera lucida*, in addition to the acicular thin-walled hairs, from the apex of the thickened unlobed hook-shaped yellow exotome aril arises a tuft of white hairs. Upon drying the thick-walled hairs stand erect and the thin-walled hairs tend to spread forming an umbrella-like structure (Verkerke 1985).

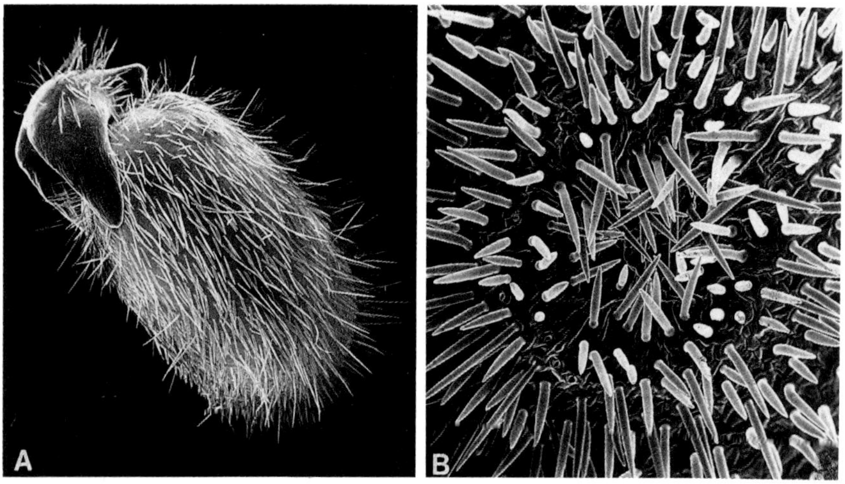

Fig. 30.13.3A, B. (SEM) *Polygala vulgaris*, **A** Seed with elaiosome and hairs. **B** Portion of chalazal region. (Courtesy Dr F. Bouman, Amsterdam)

In the mature seed the white to pale-yellow exostome aril is very prominent. It is typically three-lobed (Fig. 30.15.3A), and the lobes grow downward along the seed to one-fourth of its length. There is a loose tuft of hairs at the top, but the lobes of the aril are globose (Verkerke and Bouman 1980).

EJH Corner (1976), PK Mukherjee (1961b), A Nagaraja Rao (1964b), JH Reeves (1957), PN Schürhoff (1924), R Souèges (1941d), D Srinivasachar (1942), W Verkerke (1984, 1985), W Verkerke and F Bouman (1980), H Wirz (1910)

30.14 Taxonomic Considerations

Bentham and Hooker (1965a) include Polygalaceae, Tremendraceae and Vochysiaceae in Polygalinae, and the rest of the families in Geraniales, Engler and Diels (1936) place all the families in Geraniales, and distribute them over three suborders. Hutchinson (1973) distributes them in five suborders.

Rutaceae and Meliaceae. Rutaceae and Meliaceae are placed in Geraniales by Bentham and Hooker (1965a). Wettstein (1935) includes them together in Terebinthales, whereas Hutchinson (1973) segregated Meliaceae from Rutaceae due to the absence of gland-dotted leaves, and occurrence of a staminal tube. Embryologically, both the families are closely related. The multicellular archesporium, occasional development of Allium type of embryo sac, chalazal endosperm haustorium in *Aegle marmelos*, and the Solanad and Asterad type of embryogeny distinguish Rutaceae from Meliaceae. The seed in several members of Meliaceae is arillate, and pachychalazal. In both families the exotesta is thick-walled; it is sclerotic in Rutaceae. The tegmen is tracheidal in Rutaceae, and fibrous in Meliaceae.

Cneoraceae. Cneoraceae is represented by the genus *Cneorum*. Rau (1962) and B.M. Johri (1963b) have discussed the systematic position of Cneoraceae. Endlicher (1836–1840) includes them in Rutales near Burseraceae, while Bentham and Hooker (1965a) place *Cneorum* in Simaroubaceae. Engler and Prantl (1897) accept the family status of Cneoraceae but transfer them to Geraniales, and add that they are closely related to Zygophyllaceae. Hutchinson (1973) places them near Celastraceae in Celastrales. On the basis of embryogeny (Period I, Megarchetype II, Group II–Asterad type), Souèges (1955) relates them to Rhamnaceae, and suggests its grouping in Rhamnales. The embryogeny in the investigated taxa of Zygophyllaceae (*Tribulus terrestris* and *Zygophyllum fabago*) conforms to Period II, Series B, Megarchetype II, Group IX (Rau 1962) – Caryophyllad type.

The tribe Ruteae of Rutaceae resembles Cneoraceae in ovule and seed characters. The crassinucellate, bitegmic, primitive ovule is rather large and comparable to rutalean ovule in the massive parietal tissue, nucellar cap, and deep-seated embryo sac. The dermal initiation of integuments, initially two-layered, resembles the ovule in Rutoideae. The slightly multiplicative nature of the inner integument and the enlargement of its cells also show rutalean affinity. The seed-coat anatomy is derived and simplified (Boesewinkel 1984b). The campylotropous ovule, partially crushed exotesta, thick-walled endo- and meso-testa, and the tracheidal nature of the inner layer of inner integument of *Ruta graveolens* (Boesewinkel 1977) show affinity with Cneoraceae. Melchior (1964) has correctly placed Cneoraceae next to Rutaceae in his order Rutales.

Simaroubaceae and Burseraceae. Bentham and Hooker (1965a), Bessey (1915), and Engler and Prantl (1931) place the two embryologically allied families, Simaroubaceae and Burseraceae, together in Geraniales. Hallier (1912) includes them in Terebinthales, and Hutchinson (1973) and Takhtajan (1980) in Rutales. These two families resemble Zygophyllaceae (Sect. 29.6), Rutaceae, and Meliaceae in several respects, such as secretory anther tapetum, simultaneous cytokinesis, two-celled pollen, anatropous, bitegmic, crassinucellate ovule, Polygonum type of embryo sac, and Nuclear type of endosperm. The seeds are endospermic in both, and tracheidal exotesta gives protection; the tegmen is crushed.

Simaroubaceae can be distinguished because they are eurypalynous, and show hypostase, obturator, chalazal endosperm haustorium (*Suriana*), and perisperm as in Rutaceae. Perhaps the knowledge about the embryogeny of Burseraceae may add further to their alliance.

Malpighiaceae, Trigoniaceae and Vochysiaceae together comprise the suborder Malpighiineae.

Malpighiaceae. Malpighiaceae are included in Geraniales by Bentham and Hooker (1965a) and Hutchinson (1973), Takhtajan (1980) gives them an ordinal rank, while Cronquist (1981) treats Malpighiaceae as a primitive family in Polygalales.

Malpighiaceae are definitely distinct from the rest of the families of Rutales. Biseriate anther tapetum, nucellar beak, bisporic Allium and tetrasporic Penaea type of embryo sac development, failure of fertilization, autonomous development of the endosperm in some members, and exalbuminous seed have no parallel in Rutales. The Nuclear type of endosperm with nucleated cytoplasmic vesicles, free-nuclear chalazal haustorium in *Malpighia*, and haustorium becoming cellular in *Thryllis* are worth mentioning. Nucellar polyembryony is a common feature. The seed with thin-walled testa, and fibrous and sclerotic exo- and endotegmen resembles sapindaceous seed.

Trigoniaceae. The ovule and seed anatomy of *Trigonia cipoensis* (Boesewinkel 1987) is similar to that of *Linum* and related taxa. They share several other characters: trizonate ovule, dermal integuments, inner integument becomes thicker by repeated divisions in middle layers. The seed-coat structure of *Trigonia* also corresponds with that of *Linum*, *Erythroxylum* and *Humiria* (Boesewinkel 1980a, b, 1985a). The fibrous exotegmen in *Trigonia* is strongly lignified and forms the principal mechanical layer of the seed-coat. The outer layer of the outer integument differentiates in various ways in Trigoniaceae and families of the Linales. These similarities indicate a relationship between Trigoniaceae and Linales, also with other families, such as Vochysiaceae,

Malpighiaceae and Tremandraceae, but not with Polygalaceae. The family Polygalaceae (including Xanthophyllaceae) differs from the majority of the "Polygalales" in its ovule and seed structure (Verkerke 1985). According to Corner (1976), Trigoniaceae cannot be classified either near the Polygalaceae if Trigoniaceae have exotegmic fibres (see also under Polygalaceae).

Vochysiaceae. The Vochysiaceae (Boesewinkel and Venturelli 1987) are generally considered to be related to Polygalaceae, Trigoniaceae, Malpighiaceae, Tremandraceae and Krameriaceae. Some or all of these families have often been placed in the order Polygalales. The ovule and the endotestal seed type with an undifferentiated compressed or resorbed inner integument of Polygalaceae (Verkerke 1985) differ from those of the other families of the conventional Polygalales, which are mostly characterized by a persistent inner integument with an inner pigment layer and some form of an exotegmen. The ovules of Trigoniaceae (Boesewinkel 1987) and Tremandraceae (Mauritzon 1936b) have a thick inner integument. The ovule of *Trigonia* is similar to *Linum*. The seeds of Trigoniaceae, Tremandraceae and Malpighiaceae have a fibrous exotegmen. On the basis of seed structure, these families correspond fairly well with those of the Linaceae complex (Boesewinkel 1980a, b, 1987). The taxonomic relationships must await detailed embryological studies (see also under Polygalaceae).

Polygalinae include Tremendraceae and Polygalaceae. Wettstein (1935) placed Polygalaceae and Tremendraceae in Terebinthales. Hutchinson (1973) assigned Tremendraceae to Pittosporales, and Polygalaceae to Polygalales. As the information on Tremendraceae is scanty, their phyletic relationship cannot be discussed.

Polygalaceae. Boesewinkel (1987) points out that, except for the family Polygalaceae, he agrees with Heimsch (1942) who considered that Polygalaceae, Trigoniaceae, Tremandraceae, Malpighiaceae, Vochysiaceae and also Zygophyllaceae are interrelated and, although more specialized in general habit, floral structure and wood anatomy, these families are probably related to the Linaceae, Erythroxylaceae and Humiriaceae.

The more primitive representatives of the Polygalaceae have crassinucellar ovules with a subdermal vascularized outer integument and a two-layered inner integument. The endotesta is the most protective layer of the seed-coat.

The main embryological features of Polygalaceae are: simultaneous cytokinesis in pollen mother cell, two- or three-celled pollen, trizonate, anatropous, bitegmic, crassinucellate ovule, zigzag micropyle formed by both integuments, hypostase differentiates, Polygonum type of embryo sac, usually ephemeral antipodals, Nuclear type of endosperm which later becomes cellular, Asterad type of embryogeny, endotestal, arillate, hairy albuminous seed, and conspicuous embryo.

On morphological and embryological grounds a relationship of Polygalaceae has been suggested with Malpighiaceae, Trigoniaceae, Vochysiaceae, Tremandraceae and Zygophyllaceae. The relationships of Polygalaceae with Trigoniaceae, Vochysiaceae, Tremendraceae and Krameriaceae (all these families – except Tremendraceae – included in Polygalales by Hutchinson 1973) can be discussed with finality only when the embryology of these four families has been studied in sufficient detail.

On the basis of ovule and seed structure, Boesewinkel (1987) negates relationship of Polygalaceae with Trigoniaceae (see also Corner 1976), and Boesewinkel and Venturelli (1987) with Vochysiaceae. It has been suggested by them that the Order Polygalales (sensu Heimsch 1942) should comprise the family Polygalaceae only.

31 Sapindales

The order Sapindales comprises four suborders and ten subfamilies: Coriariineae – Coriariaceae; Anacardiineae – Anacardiaceae; Sapindineae – Aceraceae, Bretschneideraceae, Sapindaceae, Hippocastanaceae, Sabiaceae, Melianthaceae and Aextoxicaceae, and Balsamineae – Balsaminaceae.

31.1 General Features

The order Sapindales is characterized by (1) secretory anther tapetum, simultaneous cytokinesis of pollen mother cells, two-celled pollen grains, (2) anatropous, bitegmic, crassinucellate ovule, unitegmic in Sabiaceae and tenuinucellate in Balsaminaceae, parietal cell or tissue absent in Balsaminaceae, (3) Polygonum type of embryo sac, Polygonum and Allium type in Balsaminaceae, (4) Nuclear type of endosperm, Helobial in Sabiaceae, Cellular with haustorium in Balsaminaceae, (5) embryogeny of Onagrad type in Coriariaceae, Anacardiaceae, and Aceraceae, and Asterad type in Sapindaceae, Melianthaceae and Balsaminaceae, and (6) extensive simplification of seed; generally testal, tegmen gets crushed, testa or tegmen thin-walled, marked differentiation of mechanical elements absent; exalbuminous except in Melianthaceae.

On the whole, the development in the families is more or less uniform with few exceptions.

31.2 Coriariaceae

The anther is bisporangiate, and the wall comprises papillate epidermis, fibrous endothecium, two ephemeral middle layers, and secretory tapetum of one- to four-nucleate cells in *Coriaria nepalensis* (V.K. Sharma 1968a). The endothecial thickenings appear just before dehiscence, which has been contradicted by M. Garg (1980). The microspore mother cells undergo meiotic divisions, cytokinesis is simultaneous, and the tetrads are decussate and tetrahedral.

The pollen grains are three zonoporate (sometimes two- or four-zonoporate), and three-zonocolporate with perforate tectum and small spines. They are two- or three-celled at the dispersal stage; M. Garg's report of one-celled condition is incorrect. All reports of one-celled mature pollen grains require reinvestigation.

The ovule is anatropous, bitegmic and crassinucellate. The inner integument forms the micropyle. The outer integument forms a flap-like covering on either side of the funiculus. The single archesporial cell cuts off a parietal cell, the parietal tissue is only few-layered. Following meiotic divisions, the megaspore mother cell produces a linear or T-shaped tetrad. The chalazal megaspore develops into a Polygonum type of embryo sac. The antipodals are ephemeral.

Development of the endosperm is of the Nuclear type.

The embryogeny conforms to the Onagrad type, and is assigned to Period I, Megarchetype IV (V.K. Sharma 1968a). The mature embryo is straight.

The seed is exotestal, and mostly exalbuminous. The testa alone forms the seed-coat. The outer epidermis comprises thick-walled cells with tannin followed by thin-walled mesophyll and inner epidermis.

M Garg (1980), J Grimm (1912), J Mauritzon (1936b), MA Rau and VK Sharma (1970a), VK Sharma (1968a)

31.3 Anacardiaceae

The anther wall comprises the epidermis, fibrous endothecium, three ephemeral middle layers, and secretory tapetum of binucleate cells in *Pistacia chinensis* (Copeland 1955). A single middle layer develops in *Rhus mysorensis* (Kelkar 1958a). Simultaneous cytokinesis follows meiotic divisions in microspore mother cells. The tetrads are tetrahedral.

The pollen grains are tricolporate, spherical with pitted wall, and are two-celled at the dispersal stage.

The ovule is anatropous – apotropous in *Blepharocarya depauperata* (Wannan et al. 1987) – uni- or bitegmic and crassinucellar. In *P. vera* (Grundwag and Fahn 1969), the micropyle faces an outgrowth of the funiculus which may be regarded as an obturator. Both the integuments fuse to form a single structure in *P. vera* and *Anacardium occidentale* (Copeland 1955, 1961). In this connection, Grundwag and Fahn point out that in *P. vera* the outer integument is developed to various extent in different ovules. An outer integument, wholly or partially developed, is also known in other species of Anacardiaceae, e.g. in *Rhus toxicodendron* (Grimm 1912), *R. mysorensis* (Kelkar 1958a), *Lannea coromandelina* (Kelkar 1958b), and *Schinus mole* (Copeland 1959). The cells of the nucellar epidermis divide, and add to the parietal tissue. The archesporial cell cuts off a parietal cell, the latter produces a massive tissue. The megaspore mother cell undergoes meiotic divisions, and a linear or T-shaped tetrad is produced. The chalazal megaspore develops into a Polygonum type of embryo sac. The three remaining megaspores do not degenerate at the same time, and twin embryo sacs develop in *Nothopegia* (Kelkar 1974). The polars fuse before fertilization, and antipodals persist until then.

Kelkar (1958a), Copeland (1959) and Grundwag (1976) state that in some species of the tribe Rhoideae the chalazal megaspore of the linear tetrad gives rise to an eight-nucleate embryo sac of the Polygonum type. But in *Rhus lancea* (von Teichman and Robbertse 1986b) the micropylar megaspore produces an eight-nucleate embryo sac which should be considered to be of the Polygonum and not the Oenothera type. In the latter, the embryo sac is always four-nucleate – the egg apparatus and upper polar nucleus.

In *Pistacia vera* (Grundwag and Fahn 1969), the seed-set is low due to sterility of ovules, degeneration of megaspore mother cells, megaspores, and even embryo sacs.

The entry and course of pollen tube is variable in Anacardiaceae. It is strictly porogamous in *Spondias* and *Semecarpus* (Srinivasachar 1940a). According to Grundwag and Fahn (1969), in *P. vera* "germinating pollen tubes penetrate the short style, pass the locule and enter the funiculus of the ovule. In the funiculus, the pollen tubes continue to grow through the vascular bundle up to the chalaza. From the chalaza the tube enters the nucellus through the hypostase, surrounds the embryo sac and penetrates it at its micropylar end" (Fig. 31.3.1A). After pollination the pollen tube takes 24 h to reach the egg apparatus. Grundwag and Fahn observed one instance of direct penetration of a pollen tube from the hypostase to the chalazal end of the embryo sac. Copeland (1955, 1961) also reported chalazogamy in *P. chinensis* and *Anacardium occidentale*, and Grundwag (1976) in *P. atlantica*. All instances of chalazogamy require careful reinvestigation, and cleared wholemounts of ovules, longi- and transections, and dissections should be examined.

The development of the endosperm is of the Nuclear type. In *Mangifera indica* (Sachar and Chopra 1957), nuclear fusions occur (in the chalazal region) producing polyploid nuclei with a variable number of nucleoli. Nucleated cytoplasmic nodules develop in *M. indica* var. Paho (Sachar and Chopra 1957). Wall formation commences in the micropylar region along the periphery, and extends towards the centre. In *Mangifera* the lower part remains free-nuclear for a considerable time but, finally, becomes cellular. Such a free-nuclear region is absent in *Anacardium occidentale* (Copeland 1961). The endosperm is consumed by the developing embryo.

In *Pistacia vera* (Grundwag and Fahn 1969), the zygote divides 10–12 weeks after fertilization. This interval is at least partly due to delay in syngamy. If, however, there is a case of aneuspory, there is a possibility that the male gamete activates the division of the egg nucleus without fusion.

The embryogeny conforms to the Onagrad type-Euphorbia variation in *Rhus* and *Lannea* (Kelkar 1958a, 1961). The development is irregular in *P. vera* (Fig. 31.3.1B–F).

The zygotic embryo degenerates at an early stage in *Mangifera*. The nucellar cells in the micropylar region produce supernumerary embryos. Nucellar polyembryony is recorded in four varieties of mango: *Olour*, *Higgins*,

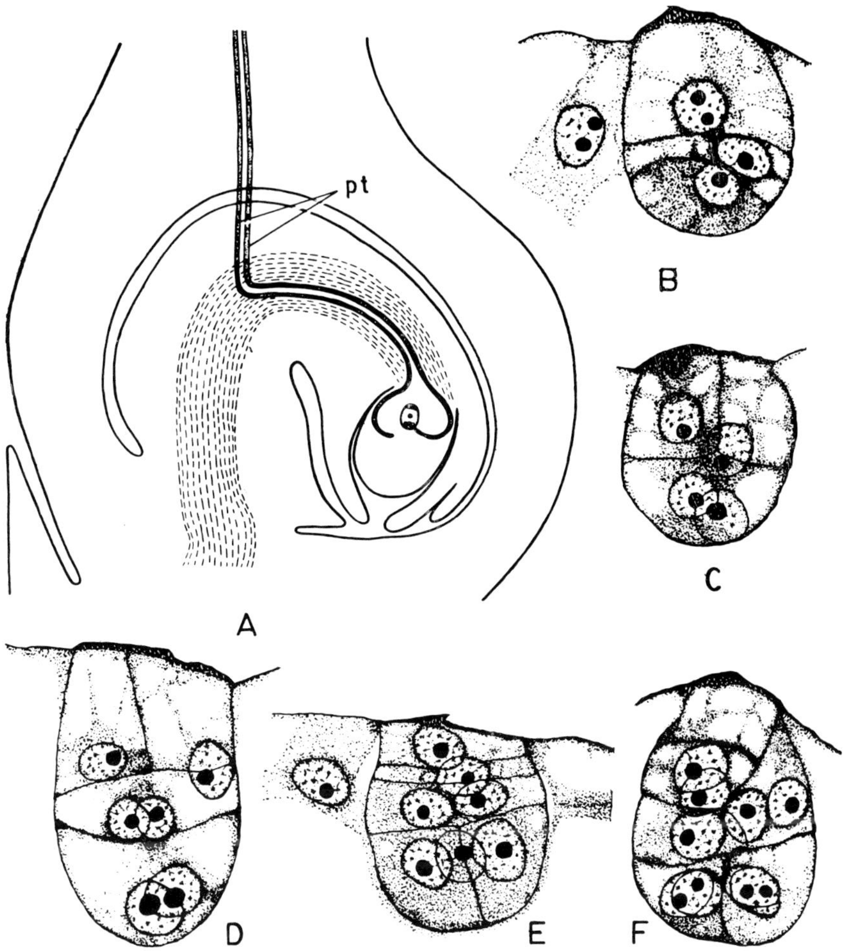

Fig. 31.3.1A–F. *Pistacia vera*. **A** Longisection of ovary. The pollen tubes (*pt*) traverse through the chalazal vascular strand to the tip of mature embryo sac. **B–F** Early embryogeny. There is an endosperm nucleus each in **B** and **E**. (See Batygina and Yakovlev 1985)

Paho and *Combodia* (Sachar and Chopra 1957). As the adventive embryos lie very close to each other, they exhibit various degrees of fusion in early stages of development. A synergid-embryo develops in *Lannea coromandelica* (Kelkar 1961).

The integuments are unspecialized, and generally get crushed except for the inner epidermis of tegmen, which is persistent, and shows bar-like thickenings on radial walls. The exotesta may become the protective layer. The endocarp is also protective.

In *Pistacia* (Copeland 1955, Grundwag and Fahn 1969), the seeds are endotegmic with abundantly pitted radial cell walls of endotegmen. The seeds of *Rhus* (von Teichman and Robbertse 1986b) are endotestal, and the radial walls of the persistent endotesta show lignified thickenings with pits. The remnants of endosperm and cotyledons contain fats and abundant proteinaceous reserve materials.

Von Teichman and Robbertse (1986a) studied the development and structure of the exo-, meso- and endocarp of the drupe of *Sclerocarya birrea* subsp. *caffra*. The mature exocarp comprises the outer epidermis with stomata and lenticels, subepidermal collenchyma, and parenchymatous layers with secretory cells. The fleshy parenchymatous mesocarp (sarcocarp) also contains secretory tissue, and develops only after the differentiation and lignification of endocarp. The endocarp (sclerocarp) consists of brachysclereids, fibres and vascular elements, and develops from the inner epidermis and adjacent tissue of the ovary wall. An "operculum" is a well-defined part of the endocarp, and it develops from the inner epidermis and some cells of the subepidermal cells of the ovary wall.

In *Rhus lancea* (von Teichman and Robbertse 1986b), the mature loose and papery exocarp consists of outer epidermis and a sclerified hypodermis. The mesocarp is not a typical sarcocarp since the ridges and inner layers are sclerenchymatous. The four-layered endocarp develops from the inner epidermis of the ovary wall.

HF Copeland (1955, 1959, 1961), EJH Corner (1976), J Grimm (1912), M Grundwag (1976), M Grundwag and A Fahn (1969), JB Juliano (1937), SS Kelkar (1954, 1958a, b, 1961, 1974), P Maheshwari (1934a), P Maheshwari and RC Sachar (1955), L Radlkofer (1879), RC Sachar and RN Chopra (1957), PK Sen and PC Malik (1940), D Srinivasachar (1940a), I von Teichman (1988), I von Teichman and PJ Robbertse (1986a, b), I von Teichman and AE van Wyk (1988), BS Wannan and CJ Quinn (1990), BS Wannan, JT Waterhouse and CJ Quinn (1987)

31.4 Aceraceae

The anther wall comprises the epidermis, fibrous endothecium, two ephemeral middle layers, and secretory tapetum of two- to three-nucleate cells; nuclear fusions occur in the tapetal cells. The microspore mother cells undergo meiosis followed by centripetal furrows and simultaneous cytokinesis. The tetrads are tetrahedral and isobilateral.

The three-colporate pollen grains are shed at the two-celled stage.

The ovule is anatropous, bitegmic and crassinucellate; the inner integument forms the micropyle. The nucellar epidermal cells divide to produce an eight to ten layers thick cap. The hypodermal archesporial cell cuts off a parietal cell, and the parietal tissue is several layers thick. The megaspore mother cell undergoes meiotic divisions, and produces a linear tetrad. The chalazal megaspore develops into a Polygonum type of embryo sac. The polar nuclei fuse before fertilization. The antipodals are ephemeral. At times, they multiply in *Acer oblongum* (Khushalani 1963b). The mature embryo sac contains starch.

The development of the endosperm is of the Nuclear type.

The zygote divides transversely. The development of the embryo appears to be of the Onagrad type in *Acer saccharinum* (Haskell and Postlethwait 1971). Suspensor polyembryony is reported in *A. platanoides.*

The seed is campylotropous with a curved embryo and flat, thick and curved, or thin and plicate, cotyledons. The thin testa alone forms the seed-coat. The outer epidermis comprises short columnar cells with thick or thin outer walls, followed by thin-walled mesophyll. The inner epidermis contains crystals.

DA Haskell and SN Postlethwait (1971), I Khushalani (1963b), CD Pigott and SJ Warr (1989)

31.5 Bretschneideraceae

The embryology has not been investigated.

H Tobe and C-I Peng (1990)

31.6 Sapindaceae

The anther wall comprises the epidermis, fibrous endothecium, two ephemeral middle layers, and secretory tapetum of uninucleate cells. Following meiotic divisions, the microspore mother cells produce tetrahedral and isobilateral tetrads. Simultaneous cytokinesis is due to centripetal furrowing.

The three-colporate pollen grains are shed at the two-celled stage. They are three-celled in *Allophylus cobbe* (S. Mathur 1980).

The ovule is ana- to campylotropous, bitegmic and crassinucellate. The inner integument forms the micropyle in *Cardiospermum halicacabum* (N.C. Nair and Joseph 1960), otherwise both the integuments are concerned. The hypodermal archesporial cell cuts off a parietal cell, and the parietal tissue is well developed. The megaspore mother cell divides by meiotic divisions, and produces a linear tetrad. The chalazal megaspore develops into a Polygonum type of embryo sac. The synergids are hooked, and the antipodals ephemeral in *Litchi chinensis* (Banerji and Chaudhuri 1944). In *Cardiospermum* (N.C. Nair and Joseph 1960), they multiply up to 14 cells and become multinucleate. A hypostase develops, and an obturator is present in *L. chinensis.*

The entry of the pollen tube is porogamous. It passes between the wall of the embryo sac and the synergid in *Cardiospermum.* Kadry (1946) describes the difference between the two male nuclei. The one which fertilizes the egg is swollen, and rounded in front and tapering behind. It is much longer than the other spherical male nucleus. Double fertilization occurs.

Development of the endosperm is of the Nuclear type, and it remains free-nuclear throughout in *Cardiospermum* (Kadry 1946).

The embryogeny conforms to the Asterad type. The mature embryo is either straight or curved with thin and spirally inrolled cotyledons, or the cotyledons may be thick and curved or straight. Parthenocarpy is reported in *Dodonea viscosa* (A.C. Joshi 1938a).

The seed is arillate, sarcotestal and exalbuminous. The aril develops, after fertilization, from the funicle and is rudimentary and vestigial in *Cardiospermum* and *Dodonea*. The testa is multiplicative. The outer epidermis comprises a palisade of columnar cells with brown walls. The mesophyll is differentiated into an outer compact thick or pulpy tissue, and an inner aerenchymatous tissue. The inner epidermis is thin-walled. The outer epidermis of tegmen alone persists, and consists of fibres and sclerotic endostome.

I Banerji and KL Chaudhuri (1944), KL Chaudhuri (1940), E David (1938), N Gulati and S Mathur (1977), AC Joshi (1938a), AER Kadry (1946), S Mathur (1980), J Muller and PW Leenhouts (1976), NC Nair and TC Joseph (1960), A Pfeiffer (1891), JE Planchon (1845b)

31.7 Hippocastanaceae

The structure and development of anther wall, embryo, and seed-coat have not been investigated.

The anther tapetum is glandular. Simultaneous cytokinesis follows meiotic divisions in microspore mother cells.

The pollen grains are two-celled at the dispersal stage.

The ovule is amphitropous (Fig. 31.7.1A), bitegmic and crassinucellar in *Aesculus woerlitzensis* (List and Steward 1965). The nucellus is massive with a nucellar cap, and the epistase consists of thick-walled refractive cells. A hypostase also develops.

The multicellular archesporium cuts off parietal cells. Development of the embryo sac is of the Polygonum type with a dilated (vesicle) chalazal region (Fig. 31.7.1B, C).

During fertilization and post-fertilization development the ovule, the nucellus, and the embryo sac grow conspicuously in the chalazal region. The nucellus can be demarcated into four regions: (1) vacuolate cells with small nuclei in the micropylar region, (2) large cells around the middle of the embryo sac, (3) smaller cells which undergo rapid divisions in the chalazal region, and (4) hypostase of radiating files of cells which are in contact with the vascular bundle at the base of the ovule.

The vesicular end of the embryo sac grows and encroaches upon the nucellus, and occupies the major portion of the ovule (Fig. 31.7.1B–D). The vesicle is filled with a fluid. The solutes enter the vesicle through the converging files of cells of the hypostase, which remain in contact with the chalazal end of the embryo sac.

Development of the endosperm is of the Nuclear type. The vesicle contains gelatinous (liquid) endosperm with numerous nuclei. The endosperm (Fig. 31.7.1E) is consumed by the developing embryo (Corner 1976).

In *A. woerlitzensis* the zygote divides transversely, vertically or obliquely. The mature embryo is dicotyledonous (Fig. 31.7.1D–J). During earlier stages

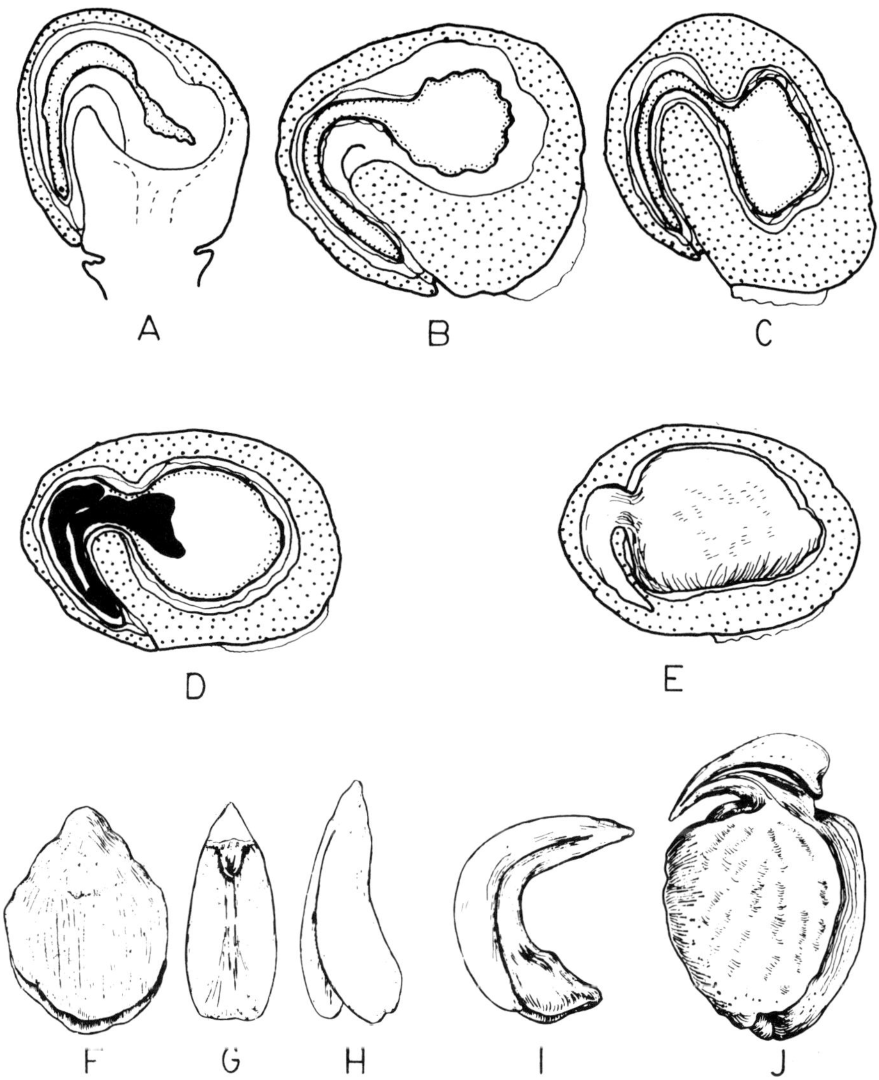

Fig. 31.7.1A–J. *Aesculus woerlitzensis.* **A** Longisection of amphitropous ovule with testa (*sparsely dotted*), tegmen, and well-developed nucellus. The embryo sac (*thinly dotted*) is at the zygote stage. **B**, **C** Later stage (endosperm and embryo not shown). The chalazal end of the embryo sac expands to form a vesicle (crushing the nucellus) which becomes filled with a fluid (liquid endosperm). **D** Longisection of ovule at dicotyledonous embryo. The larger cotyledon is suctorial, extends into the vesicle and absorbs nutritive fluid. **E** Longisection of seed with mature embryo occupying the entire vesicle; nucellus and endosperm have been absorbed. **F**, **G** Young embryo to show bilateral symmetry. **H–J** Embryos with unequal cotyledons. (After List and Steward 1965)

the growth of the two cotyledons is symmetrical (Fig. 31.7.1F, G). As the cotyledons grow up to the curved region of the embryo sac, one of them pushes into the vesicle, grows inside it and absorbs the vesicular fluid (Fig. 31.7.1D). The endosperm, and eventually the nucellus disappear. In mature seed the larger cotyledon folds in the vesicle and fills it.

The seed is pachychalazal (?) and exotestal. It is overgrown with a highly vascular wall, and tough seed-coat. The outermost layer of testa comprises palisade cells with outer and radial walls thickened. The mesophyll is aerenchymatous, and the inner epidermis is unspecialized. The thin-walled tegmen persists around the radicle.

EJH Corner (1976), GL Davis (1966), A List and FC Steward (1965), F Pax (1896b), K Schnarf (1931)

31.8 Sabiaceae

The development of pollen, embryo, and seed-coat has not been investigated.

The anther wall comprises the epidermis, fibrous endothecium, two ephemeral middle layers, and glandular tapetum with binucleate cells. Their nuclei fuse and divide.

The pollen grains are tricolporate, and two-celled when shed. The generative cell is elongated in *Meliosma arnottiana*, and lenticular in *M. wighttii* (M.V.S. Raju 1952a).

The ovule is hemianatropous, unitegmic and crassinucellate. The integument fails to organize the micropyle. The archesporial cell cuts off a parietal cell, and the parietal tissue is well developed. The megaspore mother cell divides meiotically, and produces a linear tetrad. The chalazal megaspore develops into a Polygonum type of embryo sac. The antipodals persist in *M. arnottiana* (M.V.S. Raju 1952a). The chalazal end of the embryo sac develops a diverticulum which grows towards the funiculus. The polar nuclei fuse before fertilization, and the secondary nucleus migrates to the diverticulum leaving the antipodals in situ.

Double fertilization occurs; supernumerary pollen tubes enter the embryo sac. Triple fusion takes place in the diverticulum.

The development of the endosperm is of the Helobial type. The endosperm proper develops from the micropylar chamber. The chalazal chamber becomes binucleate, and develops haustorial processes.

J Mauritzon (1936b), MVS Raju (1952a)

31.9 Melianthaceae

The anther wall comprises the persistent cutinized epidermis, endothecium, three to five middle layers, and secretory tapetum in *Melianthus major* (S. Mathur 1977). The endothecial cells are small and cutinized, and lack fibrous thickenings in the region of dehiscence; the rest of the cells develop fibrous thickenings. One or two middle layers persist. The tapetum is multiseriate towards the anterior and posterior end. Its cells are binucleate, and minute globules (Übisch) are studded on the inner tangential wall. The microspore mother cells undergo meiotic divisions. Cytokinesis occurs by furrowing, as well as cell-plate formation. When simultaneous, the tetrads are tetrahedral, isobilateral and decussate; when successive, they are linear, T-shaped and isobilateral.

The elliptical pollen grains are monad and tricolporate. They are two- or three-celled at the dispersal stage.

Sterility is very high in *M. major* (Khushalani 1963a). Due to cytological disturbances, the differentiation of generative and vegetative nuclei does not occur properly. Many pollen grains contain two to five nuclei due to supernumerary divisions.

The ovule is anatropous, bitegmic and crassinucellate. In *M. major* (S. Mathur and Gulati 1979), the inner integument organizes the micropyle and forms an operculum. In *Bersama* both the integuments form the micropyle. The epistase and hypostase are well developed. The hypodermal archesporial cell cuts off a parietal cell, and the parietal tissue is six to seven layers thick. A nucellar cap and a nucellar beak develop due to periclinal divisions of nucellar epidermal cells. After meiotic divisions, the megaspore mother cell produces a linear, T- or L-shaped, or decussate tetrad. Mostly, either the third megaspore from the micropylar end, or the chalazal megaspore functions. The mature embryo sac is of the Polygonum type. The antipodals are ephemeral. The polar nuclei fuse before or at pollination.

The entry of the pollen tube is porogamous, and it destroys one of the synergids. Double fertilization occurs, and the remnants of the pollen tube with darkly stained bodies persist for a long time.

The development of the endosperm is of the Nuclear type. It remains nuclear throughout which needs confirmation.

The embryogeny conforms to the Asterad type-Penaea variation in *M. major* (Steyn 1975).

The seed is albuminous with a short funicular aril in *Bersama*. The outer epidermis of testa is a palisade of thick-walled cells only in the region of the micropyle. It is followed by thin-walled mesophyll with styloids. The inner epidermis is thin-walled, and the tegmen gets crushed.

M Gürke (1896), I Khushalani (1963a), S Mathur (1977), S Mathur and N Gulati (1979), J Mauritzon (1936b), E Steyn (1975)

31.10 Aextoxicaceae

The development of anther, pollen, embryo sac, endosperm and embryo has not been investigated.

The ovule is anatropous, bitegmic and crassinucellar with the inner integument forming the micropyle. The nucellar beak is massive in *Aextoxicon punctatum* (Mauritzon 1936b).

In a mature seed the endosperm is ruminate, and the cotyledons orbicular-cordate (Hutchinson 1973).

J Hutchinson (1973), J Mauritzon (1936b)

31.11 Balsaminaceae

The anther wall comprises the epidermis, fibrous endothecium, two middle layers, and secretory tapetum. A peculiar feature in *Hydrocera triflora* (Venkateswarlu and Lakshminarayana 1957) and *Impatiens leschenaultii* (L.L. Narayana and Sayeeduddin 1959) is the development of trabeculate partitions of tapetal (doubtful?) cells dividing the sporogenous tissue into patches. The tapetal cells and those of trabeculae become two- to four-nucleate. In *Impatiens* (L.L. Narayana and Sayeeduddin 1959, L.L. Narayana 1963b, 1965), however, such partitions are produced due to sterilization of sporogenous tissue. The cells of septa and tapetum become binucleate. The microspore mother cells undergo meiotic divisions, and produce tetrahedral or isobilateral tetrads. Cytokinesis is simultaneous.

The pollen grains are three- to five-colporate, and shed at the two-celled stage. The generative cell is elongated. Its division occurs by a constriction in *Impatiens balsamina* (Raghavan et al. 1939). This indicates amitotic division which is, obviously, incorrect.

The ovule is anatropous, bitegmic and tenuinucellate; the inner integument forms the micropyle. The integuments are fused all along their length, except at the micropyle. They are free in *Impatiens* (L.L. Narayana 1965). The endothelium is well developed (Fig. 31.11.2F–H).

The hypodermal archesporial cell functions directly as megaspore mother cell. The embryo sac develops according to the Polygonum and Allium type. If of the Polygonum type, the tetrads may be linear or T-shaped. Polygonum type has been observed in *Impatiens inconspicua*, *I. levengi*, *I. maculata*, *I. parviflora*, *I. radiata*, *I. tenella*, *I. tripetala* (L.L. Narayana 1963b, 1965), and *I. balsamina* and *I. textori* (Takao 1966, 1968). Allium type occurs in *Hydrocera triflora* (Venkateswarlu and Lakshminarayana 1957), *I. arguta* (L.L. Narayana 1963b), *I. leschenaultii* (L.L. Narayana and Sayeeduddin 1959), *I. oliveri* (L.L. Narayana 1965) and *I. roylei* (K.V.O. Dahlgren 1934a). The polar nuclei fuse before fertilization, and the antipodals are ephemeral.

The development of the endosperm is of the Cellular type (Figs. 31.11.1A, B, E, F, 31.11.2A, B). Due to transverse partitions the embryo sac becomes four-chambered in *Hydrocera triflora* (Venkateswarlu and Lakshminarayana 1957). The micropylar chamber, with the zygote, develops into an aggressive haustorium (Fig. 31.11.2D, E, G–I). It extends beyond the micropyle, becomes vesicular, and produces branches which ramify in the funicular and placental tissue. Its nucleus becomes hypertrophied. The micropylar haustorium is multinucleate in *H. triflora* (Venkateswarlu and Lakshminarayana 1957). In *I. radiata* (L.L. Narayana 1965), the haustorium, instead of passing into the placenta, ramifies through the fused integuments along the

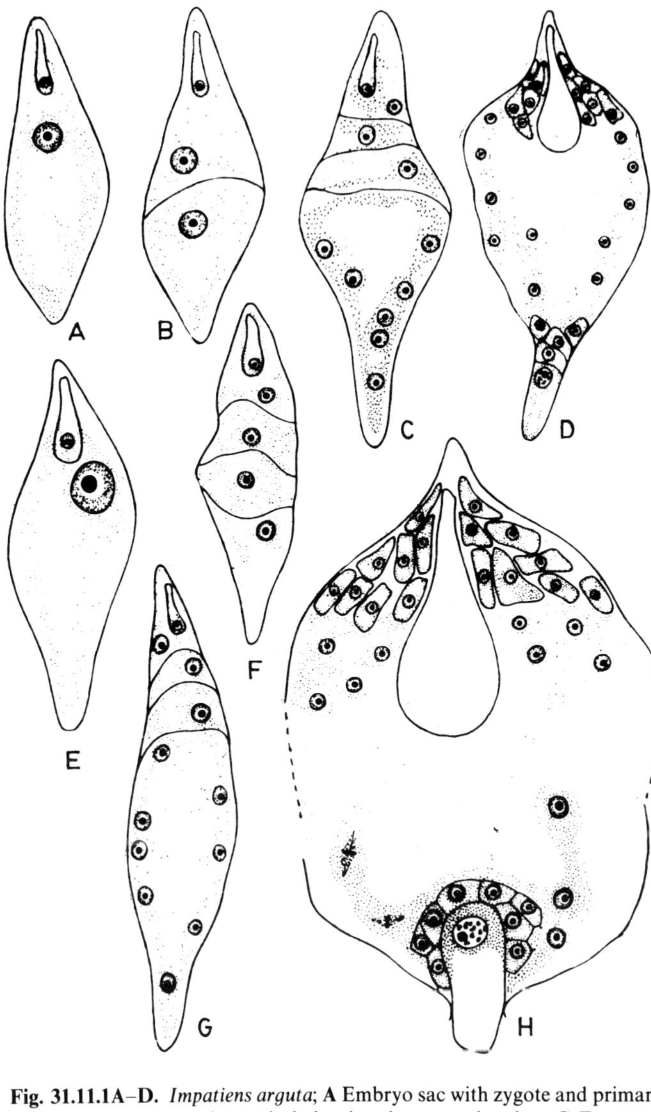

Fig. 31.11.1A–D. *Impatiens arguta*; **A** Embryo sac with zygote and primary endosperm nucleus; **B** Formation of micropylar and chalazal endosperm chamber; **C** Zygote, three-celled micropylar chamber and multinucleate chalazal chamber; **D** Globular proembryo, cellular endosperm at both poles, and free-nuclear in centre. Micropylar endosperm haustorium is absent, but a chalazal haustorium is cut off. **E–H** *I. levengi*; **E** Embryo sac with zygote and primary endosperm nucleus; **F** zygote, three-celled micropylar chamber, and chalazal chamber; **G** Zygote, three-celled micropylar chamber and multinucleate chalazal chamber; **H** Later stage, cellular endosperm at both poles, free-nuclear endosperm in centre, uninucleate chalazal haustorium, micropylar haustorium absent. (After L.L. Narayana 1965)

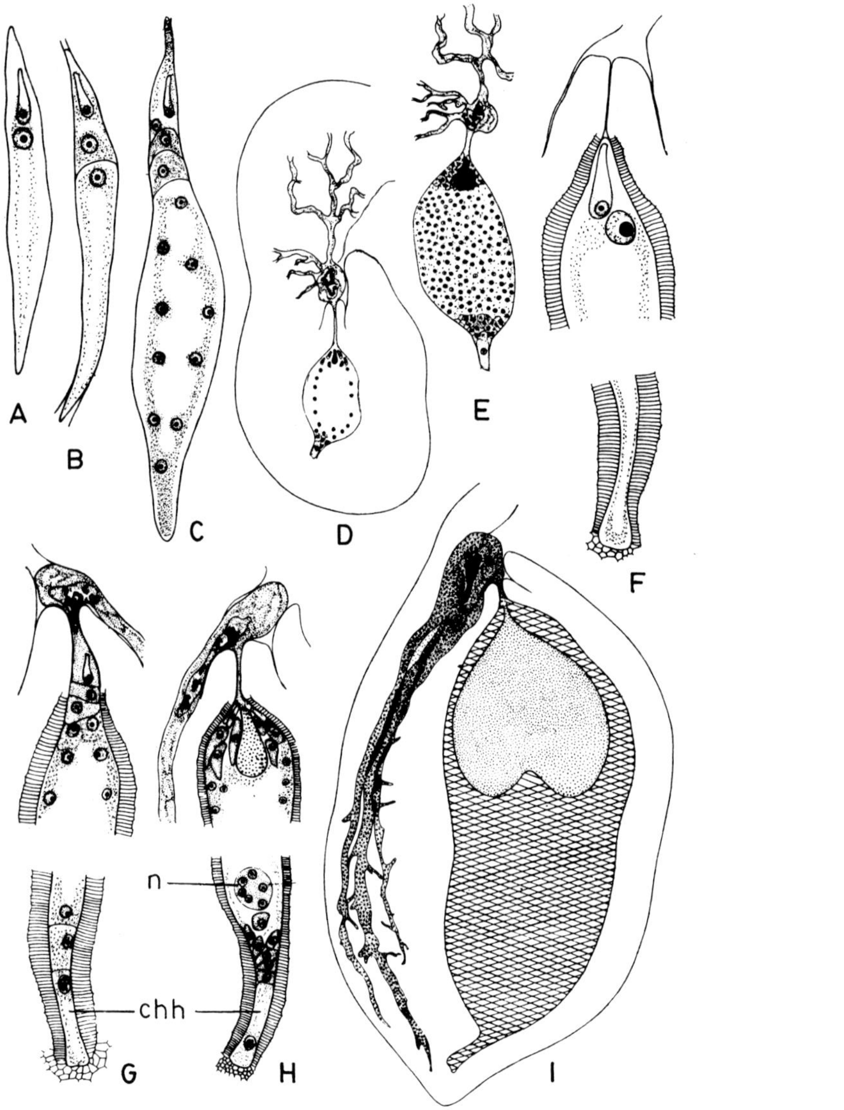

Fig. 31.11.2A–E. *Impatiens leshenaultii*; **A** Embryo sac with zygote and primary endosperm nucleus; **B** Formation of a small micropylar and a large chalazal endosperm chamber; **C** Zygote, three-celled micropylar chamber and multinucleate chalazal chamber; **D** Longisection of ovule to show micropylar and chalazal haustoria, beginning of wall formation in endosperm at micropylar and chalazal end, middle region free-nuclear. Micropylar haustorium ramifies in the placenta; **E** Same as **D**, later stage (dissected whole-mount). **F–I** *I. radiata*; **F** Embryo sac with endothelium, zygote and primary endosperm nucleus. Note the endothelium in **F–H**, **G, H** Micropylar and chalazal haustorium (*chh*), beginning of wall formation in endosperm at micropylar and chalazal end, and free-nuclear endosperm in centre. Note nucleate nodules (*n*) in free-nuclear endosperm in **H**; **I** Immature seed (cleared in chloral hydrate) to show the much-branched micropylar endosperm haustorium reaching as far as the chalaza. Cellular endosperm *hatched*, embryo *stippled*. (After L.L. Narayana 1965)

raphe. The development of haustorium remains arrested in *I. arguta* and *I. levengi* (L.L. Narayana 1965). The middle tiers develop into the endosperm proper. Free-nuclear divisions occur in the chalazal chamber, and a few cells are formed; the basal cell develops into a small haustorium (Figs. 31.11.1C, D, G, H, 31.11.2C–E, G, H). In the free-nuclear part multinucleate cytoplasmic endosperm nodules (Fig. 31.11.2H) develop in *Impatiens*.

The embryogeny conforms to the Asterad type-Erodium variation in *Hydrocera triflora*. *Impatiens textorii* (Takao 1976) exhibits polymorphism in octant formation.

The seed is exalbuminous, and mostly pachychalazal. The seed-coat is formed of two to five layers of fused integuments. In *Hydrocera* the outer region is composed of thick-walled sclerotic cells. The endothelium gets crushed. The tegmen does not persist.

KVO Dahlgren (1934a), G Heimsch (1942), J Hutchinson (1973), LL Narayana (1963b, 1965, 1970f), LL Narayana and M Sayeeduddin (1959), D Ramadevi and LL Narayana (1990), TS Raghavan, KR Venkatasubban and HD Wülff (1939), R Souèges (1945a), K Steffen (1951), S Takao (1966, 1968, 1976), JL van Went (1989), J Venkateswarlu and L Lakshminarayana (1957)

31.12 Taxonomic Considerations

Bentham and Hooker (1965a), and Engler and Diels (1936) place all the families in Sapindales. Hutchinson (1973) includes Balsaminaceae in Geraniales, and all others in Sapindales.

Coriariaceae. Coriariaceae represent the suborder Coriariineae. They are considered as an isolated member, and are in the nature of a relict family (Melchior 1964). Mauritzon (1936b) relates them to Celastrales. Takhtajan (1980) includes them in Rutales, along with Anacardiaceae. V.K. Sharma (1968a) compares the characters of Coriariaceae with Empetraceae, Anacardiaceae and Salvadoraceae, and points out that they are embryologically similar to Anacardiaceae (Table 31.12.1). It does not show any affinity with Empetraceae. The flap-like covering on either side of the funiculus develops from the outer integument, and the persistent fleshy perianth in fruit is noteworthy. The Onagrad type of embryogeny does not match with the Asterad type in Salvadoraceae. V.K. Sharma (1968a) suggests that Coriariaceae may be considered to represent an isolated family distantly related to the arborescent families included in Sapindales of Engler.

The study of seed anatomy reveals certain similarities between Coriariaceae and Anacardiaceae; the stony endocarp provides protection in Anacardiaceae. The outer epidermis of testa is protective in both. The tegmen is thin. The nucellus and endosperm are absent in the ripe seed (Mohana Rao 1974a).

Anacardiaceae. The suborder Anacardiineae includes a single family, Anacardiaceae. Anacardiaceae differ both from Coriariaceae and Aceraceae:

Table 31.12.1. Embryology of Coriariaceae (Sect. 31.2), Anacardiaceae (31.3), Salvadoraceae (33.11) and Empetraceae (42.5). (Adapted after V.K. Sharma 1968a, Maheswari Devi 1972, Corner 1976)

	Coriariaceae	Anacardiaceae	Salvadoraceae	Empetraceae
Anther	Bisporangiate	Tetrasporangiate	Tetrasporangiate	Tetrasporangiate
Wall	5-layered, endothecium fibrous	6-layered, endothecium fibrous	5-layered, endothecium fibrous	4–5-layered, endothecium non-fibrous
Tapetum	Secretory, cells 1–4-nucleate	Secretory, cells binucleate	Secretory	Secretory, cells multinucleate
Cytokinesis	Simultaneous	Simultaneous	Simultaneous	Simultaneous
Pollen	2–4-zonoporate, 3-zonocolporate, 2- or 3-celled	3-colporate, 2-celled	3-porate, 2-celled	Remain attached in pairs, individual grain 3-colpate or colporate, 2-celled
Ovule	Anatropous, bitegmic, crassinucellate, inner integument forms micropyle, outer integument forms a flap-like covering on either side of funiculus	Anatropous, sometimes apotropous, bitegmic, various degrees of fusion leading to unitegmy, crassinucellate, funicular obturator develops	Anatropous, bitegmic, crassinucellate, micropyle formed by inner integument	Anatropous, unitegmic, tenuinucellate, endothelium differentiates, funicular obturator is reported
Tetrad	Linear, T-shaped	Linear, T-shaped	Linear	Linear
Embryo sac	Polygonum type, antipodals ephemeral	Polygonum, antipodals ephemeral, twin embryo sacs may occur	Polygonum, antipodals simulate egg apparatus	Polygonum, antipodals persist up to fertilization
Endosperm	Nuclear type, becomes cellular	Nuclear, becomes cellular	Nuclear, becomes cellular	Cellular, micropylar haustorium 2- and chalazal 1-nucleate, both bulge at the extremities
Embryogeny	Onagrad type	Onagrad	Asterad	Caryophyllad
Polyembryony	Absent	Synergid, and nucellar polyembryony	Absent	Cleavage polyembryony
Seed coat	Not multiplicative, exotesta thick-walled	Unspecialized	Exotesta of thinly-cuticularized cells with mucilaginous radial and inner walls	Exotesta forms seed-coat
Seed	Exotestal, mostly exalbuminous, embryo straight	More or less exalbuminous, embryo large, stony endocarp protective	Exotestal, exalbuminous, embryo straight, cotyledons thick	Exotestal, albuminous, embryo straight

(1) bitegmic condition of ovule due to splitting of integumentary primordium, (2) hypostase with phenolic compounds, (3) occasional occurrence of chalazogamy, (4) considerable interval between fertilization and division of zygote, (5) nucellar polyembryony, and (6) protective endocarp.

Anacardiaceae also show marked similarities with Coriariaceae (Table 31.12.1) and Aceraceae: (1) secretory anther tapetum, (2) two-celled pollen grains, (3) anatropous, bitegmic, crassinucellate ovule, (4) Polygonum type of embryo sac with ephemeral antipodals, and (5) Nuclear type of endosperm. All the three families, Coriariaceae, Anacardiaceae and Aceraceae, have Onagrad type of development of embryo, and exalbuminous seed. Anacardiaceae and Coriariaceae also have exotestal seed-coat. There is no doubt that these three families are closely related.

In Anacardiaceae the unspecialized testa may degenerate to give unitegmic seeds. In some taxa the inner epidermis of the tegmen becomes thick-walled. None of these features are common to the Rutales or Sapindales and, therefore, any relationship should be ruled out (Corner 1976, see also von Teichman and Robbertse 1986b). On the basis of seed-coat and pericarp structure, there is possibly a relationship between Anacardiaceae and Julianaceae. Von Teichman and Robbertse emphasize that the seed-coat structure supports the generally accepted view that there is a strong phylogenetic affinity between Anacardiaceae and Burseraceae.

Blepharocarya. Wannan et al. (1987) discuss the affinities of *Blepharocarya*, which has been placed either in Sapindaceae or Anacardiaceae. Radlkofer (1879) had included it in the Anacardiaceae. Anacardiaceae, Burseraceae and Julianiaceae are closely related. Burseraceae have two epitropous ovules per locule, as compared to one ovule in the other two families. The tricolpate and prolate pollen grain in *B. involucrigera* is of the Rhus type (Anacardiaceae). The ovule is clearly apotropous with a dorsal raphe and a superior micropyle. These pollen and ovule characters are also common to Anacardiaceae. The secretory canals in the stem and wood clearly point to the anacardian affinity. On the basis of a detailed study of floral morphology, stem anatomy, pollen and ovule structure, Wannan et al. conclude that *Blepharocarya* should be "readmitted" into the Anacardiaceae. Final confirmation must await fuller embryological study of this taxon.

The suborder Sapindineae comprises Aceraceae, Bretschneideraceae, Sapindaceae, Hippocastanaceae, Sabiaceae, Melianthaceae, and Aextoxicaceae.

Aceraceae. The relationships of Aceraceae have already been discussed along with Anacardiaceae (Sect. 31.12).

Bretschneideraceae. *Bretschneidera*, formerly included in Hippocastanaceae (Pax 1896b), is treated as a separate family Bretschneideraceae, related to Moringaceae (Engler and Diels 1936). The embryology has not been investigated, and its taxonomic assignment cannot be discussed.

Sapindaceae. Sapindaceae can be distinguished from the other families of the suborder on the basis of: (1) anacampylotropous ovule with hypostase and obturator, (2) persistent supernumerary antipodals, and (3) arillate, sarcotestal, exotegmic seeds. The embryogeny is of the Asterad type similar to that in Melianthaceae and Balsaminaceae. In spite of these differences, Sapindaceae show embryological features in conformity with the order Sapindales.

Hippocastanaceae. The characteristic features of Hippocastanaceae are: (1) amphitropous ovule, massive nucellus, thick-walled epistase, and hypostase, (2) embryo sac with a vesicle at the chalazal end, (3) asymmetric development of cotyledons (functioning as an organ of absorption of nutrients from the vesicle), (4) insignificant development of suspensor, and (5) variability in the plane of division of the zygote (see List and Steward 1965).

Even though the above features mark out Hippocastanaceae, the other embryological features like development of anther, pollen, ovule, nucellus, and embryo sac show alliance with the other families of Sapindales. The exalbuminous, pachychalazal seed resembles that of Balsaminaceae. It follows that Hippocastanaceae are correctly placed in Sapindales.

Sabiaceae. Sabiaceae is said to be closely allied to Sapindaceae. Its unitegmic ovule with dome-shaped nucellus, formation of a diverticulum at the chalazal end of the embryo sac, and Helobial type of endosperm with haustorial processes from the chalazal chamber single it out from the rest of the families. Its position in Sapindales is inappropriate.

Melianthaceae. Melianthaceae were considered by Bentham and Hooker (1965a) and Hallier (1912) as constituents of Sapindaceae. Gürke (1896, see Lawrence 1951) separated them on morphological grounds. Embryologically also, they differ from Sapindaceae in several respects. Melianthaceae have persistent epidermis, three to five middle layers, and biseriate secretory anther tapetum. Cytokinesis in pollen mother cells is by furrowing, as well as by cell-plate formation. The pollen grains are three-nucleate at the dispersal stage. The adjacent sporangia do not coalesce, and each anther lobe dehisces independently (S. Mathur 1977).

The bitegmic, crassinucellate ovule, Polygonum type of embryo sac, and Nuclear type of endosperm occur in Melianthaceae as well as in Sapindaceae. The endosperm never becomes cellular in Melianthaceae as also in Hippocastanaceae (Khushalani 1963b, List and Steward 1965).

The seed is arillate. The endosperm is massive, horny and ameloid in *Bersama* (see Corner 1976). No other family shows endospermic seeds in Sapindales. The seed structure also differs to some extent from that of Sapindaceae. The seeds of Sapindaceae have palisade-like exotesta, and sclerotic endostome. The exotegmen is composed of longitudinally elongate lignified, or non-lignified, fibres. In Melianthaceae the exotesta is palisade-like only in the region of the micropyle, and the tegmen is almost crushed. The innermost layer of the inner integument is radially elongate at the micropyle (Khushalani 1963b). On embryological ground, Melianthaceae should be treated as an independent and specialized family.

Balsaminaceae. Balsaminaceae are treated as a subfamily of Geraniaceae by Bentham and Hooker (1965a). Engler and Prantl (1931), Hutchinson (1973) and Takhtajan (1980) considered them as an independent family in Geraniales. This family is usually placed in Geraniales, except by Engler and Prantl (1931) and Melchior (1964) who transferred Balsaminaceae to the order Sapindales, especially due to the nature of ovular attachment (pendulous). R. Dahlgren (1977a) erected a separate order Balsaminales in superorder Rutanae.

Balsaminaceae differ from Sapindales in the formation of trabeculate partitions in the pollen sac, fusion of integuments up to the tip (except in *Impatiens maculata* and *I. parviflora*), tenuinucellate ovule with endothelium, Polygonum as well as Allium type of embryo sac, Cellular type of endosperm with micropylar haustorium, and nucleated nodules in the free-nuclear part of the endosperm. Seeds are small- to medium-size, exarillate, exalbuminous, and mostly pachychalazal. The seed-coat is formed by expanded chalaza. Corner (1976) mentions that in *Impatiens* there may be a complete series of species to show the origin of the unitegmic pachychalazal seed from the bitegmic.

According to L.L. Narayana (1970f), Balsaminaceae are closely related to Oxalidaceae, Geraniaceae and Tropeaolaceae. The features of similarities are: (1) secretory anther tapetum, (2) cytokinesis by furrowing, (3) two-celled pollen grains, (4) anatropous, bitegmic ovules, (5) early fusion of polar nuclei, and ephemeral antipodals, (6) porogamous entry of pollen tube, and (7) Asterad type of embryogeny. Anatomically (Heimsch 1942) and cytologically (Warburg 1938) also Balsamineae resemble these three families. L.L. Narayana (1970f) negates the inclusion of Balsaminaceae under Sapindales, on the basis of the nature of ovular attachment as proposed by Engler and Prantl (1931).

The embryological features cited by L.L. Narayana (1970f) as affinities of these three families occur in most of the sapindaceous families. Asterad type of embryogeny occurs in Sapindaceae and Melianthaceae. These affinities do not warrant the separation of Balsaminaceae from Sapindales. In our opinion there is adequate justification to erect the order Balsaminales, as already suggested by R. Dahlgren (1977a).

32 Julianiales

The order Julianiales comprises a single family, Julianiaceae. The embryology of this family has not been investigated.

33 Celastrales

The order Celastrales comprises three suborders with 13 families: Celastrineae Cyrillaceae, Pentaphylacaceae, Aquifoliaceae, Corynocarpaceae, Pandaceae,

Celastraceae, Staphyleaceae, Hippocrateaceae, Stackhousiaceae and Salvadoraceae; Buxineae – Buxaceae; Icacineae – Icacinaceae and Cardiopteridaceae.

33.1 General Features

The order Celastrales is characterized by: (1) secretory anther tapetum, (2) simultaneous cytokinesis in pollen mother cells, (3) two-celled pollen grains except in Stackhousiaceae, (4) anatropous ovules, orthotropous in Celastraceae; variable number of integument, and nature of nucellus, (5) Polygonum type of embryo sac, (6) Nuclear type of endosperm, Cellular in Cyrillaceae and Aquifoliaceae, and both types in Buxaceae, endosperm haustoria in Cyrillaceae and Icacinaceae, and (7) probably Solanad type of embryogeny in Cyrillaceae, Solanad in Celastraceae, probably Caryophyllad type in Aquifoliaceae and Corynocarpaceae, Asterad in Stackhousiaceae and Salvadoraceae, and Onagrad in Buxaceae.

33.2 Cyrillaceae

The development of the embryo has not been investigated.

The anther wall comprises the epidermis, fibrous endothecium, two ephemeral middle layers, and secretory tapetum of binucleate cells. The epidermis, endothecium, and middle layers contain tannin in *Cyrilla racemiflora* and *Cliftonia monophylla* (Vijayaraghavan and Dhar 1978). A few cells of the connective also develop fibrous thickenings, and contain tannin and druses. Simultaneous cytokinesis follows meiotic divisions in microspore mother cells. The tetrads are tetrahedral.

The pollen grains are tricolporate, and two-celled at the dispersal stage.

The ovule is anatropous, unitegmic and tenuinucellar with a long and narrow micropyle. Druses, tannin-filled cells, and stomata occur in the funicle in *Cyrilla* (Vijayaraghavan and Dhar 1978). The inner epidermis of the integument develops into an endothelium. A single hypodermal cell functions directly as megaspore mother cell. The latter, by meiotic divisions, produces a linear tetrad. The chalazal megaspore develops into a Polygonum type of embryo sac. The polar nuclei fuse before fertilization, the secondary nucleus lies in the centre of the embryo sac. The antipodals persist up to fertilization.

The development of the endosperm is of the Cellular type (Fig. 33.2.1A–G). Due to two consecutive transverse divisions of the primary endosperm nucleus, a row of four cells is formed in *Cliftonia* (Vijayaraghavan and Dhar 1978). The two middle cells develop into the endosperm proper (Fig. 33.2.1B–F). The two terminal cells develop into haustoria. The micropylar haustorium (Fig. 33.2.1B–F) is one-celled, binucleate and aggressive, and penetrates the funicular cells. Rarely, it becomes three- to four-nucleate in *Cyrilla*. The chalazal haustorium is one-celled, binucleate (Fig. 33.2.1C–F), and persists up to the dicotyledonous stage of the embryo.

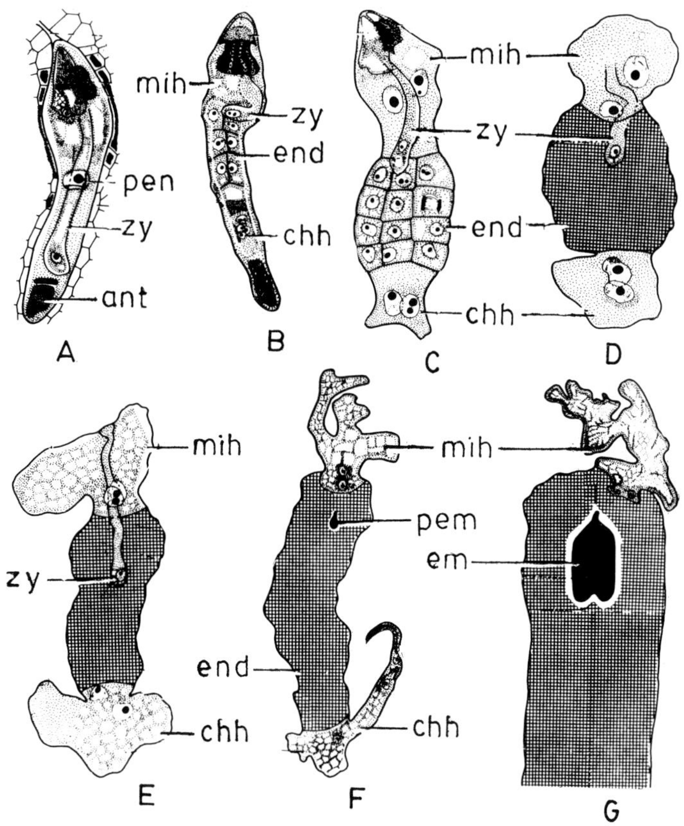

Fig. 33.2.1A–G. *Cyrilla racemiflora*. **A** Embryo sac with zygote (*zy*), degenerated synergid, remnants of pollen tube, primary endosperm nucleus (*pen*), and degenerated antipodal cells (*ant*). **B** Embryo sac with zygote, degenerated synergid, binucleate micropylar haustorium (*mih*), endosperm proper (*end*), four-nucleate chalazal haustorium (*chh*), and degenerated mass of antipodal cells. **C, D** Embryo sac with elongated tubular zygote, binucleate micropylar haustorium, cellular endosperm proper, and binucleate chalazal haustorium. **E** Bulbous endosperm haustoria. **F** Embryo sac at globular proembryo. The well-developed branched micropylar haustorium is binucleate, and the chalazal haustorium four-nucleate. **G** Upper portion of embryo sac at dicotyledonous embryo. (After Vijayaraghavan and Dhar 1978)

The degeneration of egg and zygote is frequent. The zygote elongates (Fig. 33.2.1A–E) before division, and the mature embryo is large, straight and dicotyledonous.

A noteworthy feature of the family is the occurrence of aposporic embryo sacs of Hieracium type (Battaglia 1963) in *Cliftonia monophylla* (Vijayaraghavan 1970c, Vijayaraghavan and Dhar 1978). In the chalazal region, some of the nucellar cells differentiate as megaspore mother cells and behave as embryo sac mother cells. They enlarge, undergo mitotic divisions, and develop into two- and four-nucleate embryo sacs. An eight-nucleate unorganized embryo sac was also observed. These are non-functional, and degenerate during the development of endosperm.

The integumentary cells become vacuolate and degenerate at the dicotyledonous stage of the embryo. The mature seed does not have any seed-coat (Vijayaraghavan and Dhar 1978). The multilayered pericarp is tanniniferous, and contains druses.

E Battaglia (1963), HF Copeland (1953), JL Thomas (1960, 1961), MR Vijayaraghavan (1969, 1970c), MR Vijayaraghavan and U Dhar (1978)

33.3 Pentaphylacaceae

The development of anther, pollen, embryo sac, endosperm and embryo has not been investigated.

The ovule is ana- or campylotropous, bitegmic and crassinucellar in *Pentaphylax euryoides* (Mauritzon 1936b). A long micropyle is organized by the inner integument. The parietal tissue is scanty. The embryo sac is eight-nucleate and, later, becomes curved.

Seeds are more or less winged, with scanty endosperm and horseshoe-shaped embryo (Hutchinson 1973).

A Cronquist (1981), J Hutchinson (1973), J Mauritzon (1936b)

33.4 Aquifoliaceae

The development of anther, embryo and seed-coat has not been studied in detail.

In *Ilex cornuti* (Copeland 1963), the endothecium is ribbed, the tapetum with binucleate cells is not strictly a single layer, and the three-grooved pollen grains have a papillate surface. They are two-celled at the shedding stage.

The ovule is apotropous, unitegmic and crassinucellar. One to three cells in the hypodermal region cut off parietal cells. The parietal tissue is one or two layers thick in *Ilex aquifolium* and *I. cornuti* (Copeland 1963). At times, two megaspore mother cells function. The megaspore mother cell undergoes meiotic divisions, and produces a linear tetrad. The chalazal megaspore develops into a Polygonum type of embryo sac. An endothelium and hypostase differentiate. The polars fuse before fertilization, and the secondary nucleus is prominent. The antipodals are inconspicuous.

Development of the endosperm is of the Cellular type. A transverse division of the primary endosperm nucleus results in a micropylar and a chalazal chamber. Both chambers undergo longitudinal divisions; linear cell formation (Schürhoff 1921) occurs rarely.

The embryogeny is probably of the Caryophyllad type.

According to Hu and Ferreira (1989), the ripe fruit in *Ilex* contains seeds at the early heart-shaped embryo; further development to maturity occurs later.

The seed is endospermic with a minute embryo. The cells of the integument become lignified at the micropyle. The endothelial layer accumulates tannin. The remnants of nucellus persist in mature seed.

HF Copeland (1963), JM Herr Jr (1959a, b, c, 1961), CY Hu and AG Ferreira (1989), PN Schürhoff (1921)

33.5 Corynocarpaceae

The development and structure of anther, pollen grain, embryo sac, and embryo have not been investigated.

The ovule is anatropous, bitegmic and crassinucellate in *Corynocarpus laevigata* (Pigott 1927). The inner integument forms the micropyle. The archesporial cell cuts off a parietal cell. The cells of the nucellar epidermis divide and produce three to four layers.

In a mature embryo sac, the three antipodal cells multiply.

The development of the endosperm is of the Nuclear type.

The embryogeny probably corresponds to the Caryophyllad type (see Davis 1966).

The seed is exalbuminous, and contains a large embryo with minute radicle. The massive (20–40 layers) testa is thin-walled, and gets crushed. The tegmen also gets crushed.

Possibly the seed is pachychalazal as in Anacardiaceae (Corner 1976).

EJH Corner (1976), GL Davis (1966), EM Pigott (1927)

33.6 Pandaceae

The embryology has not been investigated.

The anther is bilocular, and introse and opens longitudinally. The ovules are ortho- or anatropous, without obturator. The seeds are usually flattened, rarely ovoid, with copious fatty endosperm, and a straight embryo which has thin, flat, cordate cotyledons.

A Cronquist (1981), J Hutchinson (1973), J Vaughan and JA Rest (1969), JC Willis (1973)

33.7 Celastraceae

The anther wall comprises the epidermis, fibrous endothecium, two or three ephemeral middle layers, and secretory tapetum. The epidermal cells are papillate, and contain tannin. In *Gymnosporia* (Adatia and Gavde 1962), the endothecium becomes irregularly biseriate. The tapetal cells are multinucleate. In *Celastrus paniculata* (Adatia and Gavde 1962), they are eight-nucleate; the nuclei fuse and become polyploid. The microspore mother cells divide meiotically and produce tetrahedral and decussate tetrads. Simultaneous cytokinesis occurs by furrowing.

The pollen grains are two-celled at the shedding stage. In *Gymnosporia rothiana* they germinate in situ. Pollen grains fail to develop in *Euonymus japonica* (Copeland 1966); this species is an apomict.

The ovule is orthotropous, bitegmic and tenuinucellate in *Gymnosporia*, and weakly crassinucellate in *Celastrus* and *Elaeodendron* (Adatia and Gavde 1962). Both the integuments form the micropyle. The hypodermal archesporium is uni- to multicellular. It cuts off a parietal cell in *Celastrus* and *Elaeodendron*, otherwise it functions directly as megaspore mother cell. At times, two mother cells may be present. The megaspore mother cell undergoes meiotic divisions and produces a linear or T-shaped tetrad. The chalazal megaspore develops into a Polygonum type of embryo sac. The antipodals are

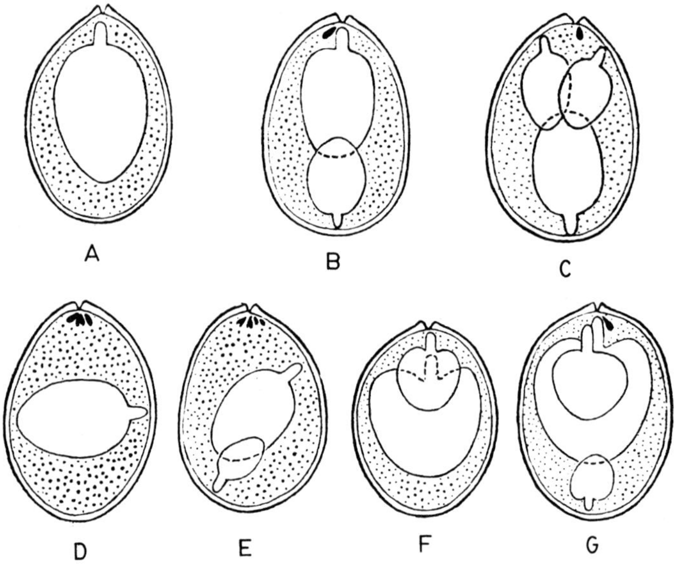

Fig. 33.7.1A–G. *Euonymus*, polyembryonate seeds. All figures diagrammatic, double seed-coat is not indicated, endosperm *stippled*. Minute "embryos" at micropylar pole *black*. **A–E** *E. americanus*; **A** Monoembryonate seed with full-sized embryo (3.0–3.5 mm); **B** Three-embryonate seed, a globular proembryo (0.25 mm) and a large embryo (2–3 mm) at micropylar pole, and a smaller embryo (1–2 mm) at chalazal pole; **C** Four-embryonate seed, a globular proembryo (0.25 mm) and two small embryos (1–2 mm) at micropylar pole, and a larger embryo (2–3 mm) at chalazal pole (mostly integumentary); **D** Two globular proembryos (0.25 mm) and a heart-shaped (0.25–0.5 mm) embryo at micropylar end, and a larger (probably integumentary) embryo (3.0–3.5 mm) lies horizontally in the middle of the embryo sac; **E** Six-embryonate seed. Two globular proembryos and two heart-shaped embryos (0.05–0.25 mm) at the micropylar pole, one smaller (1 mm) and a larger embryo (2–3 mm) lie obliquely opposite to each other. **F, G** *E. alatus* var. *apterus*; **F** Two-embryonate seed with cotyledons of smaller embryo (1–2 mm) clasping the radicle of larger one (3.0–3.5 mm); **G** Four-embryonate seed with one globular proembryo (0.25–0.5 mm) and two embryos (2–3 mm, 1–2 mm) at micropylar end, and a smaller one (under 0.25 mm) at chalazal pole. (Adapted after Brizicky 1964)

ephemeral. The polar nuclei fuse in the centre of the embryo sac. An endothelium differentiates by the time of tetrad formation. In *Euonymus japonica*, half the ovules abort, and the egg apparatus generally degenerates.

The development of endosperm is of the Nuclear type. Centripetal wall formation occurs after the formation of about 200 nuclei.

The embryogeny corresponds to the Solanad type (Adatia and Gavde 1962). It is reported to be of Caryophyllad type-Sagina variation in *Celastrus scandens* and *Euonymus europaeus* (Anderson 1931).

The endosperm in *E. japonica* (Copeland 1966) develops autonomously, and the embryo originates from a cell of the inner integument close to the micropyle.

Polyembryony occurs in *Celastrus paniculata* (Adatia and Gavde 1962). Brizicky (1964) reports integumentary (?), and nucellar embryos (Fig. 33.7.1A–G), and parthenocarpic development of fruit in *Euonymus* spp. Naumova (1970a, b) observed different patterns of embryogenesis in the integumentary and zygotic embryos.

The seed is arillate. The aril may develop from the outer integument, or funicle, or both. The outer epidermal cells of testa are thickly cuticularized, and are followed by aerenchymatous mesophyll with scattered sclerotic cells. The inner epidermis is thin-walled. The outer epidermis of tegmen comprises thick-walled, lignified, pitted, longitudinal fibres. The inner epidermis is represented by the endothelium.

RD Adatia and SG Gavde (1962), A Anderson (1931), GK Brizicky (1964), HF Copeland (1966), H Ding (1967), T Loesener (1942), J Mauritzon (1936b), TN Naumova (1970a, b), R Pilger (1915)

33.8 Staphyleaceae

The embryogeny has not been investigated.

The anther wall comprises the epidermis, fibrous endothecium, a single ephemeral middle layer, and secretory tapetum of binucleate cells. The microspore mother cells undergo meiotic divisions, and produce tetrahedral tetrads. Simultaneous cytokinesis occurs by furrowing.

The pollen grains are shed at the two-nucleate stage.

The ovule is anatropous, bitegmic and crassinucellate. The micropyle is formed by both the integuments. The hypodermal archesporial cell cuts off a parietal cell. The parietal tissue is well developed, and a nucellar cap also develops. The megaspore mother cell divides meiotically, and produces a linear tetrad. The chalazal megaspore gives rise to a Polygonum type of embryo sac. The synergids are hooked, and exhibit filiform apparatus in *Turpinia* (L.L. Narayana 1960a). The three antipodals are small.

The development of the endosperm is of the Nuclear type in *Staphylea pinnata* (Guèrin 1901) and *S. trifoliata* (Riddle 1905). The endosperm stores fatty food reserve.

The mature embryo is straight with thick flat cotyledons, and a short radicle.

The seed is large and albuminous. The testa is multiplicative. The outer epidermis comprises a palisade of thick-walled cells followed by a sclerotic mesophyll. The inner epidermis is thin-walled. The tegmen gets crushed.

P Guèrin (1901), LL Narayana (1960a), LC Riddle (1905)

33.9 Hippocrateaceae

The embryogeny and structure of seed-coat have not been studied.

The anther wall comprises the papillate persistent epidermis, fibrous endothecium, two ephemeral middle layers, and secretory tapetum of multinucleate cells. Simultaneous cytokinesis occurs by furrowing, and the tetrads are tetrahedral and decussate.

The tricolpate, rarely tetracolpate, pollen grains with reticulate exine are two-celled at the dispersal stage.

The ovule is anatropous, bitegmic and tenuinucellate with the inner integument forming the micropyle in *Hippocratea grahamii* (Adatia and Gavde 1962). The archesporial cell functions directly as megaspore mother cell. In *Salacia oblonga* the archesporial cell cuts off a parietal cell which subsequently produces a parietal tissue. The megaspore mother cell undergoes meiotic divisions. The chalazal megaspore of the linear tetrad develops into a Polygonum type of embryo sac. The antipodals are ephemeral, and polar nuclei fuse before fertilization. An endothelium develops in *Hippocratea.*

The development of the endosperm is of the Nuclear type.

The fruit is schizocarpic, and the seeds are exalbuminous, winged and exarillate.

RD Adatia and SG Gavde (1962), J Mauritzon (1936a, c), J Miers (1872)

33.10 Stackhousiaceae

The anther wall comprises the epidermis, fibrous endothecium, a single middle layer, and secretory tapetum of binucleate cells. The microspore mother cells divide meiotically, and produce tetrahedral, isobilateral and decussate tetrads. Simultaneous cytokinesis is accompanied by furrowing.

The pollen grains are shed at the three-celled stage.

The ovule is anatropous, bitegmic, tenuinucellate and develops a hypostase. The micropyle is formed by the outer integument, the inner one collapses. The hypodermal archesporial cell functions directly as megaspore mother cell. The latter, following meiotic divisions, produces linear or T-shaped tetrad. The chalazal megaspore develops into a Polygonum type of embryo sac. The embryo sac comes to lie in direct contact with the integument after the degeneration of nucellus. The three antipodals are multinucleate in *Stackhousia linariaefolia* (Narang 1953), and their number increases up to 15 in *S. monogyna* and some other species (see Mauritzon 1936c).

The development of endosperm is of the Nuclear type. The peripheral layer of cytoplasm around the central vacuole contains 60–70 free nuclei. Cytoplasmic nodules develop in the endosperm. Due to centripetal wall formation, the endosperm becomes cellular at the globular proembryo.

The embryogeny conforms to the Asterad type. The mature embryo is large and straight, with rather thick cotyledons.

The seed is albuminous. The outer epidermis of testa comprises tangentially elongate cells containing tannin. The mesophyll and inner epidermis is thin-walled. The inner integument is absorbed by the developing embryo sac, before anthesis. The funicular vascular bundle extends in the antiraphe up to the micropyle.

J Mauritzon (1936c), N Narang (1953)

33.11 Salvadoraceae

The anther wall comprises the epidermis, fibrous endothecium, two ephemeral middle layers, and secretory tapetum. The microspore mother cells undergo meiotic divisions resulting in tetrahedral and isobilateral tetrads. Simultaneous cytokinesis occurs by furrowing.

The triporate pollen grains are shed at the two-celled stage.

The ovule is anatropous, bitegmic and crassinucellar. The micropyle is organized by the inner integument in *Salvadora persica*, and by the outer integument and inner lip of the inner integument in *Azima tetracantha* (Maheswari Devi 1972). The archesporium is uni- to multicellular; the parietal tissue is well developed. The megaspore mother cell undergoes meiotic divisions, and produces a linear tetrad. The chalazal megaspore develops into a Polygonum type of embryo sac. The antipodal cells simulate egg apparatus. Sometimes twin embryo sacs are formed.

The entry of pollen tube is porogamous, and syngamy and triple fusion occur more or less simultaneously.

The development of the endosperm is of the Nuclear type. The nuclei with cytoplasm accumulate in the micropylar region around the embryo. Later, a similar accumulation of nuclei also occurs in the chalazal region where they fuse and become polyploid. Finally, the endosperm becomes cellular, and is consumed by the developing embryo.

Embryogeny corresponds to the Asterad type-Oxalis variation (Maheswari Devi 1972). The mature embryo is straight with thick flat cotyledons, and a well-developed radicle.

The seed is exotestal in *Azima*. The outer epidermis of testa comprises columnar, thinly cuticularized cells with mucilaginous radial and inner walls, and contains crystals. The mesophyll is thin-walled, except at the hilum and micropyle where it is lignified.

G Bentham (1876b), H Maheswari Devi (1972), J Mauritzon (1936b)

33.12 Buxaceae

The anther wall comprises the persistent epidermis, fibrous endothecium, two or three ephemeral middle layers, and secretory tapetum of binucleate cells. The microspore mother cells undergo meiotic divisions, cytokinesis is simultaneous, and tetrahedral and isobilateral tetrads are produced.

The pollen grains are two- or three-celled at the dispersal stage.

The ovule is anatropous, bitegmic and crassinucellate with a false endostome in *Buxus*. The archesporial cell cuts off a parietal cell which produces a thick parietal tissue. The megaspore mother cell undergoes meiotic divisions. The chalazal megaspore of a linear tetrad develops into a Polygonum type of

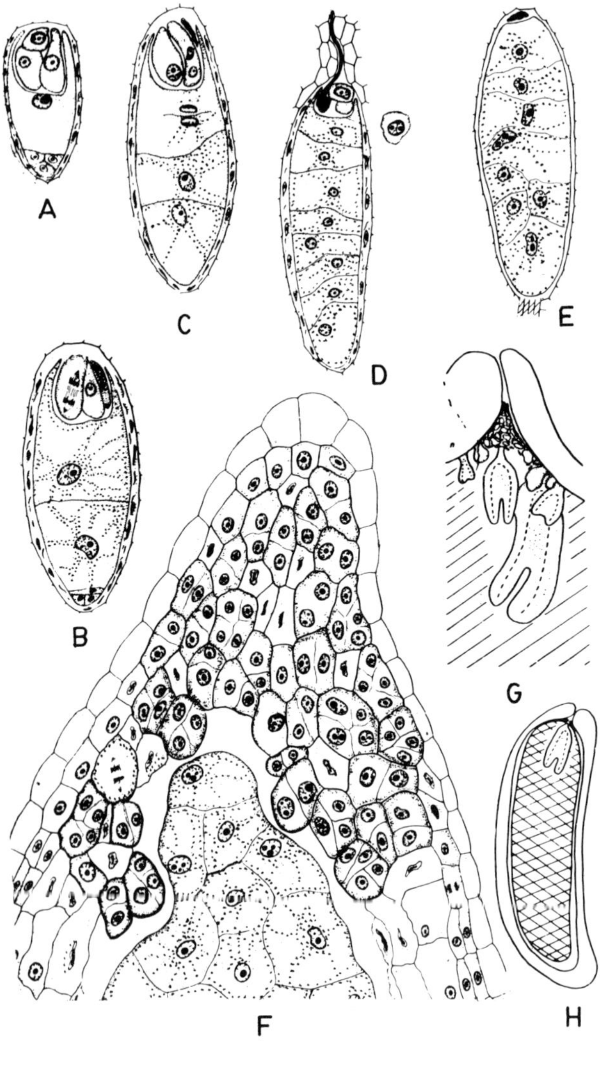

Fig. 33.12.1A–H. **A–C** *Notobuxus acuminata*, **D, H** *Buxus sempervirens*, **E–G** *Sarcococca humilis*. **A** Mature embryo sac; **B** Fertilized embryo sac, zygote in division, persistent synergid, remnants of pollen tube, and two-celled endosperm; **C** Three-celled endosperm, zygote undivided. **D** Embryo sac with zygote (*inset*), remnants of pollen tube, and multicellular endosperm. **E** Embryo sac with remnants of degenerated egg apparatus, and cellular endosperm; **F** Micropylar region of nucellus. Most of the cells function as initials of apomictic embryos; note a number of proembryos especially around the endosperm; **G** Micropylar region of ovule with a number of nucellar embryos. **H** Ovule with sexual, young dicotyledonous embryo surrounded by endosperm (*cross-hatched*). (After Naumova 1979, 1980, see Yakovlev 1981)

embryo sac (Fig. 33.12.1A). The polar nuclei fuse prior to fertilization. The antipodals are persistent, and multiply (Wiger 1935).

Simmondsia represents a young taxon which appears to have made a comparatively recent transition from insect-mediated pollination (entomophily) to anemophily. Three major alterations have occurred in this transition: (1) loss of the androecium, (2) reduction in flower number per inflorescence to one, and (3) extension of the peduncle. These are selectively advantageous for anemophily.

The development of endosperm is of the Nuclear type in *Buxus* and *Pachysandra*, and Cellular (Fig. 33.12.1B–E) in *Sarcococca* and *Simmondsia* (see Schnarf 1931, Davis 1966).

Development of the embryo is of the Onagrad type. Nucellar polyembryony (see Schnarf 1931, Davis 1966) occurs (Fig. 33.12.1 F, G) in several species of *Sarcococca* (Orr 1923), and in *S. prunifolia* at least one nucellar embryo develops autonomously while the egg degenerates (Wiger 1935).

The albuminous seed is small and exo-mesotestal with a straight or curved embryo with flat cotyledons (Fig. 33.12.1H). The outer epidermis of testa is lignified, and the mesophyll is crushed except in *Sarcococca* where one to two layers are sclerotic. The inner epidermis is palisade-like, and lignified at the micropylar end in *Buxus*. The inner epidermis of tegmen is lignified in *Sarcococca*, otherwise the tegmen is crushed. The seed of *Buxus* shows two small waxy-horny white lobes, on either side of the micropyle, as rudimentary funicular aril or caruncle.

EJH Corner (1976), GL Davis (1966), TN Naumova (1979, 1980), KJ Niklas and SL Buchman (1985), MY Orr (1923), TL Rost, AD Simper, P Schell and S Allen (1977), K Schnarf (1931), J Wiger (1935)

33.13 Icacinaceae

The embryogeny has not been studied.

The anther wall comprises the epidermis, radially elongate thick-walled endothecium, one or two ephemeral middle layers, and secretory tapetum of five- to six-nucleate cells. The microspore mother cells undergo meiotic divisions, cytokinesis is simultaneous, and the tetrads are tetrahedral.

The family is multipalynous with three-colpate and inaperturate elements (P.K.K. Nair 1970). The pollen grains are two-celled at the dispersal stage in *Gomphandra polymorpha* (Padmanabhan 1961c).

The ovule is anatropous, unitegmic and crassinucellate with a well-developed vascular supply in the integument. The integument splits (in young condition) in *Phytocrene* (Fagerlind 1945d), and the splits remain free at the micropyle. A weak obturator develops from the placenta. The archesporium is multicellular. The archesporial cells cut off parietal cells, and the parietal tissue is five to six layers thick. The megaspore mother cell undergoes meiotic divisions giving rise to a linear tetrad. The chalazal megaspore develops into a Polygonum type of embryo sac. At the four-nucleate stage, the embryo sac absorbs the nucellus, extends towards the micropyle and comes to lie between the integument and raphe of the ovule. The synergids lack vacuoles. The polar nuclei fuse before fertilization. The antipodals are ephemeral.

The development of the endosperm is of the Nuclear type. In *Gomphandra*, the zygote divides almost at the same time as the walls are laid down in the free-nuclear endosperm. The inner cells of the endosperm are large and highly vacuolate.

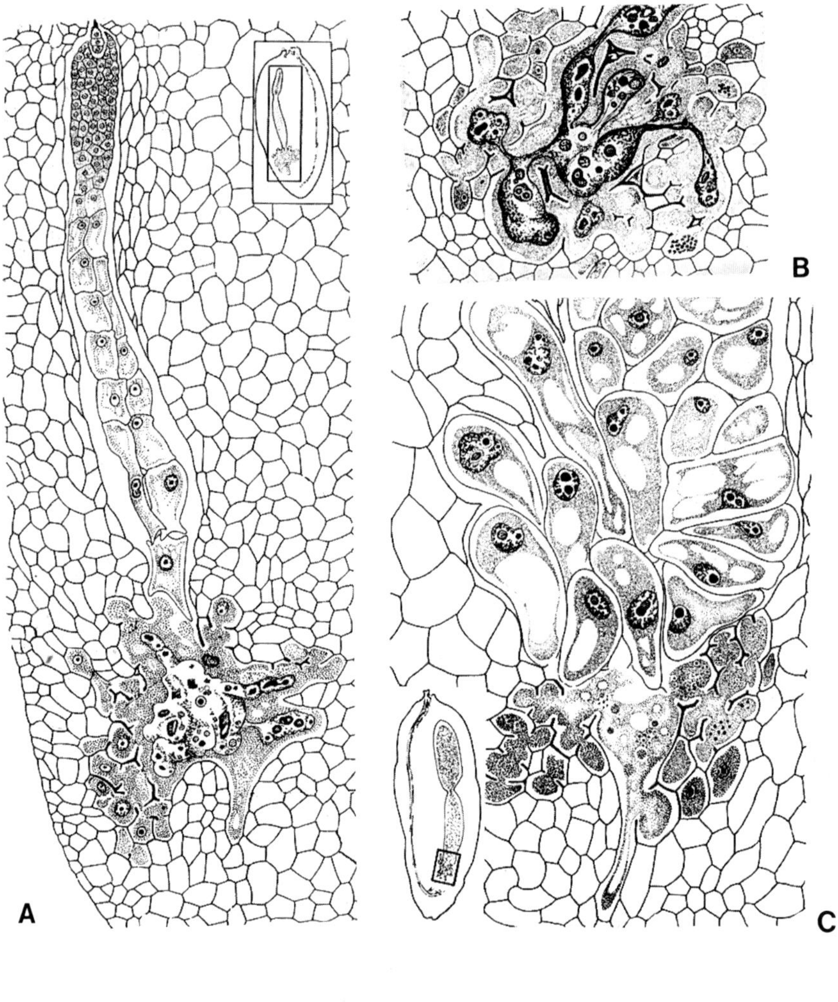

Fig. 33.13.1A–C. *Nothapodytes foetida*. **A** Longisection of part of ovule (*inset* outline diagram) to show three-celled proembryo. Cellular endosperm proper, and intracellular growth of chalazal haustorium with polyploid nucleus. **B** Nucleus of chalazal haustorium with adjoining tissue. **C** Longisection of part of ovule (*inset* outline diagram), later stage than **A**. The endosperm cells in the vicinity of disintegrated chalazal haustorium have become hypertrophied. The chalazal cells have been incorporated into the haustorium in **A–C**. (After Swamy and Ganapathy 1957b)

In *Nothapodytes foetida*, Swamy and Ganapathy (1957b) mention a very unusual type of endosperm haustorium (Fig. 33.13.1A–C). The basal cell of the endosperm proper becomes haustorial and develops a narrow, sac-like intercellular extension in the chalazal region. As the cytoplasm of the haustorial cell comes in contact with the intervening walls of adjacent cells of the chalazal tissue, an aperture develops in the wall at the point of contact, and the haustorium establishes protoplasmic connection with the affected cell. The nucleus of the affected cell becomes amoeboid, develops several nucleoli-like bodies and loses its membrane. Its chromatin becomes coarsely granular, and mixes with the cytoplasm. Finally, the cytoplasm and nucleus of the affected cell are incorporated into the haustorium. Such chain-like cytoplasmic connections develop in the chalazal tissue, and the haustorium continues to enlarge. During its aggressive growth, the haustorium affects groups of cells, and the fragments of their walls become included in the cytoplasm (Fig. 33.13.1A).

The nucleus of the haustorium becomes highly amoeboid and develops numerous nucleoli-like bodies. As the groups of protoplasts are absorbed by the haustorium, the nucleus of the haustorium extends into the affected cells, becomes lobed and dilates at the tip. The several lobes of the nucleus remain connected with the main body. The entire structure gives a very distorted appearance (Fig. 33.13.1B).

The cells of the endosperm proper exhibit distinct zonation, the upper region of smaller compact cells and the lower of larger loosely arranged cells.

When the activity of the haustorium declines, at the initiation of cotyledonary primordia in the embryo, the fragments of the cell walls of the absorbed protoplasts swell and become gelatinous. Two to three tiers of endosperm cells adjoining the haustorium become vacuolate, and develop tubular prolongations which grow intercellularly in the surrounding chalazal tissue. Their nuclei also enlarge, and give a distorted appearance (Fig. 33.13.1C). However, their haustorial behaviour is uncertain.

The seed has a small embryo. The outer layers of integument comprise thick-walled, elongate cells followed by thin-walled hexagonal cells.

F Fagerlind (1945d), J Mauritzon (1936c), PKK Nair (1970), D Padmanabhan (1961c), BGL Swamy and PM Ganapathy (1957b)

33.14 Cardiopteridaceae

The embryology has not been investigated.

33.15 Taxonomic Considerations

Engler and Diels (1936) place all the families in Sapindales. Hutchinson (1973) assigns Buxaceae to Hamamelidales, Staphyleaceae to Sapindales, and Aquifoliaceae, Salvadoraceae, Cyrillaceae, Icacinaceae, Celastraceae, Stackhousiaceae and Hippocrateaceae to Celastrales (see Table 33.15.1).

Table 33.15.1. Embryology of Cyrillaceae, Aquifoliaceae, Celastraceae, Hippocrateaceae, Stackhousiaceae, Salvadoraceae and Icacinaceae (See Vijayaraghavan and Dhar 1978)

	Cyrillaceae	Aquifoliaceae	Celastraceae	Hippocrateaceae
Tapetum	Secretory, cells binucleate	Secretory, cells binucleate	Secretory, cells multinucleate	Secretory, cells multinucleate
Tetrad	Tetrahedral	—	Tetrahedral and decussate	Tetrahedral and decussate
Pollen grains	2-celled	2-celled	2-celled	2-celled
Ovule	Unitegmic, tenuinucellate	Unitegmic, crassinucellate	Bitegmic, weakly crassinucellate, or tenuinucellate	Bitegmic, tenui- or crassinucellate
Embryo sac	Polygonum type	Polygonum	Polygonum	Polygonum
Antipodal cells	Uninucleate, persist up to fertilization	Uninucleate, inconspicuous	Uninucleate, ephemeral	Uninucleate, ephemeral
Endosperm	Cellular type	Cellular	Nuclear	Nuclear
Endosperm haustoria	Micropylar and chalazal haustoria formed	Absent	Absent	Absent
Seed-coat	Absent	Massive	Massive	Massive

Cyrillaceae. It has been suggested that the Cyrillaceae are related to Ericaceae and Celastraceae (see Thomas 1961). R. Dahlgren (1980b) places them in Ericales. Traditionally, Cyrillaceae have remained associated with Aquifoliaceae. Vijayaraghavan (1969, 1970c) compares the embryological features of Aquifoliaceae, Celastraceae, Cyrillaceae and Ericaceae. Cyrillaceae show close affinities with Ericaceae in unitegmic, tenuinucellate ovules, presence of endothelium, Polygonum type of embryo sac with a broad micropylar and narrow chalazal end, uninucleate persistent antipodals, Cellular type of endosperm with chalazal and micropylar haustoria, and probably Solanad type of embryogeny. Cyrillaceae resemble Clethraceae in (1) presence of fibrous endothecium, (2) druses in the cells of connective, and (3) pollen grains shed singly (Vijayaraghavan and Dhar 1978). Occurrence of apospory (Vijayaraghavan 1969) is a noteworthy feature in the family. Cyrillaceae differ from Celastraceae in the nature of ovule, absence of parietal tissue, type of endosperm, and seed structure. Nucellar polyembryony is absent in Cyrillaceae. Thus, embryological characters support its close association with Ericales (Clethraceae), rather than with Celastrales.

Aquifoliaceae. Several embryological features of Aquifoliaceae, such as two-celled pollen grains, unitegmic ovule, Polygonum type of embryo sac, endothelium, and Cellular type of endosperm (Copeland 1963), correspond with those of Cyrillaceae (Table 33.15.1). However, apotropous and crassinucellate condition of the ovule, inconspicuous antipodals, probably Caryophyllad type

Stackhousiaceae	Salvadoraceae	Icacinaceae
Secretory, cells binucleate	Secretory, cells binucleate	Secretory, cells multinucleate
Tetrahedral, decussate and isobilateral	Tetrahedral and isobilateral	Tetrahedral
3-celled	2-celled	2-celled
Bitegmic, tenuinucellate	Bitegmic, crassinucellate	Unitegmic, crassinucellate
Polygonum	Polygonum	Polygonum
Each cell becomes multinucleate	Uninucleate, ephemeral	Uninucleate, ephemeral
Nuclear	Nuclear	Nuclear
Absent	Absent	Absent; conspicuous chalazal endosperm haustorium reported in *Nothapodytes*
Massive	Massive	Massive

of embryogeny, and minute embryo differ from those of Cyrillaceae. Caryophyllad type of embryogeny occurs in Celastraceae, and probably in Corynocarpaceae.

Corynocarpaceae. Corynocarpaceae show affinities with Celastraceae. The common features are: bitegmic, crassinucellate ovule, parietal tissue, Nuclear type of endosperm, and Caryophyllad type of embryogeny. But anatropous ovule, absence of endothelium, secondary multiplication of antipodals, and pachychalazal exalbuminous seed with a large embryo distinguish Corynocarpaceae.

Celastraceae. Celastraceae have several special features, such as superior ovary with two ovules in each locule, orthotropous, bitegmic, tenui- or crassinucellar ovule, endothelium, ephemeral antipodals, Solanad and Caryophyllad (*Celastrus* and *Euonymus*) type of embryogeny, nucellar polyembryony, and arillate albuminous seed with sclerotic mesotesta, and fibrous exotegmen. The wings are absent. The fruit is drupaceous, capsular, or follicular.

Hippocrateaceae. Hallier (1912) combines the genus *Hippocratea* with Celastraceae. *Hippocratea* resembles Celastraceae (Table 33.15.1): (1) imbricate perianth in two whorls, each of five members, (2) glandular anther tapetum, (3) tricolpate, rarely tetracolpate, pollen grains with reticulate exine, (4) tricarpellary syncarpous ovary, (5) anatropous, bitegmic ovules, (6) Polygonum type of embryo sac, presence of an endothelium, and (7) Nuclear type of endosperm (Adatia and Gavde 1962). Adatia and Gavde suggest separation of the genus *Hippocratea*, and its placement in Hippocrateaceae. *Hippocratea* has tenuinucellate ovules with micropyle formed by the inner integument, exalbuminous, exarillate, winged seed, and schizocarpic fruit.

Staphyleaceae. Bentham and Hooker (1965a) united Staphyleaceae with Sapindaceae. Bessey (1915), Wettstein (1935) and Rendle (1959) include them in Celastrales, but Hutchinson (1973) placed them in Sapindales.

Staphyleaceae differ from Sapindaceae in having numerous ovules, copious endosperm, and straight embryo. On the other hand, Staphyleaceae have several common features with Celastraceae. Two-celled pollen grains, crassinucellate, bitegmic ovule with exo- and endostome, parietal cell, Polygonum type of embryo sac, Nuclear type of endosperm, sometimes arillate, albuminous seeds with sclerotic mesophyll, and capsular, infrequently berry-like fruit. The fibrous exotegmen of seed has been lost in this family (Corner 1976). The proper position of Staphyleaceae is in Celastrales rather than in Sapindales.

Stackhousiaceae. Narang (1953) discusses the affinities of Stackhousiaceae in relation to Rhamnaceae, Scrophulariaceae, Lobeliaceae and Celastrales. She concludes that the assignment of Stackhousiaceae to Celastrinae appears to be justified (see also Mauritzon 1936c). Stackhousiaceae closely resemble Celastraceae (Table 32.15.1) and the common features are: bitegmic, tenuinucellate ovule with a single archesporial cell, Polygonum type of embryo sac, Nuclear type of endosperm, and albuminous seeds. However, it can be distinguished due to three-celled pollen grains, supernumerary antipodals, hypostase, cytoplasmic nodules in the endosperm, Asterad type of embryogeny, and a large straight embryo (Narang 1953). The seeds are exarillate, and the inner integument is absorbed before anthesis. Supernumerary antipodals occur in Corynocarpaceae, and Asterad type of embryogeny in Salvadoraceae (Maheswari Devi 1972).

Salvadoraceae. Salvadoraceae show several features in conformity with those of the order Celastrales, and family Celastraceae (Table 32.15.1). They have anatropous bitegmic ovules, Polygonum type of embryo sac, and exarillate seeds as in Stackhousiaceae, with which it is placed. The seed is exotestal and exalbuminous, and lacks celastraceous fibrous exotegmen.

Buxaceae. Buxaceae are lone members of the suborder Buxineae. This family was included as a tribe of Euphorbiaceae by Bentham and Hooker (1965c). It has been treated as a distinct family in the order Euphorbiales by Wettstein (1935), Rendle (1959) and Cronquist (1981); Hutchinson (1973) placed Buxaceae in Hamamelidales, and Bessey (1915) and Melchior (1964) in Celastrales.

Buxaceae have three-loculed superior ovary with two pendulous collateral ovules with a dorsal raphe in each locule, and glossy black caruncular seeds as in Euphorbiaceae. Buxaceae differ in the absence of latex, anatropous ovule, fleshy endosperm, straight embryo, and loculicidal capsule, or berry-like fruit. In Euphorbiaceae latex is present, the ovule has a ventral raphe, and the fruit is schizocarpic. The development of endosperm is always of the Nuclear type, while in Buxaceae both Cellular and Nuclear types occur.

According to Dang-Van-Liem (1962), Buxaceae show affinities with Celastraceae. Bitegmic crassinucellar ovule with parietal tissue, Polygonum type of embryo sac, and Nuclear type of endosperm are common to both. Cellular endosperm occurs in Cyrillaceae and Aquifoliaceae in Celastrales. Asterad type of embryogeny is also present in Stackhousiaceae and Salvadoraceae. But Buxaceae do not resemble at all, either Euphorbiaceae or Celastraceae, in the anatomical structure of seed (see Mohana Rao 1974b).

Buxaceae resemble Hamamelidaceae in the type of pollen grain, ovule, and the development of embryo sac and endosperm. However, the embryogeny and seed structure differ.

Embryologically, Buxaceae are related more to the families of the suborder Celastrineae than to Hamamelidaceae and Euphorbiaceae.

Icacinaceae. The suborder Icacinae comprises Icacinaceae and Cardiopteridaceae. The embryology of the latter family has not been studied. In conformity with the order Celastrales, Icacinaceae have anatropous ovule, Polygonum type of embryo sac, and Nuclear endosperm. The ovules of *Gomphandra* and *Gonocarpus* are unitegmic, but in *Phytocrene* (Fagerlind 1945d) two primordia of integuments are present and the integuments are free at the micropyle. The development of endosperm is of the Nuclear type. The formation of protoplasmic connections between the chalazal nucellar cells, due to the activity of endosperm haustorium in *Nothapodytes*, is unique.

34 Rhamnales

34.1 General Features

The order Rhamnales comprises Rhamnaceae, Vitaceae and Leeaceae. These families show close resemblance in several major respects: secretory multinucleate tapetum, simultaneous cytokinesis, three-colporate pollen grains, presence of hypostase, massive parietal tissue, nucellar beak in Rhamnaceae and Vitaceae, Polygonum type of embryo sac, ephemeral antipodals, Nuclear type of endosperm, Asterad type of embryogeny, ruminate endosperm, and perichalazal seed in Vitaceae and Leeaceae.

34.2 Rhamnaceae

The anther wall comprises the epidermis, fibrous endothecium, two ephemeral middle layers, and secretory tapetum of two- to four-nucleate cells. Granular markings on the inner surface of tapetal cells are reported in *Zizyphus mauritiana* (Kajale 1944b). Only the fibrous endothecium remains intact in the mature anther. The microspore mother cells divide meiotically, and produce tetrahedral and isobilateral tetrads. Simultaneous cytokinesis occurs by centripetal furrowing.

The pollen grains are two-celled at the dispersal stage. They are three-colporate with reticulate exine.

The ovule is anatropous, bitegmic and crassinucellate. The micropyle is organized by the outer integument. The hypostase in *Gouania microcarpa* (see N.C. Nair 1970b) is massive, and projects as a cone into the nucellus. An epistase is present in *Zizyphus glabrata* (see N.C. Nair 1970b). The funicular vascular strand reaches almost up to the tip of the outer integument, except in *Berchemia polyphylla* var. *leioclada* (Kajale 1944b, see also N.C. Nair 1970b) where it terminates at the chalaza. The archesporium consists of one, two or more cells. In *Z. glabrata* there are 12–16 cells. These cells cut off parietal cells which produce a massive parietal tissue. The cells of the nucellar epidermis also divide, and add to the parietal tissue. In *Zizyphus rotundifolia* (Arora 1953), the nucellus forms a beak protruding through the micropyle. The megaspore mother cell undergoes meiotic divisions, and produces a linear or T-shaped tetrad. The chalazal megaspore develops into a Polygonum type of embryo sac in *Colletia cruciata*, *C. spinosa*, *Hovenia dulcis*, *Paliurus aculeatus* and *Rhamnus alaternus* (Dolcher 1947). Allium type of development is reported in *Zizyphus cenoplia* (see N.C. Nair 1970b), *Z. mauritiana* (Kajale 1944b), *Z. rotundifolia* (Arora 1953), and *Z. sativa* (Chiarugi 1930). The antipodals are generally ephemeral, but they persist in *Z. rotundifolia* and are coenocytic and four-nucleate in *Z. sativa* (Srinivasachar 1940b). Supernumerary antipodals are present in *Paliurus* (Dolcher 1947). The polars fuse at the time of fertilization. Multiple embryo sacs have been observed in *Colletia ferox*, *Gouania tillaefolia*, *Noltea africana*, *Ventilago bombaiensis* and *V. caluculata* (see N.C. Nair 1970b), *Hovenia dulcis* (Dolcher 1947), *Zizyphus mauritiana* (Kajale 1944b) and *Z. rotundifolia* (Arora 1953).

The remnants of pollen tubes persist in *Z. rotundifolia.*

The endosperm is of the Nuclear type. Wall formation commences from the micropylar region at an early stage of proembryo in *Zizyphus*, and the entire endosperm becomes cellular except in *Rhamnus wightii* where the nuclei in the chalazal region fuse into one or more irregularly shaped polyploid nuclei.

The embryogeny conforms to the Asterad type in *Ceanothus azureus* (Souèges 1939c) and *Rhamnus frangula* (Souèges 1941c). It is Asterad, rarely Solanad, type in *Zizyphus* (see N.C. Nair 1970b).

The embryogeny (Asterad type) of *Rhamnus* corresponds to Period I, Series B, Megarchetype II (Souèges 1941c).

Synergid polyembryony is reported in *Z. rotundifolia* (Arora 1953).

The seed is exotestal with a straight embryo. The testa is firm, smooth and multiplicative.The outer epidermis is a palisade of thick-walled cells with linea lucida separating the upper lignified part of the cell from the lower thin-walled region. The mesophyll is aerenchymatous with scattered sclerotic cells as in *Ventilago*, or is crushed. The tegmen also gets crushed, except for the inner epidermis. The rectangular cells of the inner epidermis have scalariform thickenings on radial walls.

N Arora (1953), A Chiarugi (1930), T Dolcher (1947), LB Kajale (1944b), NC Nair (1970b), EC Prichard (1955), R Souèges (1939c, 1941c), D Srinivasachar (1940b)

34.3 Vitaceae

The anther wall comprises the epidermis, fibrous endothecium, two ephemeral middle layers, and secretory tapetum. The tapetal cells are multinucleate, their nuclei fuse and become polyploid. The tapetum of binucleate cells produces a periplasmodium in *Cayratia carnosa* (Adatia et al. 1950). The microspore mother cells undergo meiotic divisions resulting in tetrahedral and, rarely, isobilateral and decussate tetrads. The cytokinesis is simultaneous.

The pollen grains are tricolporate with striate exine, and are dispersed at the three-celled stage, except in *Cayratia carnosa* and *Cyphostemma setosum* (N.C. Nair and Bajaj 1966) where they are dispersed at the two-celled stage.

The ovule is anatropous, bitegmic and crassinucellate, the inner integument forms the micropyle. In *Cissus trilobata* (N.C. Nair 1965), the much-elongated inner integument is thrown into folds. The hypostase is prominent, and well developed. In *Ampelocissus latifolia* and *Cayratia carnosa* (Kashyap 1958), the cells at the basal region of the ovary elongate and function as obturator. It develops by proliferation of the placental cells in *Cissus quadrangularis* (Goyal 1959, N.C. Nair and Mani 1960). The cells lining the ovarian cavity become richly cytoplasmic in *Cissus pallida* (N.C. Nair and Parasuraman 1962). The archesporial cell cuts off a parietal cell, the parietal tissue is well developed. The cells of the nucellar epidermis divide periclinally and produce a beak. The nucellus persists for a fairly long time. The megaspore mother cell undergoes meiotic divisions, and produces a linear or T-shaped tetrad. The chalazal megaspore develops into a Polygonum type of embryo sac. The synergids are hooked in *Ampelocissus latifolia* (Kashyap 1958) and *Cyphostemma stetosum* (N.C. Nair and Bajaj 1966), and larger than the egg. The antipodals are ephemeral. The polar nuclei fuse before fertilization. Supernumerary nuclei have been noted at the antipodal end in *Cayratia carnosa* (Adatia et al. 1950, Kashyap 1955); this may be due to the migration of nucellar cells or nuclei into the embryo sac (Kashyap 1955). Of special interest is the elongation of the tip of the embryo sac in *Cissus pallida* (N.C. Nair and Parasuraman 1954, 1962) and *C. vitiginea* (N.C. Nair unpublished), through the micropyle, until it occupies the narrow space between the ovule and ovary wall. Finally, it becomes closely appressed to the richly cytoplasmic inner epidermis of ovary wall and, probably, draws nourishment from it. In *Cissus quadrangularis* (Goyal 1959), it becomes extramicropylar and lodges itself in the obturator. Twin embryo sacs, derived from two adjacent megaspore mother cells, have been observed in *Vitis pallida* (N.C. Nair and Parasuraman 1954).

One synergid is destroyed during the passage of the pollen tube. Accessory pollen tubes are present in *C. pallida*.

Development of the endosperm is of the Nuclear type, wall formation is initiated at the chalazal end, and then at the micropylar end in *Vitis* (Kashyap 1958). The central region remains free-nuclear for a considerable period. Finally, it becomes completely cellular and stores fatty food reserve. Each cell contains a druse.

The endosperm becomes ruminate due to localized ingrowths of the integument. The formation of ruminate endosperm is of the Annona type (Periasamy 1962b). The seed becomes perichalazal after fertilization. In *Cayratia* the ingrowths are dome-shaped with two small longitudinal ridges on the inner side. In *Ampelocissus*, *Cissus*, *Tetrastigma* and *Vitis*, two ingrowths develop away from the chalaza. The ingrowths are plate-like and "hanging" in *Cissus*, ridge-like in *Ampelocissus*, narrow at the base and widening out at the free-end in *Tetrastigma*, and plate-like but not hanging in *Vitis*. In *Cyphostemma* (N.C. Nair and Bajaj 1966), the two plate-like ingrowths have lateral branches.

The embryogeny conforms to the Asterad type-Geum variation. There is natural parthenocarpy in *Vitis*. The fruit lacks the embryo and endosperm, but attains normal size due to the growth of the ovary wall.

The perichalazal (except *Cayratia*) seed is endospermic. The small embryo has flat thin cotyledons and a short radicle. The testa is multiplicative. The outer portion is thin-walled, and the mesophyll contains bundles of raphides. The inner epidermis divides periclinally, becomes two to six cells thick, and differentiates as woody endotesta (N.C. Nair 1970c). It comprises thick-walled, pitted, lignified cells with crystals. The tegmen has an outer tracheidal part, and the inner epidermis of oblong cells with thick brownish walls.

RD Adatia, BN Mulay and GR Hingorani (1950, 1953), F Gagnepain (1910), E Gilg (1896), S Goyal (1959), G Kashyap (1955, 1958), BN Mulay, NC Nair and MSRK Sastry (1953), NC Nair (1965, 1970c), NC Nair and YPS Bajaj (1966), NC Nair and KV Mani (1960), NC Nair and V Parasuraman (1954, 1962), NC Nair and PL Suri (1957), JE Planchon (1887), K Periasamy (1962a, b), K Süssenguth (1953)

34.4 Leeaceae

The anther wall comprises the epidermis, fibrous endothecium, two ephemeral middle layers, and glandular tapetum of multinucleate cells; the nuclei fuse and become polyploid. The endothecium alone persists. The microspore mother cells undergo meiotic divisions, cytokinesis is simultaneous, and tetrads are tetrahedral, isobilateral and decussate.

The pollen grains are three-colporate with smooth exine, and three-celled at the dispersal stage in *Leea sambusina* (N.C. Nair and Nambisan 1957).

The ovule is anatropous, bitegmic and crassinucellate. Both the integuments form the micropyle. The hypodermal archesporial cell cuts off a parietal cell, the parietal tissue is six to eight layers thick. The nucellar epidermis divides periclinally and produces two or three layers. A hypostase develops. The megaspore mother cell undergoes meiotic divisions, produces a linear tetrad, and the chalazal megaspore develops into a Polygonum type of embryo sac. The antipodals are ephemeral. The polar nuclei fuse before fertilization.

The development of the endosperm is of the Nuclear type. It is quiescent in the beginning but, later, becomes aggressive and destroys the nucellus. The cells become thick-walled, and store fatty substances. Each cell contains a druse. The seed shows extensive perichalazal growth. The perichalaza is pushed into the seed as a broad, forked structure. Four more ingrowths are produced due to the meristematic activity of the middle layer of the outer integument, two in the region of raphe and two laterally. They cause ruminations in the endosperm.

The development of the embryo follows the Asterad type. The embryo is dicotyledonous and small.

The seed is endotestal, and the testa is multiplicative. The outer epidermis comprises slightly thick-walled cells. The mesophyll contains tannin. The inner epidermis divides and produces four layers of cells which are lignified, and enclose a single simple crystal in each cell. The outer epidermis of tegmen is spirally thickened, while the inner contains tannin.

EJH Corner (1976), NC Nair (1968, 1970d), NC Nair and PN Nambisan (1957)

34.5 Taxonomic Considerations

Rhamnaceae. Bentham and Hooker (1965a) placed Rhamnaceae and Vitaceae (*Leea* is a genus in Vitaceae) along with Celastrineae and Stackhousiaceae in the cohort Celastrales. Wettstein (1935), Engler and Diels (1936), Rendle (1959), Lawrence (1951) and Melchior (1964) placed Rhamnaceae with Vitaceae in a distinct order, Rhamnales. Hutchinson (1973) included Heteropyxidaceae and Elaeagnaceae, along with Rhamnaceae and Vitaceae, in Rhamnales, and treated *Leea* as a genus in Vitaceae. Bessey (1915), Prichard (1955) and Ganapathy and Subramanyam (1962, see N.C. Nair 1970b) consider that Rhamnaceae are related to Vitaceae, and show many features common to Celastraceae.

Rhamnaceae are characterized by two-celled pollen grains with reticulate exine, a vascular strand traversing the outer integument up to its tip, epistase, uni- or multi-celled archesporium, and Allium type of embryo sac in the genus *Zizyphus*. The antipodal cells multiply and are coenocytic. Synergid polyembryony and, rarely, Solanad type of embryogeny occur in the family. The seed is exotestal, and has a palisade with linea lucida in the outer epidermis of the testa. None of these features are present in Vitaceae or Leeaceae.

The allied families Vitaceae (Sect. 34.3) and Leeaceae (Sect. 34.4) differ from Rhamnaceae (Sect. 34.2) in several respects. R. Dahlgren (1977a) points out: "Rhamnales should be divided. Vitaceae and Leeaceae being placed in a separate order." This is quite justifiable on embryological grounds.

How far are Rhamnaceae related to Stackhousiaceae and Celastraceae, as treated by Bentham and Hooker (1965a)? Stackhousiaceae differ from Rhamnaceae in a single middle layer in anther wall, tenuinucellate ovules on basal placenta, absence of nucellar beak, and single-celled female archesporium, the embryo sac consumes inner integument, and seed-coat is formed by the outer integument alone. Similarities between the two are extension of funicular vascular strand beyond the chalaza, Polygonum type of embryo sac, Nuclear type of endosperm, and Asterad type of embryogeny (Mauritzon 1936c, Narang 1953).

Rhamnaceae stand out from Celastraceae in crassinucellate ovules with a vascular strand passing through the integument, a prominent hypostase, Allium type of embryo sac in *Zizyphus*, fusion of polars at the time of fertilization, and in the structure of seed. Therefore, Rhamnaceae need not be placed in Celastrales along with Celastraceae and Stackhousiaceae.

Palynologically, Rhamnaceae and Elaeagnaceae are similar. Corner (1976) reports similarity between their seed-structure. Some embryological characters like the presence of a multi-nucleate secretory tapetum, tricolpate pollen grains, anatropous, bitegmic and crassinucellate ovules, and of a well-developed hypostase are features common to the two families (Rau and Sharma 1970b). Yet, the presence of a basal ovule, absence of integumentary vascular strand, parietal tissue in the ovule, polars fusing prior to fertilization, and chalazal embryo sac haustorium indicate that Elaeagnaceae are different from Rhamnaceae.

Vitaceae. Vitaceae, on the other hand, have two or three-celled pollen grains, and rarely amoeboid tapetum, the inner integument is thrown into folds, and extra-nucellar embryo sacs occur in *Cissus*. The obturator, proliferation of nucellar tissue which collapses in the seed, and parthenocarpy are common features in Vitaceae. The micropyle is organized by the inner integument in Vitaceae and Rhamnaceae. The endosperm is quiescent in the beginning in Vitaceae and Leeaceae, stores fatty substances with a druse in each cell, and is ruminate. The seed shows perichalazal growth, sarcotesta with raphides, woody endotesta, and tracheidal exotegmen.

Planchon (1887, see N.C. Nair 1970c) divided *Vitis* into a number of genera. *Cissus* was divided into three sections: *Eucissus*, *Cayratia* and *Cyphostemma*. Gagnepain (1910) removed *Cayratia* to a generic rank, and retained *Cyphostemma* under *Cissus*. In *Cissus* the embryo sac is haustorial and extends beyond the ovule by absorbing the nucellus (N.C. Nair and Parasuraman 1962). The seed structure differs from that of *Cyphostemma*. In *Cyphostemma setosum* (N.C. Nair and Bajaj 1966), eight to ten layers of nucellar cells persist at the micropyle. The ruminations take place by two median and two lateral folds. The inward projections extend throughout the major length of the dorsal side. In *Cissus* there are two median folds which are oblique, the lateral fold is absent. The perichalazal growth is along the median plane nearly up to the micropylar region on the dorsal side, whereas in *Cyphostemma* it extends throughout the major length on the dorsal side and reaches only up to the ventral side (N.C. Nair and Bajaj 1966). The seed-coat has brownish contents in *Cissus*, but not in *Cyphostemma*. In *Cayratia* the ingrowth in the seed is a single dome-shaped structure with two small longitudinal ridges on the inner side.

Planchon (1887) removed *Leea* from Vitaceae; Dumortier (see Süssenguth 1953) and Bartling (1830) gave *Leea* a family rank, and placed Leeaceae between Vitaceae and Meliaceae (see N.C. Nair 1970d).

N.C. Nair (1970c) points out that in the organization of perichalaza, basal growth of seed after fertilization, formation of ingrowths from free integumentary portion, enormous increase of nucellus after fertilization, late development of endosperm, and in the rumination of seed, Vitaceae resemble Annonaceae (Corner 1949, Periasamy 1962a, Periasamy and Swamy 1961).

Leeaceae. The three-celled pollen grains at the dispersal stage, a small nucellar cap, long persistence of nucellus, micropyle formed by both the integuments, an extensive perichalazal growth with four ingrowths in addition to two chalazal ingrowths, and absence of sarcotesta with raphides in seed distinguish Leeaceae from Vitaceae (N.C. Nair and Nambisan 1957). Concerning rumination of the endosperm, Periasamy (1962b) concludes that *Leea* differs from the other genera of Vitaceae by the large number of ingrowths and a pronounced activity of nutritive layers of the outer integument in the mature seed-coat. According to Periasamy, *Leea* need not be removed from Vitaceae and should be considered as a genus lowest in the phylogenetic series. N.C. Nair (1968), on the other hand, discusses several morphological and embryological similarities and dissimilarities, and places it close to Vitaceae under a monogeneric family Leeaceae.

35 Malvales

The Malvales comprise four suborders: Elaeocarpineae – Elaeocarpaceae; Sarcolaenineae – Sarcolaenaceae; Malvineae – Tiliaceae, Malvaceae, Bombacaceae and Sterculiaceae; and Scytopetalineae – Scytopetalaceae.

35.1 General Features

The order Malvales is characterized by: (1) secretory anther tapetum, amoeboid in Malvaceae, (2) simultaneous divisions in pollen mother cells, (3) two-celled pollen, (4) anatropous, bitegmic, crassinucellate ovule with well-developed parietal tissue, tenuinucellar in Scytopetalaceae, (5) nucellar cap, absent in Elaeocarpaceae and Scytopetalaceae, (6) Polygonum type of embryo sac with ephemeral antipodals, supernumerary in Malvaceae, (7) hypostase, (8) fusion of polar nuclei at the time of fertilization, (9) Nuclear type of endosperm, and cytokinesis at a very late stage of embryo development, (10) Asterad type of embryogeny, Onagrad in *Muntingia* (Elaeocarpaceae), (11) endospermic seed with a straight embryo, curved in Malvaceae, and (12) seed with outer epidermis of tegmen comprising lignified, longitudinal fibres.

The order is homogeneous as suggested by Venkata Rao and Sambasiva Rao (1952), and Venkata Rao (1954a, b).

35.2 Elaeocarpaceae

The anther wall comprises the epidermis, endothecium, two or three ephemeral middle layers, and secretory tapetum of binucleate cells. The epidermal cells have spinescent structures, and contain tannin in *Elaeocarpus* (Venkata Rao 1953b). In this taxon the endothecial cells develop into stone cells with thick walls and prominent pits (Venkata Rao 1953b). The endothecium has fibrous thickenings in *Muntingia* (Venkata Rao 1951a). The microspore mother cells divide meiotically, and produce tetrahedral tetrads. Cytokinesis is simultaneous.

Pollen grains are three-colporate and two-celled at the shedding stage.

In *Aristotelia, Elaeocarpus* and *Muntingia*, the ovule is anatropous, bitegmic and crassinucellate with a zigzag micropyle formed by both integuments. The archesporial cell cuts off a parietal cell, and six- to seven-layered parietal tissue develops in *Muntingia* (Venkata Rao 1952a). The megaspore mother cell undergoes meiotic divisions and produces a linear tetrad. The chalazal megaspore develops into Polygonum type of embryo sac. In *Aristotelia manqui* and *A. raceskosa* (Mauritzon 1934c), usually the subchalazal megaspore functions. The polars fuse before fertilization. The antipodals are ephemeral. An endothelium and hypostase develop in *Elaeocarpus*. A socket of thick-walled cells is present around the chalazal end of the embryo sac in *Muntingia*.

The development of the endosperm is of the Nuclear type. The nuclei accumulate in the chalazal region. The endosperm becomes cellular at the quadrant stage of proembryo, and the mature seed is endospermic.

The embryogeny conforms to the Onagrad type. The mature embryo is straight with thick cotyledons.

The seed is albuminous, and the testa and tegmen are multiplicative. The exotesta is mostly thick-walled. The outer epidermis is hairy, and the inner one contains crystals. The outer epidermis of the tegmen comprises longitudinal lignified fibres. The mesophyll is aerenchymatous (densely sclerotic in *Aristotelia* – Corner 1976), and the inner epidermis has thick brownish walls.

EJH Corner (1976), J Mauritzon (1934c), C Venkata Rao (1951a, 1952a, 1953b)

35.3 Sarcolaenaceae

The embryology has not been studied.

The anthers are bilocular with longitudinal dehiscence. The ovule is anatropous, erect to more often pendulous (Hutchinson 1973, Cronquist 1981). The seed is hairy in *Xerochlamys* (Hutchinson 1973). It contains a fleshy or horny endosperm, and the embryo is straight with flat and plaited cotyledons.

A Cronquist (1981), J Hutchinson (1973)

35.4 Tiliaceae

The anther wall comprises the epidermis, fibrous endothecium, ephemeral middle layer, and secretory tapetum of two- to four-nucleate cells. The epidermal cells contain rosette crystals in *Corchorus* and *Triumfetta* (Venkata Rao and Sambasiva Rao 1952). The microspore mother cells undergo meiotic divisions, and tetrahedral and isobilateral tetrads are produced. Simultaneous cytokinesis occurs by furrowing (Banerji 1933).

The pollen grains are two-celled at the dispersal stage. The exine is reticulate.

The ovule is anatropous, bitegmic and crassinucellate. The micropyle is organized by the outer integument, the growth of the inner integument is arrested. The nucellar epidermal cells divide periclinally and produce three layers. The archesporium is multicellular in *Corchorus capsularis* (Banerji 1932), only one cell functions. In *Triumfetta rhomboidea* four or five cells are functional. The hypodermal archesporial cell cuts off a parietal cell and the parietal tissue is five layers thick. The megaspore mother cell undergoes meiotic divisions, tetrads are linear or T-shaped, and chalazal megaspore develops into a Polygonum type of embryo sac. The synergids occupy half the length of the embryo sac in *Corchorus* (Banerji 1932). The polar nuclei do not fuse till fertilization. The antipodals are ephemeral.

In *T. rhomboidea* the obturator is hairy.

The development of endosperm is of the Nuclear type. Wall formation commences from the micropylar end at the initiation of cotyledonary primordia in the embryo, and the endosperm becomes cellular throughout.

The development of embryo is of the Asterad type-Urtica variation. Polyembryony is reported in *C. olitorius*.

Ganesan et al. (1957) state that in the cross *Corchorus olitorius* × *C. capsularis*, a hybrid embryo and Nuclear endosperm develop after normal fertilization. However, the embryo aborts at an early globular stage. They suggested that probably somatoplastic sterility is involved. Iyer (1962) has demonstrated that there is no somatoplastic sterility.

The development of endosperm and embryo has been studied in *Corchorus olitorius*, *C. capsularis*, *C. olitorius* × *C. capsularis*, and *C. capsularis* × *C. olitorius* (Iyer et al. 1961, Iyer 1962). The developmental stages in selfed *C. capsularis* are essentially similar to those of *C. olitorius*. In the latter the endosperm remains free-nuclear up to 11 days after pollination, when the proembryo is globular. Thirteen days after pollination the embryo becomes heart-shaped and the upper half of the embryo sac contains cellular endosperm; fifteen days later the entire endosperm becomes cellular and the embryo has well-differentiated cotyledons. Twentyone days after pollination the cotyledons enlarge, and 30 days after pollination the cotyledons expand and become leafy. Much of the endosperm is now consumed.

In the selfed *C. capsularis*, the embryo matures earlier by almost a week or 10 days, and is somewhat bigger.

In the cross *C. olitorius* × *C. capsularis*, fertilization and early embryogeny are comparable to the reciprocal cross, and the two parents. In *C. olitorius* × *C. capsularis*, the endosperm remains free-nuclear from seven to 30 days after cross-pollination, and the proembryo does not grow beyond the eight- to ten-celled stage. Vasculation in the proembryo occurs 11 days after cross-pollination, and the endosperm shrinks and collapses on the fifteenth day. The seeds are shrivelled with aborted embryo and remnants of endosperm.

In the reciprocal cross *C. capsularis* × *C. olitorius*, the flowers abscised and dried up. The ovules showed that the pollen tubes reached the micropyle but fertilization did not take place.

In *C. capsularis* grafted on *C. olitorius* root-stock, there was an increase in the percentage of fruit-set and marked development of hybrid embryo and endosperm. The globular proembryos were much larger than in straight crosses, but they failed to differentiate and aborted. The endosperm also remained free-nuclear until degeneration after 3 weeks.

When *C. olitorius* was pollinated with *C. capsularis* grafted on the former, occasionally late stages of heart-shaped embryos with cellular endosperm were observed.

The seed is endospermic with a straight embryo. The testa is membraneous, the inner epidermis contains crystals. The outer epidermis of the tegmen comprises a palisade of thick-walled lignified cells. The mesophyll is aerenchymatous and gets crushed. The inner epidermal cells are cuboidal with brownish walls.

I Banerji (1932, 1933), MR Chaudhury (1961), RM Datta (1954, 1955), AP Ganesan, SS Shah and MS Swaminathan (1957), RD Iyer (1962), RD Iyer, K Sulbha and MS Swaminathan (1961), J Mauritzon (1934d), C Venkata Rao and KV Sambasiva Rao (1952), R Souèges (1941b)

35.5 Malvaceae

Venkata Rao (1954a, b, 1955a) has reviewed the embryology of Malvaceae, and P. Maheshwari and Rangaswamy (1960) of *Gossypium*.

The anther is bisporangiate, and its wall comprises the epidermis, fibrous endothecium, an ephemeral middle layer, and amoeboid tapetum which forms the periplasmodium. The microspore mother cells divide meiotically, and produce isobilateral and tetrahedral tetrads. Simultaneous cytokinesis occurs by furrowing.

The mature pollen grains are two-celled at the dispersal stage, and packed with starch. They are three-colporate with reticulate or spinous markings. They contain fatty food reserve and, in later stages, starch, as in *Althaea* and *Malva* (Luxemburg 1927). Occasionally, some dyads persist up to the mature stage in *Malvastrum tricuspidatum*, and tetrads in *Hibiscus labatus* (Gaur 1979). This needs confirmation.

The ovule is campylotropous, bitegmic and crassinucellate with a zigzag micropyle formed by both the integuments. A hairy obturator of stylar origin develops in *Abutilon indicum*, and from the placenta in *Hibiscus micranthus* (Venkata Rao 1955a). A hypostase also differentiates. Numerous stomata occur on the surface of the ovary. The archesporium is uni- to multicellular in *Gossypium* (P.C. Joshi et al. 1967). The archesporial cell cuts off a parietal cell; the parietal tissue is massive, and a nucellar cap develops. The nucellus persists as perisperm. Accessory sporogenous cells differentiate from the parietal tissue, only one cell functions. The megaspore mother cell divides meiotically, and produces a linear or T-shaped tetrad. The chalazal megaspore develops into a Polygonum type of embryo sac. The antipodals undergo supernumerary divisions forming as many as 19–24 cells in *G. herbaceum* (P.C. Joshi et al. 1967) and these persist up to fertilization.

Double fertilization occurs, and is porogamous. Venkata Rao (1952a) points out that the pollen tube is persistent, and serves as a haustorial organ in *Hibiscus hirsutus*, *H. micranthus*, *H. solandra*, *Malachra* (= *Malache*) *capitata*, *Pavonia zeylanica* and *Sida veronicaefolia* (*S. humilis*). The haustorial function of the pollen tube has been questioned (Sect. 3.9), and it is only the remnants of the discharged tube that persist during post-fertilization stages, sometimes even in a mature seed (see P. Maheshwari and Johri 1950b).

The pollen tube branches in several species (Venkata Rao 1952a). Polysiphonous germination of pollen has been observed in *Hibiscus esculentus* (Purewal and Randhawa 1947) and *Malva neglecta* (Stenar 1925b), as many as 14 pollen tubes have been noted in the latter species. In *Abelmoschus esculentus* (S. Chandra and Bhatnagar 1975a, b), syngamy and triple fusion take place simultaneously.

The time-relations concerning pollination, growth of pollen tube, and fertilization have been studied in *Gossypium arboreum* and *G. hirsutum*, and their reciprocal crosses (Pundir 1972). The stigma is papillate, and the pollen tube grows through the single transmitting strand, along the placenta, and enters the zigzag micropyle. Generally, only a single pollen tube emerges from a germinating pollen grain, but polysiphonous conditions often occur in *G. arboreum*. The bulging of tubes is a common feature in *G. hirsutum*, and coiling of tube has been observed in a cross *G. hirsutum* × *G. arboreum*. The interval between pollination and fertilization is 24–48 h in *G. arboreum*, *G. hirsutum* and *G. arboreum* × *G. hirsutum*, but 72–120 h in the cross *G. hirsutum* × *G. arboreum*.

The endosperm is of the Nuclear type and several hundred nuclei are formed before wall formation is initiated at the late globular stage of proembryo in *Hibiscus micranthus*, and at the dicotyledonous stage in *Pavonia zeylanica* (Venkata Rao 1955a). A few layers of endosperm persist in *Abutilon indicum*, *Pavonia zeylanica*, *Sida cordifolia* and other taxa.

In *Gossypium arboreum*, *G. hirsutum* and *G. arboreum* × *G. hirsutum*, the primary endosperm nucleus divides 24 h after pollination but after 72–120 h in *G. hirsutum* × *G. arboreum* (Pundir 1972). In the parent species about 2000 free nuclei are produced on the seventh or eighth day after pollination, whereas in the hybrid the endosperm is 100- to 200-nucleate on the fifth day. A chalazal diverticulum develops in *G. arboreum*. The nuclei become polyploid due to endomitosis and nuclear fusions, and centripetal wall formation is initiated at the micropylar pole on the seventh or eighth day (after pollination) in the parent species. In the hybrid the endosperm nuclei become abnormally large and dumb-bell-shaped and form clumps. The cytoplasm becomes vacuolate, and the endosperm degenerates 10 to 14 days after pollination; wall formation never occurs in the hybrid endosperm. In *Gossypium* (P.C. Joshi et al. 1967), only five or six layers of endosperm persist in the seed.

In *G. arboreum* a young dicotyledonous embryo develops 15 or 16 days after pollination, and 17 or 18 days in *G. hirsutum*. The embryo aborts 12–14 days after pollination in the cross *G. arboreum* × *G. hirsutum*, and after 7 days in the reciprocal cross. The hybrid embryo remains small and shows many abnormalities (Pundir 1972).

S. Chandra and Bhatnagar (1975a, b) have made a detailed time-relation study of post-fertilization development of the ovule in *Abelmoschus esculentus*. The primary endosperm nucleus divides 10–12 h after pollination, while the zygote divides after 16–18 h. Free-nuclear divisions continue in the endosperm until about 2000 nuclei are formed, 5 days after pollination. Synchronous divisions occur only up to the 64- or 128-nucleate stage. At this time the proembryo is globular.

Five days after pollination, centripetal wall formation commences from the micropylar region, proceeds to chalaza, and continues up to 8 days after pollination. Concurrently, cotyledons are initiated, and 7 days after pollination the embryo becomes dicotyledonous, and the root cap also differentiates. The suspensor comprises seven to ten cells.

Eight to 11 days after pollination, the embryo is fully differentiated. The cotyledons become leafy. During the growth of the embryo, most of the starch-rich endosperm is consumed. Finally, in a mature seed, the embryo occupies most of the space, the cotyledons curl, and remnants of endosperm persist in the funicular region and a layer or two between the folds of the cotyledons.

The embryogeny conforms to the Asterad type. Mortier (1969) assigned the development of embryo in *Althaea officinalis* to Period I, Megarchetype II (Asterad type). The mature embryo is dicotyledonous, and the cotyledons become foliar with an inwardly curved margin in *Gossypium*. During elongation, they curl around each other and around the hypocotyl (Fig. 35.5.2A).

Polyembryony has been observed in *G. barbadense* (P.C. Joshi et al. 1967, B.M. Johri and Pundir 1976); haploid-diploid twins are also reported in this taxon (Harland 1936).

The outer epidermis of testa comprises cubical cells with thick, lignified outer walls, and contains tannin (Fig. 35.5.2B). In *Gossypium* (P. C. Joshi et al. 1967), some of the cells develop into hairs (fibres) except in the region of the micropyle. The development of hairs is in three successive crops: at anthesis, 4 to 6 days after pollination, and 8 to 12 days after pollination (Fig. 35.5.1A–I). At maturity the fibres are distinguishable into lint and fuzz. The lint fibres are longer with thinner walls, and show convolutions on the surface. The convolutions are absent on fuzz fibres. The mesophyll contains tannin cells, and is followed by a starch-containing inner epidermal layer. The tegmen is multipli-

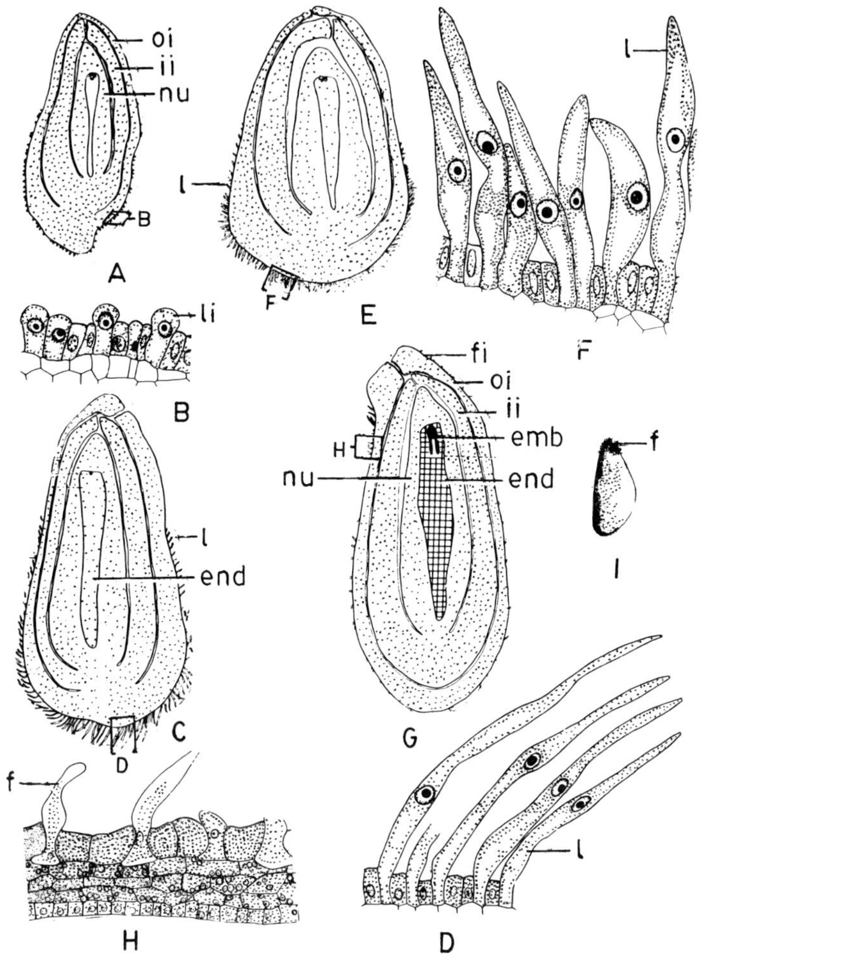

Fig. 35.5.1A–I. *Gossypium barbadense* var. Andrews, development of fibres; *emb* embryo; *end* endosperm; *f* fuzz hairs; *fi* fuzz initial; *ii* inner integument; *l* lint hair; *li* lint initial; *nu* nucellus; *oi* outer integument. **A**, **C**, **E**, **G** Outline diagrams for **B**, **D**, **F**, **H**, respectively. The ovules represent stages at anthesis (**A**), and two (**C**), five (**E**) and 10–12 (**G**) days after pollination. **B**, **D**, **F** *Inset* from **A**, **C** and **E**, development of lint hairs. **H** *Inset* from **G**, differentiation of fuzz. **I** Mature seed after the removal of lint; fuzz is mostly limited to the micropylar end. (After P.C. Joshi et al. 1967)

cative (Fig. 35.5.2B). The outer epidermis is a palisade layer of lignified pitted cells (Fig. 35.5.2B–E). In surface view the cells show radiating thickenings in the lumen. The mesophyll shows an outer tanniniferous, starchy zone of four to five layers, and an inner colourless starchy zone of nine or ten layers in *Gossypium*. The inner epidermis forms the fringe layer (Fig. 35.5.2F). Its cells do not contain nuclei and cytoplasm, and the walls develop thickenings.

In a cross *G. arboreum* × *G. hirsutum*, and in the reciprocal cross, the development of seed-coat and fibers is arrested 7 to 10 days after pollination (Pundir 1972).

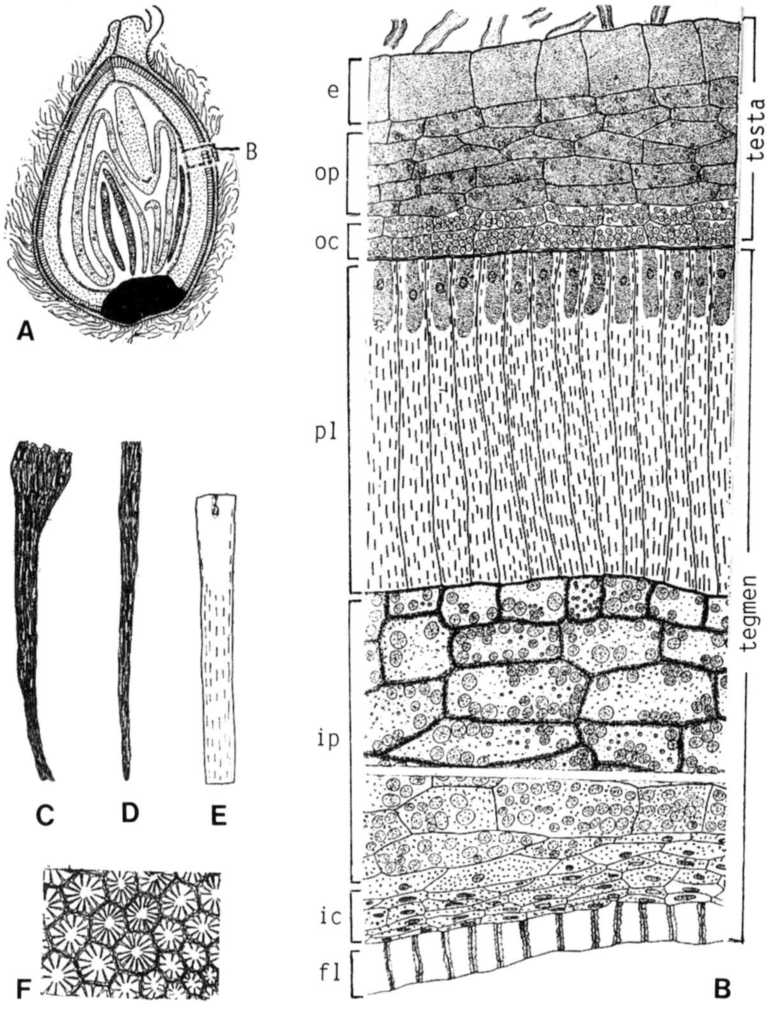

Fig. 35.5.2A–F. A, B, E, F *Gossypium herbaceum* var. *Jaydhar*, **C, D** *G. arboreum* var. *Bhoj*. **A** Longisection of mature seed studded with hairs, 40 days after pollination. Note well-developed hypostase, remnants of endosperm, and embryo with highly coiled cotyledons. **B** *Inset* **B** from **A**, to show pigmented hairy epidermis (*e*) of testa, outer pigmented zone (*op*), and outer colourless zone (*oc*). The palisade layer (*pl*) comprises exotegmen followed by inner pigmented zone (*ip*), inner colourless zone (*ic*); the inner epidermis of tegmen forms a fringe layer (*fl*) with conspicuous thickenings. The outer and inner colourless zone contain compound starch grains. (After Ramachandani et al. 1966). **C–E** Palisade cells of exotegmen (from macerated wholemount). **F** cells of fringe layer (transection) with radiating thickenings on inner wall. (After P.C. Joshi et al. 1967)

BS Ahuja (1964), T Ashley (1975a, b, c), R Balasubramanium (1931), S Chandra and SP Bhatnagar (1975a, b, 1976), HK Chaudhari (1978), PB Christensen (1986), PC Datta and A Chowdhury (1981), EJ Duncan (1970), HL Edlin (1935), GH Egley and RN Paul (1981), PA Fryxell (1968, 1969), RD Gaur (1979), UR Gore (1932), SC Harland (1936), WA Jensen and DB Fischer (1968), WA Jensen, ME Ashton and CA Beasley (1983), WA Jensen, P Schulz and ME Ashton (1977), BM Johri and NS Pundir (1976), PC Joshi, AM Wadhwani and BM Johri (1967), P Kumar and D Singh (1990), U Lantis (1912), A Luxemburg (1927), P Maheshwari and BM Johri (1950b), P Maheshwari and NS Rangaswamy (1960), J Mortier (1969), NS Pundir (1972), SS Purewal and GS Randhawa (1947), S Ramchandani, PC Joshi and NS Pundir (1966), B Singh (1967), RP Singh and MPS Chauhan (1984), R Souèges (1922b), H Stenar (1925b), M Takahashi and J Kouchi (1988), WA van Heel (1966), C Venkata Rao (1952a, 1954a, b, 1955a), DM Winter (1960), CJ Webb (1984)

35.6 Bombacaceae

The anther wall comprises the epidermis, fibrous endothecium, two or three ephemeral middle layers, and secretory tapetum of multinucleate cells, Nuclear fusions occur in the tapetal cells. Übisch granules are studded on the inner walls of tapetum in *Ochroma lagopus* (Venkata Rao 1953a). The microspore mother cells increase secondarily. They undergo meiotic divisions, simultaneous cytokinesis occurs by furrowing, and tetrahedral tetrads are produced.

Pollen grains are two-celled at the shedding stage. They are triporate, and smooth-walled.

The ovule is anatropous, bitegmic and crassinucellar with a zigzag micropyle formed by both the integuments. The hypodermal archesporial cell cuts off a parietal cell. The parietal tissue is massive, and a nucellar cap is formed. A hypostase also develops. Following meiotic divisions, the megaspore mother cell produces a T-shaped tetrad. The chalazal megaspore develops into a Polygonum type of embryo sac (Banerji 1942a), and it contains starch grains. The antipodals are ephemeral. The embryo sac becomes tubular at the chalazal end; a postament develops in *Ochroma* (Venkata Rao 1953a). In *Eriodendron anfractuosum* (Thirumalachar and Razi 1941), sometimes the mature embryo sac shows reversed polarity.

Fertilization is porogamous. The pollen tube branches and persists up to the globular proembryo. Perhaps, it may serve as a means of transport of nutrients from the integuments to the embryo (Venkata Rao 1954a, b). There is no evidence that a discharged pollen tube is concerned in short transport of metabolites. In *Eriodendron* (Thirumalachar and Razi 1941), the chalazal end of the embryo sac becomes haustorial, and penetrates into the underlying tissue.

Development of the endosperm is of the Nuclear type. It becomes cellular at the globular stage of pro-embryo.

The embryogeny conforms to the Asterad type in *Ochroma*, and Caryophyllad in *Bombacopsis glabra* (Duncan 1970). The mature embryo is straight with thin and flat, or thick and folded, cotyledons. Adventive polyembryony is reported in *Bombacopsis* (Duncan 1970).

The seeds are large with a massive seed-coat. Both testa and tegmen are multiplicative. The outer epidermis of testa comprises tangentially flattened

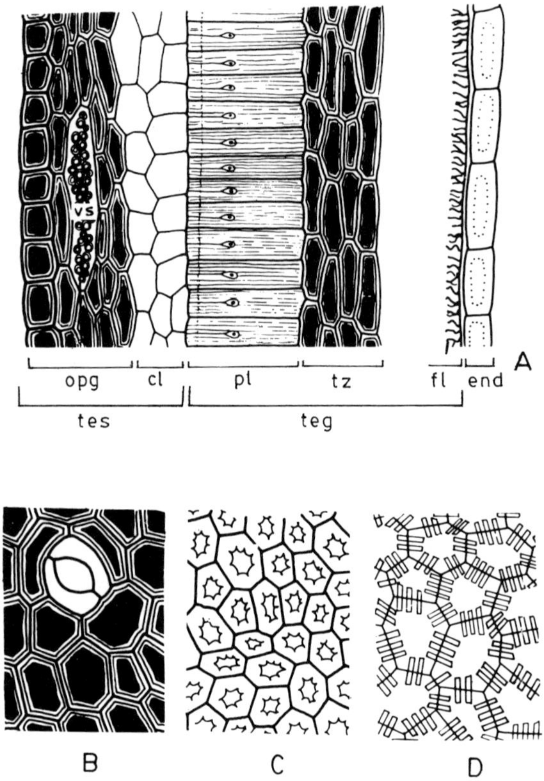

Fig. 35.6.1A–D. *Hampea nutricia.* **A** Part of seed-coat, cross-section; *cl* colourless zone; *end* endosperm; *fl* fringe layer; *opg* outer pigmented layer; *pl* palisade layer; *tes* testa; *teg* tegmen; *tz* tanniniferous zone. **B** Outer epidermis of testa with stomata, surface view. **C** Macrosclereids, cross-section. **D** Fringe layer, surface view. (After R.P. Singh and Chauhan 1984)

cells with tannin, followed by aerenchymatous mesophyll. The inner epidermis contains crystals. Hairy outgrowths develop from the outer epidermis in many genera. The tegmen consists of a palisade of radially elongate, lignified cells. The mesophyll is aerenchymatous with stellate cells. The inner epidermis is a fringe layer with irregular thickenings on the anticlinal walls, and may contain tannin (Fig. 35.6.1A–D). The chalaza is massive. The aril, when present, is either funicular or micropylar.

I Banerji (1942a), EJ Duncan (1970), BM Johri (1963b), MJ Thirumalachar and BA Razi (1941), C Venkata Rao (1953a, 1954a, b)

35.7 Sterculiaceae

The major contributions to the embryology of Sterculiaceae are by Venkata Rao (1948, 1949, 1950, 1951b, 1952b, 1953c).

The anther wall comprises the persistent epidermis, fibrous endothecium, two or three ephemeral middle layers, and secretory tapetum with multinucleate cells. The tapetum is biseriate in *Dombeya* and *Pterospermum* (Venkata Rao 1949). False periplasmodium has been noted in *Guazuma tomentosa*, *Melochia corcorifolia* (Venkata Rao 1951b) and *Waltheria indica* (Venkata Rao 1949). This needs reinvestigation. The sporogenous tissue increases secondarily. The microspore mother cells undergo meiotic divisions and produce tetrahedral tetrads. Cytokinesis is simultaneous.

Pollen grains are shed at the two-celled stage.

The ovule is anatropous, bitegmic and crassinucellate with a zigzag micropyle in *Dombeya klienhovia*, *Pentapetes* and *Sterculia* (Venkata Rao 1953c), *Melochia* (Venkata Rao 1951b), *Pterospermum* (Venkata Rao 1952b) and *Waltheria* (Venkata Rao 1950). The micropyle is produced by the outer integument alone in *Abroma*, *Buettneria* and *Guazuma* (Venkata Rao 1953c). In *Helicteres* (Venkata Rao 1953c), the growth of integuments is arrested, and the nucellus protrudes and forms a cap. The occurrence of integumentary vascular bundles is a common feature. In *Sterculia* (Venkata Rao 1953c), six to nine strands extend up to the micropylar end. The vascular supply has numerous anastomosing branches in *Pterospermum*. A hypostase develops in *Helicteres* and *Pterospermum*, and postament in *Abroma*. An obturator develops from the funicule in *Abroma angusta* and *Sterculia colarata* (Venkata Rao 1953c).

The archesporium is multicellular. As many as 100 archesporial cells have been observed in *Pterospermum suberifolium* (Venkata Rao 1952b), and ten or more embryo sacs may develop concurrently, but only two to three mature. Usually, a single archesporial cell cuts off a parietal cell which produces a parietal tissue. The cells of the nucellar epidermis also divide to form a nucellar cap. The megaspore mother cell divides meiotically and produces a linear tetrad. The chalazal megaspore develops into a Polygonum type of embryo sac. In *Pterospermum* (Venkata Rao 1949, 1952b), a fairly large number of archesporial cells develop directly into megaspore mother cells. Several tetrads are produced. Double and triple eight-nucleate embryo sacs are common. The antipodals are ephemeral. Reversed polarity in the embryo sac has been observed in *Helicteres isora*.

In *Buettneria herbacea* the obturator is hairy. The radially elongate cells of the funicule become richly cytoplasmic in *Abroma*, *Helicteres* and *Sterculia*.

Double fertilization occurs, and the entry of pollen tube is porogamous.

The endosperm is of the Nuclear type. In *Abroma* and *Pentapetes*, the nuclei in the chalazal region become large and polyploid. Later, the endosperm becomes cellular at the initiation of cotyledonary primordia. The endosperm stores starch and fatty substances.

The embryogeny conforms to the Asterad type. The mature embryo is straight with flat cotyledons. Cleavage polyembryony has been recorded in *Theobroma cacao* (Martinson 1972a). The cleavage of the zygote results in the formation of unequal diploid seedlings. Haploid seedlings are produced after spontaneous development of the egg cell following triple fusion. Martinson (1972b) observed two mature embryo sacs within the same nucellus. A case of two embryo sacs in separate nucelli, but enclosed within common integuments, has also been reported. Twin and multiple embryo sacs have been observed by Venkata Rao (1949, 1952b).

The testa and tegmen are multiplicative. The outer epidermis is of short, palisade cells, or of longitudinally elongate cells with stomata and peltate scales. The mesophyll is pulpy with mucilage sacs, or with lignified stellate cells. The inner epidermis is thin-walled. In winged seeds the wing is formed by the exotesta and mesophyll. The outer epidermis of tegmen comprises longitudinal or short, lignified malpighian cells. The mesophyll is similar to that of the testa, and gets more or less crushed. The inner epidermis may comprise thin-walled or thick-walled cells with bar-like thickenings.

EE Cheesman (1927), EJ Duncan and AW Todd (1972), VA Martinson (1972a, b), PR Mohana Rao (1976), YML Sharma (1938), C Venkata Rao (1948, 1949, 1950, 1951b, 1952b, 1953c)

35.8 Scytopetalaceae

Vijayaraghavan and Dhar (1976) have contributed to our knowledge of embryology of *Scytopetalum tieghemii*, and have studied the male and female gametophytes, seed-coat and pericarp. We have no information on the development of endosperm and embryo.

The anther wall comprises the epidermis, fibrous endothecium, one or two ephemeral middle layers, and secretory tapetum of two- or three-nucleate cells in *Scytopetalum tieghemii* (Vijayaraghavan and Dhar 1976). Druses occur in the connective. The microspore mother cells undergo meiotic divisions, cytokinesis is simultaneous, and the tetrads are tetrahedral. A thick callose wall surrounds the meiocytes, and persists up to the tetrad stage.

The triporate pollen grains are two-celled at the dispersal stage.

The ovule is anatropous, bitegmic and tenuinucellate. The inner integument is longer than the outer one, and forms the micropyle. An endothelium differentiates at the megaspore mother cell stage. It is irregularly biseriate with uninucleate cells.

The archesporial cell functions as megaspore mother cell. The latter divides meiotically and produces a linear tetrad. The chalazal megaspore develops into a Polygonum type of embryo sac. The antipodals are ephemeral.

The seeds are albuminous and exarillate. The endosperm is sometimes ruminate as in *Rhaptopetalum* and *Scytopetalum*. The seed-coat is about ten cells thick, the outer epidermis comprises large cells and the outer wall is almost hair-like in *Oubanguia*. Hairs develop along the raphe in *Brazzeia* and *Rhaptopetalum*, and the seed is completely hairy in *Pierrina*. There is post-chalazal extension of the vascular bundle almost upto the micropyle with branches to the seed-coat entering the ruminations in *Rhaptopetalum* (Corner 1976)

EJH Corner (1976), MR Vijayaraghavan and U Dhar (1976)

35.9 Taxonomic Considerations

Bentham and Hooker (1965a) include Malvaceae, Tiliaceae and Sterculiaceae in Malvales. Engler and Prantl (1931) group all the families – Elaeocarpaceae,

Tiliaceae, Malvaceae, Bombacaceae, Sterculiaceae and Scytopetalaceae – in Malvales. Hutchinson (1973) places Malvaceae in Malvales, and the rest of the families in Tiliales.

Elaeocarpaceae. The suborder Elaeocarpineae comprises a single family, Elaeocarpaceae. Cronquist (1968, 1981) treats Elaeocarpaceae as a primitive family in Malvales, with probably Thealean ancestry; R. Dahlgren (1977a, 1981) considers its position in Malvales as doubtful. This family shows close alliance with Tiliaceae and, to some extent, with Scytopetalaceae. It resembles Tiliaceae in the development of anther, pollen, embryo sac, and endosperm. Even the seed structure is more or less similar in both. Elaeocarpaceae and Scytopetalaceae develop an endothelium; absent in other families of Malvales. However, a nucellar cap, and mucilage sacs and canals in the ovary wall and ovule – common features in Malvales – are absent in both families. Also, the Onagrad type of embryogeny of Elaeocarpaceae does not occur in any other family of Malvales.

The suborder Malvineae comprises Tiliaceae, Malvaceae, Bombacaceae and Sterculiaceae.

Tiliaceae. The deep-seated archesporial cell develops directly as megaspore mother cell in *Corchorus acutangulus* (Venkata Rao and Sambasiva Rao 1952). In such cases, the ovules in the Tiliaceae may be considered as tenuinucellar, as in Scytopetalaceae. Even though a parietal cell is not cut off, there is conspicuous "parietal" tissue of nucellar origin. Therefore, it is incorrect to call it tenuinucellar; at best this condition should be considered "pseudo-crassinucellar". The synergids are exceptionally long in *Corchorus*. The pollen tube is unbranched. Cytokinesis in the Nuclear endosperm is very late as in Sterculiaceae. It occurs at the time of initiation of cotyledonary primordia in the embryo. The outer epidermis of the testa is membraneous, while the inner one forms a characteristic crystal layer. The seed is exotegmic, the inner epidermis has brown walls. Tiliaceae are closer to Elaeocarpaceae and Sterculiaceae than to Malvaceae.

Malvaceae. The Malvaceae show several features distinct from the rest of the members of Malvales. One-celled anthers, amoeboid anther tapetum, multiporate spinescent pollen grains, campylotropous ovule, multiplication of antipodals in some members, branching pollen tube, curved mature embryo with folded cotyledons, comose, albuminous, perispermous seeds, and outer epidermis of the testa of columnar cells with lignified outer walls.

Bombacaceae. Bombacaceae are treated as a tribe of Malvaceae. They differ from Malvaceae in: secondary increase of microspore mother cells, secretory anther tapetum, smooth-walled triporate pollen grains, ephemeral antipodals, presence of starch in embryo sac, massive chalaza, and arillate seeds with scanty endosperm. The embryo is straight. This family should not be treated as a tribe of Malvaceae (Venkata Rao 1954a), and erection of a separate family Bombacaceae is justified.

Hampea. Engler and Prantl (1894), Edlin (1935) and Hutchinson (1973) place the genus *Hampea* in the family Bombacaceae. However, on the basis of the structure of embryo, Fryxell (1968, 1969) places it in the tribe Gossypieae, family Malvaceae.

The pollen grains are multiporate and echinate as in Gossypieae. In Bombacaceae they are triporate and smooth (Erdtman 1952).

The single-layered endosperm fills the cotyledonary folds. The cells are thin-walled and contain food reserve.

The embryo of *Hampea nutricia* (R.P. Singh and Chauhan 1984) is nearly straight and gland-dotted. It has conduplicate, punctate, asymmetrical cotyledons and gossypol ducts.

The seed is exotegmic. The testa (eight to ten layers) is differentiated into three zones. Stomata occur in the outer epidermis, the cells show slightly thickened outer walls, and have dark-brown contents. The mesophyll (three to four layers) is pigmented and compact. The inner zone (two to three layers) is colourless.

The outer epidermis of the tegmen forms a sclerified, fibrous palisade layer which has a distinct light-line. In a cross-section these cells show a star-shaped lumen. Inner to it lie four thick-walled tanniniferous layers. The inner epidermis of the tegmen, in contact with the endosperm, is a fringe layer, as in Malvaceae. In surface view the cells of fringe tissue are polygonal and thickened along anticlinal walls with simple pits.

The seed is arillate as in *Gossypium cunninghamii* and *G. populifolium* (Fryxell 1969).

The structure of the embryo, a testa differentiated into three zones, the exotegmen as palisade layer, and the inner epidermis of tegmen as fringe layer, clearly suggest the position of *Hampea* in/near Malvaceae. Palynological data support these findings (Fryxell 1969, R.P. Singh and Chauhan 1984).

Sterculiaceae. The occurrence of integumentary vascular bundles is a common feature in Sterculiaceae. Nearly six to nine strands extend up to the micropylar tip (*Sterculia*), or anastomosing branches (*Pterospermum*) may develop. Ten or more embryo sacs develop concurrently from about 100 archesporial cells in *Pterospermum*, but only two or three mature. The winged seed contains copious endosperm, but perisperm is absent. The seed is exotegmic; the inner epidermis of the testa does not contain crystals. They are present in all the other families.

Scytopetalaceae. Edlin (1935) considers Scytopetalaceae as the most primitive family, due to the disposition of their stamens; many stamens are arranged in several rows. He compares this feature with multiseriate stamens of Magnoliaceae. The affinities of the family with Elaeocarpaceae (see earlier under Sect. 35.9) have already been dealt with. In addition, the development of anther, structure and type of pollen, and development of the embryo sac are in conformity with the other families of the order. But Scytopetalaceae show tenuinucellar ovule with the inner integument longer than the outer, and the inner integument forms the micropyle. Again, a hypostase-like strand is prominent, but a socket of thick-walled cells around the chalaza is absent (Vijayaraghavan and Dhar 1976). According to Willis (1973), Scytopetalaceae show affinities probably with Olacaceae rather than Malvales. Vijayaraghavan and Dhar (1976) negate any relationship with Olacaceae. Knowledge about the development of endosperm, embryo, and structure of seed may indicate the affinities with other members of Malvales.

36 Thymelaeales

The order Thymelaeales comprises the families Geissolomataceae, Penaeaceae, Dichapetalaceae, Thymelaeaceae and Elaeagnaceae.

36.1 General Features

The order Thymelaeales is characterized by anatropous, bitegmic, crassinucellate ovule, micropyle formed by both integuments, presence of parietal tissue, Polygonum type of embryo sac, tetrasporic Penaea type in Penaeaceae, Nuclear endosperm, Cellular in Geissolomataceae, and Asterad type of embryogeny.

36.2 Geissolomataceae

The development and the structure of anther, pollen grain, embryo sac, and seed-coat have not been studied.

The ovule is anatropous, bitegmic and crassinucellate in *Geissoloma marginatum* (Stephens 1910). Both the integuments form the micropyle. Earlier stages of development are not available, but the embryo sac is probably of the Polygonum type (Stephens 1910). The antipodals are ephemeral. The polar nuclei fuse in the chalazal region of the embryo sac prior to fertilization. Abundant starch grains occur in the mature embryo sac.

Development of the endosperm is of the Cellular type.

The embryogeny conforms to the Asterad type-Penaea variation. The suspensor is said to be absent, which requires confirmation.

Seeds have scanty endosperm. The embryo has long and linear cotyledons (Hutchinson 1973).

J Hutchinson (1973), EL Stephens (1910)

36.3 Penaeaceae

The anther is tetrasporangiate, and the anther wall comprises a persistent and hairy epidermis, ephemeral endothecium, two middle layers, and glandular tapetum with two-nucleate cells in *Penaea mucronata* and *Saltera sarcocolla* (Tobe and Raven 1984f). The development of the anther wall is of the "Basic" type (Davis 1966). The microspore mother cells undergo meiotic divisions, cytokinesis is simultaneous, and tetrads are tetrahedral and decussate, rarely isobilateral.

The pollen grain is two-celled at the shedding stage.

The ovule is anatropous, bitegmic and crassinucellate. Both the integuments are two-layered, and form the micropyle. In younger ovules, at the base the integuments are close to each other. As the chalazal tissue grows, the base of integuments become separated. The vascular supply extends in the base of integuments.

The archesporium is one-celled. It cuts off a parietal cell, the parietal tissue is three to four layers thick. In *Brachysiphon* and *Penaea* (Stephens 1909) and *Saltera* (Tobe and Raven 1984f), the megaspore mother cell develops into a tetrasporic 16-nucleate embryo sac of Penaea type. The 16 nuclei are arranged in four crosswise quartets. A nucleus from each quartet moves into the centre and the four nuclei fuse to form a tetraploid secondary nucleus. The three cells of each triad simulate egg apparatus. The micropylar triad alone is functional. The synergids are not well organized, and true antipodal cells are absent.

Fertilization is porogamous. The development of the endosperm is of the Nuclear type, and it becomes cellular at the globular proembryo (Stephens 1909).

The embryogeny is of the Onagrad type in *P. mucronata* and *Saltera sarcocolla* (Tobe and Raven 1984f). At an early stage of development the hypocotyl stores food reserve as oil.

The proembryo is massive and lacks suspensor. The two cotyledons are equal and small.

The mature seed is exalbuminous and ellipsoidal. In *P. mucronata* the seed is exotestal, endotesta is represented by crystals, whereas in *Saltera sarcocolla* the seed is endotestal and contains numerous large crystals. Exo- and endotegmen are tracheidal and fibrous.

GL Davis (1966), EL Stephens (1908, 1909), H Tobe and PH Raven (1984f)

36.4 Dichapetalaceae

Microsporogenesis, male gametophyte, and embryogeny have not been studied.

The ovule is anatropous, bitegmic and tenuinucellate in *Dichapetalum mombuttense* (Boesewinkel and Bouman 1980). The initiation of the inner integument takes place in a zone about three dermal cells broad. The outer integument is completely of dermal origin, and three to five layers thick. It is shorter than the inner one. The inner integument forms the micropyle. In *Tapura africana* (Boesewinkel and Bouman 1980), the cells of the inner layer of the inner integument differentiate into an endothelium. A funicular obturator develops in *D. mombattense*.

The hypodermal archesporial cell functions directly as megaspore mother cell. The chalazal megaspore of a linear tetrad develops into a Polygonum type of embryo sac.

Development of the endosperm is of the Nuclear type. Later it becomes cellular and only one layer is formed. The seed is exalbuminous (Hutchinson 1973).

The large straight embryo is rich in starch, and poor in lipids. In *D. cymulosum* the peripheral layers of cotyledons contain chlorophyll.

The mature seed is ellipsoid in *D. mombattense*. The seed-coat is covered with fine ramifications of the often anastomozing vascular bundles. The exotesta consists of tanniniferous cells with walls thickened with pectic substances. The cuticle is relatively smooth but spirally striate. The exotesta is followed by a layer of compressed cells with scattered groups of xylem elements of the vascular bundle. These elements show helical thickenings.

The structure of the seed-coat of *D. cymulosum*, *D. madagascariense*, *D. pallidum* and *Tapura africana* is similar to that in *D. mombattense* (Boesewinkel and Bouman 1980). In *D. bangii* the epidermis becomes multilayered. The inner epidermis of the testa consists of a fairly thick layer of flattened cells with vestiges of vascular bundles. The testa contains scattered lignified and pitted sclereids.

FD Boesewinkel and F Bouman (1980), EJH Corner (1976), J Hutchinson (1973), W Punt (1975)

36.5 Thymelaeaceae

The anther wall comprises the epidermis, fibrous endothecium, two ephemeral middle layers, and secretory tapetum. The tapetal cells are binucleate in *Daphne*, and four-nucleate in *Wikstroemia* (Venkateswarlu 1947a). The microspore mother cells undergo meiotic divisions, simultaneous cytokinesis is followed by furrowing, and the tetrads are tetrahedral and isobilateral. Polyspory (see Venkateswarlu 1947a) has been reported in *Daphne odorata* (tetrad with 5–7 cells) and *Wikstroemia indica* (tetrad with five cells). This is usually due to (a) laggard chromosomes which organize into micronuclei, or (b) supernumerary division of one or more nuclei of the tetrad.

The pollen grains are pantoporate, and three-celled at the dispersal stage. Several variations have been observed (Venkateswarlu 1945, 1947a). In *Thymelaea arvensis* one pollen grain showed two vegetative and two sperm nuclei (due to division of generative cell and vegetative nucleus), and another showed one vegetative nucleus and four sperms (due to supernumerary division of two original sperms). In *Wikstroemia canescens*, in an anther lobe with degenerated pollen, one pollen grain contained two vegetative and two generative cells.

The ovule is anatropous, bitegmic and crassinucellate with an endostome. A hypostase develops. An obturator is present in *Thymelaea*. The epidermal cells of nucellus contribute to the parietal tissue. The archesporial cell cuts off a parietal cell; the parietal tissue is nine to ten cells thick in *Daphne*, and four to five in *Wikstroemia*. The megaspore mother cell undergoes meiotic divisions, and produces a linear or T-shaped tetrad. The chalazal megaspore develops into a Polygonum type of embryo sac. The polar nuclei fuse before fertilization. There are six antipodals in *Wikstroemia*, and *Daphne* has an even larger number.

The development of the endosperm is of the Nuclear type. Wall formation is initiated at the differentiation of cotyledons in the embryo. Only a few layers of endosperm are formed so that the central region (called central vacuole by the author) is never filled up with endosperm (Venkateswarlu 1947a).

The embryogeny conforms to the Asterad type-Erodium variation. In *Daphne mezereum* it is of the Asterad type-Penaea variation (Souèges 1942a).

Nucellar polyembryony is reported in *Wikstroemia indica* (Winkler 1906), and synergid or nucellar polyembryony in *Daphne cannabina* (Venkateswarlu 1946, 1947a).

In a full-grown but immature seed of *Craterosiphon*, *Dicranolepis* and *Synaptolepis*, the nucellus shows a peripheral network of tracheids (Corner 1976).

The seed is exotegmic. The outer and the inner epidermis of the testa are thick-walled. The outer epidermis of the tegmen comprises a palisade of thick-walled, lignified, pitted columnar cells. The thin-walled mesophyll gets crushed. The inner epidermal cells are thinly lignified with reticulate thickenings (Corner 1976).

EJH Corner (1976), F Fagerlind (1940b), A Fuchs (1938), C Garagaty-Feissly (1969–70), SB Kausik (1940b), J Osawa (1913), R Souèges (1942a), K Tan (1980), J Venkateswarlu (1945, 1946, 1947a, c), H Winkler (1906)

36.6 Elaeagnaceae

The anther wall comprises the epidermis, fibrous endothecium, an ephemeral middle layer, and secretory tapetum of two- to four-nucleate cells. The microspore mother cells undergo meiotic divisions, cytokinesis is simultaneous, and the tetrads are tetrahedral and decussate.

The three-colporate pollen grains are shed at the three-celled stage.

The ovule is anatropous, bitegmic and crassinucellate. A funicular obturator of glandular cells has been observed in the micropylar region in *Elaeagnus conferta* (V.K. Sharma 1966). A prominent hypostase develops after fertilization. The archesporium consists of one to three cells, cuts off a parietal cell, and the parietal tissue is five to seven layers thick. The megaspore mother cell undergoes meiotic divisions, and the development of the embryo sac is of the Polygonum type. The antipodals are ephemeral.

The development of the endosperm is of the Nuclear type. The free-nuclear endosperm in the chalazal region exhibits aggressive haustorial activity. Wall formation is initiated in the micropylar region and proceeds towards the chalaza in *E. conferta*; however, the chalazal endosperm remains free-nuclear till its absorption (V.K. Sharma 1966).

The embryogeny corresponds to the Asterad type, and is assigned to Period I, Megarchetype II, Group I (Rau and Sharma 1970b). The mature embryo is straight with large, thick cotyledons, and a short radicle.

The seed is exalbuminous, and exotestal. The testa is highly multiplicative, and forms the seed-coat. The outer epidermis comprises radially elongate, thick-walled lignified cells with stellate lumina, and is followed by thick-walled aerenchymatous mesophyll. The tegmen is unspecialized, and gets crushed.

A Fuchs (1936), MA Rau and VK Sharma (1970b), VK Sharma (1966), B Singh (1964), R Wunderlich (1967)

36.7 Taxonomic Considerations

Engler and Diels (1936) place all the families, except Dichapetalaceae, in Myrtiflorae, under suborder Thymelaeneae. Hutchinson (1973) includes Geissolomataceae, Penaeaceae and Thymelaeaceae along with Nyctaginaceae in Thymelaeales, and Elaeagnaceae in Rhamnales.

Geissolomataceae. Geissolomataceae and Penaeaceae have plurilocular ovary, and embryogeny of the Asterad type-Penaea variation. In Geissolomataceae the embryo sac is of the Polygonum type, polar nuclei fuse in the chalazal region of the embryo sac before fertilization, and the endosperm is of the Cellular type. In Penaeaceae the development of the embryo sac is of the tetrasporic Penaea type, and endosperm of the Nuclear type. Even though Geissolomataceae are inadequately investigated, the two characters, i.e. Polygonum type of embryo sac with ephemeral antipodals and Cellular type of endosperm, justify the creation of a separate family Geissolomataceae as suggested by Stephens (1910).

Penaeaceae. Tobe and Raven (1984f) enumerate seven embryological features of Penaeaceae comparable to the order Myrtales: (1) glandular anther tapetum, (2) crassinucellate ovule, (3) two-layered inner integument, (4) micropyle formed by both integuments, (5) antipodals s. str. absent, (6) endosperm of the Nuclear type, and (7) mature seed exalbuminous. They mention that Penaeaceae is correctly placed in Myrtales sensu R. Dahlgren and Thorne (1984).

Penaeaceae have some distinctive features: (1) hairy anther epidermis, (2) ephemeral endothecium, (3) unusual development of chalaza to separate the base of the two integuments, and (4) Penaea type of tetrasporic 16-nucleate embryo sac.

On the basis of ephemeral endothecium, Tobe and Raven (1984f) compare Penaeaceae with the families Oliniaceae, Aizateaceae, and Rhyncocalycaceae.

Melchior (1964) and Hutchinson (1973) assign Penaeaceae to Thymelaeales, next to Geissolomataceae. According to recent data, Penaeaceae should be assigned to Myrtales and excluded from Thymelaeales as also mentioned by Thorne (1968), Carlquist and Debuhr (1970), Takhtajan (1969, 1980), R. Dahlgren and Thorne (1984), and Johnson and Briggs (1984).

Dichapetalaceae. Baillon (1874) includes Dichapetalaceae as a tribe Dichapetalae in Euphorbiaceae. Engler and Diels (1936) place it in Geraniales, Melchior (1964) in Thymalaeales, and Hutchinson (1973) in Rosales.

The embryological data for Dichapetalaceae are rather meagre. The ovule is tenuinucellar with Polygonum type of embryo sac, and the seed is simple, exotestal, and exalbuminous.

The ovule is crassinucellate in Euphorbiaceae, and the seed is exotegmic. The embryological data do not support relationship with Euphorbiaceae.

In Geraniales the ovules are tenuinucellate with Polygonum type of embryo sac, and the endosperm is of the Nuclear type as in Dichapetalaceae. But the seeds are exotegmic in Geraniales. The assignment of Dichapetalaceae to Geraniales cannot be confirmed unless further data are available.

The ovule, in Thymelaeales, is crassinucellate and the endosperm is of the Nuclear type except in Geissolomataceae where it is of the Cellular type. In Penaeaceae the embryo sac is of the Penaea type. The seed is exotegmic in Thymelaeales, whereas it is exotestal in Dichapetalaceae. According to Corner (1976), it is difficult to assess the significance of exotestal seed due to its simplicity.

Embryological data do not support the relationship with Rosales with crassinucellate ovule and mesotestal seed.

According to Hutchinson (1973), it is rather difficult to place Dichapetalaceae in any of these orders as they do not show any affinity with either Euphorbiaceae or Thymelaeaceae. We must await further information to discuss the taxonomic assignment of Dichapetalaceae.

Thymelaeaceae. Thymelaeaceae have secretory anther tapetum, simultaneous cytokinesis, three-nucleate pollen grains, tracheidal network in the periphery of nucellus, Polygonum type of embryo sac, secondary multiplication of antipodal cells, obturator, Nuclear endosperm, embryogeny of the Asterad type-Penaea and -Erodium variation, and nucellar or synergid polyembryony. Both the integuments take part in the formation of the seed-coat. The seed is exalbuminous, and exotegmic with the outer epidermis of the inner integument forming the palisade of sclerotic cells.

Venkateswarlu (1947a, c), while considering the systematic position of Thymelaeaceae, states that embryogeny plays a very important role in its classification. Thymelaeaceae, Geissolomataceae and Penaeaceae have Asterad type of development of embryo. The latter two have Penaea variation, as in Thymelaeaceae; *Poivrea coccinea* (Combretaceae) of Myrtiflorae also shows Erodium variation of the Asterad type, as in Thymelaeaceae. Thymelaeaceae are related to Geissolomataceae and Penaeaceae on the one hand, and Combretaceae on the other.

Thymelaeaceae (Sect. 36.5) and Nyctaginaceae (Sect. 16.5) have unilocular ovary, and indehiscent fruit. The embryogeny of Nyctaginaceae is of the

Asterad type-Polygonum variation (Kajale 1940), and it differs from the rest of the members of Centrospermales (Sect. 16). Venkateswarlu (1947b) enumerates several embryological similarities between these two families, and favours the inclusion of Nyctaginaceae with Thymelaeaceae, Geissolomataceae and Penaeaceae, as also suggested by Hutchinson (1973).

Elaeagnaceae. Elaeagnaceae show several features similar to Thymelaeaceae: secretory anther tapetum, simultaneous cytokinesis, three-celled pollen, anatropous, bitegmic, crassinucellate ovule, presence of obturator, Polygonum type of embryo sac, haustorial activity of endosperm at the chalazal end, and Asterad type of embryogeny. According to V.K. Sharma (1966), Elaeagnaceae are a highly specialized family related to Thymelaeaceae. Wunderlich (1967) refutes their alliance with Thymelaeaceae, on the basis of seed structure which is typically rhamnaceous (see Corner 1976). The relationship of Elaeagnaceae with Rhamnaceae has already been discussed (see Rhamnales, Sect. 34.5). Takhtajan (1980) treats them in a separate order Elaeagnales, close to the Proteales.

37 Violales

The order Violales comprises six suborders and 20 families: suborder Flacourtiineae – Flacourtiaceae, Peridiscaceae, Violaceae, Stachyuraceae, Scyphostegiaceae, Turneraceae, Malesherbiaceae, Passifloraceae and Achariaceae; Cistineae – Cistaceae, Bixaceae, Sphaerocepalaceae, and Cochlospermaceae; Tamaricineae – Tamaricaceae, Frankeniaceae and Elatinaceae; Caricineae – Caricaceae; Loasineae – Loasaceae; Begoniineae – Datiscaceae and Bignoniaceae.

37.1 General Features

Variations occur in this fairly large order; the following general features can be enumerated:

(1) Secretory anther tapetum, (2) simultaneous cytokinesis in pollen mother cells, (3) two-celled pollen grains, except in Frankeniaceae, (4) anatropous bitegmic ovules, unitegmic in Loasaceae and Begoniaceae, and orthotropous in Flacourtiaceae and Cistaceae, (5) presence of parietal tissue except in Bixaceae, Frankeniaceae and Loasaceae, (6) Polygonum type of embryo sac, tetrasporic in Tamaricaceae and Caricaceae, and biosporic in Stachyuraceae and Datiscaceae, (7) Nuclear endosperm, Cellular in Loasaceae, (8) ruminate endosperm in Passifloraceae, (9) micropylar haustoria develop in Frankeniaceae, and both micropylar and chalazal in Loasaceae, (10) embryogeny of Asterad type in Violaceae, Onagrad in Turneraceae, Passifloraceae, Car-

icaceae, Datiscaceae and Begoniaceae, and Solanad in Stachyuraceae, Loasaceae, Cistaceae, Tamaricaceae, Frankeniaceae and Elatinaceae, occasionally Piperad in Passifloraceae, and (11) structure of seed-coat reflects the affinities of the families of this order.

37.2 Flacourtiaceae

The staminal primordia arise irregularly on a flat, ring-like mound in *Caloncoba echinata* (Endress and Vosser 1975). The pollen sacs are transversely septate, and the sterile cells are present ab initio in the single-layered archesporium. Later, the archesporium becomes massive.

The anther wall comprises persistent epidermis, fibrous endothecium, one or two ephemeral middle layers, and secretory tapetum with two- to four-nucleate cells. The microspore mother cells undergo meiosis, cytokinesis is simultaneous, and the tetrads are tetrahedral and isobilateral.

The pollen grains are two-celled at the dispersal stage.

The ovule is orthotropous (*Casearia tomentosa* – Narayanaswami and Sawhney 1959), hemiana- (*Idesia polycarpa* – Corti 1948) or anatropous (*Kiggelaria* – see Davis 1966, *Arechavaletaia uruguayensis* – Fig. 37.2.1A), bitegmic and crassinucellate. The inner integument forms the micropyle; but

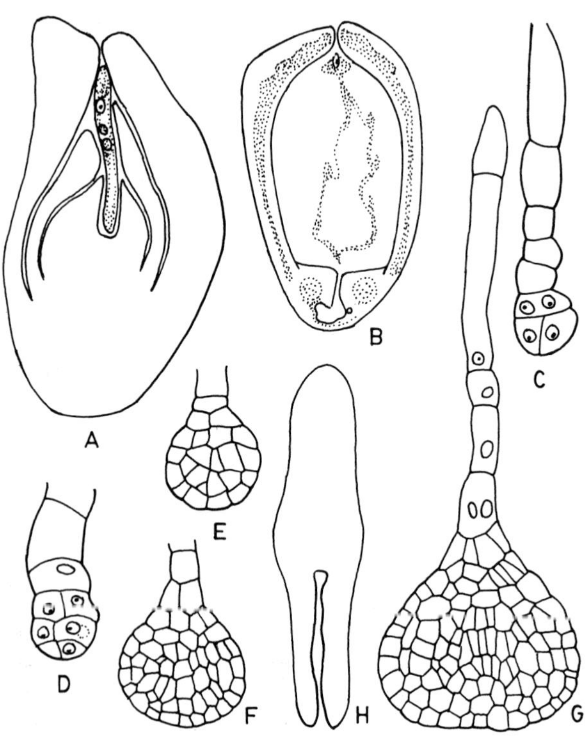

Fig. 37.2.1A. *Arechavaletaia uruguayensis*, longisection of bitegmic ovule, the tip of the mature embryo sac extends into the style (After Ventura 1938). **B** *Caloncoba glauca*, tranlongisection of young seed. Note the sclerified seed-coat, proembryo, nuclear endosperm, and vesicular chalazal caecum of embryo sac. (After van Heel 1977). **C–H** *Idesia polycarpa*, development of embryo. (After Tohda 1971)

both integuments in *A. uruguayensis*. The vascular strands reach up to the tip of the outer integument in *Hydnocarpus laurifolia* (Dathan and Singh 1979). The tip of the ovule bends like a hook after fertilization. A sclerenchymatous hypostase differentiates. The archesporial cell cuts off a parietal cell. The parietal tissue is three to seven layers thick. The megaspore mother cell undergoes meiotic divisions, and produces a linear tetrad. Variations occur in the functional behaviour of megaspores but, usually, the chalazal one develops into a Polygonum type of embryo sac in *Casearia tomentosa* (Narayanaswami and Sawhney 1959), *Flacourtia cataphracta*, *F. ramontchi* (Gopinath 1946), *F. indica* (Dathan and Singh 1973a), *Hydnocarpus laurifolia* (Dathan and Singh 1979), *Idesia polycarpa* (Corti 1948), and *Kiggelaria africana*. The embryo sac breaks through the nucellus and reaches the exostome of the micropyle in *K. africana*, and *Arechavaletaia uruguayensis* (Fig. 37.2.1A). In *F. indica* the parietal layers are crushed and the tip of the embryo sac lies below the nucellar epidermis. The antipodals are ephemeral. The polar nuclei fuse before or at the time of fertilization.

Development of the embryo is shown in Figure 37.2.1C–H. It has a conspicuous suspensor.

The development of the endosperm is of the Nuclear type (Fig. 37.2.1B). Enucleate cytoplasmic nodules occur in *Hydnocarpus*. Through centripetal wall formation, the endosperm becomes cellular at the young globular stage of proembryo in *F. indica* (Dathan and Singh 1973a). In *Hydnocarpus* the mature endosperm contains oil and calcium oxalate crystals.

In *H. laurifolia* the seeds are large, sculptured, endospermous and have a large embryo with thin cotyledons which show prominent bulgings in the position of veins. The integuments are multiplicative in *Hydnocarpus*, *Gynocardia* and *Pangium*, and non-multiplicative in *Erythrospermum*, *Flacourtia*, and *Scaphocalyx*. In the latter two genera the seeds are enclosed in a woody pseudotesta formed by the inner layers of the ovary wall (see Dathan and Singh 1979).

The seed is exotegmic, and contains one or two layers of perisperm in *F. indica* (Dathan and Singh 1973a). The outer layer of testa contains a red pigment. The cells of the other layers are lignified with numerous simple pits. The outer epidermis of the tegmen forms the main sclerenchymatous layer, the inner is tanniniferous. The middle layer may persist. The hypostase connects the main sclerenchymatous layer.

In *Hydnocarpus laurifolia* (Dathan and Singh 1979), the mature seed-coat consists of four zones: (1) outer zone of 50–90 layers of thin-walled compressed and deformed cells with patches of tanniniferous cells mostly in the peripheral region, (2) a broad sclerenchymatous zone, (3) inner parenchyma with scattered tanniniferous cells, and (4) inner zone of tanniniferous layers. The sclerenchymatous zone is of dual origin, and can be distinguished into three regions: (1) three to fourteen outer layers of brachysclereids, (2) middle region of a single layer of macrosclereids, and (3) inner region of brachysclereids with patches of thick-walled tanniniferous cells.

Corner (1976) mentions that in Flacourtiaceae the structure of seed falls in two groups:

1) Flacourtia group (Flacourtia proper) has a fibrous exotegmen as in Celastraceae. The funicular aril is vascularized in *Casearia.*

2) Hydnocarpus group has a very thick pachychalazal seed-coat without any trace of the fibrous exotegmen, even round the micropyle where the two vestigial integuments persist.

The recent discovery of a substance mangostin in the bark of *Hydnocarpus*, in contrast with mangeferin in *Flacourtia*, is significant and a pointer to the unnatural assemblage of these genera in one family (Gunasekera and Sultanbawa 1973, Corner 1976).

R Corti (1948), EJH Corner (1976), ASR Dathan and D Singh (1972b, 1973a, 1979), GL Davis (1966), PK Endress and P Vosser (1975), DM Gopinath (1946), SP Gunasekera and NU Sultanbawa (1973), S Narayanaswami and S Sawhney (1959), H Tohda (1971), WA van Heel (1973, 1974, 1977, 1979), M Ventura (1938)

37.3 Peridiscaceae

The embryology has not been investigated.

The anther is bilocular opening lengthwise, introse. Ovary unilocular with several ovules hanging from the top. Fruit unilocular and single-seeded. The seeds lack endosperm and have a curved embryo.

J Hutchinson (1973), AL Takhtajan (1966), JC Willis (1973)

37.4 Violaceae

The anther wall comprises the epidermis, fibrous endothecium, two ephemeral middle layers, and secretory tapetum of binucleate cells. There are three or four middle layers in *Viola odorata* (D. Singh and Gupta 1963, D. Singh 1966). Übisch granules are studded on the inner walls of tapetal cells in *Ionidium suffruticosum* (Mukherjee and Venugopalan 1957). The microspore mother cells undergo meiotic divisions, cytokinesis is simultaneous, and the tetrads are tetrahedral, isobilateral and decussate. Polyspory occurs in *V. odorata* (D. Singh 1966).

The pollen grains are three- to five-colporate, prolate to spheroidal, and shed at the two-celled stage. In *V. riviniana* (West 1930), the generative cell may divide in the pollen grain, or in the pollen tube. Madge (1929) reports that the vegetative nucleus increases in length. The germination of pollen takes place in situ in cleistogamous flowers of *Viola biflora*, *V. odorata* var. *paraecox* and *V. sylvatica* (Madge 1929).

The ovules are anatropous, bitegmic and crassinucellate. The outer and inner integuments form a zigzag micropyle in *Viola* and *Ionidium*, and the inner integument alone forms the micropyle in *Hybanthus concolor* (Andrews 1910). The nucellar epidermis undergoes periclinal divisions, and produces a two- to three-layered nucellar cap. The archesporium is uni- or multi-celled.

The archesporial cell cuts off a parietal cell. The megaspore mother cell divides meiotically, and forms a linear or T-shaped tetrad. The chalazal megaspore develops into a Polygonum type of embryo sac. The antipodals are small and ephemeral. In *Hybanthus* (Andrews 1910), *Viola cucullata*, *V. pedata* (Bliss 1912) and *V. odorata* (D. Singh and Gupta 1968) the antipodals are fairly large. A double hypostase is present in *Viola cinerea* var. *stocksii*, *V. tricolor* and *Ionidium suffruticosum* (D. Singh 1963, D. Singh and Gupta 1968).

The pollen tube enters the synergid and forces its way through its base (Madge 1929). This needs confirmation.

The development of endosperm is of the Nuclear type. Wall formation is centripetal, and occurs after the formation of numerous nuclei in *Ionidium* and *Viola* (D. Singh 1963). The endosperm is massive and persistent, and stores fatty food reserve.

The embryogeny corresponds to the Asterad type, and the mature embryo is erect and spatulate.

Both the integuments are three-layers thick, and contribute to the formation of testa and tegmen in *Viola*. The seed is albuminous and exotegmic with a patch of chalazal sclereids and a hypostase. It is arillate in some taxa. The cells of the outer epidermis of the testa show lamellar thickenings. They are radially-elongate in *Viola*. The middle layer contains tannin in *Viola*, and is sclerenchymatous in *Ionidium*. Each cell of the inner epidermis contains a single crystal. The exotegmen is excessively sclerified and pitted. The middle layer may persist or collapse, while the inner epidermis persists and becomes thick-walled.

FM Andrews (1910), MC Bliss (1912), ASR Dathan and D Singh (1973c), E Kuta (1990), M Madge (1929), PK Mukherjee and P Venugopalan (1957), MVS Raju (1958), D Singh (1963, 1966, 1970a), D Singh and S Gupta (1963, 1968), R Souèges (1937e), G West (1930)

37.5 Stachyuraceae

The development of anther and pollen grains has not been investigated.

The ovule is anatropous, bitegmic and crassinucellate. The zigzag micropyle is formed by both the integuments. An aril develops from the apical region of the funiculus in *Stachyurus praecox* (Sâto 1976b) and *S. chinensis* (Mathew and Chaphekar 1977), and grows towards the chalazal end of the ovule. Corner (1976) mentions that the exostome is also involved in the formation of aril. The apical cells of the nucellar epidermis divide periclinally to form a two- to three-cell thick cap. The archesporial cell cuts off a parietal cell, and the parietal tissue is three layers thick. The megaspore mother cell undergoes meiotic divisions, and produces a triad, a linear or T-shaped tetrad. A triad may develop in *S. chinensis*, the tetrad is T-shaped in *S. matsuzakii* (Sâto 1976b). The chalazal megaspore develops into a Polygonum type of embryo sac. The base of the mature embryo sac extends into the chalaza and curves convexly. The antipodals are ephemeral, and the polar nuclei fuse before fertilization.

An early stage of bisporic Endymion type of development is reported in *S. chinensis* (Mathew and Chaphekar 1977). This needs confirmation.

The development of the endosperm is of the Nuclear type. Later, it becomes cellular. It is thin-walled and stores fatty food reserve.

The embryogeny conforms to the Solanad type in *S. chinensis* (Mathew and Chaphekar 1977). The mature embryo is small and straight with flat cotyledons and a short radicle. Cleavage (Fig. 3.12.1B, C) and synergid polyembryony have been noted in *S. chinensis.*

The seed is small, arillate, and albuminous. The testa is wholly sclerotic, and the tegmen gets crushed.

EJH Corner (1976), CJ Mathew and M Chaphekar (1977), Y Sâto (1976b)

37.6 Scyphostegiaceae

The embryology has not been fully investigated.

The anther is tetralocular. Simultaneous cytokinesis occurs in the microspore mother cells.

The ovule is anatropous, bitegmic and crassinucellate. The micropyle is formed by both the integuments, the endostome is long and the exostome short. The integuments are three to four layers thick. The nucellus divides periclinally to form a nucellar cap which persists in the mature seed.

The archesporial cell cuts off a parietal cell. The megaspore mother cell undergoes meiotic divisions forming a linear tetrad, and the chalazal megaspore develops into an eight-nucleate Polygonum type of embryo sac.

Development of the endosperm is of the Nuclear type. It becomes cellular, contains starch, and persists in the mature seed.

The exotegmic seed is albuminous and perispermous. The straight embryo has oval cotyledons and occupies more than half the length of the arillate seed. The aril develops from the funiculus and outer integument, and surrounds the basal part of seed. There is an intercalary extension of the endostome and exostome to form a pseudo-funicle, in conjunction with the raphe (Corner 1976).

EJH Corner (1976), WA van Heel (1967), BGL Swamy (1953a), MS Yakovlev (1983)

37.7 Turneraceae

The anther wall comprises the epidermis, fibrous endothecium, two ephemeral middle layers, and secretory tapetum of multinucleate cells. The nuclei fuse and become polyploid. The stomata of haplocheilic type, and fibrous thickenings occur in the region of the connective. The microspore mother cells undergo meiotic divisions, cytokinesis is simultaneous, and the tetrads are tetrahedral and decussate. Polyspory is quite common.

The pollen grains are three-colporate with simple exine ornamentations. They are two-celled at the dispersal stage.

The ovule is anatropous, bitegmic and crassinucellate; the micropyle is produced by both the integuments. The hypodermal archesporium is uni- or multicelled. The archesporial cell cuts off a parietal cell. The parietal tissue is only a few cells thick. The nucellar epidermis also adds to it. The megaspore mother cell undergoes reduction divisions, gives rise to a linear or T-shaped tetrad, and the chalazal megaspore develops into a Polygonum type of embryo sac. The presence of supernumerary megaspore mother cells, dyads, triads and tetrads is a common feature. Twin embryo sacs are also present in *Turnera ulmifolia* (Vijayaraghavan and Kaur 1966). The antipodals are ephemeral. The hypostase is traversed by the funicular vascular trace.

The entry of the pollen tube is porogamous, and double fertilization occurs. The remnants of the pollen tube persist up to a late stage of the embryo.

Development of the endosperm is of the Nuclear type. The nuclei aggregate around the embryo, and in the chalazal region. Wall formation occurs at the

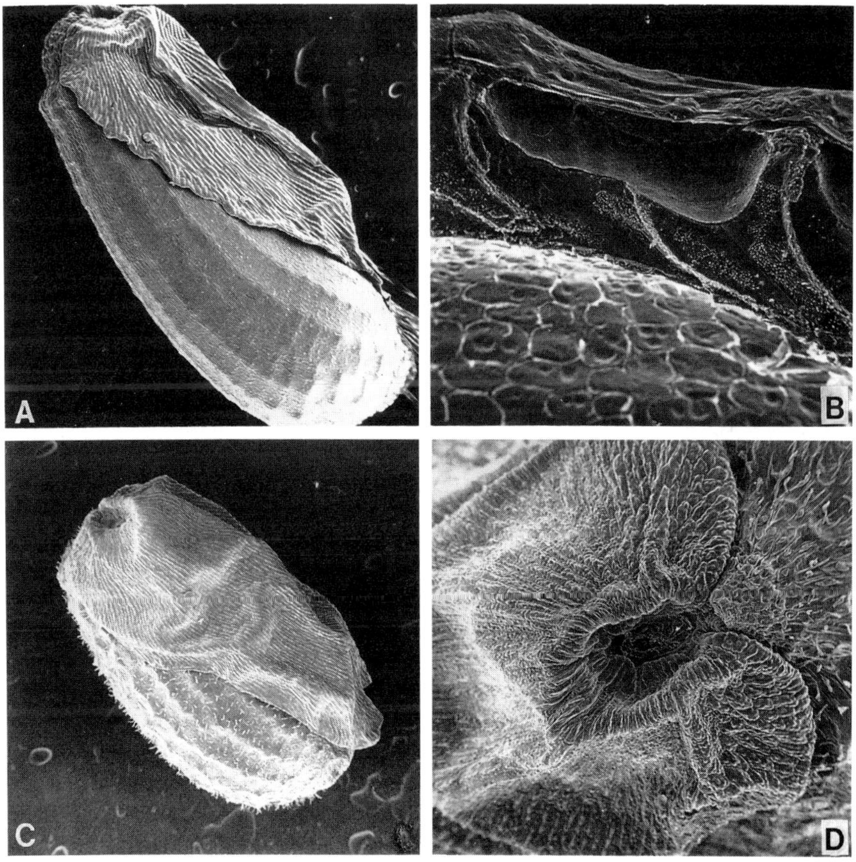

Fig. 37.7.1A–D. (SEM) **A, B** *Turnera ulmifolia.* **C, D** *Piriqueta cistoides.* **A, C** Seed with aril. **B** Cross-section of seed-coat. **D** Exostome, note aril and hilar scar. (Courtesy Dr F. Bouman, Amsterdam)

globular proembryo, and is centripetal. The cells are isodiametric, and contain starch.

The embryogeny conforms to the Onagrad type. In *Turnera ulmifolia* var. *angustifolia* (M.V.S. Raju 1956a), the embryogeny is assigned to Period I, Megarchetype I, Group I (Crucifer type). The mature embryo is dicotyledonous, and curved with a short suspensor.

The seeds are small, exotegmic and arillate (Fig. 37.7.1A, C, D). The outer epidermis of the testa contains tannin. The cells of the inner epidermis enlarge. The outer epidermis of the tegmen comprises longitudinally elongate, pitted, sclerotic cells with their ends raised to form ridges (Fig. 37.7.1B). The inner epidermis contains tannin. The aril is funicular, and arises around the apex of funicle. It is white, sickle-shaped and non-vascularized (M.V.S. Raju 1956a).

D Kaur (1970), MS Mahalingappa (1975), MVS Raju (1956a), N Ramaswamy and B Bir (1984), MR Vijayaraghavan and D Kaur (1966)

37.8 Malesherbiaceae

The embryology has not been fully investigated.

The anther is tetralocular.

The ovule is anatropous, pleurotropous, bitegmic and crassinucellate. The micropyle is formed by both the integuments. The outer integument is two layers thick, three in the micropylar region; the inner one is three-layered. The massive chalaza forms a process which resembles pecten (Fig. 37.8.1A). The seed is albuminous with aleurone grains and fat globules. The large embryo is dicotyledonous (Fig. 37.8.1B).

H Harms (1925), J Hutchinson (1973), SM Ricardi (1967)

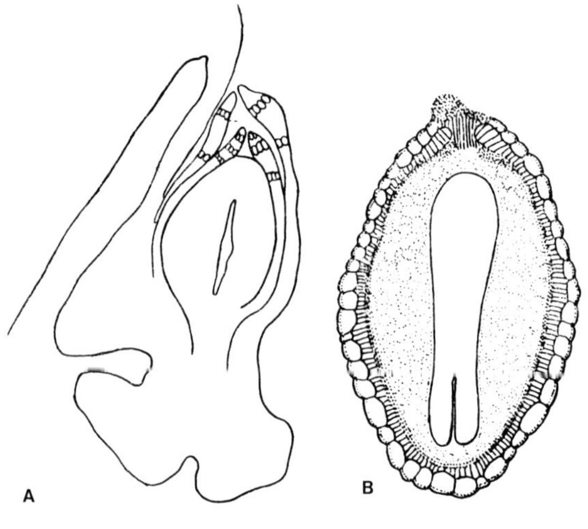

Fig. 37.8.1. A *Malesherbia scarlatiflora*, longisection of ovule. **B** *M. lirana*, longisection of seed (semidiagrammatic). (After Ricardi 1967)

37.9 Passifloraceae

The anther wall comprises the epidermis, fibrous endothecium, two or three ephemeral middle layers, and secretory tapetum of multinucleate cells. The tapetum becomes biseriate in *Passiflora calcarata* and *P. foetida* (M.V.S. Raju 1952b, 1956b). The epidermal cells are fairly large at the dehiscence slit in *P. calcarata*. The microspore mother cells undergo meiotic divisions resulting in tetrahedral and isobilateral tetrads. Cytokinesis is simultaneous.

The pollen grains are three- to twelve-colporate in *Passiflora* and the colpi are arranged in three groups; the colpi of each group coalesce at the pole. The exine is reticulate. They are two-celled at the dispersal stage.

The ovule is anatropous except in *P. calcarata* (M.V.S. Raju 1956b) where it is orthotropous. It is bitegmic and crassinucellate. Both the integuments contribute to the formation of the micropyle. The archesporium is unicelled, except in *P. foetida* (Padhye and Deshpande 1960). It cuts off a parietal cell. The parietal tissue is massive. The nucellar epidermal cells also divide and contribute to it. Following meiotic divisions, the megaspore mother cell forms a linear tetrad. The chalazal megaspore develops into a Polygonum type of embryo sac, the antipodals are ephemeral.

The entry of the pollen tube is porogamous, and double fertilization occurs. The pollen tube grows deep in the embryo sac, gets twisted, and persists for a long time. In *P. adnophylla*, Cook (1909) pointed out that the tube becomes haustorial, and absorbs all the contents. He considers that this is an instance of parasitic condition of the male gametophyte over the female one. Such a presumption is entirely incorrect since the discharged pollen tube does not show any activity.

Persistent and twisted pollen tubes have also been observed in *P. calcarata* (M.V.S. Raju 1956b), *P. foetida* (D. Singh 1962) and *P. leschenaultii* (M.V.S. Raju 1952b).

The development of the endosperm is of the Nuclear type. The nuclei are distributed around a central vacuole, and aggregate at both the poles. The chalazal tubular region behaves as a haustorium in *Passiflora* and *Tacsonia* (Dathan and Singh 1973b). Centripetal wall formation commences after the formation of a large number of nuclei. The peripheral cells simulate meristematic cells; some cells in the chalazal region are plurinucleate in *P. calcarata* (M.V.S. Raju 1956b). Enucleated cytoplasmic nodules have been observed in the chalazal region (M.V.S. Raju 1956b), and finally they merge with the general cytoplasm. The mature endosperm is ruminate.

The embryogeny in *P. foetida* (D. Singh 1962) corresponds to the Onagrad type-Myosurus variation. Padhye (1963), however, reports that in 70% of ovules the division of the zygote is vertical and the development conforms to the Piperad type. This aspect needs confirmation.

The seeds are endotegmic and arillate with a large spatulate embryo. The aril is succulent and coloured. The testa is multiplicative and pulpy. The outer epidermis is covered with a smooth cuticle in *P. foetida* (D. Singh 1962), *P. hirsuta* and *P. holoserecea* (Netolitzky 1926), and is echinulate in *P. calcarata* (M.V.S. Raju 1956b). The middle layer is parenchymatous. The thin-walled inner epidermal cells are unequally radially elongate, more in the region of ingrowths. The inner epidermis of the tegmen is a sclerotic layer with radially elongate thick-walled, lignified, pitted cells. In *Adenia* the tegmen does not become sclerotic. The inner tangential walls of the middle layer become thickened in *P. foetida* (D. Singh 1962), and the inner epidermis contains granules. The endosperm becomes ruminate due to unequal corresponding radial elongation of testal and tegmal cells (Fig. 3.10.4D). The aril arises as a collar from the funicle. It comprises thin-walled cells containing oil and starch. It is brownish-pink in *P. calcarata*, and colourless in *P. foetida*.

MT Cook (1909), ASR Dathan and D Singh (1973b), A Kloos and F Bouman (1980), F Netolitzky (1926), MD Padhye (1963), MD Padhye and BG Deshpande (1960), MVS Raju (1952b, 1956b), D Singh (1962, 1970b)

37.10 Achariaceae

The embryology has not been studied.

The anther is bilocular, opens lengthwise, and has a broad connective (Hutchinson 1973). The ovule is anatropous. The arillate seed has copious endosperm (Willis 1973), and a small straight embryo.

H Harms (1925), J Hutchinson (1973), JC Willis (1973)

37.11 Cistaceae

The anther wall comprises the epidermis, fibrous endothecium, single ephemeral middle layer, and secretory tapetum of two- to four-nucleate cells. The nuclei divide, and the daughter nuclei fuse. The microspore mother cells undergo meiotic divisions, cytokinesis is simultaneous through furrowing, and the tetrads are tetrahedral, isobilateral and decussate. Polyspory is also reported.

The pollen grains with three germ pores are two-celled at the dispersal stage; the generative cell divides in the pollen tube.

The ovule is orthotropous, bitegmic and crassinucellate. Both the integuments form the micropyle. The tip of the integuments are swollen in the micropylar region. The cells of the nucellar epidermis undergo divisions, and a nucellar cap is formed. The archesporium is multicellular. The cells cut off parietal cells; the parietal tissue is well developed. The megaspore mother cell undergoes meiotic divisions, and produces a linear tetrad. The chalazal megaspore usually develops into a Polygonum type of embryo sac; micropylar megaspore in *Helianthemum alpestre*. The third megaspore from the micropyle is reported to be functional in *H. halimifolium* (Chiarugi 1925). The synergids have a cap-like structure at the tip, and the antipodals are ephemeral. Multiple embryo sacs develop in *H. vulgare* (Kapil and Maheshwari 1964). A hypostase is present.

The development of the endosperm is of the Nuclear type. The basal prolongation is common. The endosperm is 64-nucleate at the early globular proembryo, and wall formation is initiated at the early heart-shaped stage in *H. vulgare*. The endosperm is parenchymatous, and stores starch.

The embryogeny conforms to the Solanad type, and the embryo has a uniseriate suspensor. The development of embryo in *H. guttatum* (Souèges 1937b) is assigned to Period I, Series C, Megarchetype IV (Solanad type). The mature embryo is curved.

The seed is small, albuminous and exotegmic (Kapil and R. Maheshwari 1964). The testa persists as a membraneous covering, is rich in starch and shows thick-walled, branched, septate fungal mycelium (fungus could not be identified). The outer epidermis of tegmen comprises dark-brown, thick-walled, lignified, columnar cells forming a bulge in the chalazal region. The inner epidermis contains tannin.

JG Boursnell (1950), A Chiarugi (1925), R Corpal, F Pevez-Gargia and JM Pita (1989), RN Kapil and R Maheshwari (1964), R Souèges (1937b)

37.12 Bixaceae

The epidermis of the mature anther is papillate. The endothecial thickenings are confined around the slits at the end of thecae. The two or three middle layers are ephemeral, and the secretory tapetum is irregularly two-layered, and has binucleate cells. Übisch granules are present on the inner tangential walls of the tapetum. Meiotic divisions in microspore mother cells are followed by simultaneous cytokinesis, and the tetrads are tetrahedral and decussate.

The pollen grains are tricolporate, and two- or three-celled at the dispersal stage. Pollen grains with five nuclei have been observed in *Bixa orellana* (H. Kaur 1970b).

The ovule is anatropous, bitegmic and pseudo-crassinucellate. The outer integument overgrows the inner one, and both form a zigzag micropyle. The cells of the nucellar epidermis divide and form a nucellar cap. A single cell of the uni- or bicelled archesporium functions directly as the megaspore mother cell. The latter undergoes meiotic divisions, and produces a linear or T-shaped tetrad. Twin tetrads are also formed. The chalazal megaspore develops into a Polygonum type of embryo sac. Sometimes, an epichalazal megaspore becomes binucleate. The antipodals are ephemeral, and polar nuclei lie in the middle of the embryo sac. Mature embryo sacs with supernumerary polar nuclei and egg apparatus have been observed in *Bixa orellana* (Chopra and Kaur 1965a). Twin embryo sacs also develop concurrently (Dathan and Singh 1972a).

The entry of the pollen tube is porogamous, and double fertilization occurs. The remnants of the pollen tube persist. One of the synergids persists up to an early stage of proembryo.

Development of the endosperm is of the Nuclear type. The nuclei aggregate at the micropylar end. After the formation of a large number of nuclei, wall formation is initiated at the micropylar region and proceeds to the chalaza. Except for the outer one or two layers, the endosperm stores starch.

The division of zygote is transverse; vertical walls are formed in the apical and basal cell to produce a globular proembryo. Further development is irregular (H. Kaur 1969). The mature embryo has conspicuous elongated cotyledons.

The seed is albuminous with fibrous exotegmen. The outer epidermis of the testa shows hairs and stomata. In *Bixa* the epidermal cells in the chalazal region are sclerotic. The mesophyll contains multinucleate dye cells and oil cells, followed by the inner epidermis. The outer epidermis of the tegmen consists of lignified pitted cells. The mesophyll is parenchymatous. The inner hypodermal layer comprises hour-glass-shaped cells with thickenings, followed by thin-walled cells of inner epidermis. The cells of the latter are pitted in the anticlinal walls.

RN Chopra and H Kaur (1965a), ASR Dathan and D Singh (1972a), H Kaur (1969, 1970b), J Mauritzon (1936a)

37.13 Sphaerosepalaceae

The embryology has not been studied in sufficient detail.

The anther has a broad, glandular filament with two thecae opening introsely through longitudinal slits (Hutchinson 1973).

The anatropous ovule has a massive tanniniferous chalaza. In most species of *Rhopalocarpus* (*Sphaerosepalum*), the endotegmen has a smooth surface but in *R. lucidus* (Fig. 37.13.1A, C), *R. similis* and *R. triplinervius*, the endotegmen forms conicle or cylindrical outgrowths which ruminate the endosperm.

The mature embryo (Fig. 37.13.1B, D, E) has the hypocotyl, primordial root, and leafy cotyledons which are cordate at the base and lobed along the margin (Fig. 37.13.1D, E). The cotyledons are branched in some species of *Rhopalocarpus* and *Dialyceras*. They are bilobed with fimbriate margin in *R. thouarsianus*.

R Capuron (1962), J Hutchinson (1973)

37.14 Cochlospermaceae

The development of anther, pollen, embryo sac, and embryo has not been worked out.

The ovule is anacampylotropous, bitegmic and crassinucellate with a zigzag micropyle in *Cochlospermum gossypium* (Dathan and Singh 1972a). The nucellar epidermis contributes to the massive parietal tissue.

Development of the endosperm is of the Nuclear type in *C. orinoccense* (see Schnarf 1931). Wall formation is centripetal.

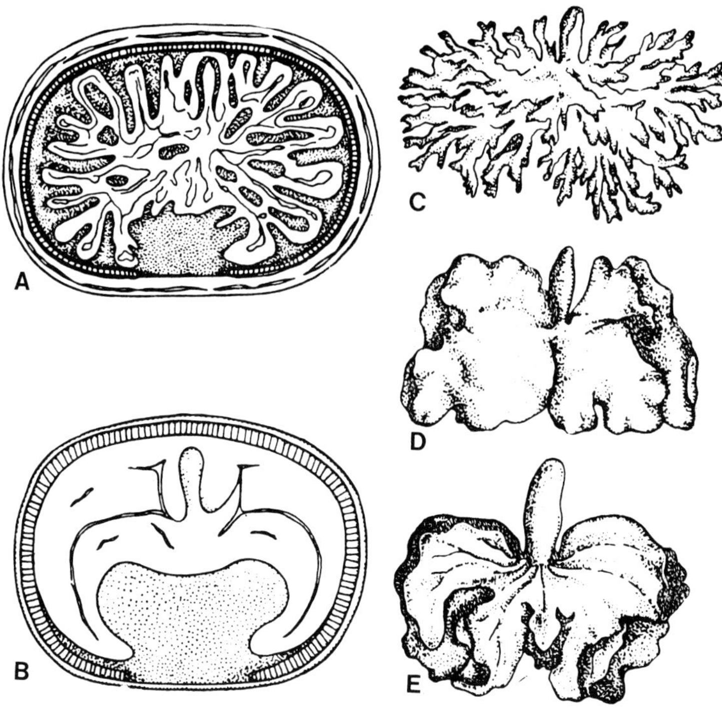

Fig. 37.13.1A–E. **A**, **C** *Rhopalocarpus lucidus*, **B**, **D** *R. thouarsianus*, **E** *Dialyceras parvifolium*. **A** Longisection of young seed with ruminate endosperm, ruminations are due to ingrowths of endotegmen. The chalaza is massive in **A** and **B**. **B** Mature seed (diagrammatic) with two bilobed cotyledons and hypocotyl. **C** Ruminate endosperm. **D**, **E** Embryo with radicle and two-lobed cotyledons. (After Capuron 1962)

The seed is reniform, exarillate and hairy along the dorsal side. It has copious endosperm, and a large curved embryo with broad cotyledons and a short radicle. The outer epidermis of the testa comprises subcuboid or transversely elongate cells. The mesophyll is aerenchymatous, the inner epidermis is of thin-walled subtabular cells. The outer epidermis of tegmen is a palisade layer. The outer hypodermis consists of large, radially elongate cells with brown and oily contents.

EJH Corner (1976), ASR Dathan and D Singh (1972a), J Mauritzon (1936a), K Schnarf (1931)

37.15 Tamaricaceae

The anther wall comprises a prominent, persistent thickly cuticularized epidermis, fibrous endothecium, two ephemeral middle layers, and secretory tapetum. The tapetal cells are large with amoeboid polyploid nuclei. Simultaneous cytokinesis follows meiotic divisions of microspore mother cells. Furrowing is centripetal, and the tetrads are isobilateral, tetrahedral and decussate.

The pollen grains are (two)-three-(four)-colpate, and two-celled at the dispersal stage.

The ovule is anatropous, bitegmic and crassinucellate (Fig. 37.15.2A). The inner integument grows faster than the outer one, and forms the micropyle. At the chalazal end, the epidermal cells elongate and grow into nucleated hairs. The hypodermal archesporial cell cuts off a parietal cell (Fig. 37.15.1A, B). The parietal tissue is single-layered. The megaspore mother cell undergoes meiotic divisions, and the development of the embryo sac is of the tetrasporic type.

Variations occur in the formation of the female gametophyte. It is of the Fritillaria type in *Myricaria germanica*, *Tamarix africana* (Battaglia 1941), *T. chinensis* (V. Puri 1939a), *T. dioica* (A.C. Joshi and Kajale 1936), *T. ericoides* (Y.M.L. Sharma 1939, 1940), *T. gallica* (Battaglia 1941), *T. odessana*, *T. parviflora* (Hjelmqvist and Grazi 1964, see also Battaglia 1988), *T. pentandra* and *T. troupii* (B.M. Johri and Kak 1954). Adoxa type has been observed in *T. aestivalis*, *T. africana*, *T. gallica* and *T. odessana* (Mauritzon 1936a), *T. parviflora* (Hjelmqvist and Grazi 1964), *T. pentandra* and *T. tetrandra* (Mauritzon 1936a) and *T. troupii* (B.M. Johri and Kak 1954). Chrysanthemum cinerariaefolium type occurs in *T. africana* and *T. gallica* (Battaglia 1941). Fritillaria, Adoxa and Chrysanthemum cinerariaefolium types occur in *T. troupii*, whereas, in addition to these three, Drusa type also occurs in *T. pentandra* (B.M. Johri and Kak 1954). Five types, i.e. Fritillaria, Chrysanthemum cinerariaefolium, Drusa, Adoxa and Plumbagella, have been recorded in *T. odessana* and *T. parviflora* (Hjelmqvist and Grazi 1964). Hjelmqvist and Grazi (1964) state that simpler reduced types are more frequent at low temperatures, whereas complicated types are frequent at high temperatures. Abnormal embryo sacs also occur in *T. pentandra.*

Since the Fritillaria type occurs more commonly in *Myricaria* and also in *Tamarix*, it may be considered to be characteristic of this family.

The pollen tube destroys one of the synergids. Sometimes, supernumerary pollen tubes have been observed in the embryo sac; one or both the synergids may be fertilized and develop into additional embryos (Figs. 37.15.2K, L, 37.15.3F).

Table 37.15.1. Distribution (%) of various types of embryo sacs in *Tamarix*

Type of embryo sac	*T. gallica*[a]	*T. africana*[a]	*T. pentandra*[b]	*T. troupii*[b]
Fritillaria	90	43	38	73
Drusa	3	10	8	0
Adoxa	2	0	48	14
Chrysanthemum *cinerariaefolium*	5	47	6	13

[a] After Battaglia (1941); [b] after B.M. Johri and Kak (1954)

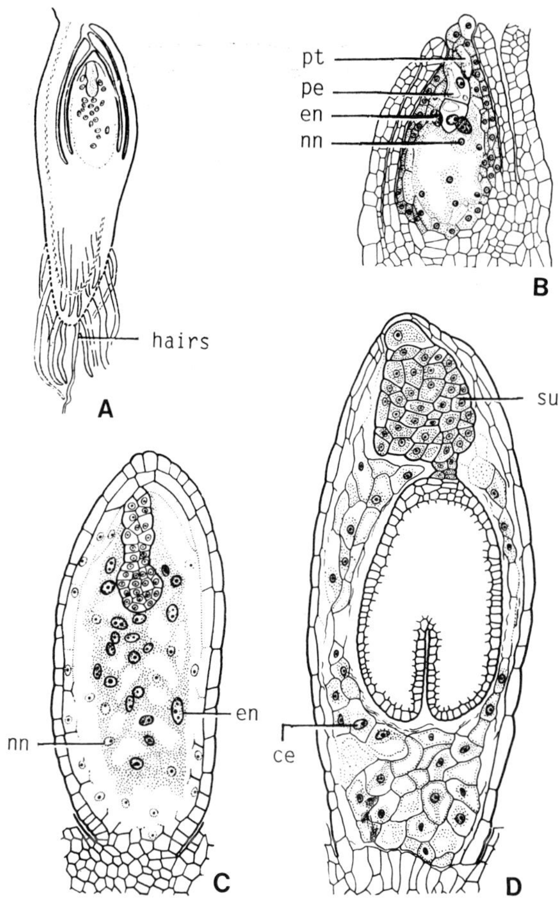

Fig. 37.15.1A–D. *Tamarix ericoides*. **A** Longisection of ovule at early globular proembryo. At the chalazal end the epidermal cells elongate and form a tuft of hairs (plume). **B** Upper part of ovule at bicelled proembryo (*pe*) and free-nuclear endosperm (*en*). The tangential walls of the innermost layer of nucellar cells break down and the nuclei (*nn*) migrate into the embryo sac. The tip of the nucellus has elongated and blocks the micropyle; *pt* pollen tube. **C** Nucellus with embryo sac from **A**; early globular proembryo and nucellar nuclei (*nn* smaller, lighter) mixed with endosperm nuclei (*en* larger, darker). **D** Dicotyledonous embryo with conspicuous suspensor (*su*), and cellular endosperm (*ce*). (After B.M. Johri and Kak 1954)

Development of the endosperm is of the Nuclear type (Fig. 37.15.1A–C). Besides the Nuclear type, Mauritzon (1936a) reports the Cellular type also in *T. tetrandra*, which is erroneous.

Wall formation in the endosperm is initiated in *T. troupii* at the globular proembryo, but in *T. ericoides* at the heart-shaped embryo. As the cotyledons differentiate, cellular endosperm fills the embryo sac (Fig. 37.15.1D). The endosperm is consumed during the maturation of seed (Fig. 37.15.3E, F); the remnants of a layer of nucellus persist (Fig. 37.15.3G).

After fertilization, the nucellar cells in contact with the gametophyte break down and their nuclei migrate into the embryo sac (Fig. 37.15.1A–C). The nucellar nuclei can be easily identified from the endosperm nuclei by their smaller size. Gradually, with the increase in the number of endosperm nuclei, the nucellar nuclei degenerate and are no more recognizable at the advanced globular proembryo. This feature has not been reported in any other species of *Tamarix*.

The development of the embryo in *T. troupii* and *T. ericoides* is of the Solanad type (Fig. 37.15.2A–J). A transverse division of the zygote generates the cells *ca* and *cb*. The division of *ca* produces the tiers *l* and *l′*, and *cb* forms *m* and *ci*. *m* divides transversely into *d* and *f* which undergo transverse and longitudinal divisions. *ci* gives rise to *n* and *n′*, and later divisions are irregular.

The derivatives of *ci* and *f* produce a massive suspensor of more or less vesicular cells which are richly cytoplasmic and contain abundant starch (Figs. 37.15.1D, 37.15.2I, J). The suspensor has a haustorial function and draws upon the reserve food from the adjacent cells of the nucellus and the endosperm. Their remnants persist in the mature seed (Fig. 37.15.3E, G).

The peripheral cells of the suspensor in *T. troupii* bulge out prominently.

The tiers *l* and *l′* divide longitudinally, twice, at right angles to each other, generating the quadrant and octant. Further divisions are followed by the delimitation of the dermatogen, and progressively the globular, heart-shaped and dicotyledonous embryo are formed.

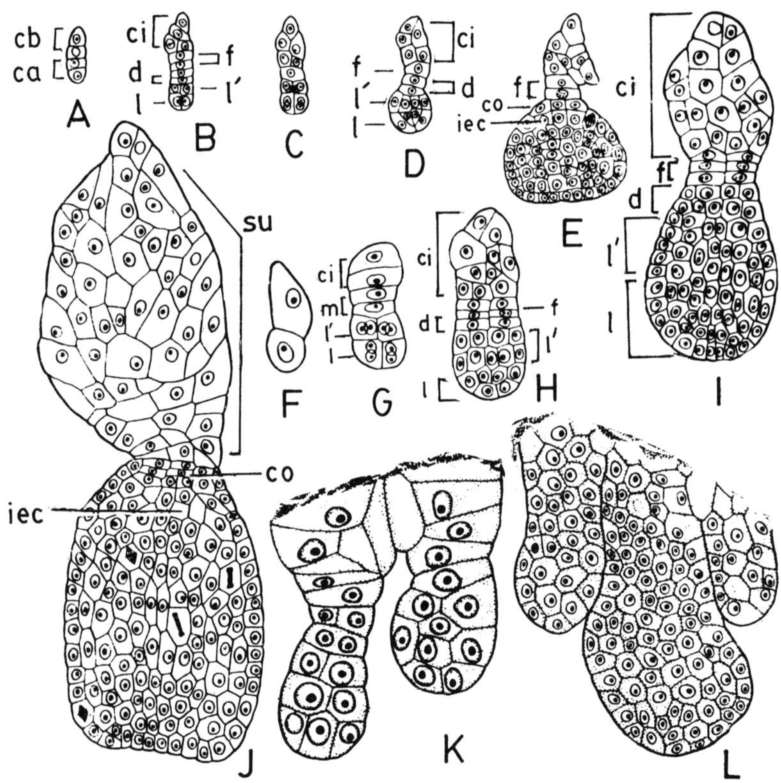

Fig. 37.15.2A–L. A–E *Tamarix troupii*. **F–L** *T. ericoides*. **A–J** Development of embryo, note prominent suspensor (*su*) in **I, J, K, L** Polyembryony, the additional embryo/s develop/s from fertilized synergid/s. (**A–J** After B.M. Johri and Kak 1954, **K, L** after Y.M.L. Sharma 1939)

The derivatives of *d* produce the tiers *co* and *iec* (Fig. 37.15.2J); *co* gives rise to root cap and *iec* to initials of the central cylinder of root.

Synergid polyembryony has been reported in *Tamarix ericoides* (Fig. 37.15.2K, L). One of the seeds showed a fully developed large embryo and an equally well developed somewhat smaller embryo (presumably from a fertilized synergid) in lateral position (Fig. 37.15.3F, G).

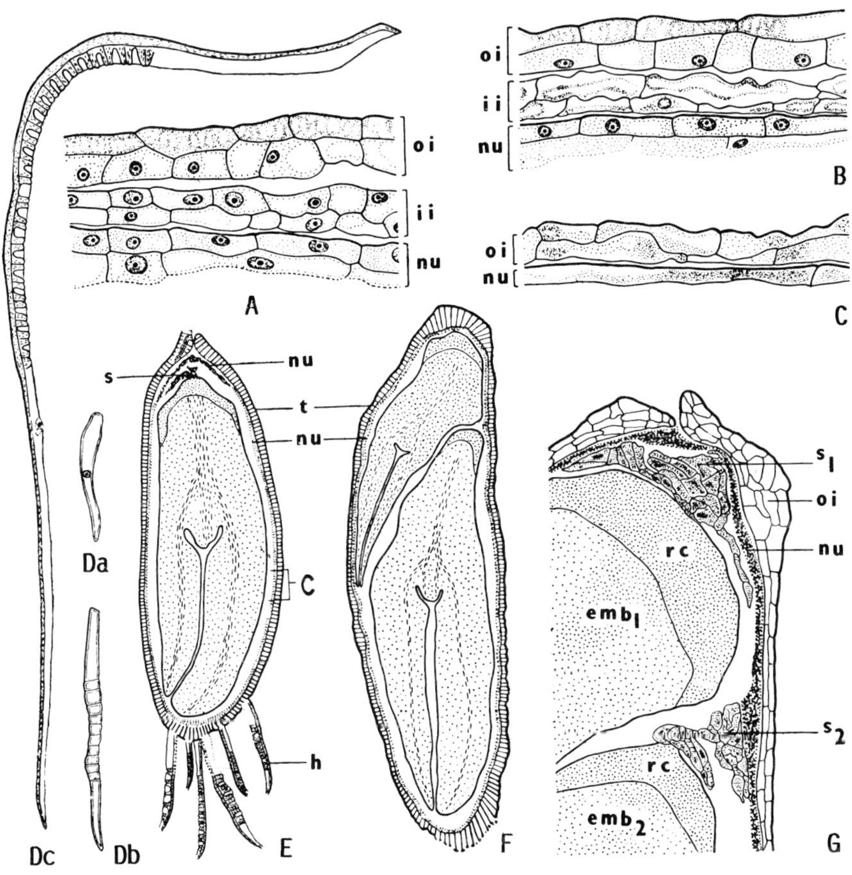

Fig. 37.15.3A–G. *Tamarix ericoides.* **A** Testa (*oi*) and tegmen (*ii*), each two-layered, and nucellus (*nu*) at early globular proembryo and free-nuclear endosperm (from **Fig. 37.15.2A**). **B** Later stage, the inner layer of nucellus has collapsed (from immature seed at the stage shown in **Fig. 35.15.2D**). **C** Testa and remnants of nucellus. **Da–Dc** Development of hair, from chalazal tuft. **E** Longisection of mature seed with chalazal thick-walled hairs (*h*), note the testa (*t*), remnants of nucellus (*nu*) and suspensor (*s*). **F** Seed with fully formed twin embryos. **G** Portion of seed (from a different successive section) shown in **F**. Note the region of micropyle, and remnants of the suspensor (S_1, S_2) of both the embryos (emb_1, emb_2) and nucellus; *oi* outer integument/testa; *rc* root cap. (After B.M. Johri and Kak 1954)

Traub (1939) observed 16 seedlings from one seed in *Myricaria cauliflora*; he did not trace the origin of polyembryony or the seedlings.

Each of the inner and outer integument is two-layered. The seed-coat comprises the longitudinally elongate thick-walled and pitted exotesta and thin-walled endotesta; the tegmen becomes crushed (Fig. 37.15.3A–C). The exotestal, exalbuminous seed has a large straight embryo, a layer of nucellus (perisperm), and a chalazal tuft of hairs (Figs. 37.15.3C–G).

Soon after fertilization, epidermal hairs develop at the elongated chalazal end of the ovule (Fig. 37.15.1A). They continue to enlarge during the growth and maturation of the seed. At first the hairs are one-celled and uninucleate. As the cotyledons differentiate, these cells repeatedly divide transversely to form a uniseriate filament. The thin-walled basal and apical cells remain uninucleate and considerably longer than the other cells which develop characteristic lignified thickenings (Fig. 37.15.3Da–Dc, E).

E Battaglia (1941, 1943, 1988), A Frisendahl (1912), H Hjelmqvist and F Grazi (1964), BM Johri and D Kak (1954), AC Joshi and LB Kajale (1936), J Mauritzon (1936a), V Puri (1939a), YML Sharma (1939, 1940), R Souèges (1924d, 1937b), HF Traub (1939)

37.16 Frankeniaceae

The anther wall comprises cutinized epidermis, endothecium, two ephemeral middle layers, and secretory tapetum of binucleate cells. The endothecium is fibrous, except in the region of stomium. The microspore mother cells undergo meiotic divisions, and produce tetrahedral, decussate and isobilateral tetrads. Cytokinesis is simultaneous.

The pollen grains are tricolporate, and two- or three-celled at the shedding stage.

The ovules are borne on a long funicle, and are anatropous, bitegmic and pseudo-crassinucellate. The inner integument forms the micropyle. An unusual feature in *Frankenia hirsuta* (see Yakovlev 1983) is the conspicuous extension of the inner integument beyond the outer integument. Consequently, the micropyle becomes very long. Walia and Kapil (1965) did not observe such a feature in the same species.

An incipient third integument originates from the outer integument in *F. hirsuta* (Walia and Kapil 1965). The nucellar epidermis may divide. The archesporial cell functions directly as megaspore mother cell. It undergoes meiotic divisions, produces a linear or T-shaped tetrad, and the chalazal

→

Fig. 37.16.1A–L. *Frankenia pulverulenta.* **A** Fertilized embryo sac. **B, C** Embryo sac with zygote and free-nuclear endosperm. **D** Later stage with two-celled proembryo. **E, F** Embryo sac at globular proembryo, endosperm nuclei aggregate in the micropylar region. **G, I, K** Outline diagrams for **H, J** and **L**. **H** Micropylar portion of embryo sac with multicellular micropylar haustorium, cellular endosperm, and part of suspensor. **J** Two-celled micropylar haustorium (each cell coenocytic), and cellular endosperm. **L** Micropylar region of embryo sac with coenocytic micropylar haustorium, from mature seed. (After Walia and Kapil 1965)

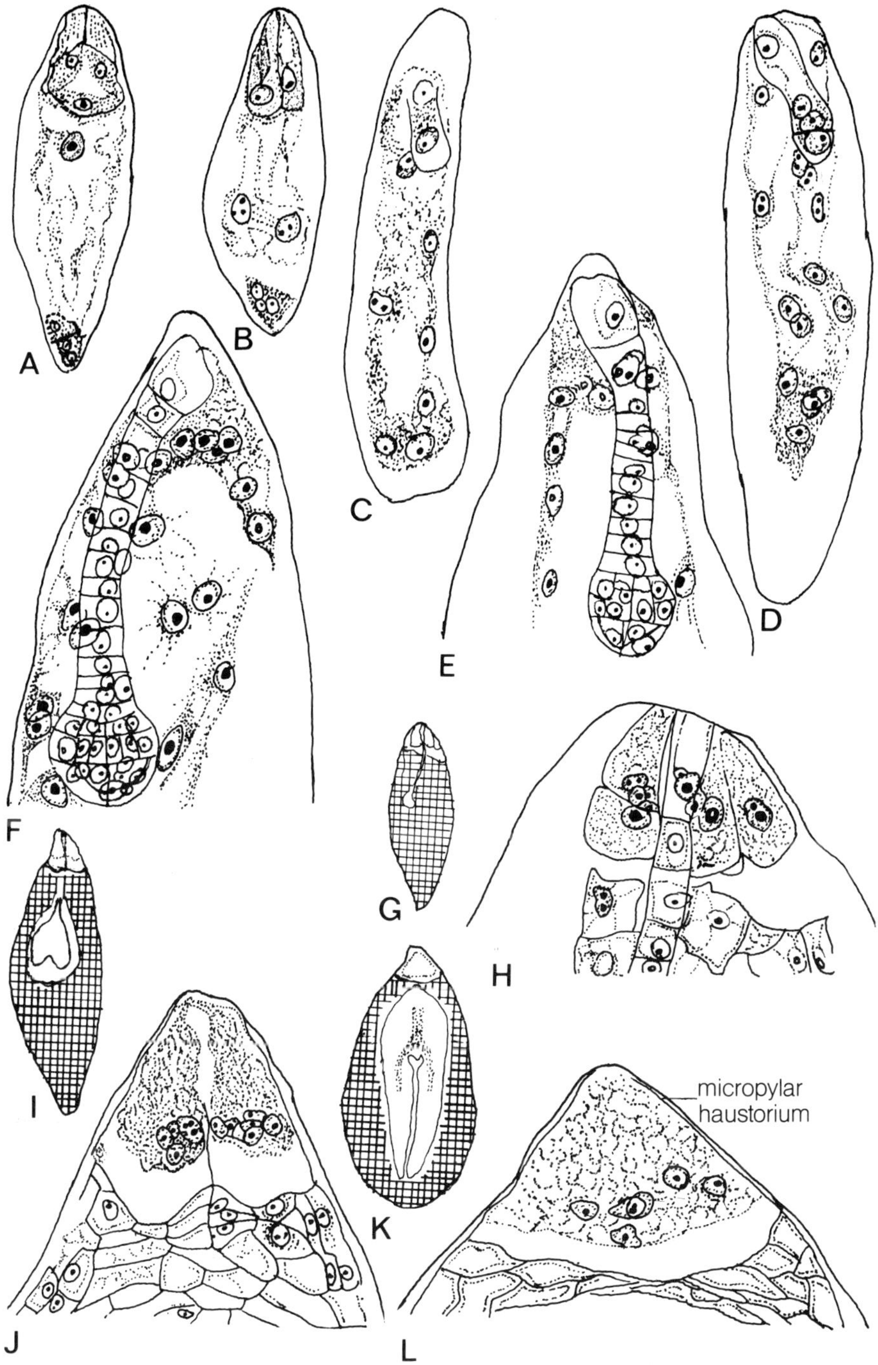
A
B
C
D
E
F
G
H
I
J
K
L
micropylar
haustorium

megaspore develops into a Polygonum type of embryo sac. The antipodals persist up to fertilization. The polar nuclei fuse near the egg.

The entry of the pollen tube is porogamous.

The development of the endosperm is of the Nuclear type (Fig. 37.16.1A–F). The nuclei aggregate in the micropylar and chalazal regions; nuclear fusions occur in the micropylar region. Wall formation starts in the chalazal region, and proceeds upward. The coenocytic micropylar part behaves as a haustorium which persists in the mature seed (Fig. 37.16.1G–L). Occasionally, one or more vertical walls are laid down in it (Fig. 37.16.1G–J). The chalazal cells are large and multinucleate. The endosperm cells contain starch and crystals.

The embryogeny conforms to the Solanad type. The mature embryo is dicotyledonous, and straight.

The outer thin-walled epidermis of the testa comprises large cells, and each cell develops a thick-walled papilla; the cells of the inner epidermis are tangentially elongate. The outer epidermis of the tegmen collapses, whereas the inner epidermal cells elongate radially, get filled with tannin, and the inner and tangential walls develop cuticle.

G Bentham (1962a), A Gundersen (1927), K Walia and RN Kapil (1965), MS Yakovlev (1983)

37.17 Elatinaceae

The anther wall comprises the epidermis, fibrous endothecium, two ephemeral middle layers, and secretory tapetum of binucleate cells. The epidermal cells contain tannin in *Bergia ammonioides* (Kajale 1939). The microspore mother cells undergo meiotic divisions. Cytokinesis is simultaneous, and the tetrads are tetrahedral and isobilateral.

The pollen grains are oblate, triangular with three germ pores, and two-celled in *Bergia*. They are three-celled in *Elatine*, and germinate in situ in the cleistogamous flowers of *E. triandra*.

The ovule is anatropous, bitegmic and crassinucellar with micropyle formed by both the integuments. The archesporium is one- to three-celled, only one is functional. It cuts off a parietal cell; the parietal tissue is massive in *B. campensis* (Raghavan and Srinivasan 1940). The divisions in nucellar epidermis also add to the parietal tissue. The megaspore mother cell undergoes meiotic divisions, and T-shaped or linear tetrads are produced. The chalazal megaspore develops into a Polygonum type of embryo sac. The antipodals persist up to fertilization, and the polar nuclei fuse prior to fertilization.

Development of the endosperm is of the Nuclear type. Centripetal wall formation commences from the micropylar pole in *Bergia* at the globular proembryo (Kajale 1939, Dathan and Singh 1971a), and from the chalazal pole upward in *Elatine* (Frisendahl 1927). Three to five layers persist in mature seed.

The embryogeny corresponds to the Solanad type. The suspensor is uniseriate, three-celled.

The seed is exotegmic, albuminous with a large embryo. Both the integuments contribute to the formation of seed-coat. The outer epidermis of the testa is thin-walled, unspecialized, and contains tannin. The inner epidermis gets crushed, or shows annular thickenings on radial walls. The outer epidermis of the tegmen comprises thick-walled, pitted, stellate cells. They may be lignified. The inner thin-walled cells contain tannin in *Bergia*.

ASR Dathan and D Singh (1971a), A Frisendahl (1927), LB Kajale (1939), R Lemesle (1929), TS Raghavan and VK Srinivasan (1940)

37.18 Caricaceae

The anther wall comprises the epidermis, fibrous endothecium, an ephemeral middle layer, and glandular tapetum. The microspore mother cells undergo meiotic divisions, and produce tetrahedral tetrads. Cytokinesis is simultaneous.

The pollen grains are three-colporate with reticulate exine; they are two-celled at the dispersal stage.

The ovule is anatropous, bitegmic and crassinucellate with micropyle formed by both the integuments. The archesporial cell cuts off a parietal cell; the parietal tissue is massive. The development of the female gametophyte is of the Polygonum type (D. Singh 1970c). Tetrasporic development is reported in *Carica candamarensis*, *C. chryopetata*, *C. papaya* and *C. pentagora* (Heilborn 1928). The antipodals are ephemeral. A hypostase is present.

The entry of the pollen tube is porogamous, and double fertilization occurs. The pollen tube is broad and its remnants persist. The pollen tube takes 10 days to reach the nucellus (under greenhouse conditions in *Carica papaya* – Foster 1943).

Development of the endosperm is of the Nuclear type. Cytoplasmic nodules are nucleate as well as enucleate in *Carica papaya* (D. Singh 1960); only enucleate nodules in *C. candamarensis* (Dathan and Singh 1976). Wall formation is centripetal in the endosperm which stores fatty food reserve.

The embryogeny conforms to the Onagrad type-Trifolium variation in *Carica* (Johansen 1950). Dathan and Singh (1976) describe it as of irregular type. The division of zygote is transverse, and a vertical division occurs in the apical cell. The basal cell divides transversely, obliquely or vertically. The mature embryo is large and spatulate. The development of the embryo requires reinvestigation.

The seed is albuminous, and exotegmic with multiplicative pulpy testa. The outer region of the testa is thin-walled (sarcotesta). The inner region comprises compact brown-walled cells. The inner epidermis contains calcium oxalate crystals. The outer epidermis of the tegmen comprises thick-walled lignified longitudinal fibres, the mesophyll may be sclerotic, and the inner epidermal cells are slightly thickened and contain tannin.

SP Agharkar and I Banerji (1930), ASR Dathan and D Singh (1970, 1976), JB Fischer (1980), LT Foster (1943), O Heilborn (1928), DA Johansen (1950), D Singh (1960, 1970c)

37.19 Loasaceae

The development of anther and pollen has not been studied.

The ovule is anatropous, tenuinucellate and unitegmic with a fairly long micropyle. The archesporial cell functions directly as megaspore mother cell. The latter undergoes meiotic divisions, and produces a linear tetrad. According to Kratzer (1918), the micropylar megaspore is functional which appears doubtful. In *Loasa bergii* (Garcia 1962a), the chalazal megaspore forms a Polygonum type of embryo sac. The antipodals persist. The embryo sac shows a constriction in the middle region separating the micropylar and chalazal regions. The endothelium is well developed, and in *Blumenbachia hieronymi* (Garcia 1962b) it surrounds only the middle region of the embryo sac.

Development of the endosperm is of the Cellular type. In *Blumenbachia hieronymi* (Garcia 1962b), the micropylar and chalazal chambers are formed following the division of primary endosperm nucleus. The micropylar chamber develops into a haustorium of two binucleate cells with hypertrophied nuclei (Fig. 37.19.1A, B). The chalazal chamber undergoes two successive transverse divisions and four cells are formed in a row. The basal cell develops into the chalazal haustorium and its branches encroach upon integumentary cells. The upper two cells divide and give rise to the cellular endosperm. The uppermost cell also contributes to the endosperm proper, but some of its daughter cells become haustorial.

The uninucleate chalazal haustorium of *Loasa burgii* is balloon-shaped (Fig. 37.19.1C–E). It remains healthy up to the initiation of cotyledons in the proembryo. The micropylar haustorium is branched, coenocytic, with hypha-like projections (Fig. 37.19.1C). The branches ramify inside the ovule and funiculus and reach up to the placenta.

The chalazal haustorium of *Cajophora silvestris* (Garcia 1962a) is elongated with a much-branched terminal portion containing a large hypertrophied nucleus (Fig. 37.19.1F, G). In *C. lateritia* (Kratzer 1918) the haustorium remains unbranched. The micropylar haustorium is short and broad with many nuclei.

In *Mentzelia laevicaulis* (Garcia 1962a) also the multicellular chalazal haustorium (with conspicuous nuclei) is unbranched. The micropylar haustorium is short and broad.

The embryogeny conforms to the Solanad type (Garcia 1962a), and it is assigned to Period I, Megarchetype V in *Loasa lateritia* (see Crètè 1946a, b). The mature embryo is straight with short cotyledons.

The exotestal seed is albuminous with thin-walled oily endosperm. The outer epidermis of the testa is variously thickened and of characteristic cells;

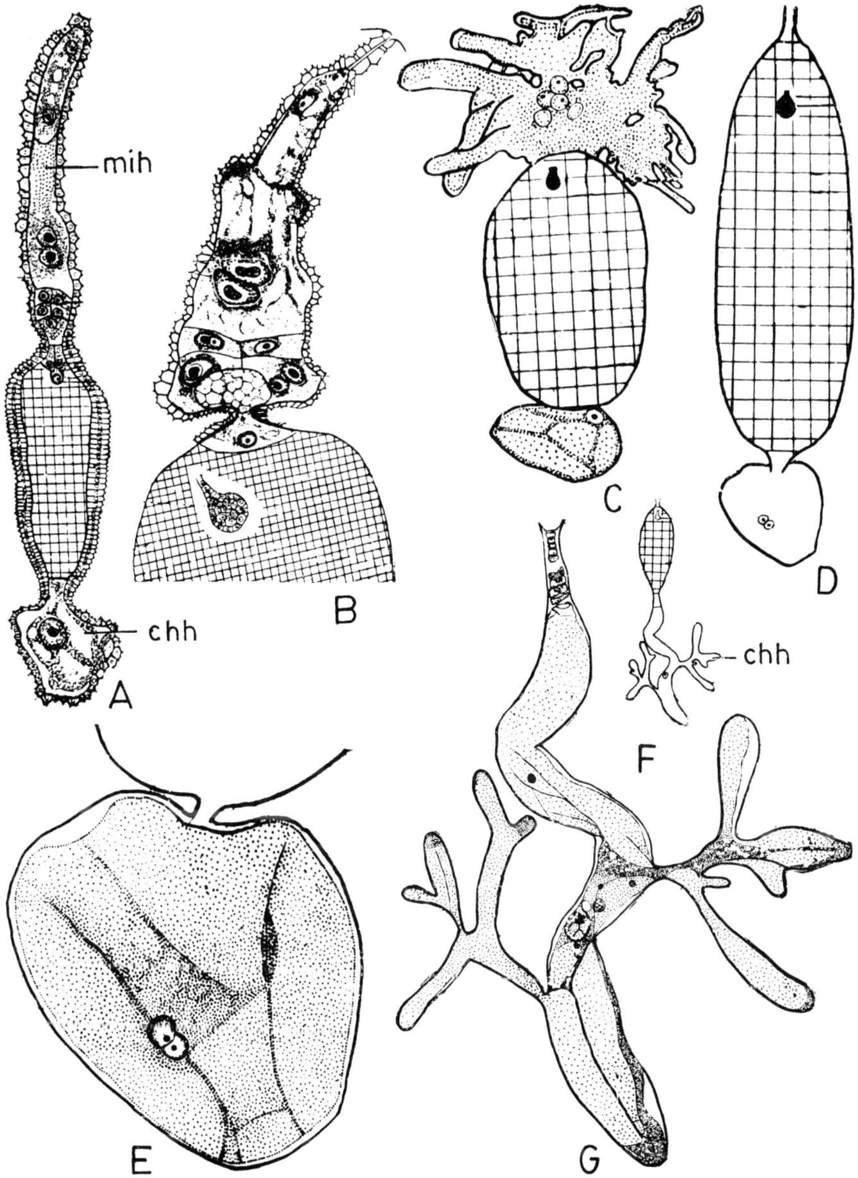

Fig. 37.19.1A–G. A, B, E–G From dissected wholemounts. **A, B** *Blumenbachia hieronymi*; **A** Embryo sac with zygote, two-celled (each cell binucleate) micropylar haustorium (*mih*), cellular endosperm proper (*cross-hatched*), and uninucleate chalazal haustorium (*chh*); **B** Upper portion of embryo sac to show globular proembryo, cellular endosperm and two-celled micropylar haustorium (After Garcia 1962b). **C–E** *Loasa bergii*; **C** Proembryo (*black* in **C, D**), cellular endosperm, and much-branched micropylar and vesicular chalazal haustorium; **D** Outline figure for **E**; **E** Chalazal haustorium from **D**. **F, G** *Cajophora silvestris*; **F** Proembryo, cellular endosperm and chalazal haustorium; **G** Chalazal haustorium from **F**. (After Garcia 1962a)

mesophyll and endothelium generally get crushed. In *Blumenbachia* the outer epidermis is composed of longitudinal fibre-like cells with strong yellowish-brown thickenings on radial walls. In *Cajophora* the outer epidermis consists of large radially elongate cells with stout brown walls, and the radial walls have reticulate thickenings (Corner 1976).

EJH Corner (1976), P Crètè (1946a, b), V Garcia (1962a, b), P Goldblatt, P Bernhardt and J Manning (1989), P Kooiman (1974), J Kratzer (1918)

37.20 Datiscaceae

The embryology has not been fully studied.

The ovule is dizonate in *Datisca cannabina* (Boesewinkel 1984c), and both the integuments are of dermal origin. They are two cell layers thick, and only

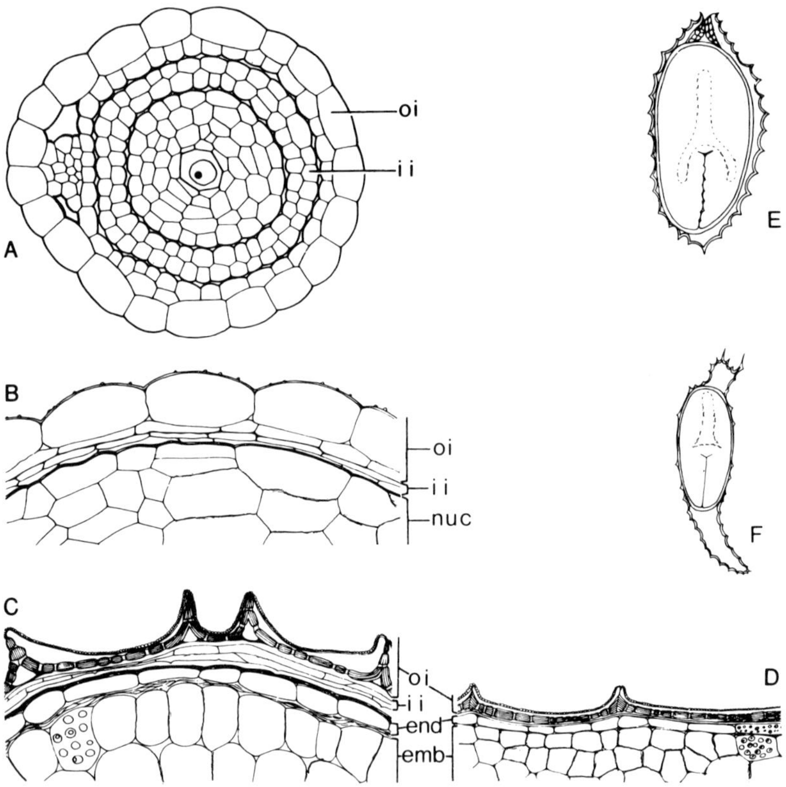

Fig. 37.20.1A–F. A–C, E *Datisca cannabina,* **D, F** *Octomeles sumatrana.* **A** Cross-section of immature ovule to show intercellular spaces next to raphe. **B** Transection of immature seed-coat, the outer epidermis of outer integument (*oi*) is cuticularized, the inner epidermis of *oi* and tegmen (*ii*) have become compressed, and the perisperm (*nuc*) persists. **C, D** Transection of mature seed-coat. The inner and radial walls of the outer epidermal cells are thick, lignified and pitted, the inner layers have become compressed. A single layer of endosperm (*end*) persists; *emb* embryonal tissue. **E, F** Longisection of mature seed. Note the chalazal air bladder in **F**. (After Boesewinkel 1984c)

the tip of the outer integument becomes multilayered. Both the integuments organize the micropyle. Before the maturation of the ovule, longitudinally oriented intercellular spaces appear on the side of the raphal bundle (Fig. 37.20.1A) and in the chalazal region.

The archesporial cell cuts off a parietal cell, the parietal tissue is well developed. The apical cells of the nucellar epidermis undergo a periclinal division. The development of the embryo sac conforms to the Allium type in *D. cannabina* (Himmelbaur 1909). The antipodals are persistent. The polar nuclei fuse prior to fertilization.

Development of the endosperm is of the Nuclear type, later it becomes cellular.

The embryogeny conforms to the Onagrad type.

The operculate (Fig. 37.20.2B) and exotestal small seed has a straight and large embryo (Fig. 37.20.1E) rich in fats and aleurone, but lacks starch. The

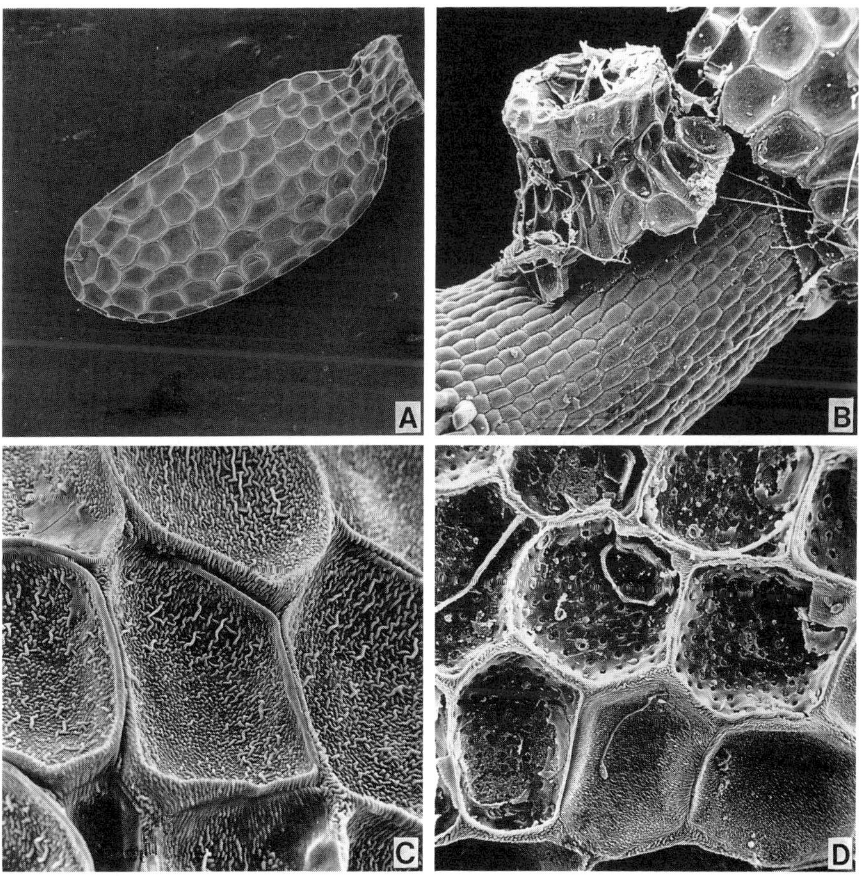

Fig. 37.20.2A–D. (SEM) *Datisca cannabina.* **A** Seed. **B** Germinated seed with seed lid. **C** Portion of seed-coat. **D** Same, outer walls partly removed to show inner pitted walls. (Courtesy Dr F. Bouman, Amsterdam)

outer epidermal cells of the testa enlarge and become cuticularized (Figs. 37.20.1B, E, F, 37.20.2A–C). The inner and the radial walls become thick, lignified and pitted (Figs. 37.20.1C, D, 37.20.2D). The tegmen becomes crushed. The perisperm persists for a while but is finally absorbed. A layer of endosperm persists in the mature seed. The intercellular spaces in the chalazal region become so prominent that an air bladder develops at the tapering chalazal region in mature seed (Fig. 37.20.1F). There is no special rupture layer to lift the operculum (Fig. 37.20.2B).

FD Boesewinkel (1984c), P Crété (1952a, b), W Himmelbaur (1909), P Leins and R Bonnery-Brachten (1977), MA Rau (1962)

37.21 Begoniaceae

The anther wall comprises the epidermis, fibrous endothecium, two or three ephemeral middle layers, and secretory tapetum of binucleate cells. The epidermal cells are protuberant and papillose. The microspore mother cells undergo meiotic divisions and produce tetrahedral tetrads. Simultaneous cytokinesis is reported in *Begonia malabarica* (Maheswari Devi and Naidu 1978).

The pollen grains are two-celled at the dispersal stage. Failure of formation of pollen grains, due to degeneration of pollen mother cells, is reported in *B. anamalayana* (Maheswari Devi and Naidu 1978).

The ovule is anatropous, bitegmic, and crassinucellate (Fig. 37.21.1A, B); weakly crassinucellate in *Begonia picta* (Madhulata 1956). Both the integuments form the zigzag micropyle. The hypodermal archesporial cell cuts off a parietal cell. The megaspore mother cell undergoes meiotic divisions and produces a linear tetrad. The chalazal megaspore develops into a Polygonum type of embryo sac. The polar nuclei fuse before fertilization, and the antipodals are ephemeral. A hypostase differentiates. The pseudo-endothelium is well developed (Maheswari Devi and Naidu 1978, see also Boesewinkel and de Lange 1983) but, at times, it accumulates reserve food which causes degeneration of the embryo sac.

Development of the endosperm is of the Nuclear type (Fig. 37.21.1C, D). The nuclei aggregate in the chalazal region with denser cytoplasm. When 30–40 free-nuclei have been formed, they arrange peripherally around the central vacuole. Wall formation results in a single layer which undergoes periclinal divisions. The inner layer divides repeatedly and gives rise to the rest of endosperm (Swamy and Parmeswaran 1960). The outer layer does not undergo any further divisions, stores aleurone grains and oil, and persists in the mature seed (Fig. 37.21.1E).

The embryogeny conforms to the Onagrad type-Lythrum variation in *Begonia crenata* (Swamy and Parmeswaran 1960). It is assigned to Period I, Series A, Megarchetype IV (Crucifer type).

The seed-coat (Boesewinkel and de Lange 1983) comprises an operculum, collar cells and the remaining testa cells (Fig. 37.21.2A, B). The entire micro-

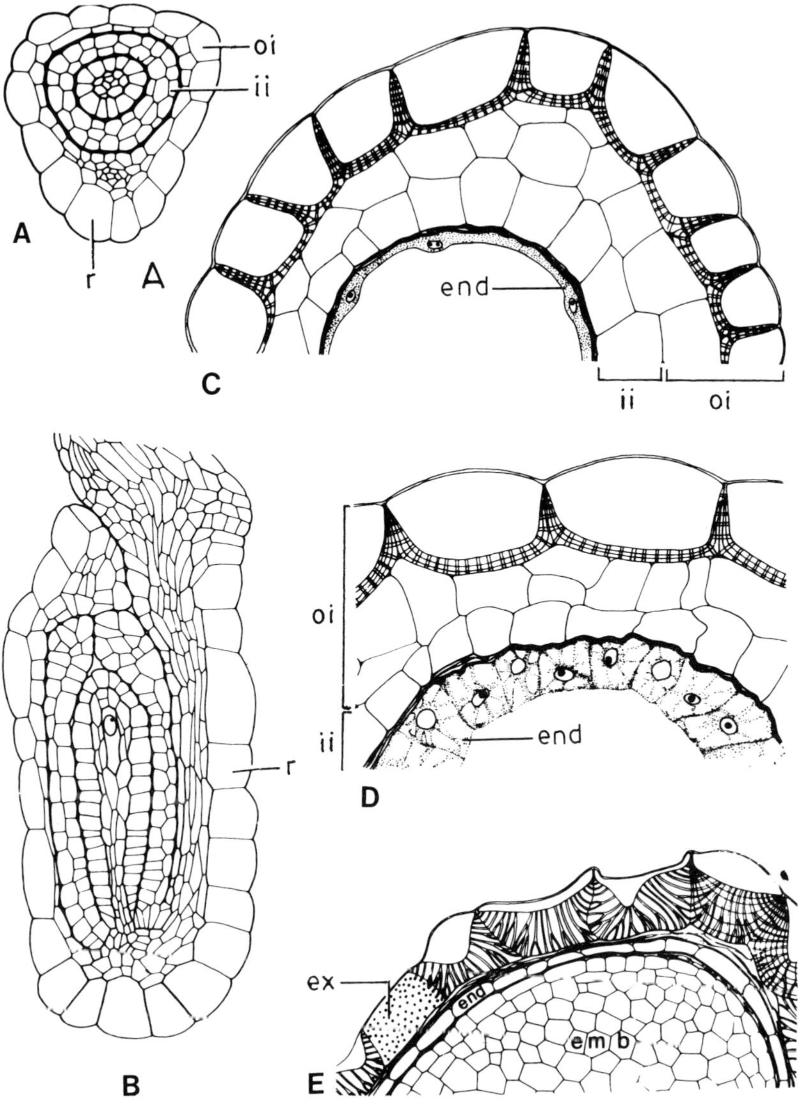

Fig. 37.21.1A–E. *Begonia squamulosa.* **A** Transection of ovule; *oi* outer integument; *ii* inner integument; *r* raphe. **B** Longisection at megaspore mother cell. **C–E** Transections, progressive development of seed-coat. Note the thickenings of exotesta (*ex*); *end* endosperm; *emb* embryo. (After Boesewinkel and Lange 1983)

pylar-hilar region of seed forms the operculum (see lid) which separates along preformed lines of rupture; the walls between the operculum and collar cells split along the middle lamella. Next to the operculum lies a ring of collar cells (Fig. 37.21.2A, B), an exclusive feature of begonian seed. The collar cells are specialized testa cells and the ring occupies nearly one-third of the seed. Sometimes, the collar cells, along with adjacent cells, become twisted.

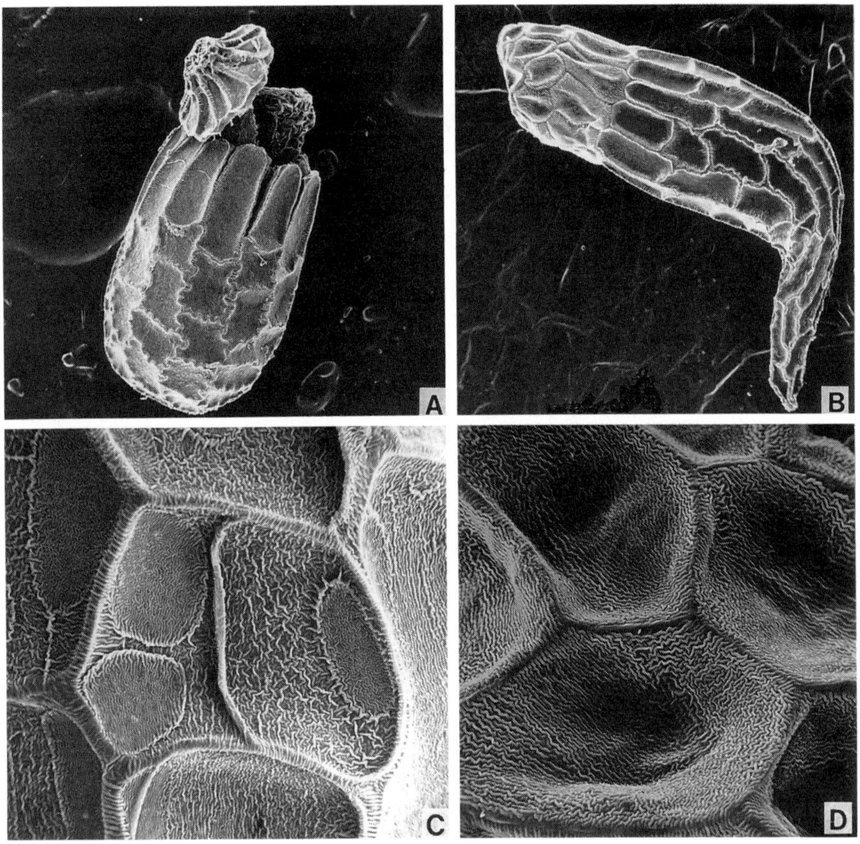

Fig. 37.21.2A–D. (SEM) **A** *Begonia leptetrida*, seed with seed-lid, after germination. **B** *B. patula*, seed. **C** *B. epipsila*, **D** *B. seychellensis*, portion of seed-coat. (Courtesy Dr F. Bouman, Amsterdam)

In *Begonia solananthera* (de Lange and Bouman 1986), the collar cells are elongated; the anticlinal walls between the cells are straight but finely undulated near their chalazal end. The cells bordering the collar are longer and form a ring. The operculum is funnel-like and sunken with hilum in *B. integerrima*, *B. procumbens* and *B. solananthera*. In *B. thomeana* the seeds show a marked adaptation to wind dispersal, and both ends of the seed consist of strongly swollen cells which function as balloons.

During germination the operculum is lifted at the top of the radicle, thus clearing the way for the emergence of seedling.

The ellipsoid seed is small, and only the testa forms the seed-coat represented by the outer epidermis (Figs. 37.21.1E, 37.21.2A–D). The cells expand in surface area. Their inner periclinal walls are strongly lignified with scattered simple pits, whereas the outer walls remain thin. They are filled with amber-coloured contents.

FD Boesewinkel and A de Lange (1983), F Bouman and A de Lange (1983), A de Lange and F Bouman (1986), Madhulata (1956), H. Maheswari Devi and KC Naidu (1978, 1982), W Preil and D Keyser (1976), R Souèges (1939a), BGL Swamy and N Parmeswaran (1960)

37.22 Taxonomic Considerations

Bentham and Hooker (1965a) distribute the families belonging to Violales in four-cohorts: Parietales, Caryophyllales, Guttiferales and Passiflorales. Engler and Diels (1936) include all the families of Violales (Melchior 1964) in Parietales under seven suborders, whereas Hutchinson (1973) places them in seven orders. According to Melchior (1964), if Guttiferales are removed from Parietales, the ·remaining taxon, i.e. Violales, forms a homogeneous and natural group.

The suborder Flacourtiineae comprises the families Flacourtiaceae, Peridiscaceae, Violaceae, Stachyuraceae, Scyphostegiaceae, Turneraceae, Malesherbiaceae, Passifloraceae and Achariaceae.

Flacourtiaceae. Flacourtiaceae closely resemble Violaceae in embryological features, yet there are important variations: a characteristic hook-like structure at the tip of the embryo sac, extra-nucellar embryo sac, variation in the functional behaviour of megaspores, and elongation of the embryo up to half the length of the embryo sac. The structure of seed-coat in Flacourtiaceae varies to such an extent that it forms two groups, Flacourtia group with fibrous exotegmen, and Hydnocarpus group with pachychalaza and without a trace of fibrous exotegmen (Corner 1976). Fibrous exotegmen occurs in Violaceae. Its embryogeny may reveal some more facts.

Hutchinson (1973) includes Flacourtiaceae in Bixales along with Bixaceae, Cistaceae, Cochlospermaceae (see Tables 37.22.1 and 37.22.2) and Peridiscaceae. Bixaceae differ from Flacourtiaceae in endothecial thickenings confined to dehiscence slits, micropyle formed by both integuments, one or two archesporial cells, absence of a parietal cell, and irregular embryogeny. The embryo does not elongate in Bixaceae and Cistaceae, and dye cells are present in the outer integument in Bixaceae. Cistaceae show a cap-like structure at the tip of synergid, Solanad type of embryogeny, and a bulge at the base of integuments in the region of attachment. Moreover, the seeds are exarillate. Therefore, Flacourtiaceae are rightly placed near Violaceae.

Violaceae. Bentham and Hooker (1965a) place Violaceae in Parietales along with Papaveraceae, Resedaceae, etc., Bessey (1915) in Guttiferales between Turneraceae and Bixaceae, and Hutchinson (1973) and R. Dahlgren (1980b, 1981) in Violales.

D. Singh (1963, 1970a) states that the structure of seed in Violaceae resembles that of Cistaceae and Bixaceae of suborder Cistineae, Caricaceae of Caricineae, and Resedaceae of the order Papaverales (Melchoir 1964). The seed is endospermic with an erect and spatulate embryo in the first three taxa. In

Resedaceae the embryo is curved, and the seed is exotegmic with tanniniferous hypostase. Both the integuments take part in the formation of seed-coat. The inner epidermis of the outer integument contains calcium oxalate crystals in Violaceae, Caricaceae and Resedaceae. The outer epidermis of the inner integument is the main mechanical layer in Violaceae, Cistaceae, Bixaceae and Resedaceae. In Violaceae and Resedaceae sclereids in the chalazal region plug the gap in sclerenchymatous layer. The sclereids are absent in the other three families. Thus, the development and structure of violaceous seed show greater similarities to those of Resedaceae. But the anatropous ovule, Asterad type of embryogeny, and massive endosperm in Violaceae differ from campylotropous ovule, Onagrad type of embryogeny, curved embryo, and scanty endosperm in Resedaceae (D. Singh 1963). While considering the structure of seeds of Capparidaceae, Resedaceae and Violaceae, D. Singh and Gupta (1968) suggest that these families are closely related, and have probably been derived from a common ancestor similar to the existing Capparidaceae, but with anatropous ovules, through parallel evolution.

Stachyuraceae. Stachyuraceae, a unigeneric family, have been included in Theaceae (Ternstroemiaceae – Bentham and Hooker 1965a), but have been separated on morphological grounds. Hutchinson (1973) retains Stachyuraceae in Hamamelidales near Buxaceae. Both Stachyuraceae and Buxaceae have superior ovary. Takhtajan (1969, 1980) places them in Violales near Flacourtiaceae. Stachyuraceae show several violaceous features: anatropous, bitegmic, crassinucellate ovule with micropyle formed by both integuments, presence of funicular aril, Polygonum type of embryo sac (Sâto 1976b), Nuclear endosperm, and Solanad type of embryogeny.

Mathew and Chaphekar (1977) and Mathew (1978) observed several important embryological similarities between Stachyuraceae and Theaceae: absence of integumentary tapetum, occasional presence of bisporic embryo sac, fusion of polar nuclei before fertilization, Solanad type of embryogeny, similar structure of suspensor, and presence of polyembryony. According to them, Stachyuraceae have affinities with Flacourtiaceae and Hamamelidaceae, the families with monosporic (Polygonum) type of embryo sac, and Theaceae, the family with bisporic type. They support the inclusion of Stachyuraceae near Theaceae in the order Theales, according to the classification of Cronquist (1968; in 1981, Stachyuraceae have been included in Violales), and negate its relationship with Hamamelidaceae. This alliance is not convincing.

In contrast to the exotegmic fibrous seed-coat of Flacourtiaceae (of Flacourtia group) and Violaceae, the tegmen in Stachyuraceae gets crushed, whereas the outer epidermis of the testa becomes sclerotic and forms the seed-coat (Mohana Rao 1974a). The absence of fibrous exotegmen negates its alliance with Flacourtiaceae (see Takhtajan 1969, 1980). Stachyuraceae show resemblances to Hamamelidaceae, where the entire testa is sclerotic and the tegmen gets crushed. The theaceous testa is rather of a different nature (Mohana Rao 1974a). It has a thin-walled epidermis followed by many layers of pitted and lignified sclerenchyma cells (mesotesta).

The foregoing discussion is clearly indicative of the relationship between Stachyuraceae and Hamamelidaceae.

Turneraceae. Turneraceae and Passifloraceae (Table 37.22.2) of suborder Flaucortiineae are embryologically allied. Turneraceae is placed in Violales by Takhtajan (1966, 1980), and in Loasales by Hutchinson (1973). Vijayaraghavan and Kaur (1966) and D. Kaur (1970) have compared the embryological features of Turneraceae, Passifloraceae, Flacourtiaceae and Loasaceae. Turneraceae show marked dissimilarities with Loasaceae in two-celled pollen grains, anatropous, bitegmic, crassinucellate ovules, ephemeral antipodal cells, Nuclear endosperm without haustoria, Onagrad type of embryogeny, and presence of aril (D. Kaur 1970). In most of these respects, it closely resembles Passifloraceae. The seed-coat is formed by a single integument in Loasaceae, whereas in Turneraceae and Passifloraceae it is formed by both integuments. The cells of the outer epidermis of the inner integument are obliquely oriented in Turneraceae and appear transitional to fibres, whereas the exotegmen is sclerotic with radially oriented cells in Passifloraceae.

Passifloraceae. Bentham and Hooker (1965a) place Passifloraceae in Passiflorales with Turneraceae (Table 37.22.2) and Cucurbitaceae. Hutchinson (1973) includes them in Passiflorales, and places the order close to Cucurbitales. Takhtajan (1980) assigns them close to Turneraceae in Violales. D. Singh (1970b) has summarized the embryological features of Cucurbitaceae in which they differ from Passiflorales: flask-shaped nucellus, equal thickness of integuments, vascular supply extending up to the tip of the outer integument, endosperm with chalazal haustorium, seed-coat from outer integument alone, formation of main sclerenchymatous layer from the outer epidermis of the outer integument, insignificant endosperm in seed, and absence of aril. The presence of ruminate endosperm, and succulent aril enclosing the seed are characteristic of Passifloraceae, otherwise this family shows close affinities with Turneraceae (Vijayaraghavan and Kaur 1966).

Cistaceae. The suborder Cistineae of Violales (Melchior 1964) comprises Cistaceae, Bixaceae, Sphaerosepalaceae and Cochlospermaceae. The embryology of Sphaerosepalaceae has not been investigated, and meagre data are available about Cochlospermaceae (see Schnarf 1931, Dathan and Singh 1972a). Bentham and Hooker (1965a) place Cistaceae and Bixaceae in Parietales, whereas Hutchinson (1973) includes them in Bixales. The three families – Cistaceae, Bixaceae and Cochlospermaceae – resemble one another (Table 37.22.1) in secretory tapetum, bitegmic, crassinucellate ovules with nucellar cap, single vascular strand terminating at the chalaza, Polygonum type of embryo sac, frequent occurrence of twin embryo sacs, Nuclear endosperm, spatulate embryo (curved in Cochlospermaceae), both integuments form seed-coat, and outer epidermis of the inner integument forms the main sclerenchymatous layer in seed.

Orthotropous ovule with a characteristic bulge of the basal cells of the outer integument in the region of attachment with the nucellus, multicellular archesporium with a massive parietal tissue, synergids with cap-like structure at the base, Solanad type of embryogeny, and association of testa with fungus (Kapil and Maheshwari 1964) are the distinguishing features of Cistaceae.

Bixaceae. Bixaceae show the presence of Übisch granules, endothecial thickenings confined to dehiscence slits, single archesporial cell, absence of parietal tissue, irregular embryogeny, presence of dye cells and oil cells in testa, and exotegmen develops into thick palisade-like cells and hypodermal layer (from inside) into hour-glass cells (H. Kaur 1969). Cochlospermaceae are closely allied to Bixaceae. The distinction between the seed structure of the two families is so meagre that Corner (1976) does not segregate them (Table 37.22.1).

The suborder Tamaricineae comprises Tamaricaceae, Frankeniaceae and Elatinaceae. Bentham and Hooker (1965a) place Frankeniaceae and Tamaricaceae along with Caryophyllaceae in Caryophyllineae, and Elatineae in Guttiferales with Hypercineae. Bessey (1915) assigns all the three families to Caryophyllales. Hutchinson (1973) includes Frankeniaceae and Tamaricaceae with Fouquieriaceae in Tamaricales, and Elatinaceae in Caryophyllales. R. Dahlgren (1980b, 1981) erected the order Tamaricales with Tamaricaceae and Frankeniaceae in superorder Violiflorae.

Tamaricaceae. The Tamaricaceae have several important features: two-celled pollen grains, nucellar parietal cells, tetrasporic (Fritillaria, Adoxa, Chrysanthemum cineraraefolium, Drusa and Plumbagella) embryo sac, Nuclear endosperm without haustoria, Solanad type of embryogeny with a massive suspensor, frequent occurrence of polyembryony, exalbuminous, exarillate, simple exotestal seed, and seed-coat derived from the outer integument only (the outer epidermis is a mechanical layer, the inner one contains crystals).

The caryophyllaceous characters, such as free-central placentation, three- to six- or 12-porate pollen grains, multicelled archesporium, Caryophyllad type of embryogeny, curved embryo, and perisperm (A.C. Joshi 1936a, Pal 1952) are absent in all the three families; Tamaricaceae, Frankeniaceae and Elatinaceae. The presence of parietal cells in the nucellus, absence of endosperm haustoria, exalbuminous seed, and seed-coat formed only by the outer integument (A.C. Joshi 1936a, Pal 1952) in Caryophyllaceae are features similar in Tamaricaceae. As several types of tetrasporic embryo sacs do not occur in any other violaceous family, B.M. Johri and Kak (1954) conclude that Hutchinson's erection of a separate order Tamaricales is justified.

Frankeniaceae. Walia and Kapil (1965) studied the embryology of *Frankenia hirsuta* and *F. pulverulenta*. In contrast to Tamaricaceae, Frankeniaceae have three-celled pollen grains, single female archesporium directly functions as megaspore mother cell, Polygonum type of embryo sac, Nuclear endosperm with two- to four-celled micropylar haustorium, albuminous seed with a straight embryo without suspensor (in mature embryo), seed-coat derived from both integuments, and finger-like projections from the outer epidermis of the testa which gives a warty appearance.

Table 37.22.1. Embryology of Cistaceae, Bixaceae and Cochlospermaceae (see Dathan and Singh 1972a)

	Cistaceae	Bixaceae	Cochlospermaceae
Pollen grains	Usually 3-colporate, suboblate-prolate, 2-celled	3-porate, or 3-colporate, oblate, 2- or 3-celled	3-colporate, prolate, spheroidal, subprolate
Gynoecium	Tricarpellary or 5–10 carpels	2-, rarely 3-carpellary	3–5-carpellary
Ovule	Orthotropous, bitegmic, crassinucellate with nucellar cap	Anatropous, bitegmic, crassinucellate with nucellar cap	Ana- or anacampylotropous, bitegmic, crassinucellate with nucellar cap
Integuments	Outer integument 2-layered, inner 4-layered	Outer integument 4-layered, inner 6–7-layered	Outer integument 4-layered, inner 6–7-layered
Vascular supply	Single, terminating at chalaza	Single, terminating at chalaza	Single, terminating at chalaza
Embryo sac	Polygonum type, twin embryo sacs occur, antipodals ephemeral	Polygonum, twin embryo sacs frequent, antipodals ephemeral	Polygonum type, twin embryo sacs frequent, antipodals ephemeral
Endosperm	Nuclear type	Nuclear	Nuclear
Embryo	Spatulate, bent, coiled or folded	Spatulate, large, cotyledons broad, incurved at the apex	Curved, radicle rather short, cotyledons ligulate
Mature seed	Ovate, endospermous	Ovate, endospermous	Reniform, endospermous
Chalazal modifications	Chalazal region distinguishable into six zones, podium and tanniniferous postament conspicuous	8 zones, podium and tanniniferous postament conspicuous	7 zones, podium and tanniniferous postament conspicuous
Seed-coat	Both integuments form seed-coat, distinguishable into 3 zones, outer epidermis of inner integument forms the sclerenchymatous layer	Both integuments, distinguishable into 8 zones, outer epidermis of inner integument forms the main sclerenchymatous layer, dye cells in outer parenchyma, inner hypodermis forms hour-glass-shaped cells	Both integuments, distinguishable into 6 zones, outer epidermis of inner integument forms mechanical layer, gum cavities in inner parenchyma

Frankeniaceae resemble Elatinaceae in secretory two-layered tapetum, simultaneous cytokinesis, Polygonum type of embryo sac, Nuclear endosperm with initiation of wall formation from the chalazal end, and Solanad type of embryo.

Walia and Kapil (1965) emphasize that the assignment of Frankeniaceae to Tamaricales of Hutchinson and Takhtajan, and to Caryophyllinae of Bentham and Hooker, is unjustifiable. There is also a remote possibility of its relationship with Fouquieriaceae or Plumbaginaceae. The proper position of Frankeniaceae is in the order Parietales, near Elatinaceae, as suggested by Wettstein (1935) and Rendle (1952).

Elatinaceae. In addition to the features mentioned along with Frankeniaceae, Elatinaceae have two-celled pollen, a zigzag micropyle, massive parietal tissue, ephemeral antipodals, and exotegmic seeds with massive perisperm and scanty endosperm. The presence of parietal cells in nucellus, and exalbuminous seed with perisperm and absence of endosperm haustoria show the similarity with Caryophyllaceae. Takhtajan (1980) places Elatinaceae in Theales. Corner (1976), on the basis of seed structure, relates it to Hypericaceae. Dathan and Singh (1971a) also confirm that the cells of the outer epidermis of the inner integument are similar to those of *Hypericum*. The massive nucellus in Elatinaceae is exceptional.

Caricaceae. The suborder Caricineae comprises the lone taxon Caricaceae. Hutchinson (1973) places Caricaceae in Cucurbitales next to Begoniaceae and Datiscaceae. D. Singh (1963) refutes their affinities with Cucurbitaceae on embryological grounds. Most embryological features are in conformity with those of the order Violales. Tetrasporic embryo sac in some species, persistent pollen tube, cytoplasmic endosperm nodules, and sarcotestal seed are its special features. The Nuclear endosperm, and Onagrad type of embryogeny resemble those of Begoniaceae and Datiscaceae. The structure of the seed is comparable to that of Violaceae.

Loasaceae. The suborder Loasineae comprises Loasaceae. Bentham and Hooker (1965a) include it in Passiflorales, Engler and Diels (1936) in Parietales, and Hutchinson (1973) in Loasales along with Turneraceae. Cronquist (1981) places it in Violales, Takhtajan (1980) in Loasales, whereas R. Dahlgren (1981) assigns it to superorder Loasiflorae, order Loasales, in sympetalous group with iridoid compounds. Vijayaraghavan and Kaur (1966) point out marked dissimilarities between Turneraceae and Loasaceae (Table 37.22.2).

The hemianatropous, unitegmic, tenuinucellate ovule, archesporial cell directly functioning as megaspore mother cell, presence of endothelium, persistent (rarely ephemeral) antipodals, and Cellular endosperm with aggressive micropylar and chalazal haustoria (Garcia 1962a) show the affinities with sympetalous families. Garcia (1962a) compares Solanad type of embryogeny with that in Solanaceae (Polemoniales), wherein iridoid compounds are absent. As the androecium is extremely variable in Loasaceae, Wagenitz (1977a, b) did not assign them to any of the existing orders in Asteridae. According to

Table 37.22.2. Embryology of Flacourtiaceae, Turneraceae, Passifloraceae and Loasaceae (See Vijayaraghavan and Kaur 1966, D. Kaur 1970)

	Flacourtiaceae	Turneraceae	Passifloraceae	Loasaceae
Anther tapetum	Secretory, cells 2–4-nucleate	Secretory, cells multinucleate	Secretory, cells multinucleate	—
Pollen grains	2-celled, colporate	2-celled, colporate	2-celled, colporate	—
Ovule	Orthotropous, bitegmic, crassinucellate, hypostase present	Anatropous, bitegmic, crassinucellate, hypostase present	Anatropous, rarely orthotropous, bitegmic, crassinucellate, hypostase absent	Anatropous, unitegmic, tenuinucellate; hypostase absent
Endothelium	Absent	Absent	Absent	Present
Embryo sac	Polygonum type, antipodal cells ephemeral	Polygonum type, antipodals ephemeral	Polygonum type, antipodals ephemeral	Polygonum type, antipodals persistent, rarely ephemeral
Endosperm	Nuclear type, haustoria absent	Nuclear, haustoria absent	Nuclear, haustoria generally absent, ruminate	Cellular, both micropylar and chalazal haustoria develop
Embryogeny	—	Onagrad type	Onagrad	Solanad
Seed	Arillate, exotegmic, seed-coat formed by both integuments	Arillate, exotegmic, seed-coat formed by both integuments	Arillate, endotegmic, seed-coat formed by both integuments	Exarillate, exotestal

R. Dahlgren (1977b), iridoids and seco-iridoids occur in Loasaceae, and they belong to "Iridoid Sympetalae" allied to Dipsacales. In Dipsacales Solanad type of embryogeny occurs in *Morina longifolia* (Vijayaraghavan and Sarveshwari 1968).

Datiscaceae and Begoniaceae. The families Datiscaceae and Begoniaceae form the suborder Begoniineae. Bentham and Hooker (1965a) place them in Passiflorales along with Loasaceae. Hutchinson (1973) includes them in Cucurbitales. The similarities between these two families are: anatropous, bitegmic ovules with micropyle formed by both integuments, single archesporium producing a parietal cell, Nuclear endosperm, Onagrad type of embryogeny, and exarillate exotestal seed. The Allium type of embryo sac and the exalbuminous seed of Datiscaceae differ from Polygonum type of embryo sac with endothelium, and persistent aleurone layer of endosperm in Begoniaceae. The collar cells of the testa, typical of a begonian seed, are absent in the seed of Datiscaceae. Both families differ from Cucurbitaceae (Sect. 38.2, 3) in several respects. Their position in Violales is quite satisfactory.

38 Cucurbitales

The family Cucurbitaceae represents the order Cucurbitales.

38.1 General Features

The order Cucurbitales has the following embryological features: (1) glandular anther tapetum, (2) two- or three-celled pollen grains, (3) anatropous, bitegmic, crassinucellate ovule with vascular supply reaching up to the tip of integument, micropyle formed by both integuments, (4) Polygonum type of embryo sac with ephemeral, rarely haustorial, antipodals, occasional presence of Allium type of embryo sac, (5) polysiphonous persistent pollen tube, (6) Nuclear type of endosperm with tubular haustorium, occurrence of enucleate cytoplasmic nodules, and coenocytic "mounds", (7) Onagrad or Solanad type of embryogeny, (8) nucellar or synergid polyembryony, and apomictic embryo sacs, (9) straight embryo with flat cotyledons, and (10) exalbuminous seed with characteristic exotestal seed-coat.

38.2 Cucurbitaceae

The development of the anther wall is of the "Dicotyledonous type" (Maheswari Devi and Naidu 1984a). It comprises the epidermis, fibrous endothecium, two ephemeral middle layers, and secretory tapetum of binucleate cells in *Bryonopsis lacinosa* (Maheswari Devi and Naidu 1984a). The tapetum forms a periplasmodium in *Luffa cylindrica*. Simultaneous cytokinesis follows meiotic divisions of microspore mother cells, and the tetrads are tetrahedral, isobilateral and decussate (Maheswari Devi and Naidu 1984a).

The family is multipalynous with various apertural forms. The pollen grains are usually shed at the two-celled stage, as in *Luffa tuberosa* (Sarada and Pullaiah 1985), but three-celled condition is reported in *B. lacinosa*.

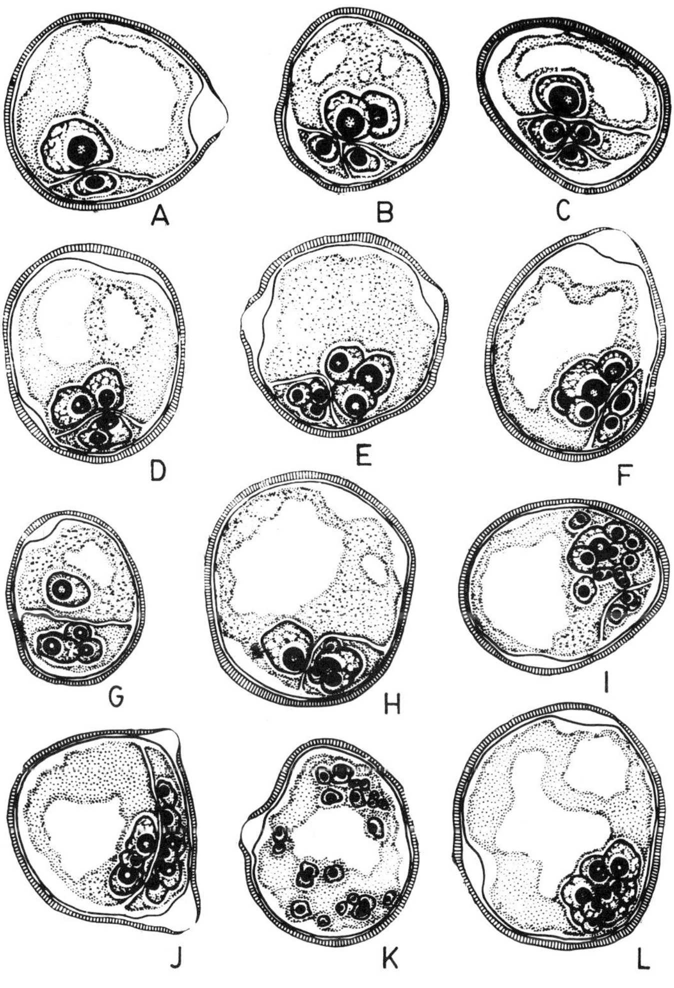

Fig. 38.2.1A–L. Variations in the development of the male gametophyte in tetraploid muskmelon (*Cucumis melo*). **A** Two-celled normal pollen grain. **B, C** Pollen grains with two (**B**) and three (**C**) generative cells, and bi- (**B**) and uninucleate (**C**) vegetative cell. **D–F** Two, three and four nuclei in vegetative cell, and two- (**D–F**) and three-nucleate generative cell. **G–I** Uni- (**G, H**) and multinucleate vegetative cell, and three- (**G, I**) and four-nucleate generative cell. **J** Three-nucleate vegetative and six-nucleate generative cell. **K, L** Pollen grains with 23 and nine free nuclei. In several pollen grains, the nuclei in vegetative and generative cell are of unequal size. (After Dzevaltosky and Litvinenko 1973, see Poddubnaya-Arnoldi 1976)

In a tetraploid variety of *Cucumis melo* (see Poddubnaya-Arnoldi 1976), the normal pollen grain is two-celled (Fig. 38.2.1A), but variations occur. The vegetative cell may be uni- (C, G, H), bi- (B, D) or multinucleate (E, F, I, J). The generative cell also shows variations. It may be bi- (B, D), tri- (C, E, G, H, I) or multinucleate (J). The pollen grain is multinucleate in K and L and does not show the delimitation into the vegetative and generative cell.

The ovules are anatropous, bitegmic and crassinucellate. The funicular bundle continues unbranched up to the tip of the outer integument. The micropyle has both exostome and endostome, or only the endostome, as in *Melothria maderaspatana* (B.M. Johri and Roy Chowdhury 1957) and *Luffa tuberosa*. The cells of the nucellar epidermis divide and produce a nucellar beak as in *M. leucocarpa* (Dathan 1974a) and *L. tuberosa*. The hypodermal archesporial cell cuts off a parietal cell. The parietal tissue is well developed. The megaspore mother cell undergoes meiotic divisions producing a linear or T-shaped tetrad. The chalazal megaspore develops into a Polygonum type of embryo sac. The development of the embryo sac from the epichalazal megaspore is reported in *Lagenaria* (Dzevaltovsky 1963). The synergids are fairly large, and the three antipodals are ephemeral. Allium type of embryo sac occurs in *Benincasa cerifera* (Chopra and Agarwal 1960) and *B. hispida* (Dzevaltovsky 1963); Polygonum type of development occurs rarely. In *B. cerifera* twin megaspore mother cells and twin embryo sacs are frequent. A hypostase develops in *Citrullus* and *Melothria*.

The germination of pollen grains is polysiphonous. The pollen tube is persistent, it may branch and dilate at the base of nucellar beak, as in *Marah macrocarpa* (Dathan and Singh 1971b). In *Citrullus vulgaris* (Gulyayev 1961), two to four pollen tubes enter the micropyle.

Development of the endosperm is of the Nuclear type. The chalazal region extends into a long tubular haustorium (Fig. 38.2.2A–I), except in *Blastania garcini* (Chopra 1953), *Ctenolepis garcinii* and *Corallocarpus conocarpus* (D. Singh 1964a). Centripetal wall formation is initiated at the micropylar region at the globular proembryo, and proceeds downward. The size of the chalazal haustorium varies. The longest (19,000 μm) and coenocytic haustorium occurs in *Sechium edule* (D. Singh 1964a). Protoplasmic streaming movement has been observed in the living haustorium of *Trichosanthes anguina* (Chopra 1955). In many genera the nuclei become polyploid. The haustorium may become completely cellular as in *Bryonia* (Chopra and Basu 1965), *Coccinia indica* (Chopra 1955) and *C. lanatus* (D. Singh 1964a), or remains partially coenocytic. The coenocytic portion is active up to the heart-shaped stage of embryo, and then disorganizes. The nuclei become amoeboid and lobulate, and then fragment (Chopra and Sachar 1963). The seed is non-endospermic.

Enucleate cytoplasmic nodules occur in several members (D. Singh 1955a, b, 1964b, Dathan and Singh 1971b). D. Singh (1964b) suggests that their formation is an active method to fill the vacuole. The peripheral cells in the chalazal region of the endosperm proper develop secondary haustoria in *Luffa* spp. (D. Singh 1964a). In *Melothria maderaspatana* coenocytic "mounds" (in the chalazal region) serve as efficient absorbing organs.

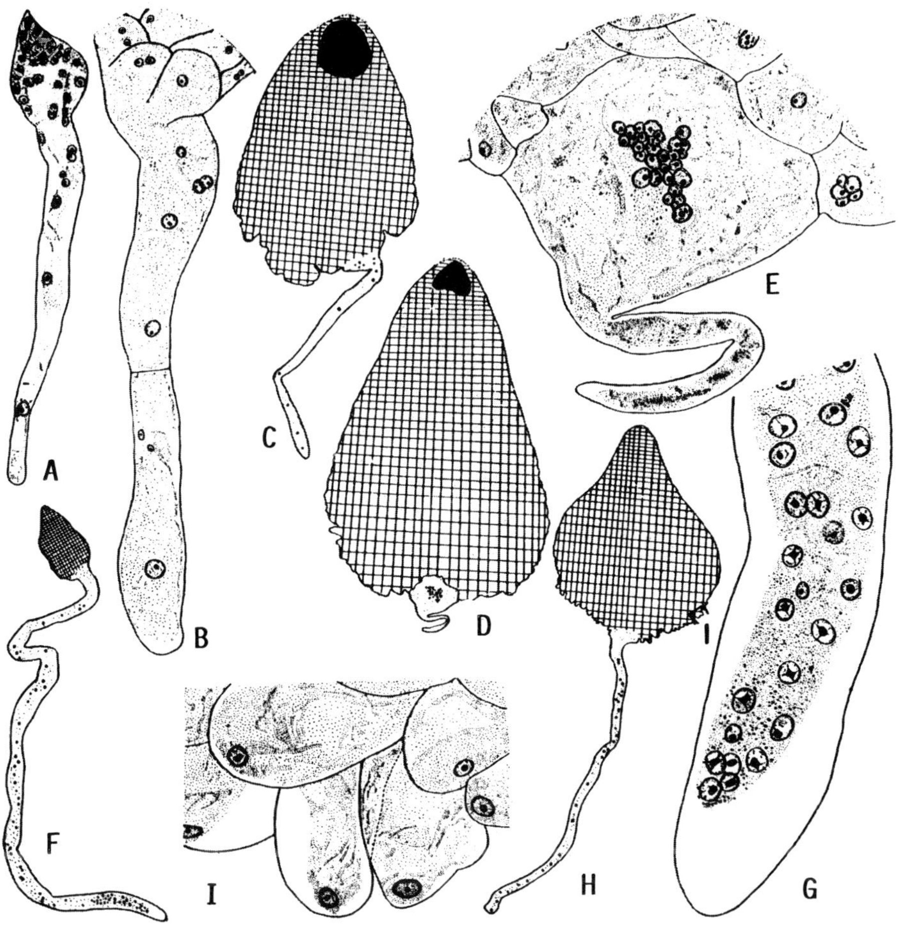

Fig. 38.2.2A–I. A–E *Citrulus vulgaris*, **F–I** *Cucurbita pepo*. **A** Free-nuclear endosperm with chalazal haustorium, and early globular proembryo. **B** Bicelled haustorium. **C** Cellular endosperm with somewhat lateral free nuclear haustorium, and pre-heart-shaped embryo. **D** Late stage with haustorium shrunk, late heart-shaped embryo. **E** Coenocytic haustorium from **D**. **F** Cellular endosperm and free-nuclear haustorium. **G** Tip of haustorium from **F**. **H** Later stage than **F**. **I** *Inset*. *I* from **H**, the peripheral cells are papillate. (After Chopra and Basu 1965)

The embryogeny corresponds to Myosurus variation of the Onagrad type in *Citrulus colocynthis*, *Dicaelospermum ritchiei*, *Melothria maderaspatana* (D. Singh 1961b), *Cucumis anguria* (Zhalatov and Dzevaltovsky 1975), *Cyclanthera explodens* (Crété 1962), *Ecballium elaterium* (Crété 1960), *Sicyos angulata* (Crété 1958a), *Marah macrocarpa* (Dathan and Singh 1971b) and *Melothria leucocarpa* (Dathan 1974a); Asterad type in *Dactyliandra welwitschii* (Dathan 1974b); Nicotiana variation of the Solanad type in *Cucumis melo* var. *pubescens*, *Fevillea cordifolia* (D. Singh 1955a) and *Luffa tuberosa* (Sarada and Pullaiah 1985).

Nucellar polyembryony has been reported in *Cucumis melo* var. *pubescens* (D. Singh 1955a) and *Momordica charantia* (J.S. Agarwal and Singh 1957), synergid embryo in *Sicyos* (Crètè 1958a), and polyembryonate seeds in *Hodgsonia* (see D. Singh 1970d). In *Marah macrocarpa* (Dathan and Singh 1971b), besides the dicotyledonous condition, some embryos with one or two additional imperfectly developed cotyledons are referred to by the authors as hemitri- or hemitetra-cotyledonary.

Mihov and Zagorcheva (1966) report apomictic embryo sacs, and synergid polyembryony in *M. macrocarpa.*

The development and structure of seed-coat has been studied in several genera (B. Singh 1952b, 1953, D. Singh 1961a, 1965a, 1967, 1970d, D. Singh and Dathan 1972a, b, 1974). The tegmen collapses and testa alone forms the seed-coat. Corner (1976) regroups the genera on the basis of the anatomical structure of the seed-coat.

Following the formation of endosperm, the nucellus increases a good deal, but most of it disintegrates and only a few layers are left around the endosperm.

Degeneration of the inner integument varies in different genera. In *Bryonia dioica, Ecballium elaterium* and *Sicyos angulata,* there is complete degeneration by the time the mature embryo sac is formed (Fig. 38.2.3A, C, E, F). In *Luffa cylindrica* (Fig. 38.2.3D), the inner integument persists in young seeds at the globular proembryo.

Soon after the radial elongation of exotestal cells, they divide tangentially (Fig. 38.2.3A, B, *ex*, ex_1, ex_2, ex_3). The inner layer (ex_2) does not divide further (Fig. 38.2.4A–H). The second division gives rise to the layers ex_1 and ex_3 (Figs. 38.2.3A–C, 38.2.4A–H). In *Echinocystis wrightii,* both ex_1 and ex_3 contribute to the formation of seed hypodermis (Fig. 38.2.4G, H). The epidermal cells at the extreme tip and base of the seeds of *B. dioica, Cyclanthera explodens, Lagenaria vulgaris* and *L. cylindrica* are always very small. The subepidermal tissue, or seed hypodermis (*hy*) lies between the epidermis and ex_2. The cells of ex_2 become sclerenchymatous, and may elongate to form a palisade-like tissue.

The structure of mature seed, in different genera, shows several similarities. The exotestal cells develop lignified or rod-shaped thickenings. The seed hypodermis is one- to many-layered and usually remains thin-walled. Next to the hypodermis is a single-layered protective covering which is heavily lignified in mature seeds. This is characteristic of the seeds of Cucurbitaceae.

The seed is exalbuminous, and has a large erect and spatulate embryo. The cotyledons contain fatty food reserve.

JS Agrawal and SP Singh (1957), JJ Asana and RD Adatia (1932), G Balasubramanium and DD Sunderaraj (1957), AK Chakravorti (1947), SVS Chauhan (1970), DP Cholakhyan (1965), RN Chopra (1953, 1955), RN Chopra and S Agarwal (1958, 1960), RN Chopra and B Basu (1965), RN Chopra and RC Sachar (1963), RN Chopra and PN Seth (1977), EJH Corner (1976), P Crété (1958a, 1960, 1962), ASR Dathan (1974a, b), ASR Dathan and D Singh (1971b), PK Deshpande, SM Bhuskute and KH Makde (1984), AK Dzevaltovsky (1963), VA Gulyayev (1961), SC Gupta and VM Rajeswari (1977), BM Johri and C Roy Chowdhury (1957), B Longo (1901, 1903), H Maheswari Devi and KC Naidu (1983, 1984a, b), A Mihov and L Zagorcheva (1966), ER Nikam

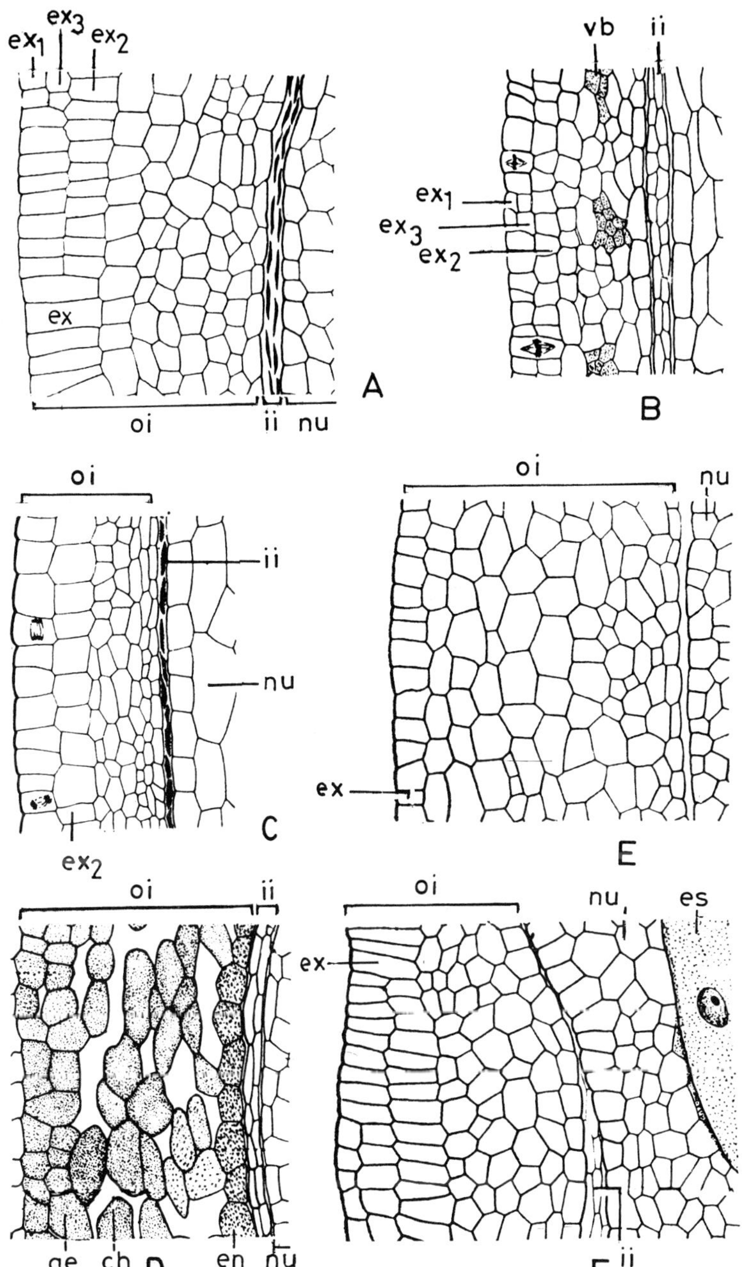

Fig. 38.2.3A–F. **A** *Bryonia dioica*, **B** *Cyclanthera explodens*, **C** *Ecballium elaterium*, **D** *Luffa cylindrica*, **E**, **F** *Sicyos angulata*. – **E** from side and **F** near the base of nucellar beak. Longisections parts of integuments; *ae* aerenchyma; *ch* chlorenchyma; *en* endotesta; *es* part of embryo sac; *ex*, ex_1; ex_2, ex_3, exotesta and its derivatives; *nu* nucellus; *oi* outer integument; *vb* vascular bundle. (After B. Singh 1953)

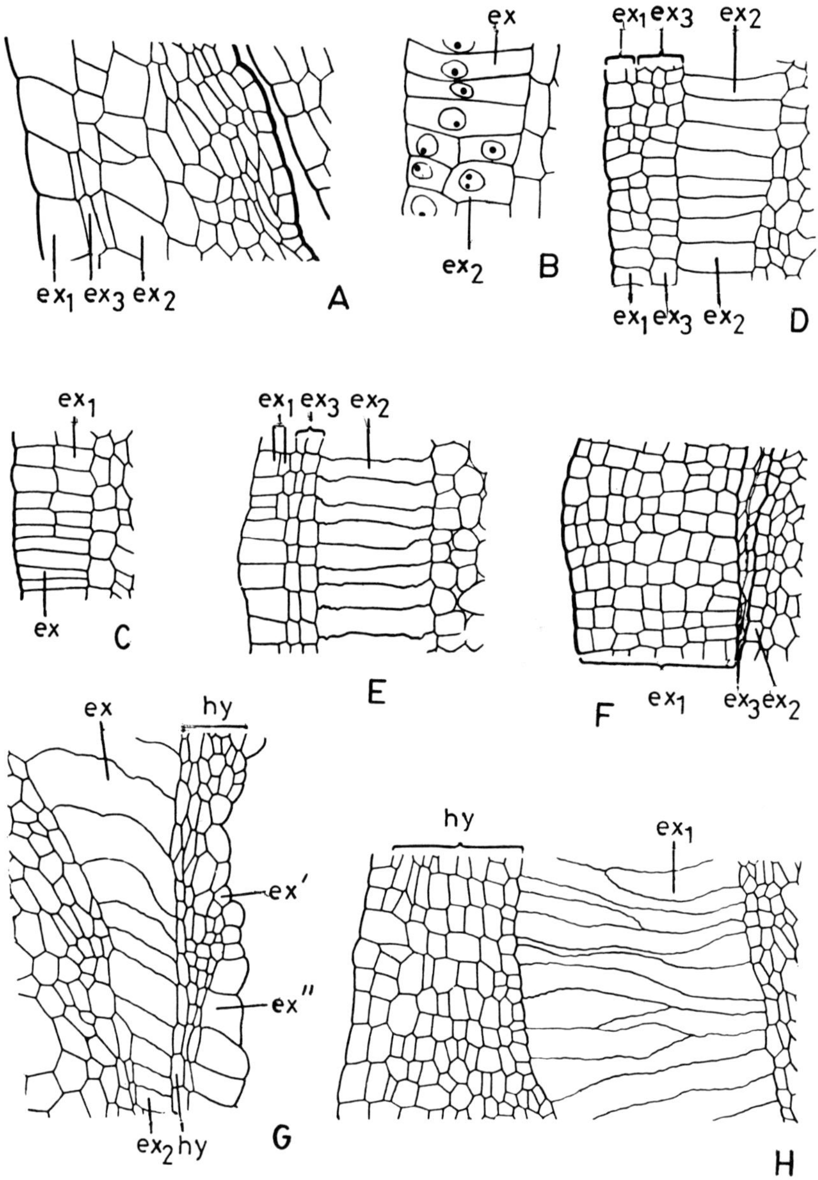

Fig. 38.2.4A–H. A *Cucurbita pepo*, **B** *Lagenaria vulgaris*, **C–E** *Luffa cylindrica*, **F–H** *Sicyos angulata*. Longisections parts of outer integument to show divisions of exotesta (*ex*) and their derivatives (ex_1, ex_2, ex_3); *ex'* small-celled exotesta; *ex"* large-celled exotesta; *hy* hypodermis. (After B. Singh 1953)

(1976), VA Poddubnaya-Arnoldi (1976), D Sarada and T Pullaiah (1985), M Sedgley and MA Blesing (1985), B Singh (1952b, 1953), D Singh (1955a, b, 1961a, b, 1964a, b, 1965a, b, 1967, 1970d), D Singh and ASR Dathan (1971, 1972a, b, 1974), Sridhar and D Singh (1986), K Turala (1969, 1971), K Turala-Szybowska (1974, 1975), IK Vasil (1960), G Zhalatov and AK Dzevaltovsky (1975)

38.3 Taxonomic Considerations

Cucurbitaceae. Bentham and Hooker (1965a) place Cucurbitaceae in Passiflorales; Wettstein (1935) and Engler and Diels (1936) allied them with Campanulaceae. Hutchinson (1973) includes them in Cucurbitales along with Begoniaceae, Datiscaceae and Caricaceae. He considered them to be derived from Passiflorales. Bessey (1915) assigns them to Loasales with begonias. Melchior (1964) and Cronquist (1968, 1981) relate and derive them from Violales. R. Dahlgren (1977a, 1981) also assigns them to Violales.

The embryological features of Campanulaceae (Sect. 51.2), such as unitegmic, tenuinucellate ovules without integumentary vascular supply, Cellular endosperm with micropylar and chalazal haustoria, endospermic seeds, and seed-coat formed by the epidermis alone or with some of the hypodermal layers of the integument, do not resemble those of Cucurbitaceae (D. Singh 1970d). D. Singh also negates their affinities with some families of Parietales. They also differ palynologically (Erdtman 1952).

Cucurbitaceae have always been allied to Passifloraceae, but the embryological features of both differ in several respects. Cucurbitaceae show similarities with Datiscaceae and Begoniaceae of Violales: anatropous, bitegmic, crassinucellate ovule with micropyle formed by both integuments, Polygonum type of embryo sac, Nuclear endosperm, Onagrad type of embryogeny, and exotestal seed with straight embryo. The occasional occurrence of Allium type of embryo sac, and exalbuminous seed are reported in Datiscaceae. But, the development of their seed-coat, which forms the basis of generic classification, has no parallel elsewhere. Their ordinal rank is fully justified, and Cucurbitales form an advanced and terminal taxon of Polypetalae (D. Singh 1970d).

39 Myrtiflorae

The order Myrtiflorae includes three suborders – Myrtineae, Hippuridineae and Cynomoriineae. Myrtineae comprise 15 families: Lythraceae, Trapaceae, Crypteroniaceae, Myrtaceae, Dialypetalanthaceae, Sonneratiaceae, Punicaceae, Lecythidaceae, Melastomaceae, Rhizophoraceae, Combretaceae, Onagraceae, Oliniaceae, Haloragaceae and Theligonaceae; Hippuridineae of Hippuridaceae; Cynomoriineae of Cynomoriaceae.

R. Dahlgren and Thorne (1984) discuss the circumscription, variation and relationships of the Myrtales, and Tobe and Raven (1983a) have attempted an analysis of the embryological characters of the various families of this order.

On this basis, Tobe and Raven (1984d) suggest a separate family Rhynchocalycaceae for *Rhynchocalyx*, and Alzateaceae (Tobe and Raven 1984e) for *Alzatea*. Tobe and Raven (1987c) consider that *Heteropyxis* should be placed in a subfamily of Myrtaceae, or in a separate but related family. Is the grouping of *Axinandra* (Tobe and Raven 1983b) with *Crypteronia* and *Dactylocladus* in Crypteroniaceae justified, or should it be grouped with *Rhyncocalyx* and *Alzatea?*

39.1 General Features

The order Myrtiflorae is characterized by (1) secretory anther tapetum, amoeboid in Lecythidaceae, (2) simultaneous cytokinesis in microspore mother cells, (3) anatropous, bitegmic and crassinucellate ovule, campylotropous in Theligonaceae, and unitegmic and tenuinucellar in Theligonaceae and Hippuridaceae, (4) marked fusion of integuments in Haloragaceae; integumentary vascular strand occurs in Myrtaceae and Rhizophoraceae, (5) development of embryo sac of the Polygonum type, Penaea type in Combretaceae, Oenothera in Onagraceae, and Peperomia in *Gunnera* (Haloragaceae), (6) development of endosperm of the Nuclear type, Cellular type in *Gunnera*, Hippuridaceae and Cynomoriaceae, and endosperm absent in Trapaceae and *Lopezia coronata* (Onagraceae), (7) embryogeny of the Onagrad type in Lythraceae, Sonneratiaceae, Melastomaceae, Onagraceae and Hippuridaceae, Solanad type in Trapaceae and Myrtaceae, Asterad type in Combretaceae, Caryophyllad type in Haloragidaceae, and Chenopodiad type in Theligonaceae, and (8) tracheidal exotegmen common in most families, except where the testa is sclerotic, or the stony endocarp gives protection to seed.

39.2 Lythraceae

The anther wall comprises the epidermis, fibrous endothecium, two ephemeral middle layers, and secretory tapetum of two- to six-nucleate cells. The microspore mother cells undergo meiosis, cytokinesis is simultaneous, and tetrads tetrahedral. The sporogenous tissue frequently degenerates in *Ammannia baccifera*, *A. pentandra*, *Lawsonia inermis* and *Nesaea myrtifolia*.

The pollen grains are three-colpate and two-celled at the dispersal stage (A.C. Joshi and Venkateswarlu 1936a).

The ovule is anatropous, bitegmic and crassinucellate; amphitropous in *Nesaea* (A.C. Joshi and Venkateswarlu 1936a). The zigzag micropyle is organized by both the integuments. A nucellar cap develops in *Pemphis acidula* (Venkateswarlu 1937c). A hypostase is also organized in most members. Frequently, two nucelli develop in an ovule and have common integuments or, at least, a common outer integument.

The archesporium is multicellular. A single archesporial cell cuts off a parietal cell, and the parietal tissue is four to five layers thick. The megaspore mother cell undergoes meiotic divisions and forms a linear tetrad, occasionally

T-shaped in *Lawsonia* (A.C. Joshi and Venkateswarlu 1935a), *Nesaea* and *Woodfordia* (A.C. Joshi and Venkateswarlu 1936a). The chalazal megaspore develops into a Polygonum type of embryo sac. Sometimes, in *Lawsonia* the epichalazal megaspore develops. The synergids develop a small vacuole at the apex, in addition to the large chalazal vacuole. The polar nuclei fuse before fertilization, and the antipodals degenerate at a very early stage. A reduction in the number of nuclei in the embryo sac, due to suppression of nuclear divisions, occurs in *Lawsonia* (A.C. Joshi and Venkateswarlu 1935a).

The chalazal end of the embryo sac is pouch-like, and penetrates the vascular strand in *Ammannia* and *Pemphis* (A.C. Joshi and Venkateswarlu 1936a).

Multiple embryo sacs occur in *Lawsonia inermis*, and inverted polarity of the embryo sac is reported in *Woodfordia floribunda* (A.C. Joshi and Venkateswarlu 1935c).

The development of the endosperm is of the Nuclear type, and the nuclei aggregate in polar regions. Wall formation is initiated as the three-histogenic layers differentiate in the embryo.

Embryogeny corresponds to the Onagrad type-Capsella variation. A hypophysis develops from the superior daughter cell of the penultimate cell of the proembryo. The suspensor is uni- or multiseriate.

The seed is exotegmic, and exalbuminous with a straight embryo and flat or convolute cotyledons. Both the integuments contribute to the formation of the seed-coat. The outer epidermis of the testa is a palisade of cuboid cells. The mesophyll persists only in *Pemphis* and is sclerotic. The inner epidermis is a crystal layer with lignified pitted cells. The outer epidermis of the tegmen comprises longitudinally elongate tracheids with different types of thickenings. The inner epidermis is thin-walled.

AC Joshi (1939b), AC Joshi and J Venkateswarlu (1935a, b, c, 1936a, b), J Mauritzon (1934e), D Oliver (1894), BB Smith and JM Herr Jr (1971), R Souèges (1925b), G Tischler (1917), H Tobe and PH Raven (1984e), J Venkateswarlu (1937c)

39.3 Trapaceae

The anther wall comprises persistent epidermis, fibrous endothecium, two or three ephemeral middle layers, and secretory tapetum of multinucleate cells. The tapetum is irregularly biseriate, and its inner wall is studded with Übisch granules (M. Ghosh 1954). The microspore mother cells undergo meiotic divisions, and produce tetrahedral and decussate tetrads. The cytokinesis is simultaneous.

The pollen grains are pyramidal with three much-folded meridional crests. They are two-celled at the shedding stage; three-celled in *Trapa astrachanica* (see Batygina and Yakovlev 1985).

The ovule is anatropous, bitegmic and crassinucellate, with a prominent hypostase. The massive nucellar beak protrudes through the micropyle. The archesporium is one- to two-celled, cuts off a parietal cell, and the parietal

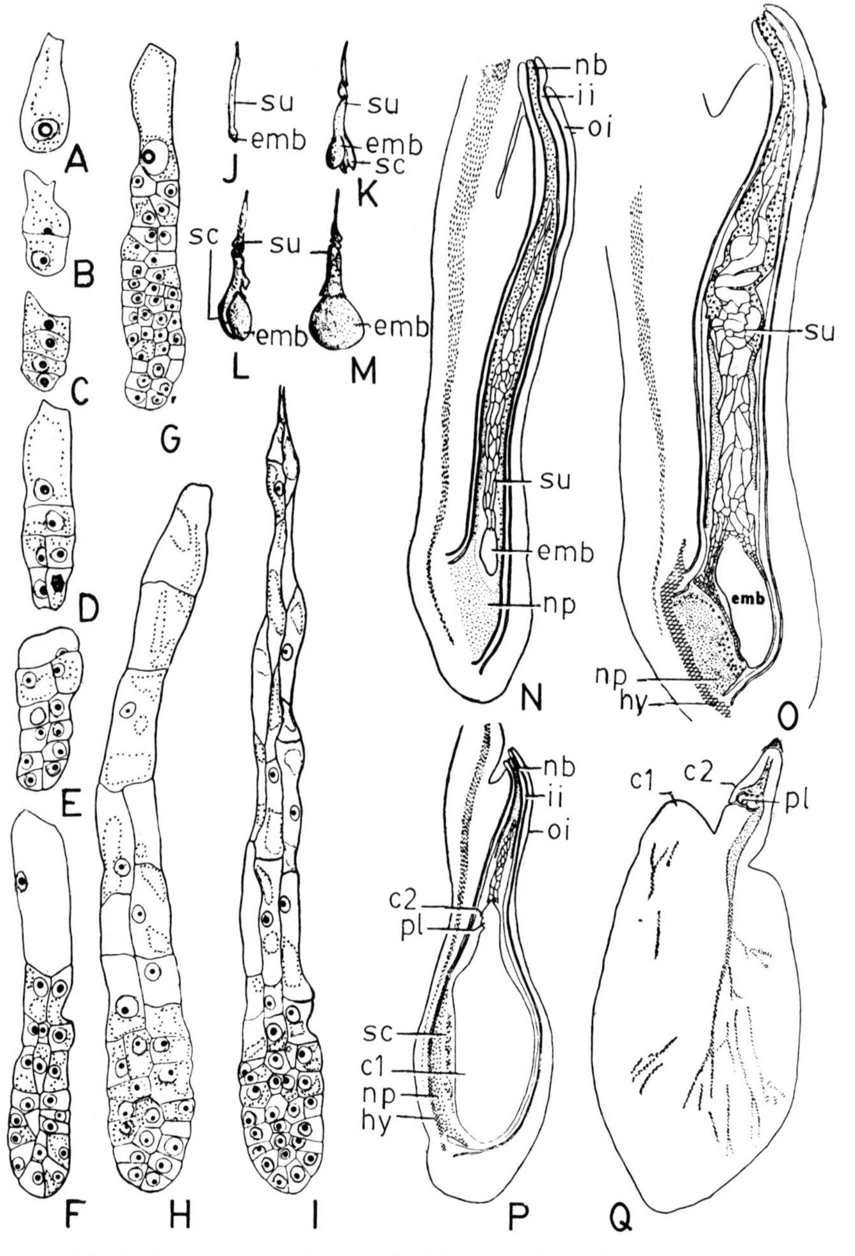

Fig. 39.3.1A–Q. *Trapa bispinosa.* **A** Zygote. **B, C** Two- and four-celled proembryo. **D–I** Further stages in the development of proembryo. **J–M** Dissected wholemount of older embryos (*emb*). Note the massive suspensor (*su*) and also the suspensor collar (*sc*) in **K** and **L**. **N, O** Longisection of ovules showing disposition of embryo. Note the nucellar beak (*nb*), inner (*ii*) and outer (*oi*) integument, nucellar pad (*np*), and hypostase (*hy*). The suspensor (*su*) is massive. **P** Longisection of seed at dicotyledonous (*c*1, *c*2) embryo; *pl* plumule. **Q** Longisection of embryo. One of the cotyledons (*c*2) is highly reduced. (After M. Ram 1956)

tissue is massive in *Trapa bispinosa* (M. Ghosh 1954, Ram 1956). A parietal cell is not cut off in *T. natans* (Trela-Sawicka 1978), which needs confirmation. Periclinal divisions occur in the nucellar epidermis. The nucellar cells undergo endomitosis and become polyploid in *T. natans*. The highest ploidy-level is 256 n (Nägl 1962). The megaspore mother cell produces a linear, sometimes isobilateral or T-shaped tetrad. The chalazal megaspore develops into a Polygonum type of embryo sac. The antipodals have not been observed in *T. bispinosa* (M. Ghosh 1954, Ram 1956). In *T. natans* (Ishikawa 1918), the antipodals are uninucleate, and become hyperthophised and probably polyploid.

The primary endosperm nucleus moves towards the chalazal region, and persists up to the formation of a several-celled proembryo. It undergoes only one or two free-nuclear divisions in *T. natans*. These nuclei finally degenerate, and the endosperm tissue is not formed. The nutritive function (for the growth of the embryo) is taken over by the nucellus and suspensor.

The division of zygote is transverse. Both the cells undergo transverse divisions and produce a linear tetrad (Fig. 39.3.1A–C). The three apical cells produce the embryo. The embryogeny corresponds to the Solanad type (Fig. 39.3.1C–I). The suspensor is long, coiled and multiseriate, and acts as a haustorium (Fig. 39.3.1I–P). Its cells exhibit endomitotic polyploidization. The suspensor forms a collar around the embryo proper (Fig. 39.3.1O). The embryo has two unequal cotyledons, one extremely reduced (Fig. 39.3.1P, Q). The embryo develops at the expense of the nucellus.

The testa is multiplicative, and can be distinguished into three zones in a young seed. The outer and inner zones are parenchymatous, and the inner zone shows reticulate thickenings on the walls. The middle zone comprises suberized cells (Corner 1976). In mature seed the testa is reduced to a thick-walled outer epidermis, and irregular tannin-containing cells of the inner epidermis. The outer epidermis of tegmen consists of narrow tracheidal cells. The inner epidermis is crushed. The fruit is a large one-seeded drupe (Ishikawa 1918).

TB Batygina and MS Yakovlev (1985), EJH Corner (1976), M Ghosh (1954), RM Ishikawa (1918), BM Johri (1963b), P Maheshwari (1950, 1964), W Nägl (1962), M Ram (1956), Z Trela-Sawicka (1978)

39.4 Crypteroniaceae

Tobe and Raven (1983b, 1987a, b) have made a critical study of the embryology and relationships of *Axinandra zeylanica, Crypteronia paniculata, C. cumingii* and *Dactylocladus stenostachys*. All these taxa share the embryological features of the Order Myrtales.

The anther is tetrasporangiate, and the wall comprises the epidermis, endothecium, one (two in *Axinandra*) middle layer, and secretory tapetum of one- or two-nucleate cells. The cells of the connective adjacent to the pollen sacs develop weak fibrous thickenings. At maturity the anther wall consists only of the persistent epidermis; the next three layers collapse.

The microspore mother cells undergo meiotic divisions, followed by simultaneous cytokinesis. The tetrads are tetrahedral (88%), decussate (10%) and isobilateral (2%) in *Axinandra* (Tobe and Raven 1983b). *Crypteronia* shows tetrahedral (66%) and often decussate (34%) tetrads (Tobe and Raven 1987a). *Dactylocladus*, on the basis of 50 selected tetrads, has tetrahedral (84%) and occasionally decussate (16%) tetrads (Tobe and Raven 1987b). These are the first comprehensive data on the quantification of various types of microspore tetrads, and this procedure should invariably be followed by all investigators (see also Schmid 1982).

The pollen grains are two-celled at the shedding stage.

The ovule is anatropous, bitegmic and crassinucellate; a hypostase does not develop. Due to degeneration of the nucellus at the four-nucleate stage of the embryo sac, an endothelium differentiates in *Axinandra*. In *Crypteronia* the chalazal tissue below the embryo sac increases but is disrupted, during post-fertilization development, due to the elongation of the embryo sac.

The hypodermal archesporial cell cuts off a parietal cell and the parietal tissue is two-layered. The megaspore mother cell undergoes meiosis, and the chalazal megaspore of the linear tetrad gives rise to a Polygonum type of embryo sac. The antipodals are ephemeral. In *Axinandra* the walls are not laid down between the nuclei but in *Crypteronia* and *Dactylocladus* antipodal cells are formed.

The development of the endosperm is of the Nuclear type. In *Axinandra* wall formation occurs at the proembryo stage when several thousand free endosperm nuclei have been formed. Cell formation in *Crypteronia* occurs at the young dicotyledonous embryo. The young seed is full of cellular endosperm but all of it is consumed by the developing embryo.

The embryogenic type is not mentioned. In *Axinandra* the zygote divides after the formation of several thousand endosperm nuclei. The suspensor is short and massive, and the mature embryo is straight and flattened; the cotyledons are not folded. In *Crypteronia* and *Dactylocladus*, the suspensor is short (but not massive). Polyembryony does not occur in any of the three taxa.

The seed is endotestal and endotegmic. In *Axinandra*, *Crypteronia* and *Dactylocladus* the exotesta is persistent and has longitudinally elongate cells, in the first taxon it collapses at places. The endotesta is persistent and cryptalliferous in the first two taxa, but degenerates in *Dactylocladus*. The exotegmen collapses. The endotegmen is persistent and tanniniferous. The seeds are winged. In *Axinandra* the membraneous wing on the chalazal side develops from the funiculus. In *Crypteronia* the flat micropylar wing is formed by the funicular tissue and apical tissue of the outer integument. The chalazal wing is narrow and long, and is probably formed by the derivatives of the epidermal cells of the funiculus. *Dactylocladus* has a flat membraneous wing along the entire margin of the seed. The wing is formed by the twofold outer epidermis and is mostly two layers thick.

J Muller (1975), JE Planchon (1845a), R Schmid (1982), H Tobe and PH Raven (1983a, b, 1987a, b), GJCM van Vliet (1975)

39.5 Myrtaceae

The anther wall comprises four to five layers. The endothecium develops fibrous thickenings; sometimes absent in *Angophora floribunda* (N. Prakash 1969d). The middle layers are ephemeral, and the tapetum is secretory with binucleate cells. It is two- to three-layered in the connective region in *Darwinia* (N. Prakash 1969c). The meiotic divisions in microspore mother cells are followed by simultaneous cytokinesis. The tetrads are tetrahedral and decussate, 84% and 16%, respectively, in *Heteropyxis dehniae* (Tobe and Raven 1987c).

The pollen grains are two-celled at the dispersal stage. Sterile one-celled pollen grains occur in *Angophora* and *Eucalyptus stellata* (see Davis 1969).

In the tetrasporangiate anther of *Heteropyxis* (Tobe and Raven 1987c), the two microsporangia in each lobe are separated by a large amount of connective tissue. At maturity, each microsporangium dehisces independently by a longitudinal slit. In other Myrtaceae, the sporangia coaleasce and dehisce by a common slit.

The ovules are ana- to campylotropous, bitegmic and crassinucellate with an endothelium and hypostase. The zigzag micropyle is formed by both the integuments. The ovules are unitegmic in *Eugenia* (S.K. Roy 1960, 1962a) and *Syzygium* (Narayanswami and Roy 1960b, S.K. Roy and Sahai 1962). A vascular bundle traverses the integument. The nucellus is massive in *Myrtus* (S.K. Roy 1962b), and a nucellar cap develops in *E. myrtifolia* (S.K. Roy 1960). The archesporial cell is subhypodermal (this is doubtful and, perhaps, earlier stages were not examined), and cuts off a parietal cell. The parietal tissue is well-developed in *Eugenia bracteata* (S.K. Roy 1955) and *Syzygium caryophyllifolium* (S.K. Roy and Sahai 1962). However, Narayanswami and Roy (1960a, b) report the absence of parietal cell in *S. cumini* and *Psidium*. The deep-seated megaspore mother cell undergoes meiotic divisions, a linear tetrad is produced, and the chalazal megaspore develops into a Polygonum type of embryo sac. The antiopodals are ephemeral. The embryo sacs are normally five-nucleate in *S. caryophyllifolium* (S.K. Roy and Sahai 1962). The chalazal end of the embryo sac develops into a haustorium, and invades the underlying tissue in *E. jambos* (S.K. Roy 1953).

S.K. Roy (1953) reports more than eight nuclei in the embryo sac, which may be due to the fusion of adjacent embryo sacs. A hypostase and postament are formed in *Psidium*.

In *Heteropyxis canescens* and *H. dehniae* (Tobe and Raven 1987c), the embryo sac is bisporic of the Allium type, the antipodals are ephemeral. This condition is not reported in any other Myrtaceae.

Several abnormalities have been observed: unorganized embryo sacs, incompletely organized embryo sacs, twin embryo sacs, and multiple embryo sacs. Sterile ovules without embryo sacs occur in *Eugenia malaccensis* (S.K. Roy 1960) and *Melaleuca* (N. Prakash 1973). Twin nucelli with embryo sacs, enveloped in a common integument, develop in *E. malaccensis*. Narayanswami and Roy (1960a) observed a case of reversed polarity, and presence of supernumerary polar nuclei in *Psidium*. Somatic nucellar aposporous embryo sacs are formed in *E. malaccensis*, *Syzygium cumini* (Narayanswami and Roy 1960b) and *S. caryophyllifolium* (S.K. Roy and Sahai 1962).

The entry of the pollen tube is porogamous, and double fertilization occurs.

Development of the endosperm is of the Nuclear type. The nuclei aggregate in the chalazal region, they fuse and form giant nuclei in *Psidium*. The endosperm remains free-nuclear up to the globular stage of proembryo in *E. bracteata* (S.K. Roy 1955). Wall formation is initiated at the micropylar end, and finally the entire endosperm becomes cellular.

In *Heteropyxis* (Tobe and Raven 1987c), wall formation probably occurs only in a small area around the proembryo/embryo, and does not extend throughout the embryo sac. The seed is exalbuminous.

The embryogeny conforms to the Solanad type in *Callistemon lanceolatus*; *Eucalyptus cinerea, E. cordata, E. deanei, E. gigentea, E. macarthuri, E. occidentalis, E. pauciflora, E. stellulata, E. myrtifolia* and *Feijoa sellowiana* (Polunina 1963) and *Heteropyxis* (Tobe and Raven 1987c), and Onagrad type in *Kunzea capitata* (N. Prakash 1969a) and *Myrtus communis* (S.K. Roy 1962b). The development of the embryo is irregular in *Angophora, Darwinia* and *Psidium*. In *Angophora* di- or tricotyledony, and a collar-like structure at the junction of hypocotyl and radicle has been observed.

The egg apparatus, or the zygote, degenerates in *Eugenia jambos, Syzygium caryophyllifolium* and *S. cumini*. The nucellar cells (in the micropylar region) proliferate, and several embryos are produced. However, the seed contains only one embryo. Polyembryonate seeds occur in *S. caryophyllifolium* (S.K. Roy and Sahai 1962). Johnson (1936) observed two to 21 embryos in the seeds of *E. hookeri*. According to van der Pijl (1934), the formation of adventive embryos begins before pollination, but they attain full size only after pollination and formation of the endosperm. The cells of the inner integument, in the vicinity of the micropyle, also give rise to adventive embryos. Haploid embryo, developing from unfertilized synergid, is reported in *Eugenia delavayi*. Polyembryony is absent in *Eugenia bracteata* (S.K. Roy 1955) and *Psidium*.

The testa is thick and extremely sclerotic. The exotesta comprises thickened lignified cells. The mesophyll is thin-walled. The inner epidermis either contains crystals or becomes strongly thickened. The tegmen generally gets crushed, or persists and contains tannin (Corner 1976, Tobe and Raven 1987c).

The crescent-shaped seed in *Heteropyxis* (Tobe and Raven 1987c) is laterally depressed with a long tapering wing on the micropylar side, and a flat membraneous wing at the chalazal end. The upper wing is formed from the derivatives of the inner and outer integument, and the lower from the derivatives of the outer integument. The wings are not known in other Myrtaceae.

Polunina (1957, 1959) studied the influence of genetic factors as against the physical factor, temperature, in *Eucalyptus*. In *E. cinerea* flowering occurs in October to December (spring in Australia, and autumn in Russia). In Italy, in *E. camaldulensis* and *E. cinerea*, sporogenesis and early stages of the embryo sac develop in summer, but in winter in Australia. The flowering cycle of *E. cinerea* does not show any alteration in adjustment to the conditions of the northern hemisphere; anthesis follows maturation of the embryo sac. But in

E. camadulensis there is an interval between differentiation of the embryo sac and anthesis. The flowering is postponed until the return of summer.

According to Davis (1968b), the embryological processes are rather slow, but regular in *Eucalyptus* spp. In *E. melliodora* in Australia, the meiosis in the megaspore mother cell occurs during winter (May–August), while in Armidale (Australia) it occurs during heavy frost. The enlargement of the egg apparatus follows the differentiation of the embryo sac, but there is a delay (period of cytological stasis) between anthesis and post-fertilization development.

Further studies of this nature and a complete follow-up of the development of seed would be very useful, especially in plants of economic importance separated by space and time.

DV Beardsell, EG William and RB Knox (1989), EJH Corner (1976), R Dahlgren and RF Thorne (1984), AN Drinnan and PY Ladiges (1989), GL Davis (1968b, 1969), E Gauba and LD Pryor (1958, 1961), AM Johnson (1936), LAS Johnson and BG Briggs (1984), S Narayanswami and SK Roy (1960a, b), L van der Pijl (1934), NN Polunina (1957, 1958a, b, 1959, 1963), N Prakash (1969a, b, c, d, 1973), SK Roy (1953, 1955, 1960, 1962a, b), SK Roy and R Sahai (1962), R Schmid (1980), NK Tiwary (1926), H Tobe and PH Raven (1983a, b, 1984e, 1987c), M Sedgley (1989), M Sedgley, FC Hand and AR Griffin (1989)

39.6 Dialypetalanthaceae

The embryology has not been investigated.

39.7 Sonneratiaceae

The anther wall comprises the epidermis, fibrous endothecium, two or three ephemeral middle layers, and secretory tapetum of binucleate cells. Cytokinesis is simultaneous.

The pollen grains are three-porate and two-celled at the dispersal stage. In *Sonneratia apetala* they contain starch (Venkateswarlu 1937a).

The ovule is anatropous, bitegmic and crassinucellate with a zigzag micropyle. The occurrence of vascular strands in the nucellus is reported in *Duabanga sonneratioides* (Venkateswarlu 1936a). Twin nucelli, within common integuments, occur in *Duabanga.*

The archesporium is uni- or multicelled. The archesporial cell cuts off a parietal cell, the parietal tissue is four to six cells thick. The megaspore mother cell undergoes meiotic divisions, and the chalazal megaspore of a linear or T-shaped tetrad develops into a Polygonum type of embryo sac. The antipodals degenerate early, and polar nuclei fuse before fertilization. The mature embryo sac is, thus, four-nucleate (egg apparatus and secondary nucleus) in *Duabanga* and *Sonneratia* (Venkateswarlu 1936a, c).

The development of the endosperm is of the Nuclear type (Venkateswarlu 1937a). The nuclei accumulate in the polar regions, and the endosperm becomes cellular at an advanced stage of proembryo. The endosperm does not fill the entire seed, and the central region remains hollow (the author prefers to call it a central vacuole in the endosperm).

The embryogeny corresponds to the Onagrad type. The suspensor is uni- or biseriate.

Seeds are filiform with an appendage at each end in *Duabanga*, and hard, thick and ellipsoid in *Sonneratia* (Corner 1976).

The exalbuminous seed shows a minute and straight embryo with flat cotyledons in *Duabanga*, and a large curved embryo with contorted cotyledons in *Sonneratia*. The outer epidermis of testa is thin-walled, mesophyll densely sclerotic, and the inner epidermis forms a sclerotic layer in *Sonneratia*, or a crystal layer in *Duabanga*. The outer epidermis of tegmen comprises narrow tracheidal cells with spiral thickenings in *Sonneratia*, or lignified fibres in *Duabanga*. The inner epidermis is thin-walled.

EJH Corner (1976), P Crété (1956g), AC Joshi (1939b), J Mauritzon (1939b), J Venkateswarlu (1936a, c, 1937a)

39.8 Punicaceae

The development and structure of anther, pollen and embryogeny has not been studied.

The ovule is anatropous, bitegmic and crassinucellate with a long funicle in *Punica granatum* (King 1947). The outer integument forms the micropyle.

The archesporial cell produces a parietal cell, and the parietal tissue is three-layered. The megaspore mother cell undergoes meiotic divisions, a linear tetrad is produced, and the chalazal megaspore develops into a Polygonum type of embryo sac. The polar nuclei fuse in the centre of the embryo sac, and the antipodals are ephemeral.

Development of the endosperm is of the Nuclear type.

The seeds are exalbuminous, and embryo straight with contorted cotyledons. The outer epidermis of testa comprises a palisade of very long, thin-walled, columnar pulpy cells (sarcotesta), mesophyll of many layers of sclerotic cells and inner layers of radially elongate cells (mesotesta), and inner epidermis of longitudinal tracheids with annular-spiral lignified thickenings (Corner 1976).

EJH Corner (1976), JR King (1947), J Mauritzon (1939b)

39.9 Lecythidaceae

The embryogeny has not been worked out.

The anther wall comprises the epidermis, fibrous endothecium, two ephemeral middle layers, and an amoeboid tapetum of binucleate cells. The microspore mother cells undergo meiotic divisions, cytokinesis is simultaneous through furrows, and tetrads are tetrahedral. A periplasmodium is formed at the tetrad stage.

The pollen grains are three-celled at the dispersal stage.

The ovule is anatropous, bitegmic and tenuinucellate. The micropyle is formed by the inner integument. The integuments are thick and the outer integument contains a single vascular trace in *Nepoleona imperialis* (Venkateswarlu 1952a, c). An endothelium develops in *N. imperialis*. In *Baringtonia acutangula* the inner integument collapses (Venkateswarlu 1952a). The archesporial cell functions directly as megaspore mother cell, undergoes meiotic divisions, the tetrad is linear or T-shaped, and the chalazal megaspore develops into a Polygonum type of embryo sac. The polar nuclei fuse at the time of fertilization, and antipodals are ephemeral. Twin embryo sacs from two megaspore mother cells, and reversed polarity of an eight-nucleate embryo sac, have been observed in *N. imperialis* (Venkateswarlu 1952c).

The development of endosperm is of the Nuclear type.

The outer epidermis of testa comprises thick-walled lignified palisade cells, followed by aerenchymatous mesophyll with scattered sclerotic cells. The inner epidermis is thin-walled. The tegmen gets crushed.

J Mauritzon (1939b), J Venkateswarlu (1952a, c)

39.10 Melastomaceae

The anther is tetralocular, and its wall comprises the epidermis, endothecium, middle layers, and secretory tapetum of uninucleate cells. The endothecium is heavily thickened in *Mouriria* (Ruys 1925), fibrous thickenings are poorly developed in *Sonerila* (Subramanyam 1944). Generally, there is a single middle layer; five to seven layers in *Melastoma* (Subramanyam 1948a) and two to three layers persist and develop thickenings. The tapetum is biseriate in *Heyneanum* and *Memecylon* (Subramanyam 1942). The microspore mother cells undergo meiotic divisions, cytokinesis is simultaneous, and tetrads tetrahedral.

The three-colporate pollen grains are three-celled at the shedding stage.

The ovule is anatropous, bitegmic and crassinucellate with a zigzag micropyle formed by both the integuments. The hypodermal archesporial cell cuts off a parietal cell, the parietal tissue is two to six cells thick. The cells of the nucellar epidermis divide and produce a nucellar cap, or the cells become conspicuously enlarged and radially arranged in *Melastoma*. The megaspore mother cell undergoes meiotic divisions, a linear or T-shaped tetrad is formed, and the chalazal megaspore gives rise to a Polygonum type of embryo sac. The antipodals rarely organize into cells except in *Oxyspora paniculata* (Subramanyam 1951a). A hypostase develops.

Etheridge and Herr (1968) report that in *Rhexia mariana*, along with the formation of a nucellar cap, a 'pore' appears at the apex of the nucellus. This pore probably results from the suppression of anticlinal divisions in the nucellar epidermis which is subsequently pulled apart with nucellar enlargement. This feature is not known in any other taxon.

In *Osbeckia rosea* (Subramanyam 1946), the micropylar megaspore of a tetrad develops up to four-nucleate embryo sac. Subramanyam (1942, 1944, 1951a) observed double tetrads in *Sonerila*, and twin embryo sacs in *Leandra* and *Oxyspora*.

Development of the endosperm is of the Nuclear type, and the nuclei accumulate in the micropylar and chalazal region. It remains free-nuclear throughout, and is absorbed by the developing embryo.

Development of the embryo is of the Onagrad type in *Leandra cordifolia*, *Memecylon heyneanum*, *Osbeckia capsularis*, *O. hispidissima*, *O. wightiana* (Subramanyam 1942), *Sonerila wallichii* (Subramanyam 1944) and *Oxyspora paniculata* (Subramanyam 1951a). The suspensor is short, multicellular and massive in *Oxyspora*.

The structure of the mature embryo is a taxonomic feature. In *Memecylon umbellatum*, the cotyledons are foliaceous and wrinkled, with the outer cotyledon crumpled over the inner one; the hypocotyl is long. The mature embryo is curved with foliaceous and rolled cotyledons in *M. arnottianum* and *M. gardneri*; the hypocotyl is short. The cotyledons are thick and fleshy in *M. rivulare*; the hypocotyl is short.

Nucellar polyembryony is reported in *Osbeckia hispidissima* (Subramanyam 1942). Due to the proliferation of suspensor cells, additional embryos are formed in *Sonerila wallichii* (Subramanyam 1944).

The seed is exalbuminous. The outer epidermis of testa comprises radially elongate thick-walled cells followed by thin-walled or sclerotic cells. The inner epidermis contains crystals, or gets crushed. The tegmen persists only at the endostome.

P Crété (1957a), AL Etheridge and JM Herr Jr (1968), T Morley (1953), JD Ruys (1925), K Subramanyam (1942, 1944, 1946, 1948a, b, 1951a), H Tobe and PH Raven (1984a), A Ziegler (1925)

39.11 Rhizophoraceae

The development of anther and pollen has not been investigated.

The ovule is anatropous, bitegmic and crassinucellate with a zigzag micropyle. The outer integument contains a vascular strand in *Rhizophora mucronata* (Carey 1934).

In *Cassipourea elliptica* (Juncosa 1984a), a non-viviparous member of Rhizophoraceae, two anatropous ovules [material from La Selva (Costa Rica) and Panama] are present in each carpel of a tricarpellary superior ovary. The material from Llorona (Costa Rica) rarely showed viable seeds. In the basal region the nucellus degenerates. The endothelium contains tannin. The micropyle is formed by the inner integument.

The archesporial cell cuts off a parietal cell, the parietal tissue is two-layered. The megaspore mother cell undergoes meiotic divisions, and chalazal megaspore of a linear tetrad develops into a Polygonum type of embryo sac. The polar nuclei fuse before fertilization; antipodals are ephemeral.

In *Polygonanthus amazonicus* (Tobe and Raven 1988a), the megaspore mother cell produces a dyad as usual. The upper smaller one fails to divide. The lower one divides, and the chalazal megaspore of the triad gives rise to a Polygonum type of embryo sac.

In *C. elliptica*, megasporogenesis is completed by the time the flower opens. The chalazal megaspore develops into a Polygonum type of embryo sac only after pollination.

The development of the endosperm is of the Nuclear type, later it becomes cellular. The endosperm protrudes through the micropyle in *Ceriops candolleana* and *Rhizophora mucronata*, and in *Ceriops* it also develops a chalazal process.

In *C. elliptica*, the nuclei aggregate at the micropylar and chalazal pole, and the central region is occupied by dense cytoplasm. Centripetal wall formation occurs at the globular proembryo. Subsequently, the central region gets filled with radial files of endosperm cells. In *Bruguiera exaristata* (Juncosa 1984b), by the time the embryo fills the seed, much of the cellular endosperm is consumed. The outer layer of the endosperm shows "transfer cells". Then there is a layer or two of large vacuolate integumentary cells encircled by the integumentary compact transfer tissue. The integumentary parenchyma is tannin-filled.

The zygote divides by a transverse (Figs. 39.11.1A, C), longitudinal (Fig. 39.11.1B), or an oblique wall (Figs. 39.11.1D, E). The orientation of the second cell generation is also variable. The proembryo derived from a transverse division of the zygote is followed by longitudinal division of the terminal

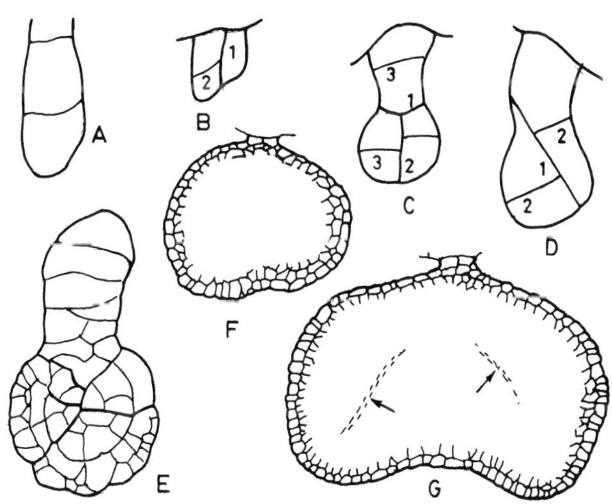

Fig. 39.11.1A–G. *Cassipourea elliptica.* **A–E** Proembryos derived from transverse (**A**, **C**), longitudinal (**B**) and oblique wall (**D**, **E**) in zygote; sequence of walls shown in **B**, **C**, **D**; protoderm demarcated in **E**, suspensor uniseriate (typical of Le Selva specimens). **F** Initiation of cotyledons. **G** Early heart-shaped embryo, provascular elements have differentiated (*arrows*), suspensor short and multiseriate (common in Llorona specimens). (Adapted after Juncosa 1984a)

cell. The protoderm differentiates in a globular proembryo (Fig. 39.11.1F). Some derivatives of the apical cell contribute to the multiseriate upper part of the suspensor. There is "delayed embryo differentiation" (see Haccius 1971). The specimens from La Selva (Costa Rica) showed a uniseriate suspensor (Fig. 39.11.1E), whereas those from Osa Peninsula (Costa Rica) had shorter multiseriate suspensor. The heart-shaped embryo shows lateral expansion (Fig. 39.11.1G). The cotyledons are initiated on the slightly longitudinally elongate embryo, and each contains a single vascular strand. *Cassipourea elliptica* has a phased pattern of growth like that of *Rhizophora*. The period of elongation of the hypocotyl overlaps the beginning of the resumption of growth in cotyledons. The growth of the hypocotyl is more rapid between *a* and *c*, and of cotyledons between *c* and *e* (Fig. 39.11.2A). Initially, there are two strands, one extending from the median bundle of each cotyledon down to the lower hypocotyl (Fig. 39.11.2B). Figure 39.11.2C shows cleared cotyledons with vasculature.

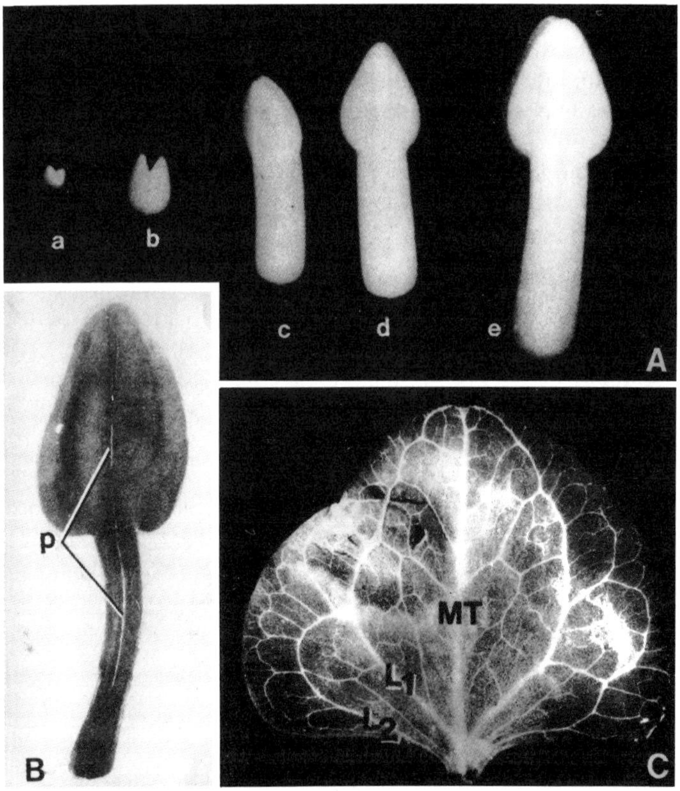

Fig. 39.11.2A–C. *Cassipourea elliptica.* **A** Successive differentiation of embryo (*a–e*). The hypocotyl is very prominent. **B** Embryo from mature seed; *p* protoxylem. **C** Cleared cotyledons; *MT* metaxylem; L_1, L_2 lateral veins. (After Juncosa 1984a)

In *Rhizophora* (Carey 1934), the mature embryo contains two almost wholly fused cotyledons.

According to Corner (1976, see also Tobe and Raven 1988b), the medium-sized seeds are albuminous or exalbuminous (*Cassipourea, Crossostylis, Weihea*), or arillate, winged in *Macarisia*. The testa is thick with a distinct exotesta (*Carallia*), or large unspecialized pellicular testa (*Anisophyllea, Rhizophora*, etc.).

In contrast to other Rhizophoraceae, in *Cassipourea elliptica* (Juncosa 1984a) the germination is epigeal with foliaceous cotyledons. In viviparous members, the cotyledons are fleshy and non-photosynthetic. During germination, the growth is diffuse and intercalary, rather than apical. Apical growth of radicle appears later, and initial vascular differentiation is alternate in the tissue produced.

In *Bruguiera exaristata* (Juncosa 1984b), from Queensland (Australia), germination occurs by the growth of hypocotyl as in non-viviparous plants. The provascular development follows cotyledonary initiation. In the embryo the radicular apex differentiates subterminally. Later, this apex becomes terminal and develops meristematic tissue protected by coleorhiza (formed by the lobes of the outer cortex).

The first two pairs of epicotyledonary leaves are decussate, but the third pair of primorida is initiated at an angle of about 70° to the previous part of the vegetative apices.

Rhizophora mangle (Juncosa 1982), tribe Rhizophorae, exhibits viviparous germination. The initiation of cotyledon and vasculature, apical anatomy of the radicle, and vascular differentiation in *R. mangle*, are highly specialized as compared to non-viviparous *Bruguiera exaristata* (Juncosa 1984b) and *Cassipourea elliptica* (Juncosa 1984a).

Development of the endosperm is of the Nuclear type, as in other members. The inner integument collapses during endosperm development. The multiseriate suspensor gets separated from the testa due to the breakdown of the basal cell followed by the intrusive growth of the endosperm (requires confirmation). Thus, the embryo moves deep into the endosperm before it begins to extrude from the micropyle. The embryo is usually extruded by expansion of the endosperm, and "germination" is effected by the growth of the endosperm rather than of the embryo. Due to the pressure of development of the endosperm, the micropyle remains wide open. The cotyledons arise as a single toroidal primordium. The apex differentiates into two or three cotyledons, i.e. only the apical one-third portion is divided. Later, these lobes fuse marginally to form a narrow tube in the centre.

The axis of the embryo connects the shoot and radicular apical meristems (Fig. 39.11.3A, B). Basipetal development of cotyledonary strands soon connect with the hypocotylary provascular cylinder. The radicular meristem arises somewhat more immersed in the embryonic tissue (Fig. 39.11.3A, B).

The diffuse intercalary meristem below the node causes the elongation of the axis. Growth takes place for several months and the mature radicle is 15–30 cm.

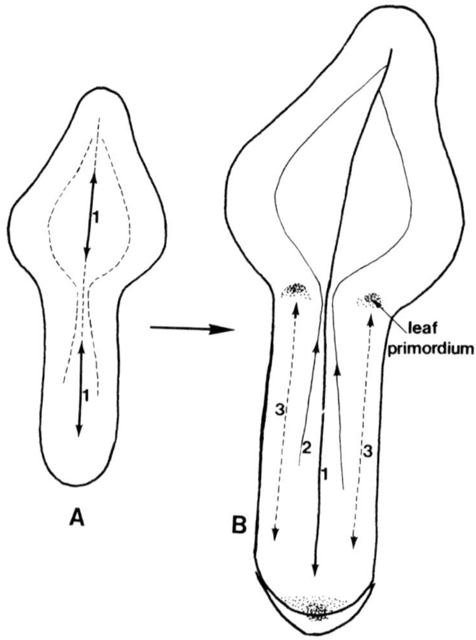

Fig. 39.11.3 A, B. *Cassipourea elliptica*, vascular differentiation in young (**A**) and old (**B**) embryo; *expanded* portion cotyledons, *thick stalk* hypocotyl. *Dotted lines* are major provascular strands. Each cotyledon is vascularized by a single median trace associated with a single provascular lacuna. *Numbers* indicate order of xylem differentiation in various separate loci. (After Juncosa 1984a)

When the apical region of cotyledons is restricted by the integument, growth ceases. Most of the endosperm is crushed or digested by this time; the outer layer of endosperm develops wall ingrowths (transfer cells).

G Carey (1934), EJH Corner (1976), B Haccius (1971), AM Juncosa (1982, 1984a, b), LSS Kumar 1942), H Tobe and PH Raven (1988a, b)

39.12 Combretaceae

The major contribution to the embryology of the family Combretaceae is by Venkateswarlu and Prakasa Rao (1972), on about 18 species and 9 genera representing all the tribes.

The anther wall comprises the epidermis, endothecium with fibrous thickenings, except in the region of stomium, one to six middle layers, and tapetum. The endothecial cells contain tannin and phenolic substances in *Lumnitzera racemosa* and *Terminalia arjuna*. The number of middle layers varies in different genera: single layer in *Guiera senegalensis*; two in *Poivrea coccinea* (Venkateswarlu 1952b), *Combretum decandrum*, *C. extensum*, *C. ovalifolium*, *Lumnitzera racemosa* and *Quisqualis indica*; three in *Anogeissus acuminata*, *A. latifolia*, *Bucida buceros*, *Calycopteris floribunda*, *Combretum grandiflorum*, *Terminalia arjuna*, *T. bellerica*, *T. catappa* and *T. paniculata* (Venkateswarlu and Prakasa Rao 1972). The cells of the connective, adjacent to middle layers, develop thickenings in *Anogeissus acuminata, A. latifolia, Lumnitzera racemosa* and *Terminalia arjuna*. The tapetum is secretory and one-, two- or multi-layered, and the cells are uni- or binucleate. It is multi-layered and three- to four-nucleate in *Combretum grandiflorum*. The cells become polyploid due to nuclear fusion. Following meiotic divisions in microspore mother cells, cytokinesis takes place by furrowing, and the tetrads are tetrahedral and isobilateral.

The pollen grains are ridged, triporate with a thick and smooth exine. They are two-celled at the dispersal stage. Sterility of sporogenous tissue and pollen is very common.

The ovule is anatropous, bitegmic and crassinucellate with a zigzag micropyle. In *Guiera senegalensis* the inner integument alone forms the micropyle. The nucellus undergoes repeated divisions to form a cap in *Bucida, Combretum, Lumnitzera, Terminalia* and *Quisqualis*. A funicular obturator is present in *Combretum, Poivrea* and *Quisqualis*. A hypostase also develops.

The archesporium is uni- to multicellular. A single archesporial cell cuts off a parietal cell, the parietal tissue is massive. The megaspore mother cell undergoes meiotic divisions, and the tetrad is linear or T-shaped. The chalazal megaspore develops into a Polygonum type of embryo sac. In *Guiera senegalensis*, due to aggressive growth at the tip, the embryo sac absorbs the parietal tissue and nucellar cap, and extends into the micropyle. The antipodals are usually ephemeral (Venkateswarlu and Prakasa Rao 1972).

Penaea type of embryo sac is reported in *Combretum paniculatum* and *C. pincianum* (Mauritzon 1939b).

In *G. senegalensis* the antipodals persist and increase up to six. Twin embryo sacs are produced in *Terminalia belerica* (Nagaraj 1955b) and *T. catappa* (M. Nagaraj 1954c), and triple gametophytes in *Calycopteris* (Venkateswarlu and Prakasa Rao 1972). The degeneration of tetrad and embryo sac has been observed.

The endosperm is of the Nuclear type. The nuclei accumulate in the micropylar and chalazal region. The number of nuclei is generally not very large. Wall formation is initiated in the micropylar region at the globular stage of proembryo. The endosperm is consumed during development of the embryo.

The embryogeny conforms to the Asterad type-Erodium variation. Nucellar polyembryony occurs in *Poivrea coccinea* (Venkateswarlu 1952b), and synergid polyembryony in *Terminalia chebula* (M. Nagaraj 1954a), *Combretum decandrum, C. extensum, C. grandiflorum, Calycopteris floribunda, Guiera senegalensis, Lumnitzera racemosa* and *Terminalia catappa*.

The outer epidermis of testa is thin-walled. The mesophyll is either sclerotic or of tracheidal cells. The inner epidermis comprises short, tangentially elongate tracheidal cells. The outer epidermis of tegmen consists of longitudinally elongate, lignified fibres with fine, close, spiral thickenings. The mesophyll and inner epidermis get crushed.

F Fagerlind (1941a), SM Jorapur and HD Noronha (1963), J Mauritzon (1939b), M Nagaraj (1954a, b, c, 1955a, b), N Pal (1951a), PS Prakasa Rao (1963), J Venkateswarlu (1947d, 1952b), J Venkateswarlu and PS Prakasa Rao (1972)

39.13 Onagraceae

The anther wall comprises the epidermis, fibrous endothecium, one or two middle layers, and secretory tapetum of binucleate cells.

Tobe and Raven (1986a) describe some unusual features. The sporogenous tissue in some species of 17 genera becomes divided by transverse septa composed only of the tapetum (Fig. 39.13.1A–C), or of both the tapetum and the parenchyma. In *Calylophus*, *Gaura*, *Clarkia*, *Heterogaura*, *Ludwigia latifolia* and *L. linearis*, the septa are transverse and mostly parenchymatous, or of tapetum only. The "packets" of sporogenous tissue appear spheroidal to ellipsoidal. In *L. linearis* each packet has its own tapetum; the pollen mother cells undergo meiosis synchronously. Finally, all the packets in an anther lobe fuse.

In *Hauya elegans*, the sporogenous tissue becomes divided by transverse as well as vertical septa (Fig. 39.13.1D–I). As many as ten packets have been observed in a transection. In a radial longisection, a large number of packets are arranged in a staggered pattern. The frequency of septa dividing the sporogenous tissue is 1.63 in *Gongylocarpus fruticulosus*.

In *Camissonia* spp. and *Gayophytum humile*, the transverse septa originate from the tapetum.

In *Gaura coccinea*, *G. longiflora* and *G. mutabilis*, by transverse parenchymatous septa, the sporogenous tissue becomes divided into cylindrical packets of varying length.

According to Tobe and Raven (1986a), the evolution of the anther has been accompanied by an increase in tapetum relative to the sporogenous tissue. The increased ratio of tapetum to sporogenous tissue relates to the distinctive surface structure of pollen grains, such as viscin threads.

The maximum number of septa (88 from four flowers with 16 anthers) occurs in *Gongylocarpus fruticulosus*. Of the 88 septa, 20 were formed by parenchymatous tissue and the rest by tapetum.

Following meiotic divisions in microspore mother cells, simultaneous cytokinesis gives rise to tetrahedral tetrads.

The three-colporate pollen grains are two-celled at the dispersal stage.

The ovule is anatropous, bitegmic and crassinucellate with micropyle formed by both the integuments (Fig. 39.13.2F). A hypostase is common. An epistase is present in *Clarkia elegans* and *Zauschneria californica* (Johansen 1928a, b), and a funicular obturator in *Stenosiphon linifolium* (Johansen 1930b). The archesporial cell cuts off a parietal cell which produces parietal tissue. The megaspore mother cell undergoes meiotic divisions, forms a linear tetrad, and the micropylar megaspore mostly develops into a four-nucleate Oenothera type of embryo sac (Fig. 39.13.2A–H).

Two megaspore mother cells occur in *Jussieua repens*, *Ludwigia octovalvis*, *Oenothera odorata* and *Taraxia ovata*, and several in *Clarkia* and *Godetia* (Täckholm 1915). T-shaped tetrads are reported in *Anogra pallida*, *L. parviflora*, *Zauschneria latifolia*, and *Fuchsia* var. *Diadem*. Twin linear tetrads, and four binucleate embryo sacs due to concurrent development of four megaspores have been observed in *O. odorata* (Subramanyam and Govindu 1948).

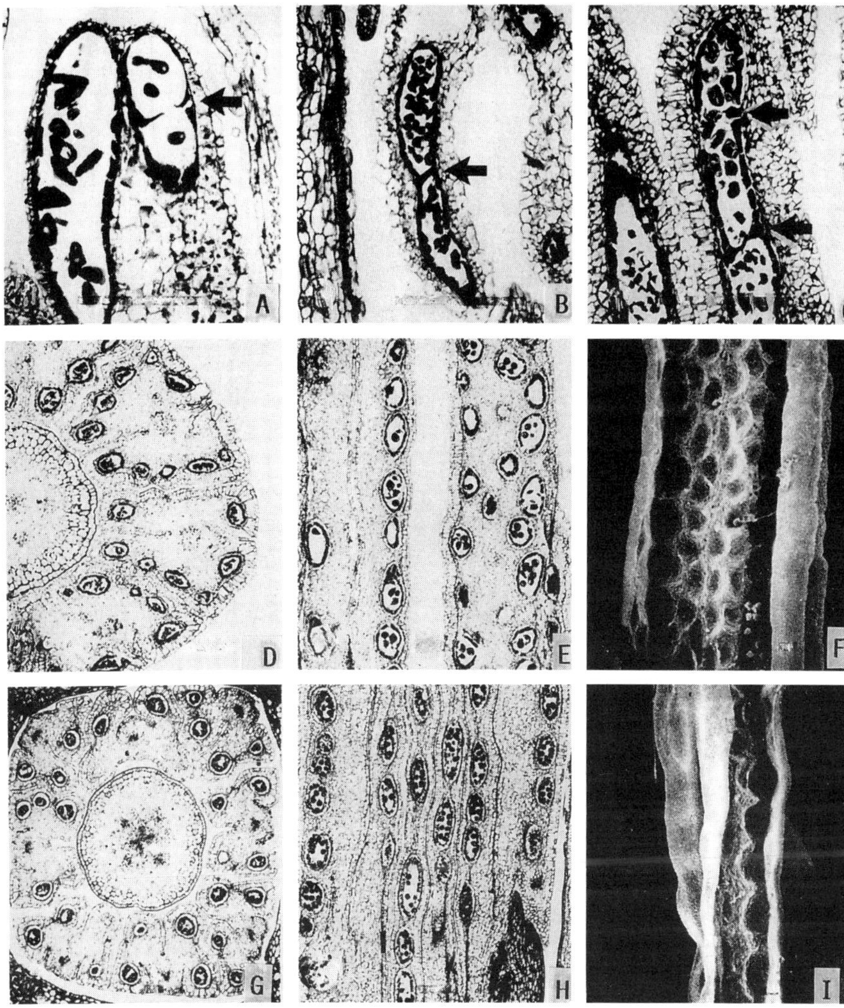

Fig. 39.13.1A–I. Polysporangiate anthers, **A–C**, **E**, **H** longisections, **D**, **G** transections, **F**, **I** SEM preparations. **A** *Lopezia racemosa*, **B** *Circea cordata*, **C** *Fuchsia jimenenzii*, **D–F** *Hauya elegans*, **G–I** *H. heydeana*. **A–C** Sporogenous tissue divided by tapetal septum (*arrow*). **D** Sporogenous tissue divided into packets (groups) by vertical parenchymatous septa (as many as ten packets in t.s.). **E** Tangential longisection of anther (*left*) with many spheroidal or ellipsoidal packets arranged linearly; in radial longisection (*right*) the packets show a scattered arrangement. **F** Remnants of parenchymatous septa on the inner surface of anther wall after dehiscence. **G** Anthers in some of which vertical septa have developed in the sporogenous tissue (not as conspicuous as in **D**). **H** Many spheroidal or elliptical packets of sporogenous cells. **I** Remnants of septa show that the packets are not as scattered as in **F**. (After Tobe and Raven 1986a)

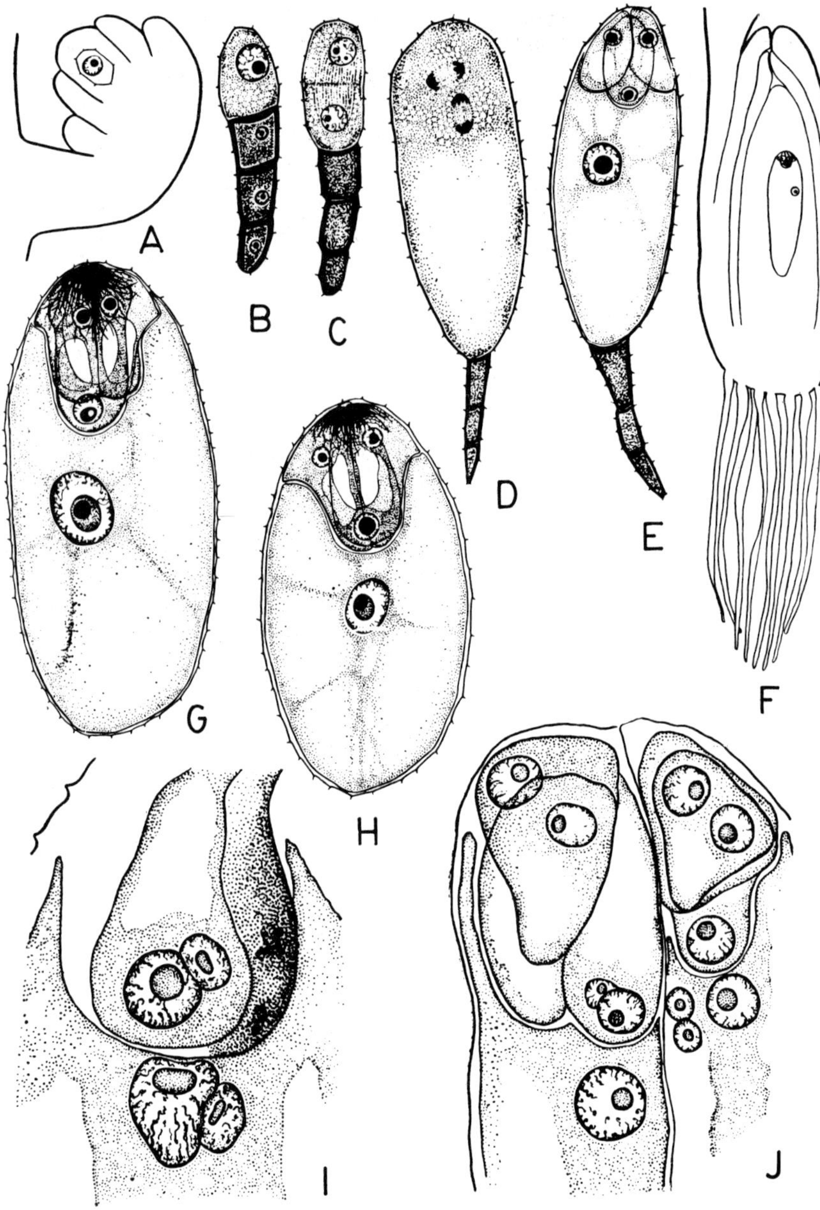

Fig. 39.13.2A–J. A–G *Chamerion dodonaei.* **A** Bitegmic ovule at megaspore mother cell; **B**, **C** Micropylar functional megaspore of linear tetrad; **D** Embryo sac with nuclei at anaphase; **E** Embryo sac with egg apparatus and single polar nucleus: Three degenerated megaspores in **B–E**; **F** Longisection of ovule at mature embryo sac, note hairy outgrowths from the base of ovule; **G** Mature gametophyte. **H** *Ludwigia palustris*, mature gametophyte. **I** *Chamerion angustifolium*, double fertilization, persistent pollen tube shows "x" bodies. **J** *Oenothera biennis*, micropylar portion of twin embryo sacs; syngamy and primary endosperm nucleus in one, in the other the two sperm nuclei have not yet effected fertilization. (See Batygina and Yakovlev 1985)

Ovular sterility is common in *Gayophytum ramosissium* (Johansen 1933). In *O. muricata* (Rodkiewicz et al. 1971), either micropylar or chalazal megaspore may develop. Abnormal three- to five- and seven-nucleate embryo sacs, and twin embryo sacs (Fig. 39.13.2J) are reported in *Jussieua repens* (Khan 1942). Normally, in *Epilobium* sp. the gametophyte is of the Oenothera type (Gachechiladze 1974a, b). However, rarely Polygonum type of development may occur in *E. hirsutum* and *E. nervosum*.

Abnormal embryo sacs also develop in *Zauschneria latifolia* (Johansen 1931b): (1) the micropylar megaspore of the tetrad does not develop further and remains uninucleate, (2) the micropylar nuclei do not organize into egg apparatus, and (3) the polar nucleus fuses with the egg nucleus. All these reports should be re-examined on a quantitative (%) basis.

Figure 39.13.2I shows double fertilization, and two X-bodies in the pollen tube. Of the twin embryo sacs in Figure 39.13.2J, one shows syngamy and primary endosperm nucleus, and in the other double fertilization has not yet taken place.

The development of the endosperm is of the Nuclear type. Khan (1942) noted ca. 900 free nuclei, but their final fate was not ascertained. In *Ludwigia octovalvis*, the endosperm becomes completely cellular; in *Zauschneria latifolia* (Johansen 1931b) the endosperm nuclei divide amitotically. Both reports require confirmation. Total absence of cell wall formation in endosperm has been observed in *Lopezia coronata*. (Täckholm 1914).

The embryogeny conforms to the Onagrad type. The suspensor is short and inconspicuous (Souèges 1920b).

Polyembryony is reported in *Jussieua repens* (Khan 1942). In one ovule two embryos were lying side by side and (in the absence of embryo sac/s) are interpreted as of nucellar origin. In *Taraxia ovata* (Johansen 1931a), "apogamous" embryos were frequently observed, probably from the continued growth of all the four megaspores of a tetrad. The embryo comprised a linear row of ten to 14 cells with a head-like cellular growth at the micropylar end. Apogamous nature is presumed due to the absence of endosperm, and filamentous growth is attributed to the activity of the tetrad, ruling out the possibility of failure of meiosis in the megaspore mother cell. The embryo is, therefore, considered to be haploid. In *Zauschneria latifolia* (Johansen 1931b), besides the perfectly normal zygotic embryos, there were embryos from fertilized synergid, embryos without any regularity in cell organization and hypophysis, and apomictic embryos. A more detailed investigation is necessary before the origin of non-zygotic embryos can be confirmed.

The seed of *Epilobium* has a conspicuous tuft of hairs at the micropylar pole (Fig. 39.13.3A, B).

The exalbuminous seed has a straight embryo with two equal cotyledons. The outer epidermis of the testa is thin-walled, papillate, or hairy. The inner epidermis is a crystal layer with inner and radial walls strongly thickened and lignified. The outer epidermis of the inner integument comprises longitudinal lignified fibres, whereas the inner one is thin-walled and gets crushed, or it may be represented by longitudinally elongate cells with subreticulate thickenings on the inner walls. The fruit is usually a coiled capsule.

Fig. 39.13.3A, B. (SEM) *Epilobium angustifolium*. **A** Seed. **B** Micropylar portion. (Courtesy Dr F. Bouman, Amsterdam)

F Bartels (1956), HD Dickinson and J Lawson (1975), MI Gachechiladze (1974a, b), RR Gates (1911), Y Heslop-Harrison (1990), RL Hulbary and A Nagaraja Rao (1959), M Ishikawa (1918), DA Johansen (1928a, b, 1930a, b, 1931a, b, 1933, 1934), R Khan (1942), A Lebègue (1948a, c), P Maheshwari (1941, 1950), P Maheshwari and BL Gupta (1934), J Modilewski (1909b), A Nagaraja Rao (1959), CE O'Neal (1923), B Rodkiewicz, J Bednara and H Pora (1971), V Seshavatharam (1970), R Souèges (1920b, 1935a), K Subramanyam and HC Govindu (1948), G Täckholm (1914, 1915), JP Tilquin, K de Brouwer, A Mathieu and M Calier (1983), H Tobe and PH Raven (1985, 1986a, b, 1988a), J Venkateswarlu (1952c)

39.14 Oliniaceae

Mauritzon (1939b) gave a brief account of the embryology of *Olinia ventosa*. The detailed studies of Tobe and Raven (1984c) are based on *O. emarginata* and *O. ventosa*.

In the tetrasporangiate anther the sporogenous tissue in a pollen sac may be divided into two groups by a tapetal partition. The Dicotyledonous type of wall (Davis 1966) comprises the epidermis, an endothecium, a middle layer, and a glandular tapetum of binucleate cells. Since all the other layers degenerate, the wall of the mature anther consists only of the persistent papillate epidermal cells. Dehiscence occurs by longitudinal slits. The cells of the connective tissue adjacent to pollen sacs are radially elongated and may develop fibrous thickenings.

The microspore mother cells undergo meiosis followed by simultaneous cytokinesis, and in *O. emarginata* the tetrads (counted on the basis of 50 selected tetrads, see Schmid 1982) are usually (84%) tetrahedral, occasionally (14%) decussate and rarely (2%) isobilateral. The heteropolar (one pole colpate and the other heterocolpate) pollen grains are two-celled at shedding.

Each locule of the five-locular ovary has two or three pendulous, campylotropous, bitegmic, crassinucellar ovules. It is neither apotropous (anatropous, Cronquist 1981) nor hemitropous (hemianatropous, Cronquist 1981). The integuments are of dermal origin, the outer one is three- and the inner two-layered. Both the integuments form the micropyle. The integuments are unvascularized, and Mauritzon's report (1939b) of a small vascular branch extending into the outer integument in *O. ventosa* is erroneous.

The archesporial cell cuts off a parietal cell, and the parietal tissue is massive. The megaspore mother cell undergoes meiosis and the chalazal megaspore of the T-shaped tetrad gives rise to the Polygonum type of embryo sac. The three antipodals are ephemeral and degenerate before fertilization. Thus, the mature embryo sac is five nucleate – egg apparatus and two polars.

The entry of the pollen tube is porogamous. A hypostase differentiates after fertilization. The development of the endosperm is of the Nuclear type; whether wall formation occurs could not be ascertained. Embryogenesis has not been studied. An embryo from a mature seed showed two equally developed unfolded cotyledons; it lacked a conspicuous suspensor.

The seed-coat appears to comprise three- or four-layered testa. The cells of the exo- and mesotesta enlarge and contain pigments. The cells of the endotesta are radially elongated. The cells of the exotegmen are extremely tangentially elongated and fibrous (but not tracheidal), those of endotegmen collapse. Remnants of nucellar epidermis persist close to the seed-coat.

The exotegmic mature seed is ovoid and ellipsoidal, and is not winged. It is exalbuminous. The embryo has a short radicle and irregularly convolute cotyledons (Hutchinson 1973, Willis 1973).

A Cronquist (1981), GL Davis (1966), J Hutchinson (1973), J Mauritzon (1939b), V Patel, JJ Skvarla and PH Raven (1983), VS Rao and R Dahlgren (1969), R Schmid (1982), H Tobe and PH Raven (1984c), JC Willis (1973)

39.15 Haloragaceae

The anther wall comprises the epidermis, fibrous endothecium, single ephemeral middle layer and glandular tapetum of uni- or binucleate cells in *Myriophyllum intermedium* (Bawa 1969a). The tapetal cells are binucleate in *Haloragis colensoi* (Kapil and Bawa 1968), and two- to five-nucleate in *Laurembergia hirsuta* (Nagaraj and Nijalingappa 1974). Due to nuclear fusion polyploidy may occur in tapetal cells. Simultaneous cytokinesis follows meiosis in microspore mother cells, and tetrahedral and decussate tetrads are produced in *G. macrophylla* (Samuels 1912).

The pollen grains are four- to five-porate and three-celled at the dispersal stage. They are two-nucleate (two-celled) in *Gunnera macrophylla* and *G. chilensis* (Modilewski 1908a).

The ovule is anatropous, bitegmic and crassinucellar. Both integuments form the micropyle. In *M. intermedium*, the integuments are closely appressed to each other, and are separate only in the micropylar region. A weak funicular obturator is present, and a hypostase differentiates in *Haloragis colensoi* (Kapil and Bawa 1968) and *H. micrantha* (Nijalingappa 1975).

The archesporial cell cuts off a parietal cell. The parietal tissue is four- to five-layered. The megaspore mother cell undergoes meiotic divisions, the tetrad is linear or, rarely, T-shaped, and the chalazal megaspore develops into a Polygonum type of embryo sac. The polar nuclei fuse before fertilization, and the secondary nucleus lies either near the antipodals, or in the middle of embryo sac. The antipodals are persistent. The development of embryo sac is of the Peperomia type in *Gunnera chilensis* and *G. macrophylla* (Samuels 1912, Modilewski 1908a).

The remnants of the pollen tube persist up to ten- to 12-celled proembryo.

Development of the endosperm is of the Nuclear as well as the Cellular type. Nuclear type occurs in *Laurembergia brevipes* (Bawa 1969b), *L. hirsuta* (M. Nagaraj and Nijalingappa 1974), *L. javanica* (Bley 1925) and *Myriophyllum alterniflorum* (Stolt 1928). Cytokinesis occurs at eight- to 16-nucleate stage. The endosperm is of the Cellular type in *Haloragis asperrima*, *H. colensoi* (Kapil and Bawa 1968), *H. micrantha* (Nijalingappa 1975) and *M. intermedium* (M. Nagaraj and Nijalingappa 1967). In *M. intermedium* (Bawa 1969a), the division of the primary endosperm nucleus and subsequent divisions are followed by transverse walls so that a linear row of eight cells is formed. In other members the division of the primary endosperm nucleus is followed either by a vertical or a transverse wall.

The embryogeny conforms to the Caryophyllad type. The basal cell of the two-celled proembryo undergoes vertical division. These two cells enlarge, become haustorial, and their nuclei become hypertrophied (Fig. 3.11.16F). The suspensor haustorium is absent in *Laurembergia hirsuta* and *Gunnera*. In *Gunnera* the mature embryo is minute and obcordate. In other members the embryo is large and cylindrical.

The seeds are small, exarillate, endosperm copious with starchy food reserve, and embryo straight (Corner 1976).

The seed-coat is simplified within the small nut or drupe. The outer epidermis and hypodermis of the outer integument form the testa. The inner integument and nucellus disorganize except for a few cells at the tip.

SB Bawa (1969a, b, 1970b), F Bley (1925), EJH Corner (1976), BM Johri (1963b), RN Kapil and SB Bawa (1968), J Modilewski (1908a), M Nagaraj and BHM Nijalingappa (1967, 1974), BHM Nijalingappa (1975), JA Samuels (1912), H Schnegg (1902), R Souèges (1940c), KAH Stolt (1928), N Virkki (1962)

39.16 Theligonaceae (Cynocrambaceae)

Kapil and Mohana Rao (1966a) have made a detailed study of *Theligonum cynocrambe*.

The anther wall comprises the papillate cutinized epidermis, fibrous endothecium, an ephemeral middle layer, and glandular tapetum of uninucleate cells. The microspore mother cells undergo meiotic divisions, cytokinesis is simultaneous, and the tetrads are tetrahedral and isobilateral.

The pollen grains are six-porate, suboblate, three-celled at the dispersal stage and contain abundant starch.

The ovule is campylotropous, unitegmic and tenuinucellate. At the four-nucleate stage of gametophyte, the nucellus begins to disorganize, and the mature embryo sac comes in direct contact with the integument. A hypodermal archesporial cell functions directly as megaspore mother cell, undergoes meiotic divisions, produces a linear tetrad, and the chalazal megaspore develops into a Polygonum type of embryo sac. The antipodals undergo mitosis and become two- to four-nucleate in *Theligonum cynocrambe* (Kapil and Mohana Rao 1966a). The polar nuclei fuse in the central region of the embryo sac.

The development of the endosperm is of the Nuclear type. The primary endosperm nucleus divides and two, four and eight free nuclei are produced. At the quadrant proembryo, the entire endosperm becomes cellular in *T. cynocrambe* (Kapil and Mohana Rao 1966a). At maturity the endosperm becomes curved, is thin-walled with highly vacuolate cells, and contains starch. Only a few cells around the embryo disintegrate, and most of the endosperm persists in the seed.

The embryogeny conforms to the Chenopodiad type. The mature embryo is dicotyledonous and horseshoe-shaped.

The single integument is multiplicative, becomes thick, and the vacuolate cells contain starch. The thin-walled testa gets crushed, and the hard shell of the endocarp gives protection to the seed.

RN Kapil and PR Mohana Rao (1966a), R Wunderlich (1971)

39.17 Hippuridaceae

The wall of the tetrasporangiate anther shows the epidermis, a conspicuous fibrous endothecium which is two-seriate close to the stomium, ephemeral middle layer, and glandular tapetum of binucleate cells. The cytokinesis in pollen mother cells is simultaneous. The four- to six-colpate pollen grains are three-celled at shedding.

The ovule is anatropous, unitegmic, tenuinucellate, or reduced (ategmic). A nucellar cap, and a feebly developed funicular obturator are present. The archesporial cell functions directly as megaspore mother cell, undergoes meiosis, and produces a linear tetrad in *Hippuris vulgaris* (Juel 1911). The chalazal megaspore develops into a Polygonum type of embryo sac. The three small antipodals are ephemeral.

The pollen tube penetrates the embryo sac laterally (see Kapil 1962). This aspect needs further study.

The endosperm is of the Cellular type, and inconspicuous micropylar and chalazal haustoria develop.

The embryogeny is of the Onagrad type. The embryo is large, and a suspensor haustorium is absent (see B.M. Johri 1963b). Most of the endosperm is consumed during the maturation of seed.

TB Batygina and MS Yakovlev (1987), SB Bawa (1970b), BM Johri (1963b), HO Juel (1910, 1911), RN Kapil (1962), R Souèges (1922a)

39.18 Cynomoriaceae

The anther wall comprises the epidermis, fibrous endothecium, ephemeral middle layer, and glandular tapetum of binucleate cells. Following reduction divisions in pollen mother cells, simultaneous cytokinesis produces tetrahedral tetrads. The mature pollen grains are two-celled in *Cynomorium songaricum* (see Batygina and Yakovlev 1987).

The pendulous ovule in an unilocular ovary is unitegmic and pseudo-crassinucellar. The integumentary cells are appressed to each other so that the micropylar canal becomes extremely narrow in *C. coccineum* (Steindl 1945) and *C. songaricum*. The cells of the nucellar epidermis undergo periclinal divisions.

The archesporial cell functions directly as megaspore mother cell, undergoes meiotic divisions, the tetrad is T-shaped, and the development of the embryo sac is of the Polygonum type in *C. songaricum*. The antipodals are not ephemeral (Fig. 39.18.1A–G).

At the time of fertilization, the cells lining the micropylar canal become mucilaginous, or glandular, and perhaps provide nutrition to the pollen tube.

The development of the endosperm is of the Cellular type, and considerable endosperm is produced by the time the proembryo becomes globular (Fig. 39.18.1A–G). The endosperm does not develop any haustoria. During the growth of the embryo, the endosperm is consumed.

When the endosperm is four-celled, the zygote divides by an oblique transverse wall (Fig. 39.18.1C). The derivatives of both apical and basal cell produce an undifferentiated suspensorless embryo (Fig. 39.18.1D–J). The development can perhaps be referred to the Asterad type. It would be interesting to study the germination of such an organless embryo (see also Kuijt 1969). The fruit is a nut (Willis 1973).

TB Batygina and MS Yakovlev (1987), HO Juel (1903, 1910), RN Kapil (1962), J Kuiji (1969), F Steindl (1945), JC Willis (1973)

39.19 Taxonomic Considerations

The suborder Myrtineae of Engler and Diels corresponds to that of Melchior (1964), except that they place the family Oliniaceae in Thymelaeineae, and Theligonaceae in Hippuridineae. Melchior (1964) assigns Oliniaceae and Theligonaceae to his suborder Myrtineae, and places Hippuridaceae in the

Fig. 39.18.1A–J. *Cynomorium songaricum.* **A** Fertilized embryo sac with zygote and primary endosperm nucleus. Note the remnants of synergid in **A–D**, and persistent antipodal cells in A–G. **B** Zygote and two-celled endosperm. **C** Zygote in division, endosperm four-celled. **D–F** Two-, four- and eight-celled proembryo, and progressive development of endosperm. **G** Delimitation of protoderm in proembryo, endosperm conspicuous. **H–J** Later stages of proembryo (**H, I**), and undifferentiated mature embryo (**J**). A few endosperm nuclei and cell walls are shown in **H** and **I**, the endosperm has been consumed in **J**. (See Batygina and Yakovlev 1987)

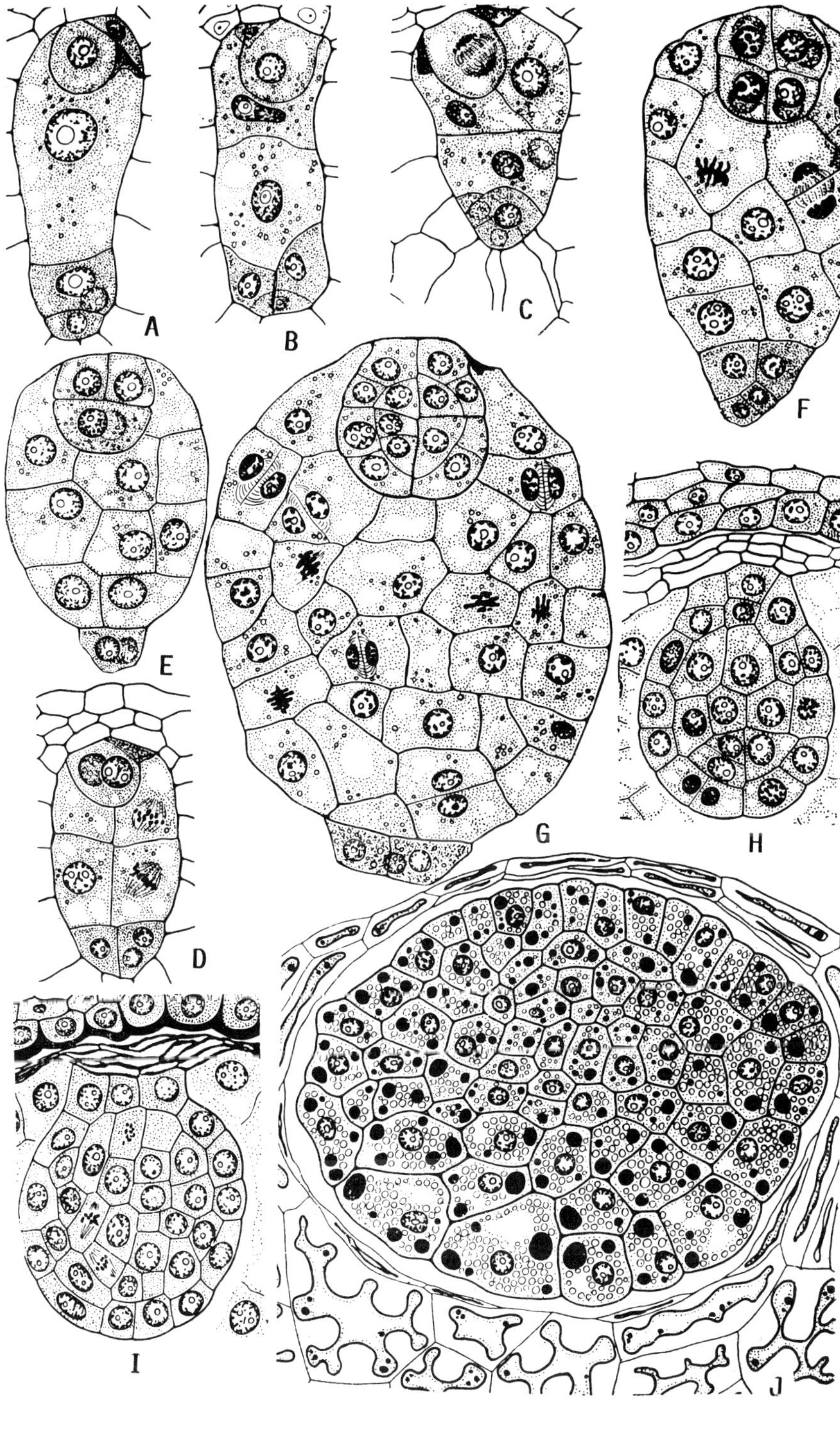
A
B
C
F
E
G
H
D
I
J

suborder Hippuridineae, and Cynomoriaceae in Cynomoriineae. Hutchinson (1973) includes Cunoniaceae and Oliniaceae in Cunoniales, *Hippuris* in Haloragidaceae, and assigns *Callitriche* to a separate family, Callitrichaceae.

Bentham and Hooker (1965a), Bessey (1915), Engler and Gilg (1924) and Wettstein (1935) group Lythraceae, Sonneratiaceae and Onagraceae along with Myrtaceae, Rhizophoraceae and Combretaceae in the order Myrtales (see also Hutchinson 1973). Hutchinson (1926) considers that Lythraceae and allied families show affinities to Caryophyllaceae, and Lythrales are related to Caryophyllales. A.C. Joshi (1939b) considered the relationship of Lythraceae and allied families with Caryophyllaceae, and negates their alliance on embryological grounds. According to him, the right place for these families is in Myrtales only.

Hutchinson (1973) transfers Lythraceae to the order Myrtales, in association with Myrtaceae, Sonneratiaceae, Punicaceae, Melastomataceae and others. According to him, the Myrtales are related to and derived from the same stock as the Guttiferales. Cronquist (1981) treats Lythraceae, Sonneratiaceae, Punicaceae and Melastomataceae, along with eight other families, in Myrtales.

Lythraceae, Sonneratiaceae, Punicaceae and Melastomataceae. Venkateswarlu (1937a, c) treats Lythraceae, Sonneratiaceae, Punicaceae and Melastomataceae as embryologically closely related. These families show similarity in the type of ovule, well-developed parietal tissue, micropyle formed by both integuments, Polygonum type of embryo sac, ephemeral antipodals (they do not even organize in Melastomataceae), Nuclear type of endosperm, Onagrad type of embryogeny, and exalbuminous seed. The endosperm is consumed in the free-nuclear state in Melastomataceae. The seed structure of Sonneratiaceae and Punicaceae is more or less comparable to that of Lythraceae. In Melastomataceae the outer epidermis of testa is sclerotic, but the inner one forms a crystal layer as in the other three families. The tegmen is present only at the endostome. Judging from these facts, Hutchinson's (1973) assignment of Lythraceae to Myrtales is fully justified.

Trapaceae. The genus *Trapa* has been treated under Onagraceae, as an appendage to Onagraceae, as an isolated member of Haloragaceae, and as a separate family Trapaceae. On an embryological basis, P. Maheshwari (1950, 1964), M. Ghosh (1954) and M. Ram (1956) favour the erection of a monogeneric family, Trapaceae (Table 39.19.1). The inferior tetralocular ovary with axile placentae bearing numerous ovules, and a loculicidal capsule in Onagraceae, differ from the semi-inferior bilocular ovary with a single ovule in each loculus, and a large one-seeded drupe of *Trapa* (P. Maheshwari 1950, 1964). In contrast to Onagraceae, *Trapa* shows irregularly biseriate anther tapetum, massive nucellar beak, Polygonum type of embryo sac, absence of endosperm, Solanad type of embryogeny, embryo with a long coiled suspensor, arrested growth of one cotyledon, and suberized mesotesta. The erection of the family Trapaceae is fully justified.

Table 39.19.1. Embryology of Trapaceae (Sect. 39.3) and Onagraceae (39.13) (See Ghosh 1954, Ram 1956, P. Maheshwari 1964, Corner 1976)

	Trapaceae	Onagraceae
Anther tapetum	Irregularly biseriate, secretory, cells multinucleate	Secretory, cells binucleate
Cytokinesis	Simultaneous	Simultaneous
Tetrads	Tetrahedral, decussate	Tetrahedral
Pollen	Pyramidal, 3 much-folded meriodonal crests, usually 2-celled, 3-celled in *Trapa astrachanica*	Tricolporate, 2-celled
Ovule	Anatropous, bitegmic, crassinucellate, true micropyle absent due to long nucellar beak, nucellar cells undergo endomitosis and become polyploid, prominent hypostase	Anatropous, bitegmic, crassinucellate, micropyle formed by both integuments, funicular obturator and epistase reported in some taxa; hypostase common
Embryo sac	Polygonum type, antipodals absent in *Trapa bispinosa*, uninucleate becoming hypertrophied and probably polyploid in *T. natans*	Oenothera, multiple megaspore mother cells, tetrads and embryo sacs develop, lower polar and antipodals are not formed
Endosperm	Primary endosperm nucleus persists up to 10–12-celled proembryo in *T. bispinosa*, 2 to 4 endosperm nuclei in *T. natans* which degenerate	Nuclear type, only a few nuclei are formed which degenerate
Embryo	Solanad type, suspensor long, coiled, multiseriate, haustorial, cells undergo endomitosis and become polyploid, one cotyledon well developed, the other highly reduced	Onagrad, suspensor short and inconspicuous, polyembryony reported in *Jussieua repens* and *Zauschneria latifolia*
Seed-coat	Testa multiplicative, except exotesta all the testal and tegmic tissue becomes crushed	Not multiplicative, inner and radial walls of endotesta strongly thickened and lignified, exotegmen comprises longitudinal lignified fibres
Seed	Exarillate, exalbuminous, enclosed in a 1-seeded drupe	Small, angular, corky, exarillate, exalbuminous, straight embryo with 2 equal cotyledons; fruit a coiled capsule with many seeds

Myrtaceae. The embryological features of Myrtaceae are in general conformity with those of the order. The integument contains a vascular strand as in Rhizophoraceae. However, the occurrence of a nucellar cap, hypostase, postament, endothelium, nucellar polyembryony, apospory, and Solanad and Onagrad type of embryogeny mark its individuality. The sclerotic testa with a crystal layer and crushed tegmen show resemblance to those of Melastomataceae, and to some extent to the mesotestal seed of Lecythidaceae.

Crypteroniaceae. Crypteroniaceae (Tobe and Raven 1987b) have been considered to be closely related to Melastomataceae (R. Dahlgren and Thorne 1984). Embryologically, these families do not differ very much from each other. The anther tapetal cells are usually two-nucleate, the endothecium degenerates much earlier, the nucellus also degenerates early, and the endotegmen persists and contributes to the seed-coat in Crypteroniaceae. In Melastomataceae the tapetal cells are uninucleate, the endothecium persists almost up to the dehiscence of the anther, the nucellus persists at least up to early post-fertilization stages, and the endotegmen is mostly crushed except near the micropyle. The available embryological data do support a relationship between Crypteroniaceae and Melastomataceae (Table 39.19.2).

Tobe and Raven (1984e) suggest that the Crypteroniaceae may also be related to Oliniaceae, Penaeaceae, Rhynchocalycaceae, and Alzateaceae because all the five families have an ephemeral endothecium. However, Crypteroniaceae differ from the other four families in the ephemeral nature of the nucellus. Crypteroniaceae do not have campylotropous ovule as in Oliniaceae; tetrasporic 16-nucleate Penaea type of embryo sac as in Penaeaceae; multicelled female archesporium as in Rhynchocalycaceae; nor is the micropyle formed by the inner integument only, and do not have multicelled female archesporium and bisporic Allium type of embryo sac as in Alzateaceae (Tobe and Raven 1984d, e). The embryological features of Crypteroniaceae are distinct from those of the related families Oliniaceae, Penaeaceae, Rhynchocalycaceae and Alzateaceae. These five families may form a monophyletic group within the order Myrtales (Tobe and Raven 1984f).

The embryological data (Tobe and Raven 1987b) indicate that *Crypteronia* and *Dactylocladus* are more closely related to one another than either is to *Axinandra* (see Sect. 39.4). *Crypteronia* has several ordinal features that characterize Myrtales: glandular anther tapetum, crassinucellate ovule, micropyle formed by both integuments, ephemeral antipodals, endosperm of the Nuclear type, and exalbuminous mature seed (Tobe and Raven 1987a). The placement of Crypteroniaceae in Myrtales is appropriate. Melchior (1964) and Cronquist (1981) also include Crypteroniaceae in Myrtiflorae near Lythraceae, Trapaceae and Myrtaceae.

Rhynchocalyx. The assignment of *Rhynchocalyx* is rather controversial. It has been related to the family Crypteroniaceae (Table 39.19.2) which includes *Rhynchocalyx* together with *Crypteronia, Dactylocladus, Axinandra* and *Alzatea.* The pollen morphology differs in each member. *Rhynchocalyx* and *Alzatea* differ from the other three genera of Crypteroniaceae sensu lato in their wood anatomy, and share characteristics with Lythraceae and Melastomataceae (van Vliet 1975). *Rhynchocalyx* is closer to Lythraceae, Oliniaceae and Melastomataceae because of nodal anatomy. Airy Shaw (1973) assigns it to Lythraceae, whereas R. Dahlgren and Thorne (1984) point out that it differs from Lythraceae in having petals and stamens on the hypanthial rim and sclereids in its leaf petioles.

Embryologically, *Rhynchocalyx* shows several Myrtalean characters: (a) glandular anther tapetum, (b) crassinucellate ovule, (c) two-layered inner integument, (d) ephemeral antipodals, (e) Nuclear type of endosperm, and (f) exalbuminous mature seed (Tobe and Raven 1984d). It differs only in that the micropyle is formed by the inner integument alone, indicating a derived position (Table 39.19.2).

The presence of ephemeral endothecium and micropyle formed by the inner integument segregates *Rhynchocalyx* from Lythraceae. In *Rhynchocalyx* (a) the septum between microsporangia on each side of the anther is persistent (septum of anther collapses in Lythraceae), nucellar subdermal cells elongate radially forming a jacket around the megaspores (radially elongated nucellar subdermal cells are absent in Lythraceae); (b) the homotypic meiotic division occurs earlier in the micropylar cell of the dyad than in the chalazal cell, starch grains are absent in the nucellus (homotypic division occurs later in the micropylar cell of dyad than in the chalazal cell, starch grains are present in the nucellus in Lythraceae); (c) the endosperm is scanty (many free nuclei aggregate in the micropylar/chalazal region in Lythraceae).

In view of the similarities of embryological characters with other Myrtales, *Rhynchocalyx* belongs to the order Myrtales. Due to several features of difference with Lythraceae, Tobe and Raven (1984d) propose a separate family Rhynchocalycaceae.

Alzatea. The assignment of the genus *Alzatea* of Myrtales is also controversial. *Alzatea* resembles Myrtales in: (a) glandular anther tapetum, (b) crassinucellate ovule, (c) two-layered inner integument, (d) ephemeral antipodals, (e) Nuclear type of endosperm, and (f) exalbuminous mature seed. *Alzatea*, like *Rhynchocalyx*, has a micropyle formed by the inner integument alone. In addition, *Alzatea* shows bisporic Allium type of embryo sac. Thus, it is an exceptional member of Myrtales (see Table 39.19.2).

Airy Shaw (1973) assigns it to Lythraceae. But the endothecium is ephemeral in *Alzatea*, the starch grains are absent in nucellus, micropyle is formed by the inner integument, the endosperm is scanty and, moreover, it has Allium type of embryo sac. Thus, *Alzatea* can be isolated from the Lythraceae. The only similarity common to *Alzatea*, *Rhynchocalyx* and Lythraceae is the multicelled ovular archesporium (Table 39.19.2). Tobe and Raven (1984e) propose erection of an independent family Alzateaceae in the order Myrtales.

Heteropyxis. Melchior (1964) assigns *Heteropyxis* to Myrtaceae. In view of its morphology, wood anatomy, floral anatomy, and palynology, Schmid (1980) also includes *Heteropyxis* in Myrtaceae. Embryologically, this taxon is closely allied to Myrtaceae except that (a) the two microsporangia of a theca are

Table 39.19.2. Embryology of Crypteroniaceae, Myrtaceae, *Alzatea*, *Rhynchocalyx*, Lythraceae and Melastomataceae (See Tobe and Raven 1983a, b, 1984a, d, e, 1987a, b, c)

	Crypteroniaceae	*Alzatea*	*Rhynchocalyx*	Myrtaceae	Lythraceae	Melastomataceae
Anther	Tetrasporangiate	Tetrasporangiate	Tetrasporangiate	Tetrasporangiate	Tetrasporangiate	Tetrasporangiate
Anther wall	—	Dicot type	Basic	Basic	Dicot	—
Epidermis	Persistent	Persistent	Persistent	Persistent	Not persistent (?)	Persistent
Endothecium	Ephemeral	Ephemeral	Ephemeral	Fibrous	Fibrous	Fibrous, except in *Melastoma* and *Oxyspora*
Middle layer/s	Ephemeral	Ephemeral	Ephemeral	Ephemeral	Ephemeral	Ephemeral
Tapetum	Glandular, cells 2-nucleate	Glandular, cells 2-nucleate	Glandular, cells 2-nucleate	Glandular, cells 2-nucleate	Glandular, cells 2–6-nucleate	Glandular, cells uninucleate
Cytokinesis	Simultaneous	—	Simultaneous	Simultaneous	Simultaneous	Simultaneous
Tetrad	Tetrahedral and decussate	Tetrahedral and decussate	Tetrahedral and decussate	Tetrahedral and decussate	Tetrahedral and isobilateral	Tetrahedral
Pollen	2-celled	Pseudocolpi absent, 2-celled	Heterocolpate, 2-celled	Tricolporate, pseudocolpi absent, 2-celled	Tricolporate, pseudocolpi may be present, 2-celled	Tricolporate, 2-celled
Ovule	Anatropous	Anatropous	Anatropous	Anatropous	Anatropous	Anatropous
Integument/s	Bitegmic, both two-layered	Bitegmic, inner longer and 2-layered, outer 2–4-layered	Bitegmic, inner longer than outer, both 2-layered	Bi- or unitegmic (*Syzygium*), inner 2-layered	Bitegmic, inner 2-layered	Bitegmic, inner 2-layered
Nucellus	Crassinucellate	Crassinucellate	Crassinucellate	Crassinucellate	Crassinucellate	Crassinucellate
Micropyle	Both integuments form micropyle	Inner integument	Inner integument	Both integuments (few exceptions)	Both integuments	Both integuments
Hypostase	Absent	Present	Absent	Present in some genera	Present	Absent

Female archesporium	1-celled, rarely 2-celled	4–8-celled, only one functions	3–5-celled, only one functions	1-celled	Multicelled, only one functions	1-celled
Tetrad	Linear	Dyad	Linear	Linear	Linear	Linear
Embryo sac	Polygonum type	Allium	Polygonum	Polygonum	Polygonum	Polygonum
Antipodals	Ephemeral	Ephemeral	Ephemeral	Ephemeral	Ephemeral	Ephemeral
Fertilization	Porogamous	Porogamous (?)	Porogamous	Porogamous	Porogamous	Porogamous
Endosperm	Nuclear type, becomes cellular	Nuclear, probably wall formation does not occur	Nuclear, few free nuclei, wall formation does not occur	Nuclear, becomes cellular	Nuclear, becomes cellular	Nuclear, wall formation does not occur
Embryogeny	—	Onagrad type	Onagrad	Solanad and Onagrad	Onagrad	Onagrad
Suspensor	Short	Short	Short	Short, or absent	Short	Short, massive
Embryo	Straight, cotyledons not folded	Cotyledons same size	Cotyledons folded inside	Cotyledons same size	Cotyledons flat or convolute	Cotyledons foliaceous, and crumpled or rolled, or thick and fleshy (*Memecylon*)
Polyembryony	Absent	Absent	Absent	Usual in *Syzygium*	Absent	Occasional
Seed	Seed-coat contributed by both integuments, wing formed on the micropylar and chalazal end, chalazal wing extremely long and narrow; exalbuminous	Exotestal, membraneous wing originates from the funicle and outer integument all round the seed; exalbuminous	Seed-coat contributed by both testa and tegmen, wing of funicular origin on the micropylar side only; exalbuminous	Exotestal, exalbuminous	Exotegmic, both testa and tegmen contribute to seed-coat, exalbuminous	Exotestal, exalbuminous

remotely separate from each other and dehisce by their respective longitudinal slits, instead of a common slit, (b) the ovule and seed have wings on both ends, and (c) the embryo sac is bisporic (Allium type).

Tobe and Raven (1987c) point out that *Heteropyxis* could be considered either as a subfamily of Myrtaceae distinct from all other members of the family, or as a separate but related family.

Lecythidaceae. Bentham and Hooker (1965a) include some genera of Lecythidaceae in Myrtaceae. Lecythidaceae are a distinct family showing amoeboid tapetum, three-celled pollen, bitegmic tenuinucellar ovule with endothelium, vascularized integument and, occasionally, reversed polarity of the embryo sac (Venkateswarlu 1952a). They show alliance with Myrtaceae due to simultaneous cytokinesis in pollen mother cells, vascularized integument, endothelium (rarely), Polygonum type of embryo sac, and Nuclear type of endosperm. The structure of seed, showing a sclerotic mesotesta and crushed tegmen, also indicates their alliance. Otherwise, they do not show similarity with the rest of the families included in Myrtiflorae. The evidence on embryogeny may help in determining their systematic position.

Rhizophoraceae. Rhizophoraceae closely resemble myrtaceous families in the type of ovule, vascularized integument, parietal tissue, Polygonum type of embryo sac with ephemeral antipodals, and Nuclear type of endosperm. The protrusion of endosperm through the micropyle in *Ceriops* and *Rhizophora*, formation of chalazal endosperm processes in *Ceriops*, and closely appressed cotyledons in the mature embryo are noteworthy features.

According to Tobe and Raven (1988b), the Rhizophoraceae are monophyletic. The vegetative, reproductive (including embryological) and chemical data suggest a close relationship to Elaeocarpaceae and Celastraceae, and possibly to Erythroxylaceae.

Combretaceae. Combretaceae are assigned a position between Melastomataceae and Rhizophoraceae, and between Onagraceae and Oliniaceae. The embryology of Combretaceae is comparable to the general pattern of development in Melastomataceae, Lythraceae, Punicaceae and Sonneratiaceae.

The genus *Guiera* differs from the rest of the members of Combretaceae in a single middle layer in the anther wall, micropyle formed by the inner integument, extension of the embryo sac at the micropylar end by absorbing parietal tissue and nucellar cap, and persistent antipodals multiplying up to six (Venkateswarlu and Prakasa Rao 1972). The separation of the genus *Guiera* from the tribe Combretae to a monogeneric tribe Guieraeae is warranted. The absorption of nucellus by the embryo sac is considered as a transitional stage in the evolution of nucellar tissue. Venkateswarlu and Prakasa Rao (1972) suggest that it may perhaps lead to tenuinucellate condition in Lecythidaceae.

The development of tetrasporic Penaea type of embryo sac has been observed only in *Combretum paniculatum* and *C. pincianum* (Mauritzon 1939b). Otherwise, the gametophyte is of the Polygonum type. To some extent, the structure of seed resembles that of Punicaceae. The presence of funicular obturator (*Poivrea*), tetrasporic embryo sac in some members, and Asterad type of embryogeny are special features of Combretaceae.

Onagraceae. Hutchinson (1973) includes Onagraceae in Onagrales. The genus *Trapa* is now removed from Onagraceae and raised to an independent family, Trapaceae. There is a close similarity between Lythraceae and Onagraceae in the development of anther, pollen, ovule, endosperm, embryo, and structure of seed. But the development of Oenothera type of embryo sac singles it out from the rest. The degeneration of antipodals occurs at a very early stage in Lythraceae. In *Lawsonia* (Lythraceae), the embryo sac shows reduction in the number of nuclei to such an extent that the Oenothera type of embryo sac can be derived from that of the Lythraceae (A.C. Joshi and Venkateswarlu 1936a).

In both families, Lythraceae and Onagraceae, the embryogeny is of the Onagrad type. A hypophysis differentiates much earlier and develops from the superior daughter cell of the cell *m* in Lythraceae, whereas in Onagraceae it develops directly from the cell *m* (A.C. Joshi and Venkateswarlu 1936a). In Lythraceae the suspensor is uni- or multiseriate, it is short and inconspicuous in Onagraceae.

Oliniaceae. Engler and Diels (1936) place Oliniaceae in Thymelaineae, whereas Hutchinson (1973) includes them in Cunnoniales. The available embryological data show their alliance with the members of Myrtineae. The Oliniaceae are presumed to be related to Thymelaeaceae (see Cronquist 1981). The similarities in ovule morphology suggest a closer relationship of Oliniaceae with the Lythraceae and Myrtaceae than with Thymelaeaceae. The relationship has been emphasized between Oliniaceae and both Crypteroniaceae and Melastomataceae (Takhtajan 1980, R. Dahlgren and Thorne 1984), or with Penaeaceae (Johnson and Briggs 1984).

Melchior (1964) places Oliniaceae in the suborder Myrtineae of Myrtiflorae. The Oliniaceae show a number of embryological characters of the Myrtales (Sect. 39.1). The ephemeral endothecium is also known in *Axinandra*, *Rhynchocalyx*, *Alzatea*, *Crypteronia*, *Dactylocladus*, as well as Penaeaceae (Tobe and Raven 1983a, b, 1984c, d, e, f, 1987a, b). The campylotropous ovule in Oliniaceae is unknown in any other Myrtales, except in *Mouiri* subg. *Taphroxylon* (Melastomataceae – Morley 1953). Oliniaceae are also unique in their pollen which is heteropolar, one pole colpate and the other heterocolpate. The pollen is said to have half-pseudocolpi (Patel et al. 1983), a condition not known in any other flowering plant.

Tobe and Raven (1984c) point out that the totality of similarities and dissimilarities in embryological characters does not suggest a close relationship of the Oliniaceae to any other group of Myrtales; rather, these features strongly support their isolated position within the order. The Oliniaceae are more closely related to Crypteroniaceae s. st., Rhynchocalycaceae, Alzateaceae and Penaeaceae than to any other. The exclusive characters of the ovule and pollen fully justify the erection of a new family Oliniaceae.

Haloragaceae. Bentham and Hooker (1965a) place the genera *Haloragis, Laurembergia, Myriophyllum, Gunnera* and *Callitriche* in Haloragaceae. Engler and Prantl, and Melchior (1964) include Haloragaceae in Myrtiflorae. Wettstein (1935) and Rendle (1952) recognize Haloragaceae as distinct from Hippuridaceae. In addition, Wettstein separates *Gunnera*, elevates it to the rank of a family – Gunneraceae, and recognizes it as an advanced taxon. Hutchinson (1973) assigns *Gunnera* and *Hippuris* to Haloragaceae (Haloragidaceae), and *Callitriche* to Callitrichaceae. Cronquist (1981) treats Haloragaceae and Gunneraceae as separate families in the order Haloragales. (This order is considered to be a reduced derivative of Myrtales.) R. Dahlgren (1981) erected an independent order Gunnerales for the family Gunneraceae, in superorder Rosiflorae. In fact, the family Haloragaceae (Bentham and Hooker 1965a) is heterogeneous, and the genera *Callitriche, Gunnera* and *Hippuris* have now been removed. According to Corner (1976), the seed structure in Haloragaceae bears no resemblance either with Lythraceae or Myrtaceae.

The embryological data support the assignment of Haloragaceae to Myrtiflorae. There is no need to erect the order Haloragales.

B.M. Johri (1963b) and Bawa (1970b) evaluate the embryological features of *Haloragis, Gunnera, Hippuris* and Callitrichaceae. The occurrence of three- to five-colpate two-celled pollen grains, bicarpellary unilocular ovary with a single ovule, micropyle formed by the inner integument alone, Peperomia type of embryo sac, and absence of suspensor haustorium distinguish *Gunnera* from other members of Haloragaceae (Table 39.19.3). Cytological data also confirm the separation of *Gunnera* (Virkki 1962). Its assignment to a family or order is a matter of choice.

Hippuridaceae. The position of *Hippuris* is debatable. Bentham and Hooker (1965a) place it in Ranales, and consider it to be derived from Nymphaeaceae. Melchior (1964) raises it to the family Hippuridaceae under suborder Hippuridineae. Thorne (1968) includes it in Cornales. R. Dahlgren (1977a, b, 1981) erects the order Hippuridales (Hippuridaceae) in superorder Lamiiflorae, close to the orders Scrophulariales and Hydrostachyales.

The aquatic habit, bisexual reduced flowers, four- to six-colpate pollen grains, monocarpellary unilocular ovary with single pendulous, unitegmic, tenuinucellate ovule (with nucellar cap), Onagrad type of embryogeny, and absence of suspensor haustorium definitely segregate Hippuridaceae from Haloragaceae, and justify their present position. Anatomical data also support this arrangement (Metcalfe and Chalk 1950).

In recent systems of classification, Hippuridaceae is placed in or near Scrophulariales, an iridoid sympetalous group (R. Dahlgren 1975a, Wagenitz 1977a). The nature of ovule (except nucellar cap), endosperm (haustoria absent), and embryogeny resembles that in Scrophulariales. The apetalous reduced flower with inferior monocarpellary ovary, single ovule, and absence of endosperm haustoria may justify their assignment to an independent order Hippuridales.

Cynomoriaceae. The suborder Cynomoriineae is represented by a single family Cynomoriaceae, and Melchior (1964) places the latter next to Hippurida-

Table 39.19.3. Floral morphology and embryology of *Haloragis*, *Hippuris* and *Gunnera* (Adapted from Kapil 1962, Hutchinson 1973, Batygina and Yakovlev 1987)

	Haloragis	*Hippuris*	*Gunnera*
Flower	Bracteate, bisexual, tetramerous, arranged in glomerules in leaf axis	Bracteate, bisexual, rarely unisexual, solitary, perianth forms a rim around the tip of ovary	Ebracteate, bisexual, bimerous, crowded in large spikes or panicles
Stamen/s	4 + 4, obdiplostamonous	Single stamen placed medianly	2, antipetalous
Anther wall	4-layered, tapetum secretory, cells 2-nucleate	4-layered, tapetum secretory, cells 2-nucleate	4-layered, tapetum secretory, cells 2-nucleate
Cytokinesis	Simultaneous	Simultaneous	Simultaneous
Pollen	4–5-porate, 3-celled	4–6-colpate, 3-celled	4–5-porate, 2-celled
Gynoecium	Tetracarpellary, tetralocular, single pendulous ovule in each locule	Monocarpellary, unilocular, single pendulous ovule	Bicarpellary, unilocular, solitary ovule
Ovule	Anatropous, bitegmic, crassinucellar, feeble funicular obturator	Anatropous, unitegmic, or ategmic, tenuinucellar, feeble funicular obturator	Anatropous, bitegmic, crassinucellar, obturator absent
Nucellus	Parietal tissue formed, nucellar cap absent	Parietal tissue not formed, nucellar cap present	Parietal tissue formed, nucellar cap absent
Embryo sac	Polygonum type, antipodals 3, persistent	Polygonum, antipodals 3, ephemeral	Peperomia, secondary nucleus formed by fusion of 6 nuclei, antipodals 6, ephemeral
Fertilization	Porogamous	Pollen tube penetrates the embryo sac laterally	Porogamous
Endosperm	Cellular type	Cellular	Cellular
Embryo	Caryophyllad type, large, cylindrical, suspensor haustorium present	Onagrad type, large, cylindrical, suspensor haustorium absent	Caryophyllad (?), minute, obcordate, without suspensor haustorium
Seed coat	Outer epidermis and hypodermis of outer integument form testa	—	—
Seed	Small, exarillate, albuminous	Exalbuminous	Albuminous

ceae (Hippuridineae). Wettstein (1935) assigns Cynomoriaceae to Santalales. As the members are root parasites, Cronquist (1981) retains *Cynomorium* in Santalales, as a genus in Balanophoraceae. Hutchinson (1973) also includes *Cynomorium* in Balanophoraceae. Kuijt (1969) states: "Although my acquaintance with Balanophoraceae is a shallow one, I am unable to support this separation" of *Cynomorium*, and assignment to an independent family Cynomoriaceae. (Initially van Tieghem split Balanophoraceae into a series of separate families including Cynomoriaceae – see Kuijt.) While *Cynomorium* and *Balanophora* are both root parasites, there are other resemblances between the two.

The similarities and dissimilarities in the embryology of Balanophoraceae (Sect. 13.2), Hippuridaceae (Sect. 39.17) and Cynomoriaceae (Sect. 39.18) are: (a) non-fibrous endothecium in Balanophoraceae and fibrous in the other two, (b) glandular tapetum of binucleate cells, (c) successive cytokinesis in Balanophoraceae, simultaneous in the other two families, (d) pollen grain three-celled in Balanophoraceae, and two-celled in the other two, (e) undifferentiated ovule in Balanophoraceae, anatropous, unitegmic, tenuinucellate ovule in Hippuridaceae and Cynomoriaceae, (f) Polygonum type of embryo sac, which becomes U- or J-shaped with reversed polarity in Balanophoraceae, (g) antipodals absent in Balanophoraceae, ephemeral in Hippuridaceae, and persistent in Cynomoriaceae, (h) endosperm of the Cellular type; a large chalazal haustorial cell in Balanophoraceae, inconspicuous micropylar and chalazal haustoria in Hippuridaceae, and none in Cynomoriaceae, (i) embryogeny of the Piperad type in Balanophoraceae, Onagrad in Hippuridaceae, and Asterad (?) in Cyanomoriaceae, (j) embryo without suspensor and undifferentiated in Balanophoraceae and Cynomoriaceae, with suspensor and typically dicotyledonous and large in Hippuridaceae, (k) seed "naked" and albuminous in Balanophoraceae, seed with testa but more or less exalbuminous in the other two families, and (1) during germination of "fruit" in Balanophoraceae, the "endosperm cells also germinate" and form tubes – *this feature is not known in any other angiosperm.*

In view of the above facts, *Cynomorium* is closer to *Balanophora* (see also Kuijt 1969) than to *Hippuris*. The families Cynomoriaceae and Balanophoraceae should be included in the Order Balanophorales, and neither in Santalales nor in Myrtiflorae.

Theligonaceae. The genus *Theligonum* is included in Urticaceae, Phytolaccaceae and Euphorbiaceae. Dumortier (1829) and Takhtajan (1980) raise it to a family rank, Theligonaceae, in the order Gentianales near Rubiaceae. Engler and Diels (1936) place Theligonaceae with Hippuridaceae in the suborder Hippuridineae of Myrtiflorae. Cronquist (1981) includes them in Rubiales along with the Rubiaceae. R. Dahlgren (1975a) and Wagenitz (1977a) include *Theligonum* in Rubiaceae. Kapil and Mohana Rao (1966a) discuss the morphology and embryology of *Theligonum cynocrambe*, and compare it with Phytolaccaceae, Urticaceae and Euphorbiaceae. They conclude that erection of the family Theligonaceae is fully justified. It does not show any relationship with either Haloragaceae (Sect. 39.15) or Hippuridaceae (Sect. 39.17). The proper place for Theligonaceae is in the order Centrospermales (Sect. 16) near the families Phytolaccaceae and Amaranthaceae. The embryological features of similarity are: (1) monocarpellary ovary with a solitary ovule, (2) secretory anther tapetum, (3) three-celled pollen grains, (4) Polygonum type of embryo sac, (5) Nuclear type of endosperm, (6) presence of raphides, and (7) unicellular, non-glandular hairs, and rubiaceous stomata (Kapil and Mohana Rao 1966a). However, the perisperm, a typical centrospermalean character, is absent in Theligonaceae.

In a very detailed and critical review, Wunderlich (1971) discusses the systematic position of *Theligonum* which has been rather problematic. Several morphological, anatomical, cytological and embryological features of *Theligonum* have been compared with those of Rubiaceae. The common embryological features are: uninucleate anther tapetum, absence of endothelium, Nuclear type of endosperm, albuminous seeds, and chromosomes n = 10 or 11.

The common features with the Anthospermeae, Spermacoceae and Gulieae (tribes of Rubiaceae) are: three-celled pollen grains, single ovule in the ovary, micropyle turned towards the base and outwards, embryogeny probably of the Solanad type, comparable development of seed, and unspecialized testa.

Theligonum has some modified rubiaceous characters, like campylotropous ovule, and horseshoe-shaped curved embryo.

Wunderlich remarks: "the relations to the Rubiaceae are largely hidden so that they had not been recognized before". She concludes that *Theligonum* certainly belongs to Sympetalae (tubular perianth of female flower, ovule tenuinucellate and unitegmic). It shows a close relationship with Rubiaceae. On account of the peculiar combination of features, it should be assigned to a separate tribe "Theligoneae", closely related to Anthospermeae, and also to Spermacoceae and Galieae.

We have some reservations in accepting the assignment proposed by Wunderlich, and would leave the question open for a future consideration.

40 Umbelliflorae

The order Umbelliflorae comprises seven families: Alangiaceae, Nyssaceae, Davidiaceae, Cornaceae, Garryaceae, Araliaceae and Umbelliferae.

40.1 General Features

The order Umbelliflorae is characterized by: (1) secretory anther tapetum, (2) simultaneous cytokinesis, (3) three-colporate two-celled pollen grains; three-celled in Umbelliferae, (4) anatropous, unitegmic ovule (5) Polygonum type of embryo sac, (6) persistent antipodals in Araliaceae and Cornaceae (rarely), and their secondary multiplication in Alangiaceae and Garryaceae, (7) Nuclear type of endosperm; Cellular in Davidiaceae and Cornaceae; both types occur in Alangiaceae, (8) Solanad type of embryogeny in Garryaceae and Umbelliferae, Onagrad or Chenopodiad in Araliaceae, and Asterad in Cornaceae, and (9) testal, albuminous seed with thin-walled seed-coat; outer epidermis lignified in Nyssaceae.

40.2 Alangiaceae

The development of pollen and embryo has not been investigated.

The anther wall comprises the epidermis, fibrous endothecium, single ephemeral middle layer, and secretory tapetum of binucleate cells. The tetrads are isobilateral.

The ovule is anatropous, unitegmic and crassinucellate, and the funicular vascular strand reaches up to the apex of integument. The nucellar epidermis divides periclinally and produces three layers. There is a hypostase of four to five layers. An endothelium develops after fertilization.

The single archesporial cell cuts off a parietal cell which produces a layer of cells. The megaspore mother cell forms a linear tetrad, and the chalazal megaspore gives rise to a Polygonum type of embryo sac. The antipodals increase up to 12 or more, and persist up to fertilization in *Alangium lamarkii* (Gopinath 1945); they are ephemeral in *A. handelii.*

The endosperm is of the Nuclear type in *A. lamarkii* (Gopinath 1945), but Cellular in *A. chinense* (Mitra and Datta 1949). Further studies on endosperm development are necessary.

Apogamous antipodal embryos are described in *A. lamarkii* (Gopinath 1943), but this is doubtful and needs reinvestigation.

The seed contains copious starchy endosperm, the embryo is straight and conspicuous, and the seed-coat is thin-walled.

DM Gopinath (1943, 1945), BM Johri (1963b), JN Mitra and RM Datta (1947, 1949)

40.3 Nyssaceae

The embryogeny has not been investigated.

The anther wall comprises the epidermis, fibrous endothecium, three ephemeral middle layers, and secretory tapetum in *Nyssa sylvatica* (Mohana Rao 1972a). Cytokinesis is simultaneous, and the tetrads are tetrahedral and decussate.

The pollen grains are tricolporate and two-celled at the shedding stage.

The ovule is anatropous, unitegmic and pseudocrassinucellar. The nucellar epidermal cells divide periclinally producing a single layer at the apex, and three layers laterally. A hypostase develops in *Nyssa biflora.* The hypodermal archesporial cell functions directly as megaspore mother cell, and forms a linear or T-shaped tetrad. The chalazal megaspore develops into a Polygonum type of embryo sac. The antipodals are ephemeral. The polar nuclei fuse before fertilization. Tandon and Herr (1971) reported Allium type of embryo sac in 20% of the ovules of *N. aquatica* and *N. biflora*.

The endosperm is of the Nuclear type, wall formation is initiated after the formation of eight free nuclei, and before the division of zygote in *N. sylvatica* (Mohana Rao 1972a).

The seed contains a large embryo, and abundant endosperm. The integument consists of four to five layers at the megaspore mother cell stage. Due to periclinal divisions the number of layers increases up to 48–50. The seed-coat is smooth and, while the outer epidermis is thick-walled and lignified, the inner layers are thin-walled.

The ovary wall is 25–28 layers thick at the megaspore mother cell stage. The thickness of wall increases to a large extent, and the pericarp differentiates into three zones: the outer zone with large cells, prominent nuclei and calcium oxalate crystals, middle zone of tangentially elongate cells with dense cytoplasm, and inner zone of 18–20 layers comprising thick-walled, elongated, tubular fibres with a narrow lumen.

RH Eyde (1966), PR Mohana Rao (1972a), SR Tandon and JM Herr Jr (1971), W Wangerin (1910a)

40.4 Davidiaceae

The structure of anther, pollen grain, embryo and seed-coat has not been fully studied (see Schnarf 1931).

The pollen grains are tricolpoidorate.

The ovule is unitegmic and crassinucellate in *Davidia involucrata* (Horne 1909). Whether bitegmic condition occurs, requires confirmation (see Corner 1976). The nucellar epidermal cells divide periclinally to produce a cap. An endothelium differentiates, but further details are not known. The archesporium is multicelled; the formation of parietal cell/s requires confirmation.

The development of the embryo sac conforms to the Polygonum type. The antipodals are conspicuous and, sometimes, their number increases and they show haustorial activity during post-fertilization stages. This aspect needs further investigation. Multiple and abnormal embryo sacs are reported.

The development of endosperm is of the Cellular type.

The seeds have fleshy endosperm, and a large embryo (Hutchinson 1973).

EJH Corner (1976), AS Horne (1909), J Hutchinson (1973), K Schnarf (1931)

40.5 Cornaceae

The anther wall comprises the epidermis, fibrous endothecium, two or three ephemeral middle layers, and secretory tapetum. The microspore mother cells produce tetrahedral tetrads, and cytokinesis is simultaneous.

The tricolporate pollen grains have a smooth exine, are oblate and spheroidal, and two-celled at the dispersal stage.

The ovule is anatropous, unitegmic and crassinucellar. An endothelium develops at the tetrad stage in *Helwingia japonica* (Sâto 1976a). The cells of the nucellar epidermis undergo divisions. The archesporium is uni- or multi-celled, cuts off a parietal cell, and the parietal tissue is three to four layers thick in *Cornus* (Chopra and Kaur 1965b). The megaspore mother cell forms a linear tetrad, and the chalazal megaspore develops into a Polygonum type of embryo sac in *Aucuba japonica* (Sâto 1971), *A. japonica* var. *borealis*, *Cornus controversa*, *Helwingia japonica* (Sâto 1976a), *C. svecica* (Håkansson 1923), *C. alba* (Fagerlind 1939c), *C. alternifolia*, *C. ammomum*, *C. macrophylla*, *C. oblonga* and *C. stolonifera* (Chopra and Kaur 1965b). In *A. japonica*, sometimes, the embryo sac develops from the lower two, or rarely the upper two, megaspores of the tetrad. In either case each megaspore becomes binucleate, the intervening wall disorganizes, and the four nuclei divide once so that an eight-nucleate embryo sac of usual organization is produced. This mode of development is essentially bisporic. Twin and multiple embryo sacs occur in *C. oblonga*. The antipodals are ephemeral. The polar nuclei fuse before fertilization. The development of the embryo sac is pseudomonosporic in *C. florida* (Morse 1907). Håkansson (1923) reported four free megaspore nuclei in *C. mas*, and Fagerlind (1939c) and D'Amato (1946b) confirmed Fritillaria type of development. D'Amato also reported five- and seven-nucleate (Pyrethrum parthenifolium type) embryo sac in *C. mas*.

B.B. Smith (1973a, 1975) studied the qualitative and quantitative development of five species of *Cornus*: *C. alternifolia*, *C. ammomum*, *C. asperifolia*, *C. florida* and *C. stricta* (Table 40.5.1). *C. florida* differs from the rest in several features, such as multicellular archesporium functioning directly as megaspore mother cells, many linear tetrads, and persistent synergid and antipodals. The maximal increase in the size of female gametophyte occurs at the four-nucleate stage in *C. florida*, in others at the eight-nucleate stage. B.B. Smith proposed the removal of *C. florida* from the genus *Cornus* to the monotypic taxon *Cynoxylon floridum* (L) Ref. (see also Small 1933).

The endosperm is of the Cellular type (Håkansson 1923). The cells are thin-walled and store starch.

The embryogeny corresponds to Period I, Megarchetype II in *Aucuba japonica* (Mestre and Guignard 1970).

The seed-coat is thin-walled, and gets more or less crushed.

RN Chopra and H Kaur (1965b), F D'Amato (1946b), H Eyde (1967), F Fagerlind (1939c), A Håkansson (1923), H Harms (1897b), H Kaur (1970c), JC Mestre and JL Guignard (1970), WC Morse (1907), Y Sâto (1971, 1976a), BB Small (1933), BB Smith (1973a, 1975), W Wangerin (1910b)

40.6 Garryaceae

The anther wall comprises the epidermis, fibrous endothecium, one or two ephemeral middle layers, and glandular tapetum of binucleate cells. The nuclei

Table 40.5.1. Embryology of five species of *Cornus* (See B.B. Smith 1973a)

	C. alternifolia	*C. ammomum*	*C. asperifolia*	*C. florida*	*C. stricta*
Nucellus	Crassinucellate	Crassinucellate	Crassinucellate	Tenuinucellate	Crassinucellate
Archesporial cells	2 or more	1	2 or more	Many	1
Parietal cell/s	1	1	2	None	1
Sporogenous cells	1	1	2	Many	1
Tetrad/s	Linear or T-shaped	Linear or T-shaped	Linear, 2	Linear, many	Linear
Degeneration of megaspore	Variable	Variable	First middle 2, then uppermost	Variable	Variable
Synergids	Degenerate rapidly	Degenerate	Persist until polar nuclei fuse	Persist until polar nuclei fuse	Persist until polar nuclei fuse
Antipodals	Persist until polar nuclei fuse	Degenerate	Persist until polar nuclei fuse	Persist until polar nuclei fuse	Persist until polar nuclei fuse

divide and nuclear fusion results in polyploid condition. The microspore mother cells undergo meiotic divisions, cytokinesis is simultaneous, and the tetrads are tetrahedral and decussate.

The tricolporate pollen grains are shed at the two-celled stage in *Garrya douglasi* (Kapil and Mohana Rao 1966b).

The ovule is anatropous, unitegmic and crassinucellate. The cells of the funiculus, adjacent to placenta, divide and form a collar around the micropylar region of the ovule. A hypostase differentiates after fertilization. The hypodermal cell cuts off a parietal cell. The parietal tissue is four to five layers thick. The megaspore mother cell produces a linear tetrad, and the chalazal megaspore develops into a Polygonum type of embryo sac. The antipodal cells divide into five to seven cells which persist till the endosperm becomes cellular (Kapil and Mohana Rao 1966b).

Development of the endosperm is of the Nuclear type, and wall formation is initiated at the micropylar and chalazal region at the 16-nucleate stage.

The embryogeny conforms to the Solanad type. Suspensor polyembryony is reported in *Garrya veatchii* (Mohana Rao 1963), and the additional embryo develops up to the globular stage.

The seed is small, and sarcotestal with a minute embryo. The endosperm stores abundant starch. The seed-coat is 12–15 layers thick, and the epidermis comprises thin-walled columnar cells. A few outer thin-walled layers persist, while the inner ones get crushed.

RH Eyde (1964), RN Kapil and PR Mohana Rao (1966b), PR Mohana Rao (1963), MF Moseley Jr and RM Beeks (1955), W Wangerin (1910c)

40.7 Araliaceae

The epidermis of the anther shows bulges at maturity, and the fibrous endothecium is biseriate in the region of the connective in *Tieghemopanax sambucifolius* (Mohana Rao 1972b). Two ephemeral middle layers are reported in *Brassia actinophylla* (Gopinath 1944), *Hedera australiana, Panax fruticosa* (D. Singh 1954) and *Tieghemopanax*. The tapetum is secretory with uninucleate cells in *H. australiana*, binucleate in *P. fruticosa* and multinucleate in *B. actinophylla* and *Tieghemopanax*. In *Tieghemopanax* nuclear fusions result in polyploid condition. The microspore mother cells undergo meiotic divisions, and produce tetrahedral and decussate tetrads. The cytokinesis is simultaneous.

The pollen grains are oblate, spheroidal, subporolate, and three-colporate, and shed at the two-celled stage.

The ovule is anatropous, unitegmic and tenui- or crassinucellar with a narrow micropyle. The nucellus disintegrates rapidly. A funicular obturator occurs in *Hedera australiana, Panax fruticosa, Polyscias pinnata* (Gopinath 1944) and *Tieghemopanax*. An endothelium and hypostase also differentiate. The archesporium is one-celled, except in *P. fruticosa* where it may be two-celled. The archesporial cell functions directly as megaspore mother cell, or may produce a parietal cell which does not divide further (Gopinath 1944). The megaspore mother cell gives rise to a linear tetrad. The chalazal megaspore develops into a Polygonum type of embryo sac. The antipodals are ephemeral, and the polars fuse before fertilization. The micropylar megaspore develops into embryo sac in *Hedera nepalensis* (D. Singh 1954). The development of more than one megaspore up to two- or four-nucleate stage is reported in *H. australiana* and *Polyscias pinnata*. Occasionally, a bisporic embryo sac occurs in *H. australiana* and *P. pinnata*; a tendency to Oenothera type is reported in *Brassia actinophylla*. The antipodals persist in *B. actinophylla* and *H. venulosum* (Gopinath 1944).

An embryo sac with reversed polarity is reported in *H. venulosum* (Gopinath 1944).

The development of the endosperm conforms to the Nuclear type, wall formation occurs at the eight-nucleate stage in *Tieghemopanax*. The endosperm is persistent, and stores abundant reserve food. It is ruminate in *Arthrophyllum diversifolium* and *Hedera helix*, due to the infoldings of the outer epidermal layer of the seed-coat (Periasamy 1962a).

The embryogeny conforms to the Onagrad type in *Aralia cachemiria*, *A. cordata*, *A. racemosa*, *Fatsia japonica* and *Hedera helix*; Chenopodiad type in *Hedera australiana* and *Panax fruticosa*.

The seed-coat consists of only the outer epidermis. Its cells are tangentially elongate, slightly thick-walled and contain tannin. It is thin and membraneous.

L Ducamp (1902), DM Gopinath (1944, 1945), PR Mohana Rao (1972b), K Periasamy (1962a), EM Pigott (1915), D Singh (1954)

40.8 Umbelliferae

The development of the anther wall is of the dicotyledonous type. It comprises the epidermis, fibrous endothecium, an ephemeral middle layer, and secretory tapetum of binucleate cells. The endothecium develops fibrous thickenings on the dorsal side in *Bupleurum tenue* (S.C. and M. Gupta 1964). The tapetum becomes biseriate near the septum, and the cells are multinucleate in some taxa (S.C. Gupta 1970). The microspore mother cells undergo meiotic divisions, cytokinesis is simultaneous, and tetrads are tetrahedral, decussate and isobilateral.

The pollen grains are shed at the three-celled stage. The male cells are located at each pole of the ellipsoidal pollen showing fixed polarity (Paliwal 1951, S.C. Gupta 1964). This aspect needs further study.

The ovule is anatropous, unitegmic and tenuinucellar. A pseudocrassinucellar ovule occurs in *Foeniculum vulgare* (S.C. Gupta 1964). A hypostase differentiates as early as the archesporial stage, and persists up to the maturation of the embryo sac. The endothelium differentiates at the megaspore tetrad stage in *Bupleurum*, and extends beyond the nucellus around the micropylar canal. An obturator also develops.

The archesporium is uni- or bi-celled. The hypodermal archesporial cell functions directly as megaspore mother cell, the chalazal megaspore of a linear tetrad develops into a Polygonum type of embryo sac. In *Pimpinella bracteata* and *P. candolleana* (Maheswari Devi and Santa Ram 1985), sometimes the tetrad is T-shaped. The occurrence of isobilateral tetrad is reported in *Peucedanum grande* (G.L. Shah 1953); this needs confirmation. Allium type of development has been observed in *Bupleurum aureum*, Drusa type in *Bowlesia tenera* and *Drusa oppositifolia*, and Penaea type in *Azorella trifurcata* (Håkansson 1923, 1927, 1952).

In most taxa, double fertilization occurs as usual. However, in *Pimpinella candolleana* and *P. monoica* (Maheswari Devi and Santa Ram 1985), syngamy is said to occur only after the formation of a large number of free endosperm nuclei. This aspect needs confirmation, also whether this is an abnormal feature limited to only a few ovules, or common one.

The endosperm is of the Nuclear type. The nuclei aggregate at the micropylar end, and enucleate nodules occur in the micropylar region in *Anethum graveolens*, *Coriandrum sativum*, *Daucus carota*, *Foeniculum vulgare* (Fig. 3.10.2F), *Heracleum sphondylium*, *Psammogeton biternatum* (S.C. Gupta 1970), *Cuminum cyminum* (Sehgal 1965), *Pseudorlaya pumilla* and *Torilis nodosa* (Al-Attar 1977) and *Pimpinella* spp. (Maheswari Devi and Santa Ram 1985). Wall formation is initiated ca. at the 64-nucleate stage in *Anethum* (Dhakre 1968), and ca. at the 256-nucleate stage in *Coriandrum* and *Foeniculum* (S.C. Gupta 1964).

The embryogeny conforms to the Solanad type. The development of a synergid proembryo has been observed in *Pimpinella candolleana* (Maheswari Devi and Santa Ram 1985). The synergid embryo is much smaller than the zygotic. Budding of suspensor is reported in *Coriandrum sativum* and *Foeniculum vulgare* (S.C. Gupta 1964). *Cuminum cyminum* (Sehgal 1965) and *P. candolleana* (Maheswari Devi and Santa Ram 1985). The origin of an additional embryo has not been traced in *Trachyspermum ammi* (Sehgal 1965).

The developing seed of *Foeniculum* is often attacked by a phytophagus chalcid fly *Systole albipennis* which causes embryolessness (S.C. Gupta 1962a, b). It lays its eggs in the mericarp with globular proembryo and partially, or completely, cellular endosperm (Fig. 40.8.1A). The eggs develop at the cost of the embryo and endosperm (Fig. 40.8.1B). Finally, in mature fruits the insect escapes by boring a hole in the fruit wall (Fig. 40.8.1C–E). A curious feature is that the larva of *Systole* may itself get parasitized by a hyperparasite, either *Liodontomerus* or *Tetrastichus* (Fig. 40.8.1F). In this case the larva of *Systole* is destroyed, and no appreciable damage is caused to the embryo or endosperm.

The seed is endospermic with a small embryo. The endosperm stores oil as reserve food. The seed-coat is almost crushed except for the outer epidermis which remains thin-walled.

The fruit is a schizocarp. In each mericarp six vittae alternate with six vascular bundles. The vittae are located in furrows. The outer epidermis of the mericarp comprises anticlinally elongate, thickly cuticularized cells. Most of the cells of the pericarp become sclerotic. The inner epidermal cells have prominent fibrous thickenings.

In *Bupleurum tenue*, stomata occur on stalks of the compound umbels and umbellets, bracts of involucre and involucel, pedicels of flowers, both surfaces

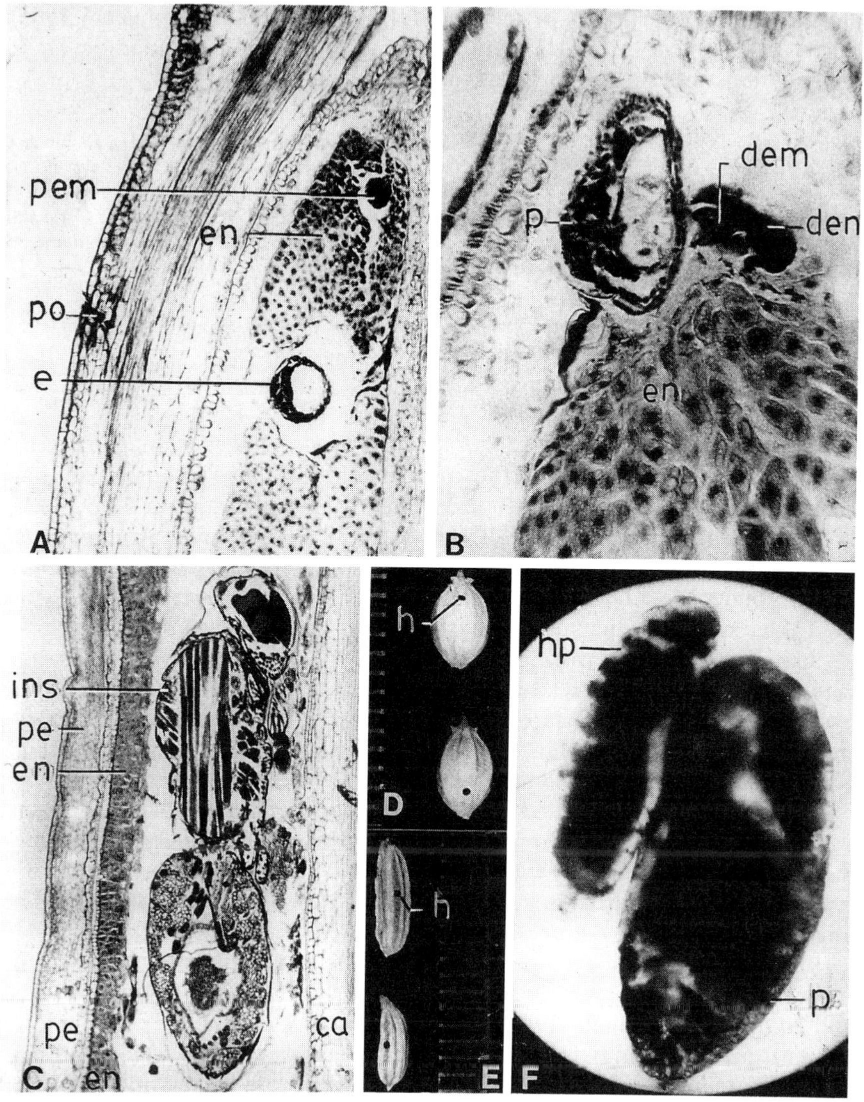

Fig. 40.8.1A–F. *Foeniculum vulgare*. **A** Longisection of immature mericarp with proembryo (*pem*), cellular healthy endosperm (*en*), and egg (*e*) of parasite (*Systole albipennis* Walk.); *po* path of ovipositor. **B** Later stage, the developing parasite (*p*) has caused degeneration of proembryo (*dem*) and adjoining endosperm (*den*). **C** Longisection of mature pericarp with adult insect (*ins*) in seed; *ca* carpophore; *pe* pericarp. **D, E** Fruits of *Coriandrum sativum* (**D**) and *Foeniculum* (**E**) with the exit holes (*h*) of parasites. **F** Hyperparasite (*hp*) larva lying adpressed to the host parasite (*p, S. albipennis*). (After S.C. Gupta 1970)

of petals, outer epidermis of pericarp, and stylopodium (S.C. Gupta et al. 1965). Their development is of the syndetochellic type, and are anomocytic in stylopodium, and anisocytic in rest of the organs.

RD Adatia and GL Shah (1952), AAA Al-Attar (1977, 1979), HA Borthwick (1931), JS Dhakre (1968), SC Gupta (1962a, b, 1964, 1970), SC and M Gupta (1964), SC Gupta, GS Paliwal and M Gupta (1965), Å Håkansson (1923, 1927, 1952), R Hegnauer (1971), H Maheswari Devi and A Santa Ram (1985), JC Mestre and JL Guignard (1971), RL Paliwal (1951), AM Pigni, G Corsi and EM Capelletti (1986), RL Rodriguez (1971), CB Sehgal (1965), GL Shah (1953), D Singh and JS Gupta (1956), R Souèges (1926a), J Szujkó-Lacza (1976, 1978)

40.9 Taxonomic Considerations

Alangiaceae. Bentham and Hooker (1965a) place *Alangium* in Cornaceae. Hutchinson (1926) recognizes the independent status of Alangiaceae, and includes them in Umbelliflorae. Subsequently, he (Hutchinson 1959, 1973) transfers Alangiaceae, along with Cornaceae and Araliaceae, to Araliales, and Umbelliferae to Umbellales. Engler and Prantl (1931) also place *Alangium* in an independent family Alangiaceae, which they include in Myrtiflorae.

Takhtajan (1980) includes Alangiaceae and Cornaceae in the Cornales.

Alangiaceae can be distinguished from Cornaceae on the basis of ovule with a vascular strand reaching up to the tip of integument, a well-developed hypostase, persistent and, rarely, supernumerary antipodals, antipodal polyembryony (which is doubtful), absence of Fritillaria type of embryo sac, and Nuclear or Cellular type of endosperm.

The embryological features justify the independent status of Alangiaceae.

The Myrtiflorae differ from Alangiaceae in an inferior ovary with numerous bitegmic ovules, several types of embryo sac – generally of the Polygonum type, tetrasporic 16-nucleate in species of *Combretum*, and Oenothera type in Onagraceae.

Alangium has a single, pendulous, anatropous, unitegmic ovule as in Umbelliferae (Håkansson 1923), Cornaceae (Morse 1907, Fagerlind 1939c), and Araliaceae (Gopinath 1944, D. Singh 1954).

B.M. Johri (1963b) rightly concludes that Alangiaceae have a closer affinity with the families included in Araliales, than with those in Myrtiflorae. Thus, the present assignment in the order Umbelliflorae along with Cornaceae, Araliaceae and Umbelliferae is in order.

Nyssaceae and Davidiaceae. Bentham and Hooker (1965a), Harms (1897b) and Hallier (1912) include *Nyssa* and *Davidia* in Cornaceae. Harms places *Camptotheca* and *Nyssa* in the subfamily Nyssoideae, and *Davidia* in Davidioideae, under Cornaceae. De Jussieu (see Takhtajan 1959) created the family Nyssaceae. Takhtajan (1959) and Melchior (1964) include Nyssaceae in Umbelliflorae along with Cornaceae, and raise the genus *Davidia* to family status Davidiaceae, next to Nyssaceae. In his later classification, Takhtajan (1980) includes the families Davidiaceae, Nyssaceae, Alangiaceae and Cornaceae, along with some other families, in Cornales. Hutchinson (1959, 1969, 1973) assigns Nyssaceae to Araliales near Cornaceae.

Some of the similarities between *Nyssa* and Cornaceae are: three-porate pollen grains, unitegmic, pendulous, anatropous ovule with a single hypodermal archesporium, unequal dyad cells, Polygonum type of embryo sac, and fusion of polar nuclei prior to fertilization (Tandon and Herr 1971). The differences with Cornaceae are: pseudo-crassinucellar ovules, absence of parietal cells, degeneration of non-functional megaspores from chalazal to micropylar, degeneration of apical region of nucellus during maturation of megagametophyte, presence of hypostase, and occasional development of bisporic embryo sac (Tandon and Herr 1971). Mohana Rao (1972a) reports Nuclear type of endosperm, and thick-walled lignified epidermis of the seed-coat. The latter feature does not occur in any other member of Umbelliflorae. Tandon and Herr relate *Nyssa* to Cornaceae, while Mohana Rao (1972a) supports its Umbellalian alliance.

The embryology of the genus *Camptotheca* has not been investigated, and its affinity cannot be discussed.

Anatropous unitegmic ovule, Polygonum type of embryo sac, albuminous seed with straight embryo, and drupaceous fruit are known in Nyssaceae and Davidiaceae. However, crassinucellar ovule, multicellular female archesporium, occasional development of supernumerary antipodals with haustorial activity, and Cellular endosperm (Horne 1909) distinguish Davidiaceae from Nyssaceae. As the embryological features show pronounced differences, both *Nyssa* and *Davidia* deserve independent family status.

Cornaceae. Engler and Diels (1936) place Cornaceae in Umbelliflorae, Hutchinson (1973) in Araliales, and Takhtajan (1959, 1980) assigns them to the order Cornales. Several genera, such as *Alangium, Garrya* and *Nyssa*, have been removed from Cornaceae, and given independent family status, which is fully supported by embryological features. No doubt, Cornaceae closely resemble Umbelallian families but have certain distinctive features: presence of uni- or multicelled female archesporium, parietal tissue, obturator, and endothelium; absence of hypostase, occasional occurrence of tetrasporic (Fritillaria) embryo sac in *Cornus mas* and pseudomonosporic in *C. florida*, Cellular type of endosperm, and nucellar apospory (Table 40.9.1). *Mastixia* of the subfamily Mastixiodeae is supposed to form a link between Cornaceae and Araliaceae, on anatomical grounds. The embryology of *Mastixia* has not been investigated. Mohana Rao (1972a) favours the segregation of Cornaceae into a separate order Cornales, on embryological grounds.

Garryaceae. Wangerin (1910b), Hallier (1912) and Bessey (1915) include the genus *Garrya* in Cornaceae. Hjelmqvist (1948) suggests ordinal rank, and points out the affinities of Garryales with the Umbelliflorae. Hutchinson (1973) includes Garryaceae in Araliales. Takhtajan (1959, 1980) and Melchior (1964) group them with Umbelliflorae (Cornales). Kapil and Mohana Rao (1966b) discuss the embryological features, and emphasize that Garryaceae differ from Salicaceae, Chloranthaceae or Piperaceae, and are allied to Alangiaceae and

Table 40.9.1. Embryological features of Cornaceae, Araliaceae and Umbelliferae (See Mohana Rao 1972a)

	Cornaceae	Araliaceae	Umbelliferae
Anther	Bilocular, anther-wall 5–6-layered, endothecium fibrous, tapetum secretory, cytokinesis simultaneous	Bilocular, anther wall 4–5-layered, endothecium fibrous, tapetum secretory, cytokinesis simultaneous	Bilocular, anther wall 4-layered, endothecium fibrous, tapetum secretory, cytokinesis simultaneous
Tetrad	Tetrahedral	Tetrahedral and decussate	Tetrahedral, decussate and isobilateral
Pollen	2-celled, oblate, spheroidal, 3-colpate	2-celled, oblate, spheroidal, subporolate, 3-colporate	3-celled, bilateral, subporolate, 3-colporate
Ovule	Anatropous, unitegmic, crassinucellate, do not abort	Anatropous, unitegmic, tenui- or crassinucellate, abortive ovules common	Anatropous, unitegmic, tenuinucellate, abortive ovules common
Embryo sac	Mono-, bi- or tetrasporic, antipodals ephemeral	Mono- or bisporic, antipodals ephemeral or persistent	Mono-, bi- or tetrasporic, antipodals persist until early stages of endosperm
Endosperm	Cellular type	Nuclear	Nuclear
Embryo	Asterad type	Onagrad or Chenopodiad	Solanad
Polyembryony	Nucellar apospory	—	Suspensor polyembryony
Seed-coat	Thin-walled, collapses	Thin and membraneous	Outer epidermis of testa remains thin-walled
Fruit	Drupe	Berry or drupe	Schizocarp

Cornaceae of the Umbelallian complex. The occurrence of oblate spheroidal three-colporate pollen grains, anatropous unitegmic crassinucellate ovule, Polygonum type of embryo sac, and berry-like fruits are similarities with Cornaceae. But a well-developed hypostase, persistent antipodals, Nuclear type of endosperm, and suspensor polyembryony distinguish *Garrya* from Cornaceae (Kapil and Mohana Rao 1966b). Solanad type of embryogeny, and suspensor polyembryony occur in Umbelliferae. Therefore, the removal of *Garrya* from Cornaceae, erection of the family Garryaceae and its inclusion in Umbelliflorae are fully justified.

Araliaceae. Bentham and Hooker (1965a), Bessey (1915), Wettstein (1935) and Engler (1964) assign Araliaceae to the order Umbelliflorae along with Umbelliferae and Cornaceae. Takhtajan (1966, 1969, 1980) groups Araliaceae and Umbelliferae (Apiaceae) under Araliales. Due to their woody habit, Hutchinson (1973) treats Araliaceae, along with Cornaceae, in a separate order Araliales. Araliaceae are embryologically allied to Umbelliferae, rather than Cornaceae (see Mohana Rao 1972b). The presence of abortive ovules, single, rarely double, female archesporium, hypostase, obturator, mono- or bisporic embryo sac, Nuclear type of endosperm, and structure of seed are similar to those of Umbelliferae. Berry (rarely drupe) type of fruit, eight-carpellate ovary with one to 15 locules, tenui- or crassinucellar ovules, and Onagrad or Chenopodiad type of embryogeny differentiate Araliaceae from Umbelliferae (Mohana Rao 1972b).

Umbelliferae. The family Umbelliferae is a natural taxon. It is characterized by schizocarpic fruit, and bilateral, subporolate, three-colporate, three-celled pollen grains at the dispersal stage. A parietal cell is absent, the ovule is pseudocrassinucellar in *Foeniculum vulgare*. A hypostase and endothelium differentiate at the megaspore mother cell stage. The endothelium grows beyond the nucellus in the micropylar region. The occasional presence of bisporic (Allium) and tetrasporic (Penaea and Drusa) types of embryo sac is also reported. The formation of enucleate endosperm vesicles is restricted to Umbelliferae. The association with parasites causes embryolessness (S.C. Gupta 1962a, b).

Cronquist (1968, 1981) maintains two distinct orders, Cornales and Umbellales (Apiales), and segregates them on morphological, anatomical, and phytochemical basis. He mentions the anatomical findings (of a Russian botanist Tamanschian) which reveal close affinities between Umbellales and Cornales.

In his new system of classification, R. Dahlgren (1977b, 1980b, 1981) places Araliaceae and Apiaceae (Umbelliferae) in the non-iridoid Sympetalous order Araliales (superorder Araliflorae), and the remaining families in the iridoid-Sympetalous order Cornales (superorder Corniflorae). A consideration under the Sympetalae is wholly unwarranted.

While discussing the new aspects of systematics of Asteridae (Sympetalae), Wagenitz (1977b) refers to two connecting lines, one between Cornaceae and Caprifoliaceae, confirmed by serological data (Hildebrand and Fairbrothers

1970a, b), and the other between Araliales and Asterales (Asteriflorae) based on chemotaxonomical investigations of Hegnauer (1971, 1973), Bohlmann et al. (1973), and Wagenitz (1977b). The polyacetylenes and sesquiterpene lactones, which are rare in other angiosperms, occur frequently in these two orders.

Cornales and Araliales have been treated as independent orders by Takhtajan (1980).

The order Umbelliflorae is a more or less natural grouping with the families Alangiaceae, Nyssaceae, Davidiaceae, Cornaceae, Garryaceae, Araliaceae and Umbelliferae. There is no justification to erect the orders Cornales and Araliales.